Introduction to Modern
Liquid Chromatography

Introduction to Modern Liquid Chromatography

Second Edition

L. R. SNYDER

Technicon Instruments Corporation
Research & Development Division
Tarrytown, New York

J. J. KIRKLAND

E. I. du Pont de Nemours & Company
Central Research & Development Department
Wilmington, Delaware

A Wiley-Interscience Publication

JOHN WILEY & SONS, INC.
New York • Chichester • Brisbane • Toronto • Singapore

Library of Congress Cataloging in Publication Data:

Snyder, Lloyd R.
   Introduction to modern liquid chromatography.

   ''A Wiley-Interscience publication.''
   Includes bibliographies and index.
   1. Liquid chromatography.   I. Kirkland, Joseph Jack,
   joint author.   II Title.

QD79.0454S58 1979   544'.924   79-4537
ISBN 0-471-03822-9

Printed in the United States of America

10 9 8

And thus the native hue of resolution
Is sicklied o'er with the pale cast of
thought.

Shakespeare
Hamlet, Scene I, Act 3

PREFACE TO THE FIRST EDITION

This book is about modern liquid chromatography. By this we mean automated, high-pressure liquid chromatography in columns, with a capability for the high-resolution separation of a wide range of sample types, within times of a few minutes to perhaps an hour. Modern liquid chromatography (LC) is now about five years old. By early 1969 it was possible to purchase equipment and high performance column packings which together largely bridged the gap between classical liquid chromatography and gas chromatography. Since that time there has been a flurry of activity on the part of companies that supply equipment and materials for LC. Within the past few years there have been further major advances in the theory and practice of LC. Finally, numerous applications of modern LC to a wide range of problems are now being reported. The technique has reached the point where the average chromatographer can achieve—by yesterday's standards—consistently spectacular results.

To get the most out of modern LC, some care is required in choosing the right technique, selecting the best separation conditions, and using the proper equipment to best advantage. In short, the practical worker must know what he is doing. Moreover, his knowledge must be a balance of theory and experience; it must include both principles and practice. Unfortunately, there has been a tendency for those in chromatography to stress either the theoretical or the "practical" side of the subject. Also, the theory of chromatography—and of modern LC—has often been represented as highly complex, with its application to real separation problems either not obvious, or impossibly tedious. We think there is a better approach.

An effective presentation of what modern LC is all about must be a blend of practical details plus down-to-earth theory. This conviction led us in 1971 to develop the American Chemical Society short course "Modern Liquid Chromatography." Within the next two years we presented the course to about 800 students, with a highly enthusiastic response. The course itself continued to evolve during this time, largely in response to the questions and comments of the students. By late 1972 it appeared worthwhile to reduce our approach to textbook form, and the present book is the result.

Our goal for this book was to retain the essential elements of the short course, and to add certain materials that could not be included in a two-day series of lectures. We did not hope to present everything of conceivable interest to LC in a book of this size, yet we were determined not to slight *any* area of significant practical importance. Where compromise between these two objectives eventually proved necessary, it has been handled by referencing other sources. The book was written to be self-sufficient in terms of the needs of the average professional or technician who plans to work with modern LC. We believe this book will prove useful in most laboratories where modern LC is practiced.

L. R. SNYDER
J. J. KIRKLAND

October 1973

PREFACE TO THE SECOND EDITION

For several years liquid chromatography (LC) with a performance fully comparable to that of gas chromatography has been available as a routine laboratory technique. In 1973, when the first edition of this book was completed, the advantages of modern LC were just coming to the attention of a wide audience. At that time there were many published examples of LC applications for a variety of sample types, and equipment and related materials were available to solve most practical problems. However, several additional techniques and instrument improvements remained to be developed. As a result, many LC applications proved challenging—and occasionally unworkable, even in the hands of experienced workers.

Six years later, in 1979, we see great changes and advances in the practical application of LC. The ubiquitous detector problem has been largely solved with the advent of spectrophotometer detectors operating down to 190 nm, making possible the sensitive detection of almost any compound type. Increased use of fluorescence and electrochemical detectors, plus off-line and on-line derivatization, has further pushed detection problems into the background, even for trace analyses of complex samples such as blood, food, and soil. Recent developments in microprocessor-controlled instrumentation also have produced greatly improved equipment performance, to permit more convenient, versatile, and precise LC separations.

Only in the past four years has the tremendous potential of small-particle, reverse-phase LC been exploited, particularly when augmented with gradient elution and special methods such as ion-pair formation. As a result, previously difficult or impossible separations of such compounds as polar dyes or basic drugs and their metabolites are now more or less routine. Also, much less time is required to develop the average LC application; successful separations within the first few tries are now common.

The exponential improvement in column packings that occurred between 1968 and 1973 has leveled off in the past four years, and most experts now agree that we are approaching a fundamental limit to further major increase in column efficiency or plate numbers. However, there has been continuing emphasis by manufacturers on reliability and reproducibility of both packings and packed columns. Consequently, today the user can expect closer agreement between packings and columns from different lots, both in terms of sample retention and column efficiency characteristics. During the past four years, packings for the separation of large, water-soluble molecules such as proteins were finally reported, and it seems likely that these and other useful new packings will become commercially available in the near future.

Problem areas such as band tailing, trace analysis, preparative separation, and so on, have received considerable attention in the past five years, and these particular problems or applications can now be approached in a relatively systematic fashion. The further development and application of special "tricks" or techniques such as column switching continues to add greatly to the potential of LC and to enable the easier handling of problem samples. The increasing volume of certain LC testing, particularly in the areas of quality control, process control, and clinical chemistry, has made full automation of these assays necessary—which has led to a number of important developments and the commercial availability of automated peripheral equipment for sampling, sample pretreatment, and so on.

Finally, superb examples of the application of LC in virtually every industry and technical area now abound, providing encouragement to the would-be practitioner, and specific examples for the chromatographer with a given job to do.

Although this second edition has been almost completely rewritten, our approach closely follows that brought to the preparation of the first edition. Aside from including new developments that have occurred over the past five years, we have expanded our treatment of many practical areas, so that the reader will have less need to chase down references to work described elsewhere. We have also intentionally added a large number of actual chromatograms of "real" samples, both because such examples are readily available and because we feel that in this case "one picture is worth a thousand words." Inevitably, all of this has meant a larger and more expensive book—for which we apologize. The final format and relative emphasis on certain areas is the result of experience gained from two American Chemical Society short courses in modern liquid chromatography—the original basic course and the problem-solving course introduced in 1974. These two courses have now been presented in over 50 sessions to more than 2000 students. Thus, the present book has benefited greatly from the questions and inputs of both beginning and experienced liquid chromatographers.

In conclusion, we want to express our appreciation to a number of people for their help in creating this second edition. Specifically, we wish to thank several of our co-workers and friends who reviewed the orginal manuscript for many helpful corrections and modifications, especially Dr. Eli Grushka of the Hebrew University, Jerusalem (Israel); Drs. Pedro A. Rodriguez, Thomas S. Turan, C. Grant Birch, Mark D. Seymour, and William J. Kozarek, all of the Proctor & Gamble Company, Cincinnati, Ohio; Dr. Dennis L. Saunders of the Union Oil Research Center, Brea, California; and Drs. John W. Dolan and J. Russell Gant of Technicon. Dr. Dolan has also contributed Section 17.2 on Sample Cleanup. We are also grateful to Mrs. Patricia C. Lyons of Du Pont for extensive assistance with the typing of the manuscript. Finally, we gratefully acknowledge the considerable support of Du Pont and Technicon in many different ways.

L. R. S.
J. J. K.

July 1979

CONTENTS

1 Introduction, 1

    1.1   Liquid versus Gas Chromatography,   2

    1.2   Modern versus Traditional LC Procedures,   3

    1.3   How Did Modern LC Arise?,   8

    1.4   The Literature of LC,   9

    1.5   About the Book,   12

    References,   13

    Bibliography,   13

2 Basic Concepts and Control of Separation, 15

    2.1   The Chromatographic Process,   16

    2.2   Retention in LC,   22

    2.3   Band Broadening,   27

    2.4   Resolution and Separation Time,   34

    2.5   The Control of Separation,   37

    References,   81

    Bibliography,   82

3 Equipment, 83

    3.1   Introduction,   84

    3.2   Extracolumn Effects,   86

    3.3   Mobile-Phase Reservoirs,   88

    3.4   Solvent Pumping (Metering) Systems,   90

    3.5   Equipment for Gradient Elution, 103

3.6   Sample Injectors, 110

3.7   Miscellaneous Hardware, 117

3.8   Integrated and Specialized Instruments, 119

3.9   Laboratory Safety, 123

References, 123

Bibliography, 124

4 Detectors, 125

4.1   Introduction, 126

4.2   Detector Characteristics, 127

4.3   UV-Visible Photometers and Spectrophotometers, 130

4.4   Differential Refractometers, 140

4.5   Fluorometers, 145

4.6   Infrared Photometers, 147

4.7   Electrochemical (Amperometric) Detectors, 153

4.8   Radioactivity Detectors, 158

4.9   Conductivity Detectors, 161

4.10   Summary of the Characteristics of Most-Used Detectors, 161

4.11   Other Detectors, 165

References, 165

Bibliography, 166

5 The Column, 168

5.1   Introduction, 169

5.2   Characteristics and Use of Different Column Packings, 173

5.3   Available Column Packings, 183

5.4   Column Packing Methods, 202

5.5   Column Evaluation and Specifications, 218

5.6   Column Techniques, 225

5.7   Reduced Parameters and Limiting Column Performance, 234

References, 243

Bibliography, 245

6 Solvents, 246

6.1   Introduction, 247

6.2   Physical Properties, 251

6.3   Intermolecular Interactions Between Sample and
       Mobile-Phase Molecules, 255

6.4   Solvent Strength and "Polarity," 257

6.5   Solvent Selectivity, 260

6.6   Purification of LC Solvents, 265

References, 267

Bibliography, 268

7 Bonded-Phase Chromatography, 269

7.1   Introduction, 270

7.2   Preparation and Properties of Bonded-Phase Packings, 272

7.3   Mobile-Phase Effects, 281

7.4   Other Separation Variables, 289

7.5   Special Problems, 294

7.6   Applications, 301

7.7   The Design of a BPC Separation, 316

References, 319

Bibliography, 321

8 Liquid-Liquid Chromatography, 323

8.1   Introduction, 323

8.2   Essential Features of LLC, 325

8.3   Column Packings, 327

8.4   The Partitioning Phases, 332

8.5   Other Separation Variables, 336

8.6   Special Problems, 336

8.7   Applications, 338

8.8   The Design of an LLC Separation, 342

References, 347

Bibliography, 348

9 Liquid-Solid Chromatography, 349

9.1   Introduction, 351

9.2   Column Packings, 361

9.3   Mobile Phases, 365

9.4   Other Separation Variables, 389

9.5   Special Problems, 391
9.6   Applications, 398
9.7   The Design of an LSC Separation, 405
References, 407
Bibliography, 409

10   Ion-Exchange Chromatography, 410

10.1   Introduction, 410
10.2   Column Packings, 414
10.3   Mobile Phases, 419
10.4   Other Separation Variables, 426
10.5   Special Problems, 427
10.6   Applications, 429
10.7   The Design of an Ion-Exchange Separation, 445
References, 450
Bibliography, 452

11   Ion-Pair Chromatography, 453

11.1   Introduction, 454
11.2   Column Packings, 457
11.3   Partitioning Phases, 458
11.4   Other Separation Variables, 470
11.5   Special Problems, 471
11.6   Applications, 473
11.7   The Design of an Ion-Pair Separation, 477
References, 481
Bibliography, 482

12   Size-Exclusion Chromatography, 483

12.1   Introduction, 484
12.2   Column Packings, 487
12.3   Mobile Phases, 500
12.4   Other Separation Variables, 503
12.5   Molecular-Weight Calibration, 509
12.6   Recycle Chromatography, 519
12.7   Problems, 522

12.8   Applications, 525

12.9   The Design of an SEC Separation, 534

References, 538

Bibliography, 540

13 Quantitative and Trace Analysis, 541

13.1   Introduction, 542

13.2   Peak-Size Measurement, 545

13.3   Calibration Methods, 549

13.4   Selection of Calibration Method, 556

13.5   Trace Analysis, 560

References, 573

14 Qualitative Analysis, 575

14.1   Retention Data for Sample Characterization, 576

14.2   Qualitative Analysis of Sample Bands from
        Analytical-Scale LC Separations, 589

14.3   On-Line Spectroscopic Analysis of LC Peaks, 603

References, 612

Bibliography, 614

15 Preparative Liquid Chromatography, 615

15.1   Introduction, 615

15.2   Separation Strategy, 617

15.3   Experimental Conditions, 623

15.4   Operating Variables, 641

15.5   Applications, 647

15.6   A Preparative Separation Example, 654

References, 661

16 Gradient Elution and Related Procedures, 662

16.1   Gradient Elution, 663

16.2   Column Switching and Stationary-Phase
        Programming, 694

16.3   Flow Programming, 712

16.4   Temperature Programming, 715

16.5   Practical Comparison of Various Programming
        and Column-Switching Procedures, 715
References, 717
Bibliography, 718

17 Sample Pretreatment and Reaction Detectors, 720

17.1   Introduction, 720
17.2   Sample Cleanup, 722
17.3   Sample Derivatization, 731
17.4   Reaction Detectors, 740
17.5   Automation of Sample Pretreatment, 746
References, 748
Bibliography, 750

18 Selecting and Developing One of the LC Methods, 752

18.1   Introduction, 753
18.2   Developing a Particular Separation, 762
18.3   Special Applications, 776
References, 779

19 Troubleshooting the Separation, 781

19.1   Troubleshooting the Equipment, 782
19.2   Separation Artifacts, 791
19.3   Troubleshooting Analytical Errors, 813
References, 823

Appendix I
Suppliers of LC Equipment, Accessories, and Columns, 824

Appendix II
Miscellaneous Tables Used by Workers in LC, 833

II.1   The Gaussian or Error Function, 833
II.2   Reduced Plate-Height Data for "Good" Columns, 836
II.3   Viscosity of Solvent Mixtures, 836
II.4   Particle Size Expressed as Mesh Size, 838
References, 839

List of Symbols, 840

List of Abbreviations, 844

Index, 847

Introduction to Modern Liquid Chromatography

ONE

INTRODUCTION

1.1 Liquid versus Gas Chromatography   2
1.2 Modern versus Traditional LC Procedures   3
1.3 How Did Modern LC Arise?   8
1.4 The Literature of LC   9
1.5 About the Book   12
References   13
Bibliography   13

Over the past 40 years the practice of chromatography has witnessed a continuing growth in almost every respect: the number of chromatographers, the amount of published work, the variety and complexity of samples being separated, separation speed and convenience, and so on. However, this growth curve has not moved smoothly upward from year to year. Rather the history of chromatography is one of periodic upward spurts that have followed some major innovation: partition and paper chromatography in the 1940s, gas and thin-layer chromatography in the 1950s, and the various gel or size-exclusion methods in the early 1960s. A few years later it was possible to foresee still another of those major developments that would revolutionize the practice of chromatography: a technique that we call *modern liquid chromatography*.

What do we mean by "modern liquid chromatography"? Liquid chromatography (LC) refers to any chromatographic procedure in which the moving phase is a liquid, in contrast to the moving gas phase of gas chromatography. Traditional column chromatography (whether adsorption, partition, or ion-exchange), thin-layer and paper chromatography, and modern LC are each examples of liquid chromatography. The difference between modern LC and these older procedures involves improvements in equipment, materials, technique, and the

1

application of theory. In terms of results, modern LC offers major advantages in convenience, accuracy, speed, and the ability to carry out difficult separations. To appreciate the unique value of modern LC it will help to draw two comparisons:

- Liquid versus gas chromatography.
- Modern versus traditional LC procedures.

1.1 LIQUID VERSUS GAS CHROMATOGRAPHY

The tremendous ability of gas chromatography (GC) to separate and analyze complex mixtures is now widely appreciated. Compared to previous chromatographic methods, GC provided separations that were both faster and better. Moreover, automatic equipment for GC was soon available for convenient, unattended operation. However, many samples simply cannot be handled by GC. Either they are insufficiently volatile and cannot pass through the column, or they are thermally unstable and decompose under the conditions of separation. It has been estimated that only 20% of known organic compounds can be satisfactorily separated by GC, without prior chemical modification of the sample.

LC, on the other hand, is not limited by sample volatility or thermal stability. Thus, LC is ideally suited for the separation of macromolecules and ionic species of biomedical interest, labile natural products, and a wide variety of other high-molecular-weight and/or less stable compounds, such as the following:

Proteins	Polysaccharides	Synthetic polymers
Nucleic acids	Plant pigments	Surfactants
Amino acids	Polar lipids	Pharmaceuticals
Dyes	Explosives	Plant and animal metabolites

Liquid chromatography enjoys certain other advantages with respect to GC. Very difficult separations are often more readily achieved by liquid than by gas chromatography, because of

- Two chromatographic phases in LC for selective interaction with sample molecules, versus only one in GC.
- A greater variety of unique column packings (stationary phases) in LC.
- Lower separation temperatures in LC.

Chromatographic separation is the result of specific interactions between sample molecules and the stationary and moving phases. These interactions are essentially absent in the moving gas phase of GC, but they are present in the liquid phase of LC—thus providing an additional variable for controlling and improving separation. A greater variety of fundamentally different stationary phases have been found useful in LC, which again allows a wider variation of these selective interactions and greater possibilities for separation. Finally, chromatographic separation is generally enhanced as the temperature is lowered, because inter-molecular interactions then become more effective. This favors procedures such as LC that are usually carried out at ambient temperature.

Liquid chromatography also offers a number of unique detectors that have so far found limited application in GC:

- Colorimeters combined with color-forming reactions of separated sample components.
- Amperometric (electrochemical) detectors.
- Refractive index detectors.
- UV-visible absorption and fluorescent detectors.

Although GC detectors are generally more sensitive and also provide unique selectivity for many sample types, in many applications the available LC detectors show to advantage. That is, LC detectors are favored for some samples, whereas GC detectors are better for others.

A final advantage of liquid versus gas chromatography is the relative ease of sample recovery. Separated fractions are easily collected in LC, simply by placing an open vessel at the end of the column. Recovery is quantitative and separated sample components are readily isolated, for identification by supplementary techniques or other use. The recovery of separated sample bands in GC is also possible but is generally less convenient and quantitative.

1.2 MODERN VERSUS TRADITIONAL LC PROCEDURES

Consider now the differences between modern LC and classical column or open-bed chromatography. These three general procedures are illustrated in Figure 1.1. In *classical LC* a column is often used only once, then discarded. Therefore, packing a column (step 1 of Figure 1.1, "bed preparation") has to be repeated for each separation, and this represents a significant expense of both manpower and material. Sample application in classical LC (step 2), if done correctly, requires some skill and time on the part of the operator. Solvent flow in classical LC (step 3) is achieved by gravity feeding of the column, and individual sample fractions are collected manually. Since typical separations require several hours in classical LC, this is a tedious, time-consuming operation. Detection and quantitation (step 4) are achieved by the manual analysis of individual fractions. Normally, many fractions are collected, and their processing requires much time and effort. Finally, the results of the separation are recorded in the form of a *chromatogram:* a bar graph of sample concentration versus fraction number.

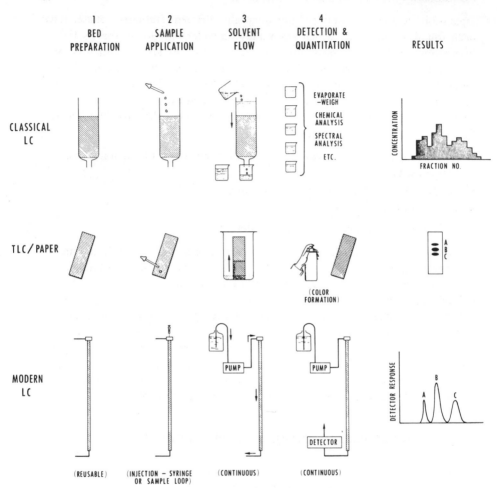

Figure 1.1 Different forms of liquid chromatography

The advent of paper chromatography in the 1940s and thin-layer chromatography (TLC) in the 1950s greatly simplified the practice of analytical liquid chromatography. This is also illustrated in Figure 1.1. Bed preparation in TLC or paper chromatography (step 1) is much cheaper and simpler than in classical LC. The paper or adsorbent-covered plates can be purchased in ready-to-use form at nominal expense. Sample application is achieved rather easily, and solvent flow is accomplished by placing the spotted paper or plate into a closed vessel with a small amount of solvent. Solvent flow up the paper or plate proceeds by capillary action, without the need for operator intervention. Finally, detection and quantitation can be achieved by spraying the dried paper or plate with some chromogenic reactant to provide a visible spot for each separated sample component.

The techniques of paper and thin-layer chromatography greatly simplified liquid chromatography and made it much more convenient. A further advantage, particularly for TLC, was that the resulting separations were much better than in classical LC and required much less time—typically 30-60 min rather than several hours. However, certain limitations were still apparent in these open-bed methods:

- Difficult quantitation and marginal reproducibility, unless special precautions are taken.
- Difficult automation.
- Longer separation times and poorer separation than in GC.
- Limited capacity for preparative separation (maximum sample sizes of a few milligrams).

Despite these limitations, TLC and paper chromatography became the techniques of choice for carrying out most LC separations.

Let us look now at modern LC. Closed, reusable columns are employed (step 1, Figure 1.1), so that hundreds of individual separations can be carried out on a given column. Since the cost of an individual column can be prorated over a large number of samples, it is possible to use more expensive column packings for high performance and to spend more time on the careful packing of a column for best results. Precise sample injection (step 2) is achieved easily and rapidly in modern LC, using either syringe injection or a sample valve. Solvent flow (step 3) is provided by high-pressure pumps. This has a decided advantage: controlled, rapid flow of solvent through relatively impermeable columns. Controlled flow results in more reproducible operation, which means greater accuracy and precision in LC analysis. High-pressure operation leads to better, faster separation, as is shown in Chapter 2. Detection and quantitation in modern LC are achieved with continuous detectors of various types. These yield a final chromatogram without intervention by the operator. The result is an accurate record of the separation with minimum effort.

Repetitive separation by modern LC can be reduced to a simple sample injection and final data reduction, although the column and/or solvent may require change for each new application. From this it should be obvious that modern LC is considerably more convenient and less operator dependent than either classical LC or TLC. The greater reproducibility and continuous, quantitative detection in modern LC also lead to higher accuracy and precision in both qualitative and quantitative analysis. As discussed in Chapter 13, quantitative analysis by modern LC can achieve a precision of better than ±0.5% (1 standard deviation or S.D.). Finally, preparative LC separations of multigram quantities of sample are now proving relatively straightforward.

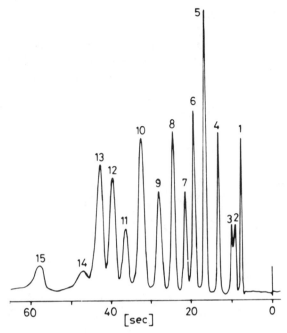

Figure 1.2 Rapid LC separation of aromatic hydrocarbons. Peaks 1, 2 are CH_2Cl_2 and $CHCl_3$, peak 3 is benzene, and peaks 4-15 are polycyclic aromatic hydrocarbons, ending with dibenzanthracene (peak 15). Column, 6.5 × 0.4 cm of porous silica (4.4 μm); mobile phase, *n*-pentane; ambient; velocity 0.93 cm/sec, $\Delta P = 1060$ psi; UV, 254 nm; 3-50 μg each compound. Reprinted from (8) with permission.

Modern LC also provides a major advance over the older LC methods in speed and separation power. In fact, LC now rivals GC in this respect. Figure 1.2 shows an example of the speed of modern LC: the separation of 15 aromatic hydrocarbons in 1 min., using a small-particle silica column. Figure 1.3 shows the separation of a urine sample into over 100 peaks in less than half an hour, using a small-particle reverse-phase column. Modern LC also features a number of new column packings that provide separations that were previously impossible. Figure 1.4 shows a chromatogram for a synthetic polymer sample, providing a rapid determination of the molecular-weight distribution of this polymer. Similar determinations by classical, physical methods required literally months of work, as compared to the 10 min. for the assay of Figure 1.4. Most important, all these advantages of modern LC are now routinely available with commercial LC equipment and supplies.

What we have called modern LC has been referred to by other names: high-performance or high-pressure LC (HPLC), high-speed LC, and simply liquid chromatography (LC). In the present book we refer to modern liquid chromato-

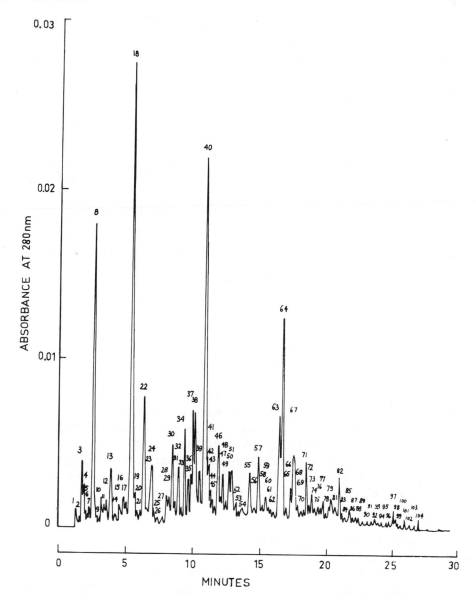

Figure 1.3 LC separation of acidified urine extract by reverse-phase small-particle column. Column, 25 × 0.46 cm LiChrosorb ODS (bonded-phase porous silica) (5 μm); mobile phase, gradient elution from 0.1 M phosphate in water (pH = 2.1) to 40 %v acetonitrile; temp., 70°; flowrate, 2.0 ml/min; detector, UV, 280 nm; sample, 10 μl injection from a 10-x concentrated urine extract. Reprinted from (9) with permission.

7

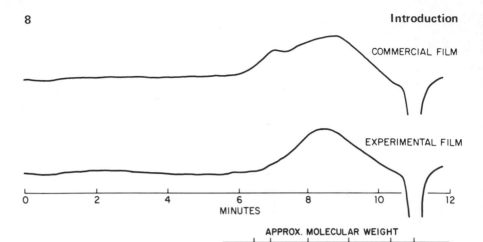

Figure 1.4 Rapid LC separation of cellulose hemi-formal for determination of molecular weight distribution (size exclusion chromatography). Columns, four 10 × 0.79 cm (in series): PSM 50S, 800 S, 1500 S and 4000 S; mobile phase, dimethyl sulfoxide; temp., 23°; flowrate, 1.0 ml/min (2000 psi); detector, refractive index, 5 × 10⁻⁵ RI units full-scale. Reprinted from (*10*) with permission.

graphy in columns as LC. Where high-pressure operation is to be contrasted with low-pressure LC, we use the term *HPLC* to define the usual technique that employs high pressure.

Recently, an improved version of TLC has been introduced (*1*), and referred to as high-performance TLC (HP-TLC). It has been implied that this new technique will displace modern LC from many of its present applications. This seems to us an overoptimistic assessment of the potential of HP-TLC (e.g., *la*, *lb*). However, TLC itself has proved to be a complementary technique that can be used effectively in conjunction with LC (see Chapter 9), and any improvement in HP-TLC will only increase the value of TLC in these applications.

1.3 HOW DID MODERN LC ARISE?

Modern LC is based on developments in several areas: equipment, special columns and column packings, and theory. High-pressure, low-dead-volume equipment with sensitive detectors plays a vital role. The new column packings that have been developed specifically for modern LC are also essential for rapid, high-performance separations. Theory has played a much more important role in the development of modern LC than for preceding chromatographic innovations. Fortunately, the theory of LC is not very different from that of GC, and a good understanding of the fundamentals of GC had evolved by the early 1960s [see, e.g., (2)]. This GC theory was readily extended to include LC, and this in turn led directly to the development of the first high-performance column packings for LC and the design of the first modern LC units. A proper understanding of how the different separation variables are selected for optimum performance is particularly important in modern LC; theory has been most useful in providing the necessary insights.

The potential advantages of modern LC first came to the attention of a wide audience in early 1969, when a special session on LC was organized as part of the Fifth International Symposium on Advances in Chromatography (3). However, modern LC had its beginnings in the late 1950s, with the introduction of automated amino acid analysis by Spackman, Stein, and Moore (4). This was followed by the pioneering work of Hamilton (5) and Giddings (2) on the fundamental theory of high-performance LC in columns, the work of C. D. Scott at Oak Ridge on high-pressure ion-exchange chromatography, and the introduction of gel permeation chromatography by J. C. Moore (6) and Waters Associates in the mid-1960s. At this point a number of workers began active research into what was to become modern LC, and their combined efforts [see (3) for a review] led to the 1969 breakthrough. Since 1969 tremendous activity has been aimed at the development of better equipment and columns and further improvements in our understanding of modern LC. These developments have now leveled off somewhat, so that it seems possible to write an account of LC that should remain useful and up-to-date for some time to come.

1.4 THE LITERATURE OF LC

The literature of LC extends back into the 1930s. Some of this information is still useful today, since the basis of retention or relative separation in LC is the same for classical LC, open-bed chromatography, and modern LC. A comprehensive review of all books on chromatography through 1969 is given in (7). However, the likelihood of finding uniquely useful information in the older literature declines each year, and the growing number of excellent review articles and books makes the search of the older literature increasingly less worthwhile.

Since 1970 over a dozen books on modern LC have appeared and they are reviewed in the Bibliography at the end of this chapter. There is no longer any "primary" journal for LC articles, since much work involving LC is now being reported in all the major chemistry journals, as well as in specialized periodicals devoted to areas such as agricultural, polymer, or biomedical chemistry. However, a high proportion of the important LC articles of general interest continue to appear in the *Journal of Chromatography*. Another journal that focuses on LC, the *Journal of Liquid Chromatography* (Marcel Dekker), made its initial appearance in 1978.

Table 1.1   Issues of the *Journal of Chromato-graphy* containing an LC bibliography section[a]

Vol. 107, No. 2 (1975)	Vol. 124, No. 1 (1976)
Vol. 108, No. 2 (1975)	Vol. 128, No. 2 (1976)
Vol. 109, No. 1 (1975)	Vol. 133, No. 2 (1977)
Vol. 110, No. 2 (1975)	Vol. 136, No. 3 (1977)
Vol. 111, No. 2 (1975)	Vol. 140, No. 2 (1977)
Vol. 114, No. 1 (1975)	Vol. 144, No. 3 (1977)
Vol. 114, No. 2 (1975)	Vol. 151, No. 1 (1978)
Vol. 121, No. 1 (1976)	Vol. 154, No. 1 (1978)

[a] These reviews are collected and organized every three years in so-called Supplementary Volumes of the *Journal of Chromatography*; for example, reviews for the period 1971-1973 are in Suppl. Vol. No. 6, 1976.

Continuing review of the LC literature is provided by the *Journal of Chromatography* in its bibliography section (see Table 1.1 for past issues containing these reviews), and by *Analytical Chemistry* in its biennial "Fundamental Reviews" (April in even years, under "Ion Exchange and Liquid Column Chromatography"). A number of abstract services for the LC literature are available and are summarized in Table 1.2.

The commercial literature offered by companies supplying LC equipment and materials is an important primary source of information, particularly for applications of LC. Several companies offering this material on a regular basis are given in Table 1.3. This literature is free and often uniquely useful.

Finally, the regular meetings of three major chromatography groups should be noted, as these meetings typically produce a substantial volume of high-quality and up-to-date material. Moreover, the papers for each meeting have been gathered into a single volume of the *Journal of Chromatography*. The three groups are

- International Symposia of Chromatography—this group is the former Gas Chromatography Discussion Group and its program still emphasizes GC rather than LC; the eleventh symposium was reported in Vol. 122 (1976) of the *Journal of Chromatography*.
- International Symposia on Advances in Chromatography—this group is similar to the preceding in emphasizing GC; the thirteenth symposium was reported in Vol. 158 (1978) of the journal.
- International Symposia on Column Liquid Chromatography—this group is dedicated entirely to LC; the third symposium was reported in Vol. 149 (1978) of the journal.

Table 1.2 LC abstract services.

#1 *CA Selects: High Speed Liquid Chromatography.* Chemical Abstracts Service, P. O. Box 3012, Columbus, Ohio (a current listing of all abstracts appearing in *Chemical Abstracts* that relate to LC; cost per year is $50; especially good for coverage of nonchromatographic journals)

Liquid Chromatography Abstracts. Science and Technology Agency, 3 Harrington Rd., S. Kensington, London, SW73ES (good coverage of out-of-the way articles, especially in European literature; cost per year is $78)

Liquid Chromatography Abstracts, Preston Technical Abstracts Co., P.O. Box 312, Niles, Ill. (tends to emphasize coverage of primary chromatographic literature and not to distinguish classical from modern LC; cost per year is $180)

Guide to the Literature of High Performance Liquid Chromatography (2nd printing with supplement) (1976), and *Scan of High Performance Liquid Chromatography 1976* (1977) D. L. B. Wetzel, Kansas State University, Manhattan, Kans., sold through Alltech Associates, Arlington Heights, Ill. (annual volumes listing abstracts and cross-referenced by area of application, column packing used, and so on; somewhat spotty coverage of literature, but useful and reasonable in price)

Liquid Chromatographic Data Compilation. AMD 41. American Society for Testing Materials, Philadelphia, Pa., 1975 (abstracts from *J. Chromatogr., J. Chromatogr. Sci., Anal. Chem.,* and *Anal. Biochem.* during period 1969-1972.

Cumulative Indexes 1969-1973, Gas and Liquid Chromatography Abstracts, C. E. Knapman, ed., Applied Science Publishers, Ripple Road, Barking, Essex, U.K. (index to LC abstracts published by Gas Chromatography Discussion group between 1969 and 1973; original abstracts somewhat limited in their coverage and discrimination between classical and modern LC articles)

Perkin-Elmer LC Bibliography Update, Perkin-Elmer Corp. Norwalk, Conn. (a key word index for various application areas, e.g., biochemistry, clinical chemistry, environmental analysis. The service costs $26 per year for each specific application area. For details, see (11) or the publisher)

Table 1.3 Some Commercial Sources of LC literature[a]

Altex (Beckman)	Spectra-Physics
Du Pont	Varian Aerograph
E M Laboratories	Waters Associates
Hewlett-Packard	Whatman
Perkin-Elmer	

[a] Addresses listed in Appendix I

These meetings are worth anticipating, either to attend or to read the papers presented.

1.5 ABOUT THE BOOK

The organization of the present book follows that of the first edition: basic principles are described first (Chapter 2), followed by equipment (Chapters 3,4), columns (Chapter 5), solvents (Chapter 6), and the individual LC methods (Chapters 7-12). Specialized techniques (Chapters 13-17), the selection and development of methods (Chapter 18), and troubleshooting (Chapter 19) are covered in the latter part of the book. We have tried to write the book in such a way that the inexperienced reader can start at the beginning and read straight through. Toward this end, advanced topics and special-interest areas are deferred until later in the book where possible. The experienced chromatographer can readily locate material of specific interest by means of the Index and Contents. In discussions of individual LC methods, some material applies equally to more than one method. To avoid repetition in these cases, the information is provided only once (usually in Chapter 7) and then referred to in later chapters. The description of LC conditions for each separation shown follows a standard format that is given in the figure caption for each chromatogram (e.g., Figure 1.2):

> Column length and internal diameter (cm), nature of packing with particle diameter in μm (see Chapter 5 for further description of packings); mobile phase composition; separation temperature ($^\circ$C); flowrate of mobile phase (ml/min) and pressure drop across column (psi); detection means (e.g., UV, fluorescent; see Chapter 4), wavelength(s) (nm) and full-scale attenuation (e.g., 0.1 abs. unit full-scale, abbreviated 0.1 AUFS); sample size (μl) and concentration.

Often all conditions for a separation are not reported by the original investigator, in which case certain items may be omitted from the description for a given chromatogram.

As in other books on chromatography, there is a general problem in this book of too many symbols. All symbols used in the present book are defined in a final List of Symbols. There, in addition to a definition of the particular term, an attempt is made to give a reference to a preceding defining equation and to give the units commonly used for that parameter. A short list of abbreviations and their meanings follows the list of symbols.

References for each chapter plus a list of recommended readings (Bibliography) are given at the end of each chapter.

REFERENCES

1. A. Zlatkis and R. E. Kaiser, eds., *High Performance Thin-layer Chromatography*, Elsevier, New York, 1976.

1a. S. Husain, P. Kunzelmann, and H. Schildknecht, *J. Chromatogr.*, **137**, 53 (1977).

1b. G. Guiochon, A. Souiffi, H. Engelhardt, and I. Halasz, *J. Chromatogr. Sci.*, **16**, 152 (1978).

2. J. C. Giddings, *Dynamics of Chromatography*, Dekker, New York, 1965.

3. A. Zlatkis, ed., *Advances in Chromatography, 1969*, Preston Technical Abstracts Co., 1969.

4. D. H. Spackman, W. N. Stein, and S. Moore, *Anal. Chem.*, **30**, 1190 (1958).

5. P. B. Hamilton, *Adv. Chromatogr.*, **2**, 3 (1966).

6. J. C. Moore, *J. Polymer Sci.*, A2, 835 (1964).

7. *J. Chromatogr. Sci.*, 8, D2 (July 1970).

8. I. Halasz, R. Endele, and J. Asshauer, *J. Chromatogr.*, **112**, 37 (1975).

9. I. Molnar and C. Horvath, *J. Chromatogr. (Biomed. App.)*, **143**, 391 (1977).

10. J. J. Kirkland, *J. Chromatogr.*, **125**, 231 (1976).

11. J. M. Attebery, R. Yost, and H. W. Major, *Am. Lab.*, 79 (October 1977).

BIBLIOGRAPHY

Basics of Liquid Chromatography, Spectra-Physics, Santa Clara, Calif., 1977 (a simple book with no unique features that recommend it over more complete textbooks on LC).

Bristow, P. A., *LC in Practice*, HETP Publ., 10 Langley Drive, Handforth, Wilmslow, Cheshire, U.K., 1976 (an excellent short book of the how-to-do-it type; contains few actual LC chromatograms).

Brown, P. R., *High Pressure Liquid Chromatography: Biochemical and Biomedical Applications*, Academic, New York, 1973 (applications oriented and now fairly out-of-date).

Charalambons, G., *Liquid Chromatographic Analysis of Foods and Beverages*, Vol. 1, Academic Press, New York, 1979.

Deyl, Z., K. Macek, and J. Janak, eds., *Liquid Column Chromatography*, Elsevier, New York, 1975 (a voluminous book covering both theory and practice, with emphasis on the older literature; important developments since 1973 are not covered).

Dixon, P. F., C. H. Gray, C. K. Lim, and M. S. Stoll, eds., *High Pressure Liquid Chromatography in Clinical Chemistry*, Academic, New York, 1976 (proceedings of an earlier conference on clinical LC; individual papers collected in this book are of uneven quality).

H. Engelhardt, *High Performance Liquid Chromatography,* Springer, Berlin, 1979.

Ettre, L. S., and C. Horvath, "Foundations of Modern Liquid Chromatography," *Anal. Chem.*, 47, 422A (1975) (a history of LC from its beginnings until the late 1950s).

Hadden, N., et al., *Basic Liquid Chromatography*, Varian Aerograph, Walnut Creek, Calif., 1971 (a very practical book on LC that is now largely out-of-date; emphasis on equipment from one manufacturer).

Hamilton, R. J., and P. A. Sewell, *Introduction to High Performance Liquid Chromatography*, Chapman & Hall, London (Wiley), 1978 (a good short book, which emphasizes applications; the latter are organized and indexed by sample type).

Johnson, E. L., and R. Stevenson, *Basic Liquid Chromatography*, 2nd ed., Varian Aerograph, Walnut Creek, Calif., 1978 (a very practical paperback on LC).

Kirkland, J. J., ed., *Modern Practice of Liquid Chromatography*, Wiley-Interscience, New York, 1971 (an excellent early book on LC that is now largely out-of-date; good mix of theory and practice).

Knox, J. N., J. N. Done, A. T. Fell, M. T. Gilbert, A. Pryde and R. A. Wall, *High-Performance Liquid Chromatography,* Edinburgh Univ. Press, Edinburgh, 1978 (a 212-page manual).

Parris, N. A., *Instrumental Liquid Chromatography: A Practical Manual*, Elsevier, New York, 1976 (a well-reviewed book aimed at the practical worker; Chap. 15 lists many LC applications under different areas: pharmaceutical, biochemical, food, pesticides, etc.).

Perry, S. G., R. Amos, and P. I. Brewer, *Practical Liquid Chromatography*, Plenum, New York, 1972 (out-of-date and contains little that is relevant today).

Pryde, A., and M. T. Gilbert, *Applications of High Performance Liquid Chromatography*, Halsted (Wiley), New York, 1978 (applications organized by compound class, e.g., lipids, carbohydrates, pesticides, etc.).

Rajcsanyi, P. M., and E. Rajcsanyi, *High Speed Liquid Chromatography*, Dekker, New York, 1975 (applications oriented but does not cover ion-exchange or size-exclusion chromatography).

Rivier, J., and R. Burgus, *Biological/Biomedical Applications of Liquid Chromatography*, Dekker, New York, 1978.

Rosset, R., M. Caude, and A. Jardy, *Manual Pratique de Chromatographie en Phase Liquide*, Varian, Orsay, 1975.

Scott, R. P. W., *Contemporary Liquid Chromatography,* Wiley-Interscience, New York, 1976 (a highly theoretical book that emphasizes the contributions of the author).

Simpson, C. F., ed., *Practical High Performance Liquid Chromatography,* Heyden, New York, 1976 (contains several excellent chapters by contributing authors, plus a final chapter that gives 21 LC experiments as a teaching aid).

Snyder, L. R., and J. J. Kirkland, *Introduction to Modern Liquid Chromatography*, 1st ed., Wiley-Interscience, New York, 1974 (Japanese translation, 1976).

Tsuji, K., and W. Morozowich, eds., *GLC and HPLC Determination of Therapeutic Agents,* Part 1, Dekker, New York, 1978 (a well-balanced and authoritative account of the determination of drugs by means of LC).

Walker, J. Q., M. T. Jackson, Jr., and J. B. Maynard, *Chromatographic Systems: Maintenance and Troubleshooting*, Academic, New York, 2d ed., 1977 (part of the book is devoted to troubleshooting LC systems and description of equipment; the remainder of the book covers GC systems).

TWO

BASIC CONCEPTS AND CONTROL OF SEPARATION

2.1	The Chromatographic Process	16
	Different LC Methods	21
2.2	Retention in LC	22
	Sample-Size Effects	26
2.3	Band Broadening	27
	Extracolumn Band Broadening	31
	Effect of Sample Size on Column Efficiency	33
2.4	Resolution and Separation Time	34
2.5	The Control of Separation	37
	Estimating R_s	38
	How Large Should R_s Be?	43
	How Many Compounds in the Sample?	43
	What Compounds Are Present?	43
	Quantitative Analysis	45
	Preparative Separation	48
	Controlling Resolution	49
	Resolution versus k'	51
	Resolution versus N	56
	A Rapid Scheme for Changing R_s via a Change in N	62
	Resolution versus α	73
	Automated Method Development	80
References		81
Bibliography		82

The successful use of LC for a given problem requires the right combination of operating conditions: the type of column packing and mobile phase, the length and diameter of the column, mobile-phase flowrate, separation temperature, sample size, and so on. Selecting the best conditions in turn requires a basic understanding of the various factors that control LC separation. In this chapter we review the essential features of liquid chromatography and apply this theory to the control of LC separation. In most cases this discussion is descriptive, rather than mathematical. The few important equations are indicated by enclosure within a box (e.g., Eq. 2.3).

2.1 THE CHROMATOGRAPHIC PROCESS

Figure 2.1 shows the hypothetical separation of a three-component mixture in an LC column. Individual molecules are represented by triangles for compound *A*, squares for compound *B*, and circles for compound *C*. Four successive stages in the separation are shown, beginning in (*a*) with application of the sample to a

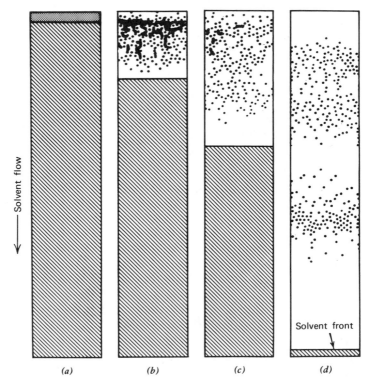

Figure 2.1 Hypothetical separation of a three-component sample: △ compound A: ■ compound B; ● compound C

dry column. In modern LC the column is always prewet prior to sample injection; however, a dry column is shown here to better illustrate the separation process (which occurs in the same manner for both wet and dry columns). In step (*b*) solvent or *mobile phase* begins to flow through the column, resulting in movement of sample molecules along the column and a partial separation of components *A*, *B*, and *C*. The movement of solvent through the column has proceeded further in step (*c*), and by step (*d*) the three compounds are essentially separated from each other.

By step (*d*) of Figure 2.1, we can recognize two characteristic features of chromatographic separation: differential migration of various compounds (solutes) in the original sample, and a spreading along the column of molecules of each solute. Differential migration in LC refers to the varying rates of movement of different compounds through a column. In Figure 2.1 compound *A* moves most rapidly and leaves the column first; compound *C* moves the slowest and leaves the column last. As a result, compounds *A* and *C* gradually become separated as they move through the column. Differential migration is the basis of separation in chromatography; without a difference in migration rates for two compounds, no separation is possible.

Differential migration in LC results from the equilibrium distribution of different compounds—such as *A*, *B*, and *C*—between particles or *stationary phase*

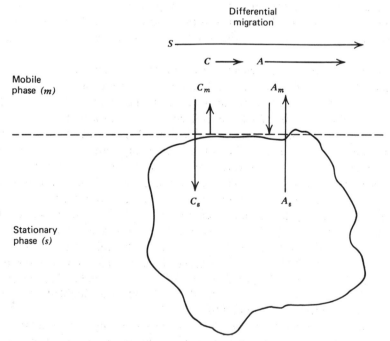

Figure 2.2 The basis of retention in LC.

and the flowing solvent or *mobile phase*. This is illustrated in Figure 2.2, for a single particle of column packing and compounds A and C. Compound C at equilibrium is present mainly in the stationary phase or particle, with only a small fraction of its molecules in the mobile phase at any given time. This situation is reversed for compound A, which is present mainly in the mobile phase. The speed with which each compound x moves through the column (u_x) is determined by the number of molecules of that compound in the moving phase at any instant, since sample molecules do not move through the column while they are in the stationary phase. Therefore, compounds such as C, whose molecules spend most of the time in the stationary phase, move through the column rather slowly. Compounds such as A, whose molecules are found in the mobile phase most of the time, move through the column more rapidly ($u_a > u_c$). Molecules of the solvent or mobile phase (S in Figure 2.2) move through the column at the fastest possible rate (except in size-exclusion chromatography; see Chapter 12).

Differential migration or the movement of individual compounds through the column depends on the equilibrium distribution of each compound between stationary and mobile phases. Therefore, differential migration is determined by those experimental variables that affect this distribution: the composition of the mobile phase, the composition of the stationary phase, and the separation temperature. When we want to alter differential migration to improve separation, we must change one of these three variables. In principle, the pressure within the column also affects equilibrium distribution and differential migration. In fact, pressure effects are negligible at the usual column pressures [i.e., 500-3000 psi; see (*1*)].

Consider next the second characteristic of chromatographic separation: the spreading of molecules along the column for a given compound, such as A in Figure 2.1. In Figure 2.1*a* molecules of A (triangles) begin as a narrow line at the top of the column. As these molecules move through the column, the initial narrow line gradually broadens, until in (*d*) molecules of A are spread over a much wider portion of the column. It is apparent, therefore, that the average migration rates of individual molecules of A are not identical. These differences in molecular migration rate for molecules A do not arise from differences in equilibrium distribution, as in Figure 2.2. Rather, this spreading of molecules A along the column is caused by physical or rate processes. The more important of these physical processes are illustrated in Figure 2.3. In Figure 2.3*a* we show a cross section of a top of the column (with individual particles numbered from 1 to 10). Sample molecules are shown as X's at the top of the column, that is, just after injection. At this point these molecules form a fairly narrow line, as measured by the vertical, double-tipped arrow alongside the X's (this is the "initial band width").

In Figure 2.3*b* we illustrate one of the processes leading to molecular spreading: *eddy diffusion* or multiple flowpaths. Eddy diffusion arises from the dif-

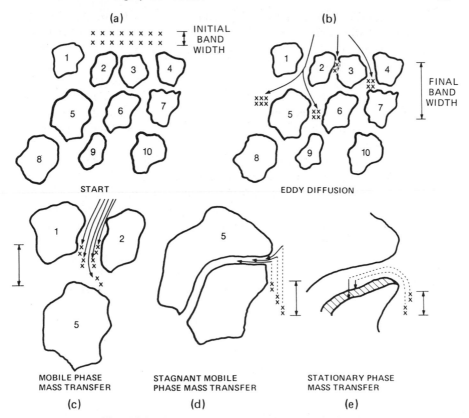

Figure 2.3 Contributions to molecular spreading in LC.

ferent microscopic flowstreams that the solvent follows between different particles within the column. As a result, sample molecules take different paths through the packed bed, depending on which flowstreams they follow. These different flowpaths are illustrated in (*b*) by the various arrows between the particles. Liquid moves faster in wide paths and slower in narrow paths. Thus the flowpath between particles 1 and 2 (or 5 and 6) is relatively wide, the solvent velocity is therefore greater, and molecules following this path will have moved a greater distance down the column in a given time. Molecules that follow the narrow path between particles 2 and 3, on the other hand, will move more slowly. These molecules therefore progress down the column a shorter distance in a given time. Thus, as a result of this eddy diffusion phenomenon in (*b*), we see a spreading of molecules from the initial narrow line in (*a*) to a broader portion of the column (see arrow in (*b*) which is the "final width"). This spreading becomes progressively greater as flow of solvent through the column continues.

A second contribution to molecular spreading is seen in Figure 2.3c: *mobile-phase mass transfer*. This refers to differing flow rates for different parts of a single flowstream or path between surrounding particles. In (c), where the flowstream between particles 1 and 2 is shown, it is seen that liquid adjacent to a particle moves slowly or not at all, whereas liquid in the center of the flowstream moves fastest. As a result, in any given time, sample molecules near the particle move a short distance and sample molecules in the middle of the flowstream move a greater distance. Again this results in a spreading of molecules along the column.

Figure 2.3d shows the contribution of *stagnant mobile-phase mass transfer* to molecular spreading. With porous column-packing particles, the mobile phase contained within the pores of the particle is stagnant or unmoving. In d we show one such pore, for particle 5. Sample molecules move into and out of these pores by diffusion. Those molecules that happen to diffuse a short distance into the pore and then diffuse out, return to the mobile phase quickly, and move a certain distance down the column. Molecules that diffuse further into the pore spend more time in the pore and less time in the external mobile phase. As a result, these molecules move a shorter distance down the column. Again there is an increase in molecular spreading.

In Figure 2.3e is shown the effect of *stationary-phase mass transfer*. After molecules of sample diffuse into a pore, they penetrate the stationary phase (cross-hatched region) or become attached to it in some fashion. If a molecule penetrates deep into the stationary phase, it spends a longer time in the particle and travels a shorter distance down the column—just as in (d). Molecules that spend only a little time moving into and out of the stationary phase return to the mobile phase sooner, and move further down the column.

Finally, there is an additional contribution to molecular spreading not illustrated in Figure 2.3: *longitudinal diffusion*. Whether the mobile phase within the column is moving or at rest, sample molecules tend to diffuse randomly in all directions. Apart from the other effects shown in Figure 2.3, this causes a further spreading of sample molecules along the column. Longitudinal diffusion is often not an important effect, but is significant at low mobile-phase flowrates for small-particle columns. Section 5.1 examines the dependence of each of these contributions to molecular spreading on experimental conditions and gives ways in which molecular spreading can be minimized in practice for improved separation.

Eventually the various compounds reach the end of the column and are carried off to the detector, where their concentrations are recorded as a function of separation time. The resulting chromatogram for the hypothetical separation of Figure 2.1 is shown in Figure 2.4. Such a chromatogram can be characterized by four features that are important in describing the resulting separation. First, each compound leaves the column in the form of a symmetrical, bell-shaped *band* or

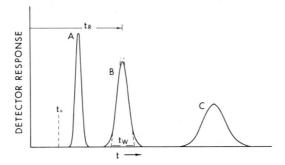

Figure 2.4 The resulting chromatogram. The time t_0 marks the elution of any unretained sample bands.

peak (a Gaussian, or standard-error, curve). Second, each band emerges from the column at a characteristic time that can be used to identify the compound, just as a melting point can be used for the qualitative analysis of an organic compound. This *retention time t_R* is measured from the time of sample injection to the time the band maximum leaves the column. A third characteristic feature is the difference in retention times between adjacent bands: for example, t_R for compound C minus t_R for compound B. The larger this difference is, the easier is the separation of the two bands. Finally, each band is characterized by a *band width t_w*, as shown for band B in Figure 2.4. Tangents are drawn to each side of the band and extended to touch the *baseline* (detector signal for zero sample concentration). Separation is better for narrower bands and smaller values of t_w.

Different LC Methods

Four separate mechanisms or processes exist for retention of sample molecules by the stationary phase. These in turn give rise to four basic LC methods: *liquid-liquid* (Chapter 8), *liquid-solid* (Chapter 9), *ion-exchange* (Chapter 10) and *size-exclusion* chromatography (Chapter 12). Liquid-liquid or *partition* chromatography involves a liquid stationary phase whose composition is different from that of the moving liquid phase. Sample molecules distribute between the mobile and stationary liquid phases, just as in liquid-liquid extraction within a separatory funnel. The moving- and stationary-phase liquids must be immiscible. Liquid-solid or *adsorption* chromatography involves high-surface-area particles, with retention of sample molecules occurring by attraction to the surface of the particle. In ion-exchange chromatography the stationary phase contains fixed ionic groups such as $-SO_3$, along with counter-ions of opposite charge (e.g., Na^+). The counter-ions are also normally present in the mobile phase in the form of a salt (e.g., NaCl). Ionic sample molecules (e.g., X^+) are retained by ion exchange:

$$X^+ + -SO_3^-Na^+ \rightleftharpoons Na^+ + -SO_3^-X^+.$$

Finally, in *size-exclusion* or gel chromatography the column packing is a porous material with pores of a certain size. Molecules that are too large are excluded from all the pores, whereas small molecules can penetrate most of the pores. Thus very large molecules move through the column quickly, and smaller molecules are retained by the packing. Usually, separation in size-exclusion chromatography is determined strictly by molecular size.

Two additional LC methods result from modification of liquid-liquid chromatography: *bonded-phase* chromatography (BPC) (Chapter 7) and *ion-pair* chromatography (IPC) (Chapter 11). Bonded-phase chromatography uses an organic stationary phase that is chemically bonded to the particle, in place of the mechanically held liquid phase used in liquid-liquid chromatography (LLC). The exact mechanism of retention of sample molecules on BPC packings is at present controversial. However, from a practical standpoint LLC and BPC can be regarded as similar in terms of retention mechanism, since similar separations are obtained with LLC and BPC, and the effect of mobile-phase composition on sample retention is similar for both methods.

Ion-pair chromatography can be regarded as a combination of LLC (or BPC) and IEC methods. The exact retention mechanism for IPC is also not yet settled, particularly as this technique can be carried out with either a mechanically held liquid stationary phase or a bonded phase. Chapters 7-12 describe the characteristic features and use of each of these six LC methods. Each method has its own advantages and disadvantages, and therefore each method has been found uniquely useful for one kind of application or another. Chapter 18 summarizes how one or more of these LC methods is selected for a given problem.

2.2 RETENTION IN LC

Up to this point our discussion of LC has been qualitative, rather than quantitative. It is possible to reduce many of the foregoing concepts to precise mathematical relationships, and we begin with band migration rate and differential migration. Consider first the average velocity within the column of molecules of solvent S (u, cm/sec) and some sample band X (u_x). From the discussion of Figure 2.2 it is seen that u_x depends on R, the fraction of molecules X in the moving phase, and upon u. Specifically,

$$u_x = uR \tag{2.1}$$

If the fraction of molecules X in the moving phase is zero ($R = 0$), no migration can occur and u_x is zero. If the fraction of molecules X in the moving phase is unity (i.e., all molecules X are in the moving phase, $R = 1$), then molecules X move through the column at the same rate as solvent molecules and $u_x = u$. Thus, R is also the relative migration rate of compound X.

R can be expressed in another manner. We begin by defining a fundamental LC parameter: the *capacity factor* k', equal to n_s/n_m. Here n_s is the total moles of X in the stationary phase and n_m is the total moles of X in the mobile phase. We can now write

$$k' + 1 = \frac{n_s}{n_m} + \frac{n_m}{n_m}$$

$$= \frac{n_s + n_m}{n_m}.$$

Since

$$R = \frac{n_m}{n_s + n_m} = \frac{1}{1 + k'}, \tag{2.2}$$

from Eq. 2.1,

$$u_x = \frac{u}{1 + k'}. \tag{2.2a}$$

The quantity u_x can be related to retention time t_R and column length L in terms of the relationship: time equals distance divided by velocity. In this case the time is that required for band X to traverse the column, t_R (sec); the distance is the column length L (cm); and the velocity is that of band X, u_x (cm/sec):

$$t_R = \frac{L}{u_x}.$$

Similarly, the time t_0 for mobile-phase (or other unretained) molecules to move from one end of the column to the other is

$$t_0 = \frac{L}{u}.$$

Eliminating L between these last two equations then gives

$$t_R = \frac{u t_0}{u_x},$$

which with Eq. 2.2a yields

$$t_R = t_0 (1 + k').$$ (2.3)

Here we have t_R expressed as a function of the fundamental column parameters t_0 and k'; t_R can vary between t_0 (for $k' = 0$) and any larger value (for $k' > 0$). Since t_0 varies inversely with solvent velocity u, so does t_R. For a given column, mobile phase, temperature, and sample component X, k' is normally constant for sufficiently small samples. Thus, t_R is defined for a given compound X by the chromatographic system, and t_R can be used to identify a compound tentatively by comparison with a t_R value for a known compound in the same LC system.

Rearrangement of Eq. 2.3 gives an expression for k':

$$k' = \frac{t_R - t_0}{t_0}.$$ (2.3a)

To plan a strategy for improving separation, it is often necessary to determine k' for one or more bands in the chromatogram. Equation 2.3a provides a simple, rapid basis for estimating k' values in these cases; k' is equal to the distance between t_0 and the band center, divided by the distance from injection to t_0. For example, for band B in Figure 2.4, $k' = 2.2$. In the same way, values of k' for bands A and C in this chromatogram can also be estimated (band A, $k' = 0.7$; band C, $k' = 5.7$). Often we require only a rough estimate of k', in which case k' can be determined by simple inspection of the chromatogram, without exact calculation.

The important column parameter t_0 can be measured in various ways. In most cases the center of the first band or baseline disturbance (following sample injection; e.g., Figure 1.2, $t_0 = 8$ sec, or Figure 2.24, $t_0 = 1.5$ min) marks t_0. If there is any doubt on the position of t_0, a *weaker* solvent (or other nonretained compound) can be injected as sample, and its t_R value will equal t_0. A weaker solvent provides larger k' values and stronger sample retention than the solvent used as mobile phase (see Chapters 6-11 for lists of solvents according to strength). For example, if chloroform is used as mobile phase in liquid-solid chromatography, then hexane (see Table 9.4) might be injected as sample to measure t_0. As long as the flowrate of mobile phase through the column remains unchanged, t_0 is the same for any mobile phase. If flowrate changes by some factor x, t_0 will change by the factor $1/x$.

Retention in LC is sometimes measured in volume units (ml), rather than in time units (sec). Thus the *retention volume* V_R is the total volume of mobile phase required to elute the center of a given band X; that is, the total solvent flow in the time between sample injection and the appearance of the band center at the detector. The retention volume V_R is equal to the retention time t_R times the volumetric flowrate F (ml/sec) of mobile phase through the column:

$$\boxed{V_R = t_R F.}$$ (2.4)

Similarly, the total volume of solvent within the column V_m is equal to F times t_0:

$$V_m = t_0 F.$$

Eliminating F between the preceding two equations gives

$$V_R = V_m \frac{t_R}{t_0}$$
$$= V_m (1 + k').$$ (2.4a)

It is sometimes preferable to record values of V_R, rather than t_R. One reason is that t_R varies with flowrate F, whereas V_R is independent of F. V_R values are thus more suitable for *permanent* column calibrations, as in size-exclusion chromatography (Chapter 12). This is particularly true where values of V_R must be known quite precisely and the flowrate provided by a given pump is not highly accurate.

So far we have ignored the effect on retention volume of the volume V_s of stationary phase within the column. The quantity k' equals n_s/n_m, where $n_s = (X)_s V_s$ and $n_m = (X)_m V_m$. Here $(X)_s$ and $(X)_m$ refer to the concentrations (moles/1) of X in the stationary and mobile phases, respectively. Therefore,

$$k' = \frac{(X)_s V_s}{(X)_m V_m}$$
$$= \frac{K V_s}{V_m}.$$ (2.5)

Here $K = (X)_s/(X)_m$ is the well-known *distribution constant*, which measures the equilibrium distribution of X between stationary and mobile phases. From Eq. 2.5, k' is proportional to V_s, and therefore in liquid-liquid chromatography the k' values of all sample bands change with the relative loading of support by stationary phase. It follows then that *pellicular* or surface-coated particles (see Figure 6.1a) provide smaller values of k', other factors being equal, because such particles have much smaller values of V_s than porous particles.

Sample-Size Effects

The general effect of varying sample size in LC is illustrated in Figure 2.5 and is further discussed in Section 7.4. For sufficiently small samples (*a*), peak height increases with increasing sample size, but retention times are not affected and relative separation remains the same (so-called *linear isotherm* retention). At some critical sample size (*b*), a noticeable decrease in retention time occurs for one or more sample bands, and for further increase in sample size (*c* and *d*) there is usually a rapid loss of separation and a further decrease in all retention-time values.

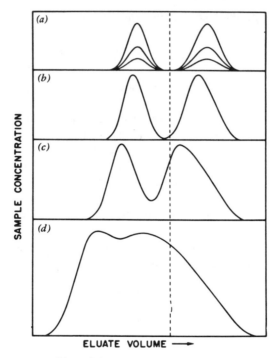

Figure 2.5   Sample size effects in LC.

The same effects are illustrated for a single sample component in Figure 2.6, where k' values are plotted versus sample size (weight sample/weight stationary phase). At a certain sample size $\theta_{0.1}$ (the *linear capacity* of the stationary phase or column) values of k' (or V_R) are reduced by 10% relative to the constant k' values observed for smaller samples. In the example of Figure 2.6 this column packing (adsorbent) has a linear capacity of 0.5 mg of sample per gram of column packing. When the linear capacity of the column is exceeded, sample retention

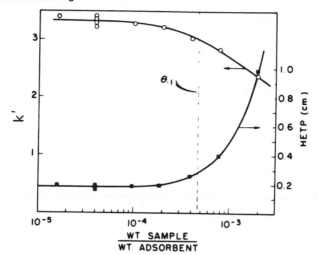

Figure 2.6 Linear capacity in LC (data for 66-μm water-deactivated silica) Reprinted from (8) with permission.

time data can no longer be used for qualitative analysis (comparison with t_R values of known compounds). Quantitative analysis is also greatly complicated, and separation can be severely degraded.

For analytical separations it is always preferable to work within a sample size range where k' values are constant; that is, sample sizes should be less than the linear capacity of the column. However, in preparative separations (see Chapter 15) the linear capacity of the column is often deliberately exceeded so as to allow maximum sample throughput (maximum weight of separated sample per unit time). Chapters 7-12 discuss the maximum allowable samples sizes for the individual LC methods. Because pellicular packings (coated glass beads) have a low concentration of stationary phase per particle, their linear capacities are normally lower by a factor of 5 or more, relative to porous particles.

As sample size exceeds the linear capacity of a given column, the resulting bands become distorted and generally asymmetrical. Band asymmetry or tailing can arise from causes other than column overloading, although such tailing should be negligible in well-designed LC systems. For a full discussion of the measurement, evaluation, and correction of band tailing see Sections 5.2 and 19.2.

2.3 BAND BROADENING

Band width t_w in LC is commonly expressed in terms of the *theoretical plate number N* of the column:

$$N = 16(\frac{t_R}{t_w})^2.$$ (2.6)

The quantity N is approximately constant for different bands in a chromato-gram, for a given set of operating conditions (a particular column and mobile phase, with mobile-phase velocity and temperature fixed). Therefore, N is a useful measure of *column efficiency*: the relative ability of a given column to provide narrow bands (small values of t_w) and improved separations.

Since N remains constant for different bands in a chromatogram, Eq. 2.6 predicts that band width will increase proportionately with t_R, and this is gener-ally found to be the case (e.g., Figures 1.2 and 2.24). Minor exceptions to the constancy of N for different bands exist (e.g., see Refs 2, 3), and in gradient e-lution chromatography (Chapter 16) all bands in the chromatogram tend to be of equal width.

Because of the widening of LC bands as retention time increases, later-eluting bands show a corresponding reduction in peak height, and eventually disappear into the baseline. This effect can be seen in Figure 2.24e for peak 5, which is barely visible because of its large t_R value. The quantity N is proportional to column length L, so that—other factors being equal—an increase in L results in an increase in N and better separation. This proportionality of N and L can be expressed in terms of the equation

$$N = \frac{L}{H},$$ (2.6a)

where H is the so-called *height equivalent of a theoretical plate* (plate height) or HETP value. The quantity H (equal to L/N) measures the efficiency of a given column (operated under a specific set of operating conditons) per unit length of column. Small H values mean more efficient columns and large N values. A central goal in LC practice is the attainment of small H values for maximum N and highest column efficiencies.

A detailed discussion of column efficiency (H and N values) as a function of different experimental variables is given in Sections 5.1 and 5.5, as well as in references cited there. In general,

- H is smaller for small particles of column packing, and for low mobile-phase flowrates.
- H is smaller for less viscous mobile phases, and for higher separation tempera-tures.
- H is smaller for small sample molecules.

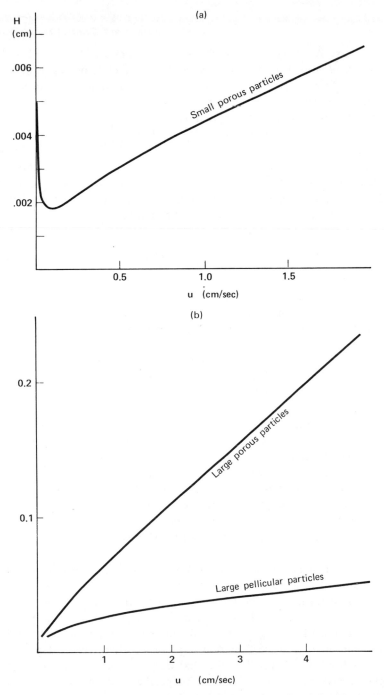

Figure 2.7 Characteristic *H* Versus *u* Plots. Calculated from Eq. 2.7 with representative values of *A*, *B*, and *C*. (*a*) 8-μm porous particles, (*b*) 40-μm porous particles and 40-μm pellicular particles.

29

Therefore, large N values and improved separation are usually favored by long columns packed with small particles, by nonviscous mobile phases flowing relatively slowly through the column, and by higher separation temperatures. Furthermore, the separation of macromolecules such as proteins is often more difficult than the separation of compounds with molecular weights below 1000.

As we see in the discussion on controlling resolution in Section 2.5, the dependence of column H values on mobile-phase velocity u is of great practical importance. A knowledge of this relationship is the key to understanding and controlling separation after a particular column packing has been selected for a given LC separation or application. It is possible to describe the dependence of H on u by means of the following equation (Section 5.1):

$$H = Au^{0.33} + \frac{B}{u} + Cu. \tag{2.7}$$

The constants A, B, and C vary from one column to another, and also depend to some extent on the sample, mobile-phase, and separation temperature. The resulting plots of H versus u exhibit a characteristic shape, which differs for (a) small, porous particles, (b) large, porous particles, and (c) large, pellicular particles. Figure 2.7a, b shows representative plots for each of these three cases, which we return to in our discussion of the control of separation in Section 2.5.

The form of the H versus u plot (as in Figure 2.7) has diagnostic value, and can be used to determine whether a given column is operating properly. Therefore, it is useful to construct such a plot for every new column before the column is placed into use. An H versus u plot also provides a reference point against which the column can be compared at some later time if it is believed that column performance has changed. To measure H versus u it is desirable to select at least two sample compounds or solutes: one unretained ($k' = 0$) and one retained ($1 < k' < 3$) for the particular mobile phase to be used. Values of H for each compound should then be determined by injecting the mixture several times, varying u after each injection. The total change in u from slowest to fastest flow of mobile phase should be at least tenfold.

The accurate, manual measurement of N values via Eq. 2.6 is somewhat difficult, as opposed to using a computer. Whatever technique is used to calculate N from a recorder tracing, the speed of the chart paper should be fast enough to give accurately measurable values of t_w and t_R. The drawing of tangents to a band for determination of t_w is an added source of error, particularly for inexperienced chromatographers. Therefore, a preferred alternative to Eq. 2.6 is to measure the band width at half-height; that is, determine the midpoint between baseline and peak maximum and draw a line parallel to the baseline through this point. If the width of the band at half-height is $t_{w\frac{1}{2}}$, then N is given as

$$N = 5.54\left(\frac{t_R}{t_{w-\frac{1}{2}}}\right)^2. \qquad (2.6b)$$

For further discussion of the measurement of N values, particularly for non-symmetrical bands, see Section 5.2.

Extracolumn Band Broadening

So far we have been discussing the band broadening that takes place within the column. In a well-designed LC system this intracolumn band broadening almost entirely determines the widths of final bands, as they pass through the detector. However, *extracolumn band broadening* takes place to some extent in every LC apparatus and can represent a serious problem in certain situations. The result of extracolumn band broadening in a particular case is illustrated in Figure 2.8. Here the separation of three compounds by the original LC system is shown in Figure 2.8*a*. It is seen in (*a*) that the bands are well separated. In Figure 2.8*b* the separation of (*a*) has been repeated after insertion of an extra length of

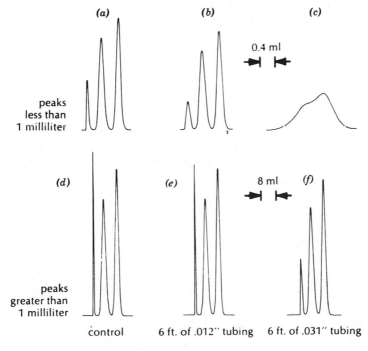

Figure 2.8 Extracolumn band broadening. Effect of added tubing between column and detector on separation. Reprinted from (9) with permission.

narrow-diameter tubing between the column and detector. Some increase in the widths of the resulting bands is apparent, but the separation is still adequate. In Figure 2.8c the length of narrow-diameter tubing is replaced by a wider-diameter tube, and now the extracolumn band broadening has increased the widths of the three bands to the point where the three compounds are poorly separated.

Extracolumn band broadening as illustrated in Figure 2.8 can arise from one of two sources: (1) increased volume in the LC system between sample injection and detector outlet (as by adding the larger-volume tube in Figure 2.8c), and (2) an increase in the volume of the injected sample. In either case the effect on the width of a final sample band can be expressed in terms of the *volume* of the sample band as a result of intracolumn broadening alone: $V_p = t'_w F$, where t'_w is the value of t_w that would be observed in the absence of extracolumn effects. If extracolumn processes lead to an analogous broadening of otherwise narrow sample bands, giving a final band volume V_{ec}, then the observed band volume in the detector ($V_w = t_w F$) is given as

$$(V_w)^2 = (V_p)^2 + (V_{ec})^2. \tag{2.8}$$

Since we are interested in the *relative* increase in V_w or t_w versus the case of no extracolumn effects, we can rewrite Eq. 2.8 as

$$\left(\frac{V_w}{V_p}\right) = [1 + (\frac{V_{ec}}{V_p})^2]^{1/2}. \tag{2.8a}$$

According to Eq. 2.8a, the relative increase in sample band width as a result of extracolumn broadening will be smaller for a band whose volume V_p (in the absence of extracolumn broadening) is larger. This has three important consequences. First, more efficient columns give narrower sample bands and smaller values of V_p. Therefore, for larger N or smaller H values (by whatever means), extracolumn effects become more important (see Section 5.5, Table 5.5). Thus, an LC system that provides no appreciable extracolumn band broadening with one column can exhibit an extracolumn problem with a more efficient column. Second, since t_w and V_p increase with t_R (Eq. 2.6), extracolumn broadening is always more severe with early-eluting bands than with later-eluting compounds. Finally, the use of wider, longer columns gives larger values of V_p (other factors being equal), which again reduces the importance of extracolumn band broadening. The latter effect is illustrated in Figure 2.8d-e, where the starting bands (d) now have greater volumes (values of V_p). In this case addition of the same length of wide-diameter tubing [in (f)] has relatively little effect on the resulting separation [compare (f) with (c)].

If there are several different contributions to extracolumn band broadening, with V_{ec} contributions V_1, V_2, and so on, then the overall effect on V_w is given as an extension of Eq. 2.8:

$$(V_w)^2 = (V_p)^2 + V_1{}^2 + V_2{}^2 + \cdots . \qquad (2.8b)$$

A further discussion of these various effects and their impact on final sample band widths is given in following chapters: tube connectors between sample injector and column or between column and detector (Section 3.2), volume of the detector cell (Section 4.2), and sample volume (Section 7.4).

The form of Eqs. 2.8 and 2.8b arises from the Gaussian curves that describe individual sample bands, and the assumption that these different band-broadening processes (intra- and extracolumn) are independent of each other. These relationships are also described in terms of so-called peak *variances* σ^2, where σ is the standard deviation of the various Gaussian curves corresponding to V_w. V_1, V_2, and so on, in Eq. 2.8b. Thus, Eq. 2.8b can be rewritten in terms of variances as

$$(\sigma_w)^2 = (\sigma_p)^2 + (\sigma_1)^2 + (\sigma_2)^2 + \cdots . \qquad (2.8c)$$

For a Gaussian curve, $\sigma = V_w/4$. However, this variance formulation is not limited to Gaussian curves alone, which makes it both general and powerful as a means of treating various chromatographic phenomena. For a more detailed discussion of extracolumn band broadening, see Refs. 2 and 4-6.

Effect of Sample Size on Column Efficiency

As the volume and/or concentration of the injected sample is increased, column plate numbers N eventually decrease, because of overloading of the column. This effect is illustrated in Figure 2.6 for a column of large-particle silica. As sample size increases beyond 3×10^{-4} g/g of column packing, values of H become significantly larger—with a resulting decrease in N. In Figure 2.6 this increase in H occurs at about $\theta_{0.1}$, the linear capacity of the column (see the discussion of sample-size effects in Section 2.2). For the more efficient, small-particle columns in use today, it is found that column N values are more sensitive to sample size than are sample t_R values; see discussion in (7). For example, the linear capacity of a small-particle silica column might be 100-300 μg/g, but N values show a significant decrease at column loadings above 5-10 μg/g. Where this potential loss in column efficiency at higher loadings is important, the user should establish the dependence of both N and t_R values on sample size. In this way a maximum sample load can be established for each LC application.

In some cases larger amounts of sample can be injected without lowering column N values, if the sample is injected as a larger volume of more dilute solution. However, this is true only if the total volume injected is less than about one-third the volume of the first peak of interest (see Section 7.4). For more details on sample size and column efficiency, see (7).

2.4 RESOLUTION AND SEPARATION TIME

The usual goal in LC is the *adequate* separation of a given sample mixture. In approaching this goal we must have some quantitative measure of the relative separation or *resolution* achieved. The resolution R_s of two adjacent bands 1 and 2 is defined equal to the distance between the two band centers, divided by average band width (see Figure 2.9):

$$R_s = \frac{(t_2 - t_1)}{(\tfrac{1}{2})(t_{w1} + t_{w2})}. \tag{2.9}$$

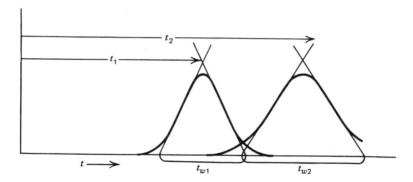

Figure 2.9   Resolution in LC.

The quantities t_1 and t_2 refer to the t_R values of bands 1 and 2, and t_{w1} and t_{w2} are their t_w values. When $R_s = 1$, as in Figure 2.9, the two bands are reasonably well separated; that is, only 2% of one band overlaps on the other. Larger values of R_s mean better separation, and smaller values of R_s poorer separation, as illustrated in Figure 2.10. For a given value of R_s, band overlap becomes more serious when one of the two bands is much smaller than the other.

The parameter R_s of Eq. 2.9 serves to *define* separation. To *control* separation or resolution we must know how R_s varies with experimental parameters such as k' and N. We next derive such a relationship for two closely spaced (i.e., overlapping) bands. Equation 2.3 gives $t_1 = t_0 (1 + k_1)$ and $t_2 = t_0 (1 + k_2)$, where k_1 and k_2 are the k' values of bands 1 and 2. Since $t_1 \approx t_2$, from Eq. 2.6

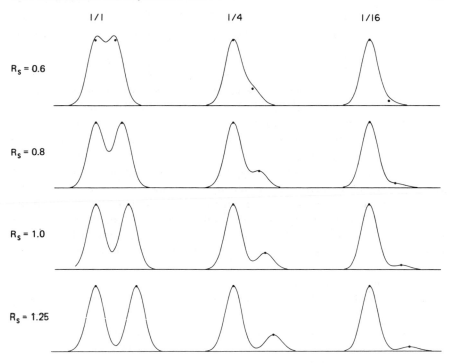

Figure 2.10 Separation as a funciton of R_S and relative band concentration (peak-height ratio).

we have $t_{w1} \approx t_{w2}$ (N assumed constant for bands 1 and 2). Inserting these relationships into Eq. 2.9 then gives

$$R_s = \frac{t_0 (k_2 - k_1)}{t_{w1}}. \tag{2.9a}$$

Similarly, Eq. 2.6 for band 1 gives $t_{w1} = 4t_1/\sqrt{N} = 4t_0 (1 + k_1)/\sqrt{N}$. Inserting this expression for t_{w1} into Eq. 2.9a gives

$$R_s = \frac{(k_2 - k_1)\sqrt{N}}{4(1 + k_1)}$$

$$= (\tfrac{1}{4}) (\frac{k_2}{k_1} - 1) \sqrt{N} (\frac{k_1}{1 + k_1}). \tag{2.9b}$$

If we define the *separation factor* $\alpha = k_2/k_1$ for bands 1 and 2, and recognize that $k_1 \approx k_2 =$ the average value k' for the two bands, Eq. 2.9b becomes

$$R_s = (\tfrac{1}{4})\,(\alpha - 1)\,\sqrt{N}\,[\frac{k'}{1 + k'}]\ .$$

$$\qquad\quad\text{(i)}\quad\ \text{(ii)}\quad\ \text{(iii)}$$

$$(2.10)$$

Equation 2.10 is a fundamental relationship in LC that allows us to control resolution by varying α, N, or k'. The three terms (i-iii) of Eq. 2.10 are roughly independent, so that we can optimize first one term, then another. *Separation selectivity* as measured by α (term i of Equation 2.10) is varied by changing the composition of the mobile and/or stationary phases. *Separation efficiency* as measured by N (term ii) is varied by changing the column length L or solvent velocity u (or by switching to a different column packing). Term iii, involving the capacity factor k', is varied by changing *solvent strength*—the ability of the mobile phase to provide large or small k' values for a given sample (see Chapters 6-11 for a listing of different solvents according to strength for each LC method). The further use of Eq. 2.10 for controlling separation is discussed in Section 2.5.

A change in the assumptions used to derive Eq. 2.10 (i.e., use of t_{w2} for t_{w1} in Eq. 2.9a) leads to a slightly different expression, in which term i is replaced by $(\alpha - 1)/\alpha$. For the usual case of closely adjacent bands ($\alpha \approx 1$), these two expressions for R_s are not sufficiently different to merit concern.

Consider now the time t required to carry out an LC separation. It is desirable that t be as small as possible, both for convenience and to allow a maximum number of separations or analyses per day. In Section 2.6 we see that separation time and resolution are interrelated, so that difficult separations require longer times. For the moment, however, let us concentrate on t apart from sample resolution. The analysis time t is approximately equal to the retention time t_R for the last band off the column. For a two-component separation as in Figure 2.9, therefore,

$$t \approx t_2 = t_0(1 + k'_2).$$

$$(2.11)$$

Since k' for the last-eluted band in LC is typically 3-20, this means that separation times are typically 4-20 times t_0. The column dead time t_0 can in turn be related to column length and mobile-phase velocity: $t_0 = L/u$. A well-known relationship gives u as a function of column conditons (see Refs. 10 and 11 for details):

$$u = \frac{K^\circ P}{\eta L}$$

$$= \frac{P d_p^{\,2}}{\phi \eta L}\ .$$

$$(2.12)$$

Here K° is the so-called column permeability and ϕ is the flow resistance factor; P is the pressure drop across the column, η is the viscosity of the mobile phase, and d_p is the particle diameter of the packing material. If all the parameters of Eq. 2.12 are expressed in cgs units (see List of Symbols), the flow resistance parameter ϕ has roughly the following values for different column packings:

pellicular* packings: $\phi = 250$
spherical porous packings: $\phi = 500$
irregular porous packings: $\phi = 1000$

This means that column permeability increases in the reverse order, with pellicular packings being the most permeable. For more commonly used units (t_0 in sec, η in cP, L in cm, P in psi, and d_p in μm), Eq. 2.12 can be rewritten

$$t_0 = \frac{15000\,L^2\eta}{Pd_p^2 f}. \qquad (2.12a)$$

Here the quantity f assumes the values 1, 2, or 4, for irregular-porous, spherical-porous, or pellicular packings, respectively.

Equation 2.12 can be used to check column permeability, which should be done for each new column. If much larger values of ϕ are found than predicted by this relationship, the cause of this discrepancy should be investigated. The column may be poorly packed, a column frit may have compacted, or the tubing between column and detector may be partially plugged by particles that have leaked out of the column. In any case the problem should be corrected immediately. A reduction in column permeability effectively reduces column performance in some applications, resulting in longer times for a given separation. Column permeability should also be checked from time to time during use, to ensure that it has not changed as a result of column plugging or other problems.

Equation 2.12a can also be used to estimate the time required for a given separation or the maximum pressure that will develop for a particular choice of separation conditions.

2.5 THE CONTROL OF SEPARATION

A strategy or approach to the design of a successful LC separation can be broken down into the following seven steps:

1. Select one of the six LC methods (Chapter 18).

*See Figure 6.1a.

2. Select or prepare a suitable column (Chapter 5).
3. Select initial experimental conditions (Chapters 7-12).
4. Carry out an initial separation.
5. Evaluate the initial chromatogram and determine what change (if any) in resolution is required.
6. Establish conditions required for the necessary final resolution.
7. If required, solve special problems or use special techniques as necessary (Chapters 13-17).

At this point we assume we are at step 5; that is, we have a given method and column type and have made an initial separation. This section describes how to carry out steps 5 and 6 by taking advantage of concepts discussed earlier in this chapter. We also present some additional equations as we proceed. However, our final strategy for achieving steps 5 and 6 is a simple (and rapid) "table-lookup" operation that avoids mathematical calculation.

Estimating R_s

The first step following the initial separation (step 4) is to evaluate the chromatogram in quantitative terms. This means we must estimate R_s for two or more

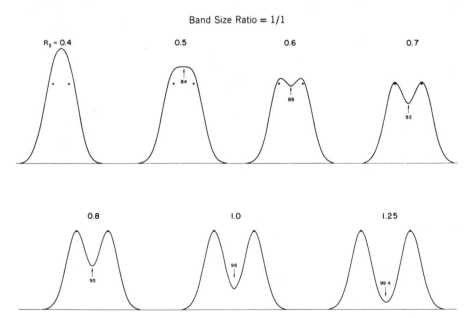

Figure 2.11 Standard resolution curves for a band-size ratio of 1/1 and R_s values of 0.4-1.25. Reprinted from (*12*) with permission.

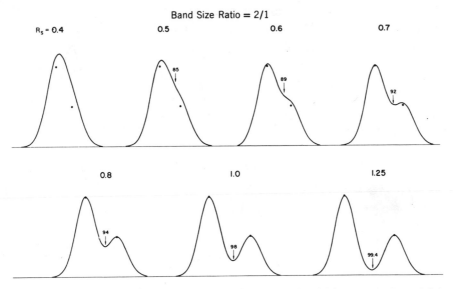

Figure 2.12 Standard resolution curves for a band-size ratio of 2/1 and R_S values of 0.4-1.25. Reprinted form (*12*) with permission.

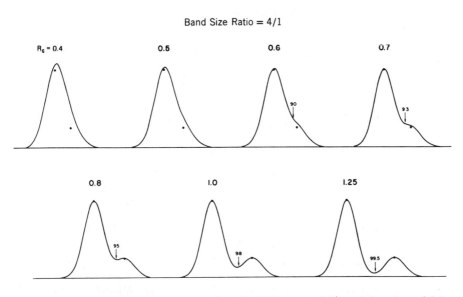

Figure 2.13 Standard resolution curves for a band-size ratio of 4/1 and R_S values of 0.4-1.25. Reprinted from (*12*) with permission.

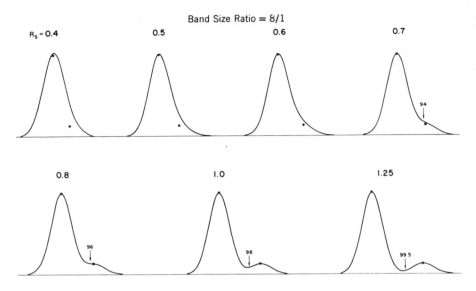

Figure 2.14 Standard resolution curves for a band-size ratio of 8/1 and R_s values of 0.4-1.25. Reprinted from (*12*) with permission.

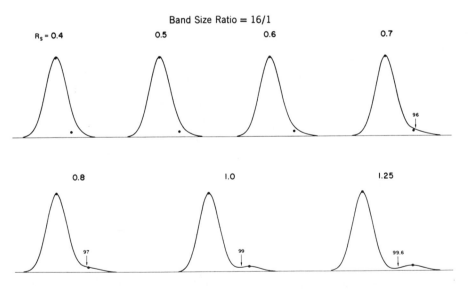

Figure 2.15 Standard resolution curves for a band-size ratio of 16/1 and R_s values of 0.4-1.25. Reprinted from (*12*) with permission.

adjacent bands. For the moment, assume that only two sample bands are involved. Shortly we will consider how to extend the R_s concept to include a complex chromatogram that contains several bands. The present approach to estimating R_s is based on comparison with a standard set of resolution curves, shown in Figures 2.11–2.16. In each case (e.g., Figure 2.11) examples are given of different values of R_s, beginning with R_s = 0.4 (almost complete overlap) through R_s = 1.25 (almost complete separation). In Figure 2.11 the two bands in each example are of equal height (1/1 ratio). In subsequent figures (2.12–2.16) the band-height ratio changes to 2/1, 4/1, 8/1, . . . , 128/1. Figures 2.11–2.16 allow R_s to be estimated for an actual pair of bands from an initial separation, by visually comparing these bands with one of the examples in these figures. Some examples are given later to show how this works. Since the essence of this approach to estimating resolution is convenience and speed, it is helpful to have

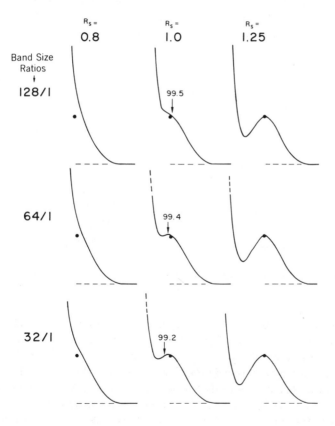

Figure 2.16 Standard resolution curves for band-size ratios of 32/1, 64/1, and 128/1 and R_s values of 0.8-1.25. Reprinted from (*12*) with permission.

enlarged copies of Figures 2.11–2.16 [see, e.g., *(12)*] mounted on the wall be-
side the LC unit. Estimates of R_s can then be made on the spot by comparing
the chromatogram on the recorder with the R_s curves on the wall.

We now estimate R_s for each of the six hypothetical chromatograms in Figure
2.17. In Figure 2.17*a* the heights of the two bands are in an approximate ratio of
2/1, which directs us to Figure 2.12. There we see a good match with the band-
pair for R_s = 0.7. Therefore, R_s = 0.7 for Figure 2.17*a*. Similarly, in Figure
2.17*b* the band-height ratio is about 8/1, and in Figure 2.14 there is a good
match for R_s = 1.0. Examples Figure 2.17*c* and *d* differ in that the smaller band
elutes first, rather than last. In these cases it is necessary to find a match with
the mirror image of the actual chromatogram; that is, to mentally reverse the
two bands. This gives peak *(c)*, 4/1 peak-height ratio and R_s = 0.7; peak *(d)*, 2/1
ratio, R_s = 0.5. Figure 2.17*e* involves two bands of widely different heights, so
much so that one of the two bands is off-scale. Matches for this case can be
found in Figure 2.16; either R_s = 1.25 for a ratio of 128/1 or R_s = 1.0 for a ratio
of 32/1. The accuracy of R_s estimates is reduced when one of the two bands is
off-scale.

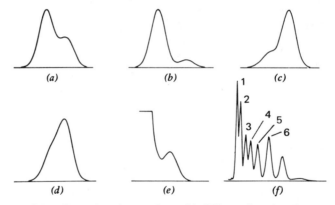

Figure 2.17   Examples of separations with different R_s values (see text).

A similar approach is used to determine the resolution of a complex chromatogram, such as in Figure 2.17*f*. The R_s value of such a chromatogram can be defined in one of two ways, depending on which bands are of interest. Where every
band in the chromatogram may be important, the two adjacent bands with the
poorest resolution (smallest R_s value) determine R_s for the entire separation. In
Figure 2.17*f* either bands 1–2 or 3–4 have the lowest value of R_s. Comparison
of either band pair with Figures 2.11 or 2.12 suggests R_s = 0.8, which is therefore the value for the entire chromatogram.

Another possibility is that not every band in a chromatogram is of interest. For example, assume that we are concerned only with band 5 in (f). In this case R_s for the chromatogram is determined by the pair of bands, 4–5 or 5–6, that is most poorly separated. For band 5 in (f) resolution is poorest for the pair 4–5 and R_s for this pair and the chromatogram is estimated at 1.0 (Figure 2.11). For further discussion of this approach to estimating R_s, see (12).

How Large Should R_s Be?

Having estimated R_s for the initial separation, the next step is to determine how large R_s must be to achieve the immediate separation goal. This in turn depends on which of several possible goals of LC separation is to be pursued.

How Many Compounds in the Sample? We may be interested only in knowing how many compounds are present in a sample, as when we examine a compound or sample for purity. Or we may wonder if a particular band in the chromatogram is really two severely overlapping bands, rather than a single compound. In either case we are faced with a band that we suspect (but do not know) is really two severely overlapping bands.

The nature of the problem is illustrated by the example for $R_s = 0.5$ in Figure 2.13. A noticeable tail appears to the right of the band center, but it is conceivable that this might represent band tailing (as in Section 19.2) rather than two overlapping bands. To differentiate between overlapping or tailing bands, R_s should be increased from 0.5 to 0.7 (see Figure 2.13). If the band in question is really two overlapping bands, this change in separation conditions will partially resolve the two bands; that is, an obvious shoulder as in $R_s = 0.7$ (Figure 2.13) will appear. If, on the other hand, a single component plus band tailing is involved, there will be little change in band shape as a result of changing R_s from 0.5 to 0.7.

Similarly, we might also question whether Figure 2.17d represents one distorted (reverse-tailing) band or two overlapping bands. Here the presence of two bands appears more likely, but we must keep in mind that peculiar band shapes are occasionally observed in LC. To be sure in this case, R_s should again be increased from 0.5 (Figure 2.12) to about 0.7.

What Compounds Are Present? More often we are interested in the identification or qualitative analysis of the various components of a sample. The simplest way of identifying an unknown sample band is to compare its t_R value with that of some known compound suspected to be present in the sample. When the retention times of the unknown and known compounds are identical, this is a good indication that the two compounds are one and the same. However, the resolution of two bands affects the accuracy with which their t_R values can be

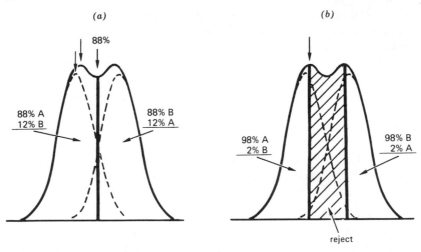

Figure 2.18 Increasing the purity of recovered fractions by rejecting a center fraction.

measured, as illustrated in Figure 2.18. Here two overlapping bands are shown (dashed curves), along with the composite curve (solid line) that would be seen in the actual chromatogram. The arrows above the first band in Figure 2.18*a* indicate the apparent and true positions of the band center, and therefore the apparent and true values of t_R for this band. From this example we see that poor resolution can result in the displacement of the apparent band center toward the overlapping band, with a resulting error in the measurement of t_R. The standard resolution curves (Figures 2.11-2.16) provide information on whether this is a problem in a given case. For each R_s example two black dots are superimposed on top of the two overlapping bands. These dots indicate the true positions of the two band centers. By comparing the horizontal position of the dot with the apparent t_R value of the overlying band, we can immediately see how close the apparent band center is to the true band center and whether R_s needs to be increased for an accurate measurement of t_R. As an example, consider Figure 2.17*c*. Comparing this example with Figure 2.13, we see that R_s must be increased from 0.7 to 0.8 for an accurate t_R measurement of the minor (second) band.

In many cases an accurate retention time measurement alone does not identify the unknown band. Either no known compound has the same t_R value or more than one compound has a retention time close to that of the unknown. For positive identification, the unknown band must be isolated in sufficient purity to allow its subsequent identification by the usual analytical techniques (e.g., infrared, mass spectrometry). Isolation and characterization are also recommended for the positive confirmation of bands whose preliminary identification has been made by comparison of t_R values (Chapter 14).

The standard resolution curves (Figures 2.11-2.16) can be used as guides in obtaining fractions of an unknown compound in some required purity. On these curves (for higher values of R_s) are seen numbered arrows between the two band centers, for example, 92 for R_s = 0.7 in Figure 2.12. The arrows in each case indicate the cutpoint that divides the two bands into fractions of *equal purity*, with the number giving the percent purity of each of the resulting two fractions. In the latter example a cutpoint at the arrow would produce two fractions of 92% purity. In Figure 2.18*a* the equal-purity cutpoint produces a fraction that is 88% *A* and 12% *B*, as well as a corresponding fraction that is 88% *B* and 12% *A*. Thus, the standard resolution curves immediately indicate whether a given separation can yield fractions of the required purity, and, if not, by now much R_s must be increased. For example, assume we require 98% pure fractions for infrared characterization, and have Figure 2.17*a* in our initial separation. Comparison with Figure 2.12 indicates R_s = 0.7 and the equal-purity cutpoint will give fractions of 92% purity. However, Figure 2.12 also shows that an increase of R_s to 1.0 will provide fractions of the necessary 98% purity.

The estimation as above of fraction purity and equal-purity cutpoints assumes that the detector response or sensitivity is equal for each of the two overlapping compounds; that is, equal quantities of each compound would give equal areas (and approximately equal peak heights) in the final chromatogram. In LC, unlike gas chromatography, *this is seldom exactly the case*. However, the detector responses for two overlapping bands are often similar, so that the error introduced by the use of Figures 2.11-2.16 for fraction purity is usually minor. In any case, in the absence of detailed information on the detector response of the two compounds, no better estimates of purity and cutpoint can be made. The equal-sensitivity assumption is generally better for refractometer detectors than for photometric detectors (cf. example given in Figure 15.13).

We should also note that the purity of recovered fractions can be increased *without* increasing R_s if yield can be sacrificed. As illustrated in Figure 2.18*b*, rejection of an intermediate fraction (cross-hatched area) increases the purity of the outside fractions from 88 to 98%.

Quantitative Analysis. Generally we are interested in how much of one or more compounds is present in the sample. The concentrations of sample components can be determined on the basis of (a) band height, (b) band area, or (c) band recovery plus weighing. Band height or area methods usually assume a proportionality between height or area and the concentration of a given sample component. This proportionality constant is determined by injecting known concentrations of the sample of interest and constructing a calibration curve (e.g., a plot of band height vs. concentration for a compound of interest). Figure 2.19 illustrates the measurement of band heights (h_1, h_2) and band areas (cross-hatched area for band 2). For overlapping bands, as in this example, it is customary to divide the areas of the two bands by means of the perpendicular from the valley (the line h_r in Figure 2.19).

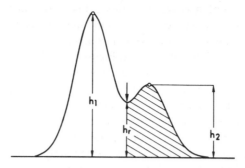

Figure 2.19 Quantitation by band height or area.

Which method we use—band height or area—depends on several factors, as discussed fully in Section 13.2. For the moment we want to focus on the effect of resolution on each type of quantitation. That is, what value of R_s is required for accurate quantitation by band height or area? The effect of resolution on the accuracy of band-height measurements is shown in the standard resolution curves (Figures 2.11-2.16). Since the dark dots in each case mark the true positions of the underlying band centers, a vertical displacement of a dot from the apparent band center is a measure of the error introduced by poor resolution. Thus in Figure 2.13 (4/1) we see that a resolution of 1.0 is required for an accurate band-height measurement; lesser values of R_s would lead to apparent band heights that are too large. As a rough guide to the overall effect of resolution on the accuracy of band-height quantitation, we can say the following: If $R_s = 1$, relative band heights can be varied from 32/1 to 1/32, with an accuracy of better than ±3%; that is, the relative concentrations of the two bands can change by an overall factor of about 1000.

Consider next how poor resolution affects the accuracy of band-area measurements. Shown in Figure 2.20 are the measured (cross-hatched) areas of the mi-

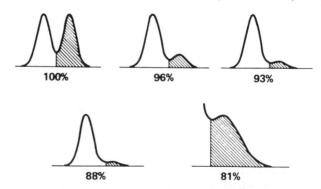

Figure 2.20 Error in quantitation by band area (100% equals no error in minor band).

nor band for R_s = 1 and several different ratios of the heights of two adjacent bands. The measured area of the minor band is expressed here as a percentage of its *true* area (i.e., as in the absence of band overlap). As the relative concentration of the minor band decreases, we see a decreasing accuracy in the measured area of the minor band (100% = no error). The error in the measured area of the major band is always positive and is never larger than 1%.

If the peak-height ratio of two overlapping bands is x ($x>1$), the errors in the measured *band areas* are related as,

$$\% \text{ error in major band} = -(\frac{1}{x})\% \text{ error in minor band.}$$

Thus, in Figure 2.20 for a peak-height ratio of 16/1, the area error of the small band is -12%. Therefore, the area error in the major band is (+12/16)% = +0.8%. Figure 2.20 and the preceding discussion together show that poor resolution has a much greater effect on band area than on band height. Thus, for R_s = 1.0 and an accuracy of at least ±3%, it is possible to vary band heights from 32/1 to 1/32. Band area, on the other hand, cannot be varied by more than 3/1 to 1/3 (Figure 2.20). This means, for comparable accuracy, that resolution must be better for quantitating by band area than by band height.

The error introduced by poor resolution in quantitation by band area can be estimated from the height of the valley h_r relative to the height of the minor band h_2, as in Figure 2.20:

h_r/h_2	Relative error in apparent area (minor band) (%)
0.25	-1
0.4	-2
0.6	-5
0.75	-10

Thus, in Figure 2.13 we can estimate the following errors in the measured areas of the minor bands: greater than 10% for R_s = 0.8, about 3% for R_s = 1.0, and less than 1% for R_s = 1.25.

When the quantity of a sample component injected onto the column is large enough, quantitation can be achieved by collecting the band, removing the solvent by evaporation, and weighing. This avoids the need for calibration, and it is particularly attractive when dealing with the one-time analysis of unknown mixtures (e.g., a competitor's product). It is obviously less attractive for the repetitive, quantitative analysis of similar samples, because it is more tedious than the band-height or band-area methods. The selection of cutpoints for band collection and the error introduced by poor resolution parallel the use of band-area quantitation. Most of what has been said concerning band-area quantitation applies equally to quantitation by recovery and weighing.

Preparative Separation. In preparative separations we are concerned with component purity and recovery. Estimating the purity of isolated fractions was discussed earlier in connection with qualitative analysis. The fractional recovery that can be expected for a given band can be estimated from the standard resolution curves plus Figure 2.21. The procedure can be illustrated by using Figure 2.17*a*. First determined the values of R_s and of the band-height ratios: For (*a*) $R_s = 0.7$ and band-height ratios are 2/1 for the major band and 1/2 for the minor band. Figure 2.12 indicates that the purity of each fraction for an equal-purity cutpoint is 92%. For this same equal-purity cutpoint, the recovery of each band can now be estimated from Figure 2.21. Thus, the 2/1 curve for a purity of 92% shows a recovery of 96%. This is the recovery of the major (2/1) band. The recovery of the minor band is about 82% (1/2 curve in Figure 2.21 and 92% puri-

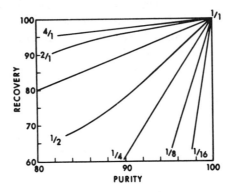

Figure 2.21 Estimating the recovery of each band (compound), using the equal-purity cutpoint.

ty). This example also illustrates a general principle: For equal purity fractions the recovery of the minor band in its fraction will be *less* than the corresponding recovery of the major band.

In determining the resolution needed for a given separation, for whatever purpose, a margin of safety is generally desirable. For example, if we decide that R_s should equal about 1.0, we might aim for an actual value of $R_s \approx 1.1$. This recognizes several contingencies: (a) Our estimate of the required change in R_s may be slightly in error, (b) our prediction (see the following discussion of controlling resolution) of the experimental conditions required for the new value of R_s will be only approximate, and (c) column performance tends to degrade with time, leading to a decrease in R_s.

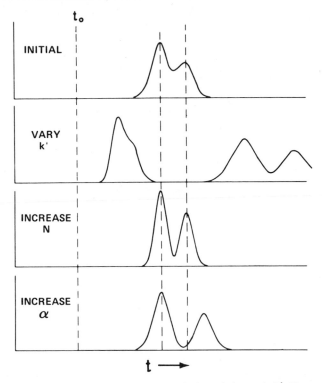

Figure 2.22   The effect on sample resolution of change in k', N or α.

Controlling Resolution

Equation 2.10 is the key to controlling resolution in LC:

$$R_s \;=\; (\tfrac{1}{4})\,(\alpha - 1)\,\sqrt{N}\,\left[\frac{k'}{(1 + k')}\right].\qquad\qquad(2.10)$$

$$\text{(i)}\quad\;\text{(ii)}\quad\;\text{(iii)}$$

Terms (i)-(iii) of Eq. 2.10 can each be varied to improve resolution, as illustrated in Figure 2.22. First, an increase in the separation factor α results in a displacement of one band center, relative to the other, and a rapid increase in R_s. Separation time and the heights of the two bands are not much changed for moderate changes in α. Second, an increase in the plate number N results in a narrowing of the two bands and an increase in band height; again separation time is not direct-

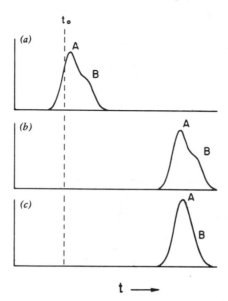

Figure 2.23 Different resolution problems, requiring different separation strategies.

ly affected.* Last, a change in k' can have a dramatic effect on separation, as shown in Figure 2.22. If k' for the initial separation falls within the range $0.5 < k' < 2$, a decrease in k' leads to a rapid worsening of separation, and an increase in k' can provide a significant increase in resolution. However, as k' is increased, band heights rapidly decrease and separation time increases.

In this section the effect on resolution of terms (i)-(iii) is examined in greater detail. We also show how each of these terms can be varied for a predictable change in R_s. First, however, consider how each of these three terms relates to an overall strategy for improving separation. Three different examples of poor resolution are illustrated in Figure 2.23. When R_s must be increased and k' for the initial separation is small (Figure 2.23a), k' should first be increased into the optimum range $1 \leqslant k' \leqslant 10$. *No other change in separation conditions provides as large an increase in R_s for as little effort.*

When k' is already within the optimum range of values, and resolution is still marginal (Figure 2.23b), the best solution is usually an increase in N. Normally,

*This assumes that the weight ratio of sample to column packing is unchanged. Also, if N is increased by change in u or L, then separation time *will* increase for increase in N (as shown later in this section).

we use a small-particle ($\leqslant 10$ μm) column, and N is increased by increasing column length L or by decreasing mobile-phase velocity u, rather than by changing to another type of column (see later discussion). In these cases separation time usually must be increased. However, the necessary change in experimental conditions is easily predicted (see below), and little effort is normally spent in achieving the required increase in N and R_s.

The final chromatogram (Figure 2.23c) shows two bands in the optimum k' range, but with R_s quite small. Here the necessary increase in N would normally require a very long separation time, and in some cases ($\alpha = 1$) the separation could not be achieved by any increase in N. In this situation an increase in α is needed. Since an increase in α can also be used to simultaneously *increase R_s* and *decrease* separation time (see end of this section), varying α is useful in any application where separation time is important (e.g., a routine assay that will be applied to a large number of samples).

Predicting the right conditions for the necessary change in α is seldom a straightforward procedure, and it can involve much effort before an adequate increase in α is achieved. Although an increase in α can provide truly optimum separations, the effort required to discover the right experimental conditions may represent a greater investment than can be justified for a given application. Thus, in the example of Figure 2.23b, a change in α would be unattractive when only a few samples are to be separated. However, an increase in α would be appropriate if many samples were to be analyzed by a given procedure.

The preceding discussion stresses the importance of resolution and separation time. The general goal is adequate resolution with minimum separation time, but also with minimum effort in the initial development of an LC separation. Later in this section we see that separation time can always be reduced if it is possible to reduce R_s. Alternatively, it is usually unwise to attempt an increase in R_s much beyond the value required for a particular application, because separation time is then unnecessarily long. These seemingly conflicting considerations involving R_s, separation time, and developmental effort require practical compromise in the design of each LC separation. Further examples in this section illustrate how to proceed in particular situations.

Resolution versus k'. The effect of k' on resolution is given by term (iii) of Eq. 2.10. Note that term (iii) $= k'/(1 + k')$ is the fraction of sample molecules X in the stationary phase (see Eq. 2.2). Differential migration requires the preferential retention of sample molecules in the stationary phase, that is, values of term (iii) different from zero. Since resolution R_s in turn depends on the extent of differential migration for two adjacent bands, it is intuitively reasonable that R_s is proportional to $k'/(1 + k')$. The latter varies with k' as follows:

k'	$k'/(1 + k')$ (proportional to R_s)
0	0
1	0.5
2	0.67
5	0.83
10	0.91
∞	1.00

When k' is initially small (< 1), R_s increases rapidly with increase in k'. For values of k' greater than 5, however, R_s increases very little with further increase in k'. Furthermore, separations that involve k' values greater than about 10 result in long separation times and excessive band broadening (Eq. 2.6) to the point where detection becomes difficult. Therefore, there is an optimum range of k' values, in terms of resolution, separation time, and band detection. We see from the preceding tabulation of values that this optimum k' range is roughly $1 \leqslant k' \leqslant 10$.

We have observed earlier that k' values in LC are controlled by means of solvent strength. When k' values must be increased, a so-called *weaker solvent* is used; *stronger solvents* are used to reduce k' values. This is illustrated in Figure 2.24 for the separation of a mixture of anthraquinones on a bonded-phase column. In this series of reverse-phase separations, solvent strength is greater for pure methanol (smaller k' values) than for pure water (larger k' values). Intermediate mixtures of these two solvents have intermediate solvent strengths. In Figure 2.24a it is seen that 70% methanol/water as mobile phase is much too strong, resulting in low k' values and poor separation. In Figure 2.24e 30% methanol/water is much too weak; that is, separation time is too long (60 min), and band 5 has widened to the point where it is barely detectable. Mobile phase consisting of 50% methanol/water (Figure 2.24c) is a good compromise; resolution of all five components is adequate, band heights are reasonable, and the separation time is less than 10 min. The approach illustrated in Figure 2.24 is the way solvent strength is usually optimized for a given separation. Two pure solvents, one too strong and the other too weak, are selected for a given application. Then various blends of the two solvents are tried as mobile phase, until k' values in the right range are obtained for the sample of interest. Chapters 7-11 discuss the relative solvent strengths of various pure liquids for each of the LC methods, thus allowing the rapid selection of two solvents that can be blended. For size-exclusion chromatography (Chapter 12), solvent composition has no effect on band migration. With this method the stationary phase must be varied to achieve changes in relative band migration.

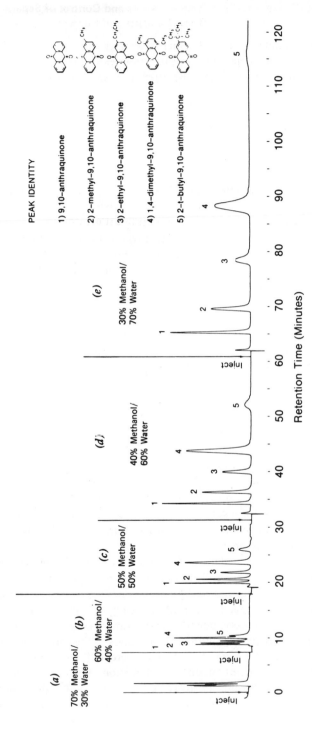

Figure 2.24 The effect of solvent strength on resolution. Conditions as stated in figure. Reprinted from (13) with permission.

PEAK IDENTITY

1) 9,10-anthraquinone
2) 2-methyl-9,10-anthraquinone
3) 2-ethyl-9,10-anthraquinone
4) 1,4-dimethyl-9,10-anthraquinone
5) 2-t-butyl-9,10-anthraquinone

53

Consider next the optimum value of k', such that resolution per unit time for a two-component sample is a maximum. In this connection it is useful to define another measure of column performance: the *effective plate number* N_{eff}.

$$N_{eff} = N(\frac{k'}{1 + k'})^2.\qquad(2.13)$$

Equation 2.10 can be rewritten in terms of N_{eff}:

$$R_s = (¼)(\alpha - 1)\sqrt{N_{eff}}.\qquad(2.13a)$$

Since α and k' are roughly independent, the optimum value of k' (for maximum R_s per unit time) can be obtained by maximizing N_{eff} with respect to k'. With this as our objective, we first substitute $L/H = N$ (Eq. 2.6a) into Eq. 2.13. If u is held constant, then H will remain constant. The column length L is in turn given (Eq. 2.11 and $t_0 = L/u$) as

$$L = \frac{tu}{1 + k'},\qquad(2.13b)$$

where t is both the value of t_R for the second band (as in Figure 2.9) and the time of separation. Combining Eqs. 2.13 and 2.13b then gives

$$N_{eff} = \left(\frac{tu}{H}\right)\left[\frac{k'^2}{(1 + k')^3}\right].$$

$$\text{(i)}\qquad\text{(ii)}$$

For a given separation with u and t constant, term (i) is constant and term (ii) and N_{eff} have a maximum value for $k' = 2$. If we allow solvent velocity u to vary, a similar analysis (see *16*) yields an optimum value of k' between 3 and 5. Since N_{eff} does not vary much with k' over the range $1 < k' < 5$, *the latter range of k' values provides optimum column performance* in terms of maximum resolution per unit time for two-component samples.

As a final example of the importance of k' in affecting resolution, consider the so-called *general elution problem* illustrated in Figure 2.25a. This shows the separation of a multicomponent mixture by normal or *isocratic* (equal-strength) elution from a silica column; that is, without changing solvent strength during the separation. Here the various components of the sample have a wide range of k' values; as a result there is poor resolution of the bands in the early part of the chromatogram (bands 1-5 in Figure 2.25a), and there is broadening of later bands (bands 10, 11) to the point of difficult detection. The problem here is that the k' values for the first bands are too small, whereas the k' values for the last bands are too large. No change of solvent (e.g., as in Figure 2.24) can solve

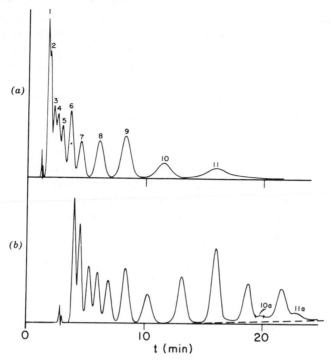

Figure 2.25 The general elution problem and the use of gradient elution. (*a*) Normal (iso-cratic) elution, (*b*) gradient elution of mixture of substituted aromatic compounds. Columns, (*a*) 100 × 0.28 cm and (*b*) 175 × 0.28 cm of Porasil 60 (35-75 μm silica); mobile phases, (*a*) *n*-pentane and (*b*) gradient); ambient; 425 psi; UV detection. Reprinted from (*14*) with permission.

this problem. The k' values of individual bands differ widely for any given mobile phase (by about 50-fold in Figure 2.25*a*), and for this sample all bands cannot elute in the optimum range of $1 < k' < 10$.

One solution to the general elution problem is to change conditions during separation so as to optimize k' for individual bands as they move through the column. This can be achieved in a number of ways, as is discussed in Chapter 16. The most common technique for solving the general elution problem is the use of *gradient elution*, as illustrated in Figure 2.25*b*. Here the same sample is eluted as in Figure 2.25*a*, from the same column, but with the composition of the mobile phase varied *during* the separation. Separation begins with a weak solvent, which elutes early bands (#1-5) with k' values near their optimum values. Mobile-phase composition is then changed during separation to successively stronger solvent mixtures, so that later bands elute with k' values that are in the optimum range. Normally, gradient elution is carried out by continuously changing the composition of a binary solvent mixture that enters the column. For example, a gradient can be formed with methanol/water to elute the compounds shown in Figure 2.24; gradient elution might begin with 30% methanol/water as mobile phase, with subsequent increase in the percentage of methanol as separation proceeds. Note that the various bands in Figure 2.25b are of roughly equal width. This is characteristic of gradient elution separations, unlike the isocratic separations of Figures 2.24 or 2.25a.

Resolution versus N. The effect of N on resolution is given by term (ii) of Eq. 2.10. The quantity N has been related to experimental conditions in Sections 2.3 and 5.5; see also (15). However, in this section we want to consider the interrelationship of N, separation time, column length L, and column pressure P. This treatment assumes that short lengths (e.g., 25-cm each) of a given column type are available and that these column lengths can be combined in series to give any desired overall length L (e.g., 50, 75, or 100-cm). In practice, this is the way that column length is normally adjusted in modern LC.

We begin with some examples of different ways in which N can be varied, and then we go on to discuss which of these techniques is appropriate for a given separation problem. Finally, we present a simple, rapid scheme for predicting the necessary change in separation conditions for a change in R_s from an initial value R_1 to a final desired value R_2. The latter scheme requires no calculations, nor even any understanding of the underlying principles on which the scheme is based.

Assume that we start with short column lengths whose performance (H vs. u) is given by the plot for the 8-μm packing shown in Figure 2.7a, page 29. The first entry in Table 2.1 gives various separation parameters that might apply in our initial separation: a 25-cm column, a pressure drop across the column of 1500 psi, a separation time of 180 sec, and a value of $u = 0.75$ cm/sec. From these data we obtain a value of $H = 0.0037$ from Figure 2.7a, which then permits $N = L/H = 25/0.0037 = 6750$ plates to be calculated. We now consider three different ways in which this initial value of N (6750) can be increased to a required final value of 10,000 plates.

The second example of Table 2.1 assumes that we increase N by simply decreasing column pressure P by the factor 0.43, holding L constant. This is equivalent to decreasing the flow of mobile phase through the column by the same 0.43-fold. Separation time t is proportional to $1/P$ in this case (Eqs. 2.11, 2.12a), so that t is increased by $1/0.43$, that is, from 180 to 420 sec. The mobile-phase velocity u is proportional to P (Eq. 2.12), and is therefore reduced 0.43-fold to a value of 0.32 cm/sec. From Figure 2.7a the corresponding value of $H = 0.0025$, from which we calculate $N = 25/0.0025 = 10,000$ plates.

In the third example of Table 2.1 we propose to increase N by increasing L while holding P constant. An increase in L to 34 cm (by 1.36-fold) leads in this case to an increase in t by $(1.36)^2$ (Eq. 2.12a), or a final value of 330 sec. The

Table 2.1 Increasing N in various ways in order to increase R_s (See Fig. 2.7a, page 29).

Procedure	L (cm)	p (psi)	F (ml/min)	t (sec)	u (cm/sec)	H (cm)	N
Initial separation	25	1,500	4.0	180	0.75	0.0037	6,750
Decrease P, L constant	25	640	1.7	420	0.32	0.0025	10,000
Increase L, P constant	34	1,500	3.0	330	0.56	0.0034	10,000
t constant, increase L, P	65	10,000	10.4	180	1.95	0.0065	10,000

mobile-phase velocity u (and the solvent flowrate F) is decreased by the same 1.36-fold (Eq. 2.12) to 0.56 cm/sec. From Figure 2.7a we determine $H = 0.0034$ and calculate $N = 34/0.0034 = 10,000$ plates. Again we have increased N to the required 10,000 plates, but now the time required (330 sec) is less than that for the previous example of decreasing column pressure (420 sec). The reason is that a higher operating pressure is used (1500 psi) for increasing L, and we will see that higher column pressures generally allow higher plate numbers if separation time is held constant (Section 5.5).

In the last example of Table 2.1 we increase N by holding t constant while increasing L and P. With L increased 2.6-fold, P is increased by $(2.6)^2 = 6.8$-fold (eq. 2.12a), to a value of 10,000 psi. Mobile-phase velocity u (and flowrate F) is increased by 2.6-fold, to a value of 1.95 cm/sec. This then yields the values of H and N shown in Table 2.1. Again we have achieved the required increase in N to 10,000 plates, but now at the price of increased column pressure. Figure 2.26 provides actual examples of the increase in N and resolution, using these three procedures: decreasing P with L constant (Figure 2.26 b,c vs. a); increasing P and L, with t constant (d vs. b); increasing L with P constant (e vs. d). Although the pressures used in Figure 2.26a-c are much lower than normally seen in high-performance LC, the principles of LC separation are the same for both low- and high-pressure separations.

What are the practical implications of the data in Table 2.1? The easiest way to increase N and resolution is by changing P (L constant) or—what is equivalent—by changing the mobile-phase flowrate F. This involves only a change in the pump setting to decrease P or F, depending on the pump type (see Section 3.4). However, in some cases the resulting increase in separation time may be considerable.

If we increase column length while holding P constant (or decrease F proportionately), some initial work is required. Thus, we will have to connect additional column lengths, and we may have to pack new columns—if additional column lengths are not on hand. But the extra effort will be repaid by shorter separation times, compared to the use of a fixed column length and decrease in P. This is especially true where a large increase in R_s is required.

Finally, if we increase both column length and pressure, resolution can be increased without any increase in separation time. Apart from the extra effort involved in increasing L, however, our LC unit may not be able to operate at the high pressures involved (e.g., 10,000 psi in the example of Table 2.1). So increasing N while holding separation time constant is an option that is not always available.

Now that we have these three options for increasing N and R_s, which procedure should we choose for a given problem? For a one-time separation, a simple decrease in column pressure (decrease in F) is attractive. The separation may take somewhat longer, but the effort required to change separation conditions is

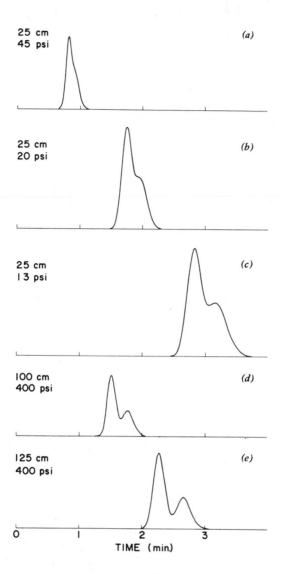

Figure 2.26 Separation of 1-nitronaphthalene ($k' = 2.4$) and methyl naphthoate ($k' = 2.8$) by liquid-solid chromatography. Column length and pressure varied as noted in parts *a-e*. Column. Porasil *A* (35-75 μm silica; mobile phase, 0.2% acetonitrile/pentane; ambient; UV detection. Reprinted from (*16*) with permission.

Table 2.2 Variation of resolution with time, pressure, and column length: *pellicular packings*

(a) **Varying pressure and time (column length constant)**

	Relative change in time or reciprocal pressure					
R_1	$R_2 \rightarrow$ 0.4	0.5	0.6	0.8	1.0	1.25
0.4	—	3.1	8	32	100	300
0.5	0.3	—	2.4	10	32	100
0.6	0.1	0.4	—	4.2	13	39
0.8	0.03	0.1	0.2	—	3.1	9
1.0	0.01	0.03	0.08	0.3	—	3.1
1.25	0.003	0.01	0.03	0.1	0.3	—

(b) **Varying column length and time (pressure constant)**

	Relative change in column length (time)				
R_1	$R_2 \rightarrow$ 0.5	0.6	0.8	1.0	1.25
0.4	(1.4, 1.9)	(1.8, 3.2)	(2.7, 7.3)	(3.7, 14)	(5.1, 26)
0.5	—	(1.3, 1.7)	(1.9, 3.6)	(2.7, 7.3)	(3.7, 14)
0.6	(0.8, 0.6)	—	(1.5, 2.3)	(2.1, 4.4)	(2.8, 8.0)
0.8	(0.5, 0.3)	(0.7, 0.4)	—	(1.4, 1.9)	(1.9, 3.6)
1.0	(0.4, 0.2)	(0.5, 0.3)	(0.7, 0.5)	—	(1.4, 1.9)
1.25	(0.3, .07)	(0.4, 0.1)	(0.5, 0.3)	(0.7, 0.5)	—

(c) **Varying column length and pressure (time constant)**

Relative change in column length (pressure)

R_1 \\ $R_2 \rightarrow$	0.5	0.6	0.8	1.0	1.25
0.4	(2.1, 4.4)	(3.9, 15)	(10, 100)	(21, 440)	(44, 1900)
0.5	—	(1.8, 3.7)	(4.8, 23)	(10, 100)	(22, 480)
0.6	(0.5, 0.3)	—	(2.6, 6.8)	(5.5, 30)	(11, 120)
0.8	(0.2, .04)	(0.4, 0.1)	—	(2.1, 4.4)	(4.4, 19)
1.0	(0.1, .01)	(0.2, .03)	(0.5, 0.2)	—	(2.1, 4.4)
1.25	(0.05, .002)	(0.1, .01)	(0.2, .05)	(0.5, 0.2)	—

minimal—just a change in the pump setting. The situation is somewhat different when it comes to developing an LC method for carrying out a large number of identical separations or analyses, as in some routine assay procedure. Here the additional work required to increase column length (with P held constant) is quite worthwhile, because the resulting saving in time per analysis (compared to separation with L held constant) is multiplied by the large number of individual separations involved. A final possibility, that of increasing both column pressure and length, occurs less often, for the reasons discussed earlier.

A Rapid Scheme for Changing R_s via a Change in N. Given the preceding three choices for varying N and R_s, we now describe a simple scheme for estimating what change in conditions will result in a change in resolution from the initial value R_1 to the final required value R_2 (*16, 17*). In the simplest case our column is filled with a *pellicular* (coated-glass-bead) packing, and Table 2.2 can be used. Assume that our starting resolution R_1 is equal to 0.8 and our desired resolution R_2 is 1.0. Table 2.2 is divided according to the three options for increasing N: (a) column length constant, (b) column pressure constant, and (c) separation time constant. Assume that we elect to increase R_s and N via a simple reduction in flowrate F and (therefore) column pressure P. In Table 2.2a we see values of R_1 plotted against values of R_2. For the present case of $R_1 = 0.8$ and $R_2 = 1.0$, the table entry is given as 3.1. This means that reciprocal pressure should be changed by this amount; that is, F or P should be changed by 1/3.1. Similarly, separation time will increase by 3.1-fold.

Alternatively, assume we decide to increase R_s by our second option: increase in column length L, holding P constant (Table 2.2b). In this case for an increase in R_s from $R_1 = 0.8$ to $R_2 = 1.0$, the table entry is (1.4, 1.9). This means that column length should be increased 1.4-fold, and separation time will be increased 1.9-fold (vs. 3.1-fold in the preceding example from Table 2.2a). Flowrate would then be decreased by 1.4-fold to maintain P constant.

Our final option is similarly covered in Table 2.2c: holding separation time constant while increasing column length and pressure together. For $R_1 = 0.8$ and $R_2 = 1.0$ the table entry is (2.1, 4.4). This means that L must be increased by 2.1-fold (as must F), and P will increase 4.4-fold. However, separation time remains unchanged.

If we are using a column that is *not* packed with pellicular particles, then a somewhat more complex approach to controlling R_s is required. However, it is still based on simple table-lookup procedures, as in Table 2.2. This alternative scheme for small-particle, porous particles is based on the so-called reduced velocity v. The reduced velocity is equal to ud_p/D_m (D_m is the diffusion coefficient of a sample molecule in the mobile phase), and is further discussed in Section 5.5. If we know the value of $v = v_1$ for the initial separation (which has $R_s = R_1$), then we can use a table (for that value of v_1) to predict what change in conditions (P, L, or P and L) will give us a final resolution equal to R_2. We do not describe the theoretical basis of this scheme here, but the interested reader can refer to (*16, 17*).

Table 2.3 summarizes values of ν_1 for various experimental conditions: column length L, particle size d_p, mobile-phase composition, and temperature. This table applies only to porous particles with $5 \leqslant d_p \leqslant 20$ μm; other cases will be discussed shortly. Once ν_1 for the initial separation has been estimated, that value defines the appropriate table relating R_1 to R_2. Tables 2.4-2.7 relate R_1 to R_2 in terms of some change in separation conditions (P, L, or P and L). Table 2.4 is for $\nu_1 = 3$, Table 2.5 is for $\nu_1 = 10$, Table 2.6 is for $\nu_1 = 30$, and Table 2.7 is for $\nu_1 = 100$. We select that table whose ν_1 value is closest to that for the initial separation.

Table 2.3 Estimating ν_1 for porous, small-particle packings from the initial separation conditions

Mobile phase and temperature	Values of ν_1 for given values of L and d_p			
	$L = 15$ cm[a]		$L = 30$ cm[a]	
	d_p=5 μm	d_p=10 μm	d_p=5 μm	d_p=10 μm
Water (or mixtures with methanol or acetonitrile), 25°	20	40	40	80
Water (or mixtures), 60°	10	20	20	40
Nonaqueous solvents, 25°	8	15	15	25

[a]For $d_p = 20$ μm, multiply ν_1 value for $d_p = 10$ by 2.

Now let us consider the use of Tables 2.3-2.7. As an example, assume initial separation is by reverse-phase, using 50% acetonitrile/water at room temperature, a 25-cm column, and a column packed with 10-μm particles. From Table 2.3 we see that for these conditions, a 15-cm column has $\nu_1 = 40$ and a 30-cm column has $\nu_1 = 80$. By interpolation we obtain $\nu_1 = 67$ for a 25-cm column. This latter value is halfway between $\nu_1 = 30$ (Table 2.6) and $\nu_1 = 100$ (Table 2.7), so we arbitrarily pick Table 2.6. Now assume $R_1 = 0.8$ and $R_2 = 1.0$. Table 2.6a describes how this change in R_s can be achieved by change in P (holding L constant). From Table 2.6a we see a value of 2.5 for $R_1 = 0.8$ and $R_2 = 1.0$, and this is the reduction in pressure or flowrate (by 1/2.5) that gives the desired increase in R_s. The separation time is increased by 2.5-fold, relative to the initial separation.

Table 2.4 Variation of resolution with time, pressure, and column length: $\nu_1 = 3$

(a) Varying pressure and time (column length constant)

Relative change in time or reciprocal pressure

R_1 \ $R_2 \rightarrow$	0.4	0.5	0.6	0.8	1.0	1.25
0.4	—	*	*	*	*	*
0.5		—	*	*	*	*
0.6			—	*	*	*
0.8				—	*	*
1.0					—	—
1.25						—

(b) Varying column length and time (pressure constant)

Relative change in column length, time

R_1 \ $R_2 \rightarrow$	0.4	0.5	0.6	0.8	1.0	1.25
0.4	—	1.7, 2.9	3.0, 9	*	*	*
0.5	0.7, 0.5	—	1.5, 2.3	4.3, 18	*	*
0.6	0.5, 0.3	0.7, 0.5	—	2.0, 4	6.0, 36	*
0.8	0.3, 0.1	0.4, 0.2	0.6, 0.4	—	1.6, 2.6	3.6, 13
1.0	0.2, 0.0	0.3, 0.1	0.4, 0.2	0.6, 0.4	—	1.6, 2.6
1.25	0.2, 0.0	0.2, 0.0	0.3, 0.1	0.5, 0.3	0.6, 0.4	—

(c) **Varying column length and pressure (time constant)**

Relative change in column length, prssure

R_1 \ $R_2 \rightarrow$	0.4	0.5	0.6	0.8	1.0	1.25
0.4	—	1.6, 2.6	2.6, 6.8	7, 49	19,360	80, 6400
0.5	0.7, 0.5	—	1.5, 2.3	3.2, 10	7, 49	19, 360
0.6	0.5, 0.3	0.7, 0.5	—	1.9, 3.6	3.7, 14	8, 64
0.8	0.3, 0.1	0.4, 0.2	0.6, 0.4	—	1.6, 2.6	3, 9
1.0	0.3, 0.1	0.3, 0.1	0.4, 0.2	0.7, 0.5	—	1.6, 2.6
1.25	0.2, 0.0	0.3, 0.1	0.3, 0.1	0.5, 0.3	0.7, 0.5	—

Note: *indicates that it is not possible to achieve a change in R_s by this procedure.

Table 2.5 Variation of resolution with time, pressure, and column length: $\nu_1 = 10$

(a) Varying pressure and time (column length constant)

			Relative change in time or reciprocal pressure			
R_1	$R_2 \rightarrow$ 0.4	0.5	0.6	0.8	1.0	1.25
0.4	—	*	*	*	*	*
0.5	0.4	—	*	*	*	*
0.6	0.2	0.4	—	*	*	*
0.8	0.1	0.1	0.3	—	*	*
1.0	0.0	0.1	0.1	0.3	—	*
1.25	0.0	0.0	0.1	0.2	0.3	—

(b) Varying column length and time (pressure constant)

			Relative change in column length, time			
R_1	$R_2 \rightarrow$ 0.4	0.5	0.6	0.8	1.0	1.25
0.4	—	1.4, 2.0	1.9, 3.6	3.2, 10	5.2, 27	11, 120
0.5	0.7, 0.5	—	1.3, 1.7	2.1, 4.4	3.2, 10	5.2, 27
0.6	0.6, 0.4	0.8, 0.6	—	1.5, 2.2	2.3, 5.3	3.5, 12
0.8	0.4, 0.2	0.5, 0.3	0.7, 0.5	—	1.4, 2.0	2.0, 4.0
1.0	0.3, 0.1	0.4, 0.2	0.5, 0.3	0.7, 0.5	—	1.4, 2.0
1.25	0.2, 0.0	0.3, 0.1	0.4, 0.2	0.5, 0.3	0.7, 0.5	—

(c) **Varying column length and pressure (time constant)**

Relative change in column length, pressure

R_1 \ $R_2 \rightarrow$	0.4	0.5	0.6	0.8	1.0	1.25
0.4	—	2.2, 4.8	4.8, 23	32, 1000	*	*
0.5	0.5, 0.3	—	1.8, 3.2	6.7, 45	32, 1000	*
0.6	0.4, 0.2	0.6, 0.4	—	2.7, 7	8.4, 71	48, 2300
0.8	0.2, 0.0	0.3, 0.1	0.5, 0.3	—	2.2, 4.8	5.8, 34
1.0	0.2, 0.0	0.2, 0.0	0.3, 0.1	0.5, 0.3	—	2.2, 4.8
1.25	0.1, 0.0	0.2, 0.0	0.2, 0.0	0.3, 0.1	0.5, 0.3	—

Note: * indicates that it is not possible to achieve a change in R_s by this procedure.

Table 2.6 Variation of resolution with time, pressure, and column length: $\nu_1 = 30$

(a) Varying pressure and time (column length constant)

R_1	$R_2 \rightarrow$ 0.4	0.5	0.6	0.8	1.0	1.25
			Relative change in time or reciprocal pressure			
0.4	—	2.6	*	*	*	*
0.5	0.5	—	2.0	*	*	*
0.6	0.3	0.5	—	3.8	*	*
0.8	0.1	0.2	0.4	—	2.5	*
1.0	0.0	0.1	0.2	0.5	—	2.5
1.25	0.0	0.0	0.1	0.2	0.5	—

(b) Varying column length and time (pressure constant)

R_1	$R_2 \rightarrow$ 0.4	0.5	0.6	0.8	1.0	1.25
			Relative change in column length, time			
0.4	—	1.4, 2.0	1.7, 2.9	2.6, 7	3.7, 14	5.0, 25
0.5	0.8, 0.6	—	1.3, 1.7	1.8, 3.2	2.6, 7	3.7, 14
0.6	0.6, 0.4	0.8, 0.6	—	1.5, 2.3	2.0, 4	2.8, 8
0.8	0.4, 0.2	0.6, 0.4	0.7, 0.5	—	1.4, 2	1.8, 3.2
1.0	0.3, 0.1	0.4, 0.2	0.6, 0.4	0.8, 0.6	—	1.4, 2
1.25	0.3, 0.1	0.3, 0.1	0.4, 0.2	0.6, 0.4	0.8, 0.6	—

(c) Varying column length and pressure (time constant)

Relative change in column length, pressure

R_1 \ $R_2 \rightarrow$	0.4	0.5	0.6	0.8	1.0	1.25
0.4	—	3.0, 9	13, 170	*	*	*
0.5	0.4, 0.2	—	2.4, 5.8	30, 900	*	*
0.6	0.3, 0.1	0.5, 0.3	—	4.5, 20	86, 7400	*
0.8	0.1, 0.0	0.2, 0.0	0.4, 0.2	—	3.0, 9	20, 400
1.0	0.1, 0.0	0.1, 0.0	0.2, 0.0	0.4, 0.2	—	3.0, 9
1.25	0.0, 0.0	0.1, 0.0	0.1, 0.0	0.2, 0.0	0.4, 0.2	—

Note: * indicates that it is not possible to achieve a change in R_s by this procedure.

Table 2.7 Variation of resolution with time, pressure, and column length: $\nu_1 = 100$

(a) Varying pressure and time (column length constant)

R_1		Relative change in time or reciprocal pressure					
	$R_2 \rightarrow$	0.4	0.5	0.6	0.8	1.0	1.25
0.4		—	2.1	3.9	18	*	*
0.5		0.5	—	1.8	5.1	18	*
0.6		0.3	0.6	—	2.5	6.3	*
0.8		0.2	0.3	0.5	—	2.0	4.7
1.0		0.1	0.2	0.3	0.5	—	2.0
1.25		0.1	0.1	0.2	0.3	0.5	—

(b) Varying column length and time (pressure constant)

R_1		Relative change in column length, time					
	$R_2 \rightarrow$	0.4	0.5	0.6	0.8	1.0	1.25
0.4		—	1.3, 1.7	1.6, 2.6	2.4, 5.8	3.1, 9.6	4.2, 18
0.5		0.8, 0.6	—	1.2, 1.4	1.8, 3.6	2.4, 5.8	3.1, 9.6
0.6		0.6, 0.4	0.8, 0.6	—	1.4, 2.0	1.9, 3.6	2.5, 6.3
0.8		0.4, 0.2	0.6, 0.4	0.7, 0.5	—	1.3, 1.7	1.7, 2.9
1.0		0.3, 0.1	0.4, 0.2	0.6, 0.4	0.8, 0.6	—	1.3, 1.7
1.25		0.3, 0.1	0.3, 0.1	0.4, 0.2	0.6, 0.4	0.8, 0.6	—

(c) **Varying column length and pressure (time constant)**

Relative change in column length, pressure

R_1 \ $R_2 \rightarrow$	0.4	0.5	0.6	0.8	1.0	1.25
0.4	—	8, 64	*	*	*	*
0.5	0.3, 0.1	—	5, 25	*	*	*
0.6	0.2, 0.0	0.4, 0.2	—	40, 1600	*	*
0.8	0.1, 0.0	0.1, 0.0	0.3, 0.1	—	8, 64	*
1.0	0.0, 0.0	0.1, 0.0	0.1, 0.0	0.3, 0.1	—	8, 64
1.25	0.0, 0.0	0.0, 0.0	0.1, 0.0	0.1, 0.0	0.3, 0.1	—

Note: * indicates that it is not possible to achieve a change in R_s by this procedure.

71

Table 2.6b similarly describes what changes in L (holding P constant) will suffice for some final value R_2. In this case we see the numbers 1.4 and 2.0 for $R_1 = 0.8$ and $R_2 = 1.0$. This means that column length L must be increased by 1.4-fold, which means an increase in separation time by 2.0-fold. To hold P constant the flowrate through the new column is reduced by 1.4-fold. As expected from our earlier discussion, the time required (2-x greater) is less than that required for the simple decrease in P (2.5-x greater) just discussed. It will sometimes prove inconvenient or impossible to increase L by some exact fraction; for example, a 1.4-fold increase for a 15-cm (initial) column would require addition of a 6-cm column. In this case the next longer column that is readily available should be used. For example, the 15-cm column might be replaced with a 25-cm column, or a second length of 15-cm (total L equal 30 cm) might be added.

Table 2.6c provides data for changing R_s via increase in P and L (separation time t constant). Here we see the values 3.0, 9 in Table 2.6c ($R_1 = 0.8, R_2 = 1.0$), which means that column length (and F) must be increased threefold and column pressure increased ninefold.

For porous particles with $d_p \geqslant 20$ μm, simply assume $\nu_1 = 100$ and use Table 2.7.

The preceding approach is somewhat approximate, and greater accuracy in predicting the necessary change in conditions can be achieved by related, but more complicated schemes described in (17). Tables 2.2-2.7 are useful in other ways than for simply estimating how one can achieve some change in R_s (either an increase, or a decrease with concomitant reduction in separation time). For example, assume we want $R_2 = 1.25$ and $R_1 = 0.8$. From Table 2.6 it is seen that this change in R_s is not even possible by means of a change in P (when $\nu_1 \approx 30$). Thus, the use of this scheme can alert us to potential problems when a particular approach for changing R_s is being considered. Similarly, one might consider a change in P and L for this same change in R_s (0.8 to 1.25). Table 2.6c shows that the necessary increase in column pressure would be 400-fold, which probably rules out this option immediately. In other cases these tables can be used to compare the different separation times for different ways of increasing R_s. Finally, these tables can provide the beginner with some feel for the practicality of increasing R_s by differing amounts and in different ways. As a general rule, increase in P and L as a means of increasing R_s is favored for small values of ν_1. On the other hand, the use of decreased P or increased L is more effective for increase in R_s at high values of ν_1. For values of ν_1 below 30, decrease in P cannot be used to increase resolution significantly.

The preceding approach and Tables 2.2-2.7 apply only to the separation of bands for pure compounds. Sometimes a group of compounds elutes as a single band whose width is greater than that for a single compound. Examples include synthetic polymers in size-exclusion chromatography (e.g., Figure 1.4), and mixtures of compounds with very similar k' values (e.g., different alkyl benzenes in liquid-solid chromatography). In these cases R_s increases much more slowly than predicted by Tables 2.2-2.7, because band width for an unresolved mixture is determined mainly by the varying retention times of the different compounds making up the band, rather than by the width of individual bands for single compounds.

Resolution versus α

The effect of α on R_s is given by term (i) of Eq. 2.10. The value of α should be as large as possible, other factors being equal. An increase in α is a powerful technique for improving resolution because R_s can be *increased* while t is *decreased*. However, determining the necessary conditions (i.e., change in compositions of moving or stationary phases, varying temperature) for a change in α is often not obvious, usually requiring a trial-and-error approach. Some guidelines for predicting changes in α are available, as discussed below and in later chapters (see especially Chapters 6-11).

When is a change in α the most likely solution to a given separation problem? First, we have already seen in Section 2.4 that no separation is possible if $\alpha = 1$. Therefore when ($\alpha - 1$) is less than about 0.1, a change in α should probably be our first goal. Second, because large values of α can yield rapid separations in certain cases, a change in α may be necessary when developing a routine assay for a large number of samples. This would be particularly true in applications such as process control, clinical analysis, and so on. Third, the separation of complex samples in routine fashion may also require that we adjust *several* α values for various adjacent bands, to give chromatograms with regularly spaced peaks (as in Figure 2.27b as compared with 2.27a); that is, similar α values for all adjacent band pairs are desirable. A particularly good example of this is seen in the amino acid analysis of Figure 10.14. Finally, large α values should be sought in preparative separations, because sample size can then be greatly increased without seriously impairing separation. This is illustrated in Figure 2.27c,d.

What is the general approach to increasing α for improved resolution? Several options are available and can be ranked in order of decreasing promise:

- Change of mobile-phase solvent.
- Change of mobile-phase pH.
- Change of stationary phase.
- Change of temperature.
- Special chemical effects.

We discuss each of these in turn. However, a general consideration is the need to maintain k' values in the optimum range ($1 \leqslant k' \leqslant 10$) while one of the following changes is made for change in α.

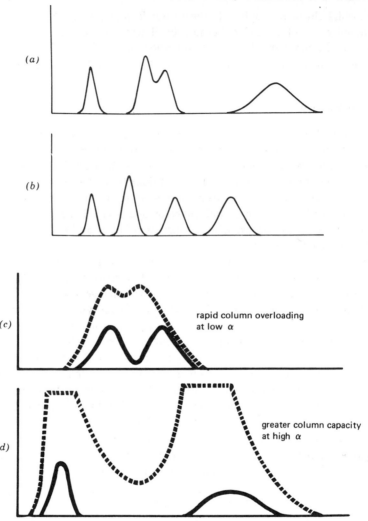

Figure 2.27 The improvement of routine LC separations by altering the α values of adjacent bands. (*a*, *b*) Analytical-scale separation, (*c*, *d*) preparative separation.

By change of the mobile-phase solvent, we mean a change to a different solvent component. An example is provided in the following separation, where solvent *A* refers to 23% CH_2Cl_2/pentane, solvent *B* is 5% pyridine/pentane, and the column packing is alumina (liquid-solid LC) (*18*):

Sample compounds	k'	
	Solvent A	Solvent B
Acetonaphthalene	5.5	2.3
Dinitronaphthalene	5.8	5.4
α	$\alpha = (1.05)$	(2.04)

With solvent A, sample k' values are in the right range, but α is small and the separation is difficult. Solvent B represents an entirely different solvent mixture (pyridine replacing methylene chloride), and therefore a change in α can be expected. The percent of pyridine in solvent B must be further adjusted (from 23 to 5%) for satisfactory solvent strength, and with this adjustment the resulting mobile-phase (solvent B) gives a much better value of α (2.04 vs. 1.05). This approach to changing α values is very powerful and widely used in LC.

Another example of the ability of the mobile-phase composition to affect separation selectivity, this time for a reverse-phase C_{18} column, is shown in the following table ([19]):

Sample compounds	k' (α)		
	50%v MeOH[a]/ water	40%v AcN[a]/ water	37%v THF[a]/ water
Aniline	1.3	1.5	1.7
	(1.23)	(1.07)	(1.35)
Phenol	1.6	1.4	2.3
Anisole	4.5	4.3	3.9
	(1.04)	(1.09)	(1.20)
Benzene	4.7	4.7	4.7
Chlorobenzene	9.2	7.7	6.5

[a]MeOH is methanol, AcN is acetonitrile, and THF is tetrahydrofuran.

The strength of the three solvents in the preceding table has been adjusted to give about the same k' values, by varying the percent of the weak solvent component water. However, the selectivity observed varies significantly. Thus, 37%v THF/water gives much larger α values for the difficultly separable pairs aniline/phenol and anisole/benzene. Furthermore, this mobile phase also gives a larger

k' value for the first band (aniline) for improved resolution, and a smaller k' value for the last band (chlorobenzene) for faster separation. Thus, this mobile phase would provide significantly better separation than the other two solvents for this particular sample.

The second most popular means of varying α values is via a change in mobile-phase pH. However, this technique is restricted to samples that include ionizable acids or bases. In such cases we vary the pH of the mobile phase and examine the resulting changes in k' and α. An example is given in Figure 2.28 for the separation of the four major mononucleotides by ion-exchange chromatography. Here a pH of 3 gives approximately equal spacing of the four peaks in the sample. The overall separation time (about 2.5 min) is also satisfactory, so no further change in separation conditions is required. When a change in pH gives optimum α values, but k' values do not fall in the right range, solvent strength can be adjusted independently of solvent selectivity by varying the *proportions* of components in the mobile phase (e.g., by adjusting salt concentration in the separation of Figure 2.28).

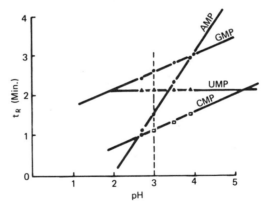

Figure 2.28 Anion-exchange separation of different mononucleotides as a function of pH. Reprinted by permission of *Applied Science Corp.*

A third means for improving α values is by changing the stationary phase. This is less convenient than a change in mobile-phase composition, and is therefore used less commonly. Usually further adjustments in the mobile-phase composition are required for the new column packing or stationary phase, for final optimization of solvent strength and k' values. An example of such a change is shown in Figure 2.29 for the separation of nine equine estrogens by normal-phase liquid-liquid chromatography. In Figure 2.29*a* partial separation of the mixture is obtained, but resolution of several bands is poor, and the poor spacing

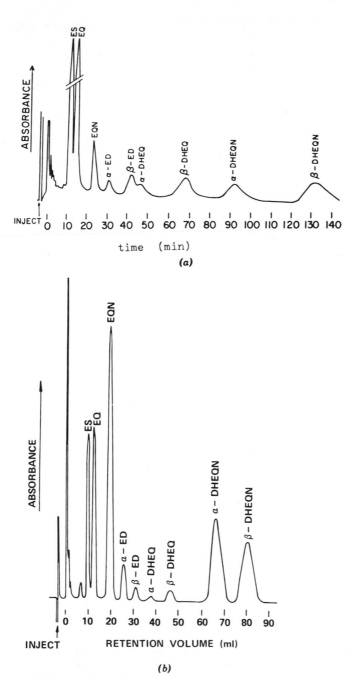

Figure 2.29 Effect of stationary phase on α values for a mixture of equine estrogens separated by liquid-liquid chormatography. Columns, (a) 200 × 0.18 cm of 1% glycol/OPN on Zipax and (b) 300 × 0.18 cm of ETH-permaphase; mobile phase, 2% THF/hexane; (a) 23° and (b) 40° ; (a) 1 ml/min and, (b) 2 ml/min; UV, 254 nm, 0.04 AUFS. Reprinted from (20) with permission

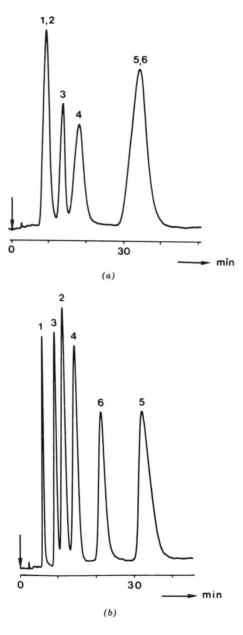

Figure 2.30 Separation of acidic compounds by ion-pair chromatography: effect of change in separation temperature from (a) 25° to (b) 43°. Compounds are (1) salicylic acid, (2) 2-naphthol-3,6-disulfonic acid, (3) 4-methyl phenol, (4) 1-naphthol-5-sulfonic acid, (5) 2,3,4-trimethylbenzenesulfonic acid, and (6) 2,5-dimethylphenol. Column, 15 × 0.3 cm Spherosil XOC with 4% added trioctyl amine; mobile phase, 0.04 M HClO$_4$ in 20%v methanol/water; (a) 25° and (b) 43°; UV, 235 nm; 10-μl sample (1 μg/ml). Reprinted from (21) with permission.

of bands within the chromatogram further results in an overall separation time of 2.5 hours. By changing from the ethylene glycol/OPN liquid stationary phase in (*a*) to the bonded-phase ETH-permaphase column in (*b*), much better α values for the various pairs of adjacent bands are obtained. As a result, good resolution of all compounds is achieved in a shorter separation time (90 min vs. 2.5 hr). In this example the same mobile phase fortuitously gave the right solvent strength for both columns.

A fourth technique for varying α values is to increase (or decrease) the separation temperature. Since an increase in temperature normally reduces all sample k' values, it is usually necessary to decrease solvent strength to compensate for this effect. A change in temperature usually has little effect on sample α values in liquid-liquid or liquid-solid chromatography, but is often important in ion-exchange and ion-pair chromatography. For this reason, change in temperature for improvement of α in ion-exchange and ion-pair chromatography is generally more promising than a change in stationary phase (and much more convenient). An example is provided in Figure 2.30 for ion-pair LC: at $25°C$ (Figure 2.30*a*) four of the six compounds in the sample are unseparated, and at $43°C$ (Figure 2.30*b*) all six compounds are well resolved.

A final means of changing α values, sometimes dramatically, is via chemical complexation effects or so-called secondary equilibria. A well-known example of this in thin-layer chromatography is the use of silver-ion-impregnated adsorbents to separate various olefinic compounds. In this case the complexation of olefin and silver ion causes dramatic changes in retention and selectivity. A similar example is shown in Figure 2.31 for the separation of vitamins D_2 and D_3 (olefinic compounds) by reverse-phase LC. In Figure 2.31*a* with the conventional 95% methanol/water mobile phase, separation is poor. In Figure 2.31*b* addition of silver nitrate to the mobile phase results in a dramatic improvement in resolution. Other chemical complexation reactions that have been used to enhance separation selectivity in LC include the borate-*vic*-diol reaction, the bisulfite-aldehyde reaction, and the reactions of various transition metal ions with amines, sulfides, and other organic ligands.

At the beginning of this section we noted that a change in α permits an increase in R_s while decreasing separation time t. How is this achieved in actual practice? A hypothetical example is used for illustration. Assume that $R_s = 0.6$ initially, a final value of $R_s = 1.0$ is desired, and α is initially equal to 1.05. Further assume we are able to vary conditions (e.g., mobile-phase composition) in such a way that k' for the second band is held constant (so that t remains constant) while α is increased to 1.105. From Eq. 2.10 this means an increase in R_s by the factor (0.105/0.05), or from $R_s = 0.6$ to about $R_s = 1.25$. Now the required value of R_s in the final separation is only 1.0. If experimental conditions at this point (with $R_s = 1.25$) are such that $\nu_1 = 30$, then from Table 2.6a we see that a change from $R_s = 1.25$ to $R_s = 1.0$ can be achieved by increasing P by $1/0.5 =$ two-fold, with a decrease in separation time by a factor of 0.5.

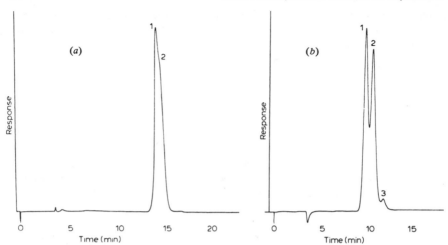

Figure 2.31 Separation of vitamin D (olefinic) compounds via reverse-phase LC and silver-ion complexation. Compounds are vitamin D_2 (#1), vitamin D_3 (#2) and impurity (#3). Column, 30 × 0.4 cm μBondapak/C_{18} (10 μm); (*a*) mobile phase, 95%v methanol/ water, and (*b*) same plus 20 g/l silver nitrate; 0.8 ml/min; UV detection; 5 μl sample (concentration 1 mg/ml). Reprinted from (*22*) with permission.

Thus, here an increase in α permits the simultaneous increase in R_s (from 0.6 to 1.0), with a 50% reduction in separation time.

The advent of small-particle columns for routine application in LC has provided relatively large values of N with little effort on the part of the user. At the same time means for increasing N further (as discussed in the preceding section) have been somewhat reduced for small-particle columns. This has placed greater emphasis on the adjustment of α as a general technique for improving the resolution of poorly separated adjacent bands.

Automated Method Development.   The step-by-step optimization of k', N, and α for a multicomponent sample can represent a substantial effort, even when guided by a good understanding of the dependence of the separation on different operating variables. Two general approaches can make method development under these conditions somewhat less tedious.

Various systematic search routines exist that allow a more or less automatic approach to the empirical selection of optimum test conditions. One of these, the so-called SIMPLEX routine, was recently adapted to the optimization of the separation of six phospholipids on a column of Corasil-II (*23*). The resulting band separation was impressive, as the following summarization shows:

Compound	t_R (min)	α
Phosphatidyl ethanol amine	5.2	4.0
Phosphatidyl serine	6.7	1.4
Phosphatidyl inositol	7.4	2.5
Phosphatidyl choline	11.5	1.6
Phosphatidic acid	15.3	1.1
Sphingomyelin	16.5	2.2
Lysophosphatidyl choline	31.2	

It should be emphasized, however, that techniques such as SIMPLEX are no substitute for a good understanding of chromatographic principles, as a careful reading of (23) amply demonstrates.

For other applications of SIMPLEX optimization as applied to LC separation, see (24, 25).

The second approach to automated method development is the use of a microprocessor-equipped LC system that allows trial separations of a particular sample to be run automatically, with solvent composition varied between runs (19, 26). Here the operator simply selects a series of solvent compositions (e.g., mixtures of three solvents A, B, and C) and enters this information into the microprocessor. The Spectra-Physics Model 8000 is one example of such a system.

REFERENCES

1. B. A. Bidlingmeyer and L. B. Rogers, *Sep. Sci.,* 7, 131 (1972).

2. J. C. Giddings, *Dynamics of Chromatography*, Dekker, New York, 1965, Chap. 2.

3. L. R. Snyder, in *Gas Chromatography. 1970*, R. Stock and S. G. Perry, eds., Institute of Petroleum, London, 1971, p. 81.

4. J. C. Sternberg, *Adv. Chromatogr.,* 2, 205 (1966).

5. M. Martin, C. Eon, and G. Guiochon, *J. Chromatogr.,* 108, 229 (1975).

6. R. P. W. Scott, *Contemporary Liquid Chromatography*, Wiley-Interscience, New York, 1976.

7. J. N. Done, *J. Chromatogr.*, 125, 43 (1976).

8. L. R. Snyder, *Anal. Chem.*, 39, 698 (1967).

9. *Basics of Liquid Chromatography*, Spectra-Physics, Santa Clara, Calif., 1977. Reprinted with permission.

10. Giddings, *op. cit.*, Chap. 5.

11. B. L. Karger, L. R. Snyder, and C. Horvath, *An Introduction to Separation Science*, Wiley-Interscience, New York, 1973, Chap. 3.

12. L. R. Snyder, *J. Chromatogr. Sci.*, 10, 200 (1972).

13. Application sheet, Du Pont Instrument Products Division.

14. L. R. Snyder, *J. Chromatogr. Sci.*, 8, 692 (1970).

15. E. Grushka, L. R. Snyder and J. H. Knox, *ibid.,* 13, 25 (1975).

16. L. R. Snyder, *ibid.,* 10, 369 (1972).

17. L. R. Snyder, *ibid.,* 15, 441 (1977).

18. L. R. Snyder, *J. Chromatogr.,* 63, 15 (1971).

19. Spectra-Physics, *Chromatogr. Rev.,* 3, No. 2 (July 1977).

20. A. G. Butterfield, B. A. Lodge, and N. J. Pound, *J. Chromatogr. Sci.,* 11, 401 (1973).

21. C. P. Terweij-Groen and J. C. Kraak, *J. Chromatogr.,* 138, 245 (1977).

22. R. J. Tscherne and G. Capitano, *ibid.,* 136, 337 (1977).

23. M. L. Rainey and W. C. Purdy, *Anal. Chim. Acta*, 93, 211 (1977).

24. M. R. Detaevernier, L. Dryon, and D. L. Massart, *J. Chromatogr*, 128, 204 (1976).

25. S. N. Deming and M. L. H. Turoff, *Anal. Chem.,* 50, 546 (1978).

26. N. A. Parris, *Amer. Lab.,* 124, October 1978.

BIBLIOGRAPHY

Giddings, J. C., *Dynamics of Chromatography*, Dekker, New York, 1965 (the definitive treatment of band-broadening processes in chromatography).

Grushka, E., L. R. Snyder, and J. H. Knox, *J. Chromatogr. Sci.*, 13, 25 (1975) (a short review of band-broadening theories and equations).

Horvath, C., and H.-J. Lin, *J. Chromatogr.*, 149, 43 (1977) (detailed theory of band spreading in LC, with emphasis on kinetic effects, i.e., slow sorption/desorption, with small particles).

Knox, J. H., *J. Chromatogr. Sci.*, 15, 352 (1977) (an excellent review of band broadening in modern LC, with specific reference to the author's important work).

Scott, R. P. W., *Contemporary Liquid Chromatography*, Wiley-Interscience, New York, 1976 (a detailed treatment of certain aspects of band-broadening theory, with emphasis on the author's particular viewpoint).

Sternberg, J. C., *Adv. Chromatogr.*, 2, 205 (1966) (definitive account of general features of extracolumn band broadening, but with specific reference to gas chromatography, rather than LC).

THREE

EQUIPMENT

3.1	Introduction	84
3.2	Extracolumn Effects	86
3.3	Mobile-Phase Reservoirs	88
3.4	Solvent Pumping (Metering) Systems	90
	Materials of Construction	90
	General Pump Specifications	91
	Constant Flowrate Pumps	92
	Reciprocating Pumps	92
	Positive-Displacement Pumps	98
	Constant-Pressure Pumps	100
	Comparison of Pumps	101
	Pump Maintenance and Care	101
3.5	Equipment for Gradient Elution	103
	High-Pressure Gradient Formers	105
	Low-Pressure Gradient Formers	106
	Comparison of Gradient Systems	109
3.6	Sample Injectors	110
	Syringe Injector	112
	Sampling Valves	113
	Automatic Injectors	116
3.7	Miscellaneous Hardware	117
	Line Filters	117
	Pressure Measurement	117
	Pulse Dampers	117
	Column Thermostats	118

Fraction Collectors   118

Flowrate Measurement   118

Data-Handling Systems   119

3.8   Integrated and Specialized Instruments   119

3.9   Laboratory Safety   123

References   123

Bibliography   124

3.1 INTRODUCTION

Modern LC is performed with equipment that is different from the relatively simple apparatus used for classical liquid chromatography. Although modest apparatus is capable of useful and sometimes elegant separations, relatively sophisticated equipment generally is needed for separating highly complex mixtures or for producing precise quantitative data. In this chapter we review the types of equipment available for modern LC and explain some of the advantages and disadvantages of the various designs. A similar discussion of detectors for modern LC is given in Chapter 4.

Appendix I lists some of the manufacturers of LC equipment. New equipment is continually appearing and often offers significant advantages in either price or performance. Therefore, current information should be obtained from various manufacturers of LC equipment before deciding on a final purchase. See also the reviews in (1, 2).

The beginner in modern LC often asks, "Which LC equipment is best?" Actually, there is no one "best" equipment, since the requirements for solving a specific problem may require different approaches. In this chapter we provide guidelines to assist the reader in selecting appropriate equipment for a particular need. Another question that often arises is, "Is it best to personally assemble apparatus from component modular units, or to obtain completely assembled, integrated instruments?" Here again the answer is dependent on the intended use of the equipment. Researchers requiring apparatus that can be easily changed or modified sometimes find the assembly of modular components to be more flexible for their separation needs. Instruction in modern LC also is often best carried out with modular or "home-made" units, to allow the student to perceive more easily the function of each component in the separation process. This also makes less likely the "black-box syndrome" which frequently occurs when neophytes are confronted with nicely assembled, integrated equipment. Equipment assembled from modular components generally is less costly, but also less versatile. More sophisticated, integrated apparatus usually is capable of better precision and is more convenient, with less operator dependency (but at higher cost). Integrated commercial equipment is particularly attractive for routine analyses and when methods are to be exchanged between laboratories.

Table 3.1 Criteria for modern LC equipment.

Performance requirements	System characteristics	Equipment needs
Versatility	Useful with different types of samples	Chemically resistant materials Variety of detectors Unique column packings
Speed	Selective, highly efficient columns High mobile-phase velocity High data output	High-pressure pump Low dead-volume fittings and detectors Fast recorder, automatic data handling
Reproducibility	Control of operational parameters	Precise control of column and detector temperature mobile-phase composition (and gradient) flowrate detector response
Sensitivity	High detector response Sharp peaks	Careful detector design for good signal-noise ratio Efficient columns

Equipment for modern LC must be designed and produced to careful specifications to obtain high-efficiency separations and precise quantitative data. An excellent, high-efficiency column placed in a poorly designed LC apparatus will produce disappointing results. Therefore, it is the *combination* of a good column and a good LC apparatus that produces high-quality separations. Table 3.1 lists the criteria generally needed for modern LC equipment.

Figure 3.1 shows a general schematic of the equipment used for modern LC. This system is composed of relatively simple equipment for routine analyses; other components may be needed for conducting particular experiments, as is described later. However, it is very difficult to design a single apparatus that will be simultaneously ideal for both analytical and preparative liquid chromatography. Equipment for analytical LC is highlighted in this chapter; the special apparatus required for preparative LC is specifically discussed in Chapter 15.

In this chapter the various components of an LC apparatus are described, roughly in the order shown in Figure 3.1. First, however, a general discussion of extracolumn effects is needed for an understanding of this important phenomenon in the selection and use of modern LC equipment.

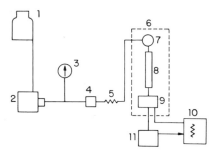

Figure 3.1 Schematic of simple LC apparatus. (1) mobile phase reservoir, (2) pump, (3) pressure gauge, (4) line filter, (5) pulse damper, (6) thermostated oven, (7) sample injector, (8) chromatographic column, (9) detector, (10) recorder, (11) data handling equipment.

3.2 EXTRACOLUMN EFFECTS

In addition to the inherent band broadening that takes place within the chromatographic column (Section 2.3), *extracolumn (ec) band broadening* occurs outside the column. This phenomenon is discussed in general terms in Section 2.3. Based on Eq. 2.8b, a good rule of thumb is that the total broadening due to the injected volume plus the other extracolumn peak volumes should be less than one-third the volume of the first peak of interest in the chromatogram. In this way, any increase in band width or band volume is limited to about 10%, which means that an unimportant loss of resolution will result.

In addition to band broadening within the fittings that attach columns, injectors, and so on, to the apparatus (Section 5.4), a common cause of band broadening within an LC instrument is the tubing that connects the injector to the column, and the column to the detector. To minimize ec band broadening, this tubing should be as short and as narrow as possible. However, in practice, longer lengths are required in most instruments, and $\geqslant 0.025$-cm i.d. tubing must normally be used since narrower bores tend to plug. The length of tubing that can be used for a maximum bandwidth increase of 5% is described by (3):

$$L = \frac{40 V_R{}^2 D_m}{\pi F d^4 N},\tag{3.1}$$

where L is the length of the column in cm, V_R is the solute retention in ml, D_m is the solute diffusivity in cm^2/sec (see Eq. 5.16), F is the volume flowrate in ml/sec, d is the diameter of the tube in cm, and N is the column plate number. The plots in Figure 3.2 show the allowable length of 0.05 cm i.d. (straight) connecting tubing with peaks of $k' = 0$ to maintain a bandwidth increase of no more than 5% for columns of various plate count and different void volumes (V_0). This plot shows that for a column with a plate count of 500 and a V_0 of 5 ml, a 0.05-cm i.d. connecting tube 270 cm long can be tolerated without adversely affecting column performance. By contrast, a 10,000 plate column with a void volume of 5 ml can tolerate only 14 cm of connecting tubing before band broadening in the tube exceeds 5% of the original bandwidth of an unretained peak. If tubing with an i.d. smaller than 0.05 cm is used, its length can be proportionately increased (as described by Eq. 3.1) with equivalent band broadening. Bending, tightly coiling, or deforming the tubing also permits a longer length of connecting tubing with the same degree of band broadening, relative to straight lengths of tubing. However, this also leads to greater back-pressure, and is rarely used except for heat exchange prior to a detector.

The peak volume V_w from a very short column of small particles (especially for a very high-efficiency size-exclusion column) can be quite small (e.g., ~ 40 μl). Not infrequently, users complain that a chromatographic column specified by the supplier with a certain plate count (e.g., 16,000) shows a much lower plate count (e.g., 9,000) when tested in their apparatus by the same procedure used by the manufacturer. Careful analysis of this apparent discrepancy often shows that both specification tests were carried out accurately, but that the larger ec band broadening associated with the user's equipment causes the apparent lowering of the manufacturer's specified plate count. Therefore, it is important for very high-efficiency columns, that extracolumn effects be kept at a minimum. Particular attention must be paid to the design of all equipment components to ensure that resolution is not degraded by band broadening.

Extracolumn effects can be identified by noting that early-eluting ($k' < 2$) peaks have a much smaller plate count than later-eluting ($k' > 2$) peaks and show

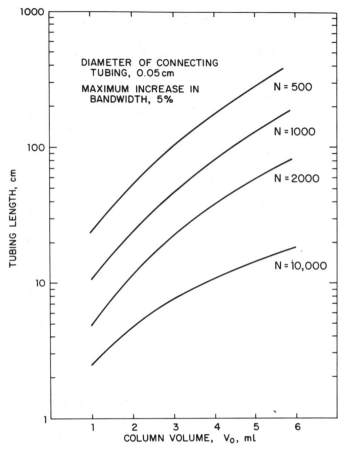

Figure 3.2 Allowable lengths of connecting-tubing for acceptable extracolumn band-broadening. Solute with $k' = 0$, $D_m = 3 \times 10^{-5}$. (Calculated from Eq. 3.1.)

greater band tailing. Alternatively, ec band broadening due to the apparatus can be recognized by connecting the first column to another "matched" column. The plate count of the two connected columns should then be nearer to that predicted by the manufacturer's data for the combination (see Section 5.6), because the volume of the connected columns is larger and the fixed volume associated with extracolumn effects then becomes a smaller fraction of the total.

3.3 MOBILE-PHASE RESERVOIRS

Solvent reservoirs should hold at least 500 ml of mobile phase for an analytical chromatograph; for preparative applications much larger volumes are required

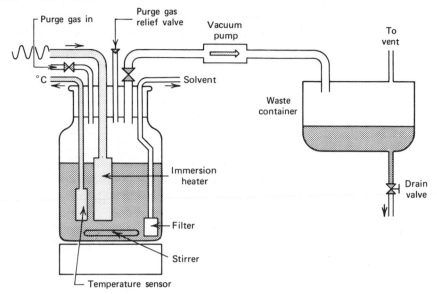

Figure 3.3 Schematic of mobile phase reservoir (Hewlett-Packard). Illustration courtesy of Hewlett-Packard.

(Chapter 15). Figure 3.3 is a schematic of a solvent reservoir and storage system used by Hewlett-Packard. This particular unit is made of glass; stainless steel can also be used. Simple systems such as a 1-l Erlenmeyer flask plus a magnetic stirrer often are adequate as "home-made" reservoirs for the solvent. The solvent feed line to the pump should be fitted with a filter (e.g., 2 μm) to prevent particulates from being drawn into the pump from the reservoir. To permit convenient in situ *solvent degassing*, reservoirs often are equipped with a heater, a stirring mechanism, and inlets for applying a vacuum and a purge of inert gas. Degassing is used to eliminate dissolved gases in the mobile phase and to reduce the possibility of bubbles forming in the pump or detector during the separation. (Degassing enhances check-valve performance by eliminating microbubbles.) Elimination of oxygen in the mobile phase is also important to prevent reaction with the mobile and/or stationary phases within the column. Degassing is more important with water and other polar solvents (especially in gradient elution), but is less needed with nonpolar solvents. To prevent the formation of bubbles in the detector, degassing can be carried out by applying a vacuum to the filled reservoir for a few minutes while stirring the mobile phase vigorously. Warming the mobile phase facilitates degassing. However, all forms of degassing must be carried out judicially so as not to change the concentration of mobile-phase mixtures or disturb the equilibrium of solvents modified with water or polar organic solvents. Any fractionation of the mobile phase as a result of handling and/or degassing can lead to variation of k' values.

It has been suggested that the benefits of degassing can be better achieved by thoroughly purging the mobile phase with helium, which has an extremely low solubility in virtually all liquids (5, 5a). After initial sparging by a fast purge of helium for a few minutes, a slow purge of helium through the solvent then is used to maintain the mobile phase. Sparging with helium (or nitrogen) also eliminates the degradation of oxidizable samples, solvents, or stationary phases by removing traces of oxygen from the mobile phase. A blanket of inert gas over the liquid in the reservoir inhibits oxygen from redissolving in the mobile phase after degassing, and also reduces the possibility of fire with flammable vapors.

Another arrangement used by Du Pont automatically vacuum-degasses the solvent in situ in a closed system (6). This approach has a substantial advantage, since dissolved gas is eliminated from the mobile phase without any other change in its composition.

In some LC pumping systems the pump is initially filled with a large enough volume of mobile phase to carry out several separations. In these systems there is no need for a separate solvent reservoir.

Because of the opportunities for change resulting from evaporation, oxidation, and so on, it is recommended that only fresh solvent be used for optimum separation reproducibility. It is poor practice to reuse solvents, or to use solvent that is left in the reservoir for long periods of time (e.g., overnight). As discussed in the chapters on individual LC methods, even minor changes in the composition of the mobile phase can result in significant changes in retention, resolution, and peak heights. Thus, it is important that the mobile-phase composition be maintained constant for optimum separation reproducibility.

3.4 SOLVENT PUMPING (METERING) SYSTEMS

The function of the solvent pumping or metering system in modern LC is to provide a constant, reproducible supply of mobile phase to the column. Because the small particles used to pack modern LC columns offer substantial resistance to flow through the column (Section 5.7), high-pressure pumps are required. The general operational requirements for pumping systems in modern LC are listed in Table 3.2. LC pumps are required to deliver solvent to the column at precise flowrates, with a relatively pulse-free output, and at pressures up to 5000 psi. The special requirements in pumps for preparative LC are discussed in Section 15.3.

Materials of Construction

Modern LC pumps should be constructed of materials resistant to chemical attack, for example, stainless steel and Teflon (polytetrafluorethylene). Pump

Table 3.2 Requirements for modern LC analytical pumping system.

The system must be made of materials chemically resistant to mobile phase

It must be capable of outputs of at least 500 psi, preferably 5000 psi

It must be pulse-free or have a pulse damper

It must have a flow delivery of at least 3 ml/min (for analysis)

It must deliver mobile phase with reproducibility of $\ll 1\%$

Desirable: Have small holdup volume for rapid solvent changes, gradient elution with external device, and/or recycle

seals are usually made from filled Teflon (e.g., graphite-filled Teflon), which will resist all solvents used in LC. In certain reciprocating pumps, pistons and check-valve balls are made of sapphire. Pumps constructed entirely of Teflon and glass may be substituted when stainless steel cannot be tolerated; however, the pressure output of these pumps generally is limited to <500 psi (35 bar).

General Pump Specifications

Pump performance specifications that are most important include:

- Resettability (or repeatability).
- Short-term precision.
- Pump noise (pulsations).
- Drift (long-term metering precision).
- Accuracy.

By *resettability* or repeatability is meant the ability to reset the pump to a particular flowrate. Thus, resettability is a measure of how reproducibly the fluid is delivered from the pump when it is set at a nominal flowrate (e.g., 2.00 ml/min) from time to time; that is, is returned to a certain set flowrate from a higher or lower flowrate. *The short-term precision* is a measure of the constancy of the volume output of the pump over a short time, usually a few minutes. Variations of this type arise from problems such as erratic check-valve leakage in reciprocating pumps. *Pump pulsation* (pump noise) is inherent in certain types of pumps, for example in reciprocating pumps, as a result of normal piston movement and check-valve operation. *Pump drift* is a measure of a continuous change (increase or decrease) in the pump output over relatively long periods (e.g., hours), often as a result of change in ambient temperature. *Pump accuracy* relates to the ability of a pump to deliver the exact flowrate indicated by a particular setting. Thus, a 2.04 ml/min delivery for a pump set nominally at 2.00 ml/min would indicate a 2% error in pumping accuracy. Pumping accuracy is usually much poorer than resettability or pumping precision, but fortunately is not so important for most applications.

A particular separation/analysis goal often dictates which pump specifications are most important: one of those discussed previously or some other performance feature. For example, pump resettability, short-term precision, and drift are critical in both qualitative and quantitative analysis, since retention-time values and peak areas are proportional to integrated (average) flowrates during a run. The determination of accurate molecular-weight-distribution data by size exclusion places even greater demands on these pump specifications (7).

If trace analysis is involved, pump noise is generally the most important specification, since minimum detectability is often limited by the detector baseline noise that arises from pump pulsations (Chapter 4). Pump noise is somewhat more critical with gradient elution (using two pumps) and with refractometer detectors. Pump noise has no adverse effect on the LC separation, although the stability of some columns is adversely affected by sharp pump pulsations. Although pump noise is a specification often stressed by manufacturers, it is by no means the only important characteristic of the pump.

If the separation goal is speed of analysis, then the maximum delivery rate and operating pressure of the pump will be of particular interest. Maximum pump flowrate is the most important characteristic of the pump for preparative separations.

The preceding considerations are important in selecting a given pump for the applications likely to be encountered in a given laboratory. The ease with which solvents can be changed during scouting studies for optimization of the mobile phase is still another factor. Finally, and among the most important considerations, are a pump's durability, operating convenience, and serviceability. Even a well-constructed pump will need occasional service, and it is important that this can be carried out quickly with a minimum of inconvenience and cost. Another feature of serviceability is the ability of the pump to operate under adverse conditions (e.g., for long periods of time, at high pressures, and with unfavorable mobile phases such as those containing salts). At the end of this section we present laboratory procedures that should be followed to ensure that pumps maintain their original performance with a minimum of servicing and repair.

Two types of pumping systems are available: those that deliver essentially constant flowrate and those that deliver a constant pressure. Each of these systems has its advantages and disadvantages, and no one pumping system is necessarily best for all types of LC operations. However, constant-flow pumps are gradually taking over in most applications of analytical LC.

Constant Flowrate Pumps

Reciprocating Pumps. Because of their generally satisfactory overall perfor-

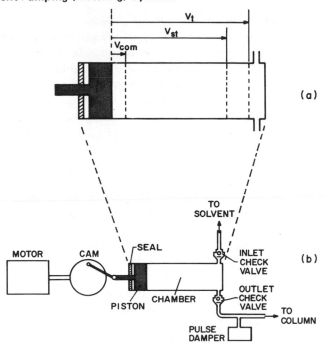

Figure 3.4 Schematic of simple reciprocating pump. (*a*) Pump chamber. (*b*) Single-head reciprocating pump.

mance, reciprocating pumps at present are most widely used in modern LC applications. These pumps use small-volume (35-400 μl) chambers with reciprocating pistons or flexible diaphrams to drive the solvent flow against the column back pressure (see Figure 3.4*b*). Several different types of reciprocating pumps are used in modern LC, including single-head, dual-head, and triple-head pumps, some with special drive mechanisms to reduce flow pulsation and some with special features for precise flow delivery. Models are available with output pressures up to about 10,000 psi; volumetric outputs of up to 28 ml/min are typical of pumps used for analytical applications. The various reciprocating pumps can be distinguished by the methods used to minimize pulsating pumping noise and to compensate for solvent compressibility and flowrate changes.

The flowrate F provided by a reciprocating pump can be changed either by varying the stroke volume V_{st} during each cycle of the pump or by varying stroke frequency. Both approaches to the control of pumping rate are used in various commercial pumps for LC. The stroke volume V_{st} is generally varied by changing the piston travel distance, as illustrated in Figure 3.4*a*; here V_{st} indicates the maximum extension of the piston into the pump chamber. Also indicated in Figure 3.4*a* is the compression of the solvent V_{com} that occurs at the beginning of the pump stroke, as the solvent pressure is raised from ambient pressure to the operating pressure of the pump (equal approximately to the column inlet pressure). Because of this solvent compression, simple reciprocating pumps deliver a slightly lower flow (smaller value of F) at higher pumping pressures—other factors being equal. Assuming that the check valves operate properly, the magnitude of this effect is determined by solvent compressibility and by the ratio of stroke volume to total chamber volume $V_t : V_{st}/V_t$. Since the check valves do not open until the compression V_{com} (Figure 3.4a) takes place, this also leads to a delay in the beginning of the pump delivery cycle (as well as to a decreased average flowrate F). These various effects illustrated in Figure 3.4a affect pumping accuracy and pulsation, but not short-term precision or drift. Resettability is affected to only a minor extent, varying with column inlet pressure. Pump pulsation, which we examine next, is affected to a much larger extent by the oscillating stroke of the pump.

The simplest reciprocating pump is the single-head type diagrammed in Figure 3.4b. In this configuration (e.g., Milton-Roy Model 196-31) the inlet and outlet check valves are actuated by liquid pressure within the pump head. Pressurization of the pump head during the delivery stroke closes the check valve on the inlet side and opens the valve on the outlet side. During the filling stroke, the depressurization of the pump head leads to closing of the outlet check valve and opening of the inlet check valve. Thus, the delivery of solvent by the pump occurs only during alternate (delivery) cycles of the pump piston, falling to zero during the filling cycle. In addition, the conversion of a rotary pump motor movement to a linear piston movement in the design of Figure 3.4b leads to a sinusoidal variation of pump flow during the delivery cycle. The resulting flow of solvent from the pump is illustrated in Figure 3.5a and is seen to be severely pulsating. Without some form of pulse damping between pump outlet and column, the pump output of Figure 3.5a would lead to severe detector noise in most cases (increasing at higher flowrates). Various approaches to pulse damping are discussed in Section 3.7.

A more sophisticated single-head pump design (e.g., Altex Model 110) allows for adjustment of the piston advance rate during each stroke and specifically provides for faster solvent delivery near the end of each delivery stroke, followed by a sudden increase in stroke frequency during the filling cycle (fast refill, 0.2 sec). This greatly smoothes the pulsation of the refill stroke and provides reduced pump noise as compared with the simple single-head pump. The Altex Model 110 provides for monitoring the motor torque, which is proportional to volume output, and returns the torque to its original value after each filling cycle. This allows for additional control over the accuracy of the average flowrate F.

Another approach to reducing pump pulsation in reciprocating pumps is the use of dual-head pumps. These consist of two identical piston-chamber units as

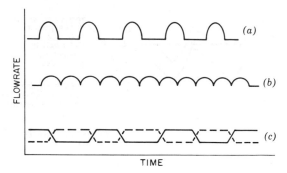

Figure 3.5 Pump output patterns (*a*) simple single-head reciprocating pump output, (*b*) dual-head circular cam pump output, (*c*) dual-head sinusoidal cam output (with changeover ramp). Reprinted from (*9*) with permission.

in Figure 3.4*b*, operated 180° out of phase. The resulting solvent delivery as a function of time, which is illustrated in Figure 3.5*b*, is seen to be much smoother than in Figure 3.5*a* for a single-head pump. This is the result of a continuous delivery by the pump, which arises from the alternate filling by one pump head during the delivery cycle of the other head. Because of the additional complexity of this pump design, dual-head pumps are significantly more expensive than are single-head units. More sophisticated dual-head pumps come with specially designed cams that reduce the sinusoidal variation of pump delivery during each stroke of the piston. Although dual-head pumps in conjunction with these special cams can produce an essentially pulse-free flow from the pump, in practice this design is subject to mismatch of the cams at the changeover points in the pumping cycle—leading to pump pulsation at the end of each piston stroke. This final problem can be overcome by arranging a piston-driving cycle slightly more than 180°, such that gradual takeover periods result between alternate delivery by each piston of the pump. The resulting flow from each pump head is shown in Figure 3.5*c*, and the total flow is seen to be constant.

The effects of using special systems to compensate for flowrate variations resulting from mobile-phase compressibility and pulsations is schematically illustrated in Figure 3.6. Output (*A*) represents the uncorrected flow output from a dual-head reciprocating pump as a function of the cam rotation driving the piston against the column back pressure; here mobile-phase pulsations are significant, and the average uncorrected flowrate is lower than the desired set value, because of solvent and pump-seal compressibility. In output (*B*) compressibility has been compensated so that the average flowrate now is at the set value; however, output is still pulsating. In trace (*C*) the pump output is additionally corrected by pressure feedback during the stroke, so pulsations are significantly decreased. Pressure- or flow-feedback systems are available in more sophisticated

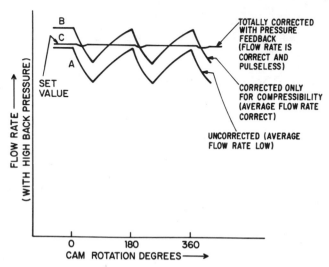

Figure 3.6 Compressibility and pressure feedback corrections in reciprocating pumps. Illustration courtesy of Spectra-Physics, Inc.

versions of digitally driven reciprocating pumps, and these devices are often useful in stabilizing pump flow output against changes due to temperature, mobilephase viscosity, column back pressure, and so on. However, accurate actuation of check valves is required in these pumps to produce a constant flow output. In one single-head pump (Varian 5000) check valves are mechanically actuated (not pressure-actuated) in proper sequence with the pumping stroke, to ensure positive check-valve operation and precise flow. An additional advantage of this approach is that the pump can be operated at low pressures in special applications, whereas other pumps (with check valves operated by pressure) require a minimum back pressure of about 50 psi to function properly.

Another method of flow control uses a "hydraulic capacitor" plus a flow resistor (the column) to produce true volumetric flowrate (Hewlett-Packard 1080 Series). This system employs twin reciprocating (diaphram) pumps, with flowrate varied by changing the effective stroke of the pump heads. The hydraulic-capacitor sensor monitors the movement of the pulsating diaphrams, and the signal generated in the hydraulic capacitor is used to control the pump stroke. The controlling signal is generated by integrating the pressure change in the capacitor over a period of several strokes during certain time fractions of the delivery cycle. Two such flow measurement systems are used to achieve independent feedback-flow control of each of the two pumps, which stabilizes mobile-phase output against changes due to temperature, mobile-phase viscosity, and column back pressure. The output of this pumping system is virtually independent of

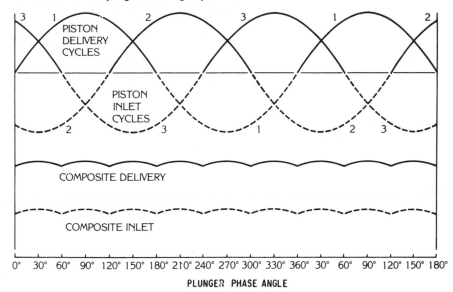

Figure 3.7 Flow profile of three-headed reciprocating pump. Solid line is composite flow-profile output; dashed line is composite pump input (suction); numbered lines represent flow profile from single pump heads 1, 2, and 3. Reprinted from (*10*) with permission.

the type of solvent pumped and there is no need for recalibration when the solvent and/or the column are changed.

Diaphragm reciprocating pumps work similarly to piston pumps, except that a flexible stainless steel or PTFE diaphragm is in contact with the mobile phase. This diaphragm is actuated by a piston working on an oil cavity, and each stroke of the piston flexes the diaphragm to produce a pulsating solvent output.

One of the more sophisticated designs for reciprocating pumps is that of Du Pont and Jasco: a three-headed, digitally driven pump. The three pistons are phased at 120° and are driven by a single-faced cam that rotates and actuates each piston in turn. This approach eliminates small differences in pump strokes that otherwise contribute to uneven flowrate output and pumping noise. In the Du Pont version of the triple-head pump, the high precision of flow output is a result of the fact that each pump head operates at full stroke, with the total pump output being digitally controlled by the speed of the pump motor. The 120° phasing allows essentially a continuous flow in the solvent output and input streams, as illustrated in Figure 3.7. The flow output of this pump is a succession of closely overlapping solvent pulses, the resultant total output having such small pulsations that pulse dampening is not required. This is advantageous in gradient elution, where low pump volumes are desirable to maintain gradient accuracy. It is particularly useful in recycle operation where all extracolumn volumes must be reduced to an absolute minimum when using a single column (Section 12.6).

A general advantage of reciprocating pumps is that solvent delivery is continuous; consequently there is no restriction on the size of the reservoir that can be used, nor on the length of time for a separation. Thus, these pumps are particularly useful in equipment to be placed on automatic operation for long periods (e.g., overnight). For gradient elution or for programmed solvent-searching experiments (Section 3.8), a specific advantage of reciprocating piston pumps is that their internal volume is small, making solvent changes rapid and accurate. Small-volume reciprocating pumps also are usually best for recycle chromatography (see Section 12.6). Additional advantages and limitations of reciprocating pumps in gradient elution are discussed in Section 3.5.

Postive-Displacement Pumps. Positive-displacement pumps take two forms: the screw-driven syringe (e.g., Perkin-Elmer, Isco) and the hydraulic amplifier (e.g., Micromeritics). A simple, single-stroke displacement pump is schematically shown in Figure 3.8. This pump is similar to a large syringe, whose plunger is

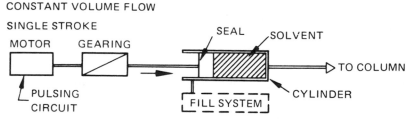

Figure 3.8 Constant displacement pump (syringe-type). Reprinted from (*8*) with permission.

actuated by a screw-feed drive through a gear box, usually actuated by a digital stepping motor. Thus, the rate of solvent delivery is controlled by changing the voltage on the motor driving the piston. Special fill systems are available on some displacement pumps to facilitate refilling of the empty pump chamber after the plunger is returned to the starting position.

Just as for the reciprocating pump, displacement pumps also provide a flow that is relatively independent of column back pressure and solvent viscosity. However, to compensate for flow changes due to compressibility effects, more sophisticated versions of this pump use a variable-restrictor valve in the output to maintain the emptying syringe at a constant pressure. Initial equilibration of a positive-displacement pump can take several minutes, unless the pump is first pressurized to the approximate operating column pressure (*11*). Another disadvantage of the displacement pump is that it has a limited solvent supply (e.g.,

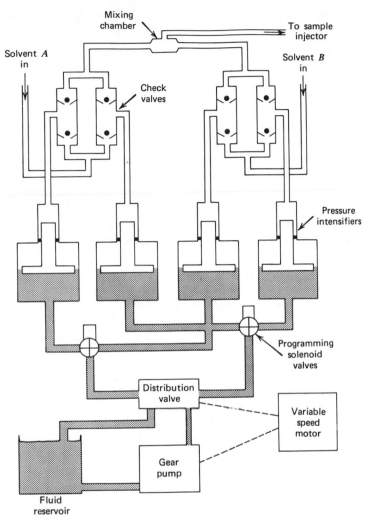

Figure 3.9 Schematic of hydraulic-amplifier pump (Micromeritics). Illustration courtesy of Micromeritics Instrument Corp.

250 ml); therefore, solvent flow must be stopped to refill the pump periodically. However, this type of pump does provide pulseless output without flow-feed-back controls.

The second type of positive-displacement pump is the hydraulic amplifier. In the Micromeritics Instruments version (Figure 3.9) a gear pump supplies oil to two hydraulic intensifiers, where the solvent pressure is increased to nine times that of the actuating oil pressure; one intensifier delivers the solvent to the column, while the other is refilling. Two sets of intensifiers are used for gradient elution, one set for each solvent A and B. The Micromeritics pump uses a pressure-feedback circuit to compensate for changes in flowrate during the switchover from one intensifier to the other. The hydraulic-amplifier pump produces a relatively constant and pulseless flow of solvent and it can operate from a limitless solvent supply.

Constant-Pressure Pumps

With constant-pressure pumps column back pressure, mobile-phase viscosity, and column temperature *must* be held constant if solvent flowrate F is to be maintained constant. A compensating advantage is that flow is completely pulseless. In the simplest form of constant-pressure pump, gas pressure is applied to the mobile phase held in a reservoir such as a coil of tubing (e.g., Du Pont, Varian). Although this pumping system is the least expensive that is so far available, it has a limited solvent supply and is relatively inconvenient to refill. Furthermore, pumping pressures are restricted to the pressure (< 2000 psi) available from the gas cylinders used to pressure the solvent. A potential problem with this type of pump is that gas dissolves at the high-pressure end, so that only about two-thirds of the reservoir volume is available for use; dissolved gas in the remaining solvent would form bubbles in the detector cell. In addition, gradient elution is impractical with these devices.

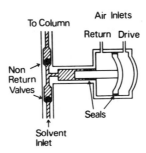

Figure 3.10 Pneumatic amplifier pump (Haskell). Reprinted from (*12*) with permission.

The pneumatic-amplifier pump (e.g., Du Pont, Haskell) (Figure 3.10) is similar to the simple gas displacement system. However, pressure amplification is obtained by using a large-area gas-driven piston to actuate a small-area liquid piston. Thus, a relatively low inlet gas pressure can be used to create high liquid-outlet pressures (to 4500 psi) without directly exposing the solvent to compressed gas (no bubble problem). The outlet pressure of this pump is directly proportional to the ratio of the piston areas and to inlet gas pressure. The pneumatic-amplifier pump of Figure 3.10 is equipped with a power-return stroke for rapid refill of the empty piston chamber by solvent. This sytem provides essentially pulseless and continuous pumping, plus (when required) relatively large flowrates for preparative applications. Also available are automatic flow-feedback systems for pneumatic-amplifier pumps (e.g., Du Pont); these continually adjust the input air supply so that a constant flowrate is maintained by a closed-loop operation. This approach has been found particularly useful in high-performance size-exclusion chromatography because of the precise flowrates obtainable; however, this system is inconvenient for gradient elution.

Comparison of Pumps

Table 3.3 compares the characteristics of various pumping systems for modern LC. All the pumping systems just described are satisfactory for some applications, but some have limitations in specific areas. Superior performance is usually obtained with the more sophisticated reciprocating pumps, which also are more expensive. In short, the user generally obtains performance in line with equipment cost.

In summary, there are two general approaches in designing pumping systems to produce high-precision flowrates with a minimum of pulsation. Both of these approaches involve relatively sophisticated hardware. The cost of such systems is substantial and the resulting performances essentially equivalent. The first approach utilizes lower-cost, less-precise and relatively "noisy" pumps with correction by flow feedback to yield a final flow that is precise and pulseless. The second approach aims at an initial pump design that provides precise, accurate, and pulseless flow inherently, without need for correction. When carried out properly, each approach is equally effective and generally satisfactory. Each pumping approach has its inherent advantages and disadvantages that must be weighed for individual applications.

Pump Maintenance and Care

Modern LC pumps are capable of very high precision; standard deviations of 0.07% in retention volumes have been obtained over 12-hr periods (9). However, to obtain this precision, special care must be taken and the system operated with the proper precautions.

To obtain and maintain pumping precision, several simple procedures need to be followed. The mobile phase should be "distilled in glass" and/or filtered through a 0.2 μm filter (e.g., Millipore), particularly if the solvent has been dried over silica gel or alumina. The filter in the pump should be regularly cleaned or changed to further ensure that particulates do not enter the pumping chamber. The pump should never be allowed to run dry while working; otherwise, abrasion between the piston and cylinder seal will eventually produce leakage. The

Table 3.3 Comparison of pump types for modern LC.

Pump characteristic	Reciprocating pumps						Positive displacement		Pneumatic		
	Simple single-head	Single-head, pulse comp.	Simple Dual-head	Dual-head compress. corr. and pulse comp.	Dual-head, closed loop flow-control	Triple-head, low-volume	Syringe-Type	Hydraulic Amplifier	Simple Amplifier	Amplifier	Amplifier with flow control
Resettability	+	+	++	++	++	++	++	++	–	–	+
Drift	+	+	+	++	++	++	++	+	–	+	+
Short-term precision ("noise")	–	+	+	++	++	++	++	++	+	+	++
Accuracy	+	+	+	+	++	++	+	+	–	–	+
Versatility and convenience	–	+	++	++	++	++	–	+	–	++	+
Serviceability	+	+	+	+	+	+	–	+	++	+	–
Durability	+	+	+	+	+	+	+	–	++	++	++
Cost	Low	Mod.	Mod.	High	Very high	Very high	Mod. to very high	Mod.	Low	Mod.	High
"Constant" flow	Yes	Yes	Yes	Yes	Yes	Yes	Yes	Yes	No	No	Yes
Constant pressure	No	No	No	Yes	No	No	No	Yes	Yes	Yes	Yes

Note: ++ = optimum, + = satisfactory, – = some deficiencies.

pump should not be used with materials known to be corrosive. Mobile phases containing dissolved salts should not be allowed to stand unused in the pump for long periods. If this is allowed to occur, a small amount of mobile phase can leak through the high-pressure seal; the liquid evaporates, leaving small salt crystals that can scratch the high-pressure seal and/or pump plunger. When using salt solutions, it is best to flush the pump thoroughly with pure water before allowing it to stand longer than overnight.

A pump should never be subjected to back pressures greater than the rated maximum for that particular pump; otherwise, the high-pressure seal can be deformed and leaks will occur. Check valves should be cleaned (e.g., with 20% HNO_3) when erratic solvent delivery is indicated; replacement may sometimes be necessary. For optimum operation some pumps require that the bore of the solvent inlet tubes should be relatively large (e.g., >2mm i.d.) to prevent "starving" of the pump for the solvent. Proper operation of some pumps also requires that the solvent reservoir be located above the pump to create a solvent head, so that the pump is properly primed. Finally, solvents should be degassed or helium-sparged (Section 3.3) for optimum pump precision.

3.5 EQUIPMENT FOR GRADIENT ELUTION

The general elution problem, which arises when a complex mixture of solutes having widely varying k' values must be separated, is most conveniently solved by the use of gradient elution (Chapter 16). Gradient elution in LC is analogous to temperature programming in gas chromatography, except that change in solute k' values during the chromatographic run is accomplished by changing the composition of the mobile phase, rather than by changing temperature. Figure 3.11 shows the separation of the same sample by gradient elution (a) versus isocratic elution (b). In gradient elution it is seen that the separation of early-eluting bands can be improved, whereas more strongly retained compounds are eluted in a shorter time. Furthermore, the peaks for later bands are greatly sharpened in gradient elution versus isocratic separation, and detection sensitivity for these bands is therefore much improved. In addition, minor components that escape detection in the isocratic run of Figure 3.11b are readily seen in the gradient run of Figure 3.11a. For another example of gradient versus isocratic elution, see the separations of Figure 16.3.

Equipments for carrying out gradient elution in LC can be divided into two distinct types: *low-pressure gradient formers* and *high-pressure gradient formers*. With either of these devices, changes in the composition of the mobile phase during gradient elution can be stepwise or continuous, depending on the mode of control. Changes in solvent composition can be programmed on a linear basis, but often some type of exponential change (convex or concave) is preferred for

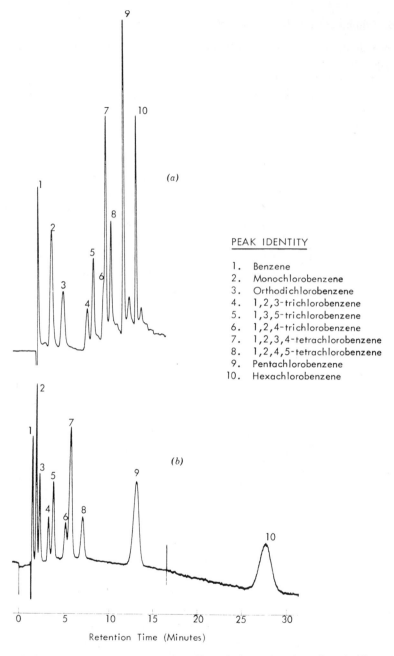

PEAK IDENTITY

1. Benzene
2. Monochlorobenzene
3. Orthodichlorobenzene
4. 1,2,3-trichlorobenzene
5. 1,3,5-trichlorobenzene
6. 1,2,4-trichlorobenzene
7. 1,2,3,4-tetrachlorobenzene
8. 1,2,4,5-tetrachlorobenzene
9. Pentachlorobenzene
10. Hexachlorobenzene

Retention Time (Minutes)

Figure 3.11 Comparison of isocratic and gradient elution separations. Sample of aromatic hydrocarbons. Column: 100 × 0.21 cm Permaphase ODS; pressure, 1200 psi; temperature, 60°C; detector, UV, 254 nm; sample, 5 µl of chlorinated benzenes in 2-propanol. (a) Gradient elution: linear gradient, 40/60 methanol-water to methanol at 8%/min, (b) Isocratic: 50/50 methanol-water, Reprinted from (*12*) with permission.

optimum separation (Section 16.1). In the most sophisticated equipment the gradient is formed with a microprocessor, so that any profile desired can be established for a particular separation. As discussed in Chapter 4, gradient elution is difficult to use with certain types of detectors (e.g., refractometers).

High-Pressure Gradient Formers

This type of gradient device involves the high-pressure mixing of two solvents to generate a mobile-phase gradient by programming the delivery from the high-pressure system. Figure 3.12 is a schematic of one form of high-pressure mixing

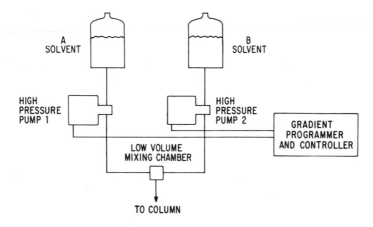

Figure 3.12 High-pressure gradient mixing with two programmed pumps.

apparatus for gradient elution. In this approach the output from two high-pressure pumps is programmed into a mixing chamber before flowing into the column. Both reciprocating and displacement-type pumps have been used with this approach. Since the output of each pump can be separately controlled (electronically), it is possible to conveniently generate almost any kind of gradient, and systems of this type are widely used in commercial instruments (e.g., Waters Associates). However, this arrangement is expensive, as two high-pressure pumps plus a gradient controller are required. One (usually minor) limitation of this approach is that the system is limited to two gradient solvents, thus restricting the k' range over which a single gradient separation can be performed (see Section 16.1).

A more important limitation of the system of Figure 3.12 for gradient elution in the high-pressure mixing mode is that reciprocating pumps operate with poor precision at low flowrates (e.g., less than ~0.1 ml/min). This is a problem at both ends of the solvent program, where a small amount of one solvent is mixed with the other. As a result of this imprecision of solvent delivery at low flowrates

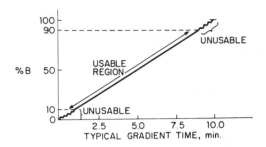

Figure 3.13 Limitation of reciprocating pumps in high-pressure-mixing gradient systems.

by reciprocating pumps, the gradient that is formed in initial and final stages of the program can be imprecise, and solute elution characteristics erratic, as in-dicated in Figure 3.13. In short, just when the highest precision of mixing is re-quired at the beginning of the solvent program (small amounts of stronger solvent *B* in the initial weaker solvent *A*), the pumping system is least able to provide the required precision, and separation reproducibility is poorest. However, impre-cision at the end of the gradient run (small amounts of *A* and large amounts of *B*) is generally less critical. Since gradient devices are often used to generate binary solvents for isocratic separation (one can dial in any desired composition), it should be noted that isocratic solvents containing less than 10% of one solvent component are also less accurate and reproducible. Some gradient systems as in Figure 3.12 correct for this effect in various ways and yield accurate gradients over the range 0-100% *B*.

One variation of the high-pressure mixing approach, which uses a single high-pressure pump, is shown in Figure 3.14. In this arrangement the weaker solvent is contained in reservoir *A* and directly feeds the high-pressure pump. The higher strength solvent *B* in reservoir *B* is first allowed to flow through valve *d* and the holding coil *c* and out valve *e*, so that the holding coil system is filled with solvent *B*. At the beginning of the gradient run, electrically actuated valves in the flow system allow the weak *A* solvent to flow from the pump through valve *a*, into the mixing chamber and then into the column. During this time, valves *b* and *f* are closed. To start the gradient, valves *b* and *f* are momentarily opened so that the flow solvent from reservoir *A* pressurizes the solvent into the holding coil. This forces solvent *B* from this holding coil through valve *d* to the mixing chamber in the column. By time-proportioning the opening and closing of the valves *a* and *b*, the solvent from the two reservoirs may be mixed in a wide variety of concentration profiles.

Low-Pressure Gradient Formers

In this approach solvent gradients are first formed by mixing two or more sol-vents at atmospheric pressure, then pumped via a single high-pressure pump. The

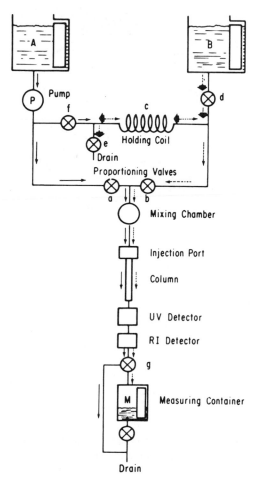

Figure 3.14 High-pressure-mixing gradient system with one high-pressure pump. Illustration courtesy of Du Pont Instrument Products Division.

simplest arrangement of the low-pressure gradient system is to add the modifying stronger solvent (e.g., from a separatory funnel) to a stirred reservoir filled with lower-strength initial solvent, which feeds the pump. Other simple gradient formers of the same type can be made from ordinary laboratory glassware, as reviewed in (*12a*). However, all such systems are generally inconvenient, often unreliable, and relatively inflexible. They are rarely used in modern LC.

A more versatile and precise approach for low-pressure gradient elution is used by several suppliers (e.g., Spectra-Physics SP-8000, Du Pont Model 850, and Varian Model 5000 liquid chromatographs). In the approach shown schematically in Figure 3.15, a series of reservoirs containing mobile phases of increasing

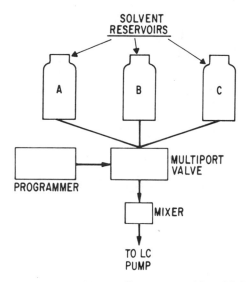

Figure 3.15 Low-pressure-mixing gradient system with multiple solvents.

solvent strength is used. Separation is started with the weakest solvent A. The gradient is produced by sequentially opening valves leading to reservoirs containing solvents of increasing strength, B and C. The concentration of these solvents can be selected by proportioning the time that the valves to the various reservoirs are opened and closed, using a microprocessor control system. After the desired solvent mixture is generated, it is mixed and fed to a single reciprocating pump for pressurization. Useful gradients can be generated with two reservoirs containing solvents of different strengths. However, by using additional solvents (e.g., up to 20), more sophisticated gradients can be generated that are capable of resolving mixtures containing compounds of very wide chemical differences (13).

It is advantageous to use low-internal-volume ($<$100 μl) reciprocating, piston pumps for low-pressure gradient systems, since then the solvent composition feeding the high-pressure pump can be varied without a significant time lag before the solvent composition reaches the column during the gradient. Generally, the lower the volume of the pumping heads, the more predictable is the gradient profile. In addition, the volume of the mixing chamber should be kept small, since this volume also significantly affects the shape of the gradient profile and, therefore, the resulting separations.

Reciprocating pumps with large chamber volumes (e.g., $>$100 μl) can cause special problems in gradient elution. Large pistons take in a relatively large volume of mixed gradient solvent of a particular composition and continue to deliver this particular composition until the next stroke. This results in a gradient with a succession of small steps at the column inlet. For example, a pump with a 400 μl piston at a flowrate of 0.8 ml/min continues to deliver the same solvent composition for 30 sec, during which time solvent strength is not changed. To eliminate this limitation, a relatively large stirred mixer can be placed after the pump, but this contributes further to the dead volume of the solvent delivering system and decreases the accuracy of the solvent gradient program.

Low-pressure gradient systems generally offer greater versatility than high-pressure systems. The reason for this is that low-pressure systems are capable of handling a series of solvents with different strengths and selectivity, whereas only two solvents are normally used in high-pressure systems. Another advantage of the low-pressure gradient-mixing system is that any solvent-volume change that occurs during mixing is completed before the solvent is pressurized and fed to the column. Therefore, flowrate changes that arise from this effect in high-pressure mixing systems are not a problem with low-pressure systems. A particular advantage of low-pressure systems is that only one pump is required, and thus they are less expensive. A potential disadvantage is that the solvents must be thoroughly degassed (or helium-sparged) before they are mixed for the gradient; otherwise, bubbles can be fed to the pump and greatly decrease flowrate precision. Built-in continuous solvent degassers are convenient to use with low-pressure mixing systems.

Because of the high precision with which gradients can be formed with low-pressure mixers, gradient separations with reproducibilities equivalent to or only slightly inferior to isocratic separations are possible. The excellent reproducibility of low-pressure gradient elution for a microprocessor-controlled LC instrument is illustrated in Table 3.4. Such systems have high repeatability because the solvent entering the pump has been accurately selected by a precise valving system, thoroughly mixed, and degassed. Most important, output is dependent on the precision of only a single pump. On the other hand, in high-pressure mixing, precision in forming gradients is dependent on the precision of two pumps to deliver solvents specified by the solvent programmer.

Automatic gradient systems are convenient for use in the isocratic (constant solvent composition) mode, since the desired composition for the mobile phase can be selected easily and rapidly. This can be carried out automatically by microprocessor-controlled solvent systems that are available in integrated LC equipments (e.g., Spectra-Physics, Du Pont, Perkin-Elmer, Varian; see Section 3.8).

Comparison of Gradient Systems

The characteristics of high- and low-pressure mixing systems for gradient elution LC are compared in Table 3.5. There is the usual compromise in performance, convenience, and cost, so that the choice between the various approaches should be based on individual need. However, there appears to be a trend at present toward low-pressure mixing, particularly when gradient reproducibility and the

Table 3.4 Reproducibility of Du Pont 850 low-pressure-mixing, gradient elution system.

Run[a] No.	Toluene		Naphthalene		Pyrene	
	Peak ret. (sec)	Peak area	Peak ret. (sec)	Peak area	Peak ret. (sec)	Peak area
1	767.3	683108	1056.2	2089019	1148.9	636621
2	777.1	673524	1058.5	2088637	1150.9	637075
3	767.0	664605	1058.0	2089744	1151.0	637810
4	768.9	677585	1058.7	2089511	1152.0	638330
5	770.0	678742	1058.0	2091480	1151.0	637744
6	769.8	694546	1058.7	2078851	1151.4	636658
7	771.1	675945	1060.0	2083292	1153.3	636350
8	768.9	683368	1058.2	2082769	1151.2	634730
9	768.7	682251	1057.3	2099405	1150.0	640320
10	767.8	685802	10583	2105269	1151.6	638731
11	769.7	677411	1056.4	2098244	1149.4	637408
12	767.5	689698	1057.1	2097557	1149.8	638853
13	767.9	678344	1057.3	2098752	1150.3	640414
14	769.0	679578	1055.9	2096177	1148.3	638338
15	767.6	692272	1057.3	2073854	1150.0	633351
16	768.1	681833	1057.7	2070294	1148.7	634388
17	768.0	675445	1054.2	2068797	1146.6	634050
18	766.2	688577	1054.5	2074871	1147.2	634645
$N=$	18	18	18	18	18	18
$\bar{X}=$	768.5	681257	10572	2087586	11500	636989
$\sigma=$	13.59	7266	15.22	10076	17.01	2090
$CV\%=$	0.18%	1.0%	0.14%	0.48%	0.15%	0.33%

[a]Column, 25 × 0.46 cm Zorbax-005; concave gradient, 50% MeOH-H$_2$O to 100% MeOH in 20 min; flow rate, 1.0 ml/min.; UV detector, 254 nm; automatic injection, computer data reduction. Taken from (6) with permission.

range of elution power are involved. Also, low-pressure gradient systems using a single high-pressure pump are somewhat less costly than two-pump high-pressure-mixing systems.

3.6 SAMPLE INJECTORS

An important factor in obtaining good column performance is proper introduc-

Table 3.5 Comparison of gradient systems (automatic equipment).

Characteristic	High-pressure mixing, two solvents	Low-pressure mixing, two solvents	Low-pressure mixing, multiple solvents
Range in elution power possible	+	+	++
Reproducibility obtainable throughout gradient	+	++	+
Cost	+	++	++
Ease of solvent changeover	+	++	+
Mechanical simplicity, dependability	+	++	+
Ease of automation	++	++	+
Sensitivity to dissolved gases	++	+	+
Gradient accuracy (sharpness)	+	++	+
Ability to program different solvent mixtures	+	+	++
Operator dependency	+	+	+
Convenience	++	++	+

Note: + = satisfactory, ++ = preferred.

tion of the sample onto an LC column. As mentioned in the discussion of sample-size effects in Section 2.2, the sample should be injected in a narrow plug so that peak broadening is neglible. Sample injection onto relatively short ($\leqslant 15$-cm), small-particle columns (particularly high-performance size-exclusion columns) can be quite critical, since peak volumes in this case are very small. Ideally, the sample injector should (a) introduce any sample into the column as a narrow plug, (b) be convenient to use, (c) be reproducible, and (d) operate at high back pressures. In addition, injection at elevated temperatures is required for some samples in size exclusion to meet sample solubility requirements (e.g., linear polyolefins).

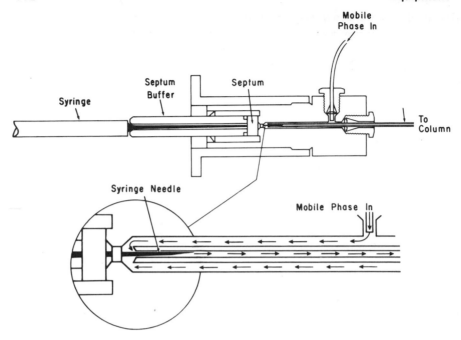

Figure 3.16 Low-volume syringe injector (Du Pont). Illustration courtesy of Du Pont Instrument Products Division.

Syringe Injector

The simplest form of sample introduction uses syringe injection as in GC, that is, into the pressurized column through a self-sealing elastomeric septum as in Figure 3.16. Injection is with a microsyringe designed to withstand pressures of up to 1500 psi. Septum materials include silicone, neoprene, and fluoro-elastomers; some have Teflon on one face to retard attack of the elastomer by certain solvents. Unfortunately, reproducibility of sample injection with a syringe is rarely better than 2%, and is often much worse. In addition, syringe injection can create practical problems as a result of pluggage of the column inlet with small particles of the elastomeric septum from repeated insertion of the syringe needle. A buildup of these elastomer particles increases the back pressure of the column, and may cause unsymmetrical peak shapes plus a marked decrease in column efficiency.

Syringe injection can be used at pressures above 1500 psi, using the *stop-flow injection* technique. In this case the pump is first turned off until the column inlet pressure becomes essentially atmospheric. The sample is then injected in the usual fashion, and the pump turned on. With this approach column efficiency is not affected, since diffusion in liquids in very slow. However, the stop-flow mode still retains the other disadvantages of syringe injection. Septum injectors also are not generally practical at higher temperatures.

Septumless-syringe injectors are available in some instruments (e.g., Hewlett-Packard) and these can eliminate the difficulties of sample leakage from syringe injection at higher pressures. Advantages of the septumless-syringe injection technique include high-pressure capability, variable injection volume, wide chemical compatability, and elimination of the problems associated with the elastomeric septums. Disadvantages include a limited sample volume range, marginal precision, and difficulties in automation. Also, if not properly designed, septumless-syringe injectors can cause significant extracolumn band broadening.

Sampling Valves

The most generally useful and widely used sampling device for modern LC is the microsampling injector valve. Because of their superior characteristics, valves are now used almost to the exclusion of syringe injection. With these sampling valves, samples can be introduced reproducibly into pressurized columns without significant interruption of flow, even at elevated temperatures. Figure 3.17 shows schematic drawings of a six-port plug-type valve from Valco (also Rheodyne and others) in which the sample fills an external loop. Compared to shorter, wider i.d. sample loops, long, narrow loops are preferred when large sample volumes

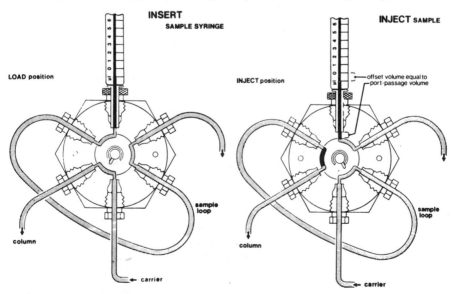

Figure 3.17 Six-port high-pressure microsampling valve (Valco). Illustration courtesy of Valco Instruments.

are required, because of lesser band-broadening effects. In Figure 3.17 the sample loop of appropriate volume is filled by flushing it thoroughly with a sample solution with an ordinary syringe, to ensure that no sample bubbles remain in the loop. (pulling the sample up into the loop usually is best.) Alternatively, a specially designed syringe may be used to inject a small volume (e.g., <10 μl) into the loop when required, although in this case the precision in the sample introduction is dependent on the precision of syringe delivery. A clockwide rotation of the valve rotor (Figure 3.17) places the sample-filled loop into the mobile-phase stream, with subsequent injection of the sample onto the top of the column through a low-volume, cleanly swept channel. Other valve types (e.g., Siemans) use an internal sample cavity consisting of an annular groove on a sliding rod that is thrust into the flowing stream.

Valve injection allows the rapid, reproducible, and essentially operator-independent delivery of large sample volumes (e.g., up to several milliliters), at pressures up to 7000 psi with less than 0.2% error. High-performance valves provide extracolumn band-broadening characteristics comparable or superior to that of syringe injection. Manually operated valves are only moderately expensive, and automated versions can be obtained at somewhat higher cost. A minor disadvantage of most valves is that the sample loop must be changed to obtain various sample volumes, but this can often be achieved in a few minutes. As shown in Figure 3.17, it is also possible with certain models to syringe-inject variable sample sizes into the loop. Another advantage of sampling valves is that they can be located within a temperature-controlled oven for use with samples that require handling at elevated temperatures (e.g., up to ~150°C). Table 3.6 compares the relative virtues of syringe and microvalve sampling; clearly, the latter is preferred for almost all applications.

A special sample valve (UK6) available from Waters Associates is illustrated by the schematic in Figure 3.18. With valve C open, the sample is syringe-injected through a loading port into a sample loop. Previous loop contents are displaced

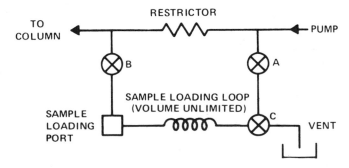

Figure 3.18 Schematic of Waters septumless valve-injector. Illustration courtesy of Waters Associates.

Table 3.6 Sampling: syringe or valve?

Parameter	Syringe (with septum)	Microsampling valve
Reproducibility	0	++
High pressure	0	++
Changing sample volume	++	+
Cost	++	+
Band spreading (optimized system)	++	++
Routine operation, automation	0	++
Delivery of large volumes ($>$ 10 ml)	++	0^a
Use with very small sample volumes	+	0^b
Useful at high temperatures	0	+

Note: 0 = deficient, or useful under some conditions, + = satisfactory, ++ = optimum or clear advantage.

[a] For very large sample volumes, use sample valve with separate pump for metering sample into column.

[b] However, sample loop can be injected with entire sample if only a very small volume is available.

by the injected sample and vented. In this step, valves A and B are closed so that the mobile phase bypasses the sample loop through a restrictor to the column. To introduce the sample into the column, the sample loading port and valve C are closed and valves A and B opened. Mobile phase from the pump then flows through the sample loading loop (in preference to flowing through the restrictor), displacing the sample into the column. This is a convenient sampling device, because very large or very small samples can be easily introduced. However, sampling precision is limited by the precision of syringe injection. Other disadvantages are that (a) the "dead" volumes can be large enough to cause extra-column band-broadening with high-efficiency columns, and (b) this device is three to four times more costly than other sample valves.

Low-volume switching valves are also available (e.g., Valco, Rheodyne, Siemans) for use in special techniques such as recycle chromatography (Section 12.6) and column switching (Section 16.2). Some of these valves can be operated at pressures up to 7000 psi, and often they can be used at elevated temperatures (but at somewhat lower maximum pressures). The more common valves can be obtained in 3-, 4-, 6-, 8-, or 10-port configurations, for use in either the manual or automated mode.

Automatic Injectors

With commercially available automatic sampling devices, large numbers of samples can be routinely analyzed by LC without operator intervention. Such equipment is popular for the analysis of routine samples (e.g., quality control of drugs), particularly when coupled with automatic data-handling systems. Automatic injectors are indispensable in unattended searching (e.g., overnight) for optimized chromatographic parameters such as solvent selectivity, flowrate, and temperature (Section 3.8). The Micromeritics automatic sample injector shown schematically in Figure 3.19 allows up to 64 samples to be pressurized consecutively into a microsampling valve for injection. In this arrangement, [1] a vial containing the sample is positioned beneath a needle and plunger assembly. [2] A motorized injection valve rotates to the "load" position, simultaneously allowing the continuing flow of mobile phase to bypass the sample loop on its way (arrows) to the column. [3] With the motorized injection valve in the "load" position, the sampling needle penetrates the polyethylene vial cap seal. [4] The needle assembly then moves the cap seal to positively displace the sample into the injector loop. To inject the sample, the motorized valve rotates [5] to allow displacement of the sample from the loop onto the column. Valve action [2] and [5] are repeated for replicate injections. [6] After the injections are completed, the needle assembly is withdrawn from the vial and a new sample vial moves

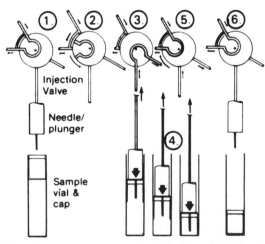

Figure 3.19 Positive displacement autosampler (Micromeritics). See text for description. Illustration courtesy of Micromeritics Instrument Corp.

beneath for the next injection. Automatic loop flushing to minimize sample carryover is part of the sampling cycle. Devices such as those in Figure 3.19 are available from several suppliers (e.g., Varian, Du Pont, Waters, Valco, Hewlett-Packard); many have build-in microprocessors for convenient and precise manipulation.

3.7 MISCELLANEOUS HARDWARE

Line Filters

A line filter should be used between the pump and the sample injector to prevent particulates from clogging the column inlet. Porous stainless steel filters having a porosity of about 2 μm are typically used in commercial instruments; however, 0.5-μm porosity filters (e.g., Alltech Associates) are desirable with columns of less than 10-μm particles. To facilitate solvent changeover, the volume of the line filters should be small.

Pressure Measurement

Pressure monitors are used in the LC equipment as diagnostic tools to optimize separation and to indicate system problems (e.g., plugging or leaks). Diaphragm or Bourdon-type gauges are simple, inexpensive, and generally robust. On the other hand, strain-gauge pressure transducers are more precise and have a smaller internal volume, which facilitates the changing of solvents. In addition, strain-gauge types are available with high- or low-pressure alarms or cutoff circuits, to protect the pump against high-pressure overload by column plugging or the instrument against fire due to solvent leaks (low-pressure alarm).

Pulse Dampers

Simple reciprocating pumps usually require pulse dampers when used with detectors that are flow sensitive (e.g., refractive index). Also, the performance of some commercial columns is seriously degraded by pump pulsations (14). Although detector output may be affected by mobile-phase pulsation, there is no effect on column efficiency. Many modular pumps and most integrated commercial instruments with reciprocating pumps are equipped with pulse-damping devices, located between the pump and the sample injector. For "home-made" equipment, a satisfactory damping system consists of a diaphragm- or Bourdon-type pressure gauge followed by 5 m of 0.25-mm i.d. capillary tubing. Pulse dampers operate analogously to an electrical capacitor-resistance circuit by storing energy in a volume (the pressure gauge, in the latter case) during the pump pressurization stroke, and releasing this energy through a restrictor during the refilling stroke. Pulse dampers of this type are used with many commercial reciprocating pumps; however, such an arrangement represents a compromise, since a volume increase in the system between the pump and the chromatographic column decreases the convenience of changing the mobile phase. A low-volume, flow-through pulse damper is especially desirable with gradient elution, to ensure an accurate gradient profile.

Column Thermostats

It often is advantageous to run ion-exchange, size-exclusion and reverse-phase columns at higher temperatures, and to precisely control the temperature of liquid-liquid columns. Therefore, column thermostats are a desirable feature in modern LC instruments. Generally, temperature variation within the LC column should be held within ±0.2°C. Maintaining a constant temperature is especially important in quantitative analysis, since changes in temperature can seriously affect peak-size measurement (Section 13.4). In LLC with mechanically held liquid stationary phases, precise control of temperature is critical, since decrease in solute k' values results from the loss of the stationary phase from the analytical column...because of the mobile-phase/stationary-phase solubility changes. It is sometimes important to be able to work at higher temperatures for the size-exclusion chromatography of some synthetic polymers because of solubility problems, but precise temperature control is not so important in this case.

High-velocity circulating air baths are most convenient in LC (as in GC). These devices usually consist of high-velocity air blowers plus electronically controlled thermostats, with configurations similar to those used in gas chromatographs. Alternatively, LC columns can be jacketed and the temperature controlled by contact heaters or by circulating fluid from a constant-temperature bath. This latter approach is practical for routine analyses, but is less convenient when columns must be changed frequently.

Fraction Collectors

Since separations are accomplished in minutes, fraction collectors are usually not needed in modern LC. Manual collection is normally used, and many commercial instruments have convenient sample-collection ports on the outlet of the detector. Fraction collectors are sometimes used in preparative chromatography and in conventional size-exclusion chromatography with larger-diameter columns, because separations are usually much slower.

Flowrate Measurement

To ensure precise retention time and peak-size measurements, it is important to know that the flowrate is held constant during the separation. Further, in size-exclusion chromatography the flowrate must be very precise, since changing flowrates can create large errors in the determination of polymer molecular-weight distributions (15). Variations in flowrate can occur as a result of the partial failure of the pumping system (e.g., a check-valve problem), or with constant-pressure pumps when there is a change in the back pressure resulting from partial pluggage of the column, a change in mobile-phase viscosity, a temperature change, and so on.

Volumetric methods are most commonly used to measure flowrates. The technique is to collect the mobile phase for a known period of time in a small volumetric flask. This technique is capable of measuring flowrates to less than 0.5%. In "flow-tube" methods, which are sometimes useful, air bubbles are introduced into the detector eluent, which then passes through a transparent, volume-calibrated tube. The bubble is timed as it travels between two volume markers on the tube for flowrate calculation. Flowrates can be measured in 20-30 sec by this approach, with a precision of about 1%.

In SEC the flowrate often is measured automatically via a syphon counter, which for high-efficiency columns should have a relatively small volume (e.g., 1 ml), since the total peak volumes also are relatively small (16). Each volume "dump" in the syphon actuates a photoelectric switch that indicates the event on the recorder trace. If the SEC column is operated at elevated temperature, the syphon also must be heated, and optimum devices use a vapor bypass tube to eliminate solvent evaporation during mobile-phase collection.

Data-Handling Systems

A good quality high-speed recorder should be used in modern LC, since many separations involve sharp peaks that require fast, accurate monitoring. Preferred characteristics of a recorder include a full-scale pen response of 1 sec. or less (equivalent to a time-constant τ of ≤ 0.3 sec), a high input impedence (e.g., greater than 1 megohm), good AC noise rejection, a "floating" input, and variable chart speeds (e.g., 10 cm/min to 10 cm/hr). As discussed in Section 13.2, digital electronic integrators and computers are valuable for precise quantitative analyses and are particularly useful for routine operations.

3.8 INTEGRATED AND SPECIALIZED INSTRUMENTS

To improve the versatility, convenience, and precision of LC, the most recent generation of integrated instruments have included a microprocessor to control operating parameters and act as a data-handling center. As illustrated by the schematic in Figure 3.20 for the Spectra-Physics Model 8000, the microprocessor

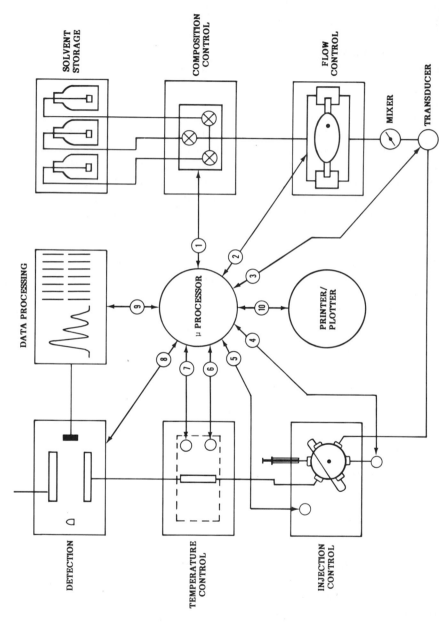

Figure 3.20 Schematic of integrated microprocessor-controlled liquid chromatograph (Spectra-Physics Model 8000). Illustration courtesy of Spectra-Physics.

120

is used to control [1] the composition of mobile phase for both isocratic and gradient elution separation, [2] and [3] pump flow output, composition of mobile phase for both isocratic and gradient elution separation, [4] and [5] sample injection, [6] and [7] column oven temperature, [8] detector output, and other functions such as [9] data processing and [10] data printout. Another advantage of the microprocessor system is that it continuously monitors all separation parameters to ensure that these are being controlled at the level set by the operator.

Operation of integrated microprocessor-controlled LC instruments takes several forms. For example, some instruments are accessed by alphanumeric keyboard systems, (e.g., Hewlett-Packard, Spectra-Physics) that are quite flexible but require considerable operator training. The Du Pont Model 850 instrument shown in Figure 3.21 uses an interactive operator-access system consisting of a series of functional touch-switches, specifically controlling each of the operating parameters. The Varian Model 5000 series uses a keyboard system and an interactive cathode-ray tube (CRT) to assist in building programs, monitoring parameters, and diagnosing problems.

Microprocessor-controlled integrated equipments generally provide instant status of the various operating parameters. These sophisticated instruments also

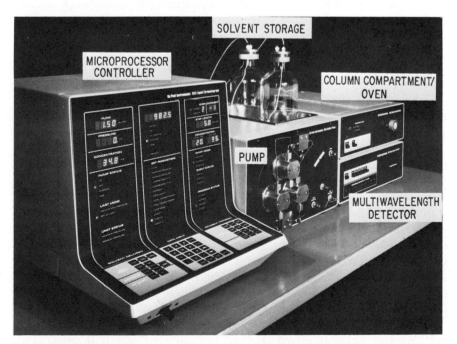

Figure 3.21 Modern integrated microprocessor-controlled liquid chromatograph (Du Pont Model 850). Illustration courtesy of Du Pont Instrument Products Division.

have additional safety features to protect continuously against overpressuring the column or fire caused by the leakage of flammable solvents. Microprocessor-controlled equipment also permits the use of very sophisticated and flexible solvent gradient programs, and with some instruments, gradient programs may be based on three or more different solvents. Some microprocessor-controlled LC instruments also can communicate with other computers, permitting the use of very sophisticated data reduction programs only available with more powerful external systems.

Most integrated, microprocessor-controlled instruments also have the capability to use internally stored parameter-sets to perform different, unattended chromatographic tasks. For example, the instruments can be operated automatically by means of user-written instructions to perform desired chromatographic runs. Method files can be held in microprocessor memory and recalled on sequential demand so that the user can mix different types of runs (i.e., different solvents, flowrates, temperature, etc.) in a single programmed sequence. Thus, a preprogrammed series of chromatographic parameters can be called up, for example, to search for optimum separation conditions involving changes in solvent composition, temperature, flowrate, and so on. With automatic sampling (Section 3.6), such an instrument can be set up for unattended overnight operation, if desired, reducing the amount of "handwork," so the operator can greatly extend his efficiency in solving LC problems. Microprocessor-controlled integrated instruments are relatively expensive and are most suited for scouting or research studies. Less-complicated and less-expensive equipments generally are used for routine, quantitative analyses.

Integrated microinstrumentation for modern LC is now commercially available (Familiac-100; Jasco) to permit analysis when only extremely small amounts of sample are available (17). The Familiac-100 uses 0.5-mm i.d. columns, which requires special instrument design to reduce extracolumn effects. For example, the UV detector cell has a volume of 0.3 μl, and optimum sample volumes of less than 1 μl are injected with a carefully designed on-column injection system. Because of the low flowrates required with the narrow i.d. columns (typically, 10 μl/min.), solvent consumption by the equipment is trivial; gradient elution is also feasible. However, the main advantage of micro-LC is the ability to utilize extremely small samples, and thus maintain solute concentrations in the column effluent many times that of conventional modern LC systems (for samples of the same size). Thus, micro-LC can be attractive for trace analysis and is particularly suited for qualitative identification by interfacing with ancillary instruments (e.g., a mass spectrometer). However, column performance of the narrow-bore column is poorer relative to conventional small-particle systems.

Finally, apparatus for on-stream or process-control LC is also being offered (e.g., Applied Automation and Du Pont Instruments Products Division). This equipment has been specifically designed for plant environments and long-term routine measurements; therefore, such apparatus should be considered if process applications are of interest (*18*). In most cases specific applications must be custom-engineered to ensure satisfactory results. The least difficulty with long-term separation reproducibility should occur with separations when SEC and BPC columns are used. Column switching (e.g., back flushing), coupled columns, and so on (Section 16.2) are techniques that are particularly suited for on-stream LC analyses.

3.9 LABORATORY SAFETY

Neophytes sometimes worry about the safety of modern LC instruments because of the operation at high pressures. However, it should be remembered that under use conditions, liquid compression is minimal (e.g., only about a 4% decrease in the volume of methanol from atmospheric to 7500 psi or 500 bar) and little energy is stored. Therefore, these pressures generally pose no real hazard. Should a leak or break occur within a high-pressure system supplied by a constant-volume pump, the pressure immediately drops to atmospheric, with leakage of only a small amount of solvent. Constant-pressure pumps will maintain pressure in the system as long as mobile phase is available to the pump, and leaks prior to the column can result in a squirt of mobile phase with considerable force and loss of solvent if they are not fixed.

The main safety consideration in LC is the handling of flammable or toxic solvents. The potential hazard of these chemicals can be eliminated by using a well-ventilated laboratory, safety glasses, and so on. Placing the LC instrument in a hood is rarely convenient, and the large volume of air that flows around the equipment can cause problems in maintaining a constant temperature. Overhead exhausts often are not adequate, since some solvent vapors are heavier than air; however, "elephant-trunk" exhaust lines can be useful in exhausting vapors from localized areas. To reduce the possibility of fire, some integrated instruments offer the option of purging the reservoir and column compartment continuously with nitrogen. Pressure-overload devices and solvent vapor-sensing systems also protect certain equipment from possible hazards due to solvent leaks.

REFERENCES

1. H. M. McNair and C. D. Chandler, *J. Chromatogr. Sci.,* **4**, 477 (1976).

2. T. Wolf, *Chromatographia,* **7**, 34 (1974).

3. R. P. W. Scott and P. Kucera, *J. Chromator. Sci.,* **9**, 641 (1971).

4. R. E. Leitch, *ibid.,* 531 (1971).

5. F. W. Karasek, *Res. Dev.,* **28**, 32 (1977).

5a. S. R. Bakalyar, M. B. T. Bradley, and R. Honganen, *J. Chromatogr.,* **158**, 277 (1978).

6. Du Pont Instruments Products Division, Bulletin E19713, 1978.

7. D. D. Bly, H. J. Stoklosa, J. J. Kirkland, and W. W. Yau, *Anal. Chem.*, **47**, 1810 (1975).

8. E. L. Johnson and R. Stevenson, *Basic Liquid Chromatography*, 2nd ed., Varian Aerograph, Walnut Creek, Calif., 1978.

8a. H. Schrenker, *Am. Lab.*, **10**, 111 (1978).

9. J. H. Knox, dir., *Manual for Intensive Course on High Performance Liquid Chromatography*, Dept. of Chemistry, University of Edinburgh, Edinburgh, Scotland, June 9-15, 1977.

10. S. Mori, K. Mochizuki, M. Watanahe, and M. Saito, *Am. Lab.*, **9**, 21 (1977).

11. M. Martin, G. Blu, C. Eon, and G. Guiochon, *J. Chromatogr.*, **112**, 339 (1975).

12. R. A. Henry, in *Modern Practice of Liquid Chromatography*, J. J. Kirkland, ed., Wiley-Interscience, New York, 1971, Chap. 2.

12a. L. R. Snyder, *Chromatogr. Rev.*, **7**, 1 (1965).

13. R. P. W. Scott and P. Kucera, *Anal. Chem.*, **45**, 749, (1973).

14. L. R. Snyder private correspondence.

15. W. W. Yau, H. L. Suchan, and C. P. Malone, *J. Polym. Sci.*, Part A2, **6**, 1349 (1968).

16. J. J. Kirkland, *J. Chromatogr.*, **125**, 231 (1976).

17. F. W. Karasek, *Res. Dev.*, **28**, 42 (1977).

18. R. A. Mowery, Jr., and L. B. Roof, *Anal. Instrum.*, **14**, 19 (1976).

BIBLIOGRAPHY

Berry, L., and B. L. Karger, *Anal. Chem.*, **45**, 819A (1973) (review of pumps and injectors for LC).

Henry, R. A., in *Modern Practice of Liquid Chromatography*, J. J. Kirkland, ed., Wiley-Interscience, New York, 1971, p. 55 (discussion of all components for LC system except detectors).

Huber, J. F. K., (ed.), *Instrumentation for High-Performance Liquid Chromatography*, Elsevier, Amsterdam, 1978 (good overall discussion of modern LC instrumentation).

International Chromatography Guide, *J. Chromatogr. Sci.*, March/April 1977 (guide to available LC equipment and accessories).

Karasek, F. W., Res. Dev., **28**, 42 (1977). (microinstrumentation for LC).

Laboratory Buyers' Guide Edition, *Am. Lab.*, November, 1977 (guide to available LC equipment and related products).

Laboratory Guide, 1978/79, *Anal. Chem.*, August 1978 (guide to available equipment and other LC products).

Martin, M., C. Eon, and G. Guiochon, *J. Chromatogr.*, **108**, 229 (1975) (discussion of problems in instrument design).

Parris, N. A., *Instrumental Liquid Chromatography*, Elsevier, New York, 1976 (practical manual with good discussion of equipment.

Schrenher, H. *Am. Lab.*, **10**, 111 (1978) (excellent review and comparison of solvent metering systems for LC).

FOUR

DETECTORS

4.1 Introduction   126
4.2 Detector Characteristics   127
4.3 UV-Visible Photometers and Spectrophotometers   130
    Photometers   131
    Spectrophotometers   135
    Characteristics of UV Detectors   139
4.4 Differential Refractometers   140
    Fresnel Refractometer   141
    Deflection Refractometer   142
    Interferometric Refractometer   143
    Characteristics of RI Detectors   143
4.5 Fluorometers   145
4.6 Infrared Photometers   147
4.7 Electrochemical (Amperometric) Detectors   153
4.8 Radioactivity Detectors   158
4.9 Conductivity Detectors   161
4.10 Summary of the Characteristics of Most-Used Detectors   161
4.11 Other Detectors   165
References   165
Bibliography   166

4.1 INTRODUCTION

A major instrumentation requirement in modern LC is a sensitive detector for continuously monitoring the column effluent. Unfortunately, the sensing of LC bands can be difficult, since the physical properties of both mobile phase and solutes are often quite similar. Various approaches to the problem of detection have been pursued in the development of modern LC:

- Differential measurement of a bulk or general property of both sample and solvent.
- Measurement of a sample property that is not possessed by the mobile phase.
- Detection after eliminating the mobile phase.

A variety of detectors have been developed for LC based on one of these approaches. An ideal LC detector would be one with the characteristics listed in Table 4.1. Unfortunately, no presently available detector possesses all these properties, nor is it likely that any such detector will be developed.

Table 4.1   Characteristics of ideal LC detector.

The ideal detector should:

    Have high sensitivity and the same predictable response

    Respond to all solutes, or else have predictable specificity

    Have a wide range of linearity

    Be unaffected by changes in temperature and mobile-phase flow

    Respond independently of the mobile phase

    Not contribute to extracolumn band broadening

    Be reliable and convenient to use

    Have a response that increases linearly with the amount of solute

    Be nondestructive of the solute

    Provide qualitative information on the detected peak

    Have a fast response

No present LC detector is as versatile or as universal as might be desired; there is no LC equivalent to the thermal conductivity or flame ionization detectors of gas chromatography. Nevertheless, presently available LC detectors allow a very wide range of applications, and only rarely is an LC application seriously limited by the detector. If sample components are very unlike and differ widely in their physical properties, it may be necessary to utilize two or more detectors in series, to ensure that each component of interest is adequately measured.

The purpose of this chapter is to acquaint the reader with the more commonly used LC detectors, including their characteristics, advantages, and limitations. Texts specifically devoted to LC detection [e.g., see (*1*)] should be consulted for more details and for additional information on less-common LC detectors.

4.2 DETECTOR CHARACTERISTICS

Two general types of LC detectors are used: (a) *bulk property* or *general detectors* measure a change in some overall physical property of the mobile phase plus the solute (e.g., refractive index); (b) *solute property* or *selective detectors* are sensitive only to some property of the solute (e.g., UV absorption). Which type of detector should be used for a particular problem depends on the characteristics of the solute, the sensitivity and selectivity required, and the convenience and versatility desired.

Specific information is required on the specifications or characteristics of detectors to allow one to be selected for a particular application. Unfortunately, the major specifications and characteristics of detectors—noise, sensitivity, response, and linearity—are not presented in a standard form by suppliers. Therefore, considerable confusion exists regarding the relative performance of some detectors. To further complicate matters, for most samples of interest there are no published reference values for many of the properties utilized by different LC detectors. Therefore, to determine whether or not a particular detector is adequate for a problem, a similar application in the literature must often be found. The application laboratories of the various equipment suppliers also have experience in determining which detector might be appropriate for a particular problem. Ultimately, the user is often required to test the detector under consideration on the sample of interest.

When evaluating detectors for LC, several features should be considered to obtain optimum use of these devices. First, *detector noise* should be known (no solvent flow). Detector noise level is defined as the maximum amplitude of the combined high-frequency noise and random baseline fluctuations arising from instrument electronics, temperature fluctuations, line voltage surges, and other effects not directly attributable to the solute. Variation of the detector signal can also result from flow changes, pump pulsations, and so on. At the top of Figure 4.1 is a representation of both high-frequency noise, which appears as a "fuzz" and widens the baseline, and *short-term noise*, which is a variation of the recorder tracing appearing as random peaks or valleys on the baseline. Detector noise makes difficult the sensing of very small peaks; it is often given as the random baseline variation in units of detector response (e.g., absorbance units for ultraviolet photometer), at a specified sensitivity.

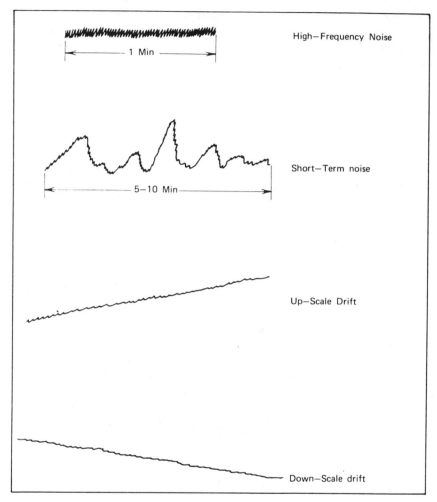

Figure 4.1   Detector noise and drift.

Also shown in Figure 4.1 is a representation of *detector drift* which is a continuous up-scale or down-scale excursion of the baseline, which tends to camouflage both noise and small peaks. Detector drift is usually defined as a change in detector response over a period of time, usually several hours. Detector noise cannot be accurately determined unless the drift is small in relation to the magnitude of the noise. Many problems with apparent detector noise and drift are a function of the total LC system (e.g., solvent impurities, temperature variations), rather than being inherent detector limitations.

Although all the characteristics listed in Table 4.1 are desirable, the need for detector sensitivity in LC must be stressed. Components of 10 ng or less in a sample must often be measured. Therefore, whether or not a particular detector has sufficient sensitivity for a desired application is of major interest. The *absolute sensitivity* of a detector is the total change in a physical parameter (e.g., absorbance for photometer) that is required for a full-scale deflection of the recorder at maximum detector sensitivity and at a specified noise level. It is important when comparing detectors that the sensitivities corresponding to equal noise levels be used. *Detection limit* or *detector sensitivity* is generally taken as the concentration or mass of solute entering the detector per unit time that will provide a signal-to-noise ratio of 2, and this is usually considered the minimum concentration or mass of a solute that can be adequately detected. The minimum detection limit often is dependent on the chromatographic system. For instance, detectability with general detectors such as a differential refractometer is very much a function of the mobile phase. For other systems the minimum detection limit is a function of other variables such as column plate count, detector response to solute and so on.

For a detector to be of use in quantitative analysis, the signal output should be linear with concentration (g/ml) for a concentration-sensitive detector and in mass (g/sec) for mass-sensitive detectors. Furthermore, the detector should have a wide *linear dynamic range* (e.g., 10^5) so that major and trace components can be determined in a single analysis over a wide concentration range.

There are other detector parameters that can be important. In particular, the cell should be designed with a minimum volume compatible with the other requirements of the detector. Peak broadening due to mixing within the detector cell is most significant when using low-volume, very efficient columns and especially so for early-eluting bands. For example, when using 5 × 0.46-cm columns of 3-μm particles, peaks with $k' < 3$ show significant broadening with 8-μl UV detector cells, relative to 1-μl cells (2). To minimize the broadening of early eluting peaks, the volume of the detector cell V_c should be less than about one-tenth the volume of the peak of interest V_p (3):

$$V_c < 0.1 V_p. \tag{4.1}$$

For example, with very high-performance LC columns, peaks of <50 μl are sometimes of interest; therefore, a detector cell with a volume of ⩽5 μl would be desirable. If the guideline in Eq. 4.1 is followed, extracolumn band broadening from the detector usually will have an insignificant effect on resolution.

Detector cell cavities should be kept smooth, with no unswept volumes (so-called dead volumes) that can cause band tailing. When placing detectors in series, the first detector should have the smallest cell volume. As discussed in Section 3.2, the tubing within the detector, connecting tubes from the column to the detector, and tubing between detectors all should have minimum internal diameter plus short coupling length, to minimize extra column band spreading.

It often is desirable that detector cells be able to operate under moderate pressure (e.g., 5-10 atm). Gas bubbles that might form in the detector can be avoided by imposing 1-2 atm of back pressure on the detector cell during flow. (Of course, prior degassing of the mobile phase also assists in eliminating gas bubbles in the detector.) Fittings on the detector and in the high-pressure flow system should be airtight to prevent diffusion of air into the mobile phase. Gases can diffuse into small holes from which pressurized liquids cannot escape; therefore, air can leak through insufficiently tightened compression fittings and dissolve in previously degassed mobile phase, eventually causing bubbles in the detector.

Separated bands as seen on the recorder can be significantly broader (and peak height reduced) relative to those actually sensed, as result of slow *detector response*. In addition to the limitation of the detector-amplifier circuit design, some detectors have an additional time constant, purposely introduced to remove high-frequency noise. A slow *recorder response* can similarly broaden and attenuate the peak. The effect of the detector (or recorder) time constant on peak distortion has been discussed by Sternberg (*3*). As a rule of thumb, the maximum detector time constant τ (sec) that can be tolerated is about one-third that of the standard deviation of the peak σ (sec). In practice this means that detectors with τ equal to 0.3 sec or less are desirable for high-efficiency small-particle columns.

Usefulness of a particular detector in a given problem can be affected by its susecptibility to mobile-phase flow and temperature changes. Precise quantitative analysis especially requires that a detector be insensitive to these factors. Finally, it is of particular importance in routine applications that the detector be trouble-free and easily operated by relatively unskilled personnel.

4.3 UV-VISIBLE PHOTOMETERS AND SPECTROPHOTOMETERS

The most widely used detectors in modern LC are photometers based on ultraviolet (UV) and visible absorption. These devices have a high sensitivity for many solutes but samples must absorb in the UV (or visible) region (e.g., 190-600 nm) to be detected. Sample concentration in the flow cell is related to the fraction of light transmitted through the cell by *Beer's law*:

$$\log\frac{I_0}{I} = \epsilon bc, \tag{4.2}$$

where I_0 is incident light intensity, I is the intensity of the transmitted light, ϵ is the molar absorptivity (or molar extinction coefficient) of the sample, b is the cell pathlength in cm, and c is the sample concentration in moles/l. Light-absorption LC detectors usually are designed to provide an output in *absorbance* that is linearly proportional to sample concentration in the flow cell,

$$A = \log\frac{I_o}{I} = \epsilon bc, \qquad (4.2a)$$

where A is the absorbance.

Properly designed UV detectors are relatively insensitive to flow and temperature changes. UV photometers with sensitivities of 0.002 absorbance units full-scale (AUFS) with ±1% noise are now commercially available. With this high level of sensitivity, solutes with relatively low absorptivities can be monitored, and it is possible to detect a few nanograms of a solute having only moderate UV absorbance. The wider linear dynamic range of UV photometric detectors ($\sim 10^5$) makes it possible to measure both trace and major components on the same chromatogram.

Photometers

Figure 4.2 is a schematic of a monochromatic UV photometer available from Du Pont; similar designs are widely used in modern LC. UV radiation at 254 nm from a low-pressure mercury lamp shines on the entrance of a flow cell (*b*) and is transmitted through this sample cell and strikes the analytical phototube (*a*). Radiation from the source also is directed through a neutral-density filter (*c*), and then to the reference photocell (*r*). Current from these photocells is fed to log amplifiers, which generate an output linear with solute concentration. The output signal to the recorder normally is attenuated with a simple bridge network.

In addition to the widely used 254-nm photometer, photometers operating at other wavelengths are available, for example, 214, 220, 280, 313, 334, and 365

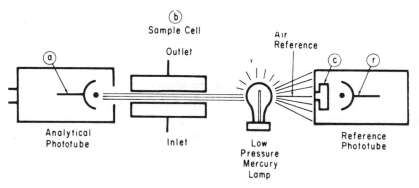

Figure 4.2 Schematic of monochromatic UV photometric detector (Du Pont). Reprinted by permission of Du Pont Instrument Products Division.

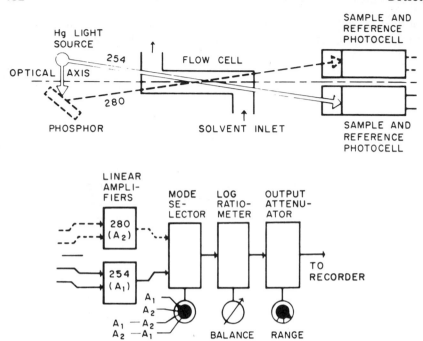

Figure 4.3 Schematic of dual-channel UV photometric detector (Altex). Reprinted by permission of Altex Scientific, Inc.

nm. Multiple-wavelength UV photometers often use a source (e.g., a medium-pressure mercury lamp) that has several of these relatively intense lines, with appropriate filters for selecting the desired wavelength. Alternatively, other wavelengths can be obtained by allowing the intense 254-nm energy from a low-pressure mercury lamp to excite a suitable phosphor.

Some UV photometers are designed to permit simultaneous monitoring at two wavelengths, as illustrated by the schematic in Figure 4.3 for an Altex detector. A 254-nm low-pressure mercury source is used to excite a phosphor that emits radiation at a second wavelength—in this case, 280 nm. Both light sources shine through the sample and reference cell, the small aperture of the flow cell passages acting as pinhole cameras to produce a physical separation between the light falling on the two photocells. With an interference filter, one pair of photocells is used to monitor at 254 nm, and a second pair of photocells with another filter is used to monitor at 280 nm. Both sets of photocells are continuously monitoring both the sample and the reference side of the flow cell, allowing for real-time monitoring at both wavelengths. Electrical output from each photocell system is fed to log amplifiers that produce a recorder output signal in linear absorbance units.

(a)   (b)

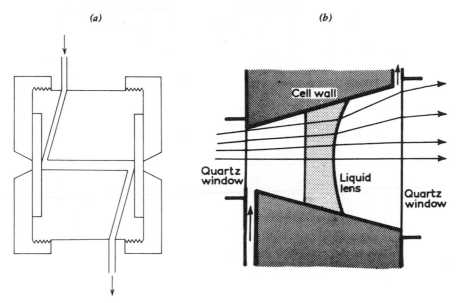

Figure 4.4 Schematics of flow cells for UV detectors. (*a*) "*Z*"-pattern flow cell, reprinted from (*3a*) with permission; (*b*) tapered flowcell, reprinted from (*3b*) with permission. "Liquid lens" refers to an eluate volume element of different refractive index.

Cells for UV photometers are typically 1 mm i.d. and 10 mm in length, with an internal volume of about 8 μl. Figure 4.4*a* shows a schematic of a conventional "*Z*"-type UV flow cell widely used in UV photometers. A potential problem in using UV detectors operated at high sensitivity is distinguishing between sample peaks and pseudopeaks due to refraction effects that affect the detector signal. If the refractive index within the cell changes (e.g., during gradient elution), the amount of energy reaching the detector can change, because of refractive effects at the cell wall. Special designs such as a tapered cell (Figure 4.4*b*, Waters Associates) reduce the magnitude of this wall effect (but can increase mixing volume within the cell if not properly designed). Alternatively, collimating and masking the light entering the cell (e.g., in the Spectra-Physics 8310) can similarly eliminate these refractive effects. Although many UV spectrophotometric cells can only be operated at pressures up to about 500 psi, some are available that can withstand pressures up to about 5000 psi, which allows the detector to be used for special applications (e.g., two-column recycle—Section 12.6).

When using the UV photometric detector, it is not necessary to work at the absorption maximum of the peak. Although the spectrum of cytidine in acidic solution shown in Figure 4.5 has an absorption maximum at 280 nm, nanogram

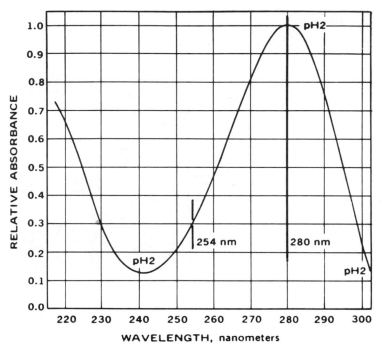

Figure 4.5 UV absorption spectrum for cytidine in acidic solution. Reprinted by permis
sion of P. L. Biochemicals.

amounts can be detected by working at 254 nm. Even though detection at the
absorption maximum provides maximum sensitivity, it is sometimes more
important to use the wavelength providing the highest selectivity, that is,
maximum freedom from interferences. In quantitative analysis, wavelengths
providing the largest signal *relative* to possible interfering substances should be
used. Figure 4.6 shows chromatograms of aflatoxins in peanut butter, obtained
with UV photometric detection at 254 and 365 nm. The aflatoxins absorb more
strongly at 365 nm, but, more importantly, many impurities that absorb and
interfere at 254 nm do not absorb at 365 nm. Thus, enhanced detection selectivity makes the longer wavelength much more desirable in this application.

Many compounds absorb fairly strongly at 254 nm; therefore, a simple photometric detector operating at this wavelength is satisfactory for many applications. The minimum structural requirement for UV detection at 254 nm is
usually a double-bonded chromophore to which is attached some structure with
unpaired electrons. Some possible minimum structures for absorption at 254
nm are shown here, but these are not limiting:

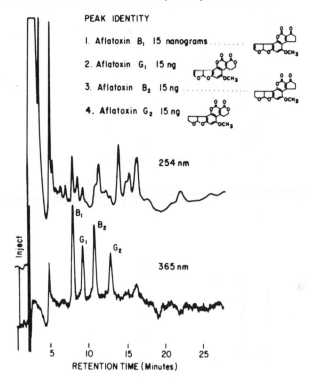

Figure 4.6 Detection of aflatoxins at two different wavelengths. Column, Zorbax-SIL, 25 × 0.21 cm; mobile phase, 60% CH_2Cl_2/40% $CHCl_3$ (both 50% H_2O-saturated)/0.1% methanol; pressure, 1500 psi; flow, 0.7 cc/min; temp., ambient; detector, UV photometer, 254 nm, 0.02 AUFS; 365 nm, 0.01 AUFS. Reprinted by permission of Du Pont Instrument Products Division.

$$-\underset{\underset{W}{|}}{X}=Y \, ,$$

Where X = C, S, P, and so on; Y = O, S, P, N, and so on; and W = UV enhancer, such as S, P, and so on.

Spectrophotometers

The single most useful and versatile detector in modern LC probably is the UV spectrophotometer, which offers a wide selection of UV and visible wavelengths. Such devices have the versatility and convenience of allowing operation at the absorption maximum of a solute, or at a wavelength that provides maximum

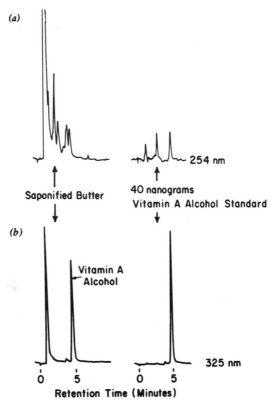

Figure 4.7 Selectivity of UV spectrophotometric detection at two wavelengths. Column, 25 × 0.21 cm, Zorbax-SIL; mobile phase, 1.6% isopropanol in hexane; flowrate, 0.7 ml/min; temp., 24°C; detector: UV, 254 nm, 0.04 AUFS; 325 mn, 0.04 AUFS. Reprinted by permission of Du Pont Instrument Products Division.

selectivity. The selectivity of a UV spectrophotometric detector versus the single-wavelength photometer is illustrated in Figure 4.7. In the chromatograms of Figure 4.7a, made with the 254-nm photometeric detector, a relatively poor response is shown for the vitamin A alcohol standard, and this component is completely obscured by other components in saponified butter. On the other hand, the Figure 4.7b chromatograms show a much larger response for the vitamin A alcohol standard with a spectrophotometric detector at 325 nm, and this compound is easily measured in saponified butter without interference.

Modern LC spectrophotometers have high-energy, continuous-spectrum sources, relatively wide band-pass monochrometers (narrow wavelengths generally are not required for LC), and stable, low-noise electronics. These factors

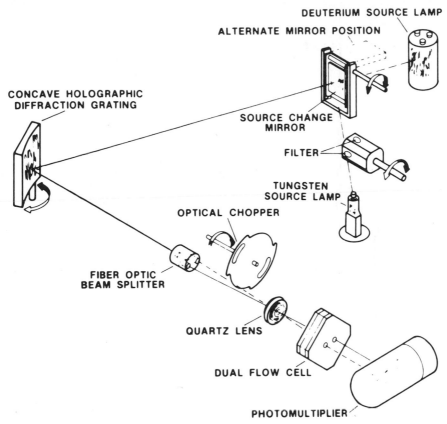

Figure 4.8 Schematic of UV spectrophotometric detector (LDC). Reprinted with permission of Laboratory Data Control.

permit signal-to-noise ratios and linearity of response almost comparable to those of UV photometric detectors. However, spectrophotometric detectors usually cost about twice as much as monochromatic photometers.

The schematic of the LDC spectrophotometric detector for LC is shown in Figure 4.8. For UV measurements an energy continuum supplied by a deuterium source lamp is directed towards a concave holographic defraction grating. By rotating the grating to the appropriate angle of incidence, the desired wavelength for a particular analysis is projected into a fiber optic beam splitter. Two separate beams of equal intensity are produced by the beam splitter and are focused onto dual flow-cells by a single quartz lens. The sample and reference cells are alternatively illuminated by interposing a rotating optical chopper disk, which permits double-beam operation for maximum detector-output stability. Energy passes through the sample cell to a photomultiplier and is compared to the

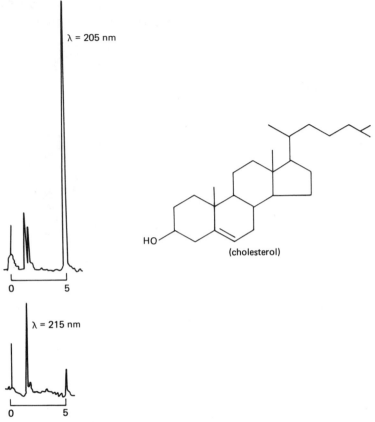

Figure 4.9 UV detection at low wavelength. Column, Partisil-PAC; mobile phase, hexane: CH$_3$CN, 98:2; flowrate, 2 ml/min; detection at wavelengths shown; sensitivity, 0.64 AUFS; sample, cholesterol in hexane solution, 5 μl. Reprinted with permission of Tracor Instruments.

energy passing through the reference cell by electronic signal-processing circuits, designed to produce an output that is linear with solute concentration. For measurements in the visible region a source-change mirror is positioned in the entrance beam to direct the energy from a tungsten source lamp toward the defraction grating.

Some UV spectrophotometric detectors (e.g., Hewlett-Packard) are designed to allow spectral scanning of chromatographic peaks during the separation, using a stop-flow technique. Depending on the equipment, the peak trapped in the cell can be manually or automatically scanned, to provide information regarding the optimum wavelength and identity of the component. The integrated, microprocessor-controlled Hewlett-Packard detector also can be programmed to select the optimum wavelength for each known peak in the chromatogram.

UV spectrophotometric detectors can be made to respond throughout a wide wavelength range (typically 190-650 nm), which makes possible the detection of a broad spectrum of compound types. Compounds with only a single double bond (e.g., carbohydrates, olefins) respond to wavelengths of less than 215 nm, and other solutes such as alkyl bromides, sulfur-containing compounds, and so on, can be detected by operating at wavelengths of less than 220 nm. Figure 4.9 illustrates the importance of being able to detect at lower UV wavelengths. Cholesterol, normally considered to be nonabsorbing in the UV, is easily detected at 205 nm, but only weakly at 215 nm.

Photometers and spectrophotometers that work in the visible range also are used as detectors in LC. These devices are potentially useful both as general and selective detectors by utilizing well-known color reactions that can be performed with the column effluent and monitored by absorption photometry (Section 17.3).

Rapid-scanning multiwavelength (RSMW) photometers have been based on oscillating mirrors (4), solid-state diode arrays (5, 6), and silicon-target Vidicon tubes (7). Each approach has its own advantages and limitations. Although RSMW devices are not yet in commercial form, it is clear that they offer substantial potential advantages. With computer support, such detectors can provide on-line spectral identification to complement retention times, thus resembling real-time use of GC/MS detection (but with less information). RSMW photometers also allow selection of wavelengths that optimize the sensitivity and resolution for each component in the chromatogram. Interferences in the chromatogram also can be tuned out, by selecting wavelengths where the response for the interferences is neglible. RSMW detectors also permit the resolving of overlapping chromatographic peaks by deconvolution techniques, and they allow the detection of impurities that are hidden within chromatographic peaks. The RSMW detector offers promise as a versatile, if relatively expensive, detector for LC. Its ultimate success depends on improvements in signal-to-noise ratios, stability, and convenience.

Characteristics of UV Detectors

Table 4.2 summarizes the general characteristics of UV detectors for modern LC. UV detectors are ideal for use with gradient elution, and many common UV-transmitting solvents of varying solvent strength are available as mobile phases. Therefore, selection of solvents for use with the UV detector is usually not a practical limitation (Table 6.1). The UV detector is very useful for the trace analysis of UV-absorbing solutes, since the UV-transmitting solvent and other non-UV-absorbing compounds in the sample are not detected and do not interfere. UV detectors are reliable and easy to operate, and are particularly suitable for use by less-skilled operators. A potential disadvantage of UV detectors is their widely varying response for different solutes.

Table 4.2 Characteristics of ultraviolet detectors.

Capable of very high sensitivity (but samples must have UV absorption)

Good linearity range ($\sim 10^5$)

Can be made with very small cell volumes (small band-broadening influence)

Relatively insensitive to mobile-phase flow and temperature changes (except at high sensitivity)

Very reliable

Easy to operate

Nondestructive to sample

Widely varying response for different solutes

Gradient elution capability

Wavelength can be selected (spectrophotometers and RSMN only)

The "background" or baseline absorbance of UV detectors can increase with continued use. This usually indicates that the cell windows have become dirty and need cleaning or replacement (Table 19.4). Although the linear-response range of UV detectors often can be as high as 2-3 Å, this range can be much reduced by dirty cell windows. The stability of UV detectors against flow and temperature changes has been improved in some instruments by thermostatting the cell. However, in some designs this is done at the expense of increasing the total detector volume, since additional capillary tubing is required prior to the cell to temperature-equilibrate the incoming mobile phase.

UV photometers are much preferred to spectrophotometers for routine applications, if wavelength requirements allow, because of the generally better signal-to-noise ratio and superior lifetime of the lamp source. In ordinary use, 254-nm UV lamps often last for a year (24-hour operation) before replacement is required because of increased noise. Deuterium sources for spectrophotometers generally have only a 500-hour operational lifetime before replacement of this relatively expensive component is required. Clearly, the choice between a photometer and a spectrophotometer depends on the versatility and the selectivity of the spectrophotometer versus the sensitivity (i.e., at 254 nm), convenience, economy, and simplicity of the simple photometer.

4.4 DIFFERENTIAL REFRACTOMETERS

The second most widely used LC detector is the differential refractometer. This device monitors the difference in refractive index (RI) between pure reference mobile phase and the column effluent. This bulk property or general detector responds to all solutes if the refractive index of the solute is significantly different from that of the mobile phase. Commercially available RI detectors operate on three different principles, each having specific advantages and disadvantages.

Fresnel Refractometer

Shown schematically in Figure 4.10 is a refractometer (e.g., LDC) based on Fresnel's law of reflection, which states that the amount of light reflected at a glass-liquid interface varies with the angle of incidence and the refractive index of the liquid. To obtain maximum sensitivity and linearity, the incident light impinges on the liquid-glass interface at an angle that is slightly subcritical. Fluctuations caused by noise and temperature changes are minimized by using a differential measurement, where the refractive indices of the sample and reference streams are continuously compared. In Figure 4.10, light from source lamp *SL* passes through a source mask *M*1, an infrared blocking filter *F*, and an aperature mask *M*2. The light is collimated by lens *L*1. Mask *M*2 yields two collimated beams, which enter the cell prism and are focused on the glass-liquid interface. Sample and reference cell cavities are formed with a Teflon gasket, which is

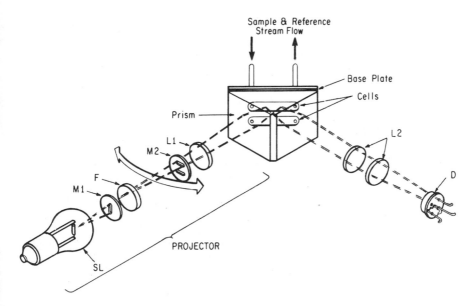

Figure 4.10 Schematic of Fresnel refractometer detector (LDC). Reprinted by permission of Laboratory Data Control.

clamped between the cell prism and a stainless base plate. All the optical components are mounted on a separate optical bench, which rotates around a pivot to allow a variation in the angle of incident light and maintain the near-critical angle of refraction. As light is transmitted through the cell interfaces, it passes through the liquid film and impinges on the surface of the cell base plate. The energy is reflected, and the lens $L2$ focuses this reflected light on the dual-element photometer-detector D, for amplification and presentation on a recorder.

Because the cells in the Fresnel RI detector can be made very small (about 3 μl) and are cleanly swept by the mobile phase, some workers prefer to use this device with very high-performance LC columns. However, the cell windows in this detector must be kept very clean for good results. This type of refractometer has a relatively limited range of linearity. In addition, two different prisms must be used to cover the useful refractive index range (about n = 1.33-1.63); however, a single prism for the lower RI range handles most situations, so this is not a significant disadvantage.

Deflection Refractometers

The most common type of RI detector is the deflection device shown schematically in Figure 4.11 (e.g., from Waters, Micromeritics). Light from source A is limited by mask B, is collimated by lens C, and passes through the detector cell D. The cell has separate reference and sample compartments, which are separated by a diagonal piece of glass. When the composition of the mobile phase changes in the sample cell, the change in refractive index causes a deflection in the final position of the light beam on the photodetector. As the incident light passes through the cell, it is first deflected, reflected from mirror E back of the

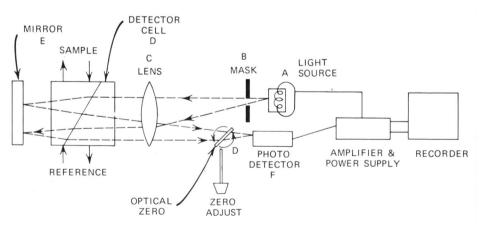

Figure 4.11 Deflection refractometer detector (Waters). Reprinted by permission of Waters Associates.

cell, and again deflected. Lens C focuses this deflected light on photodetector F, which produces an electrical signal proportional to the position of the light. The output signal is then amplified and relayed to a recorder.

Deflection RI detectors have the advantage of a wide range of linearity; also, only one cell is needed for the entire refractive index range. The cell in this detector is generally not as small nor as cleanly swept as the cells of the Fresnel type, but is less sensitive to contaminants on cell windows.

Interferometric Refractometer

A third type of refractometer (Optilab) uses the shearing-interferometer principle for measurement, as shown schematically in Figure 4.12. The light source beam (e.g., 546 nm) is divided into two parts by a beam splitter, is focused by a lens, and passes through the sample and reference cells (5 μl). The light beams are recombined by a second lens and a beam splitter, and fall on the detector. The difference in refractive index between the sample and the reference produces a difference in optical path length and this is measured by the interferometer in fractions of light wavelengths. Response of this device is linear, and the sensitivity is claimed by the manufacturer to be 10-fold greater than other RI dectectors. Under optimum operating conditions, it is possible to detect about 0.1 μg of solutes with the interferometer RI detector. Practical experience in the field with this type of detector is so far quite limited.

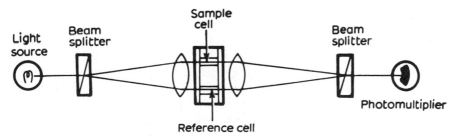

Figure 4.12 Shearing interferometric refractometer detector (Optilab). Reprinted with permission of Optilab.

Characteristics of RI Detectors

Table 4.3 summarizes the characteristics of RI detectors. Because they respond to all solutes, these devices have excellent versatility if the mobile phase is properly selected. For maximum RI detector sensitivity, the mobile phase should have a refractive index as different from the solute as possible. However, even under optimum conditions, RI detectors have only modest sensitivity. Although this detector is generally not useful for trace analysis, it is possible in optimum situations to quantitate peaks at the 0.1% concentration level. One of the severe limitations of RI detectors is that they are impractical to use with gradient elution, because it is exceedingly difficult to match the refractive in-

Table 4.3   Characteristics of refractive index detectors.

Excellent versatility—any solute can be detected

Moderate sensitivity for solutes

Generally not useful for trace analyses

Efficient heat exchanger required

Relatively insensitive to mobile-phase flow changes (if properly thermostatted)

Sensitive to temperature changes

Reliable, fairly easy to operate

Difficult to use with gradient elution

Nondestructive

dices of reference and sample streams. Despite the sensitivity limitation and impracticality in gradient elution, the differential refractometer is widely used, particularly so in size-exclusion chromatography, where sensitivity often is not important.

The sensitivity of RI detectors to temperature change also represents a severe limitation. In practice, it is difficult to maintain the temperature of the refractometer cell with a precision that permits its use at maximum theoretical sensitivity. To obtain optimum sensitivity and stability with RI detectors, efficient heat exchangers in the line between the outlet of the column and the detector cell are required. Unfortunately, these heat exchangers increase the volume of the detector flow path and can produce significant band broadening with high-efficiency analytical columns. On the other hand, flow programming of the mobile phase is feasible, because the heat exchanger effectively eliminates temperature changes in the solvent entering the detector as a result of flow changes. Refractometers are convenient and reliable, although generally not as trouble-free and easy to operate as UV photometers.

RI baseline drifts can be a problem as a result of changing from one bottle of "pure" solvent to another. Binary solvents can cause even more problems with baseline drift, and should be avoided if possible. RI-detector baseline drift can be severe when changing solvents, until the previous solvent has been completely eliminated from the equipment and the column. Change in solvent equilibrium within a column (e.g., because of temperature changes) also causes a refractive index change in the eluent. To maintain a homogeneous composition of the mobile phase during a series of runs, it is often desirable to stir the solvent continuously and slowly in the reservoir. Change in solvent composition by any means (degassing, evaporation from an open vessel, water vapor pickup, etc.) should be avoided, if maximum baseline stability and sensitivity are required in RI detection.

Maximum baseline stability with RI detectors usually is obtained by using a reference cell with a static or "captured" mobile phase identical to that used for the separation. A useful technique is to purge the reference cell frequently with the mobile phase eluting from the column (but containing no sample, and after column equilibration). The reference cell is then sealed off to prevent loss of solvent during use.

4.5 FLUOROMETERS

The fluorometer is a very sensitive and selective detector for fluorescing solutes because of its ability to measure energy emitted from certain solutes excited by UV radiation. Fluorescent derivatives of many nonfluorescing substances can also be prepared (Section 17.3), and this approach is attractive for selectively detecting certain compounds for which sensitive detection methods are otherwise not available. Compounds that are symmetrically conjugated and not strongly ionic often fluoresce. The main application areas for fluorometric detection are in the analysis of biological samples, pharmaceuticals, foods, and fossil fuels and in environmental testing.

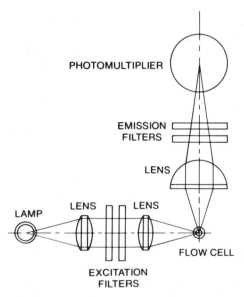

Figure 4.13 Optical diagram of right-angle filter fluorometer (Varian). Reprinted with permission of Varian Associates.

A schematic of a typical right-angle fluorometer (Varian) is shown in Figure 4.13. A tungsten-halide lamp provides strong excitation-radiation over the continuum above 280 nm. A lens plus excitation filter collects energy from the lamp, isolates the spectral band width of interest, and focuses the beam onto a 25 μl cell. Another lens collects the fluorescent energy from this cell at 90° to the exciting beam, and focuses it through emission filters onto the photomultiplier tube, which measures the radiation.

In many cases the fluorescence detector is 100-fold more sensitive than UV absorption; therefore, the fluorometer is one of the most sensitive of the LC detectors. Because of its high sensitivity, the fluorometer is particularly useful in trace analysis when either the sample size is small or the solute concentration is extremely low. Figure 4.14 compares the sensitivity of a fluorescence detector

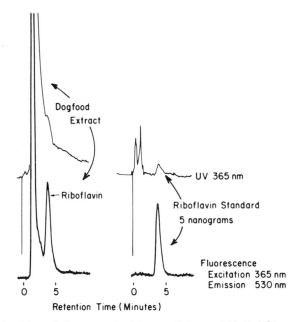

Figure 4.14 Selectivity of fluorometric detector. Column, 100 × 0.21 cm, Zipax-SCX, <37 μm; mobile phase, water; flowrate, 1.0 ml/min; temp, ambient; detector: UV, 365 nm, 0.01 AUFS; fluorescence, 530 nm, 16 namp, full-scale. Reprinted with permission of Du Pont Instrument Products Division.

with a UV spectrophotometer optimized at the absorption maximum for riboflavin. This compound is easily detected in a dog food extract with the fluorometer detector, but is not distinguishable in the UV spectrophotometer trace. Although the fluorescence detector can become markedly nonlinear at concentrations where the absorption detector is still linear in response, its linear dynamic range is more than adequate for most trace analysis applications. The dynamic range of fluorometers can be fairly large (10^4), but the *linear* dynamic range may be restricted for certain solutes to relatively narrow concentration ranges (10-fold). For all quantitative analyses the linear dynamic range should be determined over the solute concentration range of interest.

In comparison with other detection techniques, fluorescence generally offers greater sensitivity and lower dependence on instrumental instability (e.g., from temperature and pressure changes). The major disadvantage of fluorescence detection is that not all compounds fluoresce under normal LC conditions. UV absorption and fluorescence have been combined in single instruments to realize the advantages of both detection systems simultaneously (Du Pont). If solvents free of fluorescing materials are used, the fluorescence detector can be utilized with gradient elution.

The fluorometric detector is susceptible to the usual interferences that plague all fluorescence measurements: background fluorescence and quenching effects. However, in most LC applications such deleterious effects are not encountered. For quantitative applications it is highly desirable to check this effect and the response of the fluorometric detector in the actual sample to be studied. This is normally carried out by adding known amounts of the solute of interest to a sample "control" or by using the method of standard additions (Section 13.5).

Spectrofluorometers with monochrometers to select both the excitation and emitted energy wavelengths provide maximum flexibility in fluorometric detection. With these devices specific wavelengths for optimum sensitivity and/or selectivity can be conveniently selected. However, this advantage is gained only at considerably higher cost than that of fixed band-pass fluorometric detectors.

A special form of fluorometry uses a laser source and provides for extremely sensitive detection of fluorescing solutes. Figure 4.15*a*, *b* illustrates the laser-induced fluorescence detection of aflatoxins at the low-picogram level (*8*). This approach holds great promise for the future, particularly when instruments with variable-wavelength laser sources become commercially available. Figure 4.15*c* shows the similar detection of these same compounds in almonds, using conventional fluorometry. Detection levels are higher than those in Figure 4.15*a*, *b*, but they are still impressive.

In summary, the fluorescence detector is not for general use, but it is a powerful tool for specific applications when selective detection of trace components is to be carried out.

4.6 INFRARED PHOTOMETERS

Infrared (IR) absorption can be used either for selective or general detection in LC. Devices based on this principle have been successfully used for some time

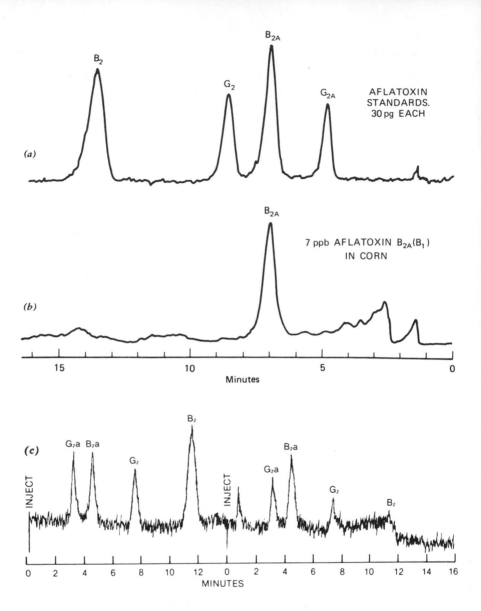

Figure 4.15 Laser fluorometer as LC detector. Aflatoxin chromatograms: (*a*) detection of the four aflatoxin standards following the addition of HCl and (*b*) detection of aflatoxin B$_1$ in corn extract by the same procedure, using a 10-μl sample injection. The chromatogram represents 45 pg of B$_1$ in the injected sample. column, 30 × 0.46-cm μ-Bondapak-C$_{18}$; mobile phase, water/ethanol (75:25 v/v). Reprinted from (*8*) with permission. (*c*) Similar analysis with conventional fluorimetric detection; Column, Lichrosorb RP18; mobile phase, water-methanol-acetonitrile (75/10/15 %v); temperature, ambient; flowrate, 2 ml/min; pressure, 2600 p.s.i.; fluorimetric detection; first sample (calibrator), 20 μl (200 pg each compound); second sample (extract of almonds, 50 g per ml), 20 μl corresponding to 30-220 parts-per-trillion various compounds in original almond sample.

Reprinted with permission of Beckman Instruments (Altex).

148

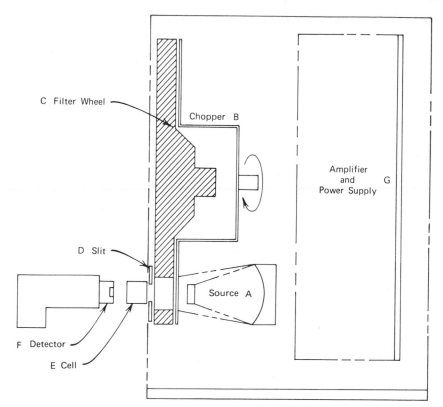

Figure 4.16 Infrared Detector for LC (Wilks). Reprinted by permission of Wilks Instruments, (Foxboro).

with larger-diameter size-exclusion chromatographic systems, but recent improvements in equipment by Du Pont and Wilks Instruments (Foxboro) now make feasible their application to separations with high-performance LC columns. Figure 4.16 shows a schematic of an infrared detector for modern LC. Energy from source A passes through chopper B and then through wheel C, which contains three gradient interference filters with ranges from 2.5-4.5, 4.5-8, and 8-14.5 μm. Positions of the filters on this wheel can be adjusted so that any wavelength in this range can be selected for detection. The monochromatic infrared energy from the source passes through slit D and then detector cell E, which is nominally 0.2 or 1.0 mm in length. Transmitted energy from the sample cell falls on a thermoelectric detector, which records the change in energy from the absorbing sample. The output of the thermoelectric detector is amplified for presentation by means of the amplifier and power supply G.

The infrared detector is limited to mobile phases that are transparent at the absorption wavelength utilized. Table 4.4 lists some common functional groups

Table 4.4 Common functional groups and wavelengths for infrared detection.

Compound Type	Functional Group	Wavelength (μm)
Alkanes	C—H	3.38-3.51
Alkenes	C—H	3.24-3.31
Alkenes	C=C	5.98-6.09
Esters	C=O	5.76-5.81
Ketones	C=O	5.73-6.01

used in LC detection by IR, along with the principal absorption wavelengths in the infrared spectrum. Other functional groups, having weaker absorption at other wavelengths, are sometimes also measured. An IR detector set at one of these wavelength ranges serves as a universal detector for any compound bearing similar functional groups. Because of its more nearly constant response with solute molecular weight change, the IR detector has a decided advantage over the differential refractometer for many applications. Figure 4.17 illustrates the constant molar response of the IR detector for a series of low-molecular-weight alkanes, contrasted to the lack of linear response for the differential refractometer. The relatively uniform molar response of the IR detector is especially useful in the analysis of macromolecules with common functional groups.

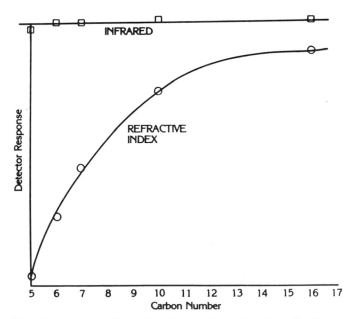

Figure 4.17 Infrared versus differential refractive index detection of *n*-alkanes. Reprinted from (*8a*) with permission.

A particular advantage of the IR detector is that it exhibits stable operation with a heated cell at elevated temperatures (up to ~150°C), making this a preferred detector for high-temperature separations of synthetic macromolecules.

The IR detector can sometimes be used for gradient elution, where the bulk-property detectors (e.g., refractometers) are not usable. For example, Figure 4.18 shows the separation of triglyceride standards by nonaqueous reverse-phase

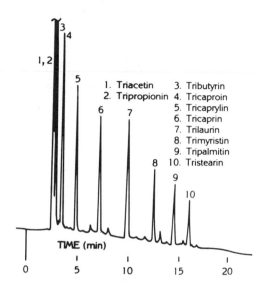

Figure 4.18 Gradient elution with infrared detector. Column, 25 × 0.46 cm, Zorbax-ODS; primary ("*A*") solvent, acetonitrile; secondary ("*B*") solvent, methylene chloride/tetrahydrofuran (47.75 + 52.25% v/v); detector, IR, 5.75 μm (carbonyl). Reprinted from (*9*) with permission.

chromatography with gradient elution and infrared detection. Because of the availability of low-volume sodium chloride or calcium fluoride infrared cells (<10 μl effective volume), the IR detector can be effectively employed with high-efficiency small-particle columns (e.g., ~5000 plates, 25 x 0.46 cm), and gives insignificant band broadening of solutes with $k' \geqslant 0.4$ (*9*).

Under optimum conditions the IR detector is capable of sensing about 1 μg of a solute (molecular weight = 300), using the C–H stretching band at approximately 3.4 μm. Carbonyl and other strongly absorbing functions provide somewhat better sensitivity. Thus, the IR detector has sensitivity that is approximately equivalent to the general refractive-index detector. A limitation of the IR detector is that the mobile phase must have sufficient light transmission at the wavelength of interest. Figure 4.19 shows the useful IR transmission windows

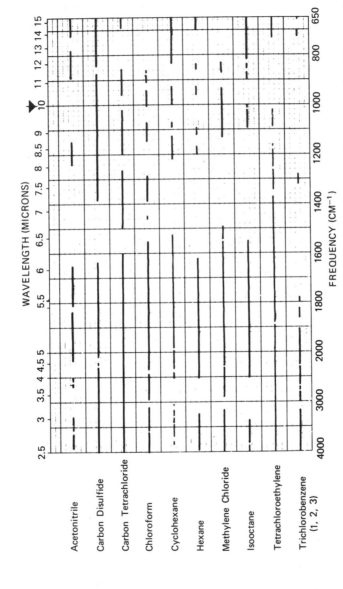

Figure 4.19 Solvent IR-transmitting windows. 1.0-mm path cell with NaCl windows. Bars indicate regions in which transmission exceeds 30%. Solvents with marginal transmission in a few wavelength regions: acetone, toluene, tetrahydrofuran. Solvents with no significant transmission above 30%: *m*-cresol, methanol, butanol, 2-propanol, *N,N'*-dimethylformamide. Reprinted from (*38*) with permission.

for a variety of LC solvents. In most cases at least 30% transmission of the solvent with the required path length cell is needed for satisfactory application.

Certain IR detectors have an advantage similar to that of some UV detectors, that is, the flow can be stopped during elution of the peak of interest and the solute scanned throughout the spectral range of interest for qualitative identification. This is generally a much more useful technique for solute identification than the stop-flow UV technique discussed earlier.

4.7 ELECTROCHEMICAL (AMPEROMETRIC) DETECTORS

Many electroreducible and electrooxidizible compounds can be detected in column efluents at very low concentrations by selective electrochemical (EC) measurements. With this approach the current between polarizable and reference electrodes is measured as a function of applied voltage. Because a constant voltage normally is imposed between the electrodes, EC detectors are more accurately termed amperometric devices. EC detectors can be made sensitive to a relatively wide variety of compound types, as illustrated in Table 4.5. Interestingly, the compounds in this list include those that also generally can be detected by UV absorption. There are, however, several compound types (e.g., mercaptans, hydroperoxides, etc.) sensed by the EC detector that either cannot be detected at all by UV absorption or can be detected only with difficulty at low wavelengths.

Table 4.5 Some compound types sensed by the electrochemical detector.

Oxidation	Reduction
Phenolics	Ketones
Oximes	Aldehydes
Dihydroxy	Oximes
Mercaptans	Conjugated acids
Peroxides	Conjugated esters
Hydroperoxides	Conjugated nitriles
Aromatic amines, diamines	Conjugated unsaturation
Purines	Activated halogens
Heterocyclic rings[a]	Aromatic halogens
	Nitro compounds
	Heterocyclic rings[a]

Note: Compound types generally not sensed include ethers, aliphatic hydrocarbons, alcohols, and carboxylic acids.
[a] Depending on structure.

The applicability of EC detectors to LC is limited by the fact that the mobile phase must be electrically conductive. However, in many cases it is feasible to add a suitable concentration of salt to the mobile phase (e.g., a buffer to an aqueous mobile phase in reverse-phase separations) without having deleterious effect on the separation. Alternatively, this problem sometimes can be circumvented by postcolumn addition of a suitable high-dielectric-constant solvent plus supporting electrolyte. An additional limitation, that the solute of interest must be electroactive, can be overcome by pre- or postcolumn derivitization (Section 17.3). The great selectivity and sensitivity of the EC detector greatly enhances its application for the trace analysis of known species in a complex matrix.

An interesting feature of the EC detector is its relatively small variation in sensitivity for various compounds to which it responds. This relatively constant molar response is due to the small number of electrons involved in EC detection

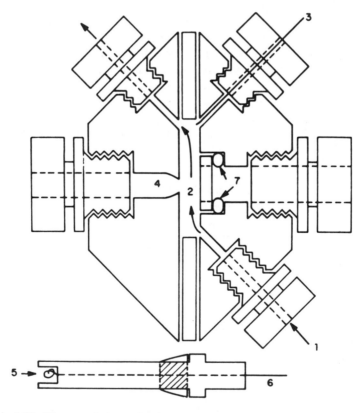

Figure 4.20 Diagram of electrochemical carbon-paste detector cell. (1) cell inlet; (2) flow cavity; (3) platinum wire auxiliary electrode; (4) reference electrode position; note that the electrode assembly (bottom) composed of (5) carbon-paste cavity and (6) platinum wire contact is held in fitting by (7) O-ring seal. Reprinted from (10) with permission.

of various compound types; that is, less that five (usually only two or three) for various structures. This limited effect on response compares to the orders of magnitude change in the response of other sensitive, selective detectors (e.g., UV and fluorometric) to various compounds. Reaction EC detectors also generally show a relatively small change in sensitivity, because the selective reactions produce derivatives that have closely similar response. This relatively constant sensitivity of the EC detector to various species is convenient in trace analysis, because the analyst can predict the sample size, concentrations, and so on, that must be used to produce the desired analytical sensitivity.

Several different electrodes have been used successfully with EC detectors, including carbon paste, glassy carbon, platinum, gold, mercury, and many of the other electrode forms commonly used in polarography. One of the most widely used is the easily formulated carbon-paste electrode, which is inexpensive, and has low residual current. However, EC detectors with this electrode must be frequently restandardized to maintain precise calibrations, because of changes that occur in the electrode surface. Therefore, carbon-paste electrodes also must be frequently replenished for satisfactory results. Figure 4.20 shows a low-volume cell that can be utilized with high-efficiency reverse-phase columns (*10*). This same cell can be used with many electrode types, and replacement of elements (e.g., carbon paste) is easily carried out.

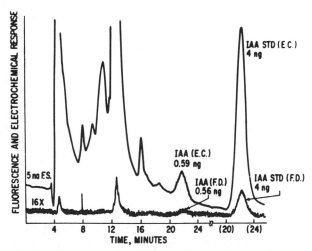

Figure 4.21 Analysis of indole-acetic acid (IAA) from a single cotton abscission zone using electrochemical (E.C.) and fluorometric detection (F.D.). Column, 25 × 0.46 cm, Partisil-SAX; mobile phase, 0.01 *M* KH$_2$PO$_4$ + 0.05 *M* NaClO$_4$; flowrate, 0.8 ml/min; detectors, electrochemical (carbon-paste) and fluorometer, sensitivities as shown; sample, 40 μl, representing 285 mg (fresh weight) of abscission zone from cotton leaf. (Note: Indoleacetic acid standard eluting at 22 min overlaps unknown sample for quantitative analysis.) Reprinted from (*10*) with permission.

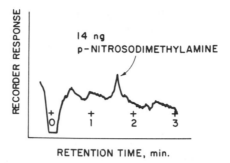

Figure 4.22 Detection of *N*-nitrosodimethylamine with mercury-drop electrochemical detector. Column, 30 × 0.46 cm, μ-Bondapak-C_{18}; mobile phase, 50% methanol/50% 0.02 *M* acetate buffer, pH 4.5; flowrate, 2.0 ml/min; amperometric detector, 500 namp, full-scale, sampled DC mode (−1.4 V. vs. Ag/AgCl). Reprinted with permission of Princeton Applied Research.

Carbon-paste electrodes are normally operated amperometrically in the range of +0.8 to −0.9 V, depending on the type and pH of the mobile phase. All solutes that oxidize or reduce at the selected electrode potential are detected. Figure 4.21 shows the analysis of indoleacetic acid from a single cotton abscission zone as determined by EC detection with a carbon-paste electrode, and a fluorometric detector. Note in this application that the EC detector is about an order of magnitude more sensitive than the fluorometric detector (FD), but the EC detector appears somewhat less selective toward other substituents in the sample. Limited studies suggest that this effect is relatively general; in many systems EC detectors are more sensitive, but fluorometric detectors are more selective.

Although carbon-paste electrode is quite effective for readily oxidizable solutes (e.g., catecholamines), the full range of electrochemical response (Table 4.5) is available only with the mercury electrode. The chromatogram in Figure 4.22 shows the determination of a trace standard by reduction at −1.4 V in a reverse-phase system. The main problems in approaching maximum sensitivity with the mercury-drop and other electrodes (in the reductive mode) include eliminating background currents from oxygen reduction and the reduction of trace metals originating mainly from the LC apparatus. Oxygen must be rigorously removed from the mobile phase, and trace metals contamination sometimes can be reduced by passifying stainless steel equipment for 1-2 hours with 20% HNO_3, or by adding ethylenediaminetetracetic acid (EDTA) complexing agent to the mobile phase.

Recent developments in cell technology have led to electrochemical detection with a versatile mercury-drop electrode (*11*). With this electrode, potentials of more than −2.0 V can be used to greatly expand the range of compound types detected. Figure 4.23 is a schematic of this device (Princeton Applied Research) for detection at a constantly renewable mercury drop. A delivery tip (*A*) directs

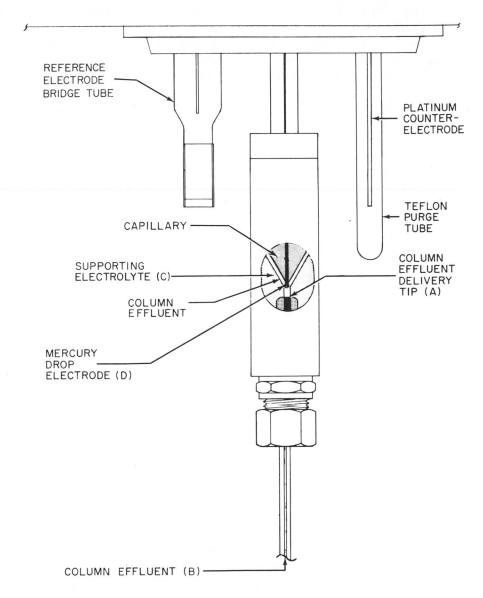

Figure 4.23 Mercury-drop electrode for electrochemical detector (PAR). Reprinted with permission of Princeton Applied Research.

the column eluate (*B*) through the supporting electrolyte (*C*) toward a mercury-drop electrode (*D*). The solute(s) of interest reacts electrochemically at this electrode, and the detected current response is recorded as a function of time. Analyses are conducted at a frequently renewed mercury drop (every second) so that the electrochemical reactions always occur at a fresh surface. The flow pattern around the drop apparently eliminates many of the inherent problems that have been experienced in previous attempts to use a mercury electrode.

EC detection is in the early stages of development; consequently, many of the experimental requirements are not fully resolved. However, it appears that for the easier oxidation reactions, carbon-paste or glassy-carbon electrodes may be preferred because of lower background currents and simpler experimental techniques. However, the mercury-drop detector will probably be required for solutes that can be reduced only at a relatively high potential. With the usual amperometric detectors, the efficiency of the EC process is normally about 1%; EC detectors are infrequently used as coulometric devices.

In addition to operating the EC detectors in the constant-voltage amperometric mode, pulse and differential-pulse systems can be employed to optimize specific experiments (*12*). For example, the virtual flowrate-independence of the differential-pulse (and pulse) currents simplifies routine analysis procedures by eliminating a need for frequent standardization. However, a disadvantage of operating in the pulse mode is the somewhat higher detection limit relative to that encountered in the amperometric mode. Fortunately, the possibility for increasing selectivity in the differential-pulse mode often can more than compensate for this difference, as illustrated in Figure 4.24. Here the conditions were chosen so that each chromatogram was optimized for the selective detection of a single component. In this case, the current collecting voltage pulse had a 51-msec duration, with a 50-mV modulation amplitude ΔE pulse differential and a 0.5-sec interval between pulses. These results clearly illustrate the increased selectivity of the differential pulse mode, optimum settings being determined by the specific compounds of interest.

4.8 RADIOACTIVITY DETECTORS

Radioactivity detectors are specific devices for monitoring radio-labeled solutes as they elute from LC columns. The continuous-flow monitoring of β radiation in LC eluents ordinarily involves the use of a scintillation technique. Depending on the method of presenting the eluent to the scintillator, this can be classified as either a heterogeneous or a homogeneous system. In the latter instance the column eluent is mixed with a liquid scintillation cocktail before passing through a flow cell positioned between the photomultiplier tubes of a liquid scintillation counter. In heterogeneous systems the eluent moves directly to the flow cell,

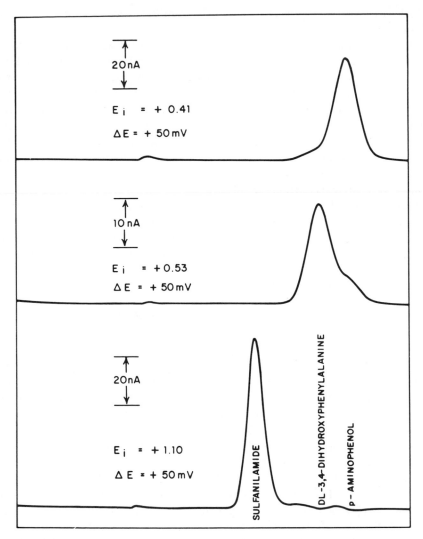

Figure 4.24 Differential-pulse electrochemical detector chromatograms, each optimized for the selective detection of one of the three components. Upper trace: *p*-aminophenol, middle: 3,4-dihydroxyphenylalanine, lower: sulfanilamide. Column, 100 × 0.21 cm, Zipax-SCX, <37 μm; mobile phase, 0.05 *M* H$_2$SO$_4$; temperature, ambient; detector, carbon-paste electrochemical; sample, 25 μl. Reprinted from (*12*) with permission.

which is packed with a finely divided solid scintillant such as anthracene or a Ce-activated lithium glass.

Homogeneous, continuous-flow detectors are best used with analytical procedures where recovery of the sample is unimportant, relative to other considerations such as sensitivity and versatility. The technique is also applicable to preparative LC when a suitable eluent splitter is used. Heterogeneous detectors are best suited to LC systems when the level of radioactivity in the sample is high and the main requirement is purification. Heterogeneous systems are relatively free of chemical quenching effects, and solutes can be easily recovered. However, this system exhibits relatively low counting efficiency for the low-energy β emitters of usual interest (i.e., ^{35}S, ^{14}C, ^{3}H, ^{32}P). For example, counting efficiency with tritium is generally well below 1%.

Radiochemical detectors have a wide response range and are insensitive to solvent changes, making them useful with gradient elution techniques. In optimized systems, radioactive peaks containing ^{14}C with as little as 100 disintegrations per minute have been detected. Chromatographic applications with stronger β, α, and γ emitters (i.e., ^{131}I, ^{210}Po, and ^{125}Sb), using Geiger counting and scintillation systems have also been reported.

With radioactivity monitors it is necessary to compromise sensitivity versus both chromatographic resolution and speed of analysis. The output of the radioactivity flow detector system can be considered to be the net result of a large number of discrete, independent measurements of samples equal in size to the volume of the flow cell V_c, averaged over a period of time equal to the transit time of the solute through the cell t_t. Thus, V_c and t_t determine the minimum volume and time period over which a change in detector response can be recorded. Although it is important that V_c should be as small as possible to reduce band spreading, any reduction in V_c is accompanied by a parallel reduction in sensitivity, because response of this detector is a function of the *total* amount of activity in the flow cell at any time.

Design of the flow cell is a particularly critical factor in radiometric detection. As with all other LC detectors, to prevent degradation of resolution the volume of the cell V_c should be no more than one-tenth the volume of the first peak of interest (Section 4.2). However, in radioactivity monitoring where resolution can be sacrificed for sensitivity and selectivity, it is often desirable to use larger flow cells. Maximum counting efficiency is obtained when the column flowrate is low, and flowrates corresponding to the minimum plate height of the column generally should be used, for combined maximum counting efficiency and detection sensitivity. In practice, peak tailing and peak broadening in a radiometric flow cell can be minimized by working with columns of larger volume. In radioactivity monitors a compromise between chromatographic resolution and detector sensitivity must be reached, the exact nature of which depends on the analytical requirements. Because of cell volume requirements, it is often not possible to use radioactivity detectors with small-particle columns for high-sensitivity detection.

Another compromise that is required in the design of radioactive monitors is that between sensitivity and speed of analysis. The precision of any estimate of radioactivity depends statistically on the number of events contributing to that measurement; therefore, the longer the measurement, the more accurate the result. However, in flow monitoring, the measuring period is determined by the transit time of the solute in the cell, and this is in turn linked to the speed of analysis by means of the flowrate. It is obvious, therefore, that detector sensitivity and column flowrate are always strongly interdependent. For a quantitative discussion of this effect and guidelines for establishing the optimum condition for radiochemical detection, the reader is referred to the excellent work of Reeve and Crozier (13), as well as the earlier study of Huber et al. (13a). Useful information on liquid scintillation counting in flowing systems also is presented in (14). Several commercial radiochemical detection systems are available for use in modern LC (e.g., Shandon Southern, Packard Instruments).

4.9 CONDUCTIVITY DETECTORS

Certain ionic solutes can be detected in aqueous mobile phases using a conductivity detector, and devices with low-volume cells suitable for use with high-performance columns are commercially available. In principle, this device should be applicable to nonaqueous systems; however, applications of conductivity cells with high-performance columns and nonaqueous media have not been adequately studied. Response of conductivity detectors is temperature dependent; consequently, temperature must be controlled carefully. Certain conductivity detector designs are also susceptible to changes in mobile-phase flowrate. The response of this detector is predictable from conductivity data, and these devices exhibit a linear response to solute concentration when properly designed. Although conductivity detectors are not generally applicable in totally organic systems, these devices are an integral part of ion chromatographs (see the discussion of ion chromatography in Section 10.6), and, as such, have proved to be extremely valuable for analyzing both inorganic and organic ionic substances in aqueous mobile phases.

4.10 SUMMARY OF THE CHARACTERISTICS OF MOST-USED DETECTORS

Table 4.6 is a summary of the general specifications for the most popular detectors in modern LC. Note that the electrochemical and fluorescence detectors

Table 4.6 Typical specifications for most-used LC detectors.

Parameter (units)	UV (Absorbance)	RI (RI units)	Radioactivity	Electrochemical (μ amp)	Infrared (absorbance)	Fluorometer	Conductivity (μMho)
Type	Selective	General	Selective	Selective	Selective	Selective	Selective
Useful with gradients	Yes	No	Yes	No	Yes	Yes	No
Upper limit of linear dynamic range	2-3	10^{-3}	N.A.[a]	2×10^{-5}	1	N.A.	1000
Linear range (max)	10^5	10^4	Large	10^6	10^4	$\sim 10^3$	2×10^4
Sensitivity at ±1% noise, full-scale	0.002	2×10^{-6}	N.A.	2×10^{-9}	0.01	0.005	0.05
Sensitivity to favorable sample	2×10^{-10} g/ml	1×10^{-7} g/ml	50 cpm ^{14}C/ml	10^{-12} g/ml	10^{-6} g/ml	10^{-11} g/ml	10^{-8} g/ml
Inherent flow sensitivity[b]	No	No	No	Yes	No	No	Yes
Temperature sensitivity	Low	10^{-4}°C	Negligible	1.5%/°C	Low	Low	2%/°C

[a] N.A., not available.
[b] Because of sensitivity to temperature changes, some detectors *appear* to be flow sensitive.

Table 4.7 Some less common LC detectors.

Detector	Selectivity	Typical detection limit, favorable sample	Inherent flow sensitivity	Useful with gradient	Linear range	Remarks	Reference
Transport (flame ionization)	All C—H containing compounds	5×10^{-10} g/sec ($\sim 2 \times 10^{-8}$ g/ml)	No	Yes	1×10^5	Solvent evaporated on transport	[15]
Transport (electron capture)[a]	Electron-capturing species	2×10^{-12} g/sec	No	Yes	5×10^2	Solvent evaporated on transport	[16]
Flame photometric	S- and P-containing compounds	2×10^{-8} g/ml P; 2×10^{-7} g/ml S	No	No	$>5 \times 10^2$	Largely useful with aqueous mobile phases	[17]
Thermal energy analyzer[a]	N-NO; N-containing compounds	$<1 \times 10^{-8}$ g/ml	N.A.	N.A.	$>10^5$	Expensive	[18]
Atomic absorption	Metal-containing compounds; Condensed phosphates	4×10^{-8} g/ml; 1×10^{-7} g/ml	No; N.A.	N.A.; N.A.	N.A.; $>2 \times 10^2$	Based on complex with Mg	[19] [20]
Spray impact	Many species	5×10^{-11} g/sec	Yes	No	1×10^4	Most sensitive to organic acids and bases	[21]
Mass spectrometry[a]	Universal	$<1 \times 10^{-9}$ g/sec	No	Yes	wide	Generally limited to volatile compounds; very expensive	[22]
Flame emission	Metal ions, metal containing compounds	1×10^{-7} g/ml (Na^+)	Yes	N.A.	$>10^2$		[23]
Chemilluminescence	Certain transition metal ions	6×10^{-12} g/ml (Cu^{+2})	N.A.	N.A.	N.A.		[24]
Photoionization	Universal	1×10^{-11} g/sec	No	Yes	10^4	Effluent must be vaporizable; ionization potential <10.2 eV	[25]
Low-angle light scattering[a]	Universal	5×10^{-4} g/ml (for MW $= 10^4$)	No	N.A.	$>10^3$	Useful only for macromolecules, particulates	[26]

163

Table 4.7 (Continued).

Detector	Selectivity	Typical detection limit, favorable sample	Inherent flow sensitivity	Useful with gradient	Linear range	Remarks	Reference
Precipitation light scattering	For solutes that can be made insoluble	$<1 \times 10^{-6}$ g/ml	Yes	Yes	N.A.		(27)
Electrolytic conductivity[a]	Halogens, P, S	5×10^{-9} g/ml Cl 1×10^{-8} g/ml P 3×10^{-7} g/ml S	No	Yes	1×10^5		(28)
Dielectric constant (permittivity)	Universal	0.1×10^{-6} g/ml	No	No	$>10^4$		(29)
Plasma chromatograph	Universal	1×10^{-11} g/sec	No	N.A.	N.A.	Column effluent splitter; solvent evaporated	(30)
Viscometer	Universal	1×10^{-6} g/ml (MW = 10^4)	Yes	No	$>10^3$	Useful only for macromolecules	(31)
Density	Universal	1×10^{-6} g/ml	Yes	No	$>10^3$	Not yet adopted for high-efficiency columns	(32)
Thermal conductivity	Universal	1×10^{-6} g/ml	Yes	No	N.A.		(33)
Density balance	Universal	1×10^{-5} g/ml	Yes	No	10		(34)
Vapor pressure	Universal	5×10^{-6} g/ml	Yes	No	N.A.	Response proportional to molality	(35)
Heat of adsorption	Universal	2×10^{-6} g/ml	Yes	No	N.A.	Adsorption and desorption peaks for solute	(36)
Plasma emission	Transition metal (complexes)	5×10^{-8} g/ml	N.A.	N.A.	$>10^3$		(37)
Piezoelectric quartz balance	Universal	$<3 \times 10^{-8}$ g/sec	No	Yes	$>10^3$	Mass scanning solvent evap., not continuous; useful for macromolecules only	(38)
Reaction detectors	See Section 17.3						

[a] Available commercially as an LC detector.

can detect less than a picogram of a favorable sample, whereas the UV detector is limited to about 1 ng in a favorable situation. The rest of the detectors listed in this table detect about 0.1-1 μg of solute under optimum conditions.

4.11 OTHER DETECTORS

Many detectors based on other solute properties (e.g., dielectric constant, density, vapor pressure, viscosity, etc.) have been reported for use in LC. Since many of these detectors are based on bulk properties, they are limited in sensitivity because of their high susceptibility to temperature changes. Consequently, many of these devices do not provide the level of sensitivity that is often needed for use with high-performance LC columns. Some of these detectors have important applications in specific instances, and it is anticipated that this list of useful devices of modern LC will grow as developments progress. Table 4.7 summarizes the properties of some of the less-used detectors for LC, with some information regarding their potential application.

REFERENCES

1. R. P. W. Scott, *Liquid Chromatography Detectors*, Elsevier, New York, 1977.
2. J. J. Kirkland, W. W. Yau, H. J. Stoklosa, and C. H. Dilks, Jr., *J. Chromatogr. Sci.*, **15**, 303 (1977).
3. J. C. Sternberg, in *Advances in Chromatography*, Vol. 2, J. C. Giddings and R. A. Keller, eds., Dekker, New York, 1966, p. 205.
3a. S. H. Byrne, Jr., in *Modern Practice of Liquid Chromatography*, J. J. Kirkland, ed., Wiley-Interscience, New York, 1971, Chap. 3.
3b. R. J. Hamilton and P. A. Sewell, *Introduction to High-Performance Liquid Chromatography*, Chapman and Hall, London, 1977, Chap. 3.
4. M. S. Denton, T. P. DeAngelis, A. M. Yacynych, W. R. Heineman, and T. W. Gilbert, *Anal. Chem.*, **48**, 20 (1976).
5. R. E. Dessey, W. G. Nunn, C. A. Titus, and W. R. Renolds, *J. Chromatogr. Sci.*, **14**, 195 (1976).
6. M. J. Milano, S. Lam, M. Savonis, D. B. Pautler, J. W. Pav, and E. Grushka, *J. Chromatogr.*, **149**, 599 (1978).
7. A. E. McDowell and H. L. Pardue, *Anal. Chem.*, **49**, 1171 (1977).
8. G. J. Diebold and R. N. Zare, *Science*, **196**, 1439 (1977).
8a. Product Bulletin E-19602, Du Pont Instrument Products Division, 1978.
9. N. A. Parris, E. I. du Pont de Nemours & Co., personal communication, April 1978.
10. P. B. Sweetser and D. G. Swartzfager, *Plant Physiol.*, **61**, 254 (1978).
11. J. L. Smith, B. N. Whitlock, and R. A. Nadolny, 29th Pittsburgh Conference on Analytical Chemistry and Applied Spectroscopy, February 27 - March 3, 1978, Paper no. 184.
12. D. G. Swartzfager, *Anal. Chem.*, **48**, 2189 (1976).

13. D. R. Reeve and A. Crozier, *J. Chromatogr.*, **137**, 271 (1977).

13a. G. B. Sieswerda, H. Poppe, and J. F. K. Huber, *Anal. Chim. Acta*, 78, 343 (1975).

14. G. B. Sieswerda and H. L. Polak, *Liquid Scintillation Counting*, Vol. II, Heyden, New York, 1972, Chap. 4.

15. O. S. Privett and W. L. Erdahl, *Anal. Biochem.*, **84**, 449 (1978).

16. F. W. Willmott and R. J. Dolphin, *J. Chromatogr. Sci.*, **12**, 695 (1974).

17. B. G. Julin, H. W. Vandenborn, and J. J. Kirkland, *J. Chromatogr.*, **112**, 443 (1975).

18. D. H. Fine, *Anal. Lett.*, **10**, 305 (1977).

19. F. E. Brinckman, W. R. Blair, K. L. Jewett, and W. P. Iverson, *J. Chromatogr. Sci.*, **15**, 493 (1977).

20. N. Yoza, K. Kouchiyama, T. Miyajima and S. Chashi, *Anal. Lett.*, 8, 641 (1975).

21. R. A. Mowery, Jr., and R. S. Juvet, Jr., *J. Chromatogr. Sci.*, **12**, 687 (1974).

22. W. H. McFadden, H. L. Schwartz, and S. Evans, *J. Chromatogr.*, **122**, 389 (1976).

23. D. J. Freed, *Anal. Chem.*, **47**, 186 (1975).

24. A. Hartkopf and R. Delumyea, *Anal. Lett.*, 7, 79 (1974).

25. J. T. Schmermund and D. C. Locke, *Anal. Lett.*, 8, 611 (1975).

26. A. C. Ouano and W. Kaye, *J. Poly. Sci., Polym. Chem. Ed.*, **12**, , 1151 (1974).

27. J. W. Jergenson, S. L. Smith, and M. Novotny, *J. Chromatogr.*, **142**, 233 (1977).

28. J. W. Dolan and J. N. Seiber, *Anal. Chem.*, **49**, 326 (1977).

29. H. Poppe and J. Kuysten, *J. Chromatogr.*, **132**, 369 (1977).

30. F. W. Karasek and D. W. Denney, *Anal. Lett.*, 6, 993 (1973).

31. A. C. Ouano, D. L. Horne, and A. R. Gregges, *J. Poly. Sci., Polym. Chem. Ed.*, **12**, 307 (1974).

32. J. Francois, M. Jacob, Z. Grubisic-Gallot, and H. Benoit, *J. Appl. Polym. Sci.*, **22**, 1159 (1978).

33. K. Ohzeki, T. Kambara, and K. Saitoh, *J. Chromatogr.*, **38**, 393 (1968).

34. R. Quillet, *J. Chromatogr. Sci.*, 8, 405 (1970).

35. R. E. Poulson and H. B. Jensen, *Anal. Chem.*, **40**, 1206 (1968).

36. J. L. Cashaw, R. Segura, and A. Zlatkis, *J. Chromatogr. Sci.*, 8, 363 (1970).

37. P. C. Uden and I. E. Bigley, *Anal. Chim. Acta*, **94**, 29 (1977).

38. Wilks Scientific Corp. Application Report No. 2, 1976.

BIBLIOGRAPHY

Byrne, S. H., Jr., in *Modern Practice of Liquid Chromatography*, J. J. Kirkland, ed., Wiley-Interscience, New York, 1971, Chap. 3 (general discussion of modern LC detectors).

Halasz, I., and P. Vogtel, *J. Chromatogr.*, **142**, 241 (1977) (some problems in quantitative analysis with concentration-sensitive detectors).

Huber, J. F. K. (ed), *Instrumentation for High-Performance Liquid Chromatography*, Elsevier, Amsterdam, 1978 (good overall discussion of modern LC detectors).

Karasek, F. W., *Res. Dev.*, March 1975, p. 34 (new detectors for LC).

Kissinger, P. T., *Anal. Chem.*, **49**, 447A, (1977) (review of amperometric and coulometric detectors).

Kissinger, P. T., L. J. Felice, D. J. Miner, C. R. Preddy, and R. E. Shoup, in *Contemporary Topics in Analytical and Clinical Chemistry*, Vol. 2, D. M. Hercules et al., eds., Plenum, New York, 1978, p. 55 (detectors for trace organic analysis by LC; principles and applications).

Maggs, R. J., in *Practical High Performance Liquid Chromatography*, C. F. Simpson, ed., Heyden, London, 1976, p. 269. (review of modern LC detectors).

Pungor, E., K. Toth, Z. Feher, G. Nagy, and M. Varadi, *Anal. Lett.*, 8, IX (1975) (review of ion-selective and voltammetric detectors).

Roberts, T. R., *Radiochromatography*, Elsevier, New York, 1978 (summary of radiotechniques for TLC, electrophoresis, and HPLC).

Scott, R. P. W., *Liquid Chromatography Detectors*, Elsevier, New York, 1977 (detailed discussion of detectors both for modern and classical LC).

Small, H., T. S. Stevens, and W. C. Bauman, *Anal. Chem.*, 47, 1801 (1975) (ion chromatography with a conductivity detector).

Steichen, J. C., *J. Chromatogr.*, 104, 39 (1975) (dual-purpose absorbance-fluorescence detector).

Swartzfager, D. G., *Anal. Chem.*, 48, 2189 (1976) (amperometric and differential pulse voltametric detection).

Wise, S. A., and W. E. May, *Res. Dev.*, October 1977, p. 54 (unusual experimental detectors for LC).

FIVE

THE COLUMN

5.1	Introduction	169
5.2	Characteristics and Use of Different Column Packings	173
	Physical Properties	173
	Separation Characteristics and Use	177
5.3	Available Column Packings	183
5.4	Column-Packing Methods	202
	Column Hardware	203
	Column Dimension and Configuration	204
	Dry-Fill Packing Procedures for Rigid Solids	206
	Wet-Fill Column-Packing Methods	207
	Packing of Rigid Solids	210
	The Down-Flow Method	212
	The Up-Flow Method	216
	Packing of Hard Gels	217
	Packing of Soft Gels	218
5.5	Column Evaluation and Specifications	218
	Column Evaluation	218
	Column Specifications	225
5.6	Column Techniques	225
	Handling of Columns	225
	Column Troubleshooting	226
	Connecting Columns	227
	Use of Guard Columns	228
	Column "Infinite-Diameter" and Wall Effects	230

5.7 Reduced Parameters and Limiting Column Performance 234

    Utility of Reduced Column Parameters 234

    Limits of Column Performance 240

References 243

Bibliography 245

5.1 INTRODUCTION

High-performance columns that provide minimum broadening of separated sample bands are the heart of the modern LC system. Both the packing put into the column and how well the column is packed are, therefore, of great practical importance. The various processes that determine relative band broadening were qualitatively discussed in Section 2.1 and illustrated in Figure 2.3. We now look more closely at each of these band-broadening phenomena and consider how their deleterious effects on column performance can be minimized.

Table 5.1 Contribution of different band-broadening processes to column plate height H.

Process	H_i	
(i) Eddy diffusion	$C_e d_p$	$\left.\begin{array}{c}\\\\\end{array}\right\} = A u^{0.33}$
(ii) Mobile-phase mass transfer	$C_m d_p{}^2 u/D_m$	
(iii) Longitudinal diffusion	$C_d D_m/u$	$= B/u$
(iv) Stagnant mobile-phase mass transfer	$C_{sm} d_p{}^2 u/Dm$	$= Cu$
(v) Stationary-phase mass transfer	$C_s d_f{}^2 u/D_s$	$= Du$

Note: C_d, C_e, C_m, C_s, C_{sm} = plate height coefficients; d_p = diameter of packing particle; d_f = thickness of stationary-phase layer; u = mobile-phase velocity; A, B, C, D = constants for a given column.

The five major contributions to band broadening are listed in Table 5.1. The effect of each of these processes on the column plate height H can be represented by a contribution H_i, that can in turn be related [see (1-3)] to such experimental variables as mobile-phase velocity u, particle diameter d_p, sample diffusion coefficient in the mobile phase D_m, the thickness d_f of the stationary-phase layer that coats the particle surface, and the sample diffusion coefficient in the stationary-phase layer D_s. Mathematical expressions of H_i for each of these five band-broadening processes also are given in Table 5.1. C_e, C_m, and so on, are constant plate-height coefficients for all well-packed columns, whereas A, B, and so on, are constants for a given column. If we add these various contributions to H, we obtain the general expression

$$H = Au^{0.33} + \frac{B}{u} + Cu + Du. \qquad (5.1)$$

Consider first the case of a well-packed column of fairly large (40-μm) porous particles, where the stationary phase thickness d_f is small and $D \approx 0$. For such a column the B/u term of Eq. 5.1 usually is small, and Eq. 5.1 reduces to

$$H \approx Au^{0.33} + Cu. \qquad (5.1a)$$

A typical H versus u plot for this particular column packing is shown as a solid curve in Figure 5.1. If the column is less well packed, so that particles of column packing are not uniformly compacted, the column constants C_e and C_m increase, which causes an increase in A. This results in larger H values and a poorer column, as shown by the dashed curve (labeled "poorly packed") in Figure 5.1. A general awareness of the importance of good column-packing procedures has prompted much research over the past 10 years on how to pack high-efficiency columns for modern LC. Section 5.4 describes in detail how different column packings are best packed.

Because many early LC column packings were not optimally designed, thick stationary-phase layers or poor stationary-phase diffusion resulted (large d_f or small D_s in H_i for process (v) of Table 5.1). Examples of these older packings include the ion-exchange resins used for years in classical column LC, as well as some of the original bonded-phase materials that were based on thick polymeric layers of stationary phase, chemically attached to a silica particle. Relatively large values of D (in Eq. 5.1) are typically observed for these packings, leading to much larger H values. This is illustrated by the dotted curve labeled "thick layer" in Figure 5.1. As discussed in Section 7.2, recent bonded-phase packings use thinner (monomolecular), more permeable layers, and H values for such packings are well represented by Eq. 5.1 with $D = 0$.

The Cu term of Eq. 5.1a can be eliminated by avoiding porous particles altogether, that is, by eliminating stagnant mobile-phase mass transfer. This is achieved by the use of surface-coated or *pellicular* packings (Figure 5.2a) consisting of a solid (glass-bead) core plus a thin external coating of stationary phase. Alternatively, in *porous-layer* or *superficially porous* beads, the solid core is coated with a thin layer of a porous support (usually silica). Comparing pellicular (or porous-layer) particles (Figure 5.2a) with porous particles (Figure 5.2c) of the same size, it can be seen that mass transfer within the porous network (stagnant mobile phase) of the particle is largely eliminated. Therefore, for pellicular particles Eq. 5.1a is further reduced to

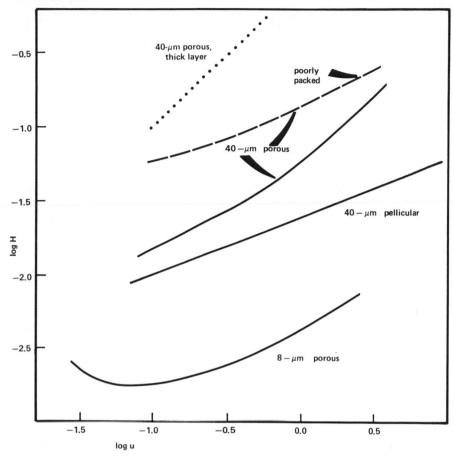

Figure 5.1 Plate height versus mobile-phase velocity plots for various columns.

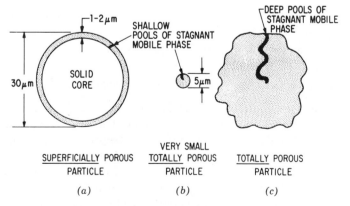

Figure 5.2 Types of particles for modern LC.

$$H \approx Au^{0.33} \tag{5.1b}$$

The resulting plot of H versus u is shown by the solid curve (labeled "pellicular") in Figure 5.1. Compared to 40-μm porous particles, a significant reduction in H values is observed for pellicular materials of the same size, particularly at higher values of u. Because of this significantly improved efficiency, the pellicular packings achieved early popularity in modern LC.

A reduction in particle size can substantially reduce column plate heights and increase column resolving power. If we look at the dependence of H_i values on particle size d_p (Table 5.1), we see that both the A and C terms of Eq. 5.1 increase for larger values of d_p. Thus, a reduction in particle size should reduce both A and C and lead to lower values of H (Eq. 5.1 with $D = 0$). This is indeed the case, as illustrated by the solid curve for the "8-μm porous" particles of Figure 5.1. Here we see a dramatic reduction in H, relative to values for 40-μm porous or pellicular packings. A major factor in these smaller H values for small particles is the virtual elimination of stagnant-mobile-phase mass transfer effects, as illustrated in Figure 5.2b. It is apparent that sufficiently small particles can give diffusion distances within the particle that are comparable to those found for larger pellicular particles (Figure 5.2a). Because of the much lower H values and greater efficiencies of porous microparticles, combined with the possibility of larger sample loadings when compared to pellicular packings (see later discussion), the porous microparticles today dominate the field of modern LC analysis.

In Figure 5.1 we see that the H-u curve for the 8-μm porous packing passes through a minimum value at about $u = 0.08$ cm/sec, and for smaller values of u the plate height H begins to increase. This increase at low u values is the result of longitudinal diffusion (B/u term of Eq. 5.1). This contribution to H does not depend on d_p (Table 5.1), so that its *relative* importance becomes greater as particle size and other contributions to H decrease. To summarize, the highest column efficiencies are obtained by attention to the following:

- Use column-packing procedures that provide a dense, tightly compacted, and uniform column bed.
- Where an added stationary-phase layer is required, use thin, permeable coatings.
- Use small particles, for example, 5-10 μm in diameter.

In Section 5.4 we describe in detail how to prepare efficient columns of different LC packings. Later chapters on the individual LC methods discuss further those aspects of column preparation that may be peculiar to a given LC method. Most column packings can be purchased in the form of prepacked columns from the various suppliers of these materials. The price of prepacked columns varies, depending on column type and size, but is generally $200-$500 per column. Because of the skills and special equipment needed to pack columns of smaller than

20-μm particles, many users find it preferable to rely on purchased, prepacked columns. However, columns of pellicular packings are easily prepared by the user and require essentially no equipment.

In a later chapter (Section 7.2) we discuss the general approach to preparing efficient bonded-phase packings that have thin, permeable surface coatings. Although most workers will not wish to synthesize their own bonded-phase packings, a knowledge of how these packings are made can be useful in their evaluation and use.

The preparation of small particles of porous silica for packings per se, or as supports for bonded-phase LC, is described elsewhere (4). In the following section we touch briefly on this area, but a detailed discussion of this subject is beyond the scope of this book.

5.2 CHARACTERISTICS AND USE OF DIFFERENT COLUMN PACKINGS

In this section we discuss the various ways in which column packings can be described or characterized, and we also look at how these particle properties affect their final use in LC. In Section 5.3 we summarize the various commercial packings that are now available for modern LC. Some further information on individual column packings is provided in Chapters 7-12.

Physical Properties

Packings for modern LC can be classified according to the following features:

- Rigid solids, hard gels, or soft gels.
- Porous versus pellicular and superficially porous particles.
- Spherical versus irregular particles.
- Particle size, d_p.

Rigid solids based on a silica matrix are the foundation of most modern LC packings in use today. Such packings can withstand the relatively high pressures (10,000-15,000 psi) used in packing stable and efficient columns of small particles (Section 5.4). These silica particles can be obtained in a variety of sizes, shapes, and varying degree of porosity. Furthermore, various functional groups or polymeric layers can readily be attached to the silica surface, thus extending the utility of these particles for applications to any individual LC method (Chapters 7-12).

Hard gels still find use in modern LC for ion-exchange and size-exclusion chromatography (SEC). However, even in these areas rigid solids are gradually replacing hard gels. Hard gels are generally based on porous particles of polystyrene cross-linked with divinylbenzene. Depending on how they are prepared, the resulting particles can vary in both rigidity and porosity over fairly wide limits. Hard gels have been used at pressures up to 5000 psi, but some of these particles (e.g., certain μ-Styragels for SEC) are limited to less than 2000 psi pressure. The major interest in the hard gels at present is for SEC applications with organic mobile phases, although they are also used in ion exchange (Chapter 10).

The soft gels, such as agarose or Sephadex for gel filtration, are still uniquely useful for the separation of large, water-soluble molecules such as proteins. However, these soft gels cannot withstand the high pressures used in modern LC, and their use is therefore not discussed in this book.

Packings for LC can be further described as either pellicular or porous, as illustrated by Figure 5.2a, and Figures 5.2b and 5.2c respectively. Pellicular particles are made from spherical glass beads, which are then coated with a thin layer of stationary phase. In one version of such particles a porous silica layer is first deposited onto the glass bead, giving a "porous-layer" or "superficially porous" particle. This porous layer can in turn be coated with some liquid stationary phase or reacted to give a bonded-phase packing. The cross-sectional surface structure of one of these superficially porous packings (Zipax) is shown in Figure 5.3. The original pellicular particles for LC were prepared by coating the glass bead with a layer of porous, polymeric stationary phase—without an intermediate layer of porous silica. These particles generally give large Du terms (Eq. 5.1) and are less efficient than the porous layer of superficially porous packings.

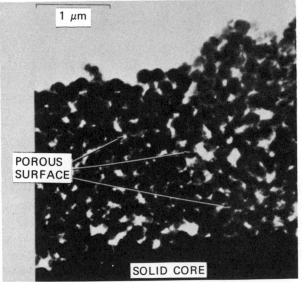

Figure 5.3 Cross-section scanning electron micrograph of the surface of a superficially porous particle (Zipax).

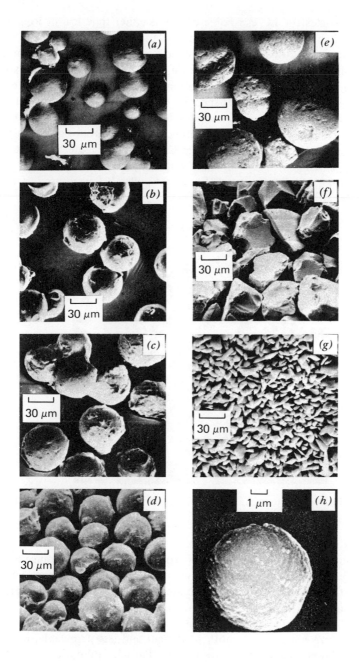

Figure 5.4 Scanning electron micrographs of some commercial LC packings. (*a*) Zipax, (*b*) Corasil-I, (*c*) Corasil-II, (*d*) Perisorb-A, (*e*) Porasil-A, (*f*) Sil-X, (*g*) LiChrosorb-60, (*h*) Zorbax-Sil.

175

Porous particles for LC can be described as either spherical or irregular, and they are available in various sizes (5 μm and larger). Irregular silicas are prepared by first precipitating silicic acid from aqueous solution, then crushing and sizing the solid silica to give particles of uniform size. Various spherical silicas are synthesized directly, usually by agglomeration and consolidation of A-size or submicron-range particles into larger spheres. By varying the conditions of aggregation for either irregular or spherical particles, the average pore size of the resulting particles can be varied over wide limits. (The implications of pore-size distribution in size exclusion are discussed in Chapter 12.) Figure 5.4 illustrates the physical appearance of several of these various types of particles. Diatomaceous earth, a natural, siliceous material (Figure 5.5), also has been used as a porous, rigid packing for LC.

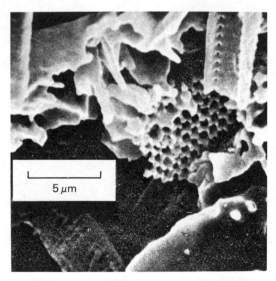

Figure 5.5 Electron micrograph of diatomaceous earth. Reprinted from (*10*) with permission.

It is generally desirable to work with particles having a narrow range of particle sizes, preferably with d_p varying by no more than 1.5-fold from smallest to largest particles. The reason is that column permeability is largely determined by the smallest particles in the column, whereas column efficiency (*H* values) is determined by the larger particles. Thus, columns packed with particles of greatly differing size are generally both inefficient and less permeable, whereas the opposite column characteristics are desired. Particles of uniform size are prepared by solvent elutriation (*5*), or, more commonly, by air classification (e.g., "zig-zag" classifiers) or cyclone separation (*6*). Some spherical particles of uniform size can be synthesized directly, without the need for a final size separation (7).

Spherical and irregular particles can each be packed to give columns of similar efficiency, that is, equal plate numbers N for columns of the same size, when packed with particles of similar d_p. In most work it has been found that the permeability of columns of spherical particles is about twice that of corresponding columns of irregular particles (Section 2.4 and Eq. 2.12), which means that spherical particles produce only half as much pressure drop as irregular particles, other factors being equal. However, a recent study has shown that in very carefully prepared columns, there is no significant difference in permeability between spherical and irregular particles (8). Additional information on this effect is needed. There is another doubling of permeability for spherical pellicular particles, because no (stagnant) mobile phase is trapped within the particles.

The structure of a bed of irregular particles is inherently less stable than for a dense, well-packed bed of spherical particles, as suggested by Figure 5.6. This

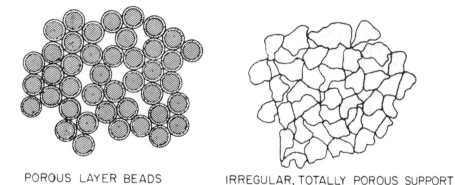

PORGUS LAYER BEADS   IRREGULAR, TOTALLY POROUS SUPPORT

Figure 5.6 Comparison of the shape of particles in packed beds.

makes the preparation of stable columns of irregular particles somewhat more difficult, which can show up as a loss in column efficiency following shipping, other rough handling, or continued use of the column at high operating pressures (see discussion in Section 5.6). The general properties of pellicular and porous column packings of varying particle size are summarized in Table 5.2.

Separation Characteristics and Use

Small, porous particles are usually preferred for most LC applications. However, no single column packing is best for every separation problem, and in this section we try to balance the opposing considerations that favor one type of packing over another. The major differences in separation characteristics of these

Table 5.2 Summary of characteristics of different packings for LC.

Property	Totally porous packings			Pellicular packings, spherical (>20 μm)
	Irregular (>20 μm)	Spherical (>20 μm)	Spherical or irregular (<20 μm)	
Efficiency	low to moderate	low to moderate	high	moderate to high
Speed	moderate	moderate	fast	fast
Packing characteristics	fair	good	fair	excellent
Sample size	large	large	large	small
Cost	low	moderate	high	high
Column permeability	high	high	low	very high

various particle types fall into one of the following areas:

- Column efficiency versus permeability or pressure.
- Sample size and detection sensitivity.
- Equipment requirements (extracolumn band broadening).
- Convenience in use.

We have already seen that as particle size decreases, column plate heights decrease (Figure 5.1) and the column becomes less permeable (Eq. 2.12). That is, for smaller values of d_p, a column of some fixed length L generates higher plate count and higher efficiency, but a greater pressure drop across the column is required for a given value of u. These trends in column N values and operating pressure P are illustrated by the typical data for porous particles shown in Figure 5.7. Comparison of these N values with the data of Figure 5.1 shows somewhat lower values of N in Figure 5.7. The reason is that the data of Figure 5.1 are for "best possible" columns, whereas the numbers of Figure 5.7 are more represen-

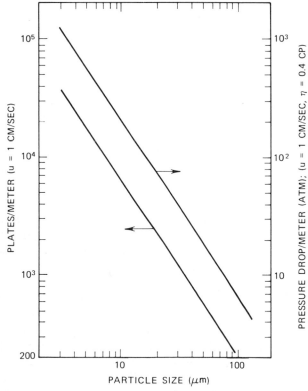

Figure 5.7 Effect of particle size on theoretical plates and column pressure drop. Reprinted by permission of D. L. Saunders.

tative of average columns in use today. Further examples of the decrease in H with decreasing d_p are shown in Figure 5.8, for silica packings (liquid-solid chromatography) of varying d_p.

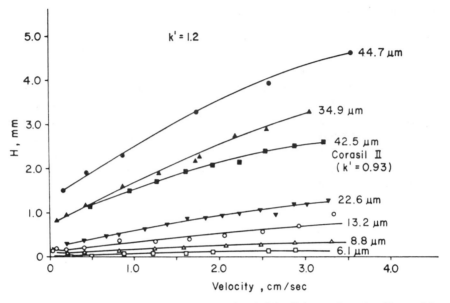

Figure 5.8 Effect of particle size on column plate height. Columns, irregular silica; mobile phase, 90:9.9:0.25 parts (v) of hexane/methylene chloride/isopropanol; sample, 1 μl of N,N'-diethyl-p-aminoazobenzene solution. Reprinted from (*11*) with permission.

A simple conclusion from Figure 5.7 is that columns of small particles are more efficient, but require higher operating pressures. However, this overlooks the fact that columns with different d_p values have different optimum values of u. When both small and large-particle columns are operated at some fixed maximum pressure P_{max}, and column length L is adjusted to give similar separation times t, then the highest N values are generally found for the small-particle columns. The dependence of N on d_p, P_{max}, and t is considered further in Section 5.7.

For columns of similar efficiency, maximum sample sizes are smaller for pellicular particles than for porous particles. The reason is that less stationary phase is available (per unit volume of column) in the case of pellicular particles, and this volume of stationary phase is more quickly overloaded by large sample injections. This is illustrated in Figure 5.9, where k' values for a given compound are plotted versus injected weight of that compound for both a porous and pellicular column. Roughly five times as much sample can be charged to the porous column before there is a significant decrease in k'. Since larger samples can be

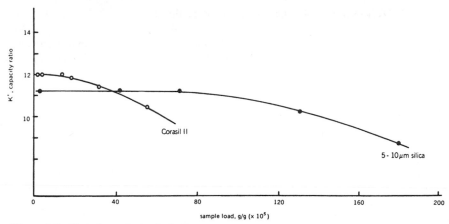

Figure 5.9 Variation of adsorbent linear capacity (maximum sample loading) with adsorbent type. Corasil II = pellicular; 5-10 μm silica = porous. Reprinted from (*12*) with permission.

charged to porous-particle columns, the resulting bands are larger and more easily detected. Therefore, detection sensitivity is usually better for porous-particle columns than for pellicular columns. With large- versus small-particle columns, the latter generally have much higher N values, and therefore give much narrower bands. Although this effect works to increase detection sensitivity, it is opposed by the fact that shorter columns generally are used with small-particle packings, and, therefore, smaller samples must be charged to these more efficient columns before H values begin to increase because of extracolumn effects (see discussion of effect of sample size on column efficiency in Section 2.3). However, on balance it appears that columns of small, porous particles give the best detection sensitivities and are preferred for trace analysis (Section 13.5). For preparative separations, where maximum sample sizes are preferred, similar reasoning suggests the use of large-particle columns rather than small-particle columns. Thus, larger columns (and more stationary phase per column) are used with larger d_p (see Table 15.4). Furthermore, as discussed in Chapter 15, column N values need not be as large in preparative separation, because band widths are determined largely by the degree of column overloading, rather than by the N values found for small-sample injections.

Although most workers have recently turned to the use of small-particle columns in modern LC, there are still areas of application in which pellicular packings show decided advantages. Table 5.3 compares the characteristics of columns made with pellicular particles with those made from 10 μm or smaller particles. Although the latter columns display superior performance, columns of pellicular particles have a decided advantage in convenience factors: ease of packing, long-term stability, and lower cost. Thus, columns of small part-

Table 5.3   Advantages: pellicular versus small particles.

Parameter	Pellicular particles	Small particles ($\leqslant$10 μm)
Efficiency and/or speed		X
Packing ease	X	
Sample capacity		X
Equipment requirements	X	
Convenience of use	X	
Long-term stability	X	
Trace analysis		X
Peak capacity		X
Cost	X	

Table 5.4   Picking the right column-packing size and type

Packing	Preliminary study	Preparative separation	Routine analysis	Other
Porous, 30-70 μm	X	XX	X	
Pellicular, 30-70 μm	X		X	Guard columns
Porous, 5-20 μm	XX	X	XX	
Porous, 20-200 μm		X		Solvent purification, precolumns

Note: X = sometimes useful, XX = often best.

icles generally are preferred for highest efficiency and speed, (a) when peaks must be isolated for identification, (b) for trace analysis, and (c) when complex mixtures requiring a large peak capacity must be analyzed. On the other hand, columns of pellicular particles are adequate for relatively simple mixtures not requiring the highest resolution.

Pellicular particles might well be preferred in some routine applications with relatively unskilled personnel, where equipment is simple and where rapidly and easily prepared, low-cost columns are important. Pellicular particles are also very useful in "guard" columns (Section 5.6). Thus, although the utility of pelliculars has diminished in recent years, these particles still remain useful in modern LC separations.

In the selection of particle size the influence of extracolumn effects must be considered. Thus, as d_p is decreased, the value of N for the column increases and column length usually is decreased in practice. Considering Eqs. 2.4 and 2.6, this means that the volume of an eluted sample band for a given value of k' decreases rapidly with decreasing d_p. As a typical example, consider the following calculated data for a 4-mm i.d. column:

d_p (μm)	L (cm)	N	Band volume [Equal to $(t_w\ F)$] $(k' = 2)$
5	15	3000	330 μl
50	100	500	5.4 ml

As discussed for sample-size effects in Section 2.3, this means that smaller-particle columns will be more prone to extracolumn band broadening, and the use of such columns requires more careful design of the LC system if serious loss in final column N values is to be avoided. This issue is further discussed in Section 5.7 and in those sections that deal with equipment design (Section 3.2 and 4.2). Table 5.4 provides a practical summary of the use of different particle types for different LC applications.

Finally, we should insert a word of caution about all packings for modern LC: Often there are significant differences in chromatographic properties among batches of nominally identical packings or packed columns (i.e., same supplier, same catalog number). The elimination of such inter-lot variations among LC columns of a given type is a major goal of column suppliers, but this goal has not yet been attained in many cases. The problem of column variability has two important implications. First, every new column from a supplier should be tested for its equivalency to previous columns of the same type. Second, for critical applications where a column is to be used over a long period of time and/or in different laboratories for a given application, consideration should be given to stockpiling a satisfactory packing. That is, a large batch of the packing should be purchased, sufficient to pack all the columns that will be required for an application. Then columns can be packed from this batch of packing as new columns are required.

5.3 AVAILABLE COLUMN PACKINGS

A large variety of column packings is available for carrying out all forms of

Table 5.5 Some microparticulate packings and prepacked columns for LSC and LLC

Type	Name	Supplier(s)[a]	Form[b]	Average particle size (μm)	Surface area (m²/g)	Columns Length (cm)	Diameter (mm)	Description
Silica, irregular	BioSil A	7, 39	B	2-10	400	—	—	Methanol extracted and activated
	Chromegasorb 60 R	13	C	10	500	30	4.6	Column filled with Li-Chrosorb
	Chrom Sep SL	41	B or C	5, 10	400	15, 30	—	Contains LiChrosorb Si60
	HiEff MicroPart	5	C	5, 10	250	15, 30	5	3 g of Silica in 30-cm column
	ICN silica	19	B	3-7, 7-12	500-600	—	—	Pore size 60 Å, pH 7
	LiChrosorb Si-60	1, 2, 12, 15, 17, 22, 24, 28, 36, 39, 41, 42	B or C	5, 10	500	25, 30	2.1-4.6	Pore size 60 Å, 100 Å also
	MicroPak Si	42	B or C	5, 10	500	25, 30	2.2, 4.0	Pore size 60 Å, contains LiChrosorb Si; Prep columns available
	Partisil	1, 2, 4, 5, 9, 15, 17, 20, 23, 24, 27, 29, 31, 41, 44	B or C	5, 10	400+	25	4.6	Pore size 50 Å; 50-cm column available for 10 μm.
	μPorasil	43	C	10	300-350	30	3.9	3000 plates/column at 1.3 cm/sec
	RSL silica	31	B or C	5, 10	>200	25	4.6	57 Å pore, iron-free ±1.5 μm distribution
	Sil 60	9, 23, 31	B or C	5, 10, 20	500	20, 25, 30	4.6	60 Å pore size, 0.75 ml/g volumn, 60D has broader distribution
	Silica A	25	B	13 ± 5	400	—	—	Acid washed, recommended for prep
	Sil-X-1	25	B or C	13 ± 5	400	50	2.7	Chemically treated surface

	Name	Ref.[a]	B or C[b]	Particle size (µm)	Surface area (m^2/g)	Length (cm)	I.D. (mm)	Remarks
Silica, spherical	Hypersil	32	B	5-7	200	—	—	
	LiChrospher Si-100	1, 2, 12, 15, 17, 22, 24, 28, 36, 39, 41	B or C	5, 10	370	25	2.1-2.6	Pore size 100 Å, larger available
	Nucleosil 50	9, 23, 28, 31	B or C	5, 10	500	20, 25, 30	4	Pore volume 0.8 ml/g, pore size 50 Å, 100 Å also; 1 ml/g pore volume
	Spherisorb SW	9, 25, 37	B or C	5, 10	220	10, 20, 25	4.6	Packing density 0.6 g/ml, pore size 80 Å, maximum pH 8
	Spherosil XOA 600	9, 30, 31, 38	B	5-8	550		—	83 Å average pore diameter, pore volume 1.2 ml/g, XOA 1000 is 860 m^2/g, 35 Å, 0.78 ml/g
	Super Microbead Si	14	B	5, 10	380	—		95 Å average pore diameter
	Vydac TP Ads.	1-5, 9, 17, 23-26, 27, 31, 32	B or C	10	100	25	3.2	Pore size 330 Å
	Zorbax Sil	11	B or C	6	350	15, 25	2.4, 4.6	70 Å pore size, pore volume 0.8 ml/g
Alumina, irregular	ALOX 60D	9, 23, 28, 31	B or C	5, 10, 20	60	20, 25, 30	4.6	60 Å pore, basic pH 9.5
	Chromasep PAA	41	B or C	5, 10	70	15, 30	—	Packed with LiChrosorb Alox T
	HiEff Micropart Al	5	C	5, 10	—	25	5	Pretested columns
	ICN Al-N	19	B	3-7, 7-12	200	—	—	Neutral, density 0.9 g/ml
	LiChrosorb ALOX T	1, 2, 12, 15, 22, 24, 28, 41, 42	B or C	5, 10	70	25, 30	2.1-4.6	Pore diameter 150 Å
	MicroPak Al	42	B or C	5, 10	70	25, 30	2.2, 4.0	Pretested columns, packed with LiChrosorb ALOX T
Alumina, spherical	Spherisorb AY	9, 26, 31, 37, 39		5, 10, 20	95	10, 20, 25	4.6	Pore diameter 150 Å, maximum pH 10, packing density 0.9 g/ml

[a]See Table 5.8.
[b]B = bulk; C = columns.

Table 5.6 Some BPC microparticulate packings for reverse-phase and ion-pair partition.

Chain length	Name	Supplier(s)[a]	Form	Functionality	Base material	Particle size (μm)	Columns L (cm)	ID(mm)	Description
Long	MicroPak CH	42	B or C	Octadecylsilane	LiChrosorb	10	25, 30	2.2, 4	Polymeric layer, 22% loading
	Partisil ODS-2	1,2,5,9,15,17, 23,24,28,31, 41,44	C	Octadecylsilane	Partisil	10	25	4.6	16% loading
	Zorbax-ODS	11	B or C	Octadecylsilane	Zorbax Sil	6	15, 25	2.4, 4.6	22% loading; monofunctional; spherical
	Nucleosil C-18	9,23,28,31	B or C	Octadecylsilane	Nucleosil 100 Å	5, 10	25	2.4, 4.6	Capacity twice that of C-8.
	Hi-Eff Micropart C-18	5	B or C	Octadecylsilane	HiEff Micropart	5	25	5	Pretested columns
	Hypersil-ODS	33	B	Octadecylsilane	Hypersil	5-7	–	–	8% carbon
	MicroPak MCH	42	B or C	Octadecylsilane	LiChrosorb	10	25,30	2.2, 4	Monomeric layer, 8% loading
	μBondaPak C$_{18}$	43	C	Octadecylsilane	μPorasil	10	30	3.9	10% loading
	Vydac RP-TP	1-5,9,17,23-26, 31,32	C	Octadecylsilane	Vydac TP silica	10	25	3.2	10% loading
	LiChrosorb RP-18	1,2,12,15,17, 24,28,39	B or C	Octadecylsilane	LiChrosorb	5,10	25	3	Monolayer, stable pH 1-9, 22% loading
	Spherisorb ODS	9,26,37	B or C	Octadecylsilane	Spherisorb	5,10	25	3	Spherical, maximum pH 8
	Partisil ODS-1	1,2,5,9,15,17, 23,24,28,31, 41,44	C	Octadecylsilane	Partisil	10	25	4.6	5% loading
	ODS-Sil-X-I	25	B or C	Octadecylsilane	Sil-X-I	13 ± 5	50	3	Irregular shape

Inter-mediate	LiChrosorb RP-8	1,2,12,15,17, 24,28,39	B or C	Octylsilane	LiChrosorb	5,10	25	3	Monolayer, stable pH 1-9, 13-14% loading
	Nucleosil C-8	9,23,28,31	B or C	Octylsilane	Nucleosil 100 Å	5,10	25,30	4.6	General-purpose reverse phase
	Zorbax-C_8	11	C	Octylsilane	Zorbax Sil	6	15,25	2.4, 4.6	For polar compounds, 15% loading; monofunctional; spherical C_{18}, C_2, C_1 also available.
	Chromegabor.J C_8	13	C	Octylsilane	LiChrosorb	10	30	4.6	
	Chromegaboud-Cyclohexane	13	C	Cyclohexane	LiChrosorb	10	30	4.6	Recommended for phenols, low polarity
	μBondaPak Phenyl	43	C	Phenyl	μPorasil	10	30	3.9	16% loading
Short	Hypersil-SAS	33	B	Short alkyl chain	Hypersil	5-7	—	—	For reverse-phase ion-pair and soap chromatography
	LiChrosorb RP-2	1,2,12,15,17, 24,28,39	B or C	Dimethylsilane	LiChrosorb	5,10	25	3	Formerly Si60 silanized, recommended for polar compounds
Organic resins	Chromex	10	B	PS-DVB	PS-DVB	11 ± 1	—	—	Available in 2,8, 12% cross-linking.
	Hamilton HN	16	B	PS-DVB	PS-DVB	7-10	—	—	Cross-linking from 2-35%, from 6-9% in increments of 0.25%.
	Hitachi Gel 3011	18	B	PS-DVB	PS-DVB	10	—	—	Stable at high pH
Specialty	μ-Bondapak-Fatty Acid	43	C	Alkylsilane	μPorasil	10	30	3.9	Free fatty acid anal., use water + acetonitrile, THF

[a]See Table 5.8.

Table 5.7 Some BPC microparticulate packings for normal phase.

Polarity (based on functionality)	Name	Supplier(s)[a]	Form	Functionality	Base material	Particle size (μm)	Columns L (cm)	Columns Diameter (mm)	Description
Weak	HiEff micropart-ester	5	C	Ester	HiEff Micropart	5	25	5	Beware of ester bond hydrolysis
	Nucleosil-NMe$_2$	9,23,28,31	B or C	Dimethylamino	Nucleosil 100 Å	5,10	25	4.6	Sil60 NMe$_2$ is I-shaped; weak anion exchanger; 6 μeq/m^2 coverage
	LiChrosorb DIOL	1,2,12,15,17, 24,28,39	B or C	Diol	LiChrosorb	10	25	3.2, 4.6	Weakly polar, tetracycline analysis
	Chromegabond DIOL	13	C	Diol	LiChrosorb	10	30	4.6	Weakly polar
	FE-Sil-X-1	25	B or C	Fluorether	Sil-X-1	13 ± 5	50	3	Carbonyl compounds
Medium	Nucleosil-NO$_2$	9,23,28,31	B or C	Nitro	Nucleosil 100 Å	5, 10	25	4.6	Sil60 NO$_2$ is I-shaped; 6 μeq/m^2 coverage
	Cyano-Sil-X-1	25	B or C	Alkyl nitrile	Sil-X-1	13 ± 5	50	3	Weaker polarity than silica
	HiEff micropart-CN	5	B or C	Nitrile	HiEff Micropart	5	25	5	Prepacked columns
	MicroPak-CN	42		Nitrile	LiChrosorb	10	25, 30	2.2, 4.0	Prep. columns available, monolayer
	μBondapak CN	43	C	Nitrile	μPorasil	10	30	3.9	9% loading; can be used in reverse-phase
	Nucleosil-CN	9,23,28,31	B or C	Nitrile	Nucleosil 100 Å	5, 10	25	4.6	Sil60 CN is I-shaped; 6 μeq/m^2 coverage
	Partisil 10 PAC	1,2,5,9,15,17, 23,24,28,31,41, 44	C	Nitrile	Partisil	10	25	4.6	Packing density 0.45 g/ml, bulk density 0.70 g/ml. Stability to 70°C. Monolayer
	Spherisorb CN	9,26,37	B or C	Nitrile	Spherisorb	5	10, 20, 25	4.6	Pore diameter 80 Å

	Refs.[a]		Functionality	Support	Particle size (μm)	Pore (nm/Å)	I.D. (mm)	Comments
Vydac Polar TP	1-5,9,17,23,24, 31,32	B or C	Nitrile	Vydac TP Ads	10	25	3.2, 4.6	Formerly an ether phase
Zorbax-CN	11	C	Nitrile	Zorbax Sil	6	25	2.4, 4.6	Also for reverse-phase. Less polar than silica; monofunctional; spherical
HPLC-Sorb Polyamide 6	23	B	aminopoly-caprolactam	Same as functionality	5-20	—	—	Irregular shaped
Chromex HEMA	10	B	Polyhydroxy-ethylmethacry-late	Same as functionality	11 ± 1	—	—	Has both hydrophilic and hydrophobic sites. Also useful in reverse-phase; 20% cross-linking
High								
Amino Sil-X-1	25	B or C	Alkylamine	Sil-X-1	13 ± 5	50	3	Selective for nitro and aromatics
Chromegabond-NH₂	13	C	Amino	LiChrosorb	10	30	4.6	Useful for peptides, carbohydrates; diamine also available
Hypersil-APS	33	B	Aminopropyl	Hypersil	5-7	—	—	Also for reverse phase
LiChrosorb NH₂	1,2,12,15,17, 24,28,39	B or C	Amino	LiChrosorb	10	25	3.2, 4.6	For sugars and peptides
MicroPak-NH₂	42		Aminopropyl	LiChrosorb	10	25, 30	2.2, 4.0	Prep. columns available, carbohydrates, nucleotides, monolayer
μBondapak NH₂	43	C	Amino	μPorasil	10	30	3.9	9% loading
Nucleosil-NH₂	9,23,28,31	B or C	Amino	Nucleosil 100 Å	5, 10	25	4.6	Sil60 NH₂ is I-shaped, 6 μeq/m² coverage
Zorbax-NH₂	11	C	Amino	Zorbax Sil	6	15, 25	2.4, 4.6	Useful for carbohydrates; weak anion exchanger; spherical
Specialty								
μBondapak-Carbohydrate	43	C	Amino	μPorasil	10	30	3.9	Use with water-acetonitrile combinations for polyhydroxy compounds, sugars
Triglyceride	43	C	PS-DVB + polar group	Poragel	10	30	7.8	Use acetonitrile-THF mixtures

[a]See Table 5.8.

Note: I-shaped = irregularly shaped

Table 5.8 Microparticulate packings for ion-exchange chromatography.

Type	Name	Supplier(s)	Particle size (μm)	Base material	Strength[a]	Functional group	Ion-exchange capacity (dry) (meq/g)	Description
Anion silica-based	Chromegabond SAX	13	10	LiChrosorb	S	$(-NMe_3)^+$	<1 (est.)	
	LiChrosorb-AN	1,2,12,23,24, 28,39	10	LiChrosorb	S	$(-NMe_3)^+$	0.55	On 100 Å silica
	Nucleosil SB	2,9,23,28,31	5,10	Nucleosil 100 Å	S	$(-NMe_3)^+(Cl)^-$	1	pH range 1-9
	Partisil SAX	1,2,5,9,15, 23,24,28,31, 39,41,44	10	Partisil	S	$(-NR_3)^+(H_2PO_4)^-$	<1 (est.)	pH range 1.5-7.5, nucleotides
	RSL Anion	31	5,10,15	RSL Silica	S	$(-NMe_3)^+$	<1 (est.)	Bulk or packed columns
	Vydac TP Anion	1-3,9,23,24, 28,29,31,32	10	Vydac Si	S	$(-NMe_3)^+(Cl)^-$	<1 (est.)	Can be used with organic
	Zorbax SAX	11	6	Zorbax Sil	S	$(-NR_3)^+$	<1 (est.)	pH 2-9; nucleotides
	-NH₂ (Table 5.7)	–	5,10	Silica	W	$-NH_2$	–	All amino phases can act as weak anion exchangers in acid solution $(-NH_3)^+$, pH 2-7
	Nucleosil NMe₂	2,9,23,28,31	5,10	Nucleosil 100 Å	W	$-NMe_2$	<1 (est.)	Weak anion exchanger; requires lower ionic strength than -SAX type
Resin based	Aminex A-series	7,39,42	A-27, 13±2 A-28, 9±2 A-29, 7±1	PS-DVB	S	$(-NH_3)^+$	3.2	All 8% cross-linked; for nucleotides, purine bases, carbohydrates; packed columns from Varian

Name	Refs	Particle size	Matrix		Functional group	Selectivity	Remarks
AN-X	9	11±1.5	PS-DVB	S	$(-NMe_3)^+(Cl)^-$	4	2,4, and 12% cross-linking; 50% moisture
Chromex	10	11±1	PS-DVB	S	$(-NMe_3)^+$	4	Cross-linking of 2,4,8, or 12X. An 8% cross-linked available with $d_p = 8\,\mu m$
Hamilton HA	5,16,31	7-10	PS-DVB	S	$(-NR_3)^+(Cl)^-$	5	Cross-linking 4,6,8 or 10%
Ionex SB	2,9,23,28,31	5-20	PS-DVB	S	$(-NR_3)^+(Cl)^-$	3	7% cross-linking. 55% moisture
Cation							
Silica-based							
LiChrosorb KAT	1,2,12,23,24,28,39	10	LiChrosorb Si	S	$(-SO_3)^-$	1.2	On 100Å silica
Nucleosil SA	2,9,23,28,31	5,10	Nucleosil 100 Å	S	$(-SO_3)^-(H)^+$	1	pH range 1-9
Partisil-10-SCX	1,2,5,9,15,23,24,28,31,39,41,44	10	Partisil	S	$(-SO_3)^-(NH_4)^+$	~1 (est.)	pH 1.5-7.5
Vydac TP cation	1-3,9,23,24,28,29,31,32	10	Vydac TP	S	$(-SO_3)^-(H)^+$	~1 (est.)	Can be used with organic solvents
Resin-based							
Aminex A series	7,39,42	A-5, 13±2 A-7, 7-11 A-8, 5-8 A-9, 11.5±0.5	PS-DVB	S	$(-SO_3)^-$	5	All 8% crosslinked; useful for amino acids; packed columns from Varian
Beckman AA	6	AA-20, 10.5 ±1 AA-15, 11±3	PS-DVB	S	$(-SO_3)^-$	5	Both 8% crosslinked; mainly for amino acid analysis
Chromex Cation	10	11±1	PS-DVB	S	$(-SO_3)^-$	5	Cross-linking 8 or 12%; general purpose; shipped in hydrated state
Durrum DC	10	-1A, 14±2 -4A, 9±.5 -5A, 6±.5 -6A, 11±1	PS-DVB	S	$(-SO_3)^-$	5	Cross-linking 8%; recommended for amino acid analysis

Table 5.8 (Continued).

Type	Name	Supplier(s)	Particle size (μm)	Base material	Strength[a]	Functional group	Ion-exchange capacity (dry) (meq/g)	Description
	Hamilton HC	5,16,31	7-10	PS-DVB	S	$(-SO_3)^-$	5.2	Cross-linking 2-35%. From 6-9% in increments of 0.25%
	Ionex SA	2 3	10±2	PS-DVB	S	$(-SO_3)^-$	3	35% moisture; 8% cross-linked

[a] S = strong; W = weak exchanger.

SUPPLIERS

1. Alltech Associates
2. Altex Scientific (Beckman)
3. Analabs
4. Applied Chromatography Systems
5. Applied Science Laboratories
6. Beckman Instruments
7. BioRad Laboratories
8. Corning (Pierce)
9. Chrompack (Holland)
10. Durrum Chemical Corp.
11. E. I. DuPont de Nemours
12. E. Merck (Darmstadt, Germany); EM Laboratories
13. ES Industries
14. Fuji-Davison Chemical Ltd. (Japan)
15. Glenco Scientific
16. Hamilton Co.
17. Hewlett Packard
18. Hitachi (Japan)
19. ICN Inc.
20. J.A. Jobling (U.K.)
21. Johns Manville
22. Kipp and Zonen (Holland)
23. Macherey-Nagel + Co. (Germany)
24. Micromeritics
25. Perkin-Elmer
26. Phase Separations (U.K.)
27. Pye Unicam
28. Rainen Instruments
29. Regis Chemical Co.
30. Rhone-Poulenc (France)
31. RSL (Belgium)
32. Separations Group
33. Shandon Southern (U.K.)
34. Shimadzu (Japan)
35. Showa Denko KK (Japan)
36. Siemens
37. Spectra-Physics
38. Supelco
39. Touzart and Matignon (France)
40. Toyo Soda Co. (Japan)
41. Tracor
42. Varian Associates, Aerograph
43. Waters Associates
44. Whatman
45. Fisher Scientific

Table 5.9 Some microparticulate packings for size-exclusion chromatography.

Name	Supplier(s)[a]	Particle size (μm)	Base material[d]	Av. Pore Size (Å)	MW Exclusion Limit, or Range (PS)	Description
ARgel	4	10	PS-DVB	n.a.[c]	10^2	Packed columns, temperature range ambient to $150°$C; nonaqueous solvents only
				n.a.	3×10^2	
				n.a.	10^3	
				n.a.	3×10^3	
				n.a.	10^4	
				n.a.	3×10^4	
				n.a.	10^5	
				n.a.	3×10^5	
				n.a.	10^6	
BioBeads S	7	10	Linear PS	12%[b]	4×10^2	Compatible with nonaqueous solvents
CPG	8	5-10	Porous glass	40Å	$1\text{-}8(\times 10^3)$	Can be used in nonaqueous or aqueous solution to pH 8 (dependent on temperature, time, volume of solution, surface area of glass). Pore volume varies from 0.1-1.5 ml/g. Glycophase G has glycerol coating
				100Å	$1\text{-}30(\times 10^3)$	
				250Å	$2.5\text{-}125(\times 10^3)$	
				550Å	$1.1\text{-}35(\times 10^4)$	
				1500Å	$1\text{-}10(\times 10^5)$	
				2500Å	$2\text{-}15(\times 10^5)$	
Hamilton HN	16	7-10	PS-DVB	HN $\times$ 7%[b]	5×10^2	Nonionic, spherical, degree of cross-linking determines exclusion limit; exclusion limit for hydrocarbon in THF solvent; sold in bulk
				HN $\times$ 4%[b]	10^3	
HSG	34	10	PS-DVB	n.a.	4×10^2	Packed columns 50 cm $\times$ 8 mm; compatible with nonaqueous solvents
				n.a.	4×10^3	
				n.a.	4×10^4	
				n.a.	2×10^5	
				n.a.	8×10^5	
				n.a.	2×10^6	
				n.a.	8×10^6 (est.)	
				n.a.	10^7 (est.)	

Table 5.9 (Continued).

Name	Supplier(s)[a]	Particle size (μm)	Base material	Av. Pore Size (Å)	MW Exclusion Limit, or Range (PS)	Description
LiChrospher	1,2,9,12 13,23,28 31,39	10	Silica	100Å 500Å 1000Å 4000Å	6×10^4 5×10^5 2×10^6 $>3.5 \times 10^6$	Must slurry pack; rigid, spherical; pore volume 1,2, 0.9, 0.8, 0.8ml/g respectively. Available in bulk. Maximum pressure 200 atm
μBondagel E	43	10	Silica gel (μPorasil)	E-125Å E-300Å E-500Å E-1000Å	$0.2\text{-}5(\times 10^4)$ $0.03\text{-}1(\times 10^5)$ $0.05\text{-}5(\times 10^5)$ $5\text{-}20(\times 10^5)$	Ether-bonded phase. Can also be used for adsorption or partition. Compatible with aqueous and nonaqueous solvents
μStyragel	43	10	PS-DVB	100Å 500Å 10^3Å 10^4Å 10^5Å 10^6Å	700 $0.05\text{-}1(\times 10^4)$ $0.1\text{-}20(\times 10^4)$ $1\text{-}20(\times 10^4)$ $1\text{-}20(\times 10^5)$ $5\text{->}10(\times 10^6)$	Prepacked columns only. Semi-rigid. Flow limit, 4 ml/min THF. Nonaqueous solvents only
SE	11	10	Silica	100Å 500Å 1000Å	$0.5\text{-}7(\times 10^4)$ $0.03\text{-}5(\times 10^5)$ $0.01\text{-}3(\times 10^6)$	Columns 25cm × 6.2mm. Compatible with aqueous and nonaqueous solvents. Must deactivate for use with polar compounds.
Shodex	35	10	PS-DVB	n.a. n.a. n.a. n.a. n.a. n.a.	1×10^3 5×10^3 7×10^4 5×10^5 5×10^6 (est) 5×10^7 (est.)	Packed columns 50cm × 8mm; nonaqueous solvents only; 8000 pls/ft

Name[a]	Degree of cross-linking[b]	Material	Pore size	MW range	Comments
Spherosil	9,30	7	Silica	80Å 140Å 360Å 680Å 1100Å	n.a. n.a. n.a. n.a. n.a.	Rigid, spherical packing; may adsorb polar molecules
TSK Gel (also MicroPak BKG)	40,42	8-10	PS-DVB	40Å 250Å 1500Å 10^4Å 10^5Å 10^6Å(est.) 10^7Å(est.) GMH	$0.5\text{-}1(\times 10^3)$ $0.05\text{-}6(\times 10^3)$ $0.1\text{-}60(\times 10^3)$ $0.5\text{-}40(\times 10^4)$ $0.2\text{-}400(\times 10^4)$ $-4\ (\times 10^7)$ $-4\ (\times 10^8)$ $1.5\times 10^3\text{-}10^7$(mixed)	8000 plates/ft guaranteed. 10000 plates/ft available. For nonaqueous solvents. Packed columns only. 30, 50, 60 cm × 8mm. Available for aqueous solution as DS type
Zorbax-PSM-60 -PSM-500 -PSM-1000	11	6	Zorbax silica	60Å 350Å 750Å	$0.01\text{-}4(\times 10^4)$ $0.1\text{-}5(\times 10^5)$ $0.03\text{-}2(\times 10^6)$	25 × 0.62 cm columns. PSM-60 and PSM-1000 combination as bimodal pore size set for linear MW calibration. Untreated and trimethylsilyl-modified spherical silica
μ-Porasil-GPC-60	43	10	Silica gel	60Å	$0.01\text{-}1(\times 10^4)$	Untreated spherical particles; primarily for small molecules
TSK Gel-SW	40,42	8-10	n.a. (hydrophilic)	n.a. n.a.	8×10^4 2×10^6	60 × 0.75 cm and 60 × 2.0 cm columns; useful in aqueous and organic solvents; use at <45°C

[a] See Table 5.8.
[b] Degree of cross-linking.
[c] n.a. = not available.
[d] PS = polystyrene; PS-DVB = polystyrene-divinylbenzene.

Table 5.10 Pellicular packings for adsorption- and liquid-liquid chromatography.

Name	Particle size (μm)	Surface area (m²/g)		Description	Supplier(s)[a]
Corasil I and II	37-50	I	7	Silica; Corasil II has a double coating of silica	43
		II	14		
Pellosil HS	37-44	HS	4	Silica; HC has a thicker coating than HS. HC means high capacity, HS high speed	1,9,15,31,39,44
HC		HC	8		
Pellumina HS	37-44	HS	4	Alumina, HC has a thicker coating than HS	1,9,15,31,39,44
		HC	8		
Perisorb A	30-40	14		Silica, pore volume 0.05 ml/g	12
Vydac	30-44	12		Silica	1-3,5,9,23,28,29,31, 32,39,41,42
Zipax	25-37	1		Silica, 800 Å pore diam., 0.025 ml/g pore volume	11

[a]See Table 5.8.
Note: all particles spherical.

Table 5.11 Pellicular packings for bonded-phase and partition LC.

Name	Particle size (μm)	Functionality	Base material	Description	Supplier(s)[a]
Bondapak C_{18}/Corasil	37–50	octadecylsilane	Corasil		43
CO:PELL ODS	37–53	octadecylsilane	Pellosil		1,44,39
Pellamidon	37–43	nylon-6	Glass bead		1,44,39
CO:PELL PAC	37–53	nitrile	Pellosil		1,44,39
ODS-Sil-X-II	35±15	octadecylsilane	Sil-X-II		25
Perisorb-RP-2	30–40	dimethylsilane	Perisorb A		12,17,39
Perisorb-RP-8	30–40	octylsilane	Perisorb A		12,17,39
Perisorb-RP-18	30–40	octadecylsilane	Perisorb A		12,17,39
Permaphase-ODS	25–37	octadecylsilane	Zipax	Loading 1% by weight; polymolecular bonded	11
Permaphase-ETH	25–37	aliphatic ether	Zipax	Loading 1% by weight; polymolecular bonded	11
Vydac SC (Reverse phase)	30–40	octadecylsilane	Vydac Adsorbent		2,5,17,23,32, 37,39,41,42
Vydac SC-Polar	30–44	ethyl nitrile	Vydac Adsorbent		2,5,17,23,32, 37,39,41,42
Bondapak Phenyl/Corasil	37–50	phenysilane	Corasil		43
Zipax-ANH	37–44	cyanoethyl silicone	Zipax	1% loading, nonbonded	11
Zipax-HCP	25–37	nonpolar hydrocarbon	Zipax	1% loading, nonbonded	11
Zipax PAM	25–37	polyamide	Zipax	1% loading, nonbonded	11

[a]See Table 5.8.
Note: all particles spherical.

Table 5.12 Pellicular packings for ion-exchange chromatography.

Type	Name	Supplier(s)[a]	Particle size (μm)	Functional group	Ion-exchange capacity (dry) (μeq/g)	Description
Anion	AE-Pellionex-SAX	1,9,15,31,39,44	44-53	$(-NH_3)^+$	10	pH 2-10, organic solvents OK, $T = 70°C$
	AL-Pellionex-WAX	1,9,15,31,39,44	44-53	$-NH_2$		pH 2-7, $T = 75°C$
	AS-Pellionex-SAX	1,9,15,31,39,44	44-53	$(-NH_3)^+$	10	pH 2-12, organic solvents OK, 8% cross-linking $T = 90°C$
	BondaPak AX/Corasil	43	37-50	$(-NMe_3)^+(Cl)^-$	10-15	pH 2-7.5, 1.3% loading
	Perisorb AN	1,2,12,17,28,39	30-40	$(-NR_3)^+$	30	Stable pH 1-9
	Permaphase-ABX	11	30	$(-NR_3)^+$	10	pH 2-9, 1% loading; organic solvents OK
	Vydac SC Anion	1-3,5,9,23,28, 29,31,32,39,41, 42	30-44	$(-NMe_3)^+(Cl)^-$	100	Organic solvents OK
	Zipax-SAX	11	25-37	$(-NR_3)^+.$	12	Lauryl methacrylate polymer coated, MW 500, pH 4-9

Cation	BondaPak CX/Corasil	43	37-50	$(-SO_3)^-(Na)^+$	30-40	Stable pH 2-7, 1.1% loading
	HC-Pellionex SCX	1,9,15,31,39,44	37-53	$(-SO_3)^-$	60	pH 2-10, organic solvents OK, $T = 40°C$
	Perisorb KAT	1,2,12,17,28,39	39-40	$(-SO_3)^-$	50	Stable pH 1-9; cationic form unspecified
	Vydac SC cation	1-3,5,9,23,28, 29,31,32,39,41, 42	30-44	$(-SO_3)^-(H)^+$	100	Can use with organic solvents
	Zipax SCX	11	25-37	$(-SO_3)^-$	5	Fluoropolymer coated, MW 1000; small amounts of organic solvent tolerated

[a]See Table 5.8.
Note: all particles spherical.

199

Table 5.13 Typical packings for preparative LC.

Method	Name	Supplier(s)[a]	Type	Size (μm)	Approx. surface area (m^2/g) (for MW 10^2-10^7)
GPC	Styragel (various pore sizes)	43	Cross-linked poly-styrene	<37	
	Spherosil[b] (various pore sizes also useful for GFC)	9,30,31,38	Porous silica beads	<40	10-400
GFC	Bio-Gel P-2	7	Polyacrylamide	37-75	(MW = 100-1800)
LSC	Spherosil-XOA-400[a]	9,30,31,38	Porous silica beads	<37	400
	Silica Gel 60	12	Irregular porous silica	37-63	400
	LiChrosorb SI 60	12	Irregular porous silica	10	500
	Woelm W-200	19	Alumina	18-30	200

LLC	Spherosil -XOBO30	9,30,31,38	Porous silica beads	<40	50
	Dia-Chrom	4	Diatomaceous earth	37-44	1-3
	Bondapak-C_{18}/Porasil -B	43	Porous silica beads with bonded C_{18} (for reverse-phase LLC)	37-75	175
Cation exchange	Rexyn 101 (Na)	45	Cross-linked poly-styrene, sulfonic acid	37-75	
Anion exchange	Rexyn 201 (Cl)	45	Cross-linked poly-styrene, amine quarternary	37-75	

[a] See Table 5.8.
[b] Also various grades of Porasil, 37-75 μm, from Waters Association.

modern LC. Tables 5.5-5.13 list some of the materials that are commercially obtainable. Data in these tables were selected from (*17*) and (*30*) and from manufacturers' literature. Although an attempt has been made to include many of the latest products, information and availability should be obtained from the various manufacturers and suppliers (Appendix I) prior to ordering. Listings of the various packings have been made in the following way:

Packing type	Table no.
Microparticulate packings and prepacked columns for LSC and LLC	5.5
Bonded-phase microparticulate packings for reverse-phase and ion-pair chromatography	5.6
Bonded-phase microparticulate packings for normal-phase chromatography	5.7
Microparticulate packings for ion-exchange chromatography	5.8
Microparticulate packings for size-exclusion chromatography	5.9
Pellicular packings for LSC and LLC	5.10
Pellicular packings for bonded-phase chromatography	5.11
Pellicular packings for ion-exchange chromatography	5.12
Typical packings for preparative LC	5.13

The relative merits of porous microparticles versus pellicular or porous-layer-bead particles were discussed in Section 5.2, and are further defined in Section 5.6. In summary, compared to pellicular packings, microparticles offer substantially improved column efficiency, sample capacity, and speed of analysis. However, they generally require more skill and equipment for the satisfactory packing of columns. A comparison of the properties of both types of packings is given in Table 5.14.

5.4 COLUMN-PACKING METHODS

The optimum or "best" procedure for packing columns is largely determined by the nature and particle size of the component particles. The column materials used (e.g., tubing blank, fittings) can have a significant effect. The goal is to pack a uniform bed with no cracks or channels, and without sizing or sorting the particles within the column. Usually, rigid solids and hard gels are packed as densely as possible, but without fracturing the particles during the packing process.

Table 5.14   Typical properties of HPLC column packings.

Property	Pellicular particles	Porous microparticles
Avg. particle size, μm	30-40	5-10
Best HETP[a] values, mm	0.2-0.4	0.01-0.03
Typical column lengths, cm	50-100	10-30
Typical column diameters, mm	2-3	2-5
Pressure drop, psi/cm[b]	2	20
Sample capacity, mg/g	0.05-0.1	1-5
Surface area (LSC), m^2/g	10-15	400-600
Bonded-phase coverage, % wt.	0.5-1.5	5-25
Ion-exchange capacity, μ equiv/g	10-40	2000-5000
Ease of packing	Easy, dry pack	Difficult, slurry pack
Cost		
Bulk packing	$4-5g (LSC) $7-9/g(BPC)	$2-3/g(LSC) $10-15/g(BPC)
Prepacked columns	$110-130 (LSC) $150-170 (BPC)	$200-250 (LSC) $225-275 (BPC)

Source: From (*30*), with some modifications.
[a]HETP = height equivalent to a theoretical plate.
[b]Columns of equal dimenions (2.1 mm i.d.) operated at flow of 1 ml/min and mobile-phase viscosity of 0.3 cP.

Column Hardware

The column blank and the associated hardware both contribute significantly to ultimate column performance. Columns must be constructed of materials that will withstand both the pressures to be used (e.g., up to 10,000 psi), and the chemical action of the mobile phase. Although most columns are made from stainless-steel tubing, heavy-wall glass columns are sometimes used for special problems. Glass blanks that withstand pressures up to about 600 psi are commercially available. Glass-lined metal columns also are available for use at high pressures for the rare case where stainless steel cannot be tolerated (*12a*).

The inside walls of the column tubing blank should be quite regular and smooth, since the condition of the column wall apparently influences the homo-

Conventional column terminator

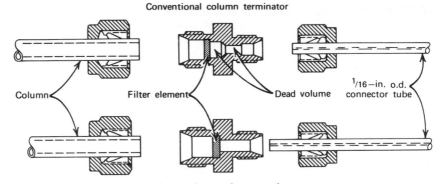

Zero dead–volume column terminator

Figure 5.10 Comparison of conventional and "zero dead volume" column terminators. Reprinted by permission of Micromeritics Instruments.

geneity of the packed bed. Satisfactory column blanks are made from materials such as LiChroma tubing (Handy & Harman Tube Co., Norristown, Pa.). The advantage of tubing with a uniform, polished interior is only significant with high-performance packings; relatively inefficient systems are not greatly affected by different types of column blanks. The effect of the column wall finish also appears to be less important as the column internal diameter increases.

Column end fittings and connectors must be designed with a minimum dead volume, having no unswept corners or pockets that can act as miniature mixing vessels and contribute significantly to extracolumn band broadening (see discussion of sample-size effects in Section 2.3). Compression fittings with "zero" dead volumes such as that shown in Figure 5.10 are now used by most LC instrument suppliers.

Porous frits or screens are used at the ends of the column to retain the packing. These elements must be homogeneous to ensure uniform flow and a minimum of unwanted broadening of sample bands. Porous stainless-steel or Hastaloy frits about 0.2 cm thick and stainless-steel screens about 0.025 cm thick have been used most commonly, with the thin stainless-steel screens apparently providing the least band broadening (*13*). The porosity of these end frits or screens should be substantially smaller than the size of the packing particles. For example, 2-μm-porosity frits are generally adequate for columns of 10-μm particles.

Column Dimension and Configuration

Straight sections of columns in lengths of 10-150 cm should be used. Some of the newer LC instruments permit maximum column lengths of only 25 cm. Thus, longer columns must be assembled from straight 25-cm lengths using low-volume capillary connectors, as described in Section 5.6.

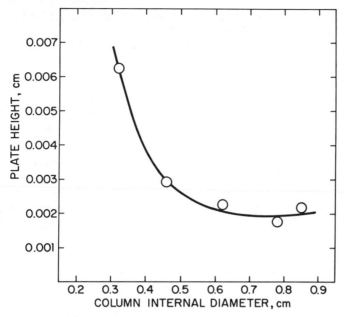

Figure 5.11 Influence of column internal diameter on plate height. Columns, 10 cm, PSM-800, 6 μm; sample, 6.7 μl, 1.0 mg/ml toluene; UV detector, 254 nm, 0.2 AUFS; mobile phase, tetrahydrofuran, 0.25 cm/sec. Reprinted from (*15*) with permission.

A complete definition of the effect of internal diameter d_c on analytical column efficiency has not yet been achieved. However, for properly packed columns of small particles (e.g., $\leqslant$10 μm) efficiency generally increases with internal-diameter [at least, up to about 0.9 cm (*14*)]. For example, we see in Figure 5.11 that lowest plate height (largest N) values for 10-cm columns of 6-μm particles occurs for $d_c \geqslant 0.6$ cm. A complex relationship between particle size, column length and internal diameter, sample injection technique, and other effects appears to determine the optimum diameter of high-performance, small-particle columns. However, columns with internal diameters of 0.4-0.5 cm represent a good compromise of performance, convenience, and the amount of mobile phase and column packing required in analytical applications. Columns of small particles of less than 0.3-cm i.d. generally show lower plate count, and these narrow-bore columns generally are reserved for those trace analyses where sample size is limited and high concentrations of the trace component are required for detection in the column effluent (see Section 13.5).

With pellicular or porous particles larger than 20 μm, columns of 0.2-0.3-cm i.d. are a good compromise between efficiency and convenience. Columns $\leqslant$0.1 cm i.d. are difficult to dry-fill, and they impose rigid limitations on extracolumn effects for all high-performance packings. Columns with i.d. greater than 0.5 cm are commonly used in analytical size exclusion, and much larger i.d.'s are used in preparative LC, as discussed in Chapter 15.

Dry-Fill Packing Procedures for Rigid Solids

The "tap-fill" procedure summarized in Table 5.15 is recommended for the dry packing of rigid particles with $d_p > 20$ μm. For both rigid solids and organic gels of $d_p \leqslant 20$ μm, the slurry-packing (wet) methods described later are preferred. Although the technique for dry-filling high-efficiency LC columns is not very different from that used to prepare high-efficiency GC columns, there are some distinct procedural differences. For example, mechanical vibration techniques should *not* be used in LC. Lateral vibration tends to size the particles across the column, producing an inhomogeneous bed in which the larger particles are near the wall and the smaller particles near the center. (This effect is more prominent in LC columns, because the relative particle size range is generally greater than that used in gas chromatography.) This particle sizing results in increased band broadening, because of the faster mobile-phase flow near the column walls, as compared to the center. The recommended "tap-fill" procedure allows a regular, dense consolidation of the packing without undue particle sizing. Vertical tapping with the right force and frequency allows a good consolidation of the column bed, but without particle sizing as result of too vigorous bouncing or vibration. Simultaneous rapping of the column along its side keeps particles from adhering to the walls during delivery of the packing. Because of their relatively high density, pellicular packings are most easily packed by this approach.

Statically charged particles cause problems, particularly with less dense, irregular porous materials. Static charge can usually be eliminated by placing the packings for several hours in a closed vessel whose atmosphere is saturated with water vapor.

With the "tap-fill" dry-packing procedure, column characteristics (e.g., plate height, permeability) usually can be reproduced within at least ±10%, and this procedure can be mastered easily by relatively unskilled operators in a reasonably short time. Column-packing machines are commercially available to dry-pack columns with greater reproducibility and efficiency and with less effort. Such devices are used by suppliers of dry-packed columns.

Although dry-packing methods have been used for some time in high-performance LC, columns of high efficiency are more difficult to produce as particle size decreases. This is due to the fact that small particles have high surface energies relative to their mass, and hence tend to clump. The result is analogous to using a packing with a wide particle-size range. Particle agglomeration causes nonuniform compaction during the packing process, which results in widely varying flow velocities along the column and poor column efficiency. Totally porous particles with $d_p < 30$ μm should not be dry-packed, unless they have a

Table 5.15 "Tap-fill" method for dry packing columns of rigid solids.

1. Degrease the inner walls of the column by washing with dichloromethane, acetone, and water, respectively.

2. Scrub interior of column tubing with hot detergent solution (using long pipe cleaner or a cloth plug tied to a nylon cord), wash with water and absolute methanol and dry.

3. Fit porous disc or screen to outlet of tubing; retain with outlet fitting.

4. Add enough packing to fill 3-5 mm of the column (100-200 mg of pellicular packing for 2 mm i.d.) via funnel to vertically held tubing.

5. Firmly tap column on floor or bench top, 2-3 times/sec, for 80-100 times, while gently rapping the side of the tube at the approximate level of the packing.

6. Discontinue rapping the side; *very gently* tap the column vertically for 15-20 sec.

7. Add another portion of packing, continue as above until column is filled (requires 15-20 min for 50 X 0.21 cm i.d. column). Rotate column slowly while filling.

8. Gently tap column vertically for additional 5 min (no rapping on the side).

9. Place rigid porous plug or screen at the inlet of the column, after carefully leveling off the packing flush with the column end. Attach fittings.

10. Place column in instrument and equilibrate with mobile phase until no bubbles are seen in column outlet.

narrow size distribution (i.e., generally less than a factor of 2). However, with care the more dense pellicular particles can be packed with this approach down to $d_p \sim 15$ μm, if particle-size distributions no greater than about a factor of 1.5 are used.

Wet-Fill Column-Packing Methods

High-pressure "wet-fill" or slurry-packing techniques are used for packing particles with $d_p \leqslant 20$ μm, because these are difficult to form into high-efficiency columns by dry filling. In the slurry technique a suitable liquid wets the particles and eliminates particle aggregation during packing. Proper selection of the liquid reduces the tendency of the particles to size-fractionate via gravitational sedimentation. Dependence of particle sizing on sedimentation is given by

$$V = \frac{2gr^2 (\rho - \rho_0)}{9\eta} \quad , \tag{5.2}$$

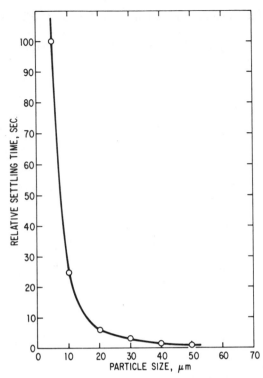

Figure 5.12 Relative settling times of various particle diameters. Time required for spherical particles of equal density to settle a given distance (roughly equal to sec/cm in methanol). Data taken from (17).

where V is the terminal velocity of the particle (cm/sec), g is the gravitational acceleration (cm/sec^2), r is the radius of the particle (cm, assumed spherical), ρ is the particle density (g/cm^3) ρ_0 is the density of the liquid (g/cm^3), and η is the viscosity of the liquid (poise).

Equation 5.2 indicates that large particles settle faster than small particles, by a factor proportional to the square of the particle radius. Figure 5.12 illustrates the relative times required for spherical particles of equal density to settle a given distance. The strong dependence of settling velocity on particle size means that the wider the distribution of particle sizes, the more quickly an initially homogeneous slurry of particles will become heterogeneous during handling. Equation 5.2 suggests that particle segregation by sedimentation decreases as the density of the slurrying liquid approaches particle density. Thus, it is sometimes desirable to use a suspending fluid that has a density equal to that of the particles, and this approach is called the *balanced-density* slurry-packing technique (18, 19). Very small particles (e.g., 5 μm) settle very slowly and seldom require the balanced-density approach in column preparation. On the other hand, particles greater than 10 μm settle much faster, and the balanced-density slurry technique is then needed to produce efficient columns, particularly if the particle-size range is greater than a factor of 2.

The suspending fluid also must be carefully chosen to maintain an adequate particle dispersion without aggregation. Because the polarity and/or charge on a particle may also be important in aggregation, media are sought that maintain discrete particles (when observed by optical microscopy) as a suspension in the proposed liquid. For example, electrolyte has been added to methanol/water slurrying liquid to improve the efficiency of slurry-packed, bonded-hydrocarbon, reverse-phase materials (20).

Table 5.16 Properties of some slurry-packing solvents.

	Density, ρ (g/ml)	Viscosity, η (cP, 20°C)
Diiodomethane (methylene iodide)	3.3	2.9
1,1,2,2-Tetrabromoethane[a]	3.0	–
Dibromomethane (methylene bromide)	2.5	1.0
Iodomethane (methyl iodide)	2.3	0.5
Tetrachloroethylene (perchloroethylene)[a]	1.6	0.9
Carbon tetrachloride[a]	1.6	1.0
Chloroform	1.5	0.6
Trichloroethylene	1.5	0.6
Bromoethane (ethyl bromide)	1.5	0.4
Dichloromethane (methylene chloride)	1.3	0.4
Ethylene glycol	1.11	1.7
Water	1.0	1.0
Pyridine	1.0	0.9
Tetrahydrofuran	0.9	0.5
n-Butanol	0.8	3.0
n-Propanol	0.8	2.3
Ethanol	0.8	1.2
Methanol	0.8	0.6
Cyclohexane	0.8	1.0
n-Heptane	0.7	0.4
Isooctane	0.7	0.5

[a] Most halogenated solvents are somewhat toxic, but these are particularly toxic.

The suspending liquid also must thoroughly wet the packing. High-surface-energy, unmodified silica or hydrophilic organic-modified silicas (e.g., diol) generally require relatively polar liquids, whereas low-surface-energy, aliphatic hydrocarbon, bonded-phase packings (e.g., C_{18}) can be packed with relatively nonpolar suspending liquids. Table 5.16 lists some solvents that have been successfully used in high-pressure slurry-packing techniques.

Equation 5.2 further suggests that particle fractionation by sedimentation decreases as the viscosity of the suspending fluid increases. Therefore, an alternative to the balanced-density technique is to employ more viscous slurrying solvents (21). Unfortunately, use of higher viscosity solvents decreases the flow through the column during the packing so that the procedure takes longer or requires much higher packing pressures. Published data suggest that the viscous-slurry technique is generally less able to produce efficient columns than the methods using low-viscosity slurrying liquids. Consequently, the viscous-slurry approach has not been widely used.

To summarize, the production of high-efficiency columns of <20-μm particles by the wet-fill procedure generally must meet these requirements: (a) the particles in the suspending liquid must be maintained without aggregation; (b) particles must be held in suspension long enough to be packed into the column before sedimentation-sizing occurs; (c) particles must be packed at modest concentrations and with fluid velocities as high as possible, to allow uniform and dense compactation into the developing bed; (d) the slurrying liquid must not react with or degrade the packing and must be easily removable after the packing process, and (e) the particles must have sufficient strength and rigidity to stand up to the mechanical stress of the packing procedure. The wet-fill technique is most useful for adsorbents and bonded-phase materials. For liquid-liquid chromatography with mechanically held stationary phases, the bare support must first be loaded into the column and then *in situ* coated with the stationary liquid (43) (see also Section 8.3).

Packing of Rigid Solids

The two different high-pressure slurry-packing procedures described in the following discussion have been used to prepare columns with approximately equivalent results. Figure 5.13 shows a typical laboratory-constructed apparatus for preparing columns by the high-pressure slurry-packing technique. Most workers use constant-pressure, pneumatic amplifier pumps for the filling process. Although high-velocity constant-flow pumps may be preferable in some cases, these devices often do not provide high enough flowrates to prepare compact bed structures. Pulsations or pump-refill cycles should be controlled during packing so as to maintain a constant flow of slurry during the filling process. The apparatus shown in Figure 5.13 can be used with either of the packing

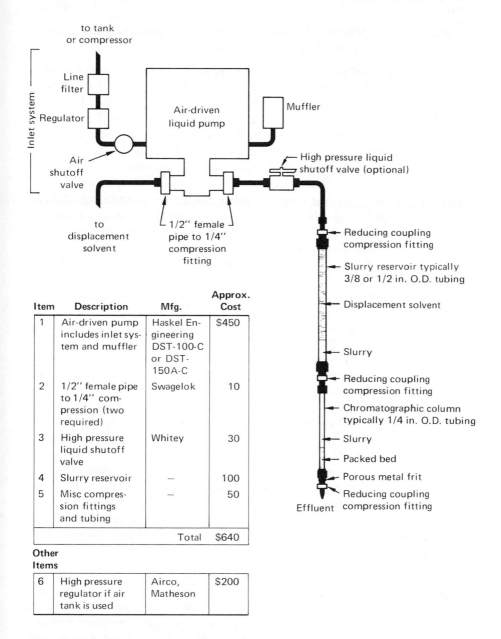

		to tank or compressor			

Inlet system

Line filter

Regulator

Air-driven liquid pump

Muffler

Air shutoff valve

High pressure liquid shutoff valve (optional)

to displacement solvent

1/2" female pipe to 1/4" compression fitting

Reducing coupling compression fitting

Slurry reservoir typically 3/8 or 1/2 in. O.D. tubing

Displacement solvent

Slurry

Reducing coupling compression fitting

Chromatographic column typically 1/4 in. O.D. tubing

Slurry

Packed bed

Porous metal frit

Reducing coupling compression fitting

Effluent

Item	Description	Mfg.	Approx. Cost
1	Air-driven pump includes inlet system and muffler	Haskel Engineering DST-100-C or DST-150A-C	$450
2	1/2" female pipe to 1/4" compression (two required)	Swagelok	10
3	High pressure liquid shutoff valve	Whitey	30
4	Slurry reservoir	–	100
5	Misc compression fittings and tubing	–	50
		Total	$640

Other Items

6	High pressure regulator if air tank is used	Airco, Matheson	$200

Figure 5.13 High-pressure slurry-packing apparatus for wet-filling columns. Reprinted from (17) with permission.

procedures to be described. Certain portions of such apparatus are also available from commercial suppliers (e.g., Alltech Assoc.).

The Down-Flow Method. The down-flow approach has been most widely used to date, and it permits the preparation of satisfactory columns of all types of microparticle packings. An apparatus such as that shown in Figure 5.13 is used, in which the slurrying mixture is rapidly forced *downward* into the column blank. With a constant-pressure pump, the slurry flowrate is dependent on the pressure used, and as the packed bed is formed the flowrate decreases. Forcing the slurry mixture into the column blank at the highest possible velocity generally produces the best column performance. This operation is carried out by pumping the slurry into the column blank at the highest pressure permitted by the compression fittings connecting the column to the slurry-packing apparatus. Strong particles are required for best results with this method. The very high initial velocities as the slurry enters the column blank may fracture weak particles, producing fines that tend to plug the column outlet and cause packing structure irregularities.

Before attempting to pack a column by the wet-fill process, fines in the packing must first be removed. This is easily accomplished by repetitively slurrying the mixture in a solvent (e.g., methanol), allowing the main fraction to settle, and pouring off the supernatant containing the fines. Aggregates also can be removed by this sedimentation procedure, because larger particles settle fastest and can be removed at the bottom of an appropriate vessel (e.g., separatory funnel).

For the wet packing procedure, slurries of 1-30% (by weight) of the packing have been used satisfactorily, but 5-15% solids generally appear best. Before packing, the slurry should be thoroughly degassed and dispersed by vigorous shaking (e.g., ultrasonic vibration) while evacuating the container containing the slurry. The liquid that is used to force a slurry into the column blank does not have to be the same as that used to prepare the slurry. Generally, a less-dense pumping fluid is preferred, to prevent unwanted mixing during the pumping.

Using the apparatus shown in Figure 5.13 as model, we now describe a typical detailed procedure for preparing a column by the down-flow slurry-packing procedure:

Procedure. Before packing, the column blank must be thoroughly cleaned. The inner walls should be washed successively with dichloromethane, acetone, and water, respectively. Then the interior of the tubing is vigorously scrubbed with a hot detergent solution, using a long pipe cleaner or a lint-free cloth tied to a nylon line (avoid scratching the tubing inner walls!). The scrubbed column blank is then rinsed with distilled water and absolute methanol and dried with purified nitrogen or air. The interior of the clean column blank should be inspected to ensure that all particulates are removed and that the column walls are smooth and clean.

To the outlet of the column blank is fitted an appropriate porous frit or screen (see earlier description), which is retained by the outlet compression fitting. An ordinary non-low-volume outlet compression fitting is used to allow maximum flowrate from the column during the packing operation. (However, this larger-bore fitting must be replaced with an appropriate low-volume or "zero-volume" fitting before use of the column in LC.) The inlet of the blank tube is then connected to a short length (3-5 cm) of connector tubing (not shown in Figure 5.13), which is then attached to the slurry reservoir. This connector tubing is useful in orienting the slurry into the column blank and also acts as a mini-reservoir to ensure that the column is completely filled at the completion of the packing procedure. The internal diameter of the connector tubing and the outlet fitting of the slurry reservoir should be identical to that of the column blank i.d. to ensure rapid, smooth delivery of the slurry without subjecting the packing particles to undue shear forces.

The chromatographic column blank and precolumn are filled with a liquid more dense than that used to make the slurry (e.g., tetrabromoethane), after closing off a stopcock attached to the column outlet. To ensure that air bubbles are not left in the column during this procedure, the column blank is rapped smartly after filling. The pump is then prepressurized with the pressurizing liquid up to the high-pressure liquid shutoff valve at the inlet of the reservoir. The packing slurry (described later) is rapidly added to the slurry reservoir, which is then closed and immediately attached to the pressurizing line. (Care should be taken to see that no air pockets remain in the top of the reservoir or the pressurizing line—these should be prefilled with slurrying liquid if required.) Immediately, the reservoir is then suddenly pressurized by opening the high-pressure shutoff valve, which delivers a pressure of 5-12,000 psi from the air-driven pump. Pump pulsations should be avoided during this filling process, if possible. Pressurization is continued for several minutes, until 50-100 ml of solvent elutes from the column; usually the actual filling of the column is completed in a fraction of a second or a few seconds. The high-pressure shutoff valve is closed and the pressure allowed to bleed through the column until no more liquid emerges from the outlet.

Experience has shown that the mechanical stability of the packed bed is enhanced by a bed consolidation or "slamming" process (22). It is convenient to use methanol for this process so as to purge the slurrying solvent and condition the packed bed for subsequent chromatography. The bed-consolidation process consists of pressurizing the pump to the highest pressure permitted (by the compression fittings), with the high-pressure shutoff valve closed. The valve is then suddenly opened and 50-100 ml of fluid forced through the column. A series of these sudden repressurizations is used to consolidate the column bed. Final stabilization of the bed is achieved when column permeability becomes constant. This is apparent when a constant flow rate of liquid from the column is obtained after repeated repressurizations at a constant pressure.

After packing, the column is removed from the apparatus, excess packing carefully leveled flush with the column inlet (e.g., with a razor blade), the face of the column end cleaned off, and the inlet frit (or screen) plus low-volume or "zero-dead volume" end assembly attached. (*Note*: Prior to assembly of the frit and end fitting, the column should *not* be touched with the bare hand; otherwise, packing will be forced out by thermal expansion, and column efficiency will be reduced.)

A variety of slurrying liquids may be employed with this procedure, as illustrated in Table 5.17. Single solvents such as methanol, isopropanol, and 0.01 M ammonium hydroxide have been used for untreated silicas of less than 7 μm with a narrow particle-size range. More satisfactory for particles of a wider size range (i.e., 5-10 μm, either unmodified silica or alumina) are 1:1 mixtures of chloroform/methanol or trichloroethylene/ethanol. These same solvent mixtures, or a 1:3 mixture of iso-octane and chloroform, have also been used satisfactorily for alkyl-silane modified silica (e.g., C_{18}). Reverse-phase packings in general can present a problem, because with very rapid flow of liquid, hydrocarbon-modified surfaces become highly charged and are difficult to pack homogeneously into stable structures. Such problems can be remedied by adding electrolyte to the slurrying liquid. For example, for hydrocarbon-modified particles, a slurrying liquid of 80% methanol/20% of 0.1% aqueous sodium acetate has been recommented (*20*).

Table 5.17 Illustrative slurry-packing systems

Packing technique	Typical slurry solvents(s)	Reference
Balanced density	Tetrabromoethane, tetra-chloroethylene, diiodo-methane (with modifiers)	18, 19, 22 44, 45
"Nonbalanced density"	Carbon tetrachloride	14, 46
	Methanol	23, 24
	Acetone	48
	Dioxane-methanol	49
	Tetrahydrofuran-water	50
	Isopropyl alcohol	51
	Chloroform-methanol	17, 23
	Methanol-water-sodium acetate	20
Ammonia-stabilized slurry	0.001 M aqueous ammonia	7
High viscosity	Cyclohexanol, polyethylene glycol 200, ethylene glycol	21

Source: From (*30*), with some modifications.

The preparation of satisfactory columns with polar organic-modified silicas (e.g., amine) has not been as widely documented. Slurrying solvents of chloroform/methanol, tetrabromoethane/tetrahydrofuran, chloroform/pyridine, and *n*-propanol all have been reported for these packings.

A balanced-density technique generally is needed if the particles have a relatively wide particle-size distribution, or if particles greater than about 10 *μ*m are involved. In this approach the relative concentrations of the solvents used for the suspending mixture are adjusted to provide a density approximately equal to that of the particles. In a proper system the particles stay suspended for at least 20 minutes. Slurries made with tetrabromoethane or methyl iodide plus relatively nonpolar modifiers such as perchloroethylene have been used for hydrocarbon-modified particles, and polar modifiers such as tetrahydrofuran and pyridine have been found satisfactory for polar bonded-phase packings.

The weights of packing needed to fill 25-cm columns of various i.d. are shown in Table 5.18. Generally, about 15% excess of this quantity is used in the slurrypacking operation to ensure that the column will be fully packed.

Table 5.18   Approximate weight of packing required in 25-cm columns (g).

Column inside diameter (cm)	Pellicular (porous-layer) beads	Porous silica microparticles			Porous alumina microparticles (Spherisorb)
		Spherisorb	LiChrospher	Zorbax	
0.21	1.4	0.5	0.5	0.7	0.9
0.32	3.2	1.0	1.0	1.5	1.9
0.40	5.1	1.7	1.6	2.2	3.1
0.48	7.3	2.4	2.3	3.3	4.5
0.62	12.2	4.0	3.8	5.3	7.4
0.78	19.2	6.3	6.0	8.0	11.5

The maximum length of column that can be prepared by the downward-flow wet-fill method depends on the size of packing particles and the packing technique. Studies with 0.6-cm i.d. columns suggest that to maintain good performance (e.g., reduced plate height of <4), maximum single-column lengths with the down-flow method should be as follows (*23*):

Particle size (*μ*m)	Maximum column length (cm)
10-12	50
7-8	25
5-6	10-15

With these lengths, expected plate count generally can be obtained; longer columns usually exhibit significantly lower column performance.

The Up-Flow Method. An alternative technique for preparing columns by wet filling is the *up-flow* approach described by Bristow (*25*). Experience with the up-flow packing method has not been extensive, but column performance results have been about equivalent to those for the down-flow procedure. On the other hand, up-flow is the only packing method that has been proposed for preparing >25-cm columns of <5-μm particles. With this method, 1-m columns of ~3-μm particles have been made with a plate count of ~120,000 (*26*). Unfortunately, the long-term stability of columns packed by up-flow has not yet been described; consequently, down-flow packing may be preferred until more information is available. The down-flow approach produces packed columns of excellent stability, providing the proper procedure is used.

The equipment used for up-flow packing is similar to that shown in Figure 5.13. However, in this case the slurry is pumped from the reservoir *up* into the column blank from a reservoir whose contents are continuously diluted by incoming pressurized liquid. The general technique is somewhat like the down-

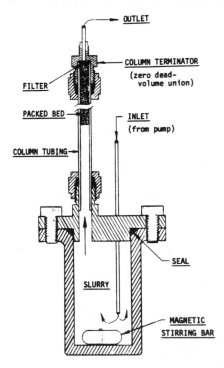

Figure 5.14 Stirred reservoir for packing columns by the up-flow method. Reprinted with permission of Micromeritics Instruments.

flow method, except for several important variables. Agglomeration of particles again must be prevented, but more dilute slurries (e.g., 1-10%w/v) are utilized. Methanol is satisfactory for suspending ≤5-μm particles, because of its low viscosity and ability to overcome interparticle forces. For particles larger than ~5 μm, a stirred slurry-packing apparatus such as that shown in Figure 5.14 is recommended, to ensure that the particles remain suspended during the up-flow packing process. n-Propanol is a preferred solvent in using the stirred reservoir.

In the up-flow approach the velocity of the particles striking the forming bed must be sufficiently high and the up-flow of the liquid sufficiently great as to prevent particles from falling from the bed back into the reservoir. Sedimentation velocities of porous particles are sufficiently low to allow columns of 3-μm particles to be packed at 2000 psi into 1-m columns with an impact velocity many times the sedimentation velocity (24). With very small particles (e.g., <5 μm), packing pressures much exceeding the pressure of intended use may not be required, because very small particles bind together strongly and are less affected by changes in forces such as mobile-phase flowrate (compared to larger particles). The relatively slow packing of the up-flow method allows the greatest opportunity for ordering of the bed to produce a densely packed structure, and on this basis dilute slurries are preferred. Up-flow packing actually causes an additional dilution to the original slurry, because the slurry is mixed with fresh liquid from the reservoir below, until particles are swept up into the column bed.

With the up-flow method, the viscosity of the slurrying liquid determines the time required to pack a column and the impact velocity of the particles, but not the maximum length of column that can be packed. By using increasingly high pressures, column lengths may also be further increased, before the final upward flow becomes too small to support the packing. Therefore, just as in the case of the down-flow method (but for different reasons), total pressure determines the total length of a packed column that may be satisfactorily prepared. The 6000 psi limit imposed by the commercial packing reservoir shown in Figure 5.14 may be adequate for many upward-flow column-packing procedures. A similar apparatus has been reported (27) with a 7500 psi limitation, and with the capability for simultaneously filling six columns by up-flow.

The wet-fill procedures described here are only illustrative of those that can be successfully used for packing columns with rigid particles. The experience of many workers has shown that other slurrying liquids, equipment, and packing procedures can be utilized for packing columns of rigid particles with excellent column performance (see Bibliography at the end of this chapter).

Packing of Hard Gels

The balanced-density slurry-packing technique (up- and down-flow) also has been widely used for preparing columns of hard gels. The same apparatus and general technique is utilized as for rigid particles. However, the handling of these gels is somewhat different than that for rigid solids. Hard gels must first be allowed to swell in the solvent in which they are to be packed. Also, since organic polymeric gels have a lower density than rigid solids, the slurry-packing technique usually is carried out with somewhat lower-density solvents (e.g., acetone/perchloroethylene mixtures). In addition, lower packing pressures must be used.

Conventional ion-exchange resins normally are wet-packed in aqueous solvents, preferably by the balanced-density technique. The procedure is to prepare a thick slurry of the swollen ion-exchange resin in a salt solution (e.g., calcium chloride for density matching). This balanced-density slurry is then forced by high pressure into the column blank, using the technique and equipment described for the packing of rigid solids. The pressure that can be utilized here depends on the strength of the resin bead, which is a function of the degree of cross-linking. Maximum pressures of about 5000 psi can be utilized with the strongest ion-exchange resin particles. Packed columns prepared by the salt-balanced-density technique must be flushed with the chromatographic mobile phase to ensure complete elimination of the salt before use.

Alternatively, conventional ion-exchange resins can be slurry-packed by a "dynamic" packing procedure (28). In this approach a very thick aqueous slurry of the swollen small-particle ion-exchange resin is prepared and rapidly pumped with high pressure into a column blank before particle sedimentation occurs. One approach is to prepare a "cartridge" column of the thick ion-exchange resin slurry, which is then connected to an empty chromatographic column containing the mobile phase. The ion-exchange resin in the cartridge is then suddenly extruded by high pressure into the blank column. Several columns may be prepared with a single cartridge.

Packing of Soft Gels

Soft gels cannot be dry-packed, nor can they be packed by the high-pressure wet-fill procedure, because most of these column packings compress at a few psi per cm. Soft gels generally are packed in the columns using a gravity slurry-sedimentation method. Optimum procedures are described in brochures furnished by the manufacturer of the particular soft gel materials.

5.5 COLUMN EVALUATION AND SPECIFICATIONS

Column Evaluation

Column performance traditionally has been defined by the plate count (methods of calculation are given later). Although useful, this parameter does not by itself provide sufficient information for properly evaluating column usefulness. In addition to plate count, other performance characteristics should be obtained on the column after it is packed, or when it is obtained from the manufacturer. Meaningful performance characteristics for a column include those given in Table 5.19. These column specifications can be measured with a standard test mixture under constant operating conditions. In some cases such data are supplied for each packed column by the manufacturer.

Table 5.19 Practical column performance specifications.

- Plate count for several solutes (e.g., k' = 0, 3, 10)
- Retention (e.g., k' values) for appropriate solutes
- Relative retention (α values) for unlike solutes
- Peak symmetry (skew or asymmetry factor)
- Column pressure drop (at a given flowrate, mobile phase, and temperature) or permeability
- Concentration of bonded organic phase

To evaluate the performance of a column completely, it has been recommended that the following (all-inclusive) data should be recorded (29): operating conditions, including column temperature, packing designation, method of packing, mobile-phase composition, composition of test sample, detection method, and injection method; properties of mobile phase and solute(s), including viscosity of mobile phase and diffusion coefficients of solute(s) in that mobile phase; geometrical parameters: column bed length, column bore, and particle size; and, finally, chromatographic parameters, including sample volume injected, peak retention times or retention volumes for unretained or retained solute(s), peak width at half-height of solutes on a time or retention volume basis, column pressure drop, and volume flowrate of mobile phase. For basic studies the following major chromatographic properties should also be considered: reduced plate height h, reduced mobile-phase velocity ν, and column flow resistance ϕ:

$$\phi = \frac{d_p{}^2}{K_0} \; , \tag{5.3}$$

where K_0 is the column specific permeability, or

$$\phi = \frac{0.1 \Delta P_i t_0 d_p{}^2}{10^9 \eta L^2} \; , \tag{5.3a}$$

where in the "practical" Eq. 5.3a, d_p equals particle size (μm), ΔP equals the column pressure drop (bar), L equals column length (m), t_0 equals the elution time for a nonretained solute (sec), and η equals the viscosity of the mobile phase (cP). Other important column parameters include the capacity ratio k', the *Knox-Parcher ratio*,

$$I = \frac{d_c^2}{Ld_p},$$ (5.4)

where in this case d_c is the internal diameter of the column bed (mm) [L and d_p as in Eq. 5.3a] and the *total porosity*,

$$\epsilon_{tot} = \frac{4F_c t_0}{\pi d_c^2 L},$$ (5.5)

where F_c is the flowrate (cm^3/sec) and the other parameters are as in Eq. 5.3a and 5.4.

Although the chromatographic properties just discussed are useful for basic studies in LC, these parameters may be too detailed for most analytical-laboratory situations. Accordingly, the practical parameters listed in Table 5.19 are usually adequate to define column performance. It is always desirable to measure these properties for a newly packed or purchased column, before placing it into use. Results should be kept for future reference, and the various tests repeated if it is suspected that a column is not performing properly for any reason. Some laboratories routinely retest columns (e.g., once weekly) when performance must be maintained constant for critical analyses. Let us now examine some of these column performance tests more closely.

The test mixture for determining column specifications preferably should be composed of at least three components, one at $k' \simeq 0$, with the other two components in the $k' = 2$-10 range. This test mixture should contain compounds that will correctly characterize the column in terms of both kinetic and equilibrium performance. For example, a test mixture for a column of untreated silica (LSC) typically should contain a neutral solute (preferably $k' \approx 0$) and at least one solute each that exhibits acidic and basic characteristics. In this case, a useful mixture of this type might contain benzene, benzyl alcohol, and dimethylaniline, respectively. The plate count of the neutral solute at $k' \approx 0$ is useful in indicating how well the column is packed. If the plate count for this solute is significantly smaller (e.g., by at least 15%) than the plate count of retained peaks, then extracolumn effects are probably significant. Plate-count levels for the retained acidic and basic test solutes indicate packing quality relating to chemical interaction effects. A significant decrease in N with k' may indicate poor quality of the packing itself.

Peak shape is also a useful column specification. For example, a significantly tailing peak for the basic test solute on a bonded-phase packing suggests interaction with free SiOH groups. If all test solute peaks are tailing (or show frontal asymmetry), a poor packing structure is likely; for a further discussion of peak shape, see the following remarks and Section 19.2.

Test solutes should be carefully selected to assess the physical and chemical characteristics of the packing. Thus, with a standard solute-solvent combination (test system), variability in k' indicates variations in packing surface area or bonded-phase concentration. Changes in plate count suggest variations in column bed-packing structure, particle size, or poor diffusion in the stationary phase. Shifts in α values for solutes of different functionality reflect changes in the chemical nature of the packing, for example, variations in the level of unreacted SiOH groups for a bonded-phase packing or varying polymerization of the organic coating.

The mobile phase used in the test system should be stable, nonviscous (e.g., $\leqslant 0.5$ cP), and easily reproduced. For instance, "dry" solvents should not be used for adsorption chromatographic packings, since their water content actually may be variable. This causes changes in adsorbent activity, with resultant variations in k' and perhaps α values. Normally, the test system should be selected to show the column at its best, that is, using mobile phases of low viscosity and simple, low-molecular-weight solutes. A test system composed of higher-molecular-weight solutes and/or higher viscosity mobile phase will show lower plate count. In some cases it is appropriate to select the test solute-solvent system to mimic actual samples. This permits a closer check over performance changes of direct interest to the analysis. For example, if solvent X is used to determine solutes Y and Z, then the test system $X/Y, Z$ provides the best possible check that the column is performing as required.

The column permeability (pressure drop) indicates whether or not the nominal particle size is correct (see Section 2.4 and Figure 5.7), or whether the column is partially plugged for any reason. The concentration of bonded organic phase, preferably described as μmoles/m^2 (see Section 7.2), shows whether the bonded organic phase has been properly reacted with the support.

To describe effectively the performance of a column, the column plate count must be accurately measured. However, as discussed later, the shape of the peak (e.g., amount of tailing) not only affects the calculated values of N, but is important in its own right. There are several methods available for measuring column plate count. The absolute method (which is applicable for tailing bands) is

$$N = \frac{(t_R)^2}{(\sigma_t)^2},\tag{5.6}$$

where N is the plate count of the column, t_R is the retention time of the peak at the apex, and σ_t^2 is the peak variance or second central moment (54). The true value σ_t^2 is too tedious to hand-calculate and is best done by computer. However, if the peaks are symmetrical (e.g., have a Gaussian shape), Eq. 5.6 can be expressed in forms that are more easily measured experimentally. The most-used approximations include the following (31):

$$N = 16 \left(\frac{t_R}{t_w} \right)^2 , \tag{5.7}$$

where t_w is the baseline width formed by the tangents of the peak intersecting the baseline (equal to 4σ; see Fig. II.1, page 834);

$$N = 5.54 \left(\frac{t_R}{w_{1/2}} \right)^2 , \tag{5.8}$$

where $w_{1/2}$ is the peak width at one-half the peak height; and

$$N = 2\pi \left(\frac{h' t_R}{A} \right)^2 , \tag{5.9}$$

where h' is the peak height and A is the peak area.

Although Eqs. 5.7-5.9 are widely used to calculate N values, this use is only valid for symmetrical bands. As discussed earlier, there are various ways of expressing the "true" efficiency of columns that give asymmetric bands, but it is generally better simply to calculate the *peak asymmetry factor* as in Figure 5.15. When the latter parameter exceeds 1.2 for a peak, the apparent N value for the column (as calculated by Eqs. 5.7-5.9) will be too high. In fact, it is a good

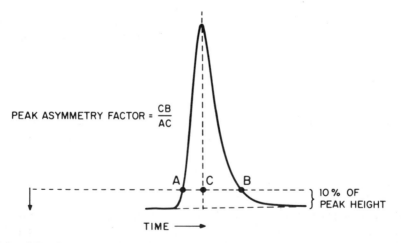

Figure 5.15 Calculation of peak asymmetry factor. Reprinted from (32) with permission.

general rule, where possible, to avoid columns (or LC systems) that give asymmetry factors greater than about 1.2. (The remainder of this section is highly specialized, and the reader may wish to skip to "Column Specifications.")

The error in the calculation of N by the simplified expressions has been quantitated in one computer-simulated study, the results of which are summarized in Table 5.20. Here we see that with increasing peak tailing (measured by peak asymmetry or peak skew, described later), serious positive errors in plate count result with the widely used hand-calculated peak-tangent method (Eq. 5.7). Moderate tailing produces positive errors of almost 100%, compared to true values, and peaks with this degree of tailing are not uncommon in the literature. Calculation of N by the other simplified plate count expressions (Eqs. 5.8 and 5.9) produce essentially the same errors.

Table 5.20 Effect of tailing peaks on plate count calculation.

Peak asymmetry, value[a]	Peak skew γ_1' [b]	Plate count		
		True value	Tangent method	% error
$\sim$1.00	0.002	7670	7330	−4
1.06	0.18	6200	6230	0
1.23	0.71	3870	4470	+15
1.62	1.15	2380	3850	+61
2.03	1.43	1550	3060	+98

Source: Data from (32, Table II).
[a] From calculations by method in Figure 5.15 (CB/AC).
[b] From Eq. 5.10.

To account for unrealistically high plate counts for asymmetric peaks, another plate-height equation has been proposed (34):

$$N_{5\sigma} = 25 \left(\frac{t_R}{W_{4.4}} \right)^2, \tag{5.9a}$$

where $N_{5\sigma}$ is the plate height (5σ) and $W_{4.4}$ is the width of the peak at 4.4% of the peak height. This approach does produce plate count values that partially compensate for the effects of peak tailing (e.g., for a tailing peak, 3300 plates for $N_{5\sigma}$ vs. 9600 for the tangent method). However, if manual calculation is used, the accuracy and precision of this method are highly variable. More significant, the correlation of $N_{5\sigma}$ values for tailing peaks with actual column performance has not been established.

Thus, the shape of the peak can significantly affect the N values that are calculated by simplified plate-count expressions. It is important, therefore, that we have means of quantitating peak shape, or the degree of distortion from the Gaussian peak model. In the study summarized in Table 5.20, the standard deviation σ of peaks was convoluted with an exponential function τ to produce increasingly tailing peaks (*32*). An exponential function is a close approximation of the peak tailing that occurs generally in chromatography, for example, extra-column effects, poor sample injection, badly packed columns, chemical effects, and so on. Peak tailing (or distortion) can be quantitated by a *peak skew* value (*33*):

$$\text{peak skew} = \gamma_1' = \frac{2(\tau/\sigma)^3}{[1 + (\tau/\sigma)^2]^{3/2}} \quad , \tag{5.10}$$

where for this model γ_1' is approximately equivalent to the peak skew from statistical moments:

$$\gamma_1 = \frac{u_3}{u_2^{3/2}} \quad , \tag{5.11}$$

where u_2 and u_3 are the second and third central moments of the peak, respectively. [See (*47*) for a discussion of statistical moments.] Although peak skew values can be determined with higher precision and better accuracy than other measures of peak shape, measurements are not convenient unless a computer is available. However, in the laboratory empirical *peak asymmetry values* can be easily hand-calculated by the procedure illustrated in Figure 5.15, and these values relate directly to the mathematical peak skew γ_1', as shown by the plot in

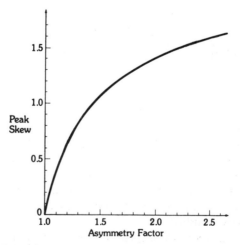

Figure 5.16 Correlation of peak skew with peak asymmetry values. Reprinted with permission of Du Pont Instrument Products Division.

Figure 5.16. Several column suppliers are using various forms of peak asymmetry or peak skew values to describe peak shape as a specification for column performance, but, unfortunately, as yet there is no uniform approach. The data in Table 5.20 suggest that column performance should only be determined on essentially symmetrical peaks (peak asymmetry values ≤1.2 or peak skew ≤0.7), or errors may be sufficient to negate any useful conclusions.

Column Specifications

What should the performance of a good column be? Experts generally agree that the reduced plate height h (Section 5.7) for a column should be about 2-3.5 for a reduced velocity of about 5, and should not exceed 20 for a reduced velocity of about 100. The column flow-resistance parameter ϕ should not exceed 1000. Thus, a 25-cm column of 10-μm particles typically would have a plate count of 10,000 at a linear velocity of about 0.08 cm/sec (h = about 2.5 and v = about 5). Similarly, a 15-cm column of 5-μm particles should exhibit a plate count of about 12,000 at $u \approx 0.15$ cm/sec. Commercial columns made in large numbers exhibit a performance slightly poorer than these values, but a degradation of more than 50% from this level should not occur. Peak asymmetry values (Figure 5.15) for columns should be ≤1.2. However, columns manufactured in volume may not meet this rigid specification, and asymmetry factors ≤1.6 may be more realistic (if less desirable). Specifications for reproducing α and k' with columns have not yet been given appropriate attention; however, variations of no more than ±5% and 10%, respectively, from the average have been proposed as a useful goal for these parameters (*32a*).

Direct comparisons of column performance may be difficult, because of the variation in extracolumn effects in various instruments. Therefore, if performance data are significantly poorer than those reported by the manufacturer of a purchased column, extracolumn effects (e.g., large-volume connectors, poorly constructed detector cells) may be at fault and should be checked first. One indication of extracolumn effects is a significant increase in N for bands with larger k' values. However, if a column does not measure up to manufacturers' specifications, it should be returned for replacement. Columns prepared in the laboratory should be repacked if they fail to meet the guidelines suggested here.

5.6 COLUMN TECHNIQUES

Handling of Columns

Certain precautions must be followed to maintain the efficiency, capacity, and permeability of columns containing particles of ≤10 μm. Such columns are readily plugged by minute particles; therefore, carefully distilled or filtered mobile phases should always be used. As discussed in Section 7.5, water is a particularly important source of contaminants. Of particular importance when using aqueous mobile phases (e.g., buffer solutions) is bacterial growth that can plug column frits or screens. This problem can be prevented by maintaining inhibitors (e.g., 0.01% NaN_3) in the aqueous mobile phase, if this does not interfere with detection. Use of in-line solvent filters (e.g., 0.5-μm porosity) and the prefiltering of samples before separation can also help eliminate the buildup of particles at the column inlet. Filtration of samples with Swinney or Swinnex filters (Millipore Corp.) is quick, efficient, and convenient.

Vibration, mechanical shock, and extremes of temperatures should be avoided in the handling of columns. Slurry-packed columns should always be used at operating pressures well below that employed in the wet-fill operation, preferably no more than one-half the pressure used during the packing process. Otherwise, a void can develop at the inlet due to the settling of the packed bed. Rapid formation of a void in the inlet of a new column usually indicates that the packing process was not carried out properly. Backflushing of columns generally should be avoided, to prevent shifting of the packed bed and performance degradation.

Loss in performance rarely occurs when columns are stored with neat organic solvents. However, storage in aqueous mobile phases outside the pH range of 2-8.5 should be avoided. For long-term storage, the best approach is to purge the column with a pure organic solvent (e.g., methanol). The drying out of columns, particularly rigid organic gels, should be avoided. For storage, closing both ends of the column with metal compression-fitting caps is desirable.

Table 5.21 summarizes the recommended practice for handling and using slurry-packed columns of $\leqslant$10-μm particles. Close adherence to these guidelines will maintain column performance and prolong the lifetime.

Table 5.21 Recommended practice with columns $\leqslant$10-μm particles.

- Use 0.5-μm filter in mobile-phase supply to column
- Use sampling valve (avoid septum inlet!)
- Filter sample with 0.5-μm filter
- Filter mobile phase (0.5-μm porosity) if not freshly distilled
- Avoid reverse-flow
- Use guard columns for "dirty" samples
- Stay within recommended pressure, pH, solvent limits
- Handle and store carefully

Column Troubleshooting

If the performance of a new column rapidly and radically degrades, and a void is found in the inlet, the column should be repacked or returned to the supplier (with a history of its use) for replacement. The performance of an initially good column rarely can be reestablished with such techniques as adding additional packing (as a slurry) or glass microbeads to the void at the inlet, although this sometimes can partially offset degraded column performance.

Plugged or partially plugged columns are evidenced by the increased pressure required to maintain a constant flow, or the indication of lower (or no) flow at a constant pressure. Broader, tailing peaks also may be due to a partially plugged column inlet. Column pluggage is most likely to develop in the narrow i.d. inlet fitting or the column inlet or outlet frits. Pluggage can arise from septum particles (syringe injection) and, to a lesser extent, from particles that arise from the wearing of microvalve seals. If a pluggage develops, the inlet connector assembly should be carefully removed, inspected for particles, and cleaned. (*Caution*: Do not disturb the packing or touch the column with the bare hand when the fitting has been removed, or degradation of column performance will occur.) If this is not the source of the problem, the inlet frit or screen should be cleaned or replaced. Outlet frits or screens are rarely plugged, but these closures also can be replaced (carefully, without disturbing the packing) if found to be dirty.

As indicated in Section 7.5, the stability of a column depends on the type of packing and the way the column has been used. No column will maintain satisfactory performance indefinitely; with care, however, a well-made unit can last for several months. Degradation of a bonded-phase column can be accelerated by certain additives (e.g., salts—see Section 7.5), and silica-based packings slowly dissolve in the presence of aqueous mobile phases. Dissolution is accelerated at high temperature and pH (e.g., above 8); after extensive use under these conditions, the column may also develop a void in the inlet. If this occurs, the column should be discarded, because the initial performance can never be restored, because of structural change in the component particles. Packing stability can be significantly enhanced by inserting a short (e.g., 3-cm), disposable precolumn of the packing in the mobile phase line prior to the sample injection system (see Section 8.6).

Connecting Columns

For separating very complex mixtures, it is necessary to have a column with a large number of theoretical plates and a high peak capacity. Long columns with large plate counts can be prepared by connecting individual columns with jumpers of short, narrow-bore tubing (e.g., 2×0.025 cm), using low-volume or "zero"-dead-volume fittings. Plate count for the individual columns should be additive if essentially symmetrical peaks are involved and the columns are joined with proper low-volume connectors. If the retention times t_R for the test solute are approximately equal for each column (t_{R_1} approximately equal to $t_{R_2} \ldots$ approximately equal to t_{R_n}), then the plate count for the connected system can be calculated according to

$$N_t = \frac{Y^2}{\Sigma(1/N_i)} \qquad (5.12)$$

where Y is the number of columns in series, N_t is the calculated total plate count, and N_i is the plate count for each individual column. As shown in Table 5.22, using Eq. 5.12, the total toluene plate count for a series of columns for high-speed gel permeation chromatography was calculated as 21,500, compared to 23,900 plates actually measured. The slightly higher observed plate count probably is due to decreased extracolumn effects for the larger-volume combined column set relative to that for the individual columns.

Table 5.22 Connected columns for high-speed GPC.

Column designation	Column length (cm)	t_R, Toluene (min)	N	σ^2
PSM-50S	10	1.27	2970	5.43×10^{-4}
PSM-300S	15	1.92	3425	10.76×10^{-4}
PSM-800S	10	1.26	5400	2.94×10^{-4}
PSM-1500S	10	1.26	3130	5.07×10^{-4}
PSM-4000S	15	1.92	8145	4.53×10^{-4}

$$\sigma_T^2 = 28.73 \times 10^{-4}$$
$$N_t = 21,450$$
$$N_{OBS} = 23,890$$

Source: From (*15*).

Sometimes the observed plate count of a series of connected columns is significantly smaller than predicted from the sum of the various individual columns. This often arises because tailing peaks are used to calculate the plate count of the individual units, causing positive errors as described in Section 5.5. Therefore, for connected columns and tailing bands it is important to use the true plate count derived from Eq. 5.6 to determine the additive plate count.

Use of Guard Columns

Continued injection of samples with strongly retained (noneluted) solutes will eventually degrade column performance. Changes can result in both retention and column plate count, as well as in column selectivity. To prolong the life of analytical columns, *guard columns* can be inserted between the sample valve and the analytical column. The guard column captures the strongly retained sample components and prevents them from contaminating the analytical column. Guard columns usually are relatively short (5-10 cm) and contain a stationary phase equivalent to (or similar to) that in the analytical column. These guard columns are replaced at required intervals (i.e., before strongly retained solutes elute into the analytical column) to ensure constant performance of the more expensive analytical column. Typically, guard columns are made with pellicular packings and, therefore, are easily dry-packed, relatively inexpensive, and conveniently replaced.

One might question whether guard columns can degrade the performance of very efficient analytical columns, because of extracolumn effects. The effect of adding the guard column to an analytical system can be predicted by considering band variances, as indicated in Table 5.23. Here we show the calculated effect of using a 5-cm guard column of 30-μm particles with a 25-cm analytical column of 10-μm porous particles. Only for the impractical (and obviously limiting) situation in which the solute shows $k' = 0$ (unretained) for both the guard and analytical columns is there a significant decrease in the apparent plate count for the analytical column. However, as k' increases for the analytical column, the volume of the peak increases and the effect of the guard column decreases to an insignificant level. Even with efficient analytical columns (e.g., 10,000 plates), the extracolumn effect of the guard column is still insignificant, providing k' for the peak of interest is sufficiently large for the analytical column. This analysis predicts that a guard column can be used with high-performance small-particle analytical columns if the peak broadening caused by the guard column is insignificant. This conclusion is verified by the experimental data in Table 5.24 with a guard column of 35-75-μm pellicular adsorbent and an analytical column of 7-μm porous silica microspheres operated at two flowrates. As predicted, plate counts significantly decrease only with the solute eluting at $k' = 0$, and no important effect occurs with solutes having larger k' values.

Table 5.23 Calculated effect of guard column performance (Guard column, 5 cm long, $\sim$30 μm, analytical column, 25 cm, 10 μm).

Peak k' of interest		Plate count of analytical column	
Guard column	Analytical column	Without guard	With guard
0	0	5,000	3750
0	1	5,000	4700
0	2	10,000	9470
2	0	5,000	1540

Source: From (*32*).

Table 5.24   Measured effect of guard column (guard column, 5 × 0.41 cm, Corasil II, 37-75 μm; analytical column, 25 × 0.46 cm, PSM-50, 7 μm).

	Plate count of aromatic alcohols (Area method, ave. of duplicate runs)			
	Flowrate, 2.0 ml/min (μ=0.3 cm/sec)		Flowrate, 4.0 ml/min (μ=0.6 cm/sec)	
k' value	Without guard column	With guard column	Without guard column	With guard column
0	7,599	6,870	5570	3430
0.3	4,310	4,130	3440	4010
0.5	11,410	11,000	7440	7650
1.2	8,180	7,370	4670	4670
2.7	7,710	7,900	5330	5290
5.4	7,620	7,290	5300	5050

Source: From (*32*).

Column "Infinite-Diameter" and Wall Effects

Careful injection of a sample at a point in the center of the column inlet can re-sult in the solute bands traversing the entire column without reaching the walls, as illustrated in Figure 5.17. This is due to the rather slow radial dispersion of the solute within the column. The theory of this so-called infinite-diameter effect has been considered in detail (*35, 36*); for a 10-cm column packed with 5-μm particles and operated at the optimum velocity, the radial spreading of the solute from a point injection amounts to no more than 0.25 cm (*20*). To predict the conditions for obtaining an "infinite-diameter" column and how this condi-tion varies with the reduced velocity, the following expression has been pro-posed (*36*):

$$\frac{(d_c - 60d_p)^2}{Ld_p} \geqslant 16\frac{1.4}{(v + 0.060)} \qquad (5.13)$$

However, most columns are not operated in the "infinite-diameter" mode. In many cases both the mobile phase and solute are introduced into the column in-let through a small capillary onto a frit or screen at the center of the column packing, as shown by the "normal" injection in Figure 5.18. With this mode of sampling, the solute does not enter as a sharp, well-defined spot in the center of column. Rather, in unfavorable circumstances it can be dispersed largely over the

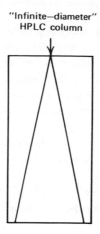

"Infinite—diameter"
HPLC column

Figure 5.17 Schematic of flow effects in "infinite-diameter" column. Injection of sample into center of "infinite-diameter" column results in molecules never approaching the column walls, with resulting sharp peak.

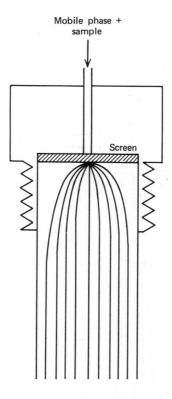

Mobile phase +
sample

Screen

Figure 5.18 "Normal" injection configuration. Reprinted from (*32*) with permission.

column inlet cross section in an unequal profile (*32*). Under these conditions a portion of sample approaches the wall area where band broadening is greater, rather than remaining in a tight sphere in the more efficient column center, thus eliminating the possibility of an "infinite-diameter" column. It has been estimated that the wall effect, where plate heights may be increased threefold, extends about 30 particles into the column (*36*). With poorly packed columns and/ or an unfortunate combination of particle size and column dimensions, use of the "normal" sample injection system shown in Figure 5.18 result in significantly tailing peaks. In particularly poor columns, peaks can exhibit "humps" on the backside, or the peaks may appear as doublets (*32*). However, use of the "normal" injection mode with well-packed columns generally does not seriously degrade performance, and allows larger sample sizes versus "infinite-diameter" injection.

Injected solutes can be forced to move down the center of the column ("infinite-diameter" mode) with point (or center) injection such as that shown in Figure 5.19. With this approach, samples are introduced into the top and center of the column onto (or just through) the inlet screen or frit, while the

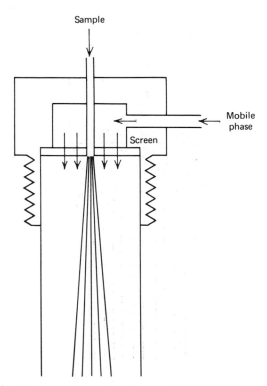

Figure 5.19 Point injection of sample. Reprinted from (*32*) with permission.

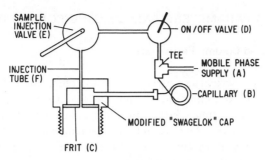

SAMPLE
INJECTION
VALVE (E)

ON /OFF VALVE (D)

INJECTION
TUBE (F)

TEE

MOBILE PHASE
SUPPLY (A)

CAPILLARY (B)

MODIFIED "SWAGELOK" CAP

FRIT (C)

Figure 5.20 Point-injector apparatus with high-pressure microsampling valve. Reprinted from (*32*) with permission.

flowing mobile phase is directed across the entire column inlet cross section. This mode of sample introduction produces a sharp sample pulse into the center of the column, and the integrity of this sharp pulse is maintained as it passes through the center of the column bed, thus ensuring a true "infinite-diameter" column effect. Point injection can be accomplished with a syringe, but generally a microsampling valve is preferred. Figure 5.20 shows a point-injection apparatus using a high-pressure valve with an external loop. The general approach is to displace a sample from the loop into the top and center of the column by means of a slight differential pressure generated by an alternate mobile-phase supply. Concurrently, the mobile phase also is flowing over the entire column inlet cross section. Basically, this technique simulates the action of a syringe, but without the disadvantages of syringe injection (Section 3.6).

With point injection, column plate count and peak symmetry often are measurably improved, particularly if the column is imperfectly packed. However, for well-packed columns the advantage of the point-injection technique over the "normal" injection technique is not large, presumably because wall effects are not so significant. However, the point-injection technique does optimize the efficiency of small-particle columns, and this technique is highly reproducible and easily automated if needed. Symmetrical peaks often are obtained even with imperfectly packed columns, leading to more accurate analyses. With point injection, column-packing techniques and special column blanks also are less critical, and the integrity of the packing in the column is more easily maintained during routine use. On the other hand, a disadvantage of point injection is the need for special hardware not now generally available. Point injection also uses only a fraction of the available column packing. Thus, column sample capacity is significantly reduced and column performance is much more susceptible to larger sample volumes (i.e., injection volumes generally should be smaller with point injection). Finally, the point-injection technique appears to be less useful for preparative separations, because high sample loadings normally are desired and all the packing should be used.

5.7 REDUCED PARAMETERS AND LIMITING COLUMN PERFORMANCE

Utility of Reduced Column Parameters

In preceding pages several concepts relating to column performance have been developed by means of more or less qualitative arguments, or in terms of specific examples. At the same time, certain questions concerning this general area have so far gone unanswered. Here we present a general and quantitative treatment of column efficiency as a function of different separation variables and provide answers to the following (and other) questions:

- In general terms, what is a "good" column?
- Is there some optimum particle size d_p that yields maximum N values in a given application?
- How does N vary with operating pressure P and separation time t, for different values of d_p?
- What effect does the molecular weight of sample components have on column N values?

The model we use is based on so-called reduced parameters introduced by Giddings (1)—the reduced plate height h and reduced mobile-phase velocity v:

$$h = \frac{H}{d_p}, \tag{5.14}$$

$$v = \frac{u d_p}{D_m}. \tag{5.14a}$$

Giddings (1) has shown that the various contributions to plate height H of Table 5.1 can be combined to give

$$H = \underbrace{\frac{1}{(1/C_e d_p) + (1/C_m d_p{}^2 u/D_m)}}_{\text{(i-ii)}} + \underbrace{\frac{C_d D_m}{u}}_{\text{(iii)}}$$

$$+ \underbrace{\frac{C_{sm} d_p{}^2 u}{D_m}}_{\text{(iv)}} + \underbrace{\frac{C_s d_f{}^2 u}{D_s}}_{\text{(v)}}. \tag{5.15}$$

In well-designed LC packings, term (v) is small—as noted in Section 5.1. If we then ignore term (v) in Eq. (5.15) and substitute Eq. 5.14 and 5.14a into Eq. 5.15, we obtain

$$h = \frac{1}{(1/C_e) + (1/C_m\nu)} + \frac{C_d}{\nu} + C_{sm}\nu \tag{5.15a}$$

The significance of Eq. 5.15a is as follows. First, h is a function of ν, but is independent of d_p. Second, the constants C_e, C_m, C_d, and C_{sm} are dependent only on how well the column has been packed, and for good columns will have constant, minimum values—regardless of particle size. As a result, a plot of h versus ν values for any "good" column should fall on the same general curve. However, because C_{sm} is essentially zero for pellicular packings, and has a finite value for porous packings, two h-ν plots for "good" columns result: one for porous columns and one for pellicular packings. Figure 5.21 illustrates these two plots for well-packed columns of "good" particles. In agreement with theory, data for "good" columns fall on top of one of these two curves, whereas data for "poor" columns lie above the appropriate curve; that is, "bad" columns have larger h values than are predicted by the curves of Figure 5.21. Thus, comparison of h versus ν data for a given column with the plots of Figure 5.21 immediately indicates whether the column is as good as the best possible LC column. Data for a "good" column fall close to the ideal curve of Figure 5.21. If the h values for a column are higher than predicted by Figure 5.21, it is also possible to assess why the column in question is "bad." High h values at low values of ν ($0 \leqslant \nu \leqslant 10$) indicate a poorly packed column. High h values at high values of ν ($100 \leqslant \nu \leqslant 1000$) indicate a poor column packing; that is, particles with large d_f or small D_s (see Table 5.1).

The two ideal curves of Figure 5.21 can be represented by the equations

$$\text{(pellicular)} \quad h = \nu^{0.33} + \frac{2}{\nu} + 0.003\nu \tag{5.15b}$$

and

$$\text{(porous)} \quad h = \nu^{0.33} + \frac{2}{\nu} + 0.05\nu \tag{5.15c}$$

Therefore, given values of L, u, d_p, and D_m for any "good" column, it is possible to calculate N for that column using Eqs. 5.14 and 5.14a and Eq. 2.6a. For example, for a 25-cm column of 10-μm porous particles tested at 0.10 cm/sec with a solute having a $D_m = 10^{-5}$ cm^2/sec, one can calculate the expected plate count as follows:

$$\nu = \frac{(0.1)(0.0010)}{10^{-5}} \quad \text{(from Eq. 5.14a)}$$

$$\nu = 10$$

and

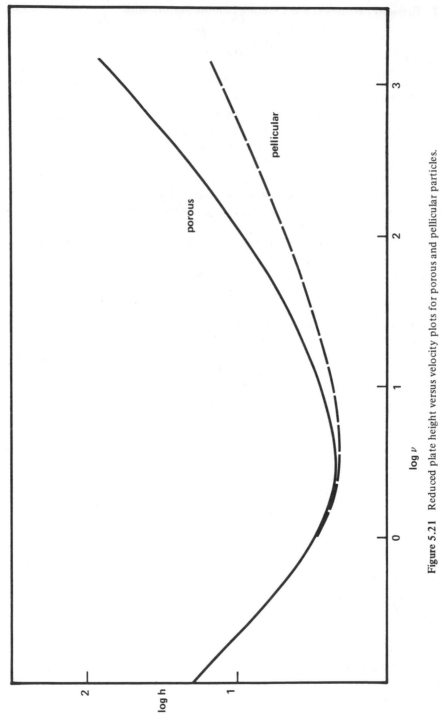

Figure 5.21 Reduced plate height versus velocity plots for porous and pellicular particles.

$$h = (10)^{0.33} + (\frac{2}{10}) + 0.05(10) \quad \text{(from Eq. 5.15c)}$$

$$h = (2.14) + (0.20) + (0.5)$$

$$h = 2.84$$

and

$$H = (2.84)(0.0010) \quad \text{(from Eq. 5.14)}$$

$$H = 0.00284 \text{ cm}$$

and, finally,

$$N = \frac{25}{0.00284} \quad \text{(from Eq. 2.6a)}$$

$$N = 8803.$$

Values of h versus ν are tabulated in Appendix II.

Alternatively, by means of Eqs. 2.11 and 2.12, it is possible to relate N for "good" columns to the variables d_p, P, and t. [Several workers have discussed the variation of N with these variables as a means of assessing the limiting performance of LC columns, e.g., (1), (2), (37), (38).] Here we simply summarize such a calculation in tabular form, for representative separation conditions, as one way of gaining insight into some of the questions raised at the beginning of this section.

Before we discuss such calculations, however, it is necessary to be able to estimate D_m for Eq. 5.14a as a function of separation conditions. The commonly used basis for calculating D_m values is the Wilke-Chang equation (39):

$$D_m = \frac{7.4 \times 10^{-8} (\psi_2 M_2)^{0.5} T}{\eta V_1^{0.6}}. \tag{5.16}$$

Here M_2 is the molecular weight of the solvent, T is the temperature in $^\circ$K, η is the solvent viscosity (cP), and V_1 is the molar volume (ml) of the sample molecule. The association factor ψ_2 is unity for most solvents, and greater than one for strongly hydrogen-bonding solvents; for example, 2.6 for water, 1.9 for methanol, 1.5 for ethanol, and so on. Equation 5.16 is not reliable for certain large molecules such as proteins, which show larger D_m values than predicted by this expression:

Protein mol. wt.	Increase in D_m vs. Eq. 5.16
1,000	1.3-fold
10,000	2-fold
100,000	3.2-fold

See also the discussion in (53).

Let us now see how the plate number of a "good" column varies with experimental conditions. In Table 5.25 we summarize column N values as a function of d_p and separation time t for two different column pressures: 500 and 5000 psi. Here separation time is arbitrarily assumed equal to $10t_0$ (maximum $k' = 9$: Eq. 2.11). Given values of P, t_0, and d_p, it is possible to calculate L from Eq. 2.12 for some fixed value of η. We assume the separation of a low-molecular-weight compound (mol. wt. 300) by reverse-phase LC at room temperature, on a porous spherical packing. As noted in Table 5.25, this means values of $\eta = 0.89$ cP, $\phi = 1000$, and $D_m = 5.5 \times 10^{-6}$.

Several interesting points emerge from a study of Table 5.25, and we will examine these in order. First, the box in the center of the table encompasses most conditions as they exist today in LC; that is, separation times between 1.5 min and 2.5 hr, particle sizes from 5 to 50 μm, and pressures from 500 to 5000 psi. We see later that there are good reasons for moving outside these limits in special cases, but the region enclosed within the box of Table 5.25 will probably include 99% of all LC separations for the foreseeable future.

Within this region of commonly used LC conditions, we can draw the following conclusions:

- The maximum value of N is about 100,000 plates; see, for example, (24) [except in size-exclusion chromatography, where for $t = t_0 = 2.5$ hr, $N \approx 550,000$ plates; see, for example, (40)].*
- The largest values of N occur for the smallest particles (5 μm), the longest separation times (2.5 hr), and the highest pressures (5000 psi).
- To obtain minimum separation time for some required value of N (e.g., 2000 plates in Table 5.25), the smallest particles (5 μm) and highest pressures (5000 psi) are again preferred (giving $t \approx 1.5$ min in Table 5.25).

Looking now at the whole of Table 5.25, let us consider first the question of optimum column pressure P. A higher operating pressure generally yields larger N values (assuming L is increased proportionately), as was noted in Section 2.5c. However, the advantage of a major increase in P (e.g., 10-fold as in Table 5.25) is only important for small values of d_p and/or large values of t. With small particles (5-10 μm) and separation times of 15 min to 2.5 hr, a 10-fold increase in P yields roughly a 2-fold increase in N. For much smaller particles and very long separation times, a 10-fold increase in P can translate into a 10-fold increase in N. However, the experimental conditions involved are totally impractical in

---

*Recently (40a) N values of 650,000 have been reported for t_0 values of about 9 hrs. (i.e., 30,000 sec) and 20-μm particles. Table 5.25 predicts N $\approx$ 300,000 for these conditions. However, a lower viscosity mobile phase was used in (40a), which would be expected to increase N somewhat (as in Fig. 5.26).

Table 5.25 Variation of column plate number N with experimental conditions.

Example: Reverse-phase separation of a low-molecular-weight compound on a "good" column of porous spherical particles. Values calculated from Eqs. 2.11, 2.12a, 5.14, 5.14a, and 5.15, with indicated values of t and d_p at two different values of P (500 and 5000 psi) and other conditions as follows: $D_m = 5.5 \times 10^{-6}$, $\eta = 0.89$ cP, $\phi = 1000$, and k' for last band = 9. L varies in each case, according to Eq. 2.12a.

$$N \times 10^{-4\ a}$$

d_p	$t =$ 10 sec / $t_o =$ 1 sec	1.5 min / 10 sec	15 min / 100 sec	2.5 hr / 1000 sec	1 day / 10^4 sec	10 days / 10^5 sec	∞ / ∞
1	0.1, 0.3	0.5, 1.5	1.0, 5.1	1.2, 9.7	1.3, 12.0	1.3, 12.6	1.3, 13
2	0.1, 0.1	0.4, 0.9	1.7, 4.4	3.6, 17	4.7, 36	5.0, 47	5.1, 51
5	0.0, 0.0	0.2, 0.3	1.1, 1.8	5.0, 11	15, 50	26, 150	32, 320
10	0.0, 0.0	0.1, 0.1	0.5, 0.7	3.0, 5.0	15, 30	51, 150	126, 1,260
20	0.0, 0.0	0.0, 0.0	0.2, 0.2	1.4, 1.9	8.5, 14	44, 85	510, 5,100
50	0.0, 0.0	0.0, 0.0	0.0, 0.0	0.3, 0.4	2.7, 3.4	18, 27	3,200, 32,000
100	0.0, 0.0	0.0, 0.0	0.0, 0.0	0.1, 0.1	0.9, 1.0	7.1, 8.8	12,600, 126,000

[a] For each column entry, N values for 500 and 5000 psi are given; e.g., for 10-μm particles and $t = 15$ min, $N = 5000$ for $P = 500$ psi and $N = 7000$ for $P = 5000$ psi. All N values must be considered approximate, because of assumption of "ideal" column (Eq. 5.15c).

that separation times are much too long, and values of d_p are nonoptimum (see following discussion).

Limits of Column Performance

(The remainder of this section is highly specialized, and the reader may wish to skip to Chap. 6).

If higher values of P can increase the maximum attainable column plate number, what practical limits on P exist? The major problem noted by previous workers [e.g., (41)] is the generation of heat within the column as a result of work done in forcing mobile phase through a column at very high pressures. At higher pressures and flowrates this heat cannot be conveniently dissipated without significant increase in the temperature within the column. The resulting inhomogeneous column temperature leads to effects that can seriously degrade column performance (i.e., lowered N values). Because it appears that the effect on N is not significant for P as large as 5000 psi, under most conditions, we will not further discuss this problem of column heating at high pressures. For P greater than 5000 psi, a variety of problems besides heating of the column must be considered, and it appears unlikely that column pressures greater than 10,000 psi will be commonly used in LC. Because an increase in P from 5000 to 20,000 psi gives only a marginal possible increase in N ($\sim$50%), pressures above 5000 psi do not appear worthwhile for most LC separations.

What is the effect of column permeability on plate number? The data of Table 5.25 assume a column flow resistance value (Eq. 5.3) of $\phi = 1000$ (see p. 37). A doubling of ϕ for a given column is exactly equivalent in Table 5.25 to a decrease in P by twofold (Eq. 2.12), so that a column with $\phi = 2000$ operated at 5000 psi is equivalent to a column with $\phi = 1000$ operated at 2500 psi. Thus columns with lower ϕ values are advantageous. However, the value of a twofold decrease in ϕ is not very important in terms of allowing larger N values or shorter analysis times.

What is the optimum particle size d_p for modern LC columns? From Table 5.25, we see for each separation time and operating pressure that there exists an optimum d_p value—for which N is a maximum. Thus, for $t = 15$ min and $P = 500$ psi, we see that a maximum value of N can be achieved ($\sim$17,000 plates for this length L) for d_p equal to about 2 μm. Although in principle column optimization for maximum N and/or minimum t can be carried out by varying d_p [as suggested by Knox (2)], in fact "good" columns of variable d_p generally are not available in the range of values ($1 < d_p < 5$ μm) required for such optimization. It is hoped that this situation will change, at which time columns with $d_p < 5$ μm should prove useful for many practical applications (see Table 5.25 and discussion below of large molecules). For separation times of 1.5-15 min and a pressure of 5000 psi, Table 5.25 indicates an optimum value of $d_p \approx$ 1-2 μm. As separation time t increases, the optimum value of d_p shifts to higher values; for example, 5 μm for a separation time of 1 day. At higher operating pressures, lower values of d_p are favored. For many applications, the ability to use "good" columns of 2-3-μm particles would mean a substantial advantage in terms of either maximum N values or minimum separation time.

The data in Table 5.25 apply only to the separation of small molecules. For larger molecules, with molecular weights above 5000, the trends shown in Table 5.25 are substantially altered. The reason is that D_m decreases with sample molecular weight (e.g., Eq. 5.16, where mol. wt. $\approx V_1$). It can be shown that the optimum value of v occurs for a given ratio of d_p/D_m (see Eq. 5.14a), so because D_m decreases with increasing sample molecular weight, the optimum value of d_p also decreases. If we have an optimum particle size of $d_p \approx 2$ μm for samples of molecular weight 300 (Table 5.25), then the following optimum d_p values can be estimated for other sample molecular weights:

	Optimum d_p (μm)	
Mol. wt.	From Eq. 5.16	For proteins
300	(2)	(2)
3,000	0.5	0.8
30,000	0.1	0.3
300,000	0.03	0.1

A differentiation of proteins from other molecules is made here, because the spherical (globular) configuration of proteins leads to larger D_m values than would be calculated from Eq. 5.16.

From the preceding data we see that submicron particles are decidedly advantageous for the separation of large molecules. Because of limitations of the pore geometry in such small particles, it is likely that only the outside surface would be involved in solute retention (i.e., nonporous particles). Furthermore, extra-column effects (see following discussion) are a major barrier to the present use of such columns, assuming that "good" columns of such particles could ever be packed.

What are the maximum N values that can be achieved, regardless of separation time? This is not a very practical question for most problems, but we see in Table 5.25 that at large values of t, N approaches a limiting, constant value for each value of P and d_p. Furthermore, the largest possible N value increases with Pd_p^2. In this case maximum N corresponds to the largest possible d_p values, which may appear surprising. However, this concept is well understood and is discussed in detail by Giddings (42).

The effect of an increase in η on the data in Table 5.25 can be understood in terms of two effects. First, an increase in mobile-phase viscosity slows down the flow of solvent through the column and is equivalent to a similar increase in flow resistance ϕ or decrease in P. Although lower viscosity mobile phases are therefore advantageous, for the same reason that large P and small ϕ is beneficial, this effect is relatively small, as long as η is less than 1 cP. A more important effect of larger η is on D_m, as seen in Eq. 5.16. In any case, lower η values are always preferred in LC, as long as the boiling point of the mobile phase is not too near the temperature of separation (see discussion in Section 6.2). To provide a more quantitative relationship for the importance of η in column performance, the boxed data of Table 5.25 are recalculated for a threefold decrease in η (other conditions the same), and these data are reported in Table 5.26.

Table 5.26 Variation of column plate number with experimental conditions: effect of increase in viscosity of mobile phase. Same conditions as Table 5.25 except η = 0.3 CP (threefold decrease). *

	$N \times 10^{-3}$		
d_p	t = 1.5 min t_o = 10 sec	15 min 100 sec	2.5 hr 1000 sec
5	3.1,5.4	14,31	42,142
10	1.5,2.1	8.7,15	43,88
20	0.6,0.7	4.0,5.5	25,40
50	0.1,0.1	1.0,1.2	7.9,10.1

* D_m is 3-fold greater because of *decrease* in η.

A final consideration of importance in optimizing column performance, particularly when considering reduction in d_p, is the effect of extracolumn band broadening. The extracolumn effects discussed under sample-size effects in Section 2.3 are more important for narrower sample bands (smaller band volumes V_p). Table 5.27 shows calculated values of V_p for the separations of Table 5.25, assuming a column internal diameter of 0.4 cm and k' = 1 (it is assumed that extracolumn broadening of peaks with $k' < 1$ is not important, because these peaks are not of general interest, except in very high-efficiency size exclusion). The V_p values of Table 5.27 are all reasonably large (200 μl or greater, in most cases), which means that columns of 5-μm and larger particles do not normally require special equipment to keep extracolumn effects at an insignificant level (see discussions in Sections 3.2 and 4.2). One exception is the use of very small particles [e.g., 5 μm as in (52)] for fast size-exclusion separations. In this case V_p values of much less than 100 μl are possible.

Table 5.27 Variation of peak volume V_p with experimental conditions as in Table 5.25. dc = 0.5 cm, k' = 1.

	V_p $(\mu l)^a$		
d_p	$t = 1.5$ min $t_o = 10$ sec	15 min 100 sec	2.5 hr 1000 sec
5	140, 365	180, 400	270, 580
10	450, 1300	530, 1400	700, 1700
20	1600, 4800	1750, 5000	2000, 5500

^{a}For example, for 10 μm, $t_o = 100$ sec, $V_p = 530$ μl at 500 psi and 1400 μl at 5000 psi.

Our focus in this section has been almost entirely on porous particles, as opposed to pellicular particles. One reason is that for small-particle columns and small-molecule separations, the advantage of pellicular particles is largely lost. As seen in Figure 5.21, in the optimum region of $\nu \approx 3$, there is relatively little advantage for pellicular particles, as their h values are not much different from the h values of porous particles. Thus, there has so far been little reason to consider the use of small pellicular particles, comparable to the 5-10-μm porous particles that are widely used. However, the separation of large molecules such as proteins and particulates (e.g., colloids) is another matter. For these compounds 5-μm particles operated under normal conditions yield ν values that are greater than 100, and for these samples 5-μm pellicular packings would be expected to yield lower values of h and higher column N values (other factors being equal). However, there are many factors that argue against the use of small-particle, pellicular packings and their future use is at present speculative.

REFERENCES

1. J. C. Giddings, *Dynamics of Chromatography*, Dekker, New York, 1965, Chap. 2.
1a. B. L. Karger, L. R. Snyder, and C. Horvath, *An Introduction to Separation Science*, Wiley-Interscience, New York, 1973, Chap. 5.
2. J. H. Knox, *J. Chromatogr. Sci.*, **15**, 352 (1977).
3. G. J. Kennedy and J. H. Knox, *ibid.*, **10**, 549 (1972).
4. K. Unger, *Porous Silica: Its Properties and Use as Support in Column Liquid Chromatography*, Elsevier, New York, 1979.
5. P. B. Hamilton, *Anal. Chem.*, **30**, 914 (1958).
6. K. Taserik and M. Necasova, *J. Chromatogr.*, **75**, 1 (1973).

7. J. J. Kirkland, *J. Chromatogr. Sci.*, **10**, 593 (1972).

8. K. Unger and W. Messer, *J. Chromatogr.*, **149**, 1 (1978).

9. J. J. Kirkland, *J. Chromatogr. Sci.*, **7**, 361 (1969).

10. W. A. Aue, C. R. Hastings, J. M. Augl, M. K. Knorr, and J. V. Larson, *J. Chromatogr.*, **56**, 295 (1971).

11. R. E. Majors, *J. Chromatogr. Sci.*, **11**, 88 (1973).

12. R. E. Majors, *Anal. Chem.*, **44**, 1722 (1972).

12a. G. Vigh, E. Gemes, and I. Inczedy, *J. Chromatogr.*, **147**, 59 (1978).

13. J. J. Kirkland, in *Gas Chromatography 1972, Montreux*, S. G. Perry, ed., Applied Science Publishers, Barking, U.K., 1973, pp. 39-56.

14. T. J. N. Webber and E. H. McKerrell, *J. Chromatogr.*, **122**, 243 (1976).

15. J. J. Kirkland, *ibid.*, **125**, 231 (1976).

16. J. J. Kirkland, *J. Chromatogr. Sci.*, **10**, 129 (1972).

17. *Basics of Liquid Chromatography*, Spectra-Physics, Santa Clara, Calif., 1976, p. 20.

18. J. J. Kirkland, *J. Chromatogr. Sci.*, **9**, 206 (1971).

19. R. E. Majors, *Anal. Chem.*, **44**, 1722 (1972).

20. J. H. Knox, *Intensive Course on High-Performance Liquid Chromatography*, Wolfson Liquid Chromatography Unit, Department of Chemistry, University of Edinburgh, 1977, p. 163.

21. R. Endele, I. Halasz, and K. Unger, *J. Chromatogr.*, **99**, 377 (1974).

22. J. J. Kirkland, *Chromatographia*, **8**, 661 (1975).

23. J. J. Kirkland and P. E. Antle, *J. Chromatogr. Sci.*, **15**, 137 (1977).

24. P. A. Bristow, *J. Chromatogr.*, **131**, 57 (1977).

25. P. A. Bristow, *LC in Practice*, HETP Publishers, Cheshire, U. K., 1977, p. 33.

26. P. A. Bristow, *J. Chromatogr.*, **149**, 13 (1978).

27. H. P. Keller, F. Erni, H. R. Lindner, and R. W. Frei, *Anal. Chem.*, **49**, 1958 (1977).

28. C. D. Scott and N. E. Lee, *J. Chromatogr.*, **42**, 263 (1969).

29. P. A. Bristow and J. H. Knox, *Chromatographia*, **10**, 279 (1977).

30. R. E. Majors, *J. Chromatogr. Sci.*, **15**, 333 (1977).

31. A. T. James and A. J. P. Martin, *Analyst (London)*, **77**, 915 (1952).

32. J. J. Kirkland, W. W. Yau, H. J. Stoklosa, and C. H. Dilks, Jr., *J. Chromatogr. Sci.*, **15**, 303 (1977).

32a. J. J. Kirkland, U. S.-Japan Seminar on Modern Techniques of Liquid Chromatography, Boulder, Colo., June 28-July 1, 1978.

33. E. Grushka, *Anal. Chem.*, **44**, 1733 (1972).

34. R. V. Vivilecchia, B. G. Lightbody, N. Z. Thimot, and H. J. Quinn, *J. Chromatogr. Sci.*, **15**, 424 (1977).

35. D. S. Horne, J. H. Knox, and L. McLaren, *Sep. Sci.*, **1**, 531 (1966).

36. J. H. Knox, G. R. Laird, and P. A. Raven, *J. Chromatogr.*, **122**, 129 (1976).

37. M. Martin, C. Eon, and G. Guiochon, *ibid.*, **110**, 213 (1975).

38. I. Halasz, H. Schmidt, and P. Vogtel, *ibid.*, **126**, 19 (1975).

39. C. R. Wilke and P. Chang, *Am. Inst. Chem. Eng. J.*, **1**, 264 (1955).

40. R. P. W. Scott and P. Kucera, *J. Chromatogr.*, **125**, 251 (1976).

40a. R. P. W. Scott and P. Kucera, *J. Chromatog.*, **169**, 51 (1979).

41. I. Halasz, R. Endele, and J. Asshauer, *ibid.*, **112**, 37 (1975).

42. J. C. Giddings, *Dynamics of Chromatography*, Dekker, New York, 1965, pp. 296-298.

43. J. J. Kirkland and C. H. Dilks, Jr., *Anal. Chem.*, **45**, 1788 (1973).

44. W. Strubert, *Chromatographia*, **6**, 50 (1974).

45. R. M. Cassidy, D. S. LeGay, and R. W. Frei, *Anal. Chem.*, **46**, 340 (1974).

46. B. Coq, C. Gonnett, and J. L. Rocca, *J. Chromatogr.*, **108**, 249 (1975).

47. R. S. Burington, *Handbook of Mathematical Tables and Formulas*, 4th ed., McGraw-Hill, New York, 1965, p. 164.

48. G. B. Cox, C. R. Liscombe, M. J. Slucutt, K. Sudgen, and J. A. Upfield, *J. Chromatogr.*, **117**, 269 (1976).

49. H. R. Linder, H. P. Keller, and R. W. Frei, *J. Chromatogr. Sci.*, **14**, 234 (1976).

50. C. J. Little, A. P. Dale, D. A. Ord, and T. R. Marten, *Anal. Chem.*, **49**, 1311 (1977).

51. S. H. Chang, K. M. Gooding, and F. E. Regnier, *J. Chromatogr.*, **125**, 103 (1976).

52. W. W. Yau, J. J. Kirkland, D. D. Bly, and H. J. Stoklosa, *ibid.*, **125**, 219 (1976).

53. L. R. Snyder, *J. Chromatogr. Sci.*, **15**, 441 (1977).

BIBLIOGRAPHY

Bakalyar, S., J. Yuen, and R. A. Henry, Spectra-Physics Chromatography Technical Bulletin TB 114-76, 1976 (practical discussion and description of wet- and dry-fill packing methods).

Bristow, P. A., *LC in Practice*, HETP Publishers, Cheshire, U. K., 1977 (brief summary of column-packing techniques, sampling).

Bristow, P. A., P. N. Brittain, C. M. Riley and B. F. Williamson, *J. Chromatogr.*, **131**, 57 (1977) (description of up-flow wet-fill column-packing method).

Bristow, P. A., and J. H. Knox, *Chromatographia*, **10**, 279 (1977) (discussion of specifications for column performance, column testing).

Coq, B., C. Gonnet, and J. L. Rocca, *J. Chromatogr.*, **106**, 249 (1975). (slurry-packing techniques; influence of column length and diameter).

Giddings, J. C., *Dynamics of Chromatography*, Dekker, New York, 1965 (discussion of fundamentals of column performance).

Knox, J. H., *J. Chromatogr. Sci.*, **15**, 352 (1977) (review of principles in column optimization).

Kraak, J. C., H. Poppe, and F. Smedes, *J. Chromatogr.*, **122**, 147 (1976) (construction of columns with large plate count; connecting columns).

Linder, H. R., H. P. Keller, and R. W. Frei, *J. Chromatogr. Sci.*, **14**, 234 (1976) slurry-packing of columns with up-flow using stirred reservoir).

Martin, M. and G. Guiochon, *Chromatographia*, **10**, 194 (1977) (review and discussion of various techniques for preparing columns).

Webber, T. J. N., and E. H. McKerrell, *J. Chromatogr.*, **122**, 243 (1976) (column-packing and sampling techniques for high-performance analytical and semipreparative columns).

SIX

SOLVENTS

6.1 Introduction 247

6.2 Physical Properties 251

    Availability 251

    Detector Compatibility 251

    Solvent Reactivity 252

    Boiling Point and Viscosity 252

    Miscibility 253

    Safety 253

6.3 Intermolecular Interactions Between Sample and Mobile-Phase
    Molecules 255

    Dispersion Interactions 256

    Dipole Interactions 256

    Hydrogen-Bonding Interactions 257

    Dielectric Interaction 257

6.4 Solvent Strength and "Polarity" 257

    Polarity of Pure Solvents 258

    Solvent Mixtures 260

6.5 Solvent Selectivity 260

    Classification of Solvent Selectivity 261

    A Systematic Approach to Selectivity Optimization 263

    Solvent Selectivity in Reverse-Phase and Adsorption LC 264

6.6 Purification of LC Solvents 265

References 267

Bibliography 268

6.1 INTRODUCTION

The design of a successful LC separation depends on matching the right mobile phase to a given column and sample. Many different properties of the solvent must be considered, including solvent strength and selectivity as discussed in Chapter 2. In view of the large number of possible solvents and solvent mixtures, a rational approach to solvent selection requires a classification of the various solvents according to their relevant properties. The present chapter discusses solvent classification and selection in general terms; Chapters 7-12, on the individual LC methods, provide additional, more specific information on solvents for each method.

Figure 6.1 illustrates the general approach to selecting an LC mobile phase. The total triangle represents all possible solvents. We first reject solvents whose physical properties (boiling point, viscosity, UV absorbance, etc.) are inappropriate for use in LC; these are represented by the lined area at the bottom of Figure 6.1. This eliminates a large number of possible solvents from further consideration. Next, we select a solvent or solvent mixture of the right chromatographic strength, so that the k' values of our sample fall in the optimum range of $2 \leqslant k' \leqslant 5$ for a two-component sample, or $0.5 \leqslant k' \leqslant 20$ for a multicomponent sample. Again this means eliminating the many solvents represented by the cross-hatched area in Figure 6.1, but many solvents of the right strength will still be available. Finally, we may find overlapping bands in the chromatogram, suggesting the need for improved solvent selectivity or increased α values for certain band pairs. In this case we can select a solvent that gives the right α values from the remaining group of solvents having the right strength. In this way we eventually arrive at an acceptable mobile phase for the particular application at hand, as illustrated by the clear area at the top of Figure 6.1.

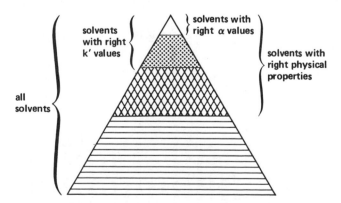

Figure 6.1   Selecting a mobile phase for an LC separation or related application.

Table 6.1 Properties of solvents for use in liquid chromatography.

Solvent[a]	Source[b]	UV cutoff[c]	R.I.[d]	Boiling point (°C)	Viscosity (cP, 25°C)	P'[e]	$\varepsilon°$[f]	Selectivity group[g]	Water solubility in solvent[h]	Dielectric constant[i] ε	$P' + 0.25\varepsilon$[j]
1.[k] FC-78 (*)		210 nm	1.267	50	0.4	<−2	−.25	—		1.88	
FC-75		210	1.276	102	0.8	<−2	−.25	—		1.86	
FC-43		210	1.291	174	2.6	<−2	−.25	—		1.9	
2. Isooctane (*) (2,2,4-trimethylpentane)	LC	197	1.389	99	0.47	0.1	0.01	—	0.011	1.94	0.1
3. n-Heptane (*)	LC	195	1.385	98	0.40	0.2	0.01	—	0.010	1.92	0.5
4. n-Hexane (*)	LC	190	1.372	69	0.30	0.1	0.01	—	0.010	1.88	0.5
5. n-Pentane (**)	LC	195	1.355	36	0.22	0.0	0.00	—	0.010	1.84	0.5
6. Cyclohexane	LC	200	1.423	81	0.90	−0.2	0.04	—	0.012	2.02	0.5
7. Cyclopentane (*)	LC	200	1.404	49	0.42	−0.2	0.05	—	0.014	1.97	0.6
8. 1-Chlorobutane (*)	LC	220	1.400	78	0.42	1.0	0.26	VIa		7.4	2.8
9. Carbon disulfide		380	1.624	46	0.34	0.3	0.15	—	0.005	2.64	1.7
10. 2-Chloropropane (**)		230	1.375	36	0.30	1.2	0.29	VIa		9.82	3.7
11. Carbon tetrachloride	LC	265	1.457	77	0.90	1.6	0.18	—	0.008	2.24	2.3
12. n-Butyl ether		220	1.397	142	0.64	2.1	0.25	I	0.19	2.8	2.4
13. Triethylamine			1.398	89	0.36	1.9	0.54	I		2.4	2.4
14. Bromoethane (*)			1.421	38	0.38	2.0	0.35	VIa		9.4	4.3
15. i-Propyl ether (*)		220	1.365	68	0.38	2.4	0.28	I	0.62	3.9	3.2
16. Toluene	LC	285	1.494	110	0.55	2.4	0.29	VII	0.046	2.4	2.9
17. p-Xylene		290	1.493	138	0.60	2.5	0.26	VII		2.3	3.0
18. Chlorobenzene			1.521	132	0.75	2.7	0.30	VII		5.6	4.1
19. Bromobenzene			1.557	156	1.04	2.7	0.32	VII		5.4	4.1
20. Iodobenzene						2.8	0.35	VII			
21. Phenyl ether			1.580	258	3.3	3.4		VII		3.7	3.7
22. Phenetole			1.505	170	1.14	3.3		VII		4.2	4.9
23. Ethyl ether (**)	LC	218	1.350	35	0.24	2.8	0.38	I	1.3	4.3	4.0
24. Benzene	LC	280	1.498	80	0.60	2.7	0.32	VII	0.058	2.3	3.6
25. Tricresyl phosphate						4.6		VIa			

248

26.	Ethyl iodide			1.510	72	0.57	2.2		VIa		7.8	4.2
27.	n-Octanol		205	1.427	195	7.3	3.4	0.5	II	3.9	10.3	5.8
28.	Fluorobenzene			1.46	85	0.55	3.1		VII		5.4	4.6
29.	Benzyl ether			1.538	288	4.5	4.1		VII			
30.	Methylene chloride (**)	LC	233	1.421	40	0.41	3.1	0.42	V	0.17	8.9	5.6
31.	Anisole			1.514	154	0.9	3.8		VII		4.3	4.6
32.	i-Pentanol			1.405	130	3.5	3.7	0.61	II	9.2	14.7	7.3
33.	1,2-Dichloroethane	LC	228	1.442	83	0.78	3.5	0.44	V	0.16	10.4	6.3
34.	t-Butanol			1.385	82	3.6	4.1	0.7	II	miscible	12.5	
35.	n-Butanol	LC	210	1.397	118	2.6	3.9	0.7	II	20.1	17.5	8.3
36.	n-Propanol	LC	240	1.385	97	1.9	4.0	0.82	II	miscible	20.3	
37.	Tetrahydrofuran (*)	LC	212	1.405	66	0.46	4.0	0.57	III	miscible	7.6	
38.	Propyl amine (*)			1.385	48	0.35	4.2		I	miscible	5.3	
39.	Ethyl acetate (*)	LC	256	1.370	77	0.43	4.4	0.58	VIa	9.8	6.0	5.8
40.	i-Propanol	LC	205	1.384	82	1.9	3.9	0.82	II	miscible	20.3	
41.	Chloroform (*)	LC	245	1.443	61	0.53	4.1	0.40	VIII	0.072	4.8	5.6
42.	Acetophenone			1.532	202	1.64	4.8		VIa		17.4	8.7
43.	Methylethyl ketone (*)	LC	329	1.376	80	0.38	4.7	0.51	VIa	23.4	18.5	9.1
44.	Cyclohexanone			1.450	156	2.0	4.7		VIa		18.3	9.1
45.	Nitrobenzene			1.550	211	1.8	4.4		VII		34.8	13.2
46.	Benzonitrile			1.526	191	1.2	4.8		VIb		25.2	10.9
47.	Dioxane	LC	215	1.420	101	1.2	4.8	0.56	VIa	miscible	2.2	
48.	Tetramethyl urea	LC	265	1.449	175		6.0		III		23.0	10.7
49.	Quinoline			1.625	237	3.4	5.0		III		9.0	7.4
50.	Pyridine			1.507	115	0.88	5.3	0.71	III	miscible	12.4	
51.	Nitroethane		380	1.390	114	0.64	5.2	0.6	VII	0.9		
52.	Acetone (*)	LC	330	1.356	56	0.30	5.1	0.56	VIa	miscible		
53.	Benzyl alcohol			1.538	205	5.5	5.7		IV		13.1	8.8
54.	Tetramethyl guanidine						6.1		I			
55.	Methoxyethanol	LC	210	1.400	125	1.60	5.5	0.88	III	miscible	16.9	
56.	Tris(cyanoethoxy) propane						6.6		VIb			
57.	Propylene carbonate	GC					6.1		VIb			
58.	Ethanol	LC	210	1.359	78	1.08	4.3	0.88	II	miscible	24.6	
59.	Oxydipropionitrile	LC					6.8		VIb			
60.	Aniline	GC		1.584	184	3.77	6.3	0.62	VIb	miscible	6.9	8.1

Table 6.1 (Continued).

Solvent[a]	Source[b]	UV cutoff[c]	R.I.[d]	Boiling point (°C)	Viscosity (cP, 25°C)	P'[e]	$\epsilon°$[f]	Selectivity group[g]	Water solubility in solvent[h]	Dielectric constant[i] ϵ	$P' + 0.25\epsilon$[j]
61. Acetic acid	LC		1.370	118	1.1	6.0		IV	miscible	6.2	
62. Acetonitrile (**)	LC	190	1.341	82	0.34	5.8	0.65	VIb	miscible	37.5	
63. N,N-dimethylaceta-mide	LC	268	1.436	166	0.78	6.5		III		37.8	
64. Dimethylformamide	LC	268	1.428	153	0.80	6.4		III		36.7	
65. Dimethylsulfoxide	LC	268	1.477	189	2.00	7.2	0.75	III	miscible	4.7	
66. N-methyl-2-pyrrolidone	LC	285	1.468	202	1.67	6.7		III		32	
67. Hexamethyl phosphoric acid triamide			1.457	233	3	7.4		I		30	
68. Methanol (**)	LC	205	1.326	65	0.54	5.1	0.95	II	miscible	32.7	
69. Nitromethane		380	1.380	101	0.61	6.0	0.64	VII	2.1		
70. m-Cresol			1.540	202	14	7.4		VIII		11.8	10.0
71. N-methylformamide			1.447	182	1.65	6.0		III	miscible	182	
72. Ethylene glycol			1.431	182	16.5	6.9	1.11	IV	miscible	37.7	
73. Formamide			1.447	210	3.3	9.6		IV	miscible	111	
74. Water	LC		1.333	100	0.89	10.2		VIII		80	

Source: From (*1*).

a (*) indicates preferred LC solvent of low viscosity (≤0.5 cP), yet with convenient boiling point (>45°); (**) indicates very low viscosity *and* low boiling point solvent.

b LC indicates that solvent can be purchased specifically for use in LC from one of following suppliers: Burdick & Jackson, Baker Chemical, Mallinkrodt Chemical, Fisher Scientific, Waters Associates, Manufacturing Chemists, Inc. (GC indicates the solvent is used as a gas chromatography stationary phase and can be purchased from companies selling GC columns and phases (these solvents are used as stationary phases in liquid-liquid LC with mechanically held phase.)

c Approximate wavelength, below which solvent is opaque.

d Refractive index at 25°C.

e Solvent polarity parameter (*8*).

f Solvent strength parameter for LSC on alumina (*10*).

g See Figure 6.4.

h %w of water dissolving in given solvent at 20°C; of interest in LSC - see Section 9.3.

i At 20°C.

j Function of P' and dielectric constant that is proportional to solvent strength in ion-pair chromatography; see Table 11.5 and related discussion.

k Fluorochemical solvents available from Minnesota Mining & Mfg. Co.

The following sections discuss in detail the various solvent properties referred to in Figure 6.1 and allow the reader to focus systematically on a satisfactory mobile phase for a given problem. However, the utility of this chapter is not limited to the simple task of finding a proper mobile phase for a particular LC separation. This discussion also provides a basis for

- Selecting a stationary phase in the case of liquid-liquid (partition) LC with mechanically held phases (Chapter 8).
- Understanding the design of optimum solvent gradients for gradient elution separations (Chapter 16).
- Selecting solvent systems for the pretreatment of difficult samples before injection onto a column (Chapter 17).
- Selecting a solvent for dissolving the sample prior to injection, when that solvent differs from the mobile phase; this is of particular interest in preparative (Chapter 15) and trace-analysis (Chapter 13) separations.

For further details on solvent selection, see (1).

6.2 PHYSICAL PROPERTIES

At a minimum, solvents used in LC should be readily available, compatible with the LC detector, safe to use, pure, and relatively unreactive. The solvent should be able to dissolve the sample, although this is seldom a problem in analytical separations when the mobile phase has the right strength. As we will see, solvents with low boiling points and low viscosities are generally preferred. Finally, it is desirable if adequately pure solvents can be obtained at a reasonable price. Solvents that largely meet these requirements and that are in relatively widespread use in LC are included in Table 6.1 in order of increasing "polarity" (see below). Preferred solvents for operation at ambient temperature are noted with an asterisk. Solvents that have found favor in elevated-temperature separations by size-exclusion chromatography are discussed separately in Chapter 12. Table 6.2 does not include a number of high-boiling liquids that have been used as stationary phases in partition LC (e.g., hydrocarbon polymer, ODPN, etc.). These are discussed in Chapter 8.

Availability

The first column of data in Table 6.1 describes whether the solvent is sold specifically for use in LC and related techniques; current suppliers are also given.

Detector Compatibility

In most cases a photometric or refractive index detector will be used, in which case it is important to know either the lowest wavelength at which the solvent transmits significant light or the refractive index of the solvent. Columns two and three in Table 6.1 list these data for the solvents shown. The minimum wavelengths shown in Table 6.1 are for highly purified solvents; traces of absorbing impurities can increase these numbers by 50-100 nm. For example, saturated hydrocarbons such as *n*-hexane are reasonably transparent down to 190 nm, but even "research-grade" solvents often contain a few tenths of a percent of olefins of similar boiling point (e.g., 1-hexene). This contamination is sufficient to make the solvent opaque at wavelengths below 260 nm.

The sensitivity of a refractive index detector for a particular compound is proportional to the difference in refractive index values of the mobile phase and sample compound. The solvent refractive index values of Table 6.1 allow the selection of a solvent for maximum detection sensitivity in a given application.

Solvent Reactivity

Many solvents are ruled out for most applications because of their tendency to react with the sample, or to undergo polymerization in the presence of certain stationary phases. Thus aldehydes, olefins, and sulfur compounds (except dimethylsulfoxide) are rarely used in modern LC, and ketones and nitrocompounds are used only sparingly. When separated sample components are to be recovered for subsequent characterization or other use, high-boiling solvents should be avoided—even those that are relatively stable. Some solvents may have to be used fresh, or purified just before use (e.g., ethers, which readily undergo oxidation to peroxides).

Boiling Point and Viscosity

These two properties of the solvent go hand in hand, low boiling solvents being generally less viscous. The fourth and fifth columns of Table 6.1 list boiling points and viscosities, respectively, for each solvent. Usually it is preferable that the solvent boil 20-50° above the temperature of separation and that its viscosity not exceed 0.5 cP at the temperature of separation. Lower-boiling solvents are difficult to use with reciprocating pumps, since they tend to form bubbles in the piston chamber, which adversely affects pumping precision and, in extreme cases, leads to loss of pump priming. Furthermore, mixtures of such solvents readily change their composition because of evaporation. Higher-boiling solvents are usually excessively viscous, which reduces separation efficiency, other factors being equal.

Relatively viscous solvents can be used as dilute solutions in some applications, because viscosity is largely determined by the viscosity of the less viscous component of a solvent mixture. Thus a typical change in mobile phase viscosity with composition is shown below, for a mixture of a less viscous solvent A with a more viscous solvent B (see also Appendix II):

% B in mixture	Viscosity (cP)
0	0.30
10	0.38
25	0.53
50	0.95
100	3.00

A further discussion of the viscosity of solvent mixtures is given in (4). Strongly associating solvent mixtures (particularly water/acetonitrile, water/methanol) show anomalous variations in viscosity with composition. For example, mixtures of water/acetonitrile show maximum viscosity at 35%v acetonitrile/water (5 and Appendix II).

Miscibility

A listing of immiscible solvent pairs is of interest for two reasons. First, partition LC with a mechanically held stationary phase requires that this phase be immiscible with the mobile phase. Second, in sample cleanup procedures prior to injection, a common separation technique is liquid-liquid extraction, where the sample of interest is shaken in a separatory funnel with two immiscible solvents. For either of these two applications, Table 6.2 catalogues a number of solvents according to whether they are miscible or immiscible toward each other. For a more complete discussion of miscibility, including practical rules for predicting the miscibility of a large number of common solvents, see (3).

Safety

Under the topic of safety we should consider the flammability and toxicity of the solvent. Certain solvents are well known for their hazardous flammability properties (e.g., CS_2) and should be avoided if possible. Unfortunately, those solvents that are preferred because of their low viscosities and low boiling points are generally quite flammable, with hydrocarbons generally being somewhat more flammable and polyhalogenated solvents being less flammable. In general it is not possible to avoid flammable solvents in LC, but reasonable precautions usually suffice to minimize any risk. Thus, well-ventilated working areas are essential, and sparking devices of any type (e.g., pumps) should not be located in spaces that tend to collect solvent vapors.

Table 6.2 Miscibility of different solvent pairs.[a]

	Phenol	Methanol	Ethanol	Ethylene glycol	Diethylene glycol	Glycerine	Acetic acid	Ethylene diamine	Formamide	Acetonitrile	Nitromethane	Water
n-Hexane	i	i	m	i	i	i	m	i	i	i	i	i
n-Heptane	i	i	m	i	i	i	m	i	i	i	i	i
Isooctane	i	i	m	i	i	i	m	i	i	i	i	i
Cyclohexane	m	i	m	i	i	i	m	i	i	i	i	i
Benzene	m	m	m	i		i	m	m	i	m	m	i
Toluene	m	m	m	i		i	m	m	i	m	i	i
m-Xylene	m	m	m	i		i	m	m	i	m	—	i
CCl$_4$	m	m	m	i	i	i	m	m	i	m	m	i
CHCl$_3$	m	m	m	m	m	i	m	m	i	m	m	i
CH$_2$Cl$_2$	i	m	m	i	m	i	m	m	i	m	m	i
1,2-Dichloroethane	—	m	m	i	m	i	m	i	i	m	m	i
CS$_2$	—	i	m	i	i	i	m	i	i	i	i	i
Ethyl ether	—	m	m	i	i	i	m	m	i	m	m	i
Isopropyl ether	—	m	m	i	i	i	m	i	i	m	m	i
Methyl, i-butyl ketone	—	m	m	i	m	i	m	i	i	m	m	i

Source: From (*2, 3*).
[a] m, miscible; i, immiscible (at room temperature); -, no data reported in (*2,3*).

254

Markedly toxic solvents should be avoided, but many of the solvents that are most useful in LC are moderately toxic. As in the case of flammable solvents, the overall risk from toxic solvents can be minimized by good ventilation. Because many common solvents have recently been found to be dangerous, all but the most innocuous solvents should be given due respect and considered toxic. It is well known that chlorinated and/or aromatic solvents are cumulative poisons, and several of these solvents are now believed to be carcinogenic (e.g., benzene, chloroform, trichloroethylene, perchloroethylene). For a recent review of solvent toxicity and explosion hazards (in French), specifically for LC, see (6). Solvent toxicity (LD_{50} values) is also discussed in the *Merck Index* (9th ed.).

6.3 INTERMOLECULAR INTERACTIONS BETWEEN SAMPLE AND MOBILE-PHASE MOLECULES

Before discussing how the solvents of Table 6.1 can be classified according to strength or selectivity, we look first at the fundamental basis of these solvent properties. A mobile phase that gives smaller k' values for a specific sample compound is a solvent that strongly interacts with molecules of that solute. As a result, such a solvent will preferentially dissolve the sample compound in question. There are four major interactions between molecules of solvent and solute that are important in LC: dispersion, dipole, hydrogen bonding, and dielectric interaction. Each of these interactions is illustrated in Figure 6.2.

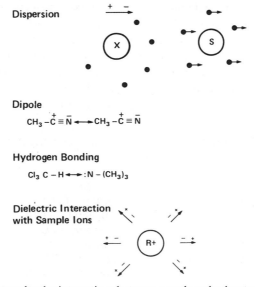

Figure 6.2 Intermolecular interactions between sample and solvent molecules in LC.

Dispersion Interactions

Dispersion interaction or attraction exists between every pair of adjacent molecules. In Figure 6.2 a sample molecule X and a solvent molecule S are shown, with the electrons that surround each molecule. Dispersion interactions arise because these electrons are in random motion and at any given instant may assume an asymmetric configuration. This is the case for molecule X in the example of Figure 6.2; as a result, X posesses at that instant a temporary dipole moment. This resulting dipole will polarize the electrons in adjacent molecule S, the temporary excess of electrons on the side of X nearer to S repelling the electrons in S, as indicated by the small arrows shown for the electrons of S in Figure 6.2. The resulting dipole created in S by the dipole in X then results in electrostatic attraction of X and S.

Dispersion interactions are stronger, for more easily polarizable electrons in the sample and solvent molecules. To a good approximation, electron polarizability increases with compound refractive index, so that dispersion interactions are stronger for sample and solvent molecules with large refractive indices. The data of Table 6.1 on solvent refractive index provide a guide to the ability of the solvent to interact selectively with sample compounds of high refractive index (e.g., aromatics, compounds with multiple substituents, or atoms from the second and higher rows of the periodic table [−Cl, −Br, −I, −S−, etc.]). Thus high-refractive-index solvents preferentially dissolve high-refractive-index samples.

Dipole Interactions

When both the solvent and sample molecule possess *permanent* dipole moments, strong interaction results by alignment of the two dipoles in a linear configuration. This is illustrated in Figure 6.2 for interaction by dipole orientation of two molecules of acetonitrile. These dipole interactions usually occur between individual functional groups of the two molecules, so that solvent and sample molecules having polar functional groups will selectively interact, the interaction being stronger for groups with larger dipole moments. A few examples of group dipole moments are given below:

Functional Group Dipole Moments (in debyes)

amine, −N=	(0.8-1.4)	halogen, −F, −Cl,	
ether, −O−	(1.2)	−Br, −I	(1.6-1.8)
sulfide, −S−	(1.4)	ester, −COO−	(1.8)
thiol, −SH	(1.4)	aldehyde, −CHO	(2.5)
carboxylic		ketone, −CO−	(2.7)
acid, −COOH	(1.7)	nitro, −NO$_2$	(3.2)
hydroxy, −OH	(1.7)	nitrile, −C≡N	(3.5)
		sulfoxide, −SO−	(3.5)

For additional examples, see (*6a*).

Hydrogen-Bonding Interactions

Hydrogen-bonding interactions between a proton donor and proton acceptor molecule are well known to chemists and are illustrated by the donor chloroform and acceptor trimethylamine in Figure 6.2. Hydrogen bonding becomes stronger as the donor is better able to give up a proton and as the acceptor is better able to accept a proton. That is, more acidic donors and more basic acceptors favor hydrogen bonding. Alcohols, carboxylic acids, and phenols are all strong donors, as is chloroform. Amides with N—H groups are intermediate in their ability to serve as proton donors, and primary and secondary amines are weak donors. Other compounds such as nitro-compounds, sulfones and sulfoxides, esters, and so on, can be ignored as possible H-donors. Proton acceptors can be classified according to basicity or acceptor strength by experimental measurement [e.g., see (6b)]. One such study gives the following relative strengths:

Compound type	Acceptor strength
Aromatic hydrocarbons	0.0-0.3
Ethers	0.7-1.2
Esters	0.8-1.3
Amines	1.5-2.1
Amides	2.1-2.4
Sulfoxides	2.2-2.6
Alcohols	large

Strong donor solvents preferentially interact with and dissolve strong acceptor sample compounds, and vice versa.

Dielectric Interaction

Dielectric interaction refers primarily to the interaction of sample ions with liquids of high dielectric constant ϵ (e.g., water, the lower alcohols, etc. [see Table 6.1]). As illustrated in Figure 6.2, a charged sample ion R+ polarizes molecules of the surrounding solvent, resulting in electrostatic attraction of R+ and solvent molecules. As is well known to chemists, these interactions are quite strong and very much favor the preferential dissolution of ionic or ionizable sample molecules in polar phases such as water, methanol, and so on.

6.4 SOLVENT STRENGTH AND "POLARITY"

The total interaction of a solvent molecule with a sample molecule is the result of the four interactions shown in Figure 6.2. The larger these dispersion, dipole, hydrogen bonding, and dielectric interactions are *in combination*, the stronger is the attraction of solvent and solute molecules. The ability of a sample or solvent molecule to interact in all four of these ways is referred to as the "polarity" of the compound. Thus "polar" solvents preferentially attract and dissolve "polar" solute molecules. In the same fashion, the strength of a solvent is directly related to its polarity. Solvent strength increases with solvent polarity in normal-phase partition LC and in adsorption LC, whereas in reverse-phase LC the reverse is true: solvent strength decreases with increasing polarity.

Polarity of Pure Solvents

Solvent polarity has been defined in various ways [e.g., (*1*)], but here we will use the parameter P' based on experimental solubility data reported by Rohrschneider (*7, 8*). There are several reasons for believing that P' is a better overall index of polarity than other solvent properties. Table 6.1 lists values of P' for each solvent (sixth column), and these are seen to vary from less than -2 for the nonpolar 3 *M* Fluorochemicals to a value of 10.2 for the very polar solvent water. A change in P' by two units causes (very roughly) a 10-fold change in k' values, which translates in the general case to

$$\frac{k_2}{k_1} = 10^{(P_1 - P_2)/2}. \tag{6.1}$$

for normal-phase partition LC. Here k_1 and k_2 are k' values for a given compound initially and finally, and P_1 and P_2 refer to solvent P' values initially and finally. For reverse-phase LC, Eq. 6.1 becomes

$$\frac{k_2}{k_1} = 10^{(P_2 - P_1)/2}. \tag{6.1a}$$

As an example, one study (*9*) shows sample k' values in reverse-phase LC varying by a factor of about 10^3 in going from water as solvent to methanol. Since the P' values of methanol and water in Table 6.1 are seen equal to 5.1 and 10.2, respectively, the predicted change in k' values (Eq. 6.1a) would be $10^{2.6}$, in approximate agreement with the experimental values.

Because P' is based on actual solubility data, and sample solubility in the mobile and stationary phases of partition LC (LLC) determines k', it can be expected that P' will provide a reasonably accurate measure of solvent strength in partition LC. P' is also a rough measure of solvent strength for adsorption LC, but a better index of solvent strength in this case is afforded by the experimental adsorption solvent strength parameter ϵ° (*10* and Section 9.3). Figure 6.3 shows

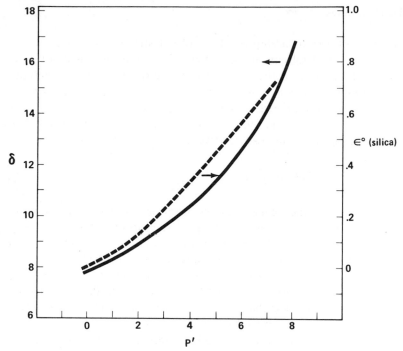

Figure 6.3 Relationship of Hildebrand solubility parameter (solid line) and adsorption solvent strength parameter (dashed line) to solvent polarity parameter P.

the rough correlation between values of ϵ° and P' for various pure solvents. Values of ϵ° for the solvents of Table 6.1 are listed in the seventh column.

Many chemists are more familiar with another way of defining solvent polarity: the *Hildebrand solubility parameter* δ [see (1)]. For most solvents, P' and δ are related as shown in Figure 6.3. Therefore, P' values for solvents not listed in Table 6.1 can be estimated from their δ values (which are known for most solvents). The major exception to the relationship between P' and δ shown in Figure 6.3 is for strong-acceptor (basic) solvents that do not possess donor properties (e.g., diethyl ether, triethylamine, etc.) Such solvents have P' values about 1.5 units greater than predicted from their δ values via Figure 6.3.

Because the P' values of Table 6.1 vary over a range of 11+ units, a change in mobile phase in partition LC can in principle result in a change of k' values by a factor of about 10^5, which is far more than is ever required in actual practice. Thus, the k' values of most compounds can be adequately controlled for optimum separation by changing the mobile-phase solvent. As we see later, however, the use of solvent mixtures is more convenient for this purpose than are pure solvents.

Solvent Mixtures

In either partition or adsorption LC, solvent strength is commonly adjusted through the use of solvent mixtures. Typically, a solvent A that is too weak is blended with a solvent B that is too strong, and some intermediate mixture will then have exactly the right strength (e.g., Figure 2.25). The polarity P' of a solvent mixture is the arithmetic average of the P' values of the pure solvents in the mixture, weighted according to the volume fraction of each solvent. For a binary solvent mixture composed of solvents A and B, P' for the mixture is given as

$$P' = \phi_a P_a + \phi_b P_b. \tag{6.2}$$

Here ϕ_a and ϕ_b are the volume fractions of solvents A and B in the mixture, and P_a and P_b refer to the P' values of the pure solvents. A 10%v change in solvent composition (e.g., from 30%v to 40%v methanol/water) will change P' of the mixture by 0.1 $(P_b - P_a)$. For reverse-phase LC with methanol/water as mobile phase, P' for methanol and water equal 5.1 and 10.2, respectively (Table 6.1), so that a 10%v change in solvent composition represents an 0.5 unit change in P'. From Eq. 6.1a, this means about a 1.7-fold change in k', a value that agrees with the 1.5-2.5-fold change reported by several workers [e.g., (9, 11-14)].

In adsorption LC, the solvent strength ϵ° does not vary linearly with solvent composition, as in Eq. 6.2. Rather, ϵ° increases rapidly for small additions of a stronger solvent B to a weaker solvent A, then asymptotically approaches the value of pure B. This is illustrated below for mixtures of pentane (A) and diethyl ether (B) for silica as adsorbent:

%v Ether in pentane	ϵ°
0	0.00
10	0.13
20	0.20
30	0.24
50	0.30
100	0.38

Values of ϵ° for solvent mixtures can be calculated from the ϵ° values of the constituent pure solvents (Eq. 9.1 and 10). Some examples of the ϵ° values of solvent mixtures are given in Figure 9.7.

6.5 SOLVENT SELECTIVITY

Once we have selected a mobile phase that gives sample k' values of 1-5 (solvent strength is right for the separation), most separations can be achieved by optimizing column efficiency as outlined in the discussion on controlling resolution: Section 2.5. In some cases, however, two or more sample bands will overlap badly, and a change in separation selectivity is indicated. This is most readily achieved by a change in mobile-phase selectivity, which in turn is achieved by holding solvent strength constant while varying the composition of the mobile phase. Let us illustrate with an example from normal-phase partition LC.

Assume that we have blended a nonpolar solvent A ($P_a \leqslant 1$) with a polar solvent B to give a mixture A/B whose P' value gives the right solvent strength and k' values for a particular separation. Now we desire to change solvent selectivity, and we do this by choosing another polar solvent C in place of B. Again the composition of the mixture A/C can be adjusted to give the correct solvent strength (i.e., P' equal to that of the first mobile phase A/B). Since $P_a \approx 0$, from Eq. 6.2 it is seen that the compositions of the first (A/B) and second (A/C) mobile phases are related as

$$\phi_c = \phi_b \left(\frac{P_b}{P_c}\right). \tag{6.3}$$

For example, assume A/B is 25%v $CHCl_3$/hexane, and A/C is to consist of a mixture of ethyl ether/hexane. Here $P_b = 4.4$ ($CHCl_3$) and $P_c = 2.9$ (ethyl ether), so ϕ_c is 0.25 (4.4/2.9) = 38%v ether/hexane.

Equation 6.3 allows us to rapidly substitute mobile phases of different composition but similar strength for the initial solvent A/B. Since Eq. 6.3 is at best approximate, further small adjustments in the concentration of polar solvent in the mobile-phase mixture will often be required to maintain k' values in the optimum range $1 \leqslant k' \leqslant 5$.

Classification of Solvent Selectivity

The greatest change in mobile-phase selectivity results when the relative importance of the various intermolecular interactions between solvent and sample molecules is markedly changed. Thus, the substitution of one polar solvent B (e.g., methanol) by its homolog C (e.g., propanol) should result in little change in these interactions or in solvent selectivity, because in both cases (methanol and propanol mixtures) a donor solvent is present. A much greater change in selectivity would be expected by using a solvent C that either is a pure acceptor (e.g., ethyl ether) or has a large-dipole moment (e.g., methylene chloride). With a pure-acceptor solvent C like diethyl ether, donor sample molecules will be preferentially held in the mobile phase. With a large-dipole solvent C, sample molecules with strong-dipole groups will be preferentially retained in the mobile phase. And with the original donor solvent methanol, acceptor sample molecules

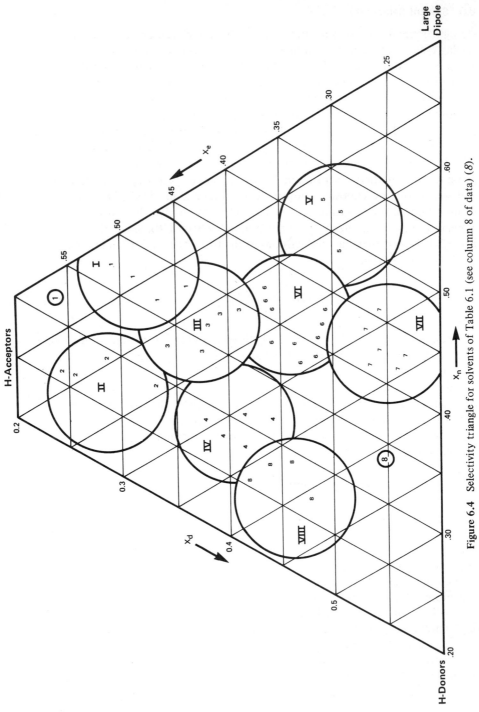

Figure 6.4 Selectivity triangle for solvents of Table 6.1 (see column 8 of data) (8).

would prefer the mobile phase. In each case preferential retention of a particular type of sample molecule in the mobile phase would reduce its k' value relative to other sample bands, possibly resulting in a desired change in the positioning of sample bands within the final chromatogram.

It should be apparent that our discussion of intermolecular interactions in Section 6.3 can be used to guide the selection of a solvent C that will yield maximum changes in solvent selectivity. Generally this will allow us to achieve the necessary change in solvent selectivity much more quickly than if we try various polar solvents C at random. An alternative, more convenient approach is simply to classify the common solvents according to selectivity—without considering specific intermolecular effects. The data of Rohrschneider (7) used to define solvent polarity values P' also allow such a classification according to solvent selectivity (8): x_d, x_e, and x_n, respectively. When these latter values for each solvent are plotted on a triangular diagram, as in Figure 6.4, it is found that various solvents are grouped into clusters of similar selectivity. Thus, group I includes pure acceptors such as ethers and amines, group II includes donor-acceptors such as the alcohols, and group VIII consists of pure donor solvents such as chloroform. The various solvents of Table 6.1 are classified as in Figure 6.4 under the heading "selectivity group." Inspection of the latter values shows that homologous solvents fall within the same selectivity group, as we would expect from the preceding discussion.

A Systematic Approach to Selectivity Optimization

Selectivity optimization is best illustrated with a hypothetical example. Assume that we carry out an initial separation by normal-phase LC, using solvent mixtures of hexane (A) and chloroform (B). Further assume that the correct solvent strength is found for 20%v chloroform/hexane as mobile phase. From Table 6.1 and Eq. 6.2 we calculate P' for the correct solvent strength equal to (0.8 x 0.0) + (0.2 x 4.4), or $P' = 0.9$. Now assume we want to change separation selectivity. This is most readily achieved by a major change in the selectivity group of the polar solvent B. From Table 6.1, we see that chloroform belongs to selectivity group VIII. By reference to Figure 6.4, we see that groups I and V are most distant from group VIII, and, therefore, would be most likely to yield a large change in separation selectivity. Let us assume we try a solvent from group V. Reference to Table 6.1 (eighth column) shows that methylene chloride and ethylene chloride belong to group V. Selecting methylene chloride for its lower volatility and ready availability, we calculate from Eq. 6.3 that its concentration in the final mobile phase should equal 0.2 (4.4/3.4), or 26%v CH_2Cl_2/hexane (for which P' also equals 0.9). If the latter mobile phase does not give acceptable separation selectivity, we can repeat the process with a polar solvent from group I (e.g., triethylamine, diethyl ether) or from any other selectivity group. However, if a particular polar solvent does not give good selectivity, it is unlikely that another solvent from the same selectivity group will be any better.

What about changing the nonpolar solvent A, in order to change solvent selectivity? Normally the strong interactions associated with the polar solvent will overpower similar interactions by the nonpolar solvent, and a change in the nonpolar solvent A will seldom be profitable. An occasional exception is noted for samples with nonpolar compounds of widely varying refractive indices (meaning differences in dispersion interactions). In this case the substitution of a nonpolar solvent of differing refractive index n for the original solvent A may provide a useful change in selectivity. For example, cyclopentane (n = 1.404) or carbon disulfide (n = 1.624) might be substituted for n-hexane (n = 1.372); however, note the discussion of (8).

Solvent Selectivity in Reverse-Phase and Adsorption LC

In reverse-phase LC, water is the polar solvent B and any water-miscible less-polar solvent A can be used. Common examples of the second solvent include methanol, acetonitrile, and tetrahydrofuran. We saw earlier that a change in the less-polar solvent (e.g., from methanol to acetonitrile) is generally less effective in changing solvent selectivity, so that reverse-phase separations provide more limited opportunity for change in separation selectivity. Nevertheless, significant changes in reverse-phase selectivities have been observed on changing the second solvent from methanol to propanol to acetonitrile (see Reference 9 and examples given in the discussion on controlling resolution in Section 2.5), so this possibility should not be ruled out. See also the discussion of (19).

In reverse-phase LC, solvent strength should be held constant while the less-polar constituent is varied for selectivity change. A variation on Eq. 6.3 can be used in this regard:

$$\phi_c = \frac{\phi_b(P_w - P_b)}{(P_w - P_c)}. \tag{6.3a}$$

Here P_w refers to P' for water (solvent A), equal to 10.2. Alternatively, we can define a more precise solvent strength parameter S for reverse-phase systems, such that

$$\phi_c = \phi_b \left(\frac{S_b}{S_c}\right). \tag{6.3b}$$

for a solvent of interest. Experimental values of S have been reported (15):

Solvent strength (reverse-phase LC)	
water (0.0)	dioxane (3.5)
methanol (3.0)	ethanol (3.6)
acetonitrile (3.1)	i-propanol (4.2)
acetone (3.4)	tetrahydrofuran (4.4)

Thus, if the starting mobile phase is 50%v methanol/water and it is desired to change to some mixture of tetrahydrofuran/water for a change in selectivity, the volume fraction of tetrahydrofuran that will give about the same solvent strength is obtainable from Eq. 6.3b as

$$\phi_c = \frac{0.5(3.0)}{(4.4)}$$
$$= 0.34.$$

Thus, 34%v tetrahydrofuran/water should be used, assuming k' values for 50%v methanol/water were already in the right range.

Solvent selectivity in adsorption LC also varies from one selectivity group to another, as in the preceding examples for partition LC. However, predictable changes in solvent selectivity are more readily made by use of a somewhat different approach, as discussed in detail in Section 9.3.

Ternary mixtures of two strong solvents B and C with a weaker solvent A are sometimes useful in fine-tuning solvent selectivity to give separations intermediate between those provided by the solvent systems A/B and A/C. For an example, see the gradient elution separation shown in Figure 16.13, as well as (*19*). Note that it is possible with ternary mixtures to continuously vary *both* solvent strength and selectivity. Thus, selectivity is controlled by the ratio of concentrations of B to C, whereas solvent strength can be controlled by changing the concentration of A.

6.6   PURIFICATION OF LC SOLVENTS

A detailed discussion of solvent purification for LC is beyond the scope of the present book, as most solvents used in LC can be purchased in adequate purity (see source column of Table 6.1). However, a few general remarks may prove useful.

Normally the *water* used for reverse-phase partition LC and ion-exchange LC will not be purchased, but will be prepared by the user. Depending on the detection sensitivity required and the wavelength used, as well as on the starting purity of water from the tap, varying levels of solvent purification may be required. In some cases, ordinary distilled water can be used. More commonly, it is desirable to use deionized water that has subsequently been distilled. In very demanding applications, a final treatment with charcoal may be necessary.

The Millipore Corp. (Worthington Diagnostics Division) sells a water purification system for about $1500 that combines filtration, deionization, and charcoal treatment in a convenient, high-volume unit. Many LC users have found that water purified in this way is completely satisfactory for all LC applications [e.g. (*16*)].

Aliphatic hydrocarbon solvents, unless specially purified for LC, will usually contain olefinic impurities that prevent the use of the solvent with UV detectors at wavelengths below 260 nm. Such solvents can easily be freed of these olefins by an adsorption separation on silver-nitrate-impregnated silica. For example, Davison Code 12 silica (Grace Chemical Co.) can be slurried with 10%w silver nitrate dissolved in water, mixture can then be dried at 125°C and used to pack a large glass column (e.g., 48 × 2 in.). The solvent to be purified is then percolated by gravity through the column, with fractions of 250 ml being collected and checked for their UV absorbance. Fractions with adequate UV transparency are retained for use as LC solvents.

Reagent-grade ethers (notably tetrahydrofuran) may come with the UV-absorbing antioxidant BHT as stabilizer. In this case a simple one-stage distillation in glass suffices to remove the less volatile BHT from the solvent, which should then be used immediately. Other solvents (e.g., 2-chloropropane) may have excessive UV absorbance that can also be eliminated by distillation.

Some solvents contain impurities that are more polar and would therefore interfere in the use of the solvent as mobile phase in most LC applications (e.g., chloroform is often stabilized by small concentrations of ethanol). If the solvent is water-immiscible (e.g., chloroform), a simple extraction with water followed by drying the solvent over anhydrous sodium sulfate may prove adequate. Where the polarity of solvent and impurity are quite different (and the impurity is more polar), passage of the solvent by gravity through a large glass column of low-cost alumina or silica (Table 5.13) may be successful.

In liquid-solid (adsorption) LC, control of the solvent water content is important (see Section 9.3). In some cases it is desired to mix known amounts of dry (water-free) and water-saturated solvent together to obtain a known "percent-water-saturation." Normally reagent-grade solvents are adequately dry as they leave the bottle. However, addition of 10-20 g of Linde 4Å activated molecular sieve beads to the bottle, followed by occasional mixing of the stoppered bottle, will ensure total dryness of the final solvent. The separation of easily oxidized samples may require that the mobile phase used be oxygen-free. This is easily achieved by bubbling helium (for 15-30 min) through the individual solvents making up the mobile phase before they are mixed to form the final mixture. The recovery of trace sample constituents for characterization (e.g., in residue analysis) requires special consideration of possible solvent impurities. Distillation of the solvent before use usually suffices to remove involatile impurities that could contaminate recovered sample fractions.

Halogenated solvents may contain traces of HCl, HBr, and so on, which in turn can react with the stainless steel of pumps, columns, and so on. Aside from the potentially corrosive nature of such solvents, they can also give problems as a result of solubilization of metal ions in this way. The latter can interfere with the LC separation, react with sample components, and so on. Solvents can be freed of HCl, HBr, and so on, by being passed over basic alumina. Storage of halogenated solvents in mixtures with water should be avoided, as the decomposition of these solvents is thereby accelerated. Mixtures of halogenated solvents (particularly CCl_4 and $CHCl_3$) with various ethers (diethyl ether, diisopropyl ether, tetrahydrofuran) react to form products that are particularly corrosive for 316 stainless steel (*17, 18*); such mobile phases should therefore be used sparingly and made up fresh before use. Halogenated solvents such as CH_2Cl_2 react with other reactive organic solvents (e.g., acetonitrile) on standing, to form crystalline products. With the exception of dry mixtures of halogenated solvents and saturated hydrocarbons, the general rule seems to be to make up all solutions of halogenated solvents fresh before use and to avoid their use where possible.

REFERENCES

1. L. R. Snyder, in *Techniques of Chemistry*, 2nd ed., Vol. III, Part I, A. Weissberger and E. S. Perry, eds., Wiley-Interscience, New York, 1978, Chap. 2.

2. F. A. V. Metsch, *Angew. Chem.*, **65**, 586 (1953).

3. N. B. Godfrey, *Chemtech*, **360** (1972).

4. W. R. Gambill, *Chem. Eng.*, 151 (1959).

5. S. R. Abbott, J. R. Berg, P. Achener, and R. L. Stephenson, *J. Chromatogr.*, **126**, 421 (1976).

6. H. Forestier and L. Truffert, *Analysis*, **3**, 30 (1975).

6a. C. P. Smyth, *Dielectric Behavior and Structure*, McGraw-Hill, New York, 1955.

6b. R. W. Taft, D. Gurka, L. Joris, P. von R. Schleyer, and J. W. Rakshys, *J. Am. Chem. Soc.*, **9**, 4801 (1969).

7. L. Rohrschneider, *Anal. Chem.*, **45**, 1241 (1973).

8. L. R. Snyder, *J. Chromatogr.*, **92**, 223 (1974); *J. Chromatogr. Sci.*, **16**, 223 (1978).

9. B. L. Karger, J. R. Gant, A. Hartkopf, and P. H. Wiener, *J. Chromatogr.*, **128**, 65 (1976).

10. L. R. Snyder, *Principles of Adsorption Chromatography*, Dekker, New York, 1976, Chap. 8.

11. P. J. Twitchett and A. C. Moffat, *J. Chromatogr.*, **111**, 149 (1975).

12. K. Karch, I. Sebestian, I. Halasz, and H. Engelhardt, ibid., **122**, 171 (1976).

13. C. Horvath, W. Melander, and I. Molnar, ibid., **125**, 129 (1976).

14. M. Lafosse, G. Keravis, and M. H. Durand, ibid., **118**, 283 (1976).

15. L. R. Snyder, J. W. Dolan, and J. R. Gant, *ibid.*, **165**, 3 (1979).

16. R. L. Sampson, *Am. Lab.*, 109 (May, 1977).

17. A. Y. Ku and D. H. Freeman, *Anal. Chem.*, **49**, 1637 (1977).

18. W. E. F. Gerbacia, *J. Chromatogr.*, **166**, 261 (1978).

19. N. Tanaka. H. Goodell and B. L. Karger, *ibid.*, **158**, 233 (1978).

BIBLIOGRAPHY

L. R. Snyder, in *Techniques of Chemistry*, 2nd ed., Vol. III, Part I, A. Weissberger and E. S. Perry, eds., Wiley-Interscience, New York, 1978, chap. 2.

B. L. Karger, L. R. Snyder, and C. Horvath, *An Introduction to Separation Science,* Wiley-Interscience, New York, 1973.

J. A. Riddick and W. B. Bunger, *Organic Solvents,* 3rd ed., Wiley-Interscience, New York, 1970.

S. R. Bakalyar, *"Mobile Phases for High Performance Liquid Chromatography,"* Am. Lab. 43 (June 1978) (a good practical discussion of solvents for LC).

SEVEN

BONDED-PHASE CHROMATOGRAPHY

7.1 Introduction   270
7.2 Preparation and Properties of Bonded-Phase Packings   272
     General   272
     Preparation of BPC Packings   276
     Properties of the Bonded Stationary Phases   278
7.3 Mobile-Phase Effects   281
     Retention Mechanism   281
     Stability of Bonded Phases to Solvents   281
     Solvent Strength and Selectivity   281
         Mobile Phases in Normal-Phase BPC   284
         Mobile Phases in Reverse-Phase BPC   285
7.4 Other Separation Variables   289
     Sample Size   289
     Effect of Temperature   292
7.5 Special Problems   294
7.6 Applications   301
7.7 The Design of a BPC Separation   316
     Selecting the Column   316
     Selecting the Mobile Phase   317
     Operating Variables   319
References   319
Bibliography   321

7.1 INTRODUCTION

The most widely used column packings for modern LC are those with surface-reacted (chemically bonded) organic stationary phases. As of 1978, many laboratories reported that over three-fourths of their LC separations were carried out on these bonded-phase chromatography (BPC) columns. BPC columns were originally developed to eliminate the disadvantages of LLC columns based on mechanically held stationary liquids (Chapter 8). However, the unique features of these BPC packings now permit applications well beyond this initial intent. Because of the many practical differences in the use of BPC columns versus the older LLC columns and because the basis of separation (retention mechanism) with the BPC columns may differ significantly from that for LLC columns, we have provided a separate discussion of bonded-phase chromatography in this chapter.

In contrast to LLC, the main advantage of BPC packings is that they are quite stable, because the stationary phase is chemically bound to the support and cannot be easily removed or lost during use. Therefore, precolumns and/or presaturation of the two phases is not required. As a result, BPC is especially suited for samples containing components with widely varying k' values (which require gradient elution). The availability of a wide variety of functional groups in BPC packings allows both normal- and reverse-phase chromatography to be carried out in a relatively simple, straightforward manner.

One disadvantage of BPC is the poor reproducibility of some commercial packings (Section 5.3). Also, BPC packings with thick polymeric stationary phases exhibit lower column efficiencies than LLC packings. Nevertheless, BPC is normally the preferred approach, if for no other reason than the greater convenience it affords. Additionally, the choices among BPC packings and the mobile phases that can be used with these packings are sufficiently flexible to permit almost any separation to be achieved by this method.

Polar BPC packings are used for normal-phase separations in much the same manner as adsorbents for LSC (e.g., silica, Section 9.2). Samples of moderate to strong polarity are usually well separated on the polar BPC packings. Reverse-phase BPC normally involves a relatively nonpolar stationary phase (e.g., C_8 or C_{18} hydrocarbon) used in conjunction with very polar (e.g., aqueous) mobile phases to separate a wide variety of less polar solutes. BPC packings also can be used to separate ionic species via ion-pair chromatography (Chapter 11). As with LLC (Section 8.2), BPC generally separates on the basis of the type and number of functional groups within the solute molecule. BPC is particularly useful for separating compounds that differ in molecular weight (e.g., homologs, benzologs, oligomeric species), as long as molecular weight does not exceed about 3000. Table 7.1 provides some examples of such separations, including several related LLC systems. Reverse-phase BPC with alkyl-substituted packings comes

Table 7.1 Separation selectivity in bonded-phase and liquid-liquid chromatography.

Separation By	LC (stationary phase/mobile phase)	Compounds separated	k'	Ref.
Compound type	ODS-Permaphase/25% methanol in water (reversed-phase BPC)	1-naphthylamine	1.1	(1)
		1-naphthonitrile	2.3	
		1-nitronaphthalene	3.5	
		naphthalene	6.6	
		1-methoxynaphthalene	9.6	
		1-methylnaphthalene	15.7	
Homologous series (molecular weight)	tris-(2-cyanoethoxy)propane/hexane (conventional LLC)	3-methylpentanal-DNPH^a (C_6)	0.2	(2)
		n-pentanal-DNPH (C_5)	0.5	
		n-butanal-DNPH (C_4)	0.8	
		propanal-DNPH (C_3)	1.3	
		crotonaldehyde-DNPH ($C_3=$)	1.6	
		acrolein-DNPH ($C_2=$)	2.5	
Number of functional groups (oligomers)	polyethylene glycol-400/iso-octane-CCl$_4$, 2:1 (v/v) (conventional LLC)	ethylene oxide oligomers (7-14 EO units)	0.4-21	(3)

^{a}DNPH = 2,4-dinitrophenylhydrazine derivatives.

the closest to a universal system for modern LC, and "scouting" separations of new samples are commonly first done on these columns.

7.2 PREPARATION AND PROPERTIES OF BONDED-PHASE PACKINGS

General

Bonded-phase packings are prepared by a variety of synthetic methods. Some of the products are more useful than others, and commercial materials generally are those that have the best combination of desirable properties. Almost all the reported BPC packings have used rigid silica or silica-based supports that provide high column efficiency and excellent mechanical stability under high pressures (Section 5.3).

All the various methods for attaching a bonded phase to a siliceous support rely on the reaction of surface silanol groups. Fully hydrolyzed silica contains a concentration of about 8 μmoles of silanol groups per m^2 of surface. At best, because of steric considerations, a maximum of about 4.5 μmoles of silanol groups per m^2 can be reacted (4), the remainder of the silanol groups being shielded by the reacted groups. Thus, fully reacted BPC packings based on low-

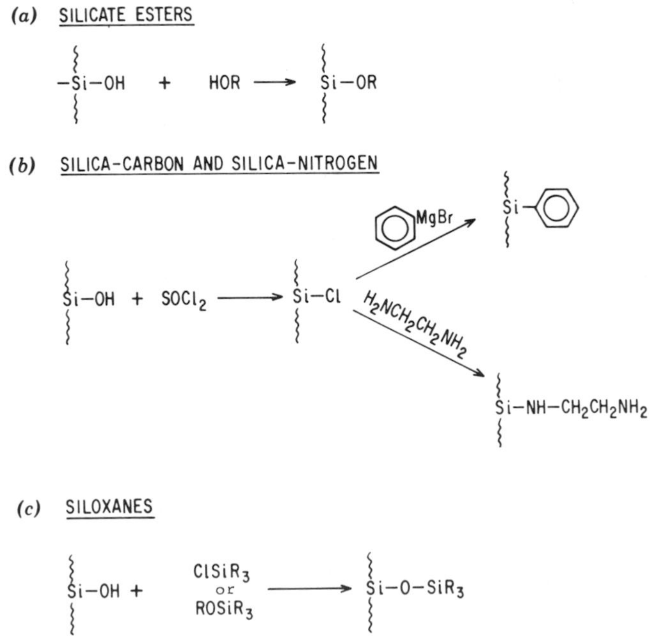

(a) SILICATE ESTERS

$$-\overset{\overset{\displaystyle\mid}{|}}{\underset{\underset{\displaystyle\mid}{|}}{Si}}-OH \quad + \quad HOR \quad \longrightarrow \quad \overset{\overset{\displaystyle\mid}{|}}{\underset{\underset{\displaystyle\mid}{|}}{Si}}-OR$$

(b) SILICA–CARBON AND SILICA–NITROGEN

(c) SILOXANES

$$\overset{\overset{\displaystyle\mid}{|}}{\underset{\underset{\displaystyle\mid}{|}}{Si}}-OH \ + \quad \begin{matrix} ClSiR_3 \\ or \\ ROSiR_3 \end{matrix} \quad \longrightarrow \quad \overset{\overset{\displaystyle\mid}{|}}{\underset{\underset{\displaystyle\mid}{|}}{Si}}-O-SiR_3$$

Figure 7.1   Reactions for preparing bonded-phase packings.

surface-area pellicular supports (e.g., Corasil and Zipax) have a relatively low concentration of organic coverage, whereas high surface area porous supports (e.g., LiChrosorb SI-60 and Zorbax-Sil) yield BPC packings with relatively high organic concentrations.

The reactions that have been used to prepare bonded-phase packings are summarized in Figure 7.1. So-called *silicate esters* (7.1a) were among the first reported for LC use (*3*). These are prepared by direct esterification of silanol groups ($\equiv$Si—OH) with an alcohol (R—OH); for example, 3-hydroxypropionitrile or Carbowax 400 (to give $\equiv$Si—O—R). Alternatively, silicate esters can be prepared by first chlorinating the siliceous support with thionyl chloride, as shown in the initial reaction of Figure 7.1b. Subsequent reaction with an alcohol produces the desired silicate ester. The surface of these monomolecular-layered particles may be pictured as a forest of organic groupings standing on end—thus, the term *brushes,* which is sometimes used to describe such materials. The organic surface of these bonded-phase materials is monomeric, and therefore, good mass transfer characteristics and high column efficiencies are found. Table 5.7 lists some of the silicate-ester-type packings that are commercially available. Unfortunately, these esterified siliceous packings are not hydrolytically or thermally stable and thus are restricted to applications with mobile phases that do not contain water or alcohol. This presents a severe limitation for many potential applications.

Other reactions result in covalent Si—C or Si—N bonds being formed (Figure 7.1b). These materials are prepared by first chlorinating the silanol groups of the support with thionyl chloride. This chlorinated silica can then be reacted with a Grignard reagent (also Wurtz and other similar reactions) to produce Si—C bonds (*5, 6*) or with amines to give Si—N bonds (*7, 8*). In both reactions the organic is bonded directly to the silica. Further modifications to the organic can be made, if desired. For example, the aromatic ring in Figure 7.1b can be sulfonated to produce a strong cation exchanger (*6*). Hydrolytic and thermal stability are markedly superior for the Si—C and Si—N bonded phases, versus the silicate esters. However, Grignard reactions are inconvenient, sometimes produce a relatively low concentration of surface organic, and often leave undesired residues. The Si—N bonded phases are more stable than silicate esters but are restricted to the pH 4-8 range when aqueous solvents are used.

By far the most useful and most widely available BPC packings are those based on siloxanes. Figure 7.1c shows that siloxane-type (Si—O—Si—C) bonded-phase packings typically are prepared by reacting the silanol groups of the support with organochlorosilane or organoalkoxysilane reagents (depending on the *R* group desired). Bonded phases of this type are hydrolytically stable throughout the pH range 2-8.5. Siloxane bonded-phase packings are available with pellicular or totally porous supports, and representative materials of both types are listed in Tables 5.6, 5.7, and 5.11.

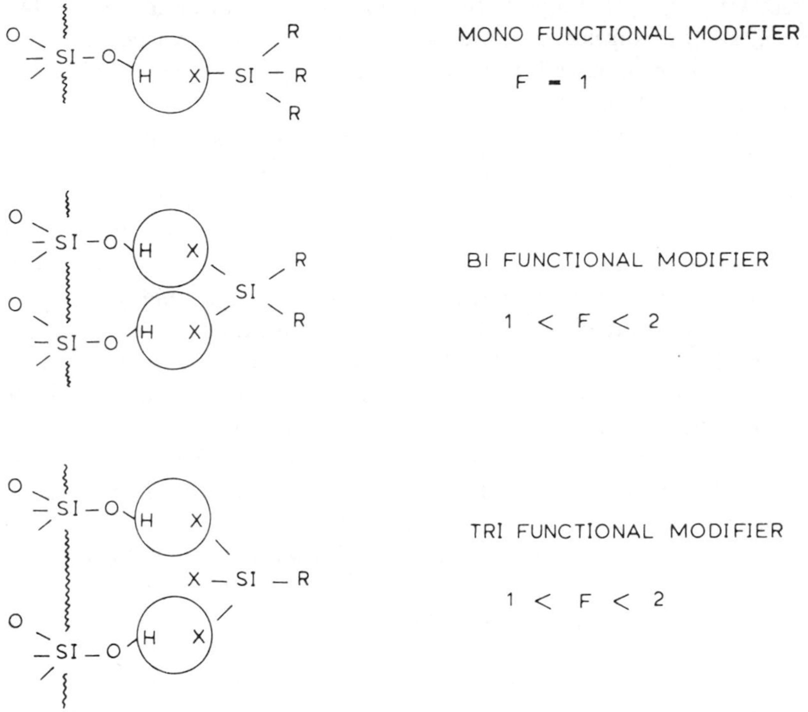

Figure 7.2 Stoichiometry of siloxane surface reactions. Reaction between hydroxyl groups at the silica surface and modifiers R_nSiX_{4-n} ($1 \leqslant n \leqslant 3$), which differ in the number of reactive Si-X groups. Reprinted from (4) with permission.

The organic bonded-phase siloxane coating can be made as a monomolecular layer (e.g., μ-BondaPak-C$_{18}$, Zorbax-CN) or as a polymerized multilayer coating (e.g., Permaphase ETH). BPC packings with so-called monomolecular organic layers are normally prepared by reacting the surface silanol groups of porous siliceous-base particles with mono-, di-, or tri-functional chloro-, dimethyl-amino-, siloxy-, or alkoxy-silanes (Figure 7.2). Usually in these reactions, $X = $ Cl, OH, OCH$_3$, or OC$_2$H$_5$, and $R = $ an organic radical. The stoichiometry of surface reactions with various modifiers has been studied extensively, with the results indicated in Figure 7.2. With a monofunctional reactant (i.e., only one reactive Si$-X$ group per molecule), a one-to-one reaction must take place, giving a value of 1.0 for F, the ratio of (number of moles of silanol groups reacted) to (number of moles of modifier reacted). For bi- and trifunctional modifiers, F values of 1-2 have been found. Values of $F = 3$ have never been found experimentally, probably because of steric factors. Therefore, using bi- and trifunctional reactants such as R_2SiX_2 and $RSiX_3$ for the surface modifications, up to two Si-X groups per bonded functional group remain unreacted. After treatment with water, hydrolysis of these unreacted groups takes place, and additional silanol groups are formed in about the same concentration as the bonded organic functional groups present in the packing. These acidic organo-silanol groups can significantly affect the retention behavior of solutes and adversely influence the stability of the packing in aqueous solutions at pH > 7 (4).

Thus, incomplete reaction of the surface with the silane reagent, or the formation of new Si—OH groups from using bi- or trifunctional modifiers, can result in a population of residual acidic Si—OH groups that are readily accessible to molecules of the mobile phase or sample. These residual silanol groups can cause tailing of chromatographic peaks (particularly for basic solutes) and a low sample capacity for the column (see the discussion of sample-size effects in Section 2.2). Therefore, the recent trend is toward (a) a dense monolayer of functional groups instead of a partial coverage and (b) the use of monofunctional dimethylsilanes [X-Si(CH$_3$)$_2$-R] to provide a homgeneous organic coating with a minimum possibility of residual Si-OH groups (4, 9). Monochlorosilane reagents are preferred, if the required organic functionality can be prepared. If two of the R groups in the monofunctional modifier (Figure 7.2) are methyl (preferred!) surface coverage can be as high as about 4 μmoles/m^2 of organic (based on carbon analysis). In the latter case, residual Si—OH groups on the silica surface are unavailable for chromatographic interactions with most solutes, because of steric shielding (4).

The reaction of organosilanols (e.g., HO—Si—R_3), organodimethylamine- [e.g., (CH$_3$)$_2$N-Si-R_3], or organoalkoxy- (e.g., RO-Si-R_3) silanes with high surface area silica supports without polymerization can also produce good BPC packings. These reactions are relatively reproducible, provided traces of water or other reactive species are absent. Unreacted, accessible silanols may be left after the initial reaction, but these can be removed by a further treatment ("capping") of the packing with chlorotrimethylsilane (providing the R groups do not react with the latter silane).

To achieve a sufficient volume of bonded-phase coating on low-surface-area pellicular particles, for adequate sample retention and capacity, polymerization of the organosilane reactants may be required (10). Since bi- or trifunctional organosilanes normally are used with these reactions, both cross-linking and linear polymerization reactions probably occur. Figure 7.3 is a schematic representation of a polymolecular bonded-phase packing made by this approach. Such reactions must be carefully controlled, since too thick a polymeric network will result in low chromatographic efficiency due to slow stationary-phase mass transfer (11). For the Permaphase packings (prepared from a ~1 m^2/ g superficially-porous support), 1-1.5% of organic phase appears to be optimum.

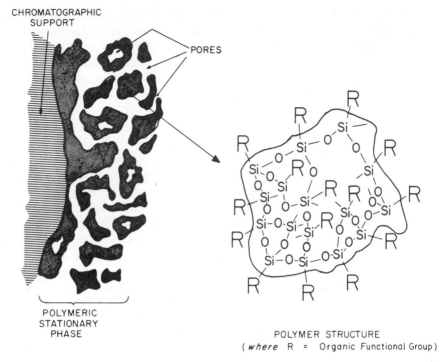

Figure 7.3 Schematic cross section of pellicular support with a chemically bonded polymolecular organic stationary phase.

Preparation of BPC Packings

In this and the following subsection we discuss the preparation and properties of BPC packings. These sections may be of less interest for those who purchase BPC packings or columns; those readers may want to skip to Section 7.3 because the rest of this section is not essential to the use of BPC packings. However, this section should be of interest to those who plan to make their own BPC packings, particularly packings for preparative LC (where the cost of commercial columns is relatively high).

Before carrying out reactions such as those shown in Figure 7.2, the silica support must be acid-hydrolyzed or hydrated (e.g., in 0.1 M HCl at 90°C for at least 24 hours), so that its surface has a maximum number of Si-OH groups per unit surface area. Physically adsorbed water must then be quantitatively removed from the surface, or else the silane-modifier may be hydrolyzed; hydrolyzed di- and trifunctional silanes can in turn condense to produce an organosilicon polymer. Heating of hydrated silicas at 200°C under vacuum for 8-16 hr reduces the concentration of physically adsorbed water to required levels.

To achieve maximum conversion of the silanol groups on the silica surface, a large excess (e.g., 2-X stoichiometric) of the silane modifier should be employed (12). The reaction rate generally is controlled by diffusion of the reactant to the active surface sites within porous particles. Therefore, the reaction temperature and time should be adequate for maximum conversion. However, reaction temperature should not exceed 250°C, or dehydration of the silica surface takes place with formation of unreactive siloxane bonds (4):

$$2(\equiv Si-OH) \rightarrow Si-O-Si + H_2O \uparrow$$

The reaction conditions used to prepare BPC packings depend on the functionality (i.e., R-groups) required. For reverse-phase packings the reaction almost invariably is carried out with chlorosilanes. Reaction of surface Si-OH groups with chlorosilanes reagents to the desired high yields is promoted by (a) using a large excess of reagent (at least 2-X stoichiometric); (b) conducting the reaction in the neat liquid reactant or in a dry, high-boiling organic solvent (e.g., toluene); (c) mechanically removing the volatile reaction product (HCl) during the reaction (10); or (d) using an appropriate acid-acceptor catalyst (e.g., pyridine) or an acid-accepting solvent (13). A typical preparation of monomolecular BPC packings involves refluxing 10 g. of dry, hydrolyzed silica support in a mixture of 80 ml of dry pyridine plus the chlorosilane for 16-24 hours, in an apparatus protected from water vapor. For example, for a 300 m²/g silica (~10 nm pores), about 9 g of dimethyl-n-octylchlorosilane would be added. Following this reaction, the product should be filtered, washed with methanol, refluxed with about 150 ml of methanol for 30 min. To ensure optimum surface coverage by organic groups, this reacted material should be "capped" by a following similar reaction with chlorotrimethylsilane. The final product is carefully dried before packing into a column.

To prepare polar bonded phases for normal-phase chromatography (or to cover the silica surface with a hydrophilic layer for gel filtration chromatography—see Section 12.2), reactions are normally carried out with alkoxysilanes or organosilanols. However, there are some notable exceptions (e.g., nitrile bonded phases) where chlorosilanes are preferred. The conditions for preparing polar bonded phases using alkoxysilanes and organosilanes have not been described in detail for every R group, and also depend on the individual situation. However, (12), (14), and (15) provide further information.

Hydrocarbon bonded-phase packings also can be prepared by the in situ treatment of prepacked silica columns (16, 17). A satisfactory procedure is slowly to pump the silanizing mixture through the hydrolyzed, dry silica column at elevated temperature (e.g., 2% dimethyl-n-octylchlorosilane in *dry* pyridine at 100°C for a C_8 packing) for 16-24 hours. This approach is not as convenient as simply packing a column of the originally bonded-phase material, but it appears to be useful in restoring the coating on a silanized packing that has been partially removed by extensive use (*17*). Such procedures are, however, rarely used.

Properties of the Bonded Stationary Phases

Bonded organic groups can be characterized by transmission infrared spectroscopy, attenuated total reflectance (*18*), or pyrolysis mass spectrometry (*19*). Elemental analysis also can be used to measure molar ratios for comparison with the expected ratio. However, by itself the carbon content of a bonded-phase packing is insufficient. A preferred parameter is the surface concentration of bonded groups, a_{exp}, calculated according to (*4*)

$$a_{exp} \text{ (mole/m}^2) = \frac{W}{(M)(S_{bet})} \quad , \tag{7.1}$$

where W equals weight of functional groups as grams per gram of adsorbent, M = molar weight of the bonded functional group (g/mole), and S_{bet} equals specific surface area of the support, corrected for the weight increase due to the modification (m^2/g). The concentration of silanol groups on the original and modified silicas is best estimated by means of isotopic exchange with tritium-labeled water (*20*).

The a_{exp} value for a particular bonded-phase functional group can vary ±10% for different samples of supposedly identical silica supports. Under optimum conditions a surface concentration of the smallest modifier group available, trimethylsilyl-, is about 4.7 μmoles/m^2. With this concentration only about 60% of the total silanol groups are reacted; the remaining silanol groups are shielded by the dense layer of trimethylsilyl groups (*4, 9, 12*).

The bulk volume and chain length of the bonded-phase modifier influences the maximum surface concentration obtainable. For example, for bonded *n*-alkyl groups [-Si-(CH$_3$)$_2$R], a_{exp} changes only slightly with increasing chain length, with the average value for n = 4-16 being about 3.5-4.1 μmoles/m^2 (*4, 12*). However, the presence of bulky groups (e.g., phenyl) on the silane drastically reduces the possible surface coverage (e.g., as low as $\sim$2.5 μmoles/m^2).

Whether the desired dense coverage of organic functional groups has been obtained for a given BPC packing can be readily determined. For example, reverse-phase packings can be tested as an adsorbent with dry *n*-heptane as the mobile phase. If the packing has a dense coverage, the capacity factors of polar solutes (e.g., methanol, benzyl alcohol, acetone) will show symmetrical peaks and $k' \ll 0.5$. Also, an additional treatment of the modified packing with trimethylchlorosilane ("capping") should not increase the carbon content of the reverse-phase packing or change its chromatographic properties according to the test just described.

Attaching a chemically reacted, monomolecular layer to a porous silica reduces the mean pore diameter by about twice the thickness of the layer. The specific surface area and pore volume are likewise decreased. The longer the chain length of the modifying group, the more pronounced the effect [e.g., specific surface areas = 201 and 139 m^2/g, and specific pore volume = 0.82 and 0.52 ml/g, respectively, for C_8 and C_{18} –BPC packings prepared from the same support (4)]. However, as alkyl chain length is increased, sample retention likewise increases, suggesting that the volume of bonded phase (rather than surface area) is the critical factor in determining retention on BPC packings.

Further investigation is needed on the effect of surface bonding and pore structure and the resulting influence in the solute mass transfer and column efficiency. Some studies have suggested that supports with somewhat larger mean pore diameters (e.g., 10 nm) give more efficient columns (21), and many workers prefer BPC supports with 10-nm rather than 6-nm pores. Intuitively, one would expect supports with small pores to be more easily blocked by bonded stationary phase, leading to less-rapid solute equilibration. Therefore, bonded-phase chain length can affect column separating efficiency. In general, for the same support, a decrease in chain length appears to increase column efficiency, particularly for solutes with larger k' values (22). However, this effect may be influenced by the amount of organic modifier in the mobile phase (wetting of the packing; see Section 7.6), the column packing method, and the size and geometry of the pores for the bonded-phase packing support.

Effect of the length of bonded-alkyl groups (reverse-phase BPC) on solute retention is illustrated in Figure 7.4 for a series of phenols. Although relative retentions are similar, k' values for a given solute increase with chain length (or % carbon) [the anomalous behavior of the Si-60/C_{18} column (broken line) may be due to the fact that some or all of the pore volume of this small-pore packing is not readily accessible to the mobile phase because of blockage by the bonded phase]. There appears to be a linear relationship between k' values and the %w concentration of organic coverage, independent of chain length (23).

Sample capacity or loadability for reverse-phase BPC also is influenced by chain length, as illustrated in Table 7.2. For this system, increasing the chain length from C_4 to C_{18} roughly doubled the column loadability. Although maximum sample size increases with stationary phase carbon content, the relationship is not linear. For this example, the maximum sample size for the C_{18} packing is about 2 mg of sample/g stationary phase, which is comparable to that reported for bare silica (Figure 9.9). Sample loadability also generally improves with increases in the concentration of organic in the mobile phase and with increased temperature.

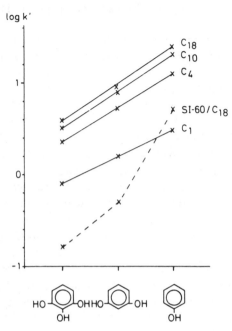

Figure 7.4 Influence of chain length on reverse-phase retention. Packings, SI-100 (and SI-60) reacted with methyl- (C$_1$), butyl- (C$_4$), decyl- (C$_{10}$), and octadecyl- (C$_{18}$-22% carbon); mobile phase, water; samples, aromatic phenols as shown. Reprinted from (21) with permission.

Table 7.2 Sample loadability for reversed-phase systems.[a]

| | k' value | | Maximum sample size[b] (10^{-4} g/g) | |
Sample	C$_4$	C$_{18}$	C$_4$	C$_{18}$
Hexanol	0.3	0.5	7.1	10.5
m-Xylenol	0.2	0.5	9.8	20
Gammexane	0.7	2.1	11.5	18.5
Lanatosid A	0.3	0.6	9.5	21

Source: From (21).
[a]C$_4$- and C$_{18}$-, silica, SI-100; d_p = 10 μm; column dimensions, 30 × 0.42 cm; eluent, water/methanol (1:4 v/v).
[b]Sample size causing k' to decrease by 10% from constant value at small samples sizes (value of $\theta_{0.1}$).

7.3 MOBILE-PHASE EFFECTS

Retention Mechanism

The mechanism of retention onto BPC packings has not been definitely estabblished. It has been variously proposed that sample molecules adsorb onto the surface defined by the organic coating (which covers the silica substrate) or that partitioning of sample molecules occurs into a "liquid" phase defined by the organic coating plus associated molecules of the mobile phase. In the case of an adsorption mechanism, sample and solvent molecules presumably compete for a place on the organic surface (much as in LSC, Figure 9.5a) (see 31). Alternately, it has been suggested that dimethylalkyl substituents in reverse-phase packings may act like liquid crystals [i.e., forming an ordered liquid phase (22)]. Still another suggestion is that the organic coating inbibes certain components of the mobile phase to form a more or less conventional liquid stationary phase, as in LLC (11, 22a).

Whatever the actual retention mechanism in BPC systems, it is convenient to consider the organic bonded phase as if it were an equivalent mechanically held liquid phase. This leads to reasonable predictions of retention as a function of solute structure (e.g., Figure 14.4) and mobile-phase composition.

Stability of Bonded Phases to Solvents

The stability of bonded-phase packings is largely determined by the silica support. Consequently, most solvents can be used if they are within about pH 2-8.5. Some bonded-phase materials can be used at temperatures up to 80°C, and 50-60°C is commonly employed with aqueous mobile phases in reverse-phase BPC. The same bonded-phase packing can sometimes be used in both normal-phase and reverse-phase chromatography. In Figure 7.5a we see the normal-phase separation of a mixture of urea-herbicides on an intermediate-polarity Permaphase ETH (aliphatic ether) column using a relatively nonpolar mobile phase. A similar mixture was separated (with approximate reversal of elution order) on the same column (Figure 7.5b) by reverse-phase LC, using an aqueous mobile phase.

Additional discussion of the stability of BPC packings is given in "Special Problems" in Section 7.6.

Solvent Strength and Selectivity

Selection of mobile phases in BPC generally parallels that found for LLC systems. Therefore, the same general approaches described here also are used in LLC separations (Section 8.4).

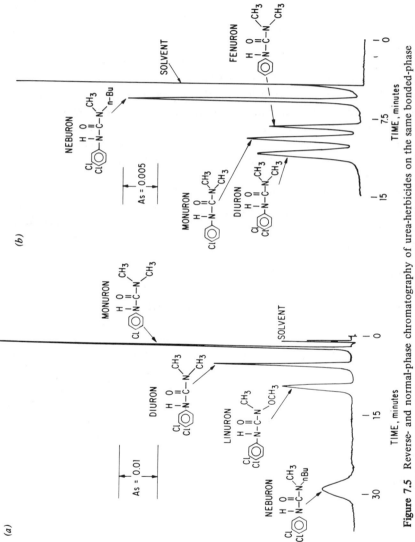

Figure 7.5 Reverse- and normal-phase chromatography of urea-herbicides on the same bonded-phase column. Column, 100 × 0.21 cm, Permaphase ETH, < 38 μm. (*a*) Mobile phase, 35%v methanol in water; flowrate, 2.5 ml/min; temp, 50°C; (*b*) mobile phase, 1%v dioxane in hexane; flowrate, 1.0 ml/min; temp, 27°C; detector. UV, 254 nm. Reprinted from (24) with permission.

How do we evaluate solvent strength so that the mobile phase can be systematically varied to optimize k' values? A useful system is one based on the P' parameter derived from experimental solubility data and discussed in detail in Section 6.5. Table 6.1 lists various possible solvents for LLC in order of increasing P' values (values of P' are in the sixth column). Nonpolar solvents have low values of P', whereas polar solvents have large values. In normal-phase BPC (more polar stationary phase), the strength of the mobile phase increases with increasing solvent polarity. Thus, in normal-phase BPC, by increasing the solvent P' value, we increase solvent strength and decrease sample k' values. The dependence of k' on solvent strength P' in normal-phase LC is shown in Figure 7.6. Here it appears that k' is decreased about 2.2-fold for each increase in P' by 1.0 unit. Table 7.3 lists a few preferred solvents for normal-phase BPC, in order of increasing strength. On the other hand, in reverse-phase LLC, as solvent P' values are increased, solvent strength decreases and sample k' values increase.

Table 7.3 Preferred solvents for normal-phase BPC.

Solvent	P'	Reduction in k' for 20%v addition of solvent to hexane[b]
Hexane, heptane	0.1	—
1-Chlorobutane	1.0	1.2-fold
i-Propyl ether (I)[a,c]	2.4	1.5
CH_2Cl_2 (V)[a]	3.1	1.7
Tetrahydrofuran	4.0	2.0
$CHCl_3$ (VIII)[a]	4.1	2.2
Ethyl acetate	4.4	2.0
Ethanol	4.3	2.0
Acetonitrile	5.8	2.6
Methanol (I)[a]	5.1	2.3

[a] Preferred solvents for selectivity change.
[b] Very approximate values.
[c] Methyl t-butyl ether a desirable alternative—less suscepible to oxidation.

Solvent selectivity is controlled by the selectivity group of Figure 6.4, reflecting contributions from donor, acceptor, and dipole characteristics of the solvent, respectively (Section 6.5). Thus, solvents from group VIII are good proton donors, and they interact preferentially with basic samples (e.g., amines, sulfoxides, etc.); basic solutes will preferentially interact with solvents from group VIII. Solvents from group I are good proton acceptors and tend to interact preferentially with hydroxylated molecules (e.g., acids, phenols). Solvents from group V tend to interact preferentially with sample molecules having large

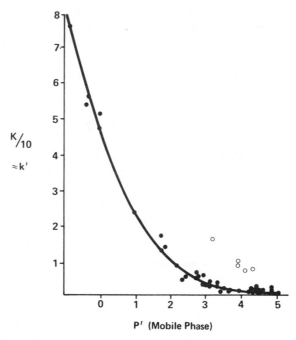

Figure 7.6 Dependence of solvent strength on P' values in normal bonded-phase and liquid-liquid chromatography. Calculated assuming β, β'-oxydipropionitrile as stationary phase. Open circles are alcohols as mobile phase. Nitromethane as solute.

dipole moments (e.g., nitro-compounds, nitriles, sulfoxides, amines). In setting up a separation, sample k' values first are adjusted into the optimum k' range (1-10) by choosing a solvent of right P' value. Then selectivity is varied by choosing another solvent of similar P', but from a different selectivity group. See Section 6.5 for further details

Mobile Phases in Normal-Phase BPC. The commonly employed mobile phases for BPC are similar to those used for LLC systems (see Table 8.2). Typically, solvent mixtures are used, consisting of a hydrocarbon solvent such as hexane, heptane, or isooctane, plus small amounts of a more polar solvent. Typical polar solvents used in this fashion are listed in Table 7.3. Solvent strength is systematically varied for a given application by varying the concentration of the more polar solvent component. For changing of solvent selectivity to provide better spacing of bands within the chromatogram, it has been found convenient to choose one of the four extreme-selectivity groups defined in Section 6.5 (see Figure 6.4); I, II, V, or VIII. In Table 7.3 we have chosen *i*-propyl ether (I), methanol (II), methylene chloride (V), and chloroform (VIII) as representing

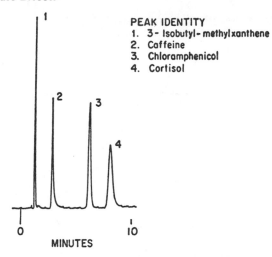

PEAK IDENTITY
1. 3-Isobutyl-methylxanthene
2. Caffeine
3. Chloramphenicol
4. Cortisol

Figure 7.7 Normal-phase separation of drug mixture with polar bonded-phase column. Column, 25 × 0.46 cm, Zorbax-CN; mobile phase, 15%v isopropanol in hexane; pressure, 750 psi; temp., ambient; detector, UV, 254 nm, 0.32 AUFS. Reprinted from (25) with permission.

convenient and effective polar solvents for the selectivity extremes. If the right solvent selectivity is not obtained with mixtures of one of these solvents plus, for example, hexane, then ternary mobile phases based on *two* of these "extreme-selectivity" solvents plus hexane may be of help (see example of Figure 11.4). If this approach is not successful, then some other variable (e.g., temperature) will probably be required to improve selectivity and band spacing. An example of a normal-phase BPC separation is shown in Figure 7.7. In this case a "nitrile" bonded-phase packing was used as the stationary phase and hexane modified with isopropanol was the mobile phase. To elute very polar solutes, stronger base-solvents (e.g., butyl chloride or dichloromethane instead of hexane) can be used with polar modifiers such as methanol.

Mobile Phases in Reverse-Phase BPC. The dependence of solvent strength and selectivity in reverse-phase BPC is similar to the case of reverse-phase LLC (Section 8.4). Usually water is used as base solvent, to which varying concentrations of miscible organics are added. Table 7.4 summarizes the relative strengths (less polar solvents with smaller P' values are stronger) of some solvents that have been used in reverse-phase BPC. Solvent strength is usually adjusted by varying the composition of the solvent mixture, and the data of Table 7.4 provide information on how k' changes with change in solvent composition. Methanol is the most commonly used organic solvent, since it meets the various requirements for a BPC solvent and is relatively cheap. Acetonitrile and tetrahydrofuran are the next most commonly used solvents, in that order.

Table 7.4   Solvent strength in reverse-phase LC.

Solvent	P'	Decrease in k' for each 10%v addition of solvent to water[a]
Water	10.2	—
Dimethyl sulfoxide	7.2	1.5-fold
Ethylene glycol	6.9	1.5
Acetonitrile	5.8	2.0
Methanol	5.1	2.0
Acetone	5.1	2.2
Dioxane	4.8	2.2
Ethanol	4.3	2.3
Tetrahydrofuran	4.0	2.8
i-Propanol	3.9	3.0

[a] Rough values that are averages calculated as in Chapter 6, using experimental data of (26a).

Changing selectivity in reverse-phase BPC is less easily accomplished, compared to normal-phase operation, because water dominates the sample mobile-phase/stationary-phase interactions. However, as seen in Figure 7.8, significant differences in separation selectivity can be achieved by replacing the organic solvent by a different solvent. Here acetonitrile (Fig. 7.8a) is replaced by methanol (Fig. 7.8b), and the elution order of these steroids is markedly altered. Methanol, which, like water, is both a proton donor and an acceptor, appears not to alter the selectivity of water, until very high concentrations of methanol are used (27). Recent studies have also shown that ternary aqueous mobile phases (e.g., methanol/dioxane/water) can provide unique selectivity for certain separations [(28, 28a, 28b); see also the practical examples of selectivity change in (28c)].

Changes in pH can change the separation selectivity for ionized or ionizable solutes, since charged molecules are distributed preferentially into the aqueous or more polar phase. For example, in Figure 7.9 we see the effect of pH on the reverse-phase elution of some drugs. A pH of about 5 permits the rapid separation of all the compounds. However, at intermediate or higher pH values, overlap of various components occurs or else a very long separation time is required. Variations in pH normally are not effective in obtaining desired changes in selectivity with solutes that do not ionize. In Figure 7.9 a strong acid such as salicylic acid (1, pK_a = 3.0) shows a marked decrease in retention volume when the pH rises to about 3. By contrast, a weak acid (e.g., curve 2, phenobarbitone, pK_a = 7.4) has a relatively constant elution volume with increase in pH until the pH exceeds the pK_a of the drug; then the elution volume decreases. This effect is anticipated, as only the nonionized form of these compounds is expected to

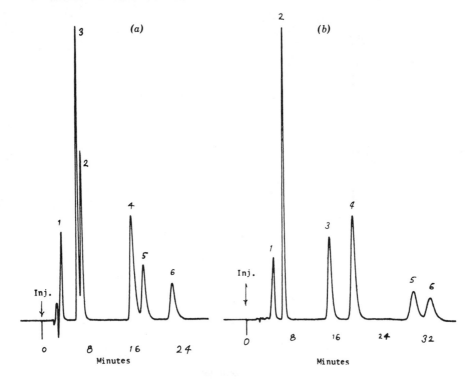

Peak Identification
1. Hydrocortisone Hemisuccinate
2. Methyl Paraben
3. Hydrocortisone
4. Propyl Paraben
5. Hydrocortisone Acetate
6. Cortisone Acetate

Figure 7.8 Variation of selectivity in reverse-phase BPC by changing organic mobile-phase modifier. Column, 30 × 0.46 cm, μ-Bondapak-C$_{18}$; flowrate, 1.5 ml/min; temp., ambient; detector, UV, 254 nm, 0.32 AUFS; sample, 10 μg each compound. (a) Mobile phase, 30% acetonitrile/70% water; (b) mobile phase, 45% methanol/55% water. Reprinted from (26) with permission.

partition into the hydrocarbon stationary phase. In contrast to these acids, the retention of basic drugs (e.g., curve 4, nicotine and curve 5, methylamphetamine, pK_a = 8.0 and 10.1, respectively) increases as pH increases. A neutral compound (curve 3, phenacetin) shows little change in retention with variation in pH, because it does not contain ionizable groups.

Occasionally, various salts are added to the aqueous mobile phase in reverse-phase BPC to vary solute retention or selectivity. In Table 7.5 we see that the concentration of phosphate in the mobile phase significantly influences not only

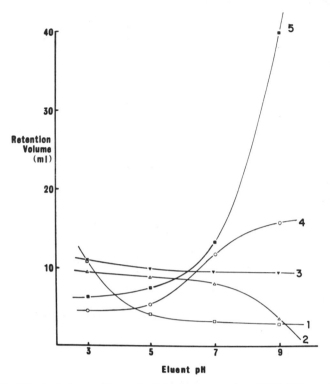

Figure 7.9 Effect of eluent pH on selectivity in reverse-phase *BPC*. Column, 30 × 0.4 cm, μ-Bondapak-C$_{18}$, 10μm; mobile phase, 0.025 *M* NaH$_2$PO$_4$–Na$_2$HPO$_4$ with 40%v methanol; flowrate, 2 ml/min; temp., ambient; detector, UV, 220 nm; samples, 1–10 μg. (1) Salicylic acid; (2) phenobarbitone; (3) phenacetin; (4) nicotine; (5) methylamphetamine. Reprinted from (*29*) with permission.

k' values for a series of nonapeptides but the selectivity of a particular solute pair. The data in this table apparently can be explained by a change in the solubility of the peptides as a result of increasing concentration of salt in the mobile phase. At low ionic strength, a strong increase in the solubility of the peptide occurs with increasing salt concentration (salting-in effect). At higher concentration, this increase reaches a maximum and then decreases (salting-out effect). Nonlinear increases in retention with increased salt concentration have been reported for other systems containing aromatic acids and bases (*31*). These seemingly conflicting results have not been resolved. However, it is likely that when the addition of salt increases the solubility of the solute in the mobile phase, retention decreases. Conversely, an increase in retention will occur when addition of salt decreases the solubility of the solute in the aqueous mobile phase. See also the discussion of (*32*).

Table 7.5 Influence of salt concentration on k' values for nonapeptides.[a]

Phosphate concentration in the mobile phase (10^{-2} M)	k'						$\alpha(Ox/Lyp)$
	Orp[b]	Lyp	Ox	Fely	Demox	Trichlorobutanol	
0.67	6.3	6.3	10.1	28.5	29.0	18.6	1.60
2.67	3.0	3.0	8.0	13.0	23.8	18.1	2.67
4.00	2.4	2.4	7.1	10.4	20.5	17.6	2.96
5.33	2.3	2.3	7.3	10.0	21.1	17.9	3.17

Source: From (*30*).
[a]Mobile phase, phosphate buffer pH 7-water/acetonitrile (x:80-x:20, where x = 10, 40, 60, or 80 ml); column, Nucleosil C_{18}, 10-μm particle size.
[b]Orp = ornipressine, Lyp = lypressine, Ox = oxytocine, Fely = felypressine, Demox = demoxytocine.

7.4 OTHER SEPARATION VARIABLES

Sample Size

Both the volume of the injected sample and the total weight of components in the sample solution affect final resolution. Consider first the importance of sample volume. Let the volume of injected sample be V_s, let the volume of a sample band leaving the column be V_p (for a small sample), and let the observed peak volume be V_w for some actual (finite-size) sample (Eq. 2.8b). A well-known relationship (*33*) then produces the expression

$$V_w^2 = V_p^2 + \frac{4}{3}\ V_s^2. \tag{7.2}$$

Here band volume V_p is measured as in Figure 7.10 and is equal to the baseline band width expressed in volume units. For the condition that V_w should not exceed V_p by more than 5% (which means R_s is not increased by more than 5% because of the large sample volume), a maximum value of V_s is equal to one-fourth V_p. In many cases, a 5% loss in resolution is too restrictive, and larger sample volumes can be used to obtain higher sensitivity (but with increasing loss of resolution). With Eq. 7.2 one can estimate the maximum volume of sample (based on the volume of the peak of interest) and the resultant resolution loss as

$V_s =$	% Resolution loss
0.25 V_p	5
0.33	8
0.40	10
0.57	20

An example of the calculation of maximum sample volume is shown in the hypothetical example of Figure 7.10, assuming that R_s should not be decreased by more than 8%. If the first band of interest has a t_w of 20 sec and the flowrate is 2 ml/min, the value of F(ml/sec) is 2/60, and the maximum sample volume from the preceding Table is (0.33 x 20 x 2/60), or 200 μl. If the second peak of Figure 7.10 is the first band of interest, a similar calculation yields a maximum sample volume of 400 μl.

The similar estimation of maximum sample volume in a real application is illustrated in Figure 7.11. In this case the volume V_p of the peak of interest (phenol) in this trace analysis was 190 μl (0.47 min. x 0.4 ml/min). For an arbitrary 20% loss in resolution, the maximum sample volume V_s would equal 0.57 V_p or (0.57 x 190 μl) or 108 μl. Therefore, for the separation in Figure 7.11, a 100-μl sample was utilized to obtain maximum sensitivity for phenol, still with adequate resolution from possible neighboring interferences. In this example, larger sample volumes gave broadening of the phenol peak.

Equation 7.2 assumes (a) that no more than a certain loss in R_s (e.g., 8%) as a result of sample volume can be tolerated, and (b) that the sample is dissolved

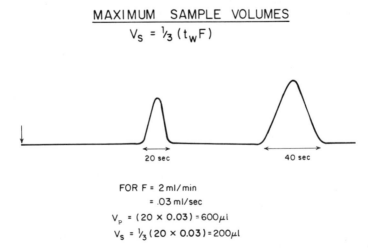

$$\text{MAXIMUM\ \ SAMPLE\ \ VOLUMES}$$
$$V_s = \tfrac{1}{3}(t_w F)$$

20 sec   40 sec

FOR F = 2 ml/min
= .03 ml/sec
V_p = (20 x 0.03) = 600μl
V_s = $\tfrac{1}{3}$(20 x 0.03) = 200μl

Figure 7.10 Maximum sample injection volumes.

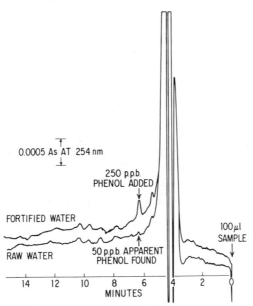

Figure 7.11 Maximum sample volume in trace analysis of phenol in surface water. Column, 50 × 0.32 cm, C$_{18}$–porous-silica microspheres, 7 μm; mobile phase, 60%v methanol in water; flowrate, 0.40 ml/ min; temp., 50°C; detector, UV, 254 nm, 0.05 AUFS; sample, 100 μl of Brandywine Creek, Brecks Mill, February 4, 1974. Reprinted from (9) with permission.

in a solvent of the same strength as the mobile phase. If a larger loss in R_s can be tolerated, then correspondingly larger V_s values are allowed. If the sample is dissolved in a solvent that is much weaker than the mobile phase, a corresponding increase in V_s without loss in resolution then becomes possible. The reason is that solute k' values are much larger in the weak mobile phase, which results in a compression of the sample volume for retained components as the sample enters the column (Section 13.5).

Typically, sample volumes for high-efficiency LC columns are in the 25-250 μl range, although very short, high-efficiency columns may require smaller sample volumes, as predicted by Equation 7.2.

Apart from the volume of sample injected, the *weight* of retained sample components is also important. Sample-size effects in BPC are similar to those of other LC methods (e.g., Figures 2.5, 2.6). With sufficient increase in sample weight, retention times begin to decrease and band plate-heights H begin to increase. For very small samples the plate heights of similar LLC, BPC, and LSC columns are generally comparable. However, as shown in Figure 7.12, BPC columns generally show a slower decrease in column efficiency (increase in H) with overloading of the column by sample (increasing sample size). However,

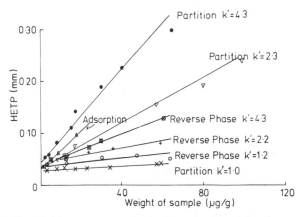

Figure 7.12 Plate height versus sample load curves for adsorption, reverse-phase, and partition modes. Column, 14.5 × 0.5 cm, Spherisorb-ODS; temp., ambient; detector, UV, 254 or 320 nm. Adsorption: mobile phase, 5%v methylene chloride in *n*-hexane (50% saturated with water); solute, nitrobenzene. Reverse-phase: mobile phase, 40%v methanol in water; solutes, phenol (k' = 1.2), *p*-cresol (k' = 2.2), and 3,5-xylenol (k' = 4.3). Partition: mobile phase, β,β'-oxydipropionitrile-saturated *n*-hexane; solutes, methylbenzoate (k' = 1.0), acetophenone (k' = 2.3), and dibutylphthalate (k' = 4.3). Reprinted from (*34*) with permission.

for larger samples, and particularly for less efficient columns, LLC columns with maximum loading of the packing by nonviscous stationary phase generally allow larger sample sizes before severe loss in column efficiency results. For further discussion, see (*21*).

Samples generally should be injected as larger volumes of a more dilute solution (using the maximum sample-volume limits discussed earlier), rather than as smaller volumes of more concentrated solutions. As sample concentration is increased, overloading of the column inlet eventually occurs, and this effect can be more important than that of sample volume per se. Column overloading results in unsymmetrical, broader bands and decreased column resolution, particularly for solutes with small k' values. In this case, injection of the sample as a larger volume of more dilute solution can greatly improve separation efficiency and resolution, particularly where V_s is maintained less than $(¼)V_p$.

Effect of Temperature

Retention in all LC separations generally decreases with increase in temperature T, and plots of log k' versus $1/T$ are generally linear. An example for a reverse-phase LC system is shown in Figure 7.13. A rough rule of thumb is that a 30°

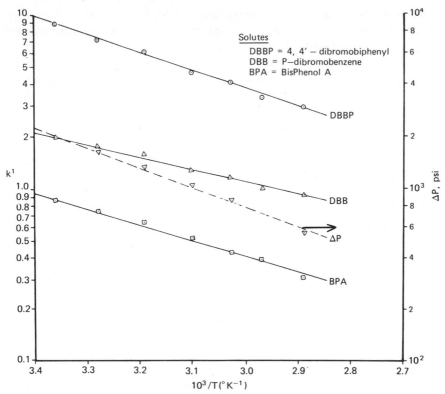

Figure 7.13   Effect of temperature on k' and column back pressure. Column, 25 cm, Micro-Pak-CH; mobile phase, 60% methanol in water; flowrate, 1.67 ml/min; temp., ambient; solutes: DBBP = 4,4-dibromobiphenyl; DBB, p-dibromobenzene; BPA = bisphenol A. Reprinted from (*35*) with permission.

increase in T leads to a two-fold decrease in k'. A change in T for BPC separations can be used to (a) decrease the viscosity of the mobile phase to improve separation efficiency; (b) increase the ability of the mobile phase to dissolve larger amounts of sample, particularly for preparative separations of less-soluble materials (Section 15.3); and (c) alter separation selectivity.

Most separations with normal-phase BPC packings are carried out at room temperature, using mobile phases of low viscosity for improved column efficiency. Reverse-phase BPC separations, on the other hand, use aqueous mobile phases that are relatively viscous. In this case, it is advantageous to operate at higher temperatures to decrease mobile phase viscosity. Hydrocarbon (e.g., C_{18}) BPC packings have been used with aqueous mobile phases at temperatures up to 80°C for extended periods without problems. A temperature of 50-60 C appears to be convenient in reverse-phase BPC, when sample components permit. Compared to ambient conditions, operation at these temperatures will usually about double column plate-count.

Changes in column temperature can affect selectivity in reverse-phase separations, particularly when separating solutes with different functionalities. Unfortunately, the direction and extent of this effect is not easily predicted and varying temperature is not generally used as the initial means of adjusting selectivity. For further discussion, see (*35a*).

7.5 SPECIAL PROBLEMS

Retention and selectivity characteristics of BPC packings vary from lot to lot and manufacturer to manufacturer, and this has been a particular problem with this LC method since its inception. One of the main sources of this irreproducibility is·the lack of complete coverage of the surface of the particles by the bonded stationary phase. It has been shown that residual SiOH groups on partially reacted BPC packings can be a major source of retention (*36*). These residual SiOH groups are available for interaction with solutes in a mode different from the bonded stationary phase, which often leads to tailing peaks. Residual SiOH groups are a particular problem when separating basic compounds that can interact by an ion exchange process (see the following discussion on the use of salts to minimize such effects).

Variability also arises from the fact that commercial BPC packings with the same functionality either can be "brushes" formed by reaction with monofunctional silanes or can be partially polymerized coatings derived from bi- or trifunctional reactants.

Some suppliers are now producing bonded-phase packings whose surfaces have been fully reacted with monofunctional silanes, so that SiOH groups are not available for unwanted interactions, and the same reproducible surface coating is formed from lot to lot. To obtain reproducible separations with BPC, it is strongly recommended that columns of such materials be employed wherever possible. As discussed in Section 7.3, surface coverages of at least 3.0-3.5 μmoles/m^2 are often required to eliminate unwanted residual SiOH interactions.

Buffering the aqueous mobile phase at a particular pH to suppress solute ionization (so-called *ion suppression*) is a technique often used to reduce band tailing in reverse-phase BPC. Figure 7.14 shows the reverse-phase separation of some weak acids, in which an acidic (pH = 2) mobile phase was used to inhibit ionization. In this case, separation was made with the free acids to maintain good peak symmetry. Addition of mineral acids (e.g., sulfuric or phosphoric) or acetic or formic acid to the mobile phase is used to produce symmetrical peaks of the acidic sample compounds in reverse-phase BPC. Figure 7.15 provides

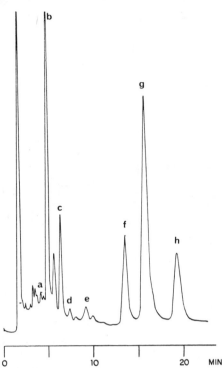

Figure 7.14 Suppression of solute ionization in reverse-phase BPC in separation of weak acids. Column, 25 × 0.49 cm, ODS-modified Partisil 5 (15% ODS); mobile phase, methanol/ 0.02 N sulfuric acid (4:1); flowrate, 2 ml/min; detector, UV, 254 nm; sample: (a) cannabidiol; (b) cannabidiolic acid; (c) cannabinol; (d) Δ^9-tetrahydrocannabinol; (e) cannabichromene; (f) cannabinolic acid; (g) Δ^9-tetrahydrocannabinolic acid; (h) cannabichromenic acid. Reprinted from (37) with permission.

specific example of this effect. Here dodecanoic acid as sample is separated in a reverse-phase system. With either methylene chloride (a) or water (b) added to the acetonitrile mobile phase, band tailing is quite pronounced. However, addition of acetic acid to the mobile phase (c) yields a symmetrical band, because of ion suppression.

Various salts are also added to the aqueous mobile phase in reverse-phase BPC to reduce band tailing, and this approach is especially effective with packings that are not fully silanized. Typically, 0.1% of a salt such as ammonium carbonate is used to improve band symmetry. Tetraalkylammonium salts (e.g., tetramethylammonium nitrate) and various organic anions (e.g., ammonium formate) also have been used for this purpose. Presumably, the salt reduces the tendency of susceptible solutes to interact by ion exchange with residual SiOH groups on the surface of the packing.

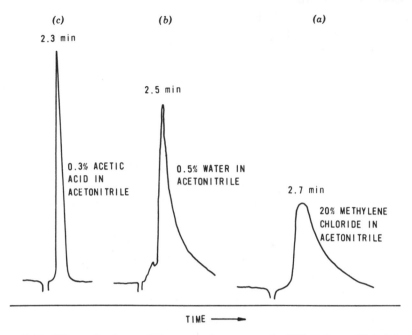

Figure 7.15 Effect of polar modifiers on peak symmetry in BPC. Column, 30 × 0.4 cm, μ-Bondapak-C_{18}; flow rate, 2.0 ml/min; detector, refractive index; sample, 100 μl of ~1 mg/ml solution of dodecanoic acid in mobile phase. Reprinted from (*38*) with permission.

The lifetime of a BPC column depends on the coverage and functionality of the stationary phase, the mobile phase, and the care with which the column is used. Reverse-phase packings with bonded hydrocarbon phases are most stable and generally give 3-6 months of satisfactory performance with proper use. A very wide range of aqueous and nonaqueous mobile phases can be used with these packings, providing the pH is within the 2-8.5 range. (Alkali- and tetra-alkyl-hydroxides *must* be avoided.) The stability of reverse-phase packings appears to be poorer with the use of salts in the mobile phase (e.g., in ion-pair chromatography), and fully reacted packings generally are more stable than partially reacted materials (*39*). Phosphates appear to degrade silane-modified packings more rapidly than mobile phases containing anions such as acetates, sulfates, nitrates, and borates.

The stability of normal-phase BPC columns with potentially reactive functional groups is generally less than that of hydrocarbon BPC packings for reverse-phase separation. For example, columns of bonded primary amines are quickly degraded at pH $\geq$ 7, or if the sample contains aldehydes, because of the formation of Schiff bases. Other unwanted chemical reactions can similarly alter the bonded phase, leading to different retention characteristics. Column degradation by unwanted reactions often can be controlled by utilizing an appropriate guard column (Section 5.6). For example, if a small-particle, primary-amine BPC column is reacting with sample components, a guard column of a pellicular primary-amine packing can be useful.

Gradual deterioration of BPC columns sometimes occurs as a result of the slow reaction of the bonded organic phase with traces of oxygen. This effect can be retarded by degassing the mobile phase or by carefully sparging it with an inert gas. Recent studies (40) suggest that helium is particularly effective in this regard, because of its very low solubility in the mobile phase. As a result, formation of microbubbles of gas in the detector is also reduced, with better detector noise levels.

BPC columns also can be altered by continued injection of samples with strongly retained (noneluted) components. Buildup of such materials on the column inlet can lead to unsymmetrical or broadened peaks, variation in solute k' values, and decreased column resolution. This process can again be controlled by using an appropriate guard column (Section 5.6). In some cases, analytical columns can be regenerated by purging with a very strong solvent for several hours. Short-chain alcohols normally are used to regenerate normal-phase BPC packings, while a 1:1 methanol/chloroform mixture can be beneficial with reverse-phase columns. Occasionally, solvents such as acetone, dimethyl-formamide, and aqueous mineral acids ($\sim$0.01 N) have been helpful with columns not responding to these regeneration treatments. If regeneration is unsuccessful, the affected column should be replaced.

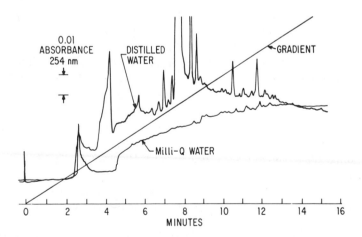

Figure 7.16 Comparison of distilled and Milli-Q-treated water by reverse-phase gradient elution. Column, 30 × 0.4 cm. μ-Bondapak-C$_{18}$; mobile phase, acetonitrile; gradient profile, linear as shown by straight line in figure; flowrate, 4 ml/min; detector, UV, 254 nm, 0.1 AUFS; sample, 160 ml of water. Reprinted from (43a) with permission.

Solvents used as mobile phases in BPC should be purified (Section 6.6); sources of adequately pure commercial solvents are listed in Table 6.1. A particular problem in reverse-phase BPC is the purity of water used in gradient elution separations. Ordinary distilled water often contains trace organics and develops microorganism growth on standing. These trace contaminants are concentrated at the column inlet when water or water-rich mobile phases are used. If the concentration of organic solvent (e.g., acetonitrile) is increased (e.g., in gradient elution), these trace materials are eluted as a series of peaks throughout the chromatogram. The intensity of these peaks depends on the volume of impure water passed through the column prior to the start of the gradient. A satisfactory solution to this problem is to use highly purified water [e.g., that provided by the Milli-Q system (Millipore Corp., Bedford, Mass) (Section 6.6)]. Figure 7.16 illustrates the advantage of using Milli-Q-treated water, as opposed to ordinary distilled water, in blank (no sample) runs by reverse-phase gradient elution BPC.

If samples are injected as solutions in a solvent that is stronger than the mobile phase, the shape and resolution of early-eluting bands are usually adversely affected. The reason is that the solvent mixes with mobile phase and effectively increases its strength over a small region of the column. This "plug" of enriched mobile phase then carries the early-eluting bands through the column in an effectively stronger solvent, and decreases the average k' values of the sample bands during their elution. Therefore, injected sample solutions preferably should be prepared in the initial mobile phase used for the separation, or in a solvent that is weaker than the mobile phase. In some cases, small (e.g., <10 μl)) volumes of sample in a stronger solvent can be injected, particularly if bands with $k' < 1$ are of no interest. However, this approach must be used with caution, and it should be confirmed that the chromatogram does not change significantly for still smaller-volume injections (apart from decrease in peak heights).

Poorly soluble samples sometime present a problem for all forms of LC, in that it is difficult to inject large enough quantities of the sample for adequate detection sensitivity. One solution to this problem is to select a weaker solvent, then inject a relatively large volume of dilute sample solution. For example, a particular sample may not be completely soluble at the 1 mg/ml concentration in the initial mobile phase (intended 25 μl injection). However, it may be possible to prepare a 0.1 mg/ml (true) solution and inject 250 μl sample with no loss in resolution (Section 7.4). Fairly large sample volumes (e.g., > 1 ml) can often be injected when the solvent used is much weaker than the mobile phase (Section 13.5).

In reverse-phase BPC, use of mobile phases with low concentrations (0-10%) of organic solvents often produces broad peaks. Presumably, this is because of the poor wetting of the packing, or the slow equilibrium across the interface of the two highly dissimilar (hydrocarbon and aqueous) phases. The bonded hydro-

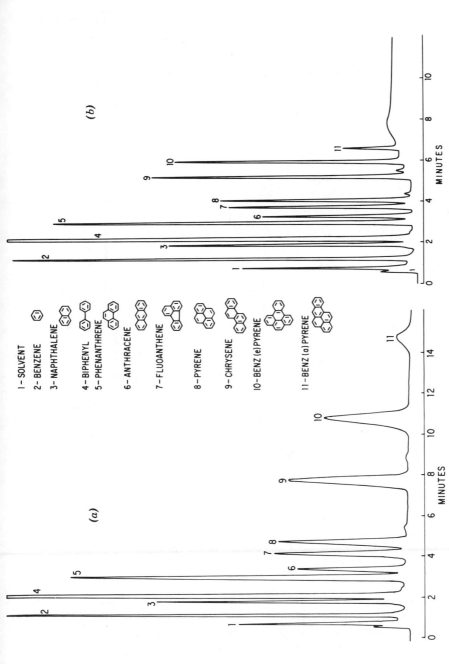

Figure 7.17 Reverse-phase separation of fused-ring aromatics. Column, 25 × 0.32 cm, C$_{18}$-modified porous-silica microspheres (18% carbon), 7 μm; flowrate, 1.4 ml/min; temp., 50°C; detector, UV, 254 nm, 0.32 AUFS; samples, 10 μl of test mixture in methanol. (*a*) Isocratic: mobile phase, 87.5% methanol/water (*b*) Gradient elution: mobile phase, linear gradient, 62.5% methanol/water to 100% methanol at 6%/min. Reprinted from (*9*) with permission.

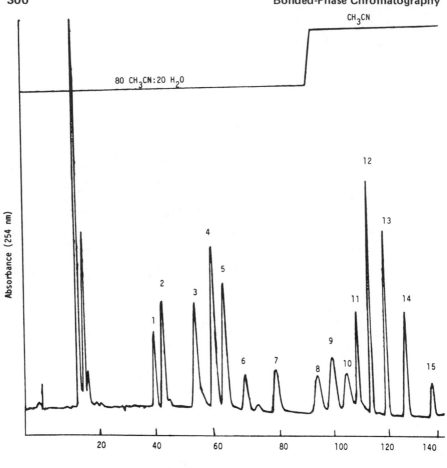

Figure 7.18 Reverse-phase separation of fatty acid phenacyl esters. Column, 90 × 0.64 cm, μ-Bondapak-C_{18}; mobile phase, acetonitrile/water; flowrate, 2.0 ml/min; detector, UV, 254 nm. Peak (1) lauric (12:0), (2) myristoleic (14:1); (3) linolenic (18:3); (4) myristic (14:0); (5) archidionic (20:4); (6) linoleic (18:2); (7) eicosatrienoic (20:3); (8) palmitic (16:0); (9) oleic (18:1, Δ^9); (10) petroselinic (18:1, Δ^6); (11) eicosadienoic (20:2); (12) stearic (18:0); (13) arachidic (20:0); (14) behenic (20:0); (15) lignoceric (24:0). Reprinted from (42) with permission.

carbon packings, particularly fully reacted C_{18} materials, are especially hydrophobic. Consequently, in using such packings in reverse-phase separations (particularly at the beginning of gradient elution), it is desirable to use aqueous mobile phases that contain some minimum concentration of organic solvent to improve wetting of the packing. Recent studies indicate that as much as 40-50%v of organic solvent may be required to allow some reverse-phase packings to be

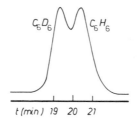

Figure 7.19 Separation of toluene-deuterotoluene by reverse-phase BPC. Column, 100 ×
0.21 cm, ODS-Permaphase; mobile phase, water/methanol (9:1); flowrate, 0.2 ml/min; temp.,
20°C; detector, UV, 254 nm. Reprinted from (43) with permission.

operated at highest efficiency (41). At higher temperatures, these same columns
tolerate less organic solvent in the mobile phase, with a minimum concentration
of 10%v suggested (9).

7.6 APPLICATIONS

BPC is presently the most widely used form of LC and can be applied to a very
wide range of sample types. The convenience of BPC columns, particularly in the
reverse-phase mode, has made this particular technique virtually a standardized
form of LC, which is used first, before other LC methods are tried. The original
applications of reverse-phase BPC included the separation of such nonpolar compounds as hydrocarbons. Figure 7.17 illustrates the use of a bonded-C_{18} column
for separating a complex mixture of fused-ring aromatics by both isocratic and
gradient elution operation. Traces of such compounds can be determined (e.g.,
as atmospheric contaminants) using this approach with a highly selective (e.g.,
fluorometric) detector. Lipophilic compounds (e.g., triglycerides, alkenes) that
have very poor solubility in aqueous reverse-phase solvents often can be separated by nonaqueous reverse-phase (NARP) chromatography (41a). With this
approach C_{18}-BPC columns are used with polar organic solvents such as acetonitrile or tetrahydrofuran (see Figure 4.18).

Reverse-phase columns also provide excellent resolution for a number of dif-
ficult-to-separate fatty acids, as shown in Figure 7.18. Here UV-absorbing
phenacylester derivatives have been used to obtain detection sensitivity in the
nanogram range. In this separation a step-gradient was employed, as indicated at
the top of the chromatogram. Reverse-phase BPC has also been used to separate
hydrocarbons with varying degrees of isotopic substitution (Figure 7.19). Separation factors for compounds of this type are usually greater than observed in
the gas chromatographic separations of the same species.

BPC has had particular impact in the area of pharmaceutical and drug analysis,
where it is used routinely for a wide variety of assays. Figure 7.20 shows a

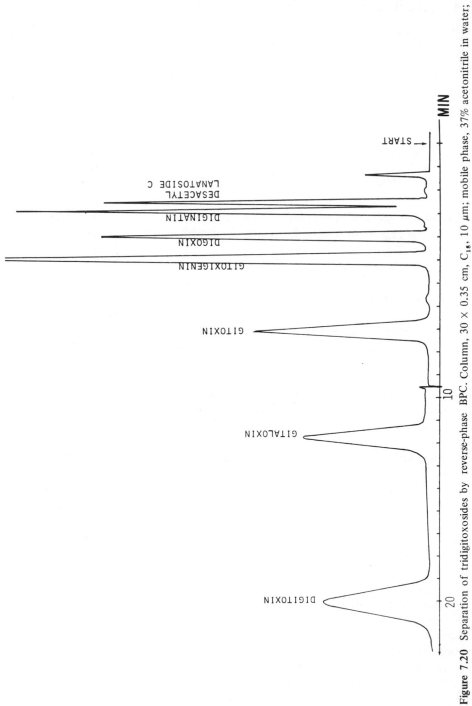

Figure 7.20 Separation of tridigitoxosides by reverse-phase BPC. Column, 30 × 0.35 cm, C₁₈, 10 μm; mobile phase, 37% acetonitrile in water; flowrate, 1.4 ml/min; detector, UV, 220 nm; sample 25 μl of solution. Reprinted from (44) with permission.

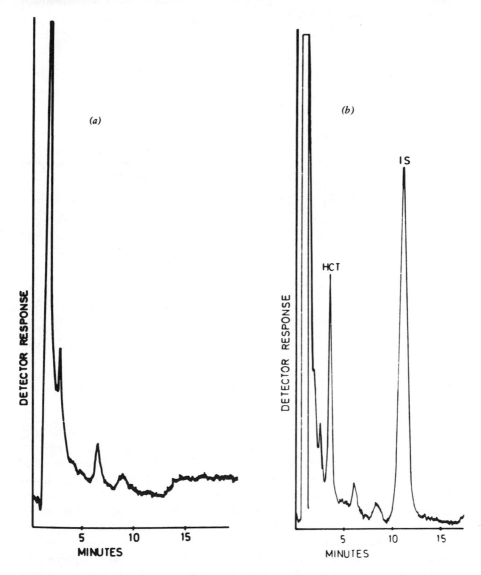

Figure 7.21 Analysis of blood serum for HCT, hydrochlorothiazide. Column, 25 × 0.3 cm, Spherisorb ODS, 10 μm; mobile phase, 15% methanol in water; flowrate, 2.1 ml/min; temp., ambient; detector, UV, 280 nm. (*a*) Serum blank (control); (*b*) patient serum containing 458 ng of hydrochlorothiazide per ml (3 hr after dosage); (IS) internal standard. Reprinted from (*45*) with permission.

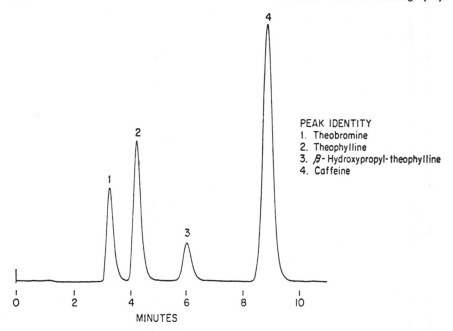

Figure 7.22 Separation of theophylline and related compounds by reverse-phase BPC. Column, 25 × 0.46 cm, Zorbax-ODS; mobile phase, 20% acetonitrile−1% tetrahydrofuran−0.1% H_3PO_4/80% water; column pressure, 1200 psi; temp., ambient; detector, UV, 254 nm, 0.08 AUFS. Reprinted from (*25*) with permission.

separation of a series of tridigitoxosides, using reverse-phase chromatography with a C_{18}-microparticle column. Reverse-phase BPC is particularly useful for applications involving biological fluids. Figure 7.21 gives chromatograms of (*a*) blank serum and (*b*) the serum of a patient treated with the drug hydrochlorothiazide. An internal standard was added to the serum for quantitative measurement (Section 13.3). In this case the drug was first extracted from the serum with ethyl acetate; the extract was then evaporated and redissolved for injection. In other applications it sometimes is possible to inject serum samples directly without prior solvent extraction, providing a guard column is used to protect the analytical system (Section 5.6). Such an approach has been used to analyze theophylline and related compounds for therapeutic monitoring. The separation of a synthetic mixture of such materials is shown in Figure 7.22.

Polar bonded-phase columns also have been used to analyze a wide variety of biologically important materials. These BPC packings are used in the normal-phase mode in much the same manner as LSC columns, but generally with more convenience; control of polar modifiers (especially water) in the mobile phase

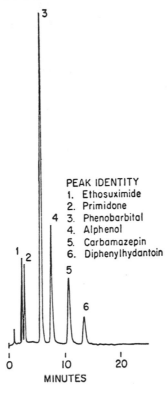

Figure 7.23 Separation of anticonvulsants by reverse-phase with a polar bonded-phase column. Column, 25 × 0.46 cm, Zorbax-CN; mobile phase, 20% acetonitrile–0.1% NaH_2PO_4/80% water; pressure, 1800 psi; temp., ambient; detector, UV, 254 nm, 0.08 AUFS. Reprinted from (25) with permission.

is less critical. Figure 7.7 shows the normal-phase separation of a mixture of drugs on a polar "nitrile" bonded-phase column. Such polar bonded-phase packings also can be used in a reverse-phase mode for separating polar compounds. For example, in Figure 7.23 we see a series of anticonvulsant drugs separated with a "nitrile" bonded-phase column in the reverse-phase mode, using a water-acetonitrile mobile phase. Pellicular normal-phase packings also have been used to separate biologically important materials. In Figure 7.24 a series of estrogen standards is separated on a pellicular Permaphase-ETH column. Elevated column temperature was used here to improve separation efficiency.

A large number of naturally occurring samples have been separated by BPC. Figure 7.25 shows the separation of phenolic constituents of a flue-cured tobacco on a reverse-phase column, using gradient elution. A number of the

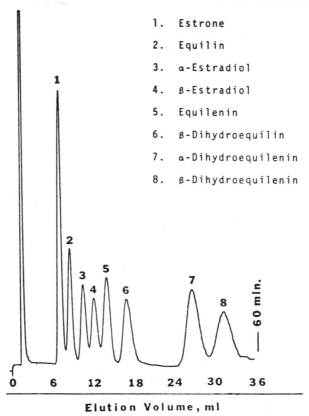

1. Estrone
2. Equilin
3. α-Estradiol
4. β-Estradiol
5. Equilenin
6. β-Dihydroequilin
7. α-Dihydroequilenin
8. β-Dihydroequilenin

Figure 7.24 Separation of estrogen standards on a pellicular normal-bonded-phase column. Column, 100 × 0.21 cm, Permaphase ETH; mobile phase, 1% 2-propanol in *n*-heptane; flowrate, 0.6 ml/min; temp., 75°C; detector, UV, 280 nm; 0.16 AUFS. Reprinted from (*46*) with permission.

component peaks were isolated and identified in this study. For instance, chlorogenic acid and rutin (peaks 2 and 18, respectively) were found to be the major phenolic constituents.

Simple extraction techniques in combination with reverse-phase BPC often can solve rather complex problems, as illustrated in Figure 7.26. In chromatogram *a* we see the separation of an aqueous opium extract that had been washed with methylene chloride under basic conditions to remove all the extractable opium alkaloids. Figure 7.26*b* shows a chromatogram of a methylene chloride extract after the solvent was evaporated and the residue redissolved in the mobile phase for reverse-phase chromatography. In this problem it was not pos-

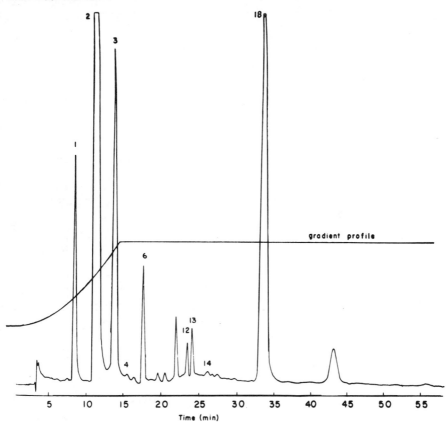

Figure 7.25 Separation of phenolic constituents from flue-cured tobacco. Column, 30 × 0.4 cm, μ-Bondapak-C$_{18}$; mobile phase, 0.1 N KH$_2$PO$_4$-methanol gradient; flowrate, 1.0 ml/min; detector, UV, 350 nm, 0.2 AUFS; sample 5 μl. Reprinted from (47) with permission.

sible to inject the crude opium extract directly without serious band overlap. Separating a mixture into two component parts made this rather complex problem relatively simple.

With reverse-phase BPC, it is often possible to inject aqueous samples without further treatment (except filtration). For example, in Figure 7.27 we see a chromatogram of an artificially sweetened soft drink, the components of which were easily quantitated without interference and with no sample pretreatment.

The availability of polar BPC packings has greatly simplified the analysis of certain highly polar samples. For example, in Figure 7.28 we see the separation of the sugars from a cellulose hydrolysate, using a primary-amine BPC column.

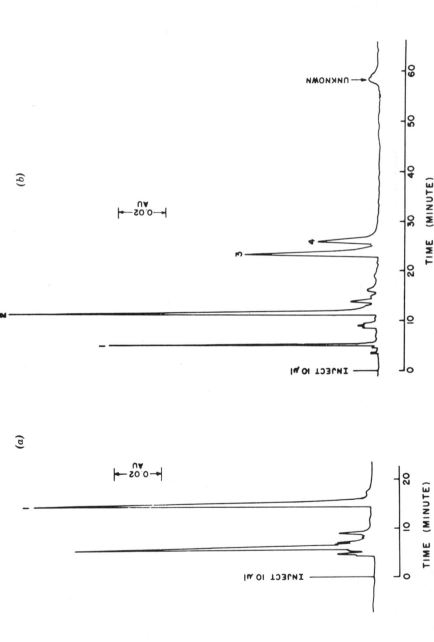

Figure 7.26 Separation of opium extracts. Column, 60 × 0.4 cm, μ-Bondapak-C$_{18}$; mobile phase, 0.1 M NaH$_2$PO$_4$ in 25% acetonitrile/water, pH 4.8; flowrate, 1.25 ml/min; temp., ambient; detector, UV, 286 nm. (*a*) Aqueous Indian gum opium extract containing the equivalent of 0.6 g opium in 25 ml. Peak identity: (1) morphine. (*b*) Methylene chloride extract of Indian gum opium containing the equivalent of 0.24 g opium in 10 ml. Peak identity: (1) codeine, (2) thebaine, (3) papaverine and (4) noscapine. Reprinted from (*48*) with permission.

308

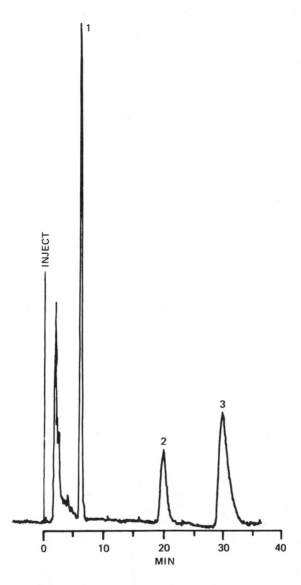

Figure 7.27 Analysis of an artificially sweetened soft drink. Column, 30 × 0.4 cm, μ-Bondapak-C$_{18}$; mobile phase, 5% acetic acid solution; flowrate, 2.0 ml/min; temp., ambient; detector, UV, 250 nm, 0.02 AUFS; sample, 10 μl; components: (1) saccharin, (2) benzoic acid, (3) caffeine. Reprinted from (49) with permission.

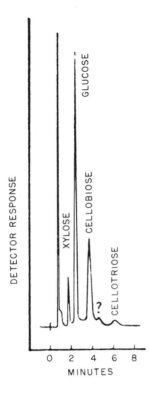

Figure 7.28 Separation of sugars in a cellulose hydroly-sate. Column, 30 × 0.4 cm, μ-Bondapak-Carbohydrate; mobile phase, water/acetonitrile (25:75); flowrate, 3.0 ml/min; temp., 25°C; detector, refractive index; sample, 20 μl. Reprinted from (*50*) with permission.

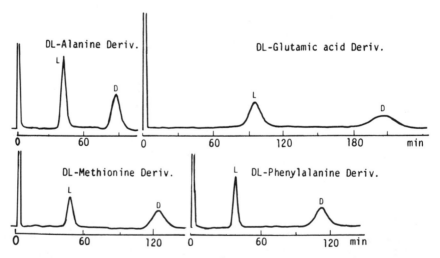

Figure 7.29 Normal bonded-phase separation of diastereomers of *N-d*-10-camphorsul-phonyl *p*-nitrobenzyl amino acids. Column, 25 × 0.22 cm, MicroPak-NH$_2$, 10 μm; mobile phase, dichloromethane; flowrate, 0.4 ml/min; temp., ambient; detector, UV, 254 nm. Reprinted from (*51*) with permission.

310

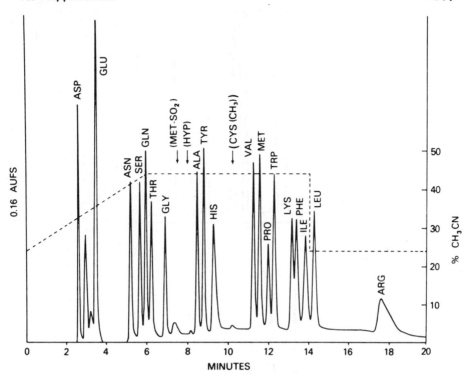

Figure 7.30 Gradient elution of PTH-amino acids by reverse-phase BPC. Column, 25 × 0.46 cm, Zorbax-ODS; mobile phase: gradient elution, 0.01 N sodium acetate (pH 4.5)/ acetonitrile; initial flowrate, 1.0 ml/min; temp., 62°C; detector, UV, 254 nm; sample, 1–2 nmoles of each PTH-amino acid. Reprinted from (52) with permission.

Such an approach has been useful in determining the role of individual carbohydrates in animal and human nutrition. In many separations of this type the only sample pretreatment required is filtration prior to injection. However, a guard column generally should be used to protect this rather reactive bonded-phase column. Similar NH_2-bonded-phase packings have been used to separate diasteromeric D- and L-amino acids, as illustrated in Figure 7.29. In this case the amino acids were reacted first with d_{10}-camphorsulphonyl chloride to form the diastereomers, and then with p-nitrobenzylbromide to introduce the chromophoric p-nitrobenzyl group. The resulting derivatives were chromatographed in the normal-phase mode.

Samples that can be separated by ion-exchange or ion-pair chromatography can often also be separated by BPC, and usually more conveniently. For example, in Figure 7.30 we see the gradient elution separation of the

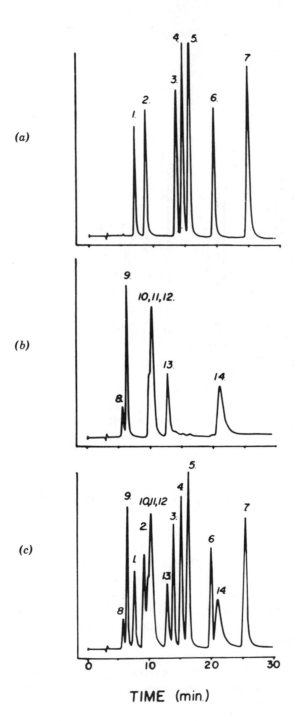

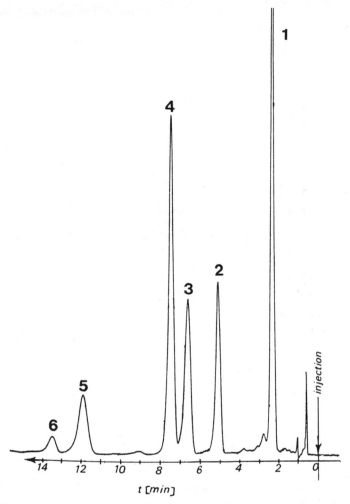

Figure 7.32 Separation of nonapeptides by reverse-phase BPC. Column, 15 × 0.4 cm, Nucleosil C$_8$, 5 μm; mobile phase, 20% acetonitrile in phosphate buffer (pH 7); flowrate, 2.0 ml/min; temp., ambient; detector, UV, 210 nm; peaks (1) ornipressine and lypressine (not separated) (2) oxytocine, (3) felypressine, (4) nipagine M, (5) demoxytocine, (6) trichlorobutanol. Reprinted from (*30*) with permission.

Figure 7.31 Separation of nucleosides and corresponding bases by reverse-phase BPC. Column, 30 × 0.4 cm μ-Bondapak-C$_{18}$; mobile phase, 0.01 M KH$_2$PO$_4$, pH 5.5/methanol-water (80:20), linear gradient, 0.25% of methanol/water in 30 min; flowrate, 1.5 ml/min; temp., ambient; detector, UV, 254 nm, 0.02 AUFS. (*a*) Separation of 5 nmoles each of (1) cytidine, (2) uridine, (3) xanthosine, (4) inosine, (5) guanosine, (6) thymidine, (7) adenosine. (*b*) Separation of 5 nmoles each of (8) cytosine, (9) uracil, (10) hypoxanthine, (11) guanine, (12) xanthine, (13) thymine, (14) adenine. (*c*) Separation of a mixture of the nucleosides and the bases. Reprinted from (*53*) with permission.

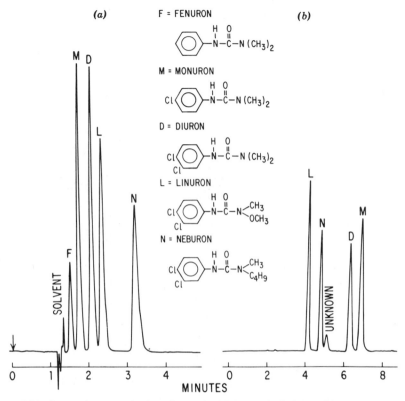

Figure 7.33 Separation of substituted-urea herbicides. (*a*) Column, 23 × 0.46 cm, C$_8$-modified porous-silica microspheres, 8.4 μm; mobile phase, 75% methanol/25% water; flowrate, 2.0 ml/min; temp., 50°C; detector, UV, 254 nm, 0.16 AUFS; sample, 25 μl, 0.1 mg/ml each. (*b*) Column, 25 × 0.46 cm, Zorbax-CN; mobile phase, 20% tetrahydrofuran/80% hexane; pressure, 400 psi; temp., ambient; detector, UV, 254 nm, 0.64 AUFS. Reprinted from (*25*) with permission.

phenythiohydantoin derivatives of the 20 amino-acids in less than 20 min using a reverse-phase system. At least three analyses can be performed per hour, making it possible to keep pace with an automated Edman protein sequencer. Nucleic acids and nucleosides also can be analyzed by reverse-phase BPC, as illustrated in Figure 7.31. Chromatogram (*a*) shows a separation of several nucleosides, chromatogram (*b*) a series of corresponding bases, and chromatogram (*c*) a mixture of both the nucleosides and the bases. Peptides also can be separated by reverse-phase BPC, as illustrated in Figure 7.32. Here we see that nipagine *M* and trichlorobutanol (peaks 1 and 6, respectively), often present as preservatives in liquid formulations, can be readily separated from several nonapeptides.

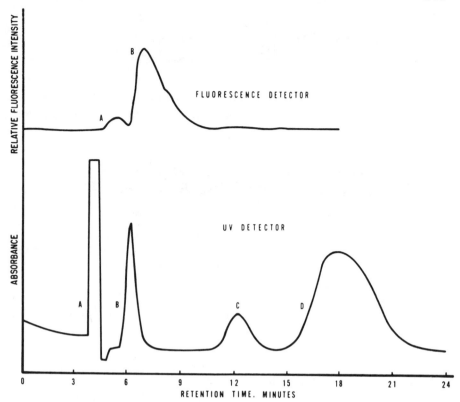

Figure 7.34 Reverse-phase determination of selenium and nitrite. Column, 30 × 0.4 cm, μ-Bondapak-C_{18}; mobile phase, ethanol (95%)–water (1.25:2); flowrate, 0.6 ml/min; temp., ambient; detectors, UV, 254 nm and fluorimeter; peaks: (*A*) sample matrix; (*B*) triazole, 1 mg/1; (*C*) DANSe, 3 mg/1; (*D*) excess DAN. Reprinted from (*54*) with permission.

Many pesticides also have been handled successfully by BPC, both for high precision assay and for trace analysis in environmental studies. In Figure 7.33 chromatogram (*a*) is a synthetic mixture of urea herbicides separated on C_8 reverse-phase column. Chromatogram (*b*) is a similar mixture separated by normal phase using a polar bonded-phase packing. The reversed elution order of the compounds in these chromatograms illustrate the great versatility of BPC, utilizing in this case changes in both stationary and mobile phases.

Reverse-phase BPC has been used to measure both cations and anions, by first forming derivatives with certain reagents. Selenium (as selenous acid) reacts with 2,3-diaminonaphthalene (DAN) to produce the highly UV-absorbing and fluorescing product, naphtho-[2,3,-d]-2-selena-1,3-diazole (DANSe):

DAN also can be used as a spectrophotometric and fluorometric reagent for the analysis of nitrite ion, based on the formation of 2,3-naphthotriazole ("triazole"):

Figure 7.34 shows the fluorescence and ultraviolet signals for the separation of a sample of DAN-reacted nitrite and selenium (IV) (1 mg/l each). In this case the amount of nitrite present is determined by measuring the "triazole" peak (*B*), and the amount of selenium by the DANSe peak (*C*).

Reverse-phase BPC is useful for characterizing certain oligomeric mixtures, as illustrated in Figure 7.35. Gradient elution (Section 16.1) is usually required for the adequate resolution of such samples.

7.7 THE DESIGN OF A BPC SEPARATION

Selecting the Column

When attempting a separation, a reverse-phase, bonded-hydrocarbon (i.e., dimethylalkyl) column should be selected first, because such columns have the widest applicability. Fully reacted monomeric bonded C_8 packings represent a good compromise for reverse-phase separations, because these materials have moderate retention, good efficiency and stability, and a useful k' range for a wide variety of samples. C_{18} can be used for applications in which maximum retention and sample size is required. C_{18} packings sometimes also exhibit superior characteristics for compounds that have higher water solubility. Shorter-chain bonded hydrocarbon phases (e.g., SAS $\equiv C_1$) are useful in applications involving very strongly retained solutes, or to improve selectivity by the use of the higher concentrations of water required in the mobile phase for these packings.

As in LLC with mechanically held stationary liquids, retention in normal-phase BPC increases with the polarity of the bonded-stationary phase. Depending on the organic functionality, polar BPC packings show significant selectivity differences when compared to bare silica packings and to each other,

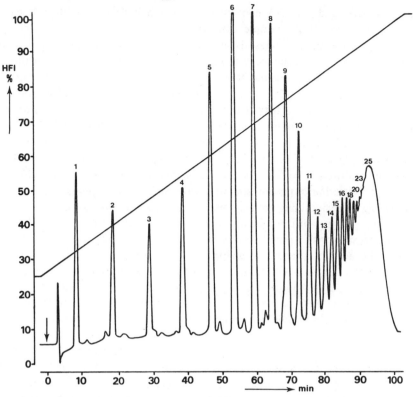

Figure 7.35 Reverse-phase separation of poly(ethylene terephthalate) oligomers. Column, 30 × 0.39 cm, μ-Bondapak-C_{18}; mobile phase, solvent gradient (straight line) from 25% to 98% hexafluoroisopropanol in water; flowrate, 0.8 ml/min; detector, UV at 270 nm, 0.5 AUFS; sample, 200 μg in hexafluoroisopropanol. Reprinted from (55) with permission.

and some polar bonded-phase packings are actually more retentive than bare silica (4).

Normal-phase BPC columns can be used as an alternative to adsorption chromatography, and one of the most versatile of these materials is the nitrile (CN or cyano-) bonded-phase materials. Separation of very polar (including water-soluble) samples may require packings such as diol- or amino-BPC packings. However, if the sample itself is aqueous, a reverse-phase system should be attempted first, because this approach offers greater sampling convenience.

Selecting the Mobile Phase

Water/methanol and water/acetonitrile mixtures have been most popular for reverse-phase BPC. The optimum mobile-phase composition generally must be found by trial and error. A convenient approach is to start with 1:1 water/methanol mixture. If the sample components elute at or near t_o, a lower concentration of the organic phase is indicated. Conversely, if the sample is too strongly retained, the organic solvent concentration must be increased. Although methanol is a preferred organic component for reverse-phase BPC, changing to acetonitrile, dioxane, or mixtures of dioxane/methanol, acetonitrile/isopropanol, and so on, can alter selectivity. Different concentrations of the various organic solvents are required to maintain constant solvent strength (Section 6.5).

Separation selectivity in reverse-phase BPC can be drastically altered by varying the pH of the mobile phase, if the sample components are acids or bases. The addition of salts to the mobile phase also can influence retention and selectivity. Buffering or pH adjustment is also used for ion suppression to reduce band tailing by maintaining sample components in a nonionized form.

A general scheme for optimizing the mobile phase in normal-phase BPC (e.g., nitrile) is to start with a pure-hydrocarbon mobile phase. Heptane is a good compromise, because it is readily available, cheap, of low viscosity, and not highly volatile. If the sample is unretained in this solvent, as evidenced by a large peak at t_0, a polar bonded-phase packing is not a good candidate for this sample, and reverse-phase BPC is indicated. If the sample is strongly retained, with no sign of the main sample components after $k' = 15$-20, a more polar mobile phase should be employed. This is accomplished by adding the appropriate concentration of polar modifiers (e.g., methanol, chloroform, dioxane) to obtain elution of the desired sample components in the preferred $k' = 1$-10 range. For highly basic samples, addition of basic modifiers to the mobile phase may be required to eliminate band tailing. Tertiary amines such as triethylamine are preferred over primary or secondary amines. Conversely, organic acids such as trichloroacetic, formic, and acetic may be required as mobile-phase modifiers to eliminate the tailing of bands for acidic compounds.

If a variation in separation selectivity is needed, the polar modifying solvent (constituent B) in the mobile phase must be changed. In many cases, modifying the nonpolar mobile phase with either diisopropylether, methylene chloride, or chloroform, representing typical B-solvents from the apex of the solvent triangle (Figure 6.4, groups I, V, VII, respectively) produces the desired change in selectivity. Changes in the polar solvent B also require adjustment of solvent composition (%vB) to maintain optimum k' values (Section 6.5).

A gradient elution separation may be required for samples that have components with widely differing k' values. Because of the great stability of the bonded-phase columns, gradient elution can be used with both reverse-phase and normal-phase systems. If a routine quantitative analysis is subsequently to be performed, the gradient elution run can be used first to produce qualitative information on the sample. The strength and selectivity of the mobile phase then can be adjusted to optimize the separation of the components of interest for subsequent routine isocratic quantitative analyses. If selectivity has been optimized and further increase in resolution is required, column efficiency can be increased using the approach described in Section 2.5.

Operating Variables

For reverse-phase BPC systems, elevated temperatures (50-60°C) should be used if the stability of the sample components allows. Normal-phase separations are usually carried out at ambient temperatures.

In LC it is generally desirable that samples not be injected in solvents stronger than the mobile phase. However, small volumes of a somewhat stronger solvent sometimes can be used for poorly soluble samples, at least for bands with k' values greater than about unity.

Optimum mobile-phase flowrate in BPC depends on the column internal diameter, the particular packing, operating temperature, mobile phase, and the separation goal (Section 2.5). However, convenient flowrates are often about 1-2 ml/min for 0.4-0.5-cm i.d. columns of 5-10 μm porous particles and 0.2-0.3-cm i.d. columns of pellicular particles.

Sample weights permitted in BPC generally are similar to those used in LLC and LSC separations. For typical analytical applications, less than 50 μg samples are used on columns of pellicular BPC packings; up to 0.5 mg can be used on BPC columns with porous packings. As with the other forms of LC, the volume of sample injected should not exceed one-third the volume of the first peak of interest. However, larger volumes can be tolerated if the strength of the solvent in which the sample is dissolved is less than that of the initial mobile phase.

REFERENCES

1. R. A. Henry, J. A. Schmit, and R. C. Williams, paper presented at the 163rd National American Chemical Society Meeting, Boston, Mass., April 1972.
2. L. J. Papa and L. P. Turner, *J. Chromatogr. Sci.*, **10**, 747 (1972).
3. J. F. K. Huber, F. F. M. Kolder, and J. M. Miller, *Anal. Chem.*, **44**, 105 (1972).
4. K. K. Unger, N. Becker, and P. Roumeliotis, *J. Chromatogr.*, **125**, 115 (1976).
5. D. C. Locke, J. T. Schmermund, and B. Banner, *Anal. Chem.*, **44**, 90 (1972).
6. P. Magidman, D. H. Saunders, R. A. Barford, L. T. Olszewski, and H. L. Rothbart, *ibid.*, **46**, 834 (1974).
7. I. Sebestian, O. E. Brust, and I. Halasz, in *Gas Chromatography, Montreux*, S. G. Perry, ed., Applied Science Publishers, Barking, U.K., 1972, p. 281.
8. O. E. Brust, I. Sebestian, and I. Halasz, *J. Chromatogr.*, **83**, 15 (1973).
9. J. J. Kirkland, *Chromatographia*, 8, 661 (1975).

10. J. J. Kirkland and P. C. Yates, U. S. Patent 3,722,181, March 27, 1973; U. S. Patent 3,795,313, March 5, 1974.

11. J. J. Kirkland, *J. Chromatogr. Sci.*, 9, 206 (1971).

12. L. Boksanyi, O. Liardon, and E.sz. Kovats, *J. Adv. Colloid Interface Sci.*, 6, 95 (1976).

13. I. Halasz and I. Sebestian, *Chromatographia*, 7, 371 (1974).

14. F. E. Regnier, U. S. Patent 3,983,299, September 28, 1976.

15. K. Unger, N. Becker, and E. Kramer, *Chromatographia*, 8, 283 (1975).

16. R. K. Gilpin, D. J. Camillo, and C. A. Janicki, *J. Chromatogr.*, 121, 13 (1976).

17. J. J. Kirkland and P. E. Antle, *J. Chromatogr. Sci.*, 15, 137 (1977).

18. A. Ahmed, E. Gallei, and K. Unger, *Ber. Bunsenges. Phys. Chem.*, 79, 66 (1975).

19. L. T. Zhuravlev, A. V. Kiselev, and V. P. Naidina, *Russ. J. Phys. Chem.*, 42, 1200 (1968).

20. K. Unger and E. Gallei, *Kolloid-Z. Z. Polym.*, 237, 358 (1970).

21. K. Karch, I. Sebestian, and I. Halasz, *J. Chromatogr.*, 122, 3 (1976).

22. F. Riedo, M. Czencz, O. Liardon, and E.sz. Kovats, *Helv. Chim. Acta*, 61, 1912 (1978).

22a. J. H. Knox and A. Pryde, *ibid.*, 112, 171 (1975).

23. H. Hemetsberger, W. Maasfeld, and H. Ricken, *Chromatographia*, 9, 303 (1976).

24. J. J. Kirkland, *Anal. Chem.*, 43, 36A (1971).

25. J. J. DeStefano, E. I. du Pont de Nemours & Co., Wilmington, Del., unpublished results, 1977.

26. N. W. Tymes, *J. Chromatogr. Sci.*, 15, 151 (1977).

26a. L. R. Snyder, J. W. Dolan, and J. R. Gant, *J. Chromatogr.*, 165, 3 (1979).

27. B. L. Karger, J. R. Gant, A. Hartkopf, and P. H. Weiner, *ibid.*, 128, 65 (1976).

28. S. R. Bakalyar, R. McIlwrick, and E. Roggendorf, *ibid.*, 142, 353 (1977).

28a. N. Tanaka, H. Goodell, and B. L. Karger, *ibid.*, 158, 233 (1978).

28b. P. J. Schoenmakers, H. A. H. Billiet, R. Tjissen, and L. de Galen, *ibid.*, in press.

28c. S. van der Wal and J. F. K. Huber, *ibid.*, 149, 431 (1978).

29. P. J. Twitchett and A. C. Moffat, *ibid.*, 111, 149 (1975).

30. K. Krummen and R. W. Frei, *ibid.*, 132, 27 (1977).

31. C. Horvath, W. Melander, and I. Molnar, *ibid.*, 125, 129 (1976).

32. C. Horvath and W. Melander, *J. Chromatogr. Sci.*, 15, 393 (1977).

33. J. C. Sternberg, in *Advances in Chromatography*, Vol. 2, J. C. Giddings and R. A. Keller, eds., Dekker, New York, 1966, p. 205.

34. J. N. Done, *J. Chromatogr.*, 125, 43 (1976).

35. R. E. Majors, in *Bonded Stationary Phases in Chromatography*, E. Grushka, ed., Ann Arbor Science Publishers, Ann Arbor, Mich., 1974, p. 146.

35a. H. Colin, J. C. Diez-Masa, G. Guiochon, T. Czakowska, and I. Miedziak, *J. Chromatogr.*, 167, 41 (1978).

36. B. L. Karger and E. Sibly, *Anal. Chem.*, 45, 740 (1973).

37. B. B. Wheals, *J. Chromatogr.*, 122, 85 (1976).

38. R. M. Cassidy and C. M. Niro, *ibid.*, 126, 787 (1976).

39. R. P. W. Scott, Hoffman-La Roche, Nutley, N. J., personal communication, June 1977.

40. F. W. Karasek, *Res. Dev.*, 28, 32 (1977).

41. R. P. W. Scott and P. Kucera, *J. Chromatogr.*, **142**, 213 (1977).

41a. N. A. Parris, *ibid.*, **149**, 615 (1978).

42. R. F. Borch, *Anal. Chem.*, **47**, 2437 (1975).

43. G. P. Cartoni and I. Ferretti, *J. Chromatogr.*, **122**, 287 (1976).

43a. Millipore Corporation, Milli-Q advertisement, *Anal. Chem.*, June 1977; See also R. L. Sampson, in *Liquid Chromatography of Polymers and Related Materials*, J. Cazes, ed., Dekker, New York, 1976, p. 149.

44. F. Erni and R. W. Frei, *J. Chromatogr.*, **130**, 169 (1977).

45. A. S. Christophersen, K. E. Rasmussen, and B. Salvesen, *ibid.*, **132**, 91 (1977).

46. R. W. Roos, *J. Chromatogr. Sci.*, **14**, 505 (1976).

47. W. A. Court, *J. Chromatogr.*, **130**, 287 (1977).

48. C. Y. Wu and J. J. Wittick, *Anal. Chem.*, **49**, 359 (1977).

49. D. S. Smyly, B. B. Woodward, and E. C. Conrad, *J. Assoc. Off. Anal. Chem.*, **59**, 14 (1976).

50. J. K. Palmer, *Anal. Lett.*, **8**, 215 (1975).

51. H. Furukawa, Y. Mori, Y. Takeuchi, and K. Ito, *J. Chromatogr.*, **136**, 428 (1977).

52. C. L. Zimmerman, E. Appella, and J. J. Pisano, *Anal. Biochem.*, **77**, 569 (1977).

53. R. A. Hartwick and P. R. Brown, *J. Chromatogr.*, **126**, 679 (1976).

54. G. L. Wheeler and P. F. Lott, *Microchem. J.*, **19**, 390 (1974).

55. F. P. B. van der Maeden, M. E. F. Biemond, and P. C. G. M. Janssen, *J. Chromatogr.*, **149**, 539 (1978).

BIBLIOGRAPHY

Bristow, P. A., *LC in Practice*, HETP Publishers, Handforth, Wilmslow, Cheshire, U.K., 1976, p. 62 (brief survey of BPC technology).

Colin, H., and G. Guiochon, *J. Chromatogr.*, **141**, 289 (1977) (an excellent review covering the preparation, use and structure of bonded phases).

Grushka, E., *Bonded Stationary Phases in Chromatography*, Ann Arbor Science Publishers, Inc., Ann Arbor, Mich., 1974 (collection of research papers on bonded-phase technology).

Hemetsberger, F., W. Maasfeld, and H. Ricken, *Chromatographia*, **9**, 303 (1976) (effect of chain length of bonded organic phases in reversed-phase BPC).

Horvath, C., and W. Melander, *J. Chromatogr. Sci.*, **15**, 393 (1977) (theory of interaction between solutes and reverse-phase BPC packings).

Karger, B. L., and E. Sibley, *Anal. Chem.*, **45**, 740 (1973) (effect of unreacted surface hydroxyls on solute retention).

Locke, D. C., *J. Chromatogr. Sci.*, **11**, 120 (1973) (review of various methods for preparing BPC packings in addition to silane reactions).

Majors, R. E., in *Practical High-Performance Liquid Chromatography*, C. F. Simpson, ed., Heyden, London, 1976, Chap. 7 (review of BPC).

Pryde, A., *J. Chromatogr. Sci.*, **12**, 486 (1974) (review of methods for preparing BPC packings).

Regnier, F. E., and R. Noel, *J. Chromatogr. Sci.*, 4, 316 (1976) (polar bonded-phase packings for proteins and nucleic acids).

Unger, K., N. Becker, and E. Kramer, *Chromatographia*, 8, 283 (1975) (modifying of silica
with alcohol groups).

Unger, K. K., *Porous Silica: Its Properties and Use as a Support in Column Liquid Chromatography*, Elsevier, Amsterdam, 1979 (definitive text on the role of silica supports in
BPC).

EIGHT

LIQUID-LIQUID CHROMATOGRAPHY

8.1	Introduction	323
8.2	Essential Features of LLC	325
8.3	Column Packings	327
	Characteristics	327
	Preparation of Coated Supports	328
8.4	The Partitioning Phases	332
8.5	Other Separation Variables	336
	Sample Size	336
	Column Temperature	336
8.6	Special Problems	336
8.7	Applications	338
8.8	The Design of an LLC Separation	342
	Selecting the Column	342
	Selecting the Mobile Phase	344
	Operating Variables	345
References		347
Bibliography		348

8.1 INTRODUCTION

In 1941 Martin and Synge described a new separation technique: liquid-liquid chromatography (LLC), sometimes called liquid-partition chromatography (*1*). Subsequent development of this method led to its application to a wide variety of sample types, including water-soluble and oil-soluble samples and ionic and nonionic compounds. The basics of modern LLC are little different from those

323

of the original method, although improvements in packings, column preparation, and equipment have produced a major increase in separation speed and resolution. In LLC the solute molecules are distributed between two immiscible liquids according to their relative solubilities. One liquid is the *mobile phase* (sometimes called the carrier); the other liquid (*stationary phase*) is dispersed onto a finely divided, usually inert, support. In conventional LLC the stationary phase is a bulk liquid, mechanically held to the support by adsorption. Organic phases that are chemically bonded to the support have become more popular in recent years, leading to a separate LC method: bonded-phase chromatography (BPC), as discussed in Chapter 7. Bonded-phase LC often is more convenient than LLC with a mechanically held stationary phase, and many of the types of separations previously made by LLC now are carried out by BPC. However, LLC remains an important separation approach because LLC columns have certain unique advantages over columns for other LC methods:

- Convenient renewal—packed columns can be easily stripped of stationary phase and recoated *in situ* many times, for longer overall column life—very useful for routine separations.
- High loadability—the pores of the support can be completely filled with a low-viscosity stationary phase, thus providing a maximum stationary-phase volume and allowing maximum sample sizes with highest column efficiencies.
- Broad range of selectivity—resulting from the large number of stationary-phase liquids available (but mobile- and stationary-phase liquids must be immiscible).
- Reproducible columns—because the column support itself plays little role in the retention characteristics of the stationary phase.

In addition, the basic LLC method has been further adapted to yield important special forms of LLC, namely, extraction chromatography and normal-phase ion-pair chromatography (Chapters 10, 11). For these reasons it is important to include a discussion of LLC with mechanically held stationary phases.

The process of LLC resembles simple extraction between two immiscible liquids in a separatory funnel. A successive series of such extractions forms the basis of countercurrent distribution, which is more efficient than a simple one-stage extraction. However, LLC is many times faster and more efficient than countercurrent extraction. This results from the large interface between the moving and stationary phases, for a rapid equilibrium-distribution of sample components. As in extraction or countercurrent distribution, separation of a sample in LLC results from the differing distribution of the various solutes between the two liquid phases. Because of this common basis for both solvent extraction and LLC, extraction data (K values) can be used to predict LLC retention (k' values) (2). Retention in LLC is defined by Eq. 2.5; this shows that for larger k' values (increased retention), either a system has to be selected in which there is greater solubility of the solute in the stationary phase (larger K value) or the volume of the stationary phase must be increased compared to that of the mobile phase.

8.2 ESSENTIAL FEATURES OF LLC

LLC offers unique selectivity for various samples, because a very wide range of liquids can be selected for the stationary phase. Two types of LLC are possible, based on the relative polarities of the mobile and stationary phases. In *normal-phase LLC* the support is coated with a polar stationary phase, whereas a relatively nonpolar solvent is used as the mobile phase. Normal-phase LLC is used for more polar, water-soluble samples, and solute elution order is similar to that observed in adsorption chromatography. Nonpolar solutes prefer the moving

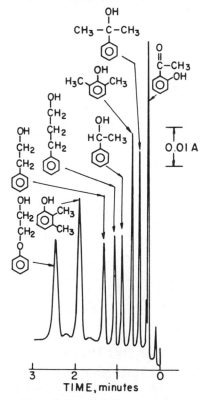

Figure 8.1 Normal-phase liquid-liquid chromatography. Column, 25 × 0.32 cm, 5-μm porous-silica microspheres (~350 Å), with 30% β,β'-oxydipropionitrile by weight; mobile phase, hexane, 27°C; flow, 7.7 ml/min; sample, 15 μl of solution. Reprinted from (*3*) by permission.

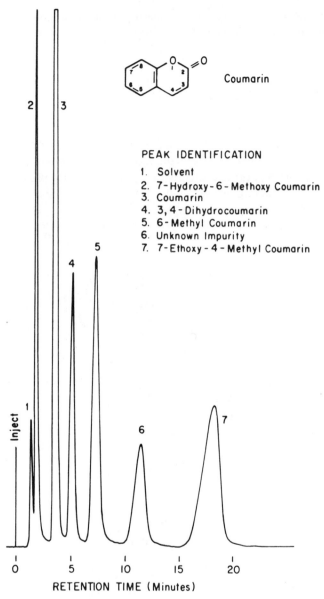

Figure 8.2 Reverse-phase liquid-liquid chromatography. Column, 100 × 0.21 cm, Zipax ANH (cyanoethylsilicone), 38 μm; mobile phase, 1% isopropanol in water; flowrate, 1.0 ml/min; temp., ambient; detector, UV, 254 nm. Reprinted from (4) with permission.

phase and elute first with low k' values; polar solutes prefer the polar stationary phase and elute later with higher k' values. The separation shown in Figure 8.1 illustrates the normal-phase mode of LLC. Retention of these hydroxylated aromatics generally increases with increasing solute polarity.

The two LLC phases can be interchanged so that the less polar liquid is now the stationary phase and the polar liquid is the mobile phase. This form of separation is referred to as *reverse-phase LLC* and is generally used to separate samples with poor water solubility. Usually the elution order of solutes is the reverse of that observed in normal-phase LLC, with polar compounds eluting first and nonpolar solutes eluting later. A reverse-phase LLC separation is illustrated in Figure 8.2 for a mixture of coumarin and derivatives. In this case a relatively nonpolar stationary phase, cyanoethysilicone, was coated onto hydrophobized (silanized) silica and a polar acetonitrile/water mixture was used as the mobile phase. Compounds elute essentially in the reverse order of water solubility.

As with bonded-phase chromatography (Section 7.1), LLC generally separates on the basis of the type (and number) of substituent groups, and by molecular-weight differences (up to about 2000 molecular weight). Therefore, this method is useful for the separation of homologs and mixtures of compounds of different functionality (Table 7.1).

An important practical advantage of LLC is the ability to prepare columns that can allow reproducible separations over long operating periods and for many replicate columns. The reason for this is that retention is determined by the bulk liquid phase. As long as a relatively inert support is employed, the partitioning system and the resulting separation should remain constant. This situation is in contrast to some of the other LC methods (e.g., ion-exchange chromatography), where the reproducibility of columns from different batches of packing material often is poor. LLC has been used widely for the precise quantitative analysis of many mixtures. Both high-precision assays in the 90-100% purity range (5) and trace analyses at the ppm level (6) have been carried out successfully.

8.3 COLUMN PACKINGS

Characteristics

In LLC it is desirable to use supports that are inert. Such materials normally have relatively large pores and low surface areas so as to minimize adsorption. With high-surface-area supports, broad, tailing peaks and irreproducible separations can result when relatively low liquid loadings are used.

Silicas with 200-500Å pores and diatomaceous earth supports appear to be best suited for LLC. Pores should be large enough to allow ready access of the solute molecules to the stationary phase contained within the porous structure, but small enough to resist removal of the stationary liquid by mechanical shear of the mobile phase. Both totally porous and superficially porous supports can be used for preparing LLC columns. Some of the commercially available supports for LLC are listed in Tables 5.5 and 5.10.

Preparation of Coated Supports

There are several different techniques for preparing liquid-coated packings for LLC, each having advantages in certain applications. The *solvent evaporation* procedure (also widely used for preparing packings for gas-liquid chromatography) is most commonly used for larger than 20-μm solid supports that are to be dry-packed. In this technique a nonvolatile stationary liquid is deposited onto the support from a volatile solvent. A known weight of the dry support is placed in a shallow evaporating dish, and to this is added an excess volume of a volatile solvent (e.g., dichloromethane) containing the desired weight of stationary phase. The volatile solvent is slowly evaporated from the packing under a stream of nitrogen, with gentle stirring and careful heating by an infrared lamp, until a dry, free-flowing powder is obtained. With mechanically stable pellicular supports, the solvent can be removed in a rotary evaporator.

Efficient LLC columns also can be prepared by *in situ* coating. This approach must be used for columns of smaller than 10-μm particles that are slurry-packed (Section 5.4). Regardless of particle size, when columns are prepared by slurry packing, the bare support is first loaded into the column blank and then the stationary liquid phase is coated onto the support.

In situ coating can be carried out by any technique that results in a uniform layer of the stationary-phase liquid. If the stationary phase has relatively low viscosity, it can be loaded directly into the column of prepacked bare support. This technique has been effective in preparing columns with mixtures of water, ethanol, and 2,2,4-trimethylpentane as stationary phase (7). In one variation of this procedure, the mobile phase (that has previously been saturated with the stationary phase) is pumped through the prepacked dry column to displace all air. Successive small volumes of the stationary phase then are injected into the column. When the stationary phase begins to elute from the column, the loading of the packing with the stationary phase is complete. In another variation, the stationary phase can be pumped through the dry-packed bed. The column then is flushed with mobile phase (preequilibrated with the stationary phase) to displace excess stationary liquid from between the particles. These two procedures are effective with stationary liquids of low viscosity that wet the support; they are less satisfactory with more viscous stationary liquids ($\eta \geqslant 2$ cP).

Another *in situ* procedure has been especially useful with stationary phases of higher viscosity (8). The stationary phase is first dissolved in a good solvent at high concentration (e.g., 20-40% wt.) and this mixture is pumped through the column of prepacked uncoated support. Pumping of the mixture continues until column equilibrium is achieved (e.g., constant detector reading). Next, a second solvent, which is miscible with the first mobile phase but immiscible with the stationary liquid, is slowly pumped through the column bed. Under these conditions the stationary phase is precipitated into the pores of the support. If high concentrations of the stationary phase are used in the initial solvent, the pores of the support are almost completely filled with stationary phase.

With viscous stationary phases it sometimes is necessary to pass mobile phase through the *in situ*-coated columns for several hours to eliminate excess liquid. Increasing the mobile-phase flowrate assists in this process. If the support consists of a very wide pore structure (e.g., greater than ~500 Å), high mobile-phase velocities can cause some loss of stationary phase from the support pores. Because of this it is desirable that all columns be conditioned at flowrates higher than those used for actual separations. An additional advantage of *in situ* coating is that the stationary liquid can be stripped with an appropriate solvent and the packed bed recoated many times without altering the packing structure or subsequent column efficiency. Thus, repeatable columns of the same stationary phase can easily be prepared. Alternatively, columns with different stationary liquids can be made from the same packed column of bare support.

The *equilibration technique* is a modification of the *in situ* method and is based on the fact that an equilibrium concentration of the stationary phase will be adsorbed onto a support from the mobile phase saturated with the stationary phase (*9*). However, caution should be observed when attempting this approach. If the concentration of liquid phase loaded onto the support is less than the equilibrium concentration, tailing peaks can result from the combination of liquid-liquid partition retention, plus adsorption.

Table 8.1 suggests preferred liquid-phase loadings for some LLC packings in common use. The amount of stationary phase on the packing can vary from a monomolecular layer to complete filling of the pores with stationary phase. Generally, higher liquid loadings (of nonviscous phases) are preferred, because sample capacity increases with the amount of stationary phase present. With nonviscous stationary phases there is no disadvantage in the use of maximum liquid loadings. However, more viscous stationary phases can give lower column efficiencies for two reasons. First, mass transfer in viscous liquids is slower, and when this is combined with complete filling of particle pores by a viscous stationary phase, the result can be a drastic decrease in plate number. In Figure 8.3 we see that some variation in liquid loading is possible, even for relatively viscous stationary phases, without adverse effect on column efficiency. In this case, a precoated column of pellicular packing with 1% (wt.) stationary liquid was tested, the phase removed by purging the column with dichloromethane, and the support recoated *in situ* with 3% (wt.) stationary phase. On the other hand, the same support with increasing concentrations of the same stationary phase can

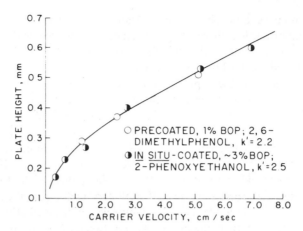

Figure 8.3 Effect of stationary-phase concentration. Column, 100 × 0.21 cm i.d., Zipax, <38 μm; stationary phase, β,β′-oxydiproprionitrile; mobile phase, hexane, 27°C; detector, UV, 254 nm. Reprinted from (8) with premission.

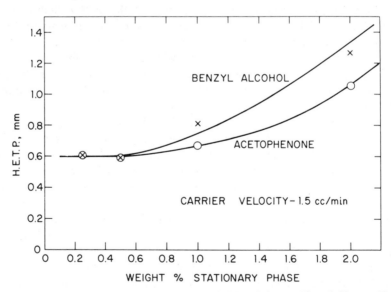

Figure 8.4 Effect of liquid loading on plate height. Column, 50 × 0.32 cm, 29-37 μm controlled surface porosity support; mobile phase, hexane, 27°C; flowrate, 0.45 ml/min; sample, 5 μl, hexane solution with 1 mg/ml acetophenone and 5 mg/ml benzyl alcohol; detector, UV, 254 nm. Reprinted from (10) with permission.

exhibit decreased column efficiency (increased plate heights) when dry-packed, as shown in Figure 8.4. Here the particles become sticky as the level of liquid coating is increased, and it is more difficult to dry-pack the precoated packing homogeneously. Stationary-phase concentration must be maintained below a level such that the coated support becomes sticky and dry-packing is impaired.

Table 8.1 Illustrative stationary-phase loadings for LLC.

Support type	Name[a]	Surface area (m^2/g)	Particle size (μm)	Suggested liquid loading range (% wt)
Porous supports				
Diatomaceous earth	Chromosorb LC-1	10	37-44	5-30
Porous silica beads	Porasil C	50-100	37-75	5-15
	Spherosil XOA-075	50-100	<40	5-15
	Zorbax-PSM-1000	15	6	10-30
Silica gel	LiChrosorb SI-100	250	5, 10, 30	10-30
Pellicular supports				
"Inactive silica"	Zipax	1	25-37	0.5-2
	Liqua-Chrom	<10	44-53	0.6-1.8
	Corasil I	7	37-50	0.5-1.5
"Active silica"	Perisorb A	10	30-40	0.5-1.5
	Vydac Adsorbent	12	30-44	1.5-3

[a] See Tables 5.5 and 5.10 for other supports and suppliers.

Equation 2.5 suggests that a linear relationship should exist between k' or retention volume and the concentration of liquid stationary phase, and this relationship is verified by the data in Figure 8.5. These data also indicate that the solvent evaporation technique of coating a support with a stationary phase is relatively precise. The plots suggest minor retention by uncoated Zipax support at zero concentration of liquid phase.

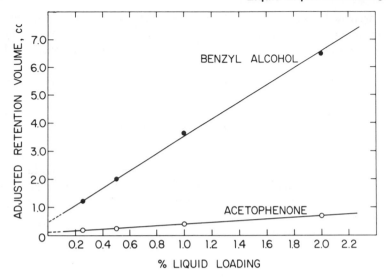

Figure 8.5 Effect of stationary-liquid loading on retention volume. Conditions same as for Figure 8.4. Reprinted from (*10*) with permission.

8.4 THE PARTITIONING PHASES

When designing an LLC separation, one should consider some basic characteristics of the two partitioning phases. First, the mobile-phase/stationary-phase combination must be an immiscible pair. Both phases, but especially the mobile phase, should have relatively low viscosity (e.g., 0.5 cP) to promote high column efficiency and permeability. Also important is the choice of a selective liquid-liquid pair for a particular sample. The selection of LLC phases is somewhat empirical, although some useful guidelines are presented below. A general discussion of solvents in LC is given in Chapter 6.

As with the other LC methods, in LLC the k' values for solutes are generally controlled by changing the strength of the mobile phase. Although it is possible to vary k' values via the stationary phase, this is generally less convenient. However, it is the combination of mobile and stationary phases that controls the overall range of solute k' values for a particular LLC system.

The mobile phase in LLC can be systematically varied to optimize both k' and selectivity by using the P' system described in Sections 6.5 and 7.3. Although the retention mechanism probably is not identical for LLC and BPC, the experimental approach in manipulating solvents to vary k' and selectivity is essentially the same.

In LLC with an aqueous phase (either mobile or stationary), it often is advantageous to buffer this sytem or adjust the pH for ion suppression, as described for reverse-phase BPC in Section 7.5. Ion suppression ensures that un-ionized solutes (e.g., free carboxylic acids at pH = 2) are available for partitioning; otherwise, tailing peaks can result.

Illustrative binary liquid-liquid systems suitable for obtaining a wide range of separation selectivity are given in Table 8.2. Many separations by normal LLC have been performed using either β, β'-oxydipropionitrile (BOP), or one of the various polyethylene glycols as stationary phase. Both liquids are useful for the same types of compounds. However, polyethylene glycols may be more selective for alcohols, and BOP often is preferred for amines and other basic compounds. As might be predicted from viscosity data, columns with lower molecular weight stationary phases (e.g., Carbowax 400) generally produce more efficient columns than comparable higher-molecular-weight stationary phases (Carbowax

Table 8.2 Some representative liquid-liquid chromatographic systems.

Stationary phase	Mobile phase
Normal-phase LLC	
β,β'-Oxydipropionitrile	Pentane, cyclopentane, hexane, hep-
1,2,3-Tris(2-cyanoethoxy)propane	tane, isooctane
Carbowax 600	Same, modified with up to 10-20%
Triethylene glycol	chloroform, dichloromethane, tetrahydrofuran, acetonitrile,
Trimethylene glycol	dioxane, etc.
Ethylene glycol	Di-n-butyl ether
Dimethylsulfoxide	Isooctane
Water/ethylene glycol	Hexane/CCl_4
Ethylenediamine	Hexane
Water	n-Butanol
Ethylene glycol	Nitromethane
Nitromethane	CCl_4/hexane
Reversed-phase LLC	
Cyanoethylsilicone	Methanol/water
Dimethylpolysiloxane	Acetonitrile/water
Heptane	Aqueous methanol
Hydrocarbon polymer	Methanol/water

4000), especially at higher loading. Lower-molecular-weight stationary phases often are more retentive and selective than the higher-molecular-weight equivalents (11).

Squalane, cyanoethylsilicone, and hydrocarbon polymers have found use as stationary phases in reverse-phase LLC, but these materials have not achieved popularity in high-performance applications with small particles (<10 μm). The advent of stable bonded-hydrocarbon stationary phases (Chapter 7) for reverse-phase separations has reduced the need for coated supports. Nevertheless, mechanically held nonpolar stationary phases do have an advantage for certain operations [e.g., extraction chromatography with highly radioactive species (12)], because of high sample capacity, separation selectivity, and excellent column repeatability. When they are used in the reverse-phase mode, nonpolar stationary phases should be coated onto hydrophobic supports (e.g., trimethylsilyl-modified; see Section 7.2) for proper wetting of the support and to prevent desorption and loss of the stationary phase as a result of preferential adsorption of water on the support surface.

In addition to the normal binary-liquid systems for LLC, immiscible ternary liquid-liquid systems have been used. These are obtained from immiscible binary-liquid systems by adding a third component (co-solvent) that is miscible with both phases. With ternary systems, either of the two resulting phases can be used as a stationary liquid, with the other phase as a mobile phase. Thus, normal- or reverse-phase LLC can be carried out by using the appropriate phase of a ternary system as the stationary liquid. The water/ethanol/2,2,4-trimethylpentane ternary is a versatile choice for separating a wide range of compounds by LLC, including pesticides, steroids, and metal chelates (13). With systems of this type it is possible to establish a range of two-phase systems using three simple solvents. Differences in solvent strength may be large or small, depending on the relative concentrations of the individual solvents in the mixture. Generally, as the concentration of the co-solvent (ethanol in the water/ethanol/2,2,4-trimethylpentane ternary) is increased, the distribution coefficients of all solutes approach unity. Figure 8.6 shows the separation of a mixture of corticosteroids using a ternary liquid-liquid partitioning system. There are many combinations of solvents that can be used in ternary liquid systems [e.g., see (14-16)] to obtain a highly selective pair of phases for LLC separations.

Correlation of chromatographic retention with partition coefficients obtained from simple extraction is possible as predicted by Eq. 2.5. Therefore, simple extraction studies can quickly predict whether the organic- or aqueous-rich phase should be the stationary phase for a particular system. If the stationary phase is hydrophilic, it should be coated onto untreated support. Conversely, if the stationary phase is hydrophobic, it should be coated onto a hydrophobic (silanized) support.

LLC liquid pairs that have a relatively high degree of mutual solubility tend to give unstable columns and require special experimental precautions (thermo-

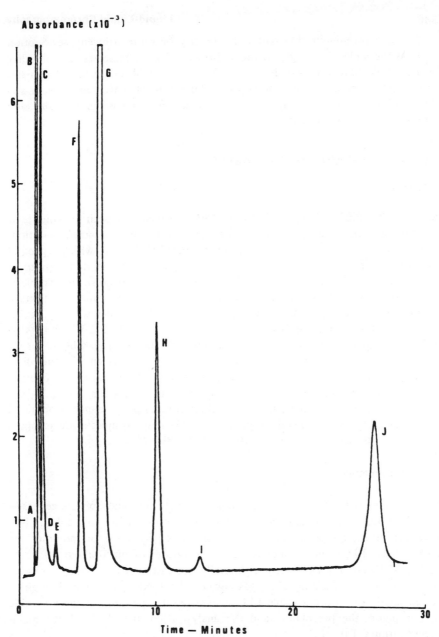

Figure 8.6 Separation of corticosteroids using a ternary liquid-liquid partitioning system. Column, 25 × 0.21 cm, Zorbax-Sil, equilibrated dichloromethane-methanol-water (970 ml:10 ml:20 ml), organic layer as mobile phase; flowrate, 0.8 ml/min; temp. 25°C; detector, UV, 254 nm; components: (A) unknown, (B) progesterone, (C) 11-deoxycorticosteroid, (D) unknown, (E) unknown, (F) 11-deoxycortisol, (G) corticosteroid, (H) cortisone, (I) prednisone, (J) cortisol. Reprinted from (14) with permission.

statting and preequilibration requirements may be quite stringent; see Section 8.6). As the molecular weight of the stationary phase is increased, its solubility in the mobile phase generally decreases, giving more stable columns and a greater choice among potential (immiscible) mobile phases. However, increase in stationary-phase molecular weight is accompanied by increase in viscosity and generally poorer column efficiency.

8.5 OTHER SEPARATION VARIABLES

Sample Size

As with BPC, final column resolution in LLC is affected by both the volume of the injected sample and the weight of components in the sample solution. Maximum sample volume for LLC is again predicted by Eq. 7.2, which is general for each of the LC methods. A much larger volume can be tolerated in preparative separations (Section 15.4), or if the solvent used to dissolve the sample is weaker than the mobile phase used for the separation. The amount of stationary liquid largely determines the sample weight that can be handled without significantly affecting column efficiency. For example, 0.2-0.3-cm i.d. columns of pellicular supports containing 1-3% w. of stationary liquid often will accept up to 0.2 mg of a solute before significant band broadening is evident. On the other hand, 1-2 mg of a solute generally can be used without serious loss in column efficiency for a 0.4-cm i.d. column of totally porous packing containing 25-50% w of stationary liquid. In both cases, highest column performance is obtained with small sample volumes and sample weights. Overlarge samples cause asymmetrical peaks, decreased column resolution, and variable k' values.

Column Temperature

LLC columns with low-viscosity, mechanically held stationary phases generally should be operated near ambient temperature to ensure satisfactory column life. Columns of some viscous stationary phases (e.g., hydrocarbon polymer for reverse-phase LLC) can be operated at temperatures of up to about 50°C. As with BPC (Section 7.4), retention generally decreases with increasing temperature, and column efficiency improves because the viscosity of the mobile phase is decreased. To ensure a stable LLC system, the temperature of the incoming mobile phase, the precolumn, and the analytical column all should be maintained within 0.1-0.2°C.

8.6 SPECIAL PROBLEMS

To prevent the loss of mechanically-held liquid stationary phases from LLC columns, it is necessary to *presaturate* the mobile phase with the stationary liquid. This precaution is particularly important if column performance must be maintained constant over an extended period, or if the two LLC phases have significant mutual solubility. Liquids that have relatively poor mutual solubility are difficult to equilibrate by simple shaking in a separatory funnel. In these cases equilibrium is achieved by placing a 50-100-fold excess of the mobile phase in a large bottle with the stationary liquid. After thorough sparging of these phases with helium or nitrogen to eliminate dissolved oxygen, the mixture is rapidly mixed in the closed container for several hours by magnetic stirring under nitrogen. This stirring should be sufficiently vigorous to produce a turbulent vortex in the bottle. The phases are then allowed to separate, and the equilibrated mobile phase is removed for use.

In LLC, as discussed in Section 7.5 for BPC, degassing the mobile phase or sparging it with helium can eliminate problems from oxygen reacting with the partitioning phases; it also reduces the possibility of small bubbles forming in the detector.

With LLC columns, a *precolumn* should be placed between the pump and the sample inlet to ensure true equilibration of the two partitioning phases. A precolumn is particularly important when the two partitioning phases have significant mutual solubility (e.g., >50%). A convenient precolumn consists of a 50-cm length of 0.46-cm i.d. tubing packed with 20-30% by weight of the stationary liquid impregnated on a relatively coarse (e.g., 120-140 mesh) diatomaceous earth support of the type commonly used in gas chromatography. This precolumn does not need to be carefully packed, because it does not participate directly in the separation. Precolumns should be replaced frequently to ensure complete protection of the analytical column.

As with the other LC methods, it is important that purified solvents be used in LLC. Commonly used solvents are commercially available in satisfactory purity (Table 6.1). Solvents may also be purified as described in Section 6.6.

Sample solutions injected into LLC columns should be prepared in the mobile phase, preferably saturated with the stationary liquid. If a solvent stronger than the mobile phase is used to dissolve the sample, the stationary phase may be stripped from the column, resulting in decreased solute retention and resolution. In addition, the shape and resolution of early-eluting peaks also can be affected by upset of the mobile-phase/stationary-phase equilibrium.

Column deactivation occurs in LLC with continued injection of samples containing strongly retained (noneluted) components, just as in the case of LSC, BPC, IEC, and so on, and can be controlled by utilizing an appropriate guard column (Section 5.6).

Gradient elution generally cannot be used in LLC, because dissolution of the stationary liquid causes loss of column performance. Therefore, if samples having components of a wide k' range must be separated and gradient elution subsequently is required, packings having chemically bonded stationary phases (Chapter 7) must be used.

In the recovery of LLC fractions for subsequent identification or other use, it should be recalled that the mobile phase is saturated with the stationary liquid. Therefore, high-boiling stationary phases may be difficult to remove from separated fractions, and volatile stationary phases are preferred.

Alternative liquid phases should be investigated if a particular mobile-phase/stationary-phase combination provides inadequate resolution of sample components eluting in the optimum k' range (1-10). The simplest procedure is to modify the mobile phase of the original column. However, before using a new mobile phase, it must be carefully presaturated with the stationary phase. These manipulations can prove to be somewhat time-consuming, but may be justified for a particular separation if a large number of routine samples are to be analyzed.

The lifetime of an LLC column depends on the particular partitioning system and the care with which it is used. For many columns, 3-6 months of satisfactory performance is typical. Generally, the higher the mutual solubility of the partitioning phase, the more difficult it is to maintain true equilibration, which usually results in shorter column life. One advantage of LLC systems with mechanically held stationary phases is that if the performance of the column becomes unsatisfactory, the stationary phase can be eluted with a strong solvent and the support recoated by the *in situ* procedure (Section 8.3). Good column lifetime in LLC is promoted by using purified solvents (both the stationary and mobile phases), by excluding oxygen from the system to prevent solvent degradation, by using presaturated solvents and an appropriate precolumn, and by carefully thermostatting the entire system.

8.7 APPLICATIONS

Modern LLC has been used to separate a wide range of sample types, some of which are illustrated in Table 8.3. Despite the experimental limitations in Section 8.6, LLC with simple liquid stationary phases is a useful method for many applications.

Several important separations of compounds containing metal atoms have been made, including isomers of cobalt complexes involved in the synthesis of vitamin B-12 (22) and metal-β-diketonates (Figure 8.7). Extraction-LLC also has been used for separating a variety of radioactive metals (see Figure 10.7).

Many biologically important natural and synthetic compounds have been separated by LLC, including cortisol (6), derivatized urinary 17-ketosteroids (24), and free underivatized steroids (19). In Figure 8.8, chromatogram (a) is the separation of a placebo cream and (b) is a cream containing 0.03% flumethasone pivalate, a synthetic steroid.

Table 8.3 Typical applications of modern liquid-liquid chromatography

Stationary phase	Mobile phase	Systems separated	Ref.
Ethylene glycol	3% CHCl$_3$/heptane	steroids	(17)
1,2,3-tris(2-cyanoethoxy)propane	2,2,4-trimethylpentane	phenols	(18)
β,β'-oxydipropionitrile	7% CHCl$_3$/hexane	insecticides	(5)
Trimethylene glycol	n-hexane	pesticide metabolites	(19)
Carbowax 400	hexane	nonionic detergents	(19)
Squalane	20% H$_2$O/CH$_3$CN	hydrocarbons	(19)
Cyanoethylsilicone	water	coumarins	(19)
Polyamide	5% hexane/ethanol	pencillin esters	(19)
Hydrocarbon polymer	3% acetonitrile, 0.001 M EDTA in pH 4.5, 0.05 M phosphate buffer	tetracyclines	(20)
Water/ethanol/2,2,4-trimethyl-pentane (ternary)	water-rich phase	steroids	(21)

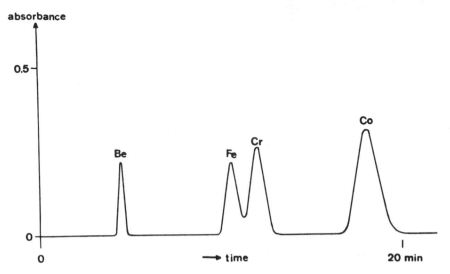

Figure 8.7   Separation of metal-β-diketonate complexes by LLC. Column, 50 × 0.27 cm, 5-10 μm, water-ethanol-2,2,4-trimethylpentane ternary with less-polar liquid as mobile phase; mobile-phase velocity, 0.3 cm/sec; detector, UV, 254 nm. Reprinted from (23) with permission.

Various pesticides have been separated by LLC, and the technique has proved useful for the trace analysis of these compounds in environmental samples. Studies have included the analysis of phosphorus-containing insecticides in crop tissues (26) and a mosquito larvacide in salt ponds, as shown in Figure 8.9. In this case water from a series of ponds was acidified, extracted with chloroform, and the chloroform extract concentrated by evaporation to enhance sensitivity about 100-fold.

Modern LLC has been used extensively to analyze drugs and various food products [e.g., methyl prednisolone in milk (28), rifampin (29), and alkaloids (30)]. Various esters of penicillin have been separated as in Figure 8.10, using a polymeric (polyamide) stationary phase on a pellicular support. A relatively polar mobile phase was required to dissolve and to elute the highly polar solute properly. N-Nitrosamines have been determined in food products, as illustrated in Figure 8.11. Part (a) shows a separation of a synthetic mixture of nitrosamines using normal-phase LLC with a pellicular support, and (b) an extract of pork. Many of the compounds separated in (a) were not detected in the pork extract.

Reverse-phase LLC has been used to separate mixtures of aromatics, paraffins, olefins, and diolefins (19); chlorinated benzene (19); and fused-ring aromatics (32). Reverse-phase LLC also has been used to separate mixtures of fat-soluble

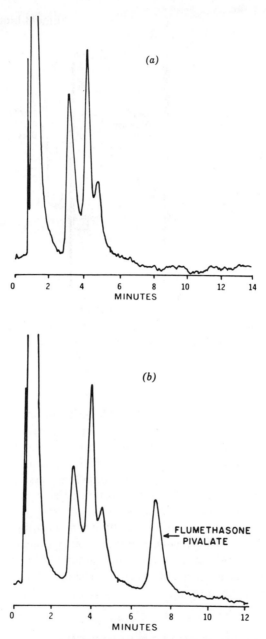

Figure 8.8 Determination of flumethasone pivalate in topical cream. Column, 1% β,β'-oxydipropionitrile on Zipax, 100 × 0.21 cm; mobile phase, 5% ethyl acetate, 0.2% acetonitrile in hexane; flowrate, 1 ml/min; temp., ambient; sample, 3 µl of extract; detector, UV, 254 nm; (a) placebo cream, (b) 0.03% flumethasone pivalate cream. Reprinted from (25) with permission.

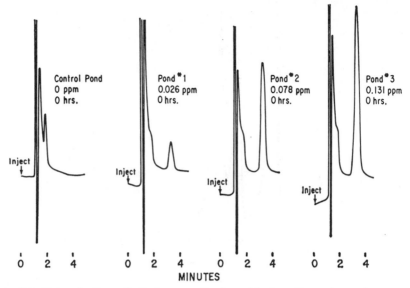

Figure 8.9 Determination of Abate mosquito larvacide in salt marsh pond extracts. Column, 100 cm × 0.21 cm, 1% β,β′-oxydipropionitrile on Zipax; mobile phase, heptane; flowrate, 1.0 ml/min; detector, UV, 254 nm; sample, 5 μl of heptane extract. Reprinted from (27) with permission.

vitamins, as illustrated in Figure 8.12. The gelatin-coated tablet was ground and shaken with warm dimethylsulfoxide, which dissolved the gelatin and released the tablet components, which were then extracted into hexane. The hexane was evaporated to dryness and redissolved in methanol/water for reverse-phase LLC on a pellicular packing with a hydrocarbon polymer stationary phase.

A large variety of commercially important organic chemicals have been separated by modern LLC, including antioxidants and plasticizers (34), aliphatic carbonyl compounds (35), and a variety of nonionic organic dyes. As illustrated in Figure 8.13, similar compounds of this latter type are easily separated by LLC because of the high selectivity of this method. Also, the symmetrical peaks obtained for these compounds permit accurate peak size measurement for routine quantitative analysis.

8.8   THE DESIGN OF AN LLC SEPARATION

Selecting the Column

For polar or water-soluble samples, normal-phase LLC is generally preferred. (Nonpolar and organic-soluble compounds are best separated by reverse-phase

Figure 8.10 Separation of penicillin, methyl-benzyl ester t_R = 3.0 min, from minor impurities. Column, 100 × 0.21 cm, polyamide on Zipax; mobile phase, 5% *n*-hexane in ethanol; flowrate, 0.9 ml/min; detector, UV, 254 nm. Reprinted from (*19*) with permission.

RETENTION TIME (MINUTES)

LLC.) For the analytical separation of mixtures by normal-phase LLC (i.e., relatively polar solutes), the best initial choice of a column support is an untreated porous-silica microparticle such as Zorbax-PSM-1000 or LiChrospher 500. or, alternatively, a pellicular support such as Zipax. Uncoated small particles (<20 μm) must first be loaded into the column by high-pressure slurry packing and the stationary phase subsequently introduced *in situ*. With pellicular supports (~30 μm) the liquid phase may be coated by the evaporation procedure and the coated packing dry-packed into the column as described in Chapter 5. Table 5.10 lists a number of acceptable supports for each type.

Either β,β'-oxydipropionitrile or Carbowax 400 is a reasonable initial choice of a stationary liquid for a normal-phase LLC system. If the sample is aqueous, use of a partitioning system having a significant concentration of water is indicated, and reverse phase may be a better approach. Alternatively, the 2,2,4-trimethylpentane/ethanol/water ternary can be used.

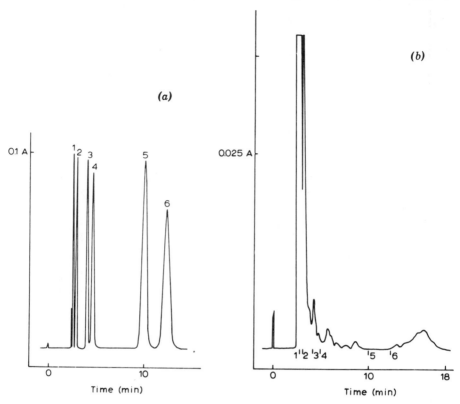

Figure 8.11 Determination of *N*-nitrosoamines in food products. Column, 100 × 0.21 cm, di(2-cyanoethyl)ether on Zipax; mobile phase, 2,2,4-trimethylpentane; flowrate, 9.6 ml/min; detector, UV, 254 nm. (*a*), Separation of (1) *N*-nitrosodibutylamine, (2) *N*-nitrosodipropylamine, (3) *N*-nitrosodiethylamine, (4) *N*-nitrosopiperidine, (5) *N*-nitrosodimethylamine, and (6) *N*-nitrosopyrrolidine; (*b*), extract of pork. Reprinted from (*31*) by permission.

Reverse-phase LLC systems generally require a hydrophobic (silanized) support, either porous microparticle or pellicular. Cyanoethylsilicone (50% cyano) often is a satisfactory initial choice as stationary phase, and the water-poor phase of the 2,2,4-trimethylpentane/ethanol/water system also can be used for many separations.

Selecting the Mobile Phase

Selection of mobile phases for optimum k' and selectivity for an LLC separation are essentially the same as described for BPC in Section 7.3. Also, the general aspects of solvent strength and selectivity in LLC are discussed in Section 6.5. Therefore, we will make only a few supplementary comments.

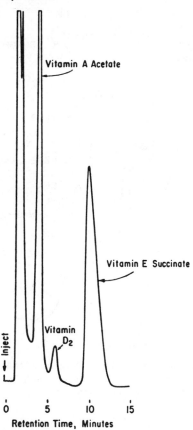

Figure 8.12   Determination of fat-soluble vitamins in a tablet extract. Column, 100 X 0.21 cm Zipax-HCP; mobile phase, 79% methanol, 21% water, and 0.1% H_3PO_4; temp., ambient; pressure, 1500 psi; detector, UV, 254 nm. Reprinted from (*33*) with permission.

 If it is apparent that the sample of interest has solutes with widely differing k' values, gradient elution may be necessary. Since gradient elution cannot be used with LLC, chemically-bonded stationary phases are indicated (Chapter 7). When the selectivity of the LLC system has been optimized and further increase in sample resolution is required, column efficiency can be increased according to the concepts discussed in Section 2.5.

Operating Variables

To ensure satisfactory LLC column life, the partitioning phases must be pre-equilibrated and a precolumn should be used. Oxygen generally should be excluded from the phases, and the temperature of the incoming liquid, pre-

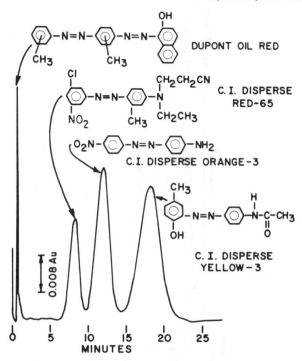

Figure 8.13 Separation of azo dyes by LLC. Column, 100 × 0.21 cm; β,β'-oxydipropioni-trile on Zipax; mobile phase, hexane; flowrate, 1.5 ml/min; temp., ambient; detector, UV, 254 nm. Reprinted from (*36*) with permission.

column, and analytical column must be maintained constant to ensure maximum column stability. Near-ambient temperatures normally are recommended for LLC. With mechanically-held stationary liquids, it is important that the velocity of the mobile phase not be increased to the point where shear forces remove the stationary phase from the support. This is unlikely to occur as long as velocities are maintained in the commonly used 0.5-2 cm/sec range. However, if the velocity exceeds this level, loss of stationary liquid may be noticeable. At very high velocities, increase in column temperature due to frictional forces also can remove stationary liquid. Bleed of the stationary liquid from the support in spite of recommended mobile phase velocities indicates that the level of stationary phase is too high for the support used, or that the temperature of the system or the composition of the mobile phase is not constant.

Samples should not be injected in solvents stronger than the mobile phase. In rare cases where it is not possible to prepare samples in the mobile phase, it is essential to use solvents that are miscible with the mobile phase. For example, injecting water or methanol samples into a β,β'-oxydipropinitrile column with a heptane mobile phase can result in poor separations and spurious peaks.

The maximum sample weight permitted in LLC depends on the particular system. Typically, less than 200 μg of sample are used on columns of pellicular packings; up to 2 mg can be used on columns with porous packings. Generally, the volume of sample injected should not exceed one-third of the volume of the first peak of interest. Larger volumes can be tolerated if the strength of the solvent in which a sample is dissolved is less than that of the mobile phase.

REFERENCES

1. A. J. P. Martin and R. L. M. Synge, *Biochem. J.*, **35**, 1358 (1941).
2. D. C. Locke, in *Advances in Chromatography*, Vol. 8, J. C. Giddings and R. C. Keller, eds., Dekker, New York, 1969.
3. J. J. Kirkland, *J. Chromatogr. Sci.*, **10**, 593 (1972).
4. Du Pont Liquid Chromatography Applications Lab Report, No. 72-06, 1972.
5. R. E. Leich, *J. Chromatogr. Sci.*, **9**, 531 (1971).
6. C. A. M. Meijers, J. A. R. J. Hulsman, and J. F. K. Huber, *Z. Anal. Chem.*, **261**, 347 (1972).
7. J. F. K. Huber, C. A. M. Meijers, and J. A. R. J. Hulsman, in *Advances in Chromatography, 1971*, A. Zlatkis, ed., Chromatography Symposium, Department of Chemistry, University of Houston, Houston, Tex., 1971, p. 230.
8. J. J. Kirkland and C. H. Dilks, Jr., *Anal. Chem.*, **45**, 178 (1973).
9. B. L. Karger, H. Engelhardt, and K. Conroe, in *Gas Chromatography 1970*, R. Stock and S. G. Perry, eds., Institute of Petroleum (Elsevier), London, 1971, p. 124.
10. J. J. Kirkland, *J. Chromatogr. Sci.*, **7**, 7 (1969).
11. J. A. Schmit, in *Modern Practice of Liquid Chromatography*, J. J. Kirkland, ed., Wiley-Interscience, New York, 1971, p. 378.
12. E. P. Horwitz, C. A. Bloomquist, and W. H. Delphin, *J. Chromatogr. Sci.*, **15**, 41 (1977).
13. J. F. K. Huber, *ibid.*, **9**, 72 (1971).
14. N. A. Parris, *ibid.*, **12**, 753 (1974).
15. J. F. K. Huber, C. A. M. Meijers, and J. A. R. J. Hulsman, *Anal. Chem.*, **44**, 111 (1972).
16. C. Hesse and W. Hovermann, *Chromatographia*, **6**, 345 (1973).
17. R. A. Henry, J. A. Schmit, and J. F. Dieckman, *J. Chromatogr. Sci.*, **9**, 513 (1971).
18. J. F. K. Huber, 5th International Symposium, Column Chromatography, Lausanne, 1969, published as supplement to *Chimia*, **24**, (1970).
19. J. A. Schmit, in *Modern Practice of Liquid Chromatography*, J. J. Kirkland, ed., Wiley-Interscience, New York, 1971, Chap. 11.
20. K. Tsuji, J. H. Robertson, and W. F. Beyer, *Anal. Chem.*, **46**, 539 (1974).
21. J. F. K. Huber, C. A. M. Meijers, and J. A. R. J. Hulsman, in *Advances in Chromato-*

graphy, 1971, A. Zlatkis, ed., Chromatography Symposium, Department of Chemistry, University of Houston, Houston, Tex., 1971, p. 230.

22. J. Schreiber, *Chimia*, **25**, 405 (1971).

23. J. F. K. Huber, J. C. Kraak, and H. Venning, *Anal. Chem.*, **44**, 1554 (1972).

24. F. A. Fitzpatrick, S. Siggia, and J. Dingman, *ibid.*, **44**, 2211 (1972).

25. J. A. Mollica and R. F. Strusz, *J. Pharm. Sci.*, **61**, 444 (1972).

26. J. G. Koen and J. F. K. Huber, *Anal. Chim. Acta*, **51**, 303 (1970).

27. R. A. Henry, J. A. Schmit, J. F. Dieckman, and F. J. Murphey, *Anal. Chem.*, **43**, 1053 (1971).

28. L. F. Krzeminski, B. L. Cox, P. N. Perrel, and R. A. Schlitz, *J. Agric. Food Chem.*, **20**, 970 (1972).

29. J. A. Schmit, E. I. du Pont de Nemours Co., unpublished studies, 1971.

30. C. Wu and S. Siggia, *Anal. Chem.*, **44**, 1499 (1972).

31. G. B. Cox, *J. Chromatogr.*, **83**, 471 (1973).

32. J. A. Schmit, *Chromatographic Methods*, 820M4, E. I. du Pont de Nemours Co., Instrument Products Division, March 30, 1970.

33. Du Pont Liquid Chromatography Methods Bulletin, 820M10, March 23, 1972, E. I. du Pont de Nemours Co., Instrument Products Division, Wilmington, Del.

34. J. A. Schmit, R. A. Henry, R. C. Williams, and J. F. Dieckman, *J. Chromatogr. Sci.*, **9**, 645 (1971).

35. R. C. Williams, J. A. Schmit, and R. A. Henry, *ibid.*, **10**, 494 (1972).

36. R. J. Passarelli and E. S. Jacobs, *ibid.*, **13**, 153 (1975).

BIBLIOGRAPHY

Keller, R. A., and J. C. Giddings, in *Chromatography*, 3rd ed., E. Heftmann, ed., Van Nostrand Reinhold, New York, 1975, Chap. 6 (theoretical basis of partition chromatography).

Kirkland, J. J., in *Modern Practice of Liquid Chromatography*, J. J. Kirkland, ed., Wiley-Interscience, New York, 1971, Chap. 5 (practice of modern LLC).

Locke, D. C., in *Advances in Chromatography*, Vol. 8, J. C. Giddings and R. A. Keller, eds., Dekker, New York, 1969 (thermodynamics of LLC).

Majors, R. E., in *Practical High-Performance, Liquid Chromatography*, C. F. Simpson, ed., Heyden, London, 1976, Chap. 6 (review of modern LLC).

Metzsch, F. A. V., *Angew. Chem.*, **65**, 586 (1953) (extensive list of solvent pairs for liquid-liquid partition).

Parris, N. A., *Instrumental Liquid Chromatography*, Elsevier, New York, 1976, Chap. 8 (practice and techniques of modern LLC).

Perry, S. G., R. Amos, and P. I. Brewer, *Practical Liquid Chromatography*, Plenum, London, 1972, Chap. 4 (mechanism and materials for LLC).

Snyder, L. R., in *Techniques of Chemistry*, 2nd ed., Vol. III, Part I, A. Weissberger and E. S. Perry, eds., Wiley-Interscience, New York, 1977 (solvent selection for separation processes, including LLC).

NINE

LIQUID-SOLID CHROMATOGRAPHY

9.1	Introduction	351
9.2	Column Packings	361
9.3	Mobile Phases	365
	Solvent Strength	365
	Solvent Selectivity	370
	Addition of Water or Polar Modifiers to the Mobile Phase	374
	The Value of Adding Water	374
	Isohydric Solvents	377
	Preparation of Isohydric Solvents	380
	Use of Organic Modifiers instead of Water	381
	Thin-Layer Chromatography to Pilot	
	LSC Separation	383
9.4	Other Separation Variables	389
9.5	Special Problems	391
9.6	Applications	398
9.7	The Design of an LSC Separation	405
	Selecting the Column	405
	Selecting the Mobile Phase	406
	Temperature	407
	Sample Size	407
	Other Considerations	407
References		407
Bibliography		409

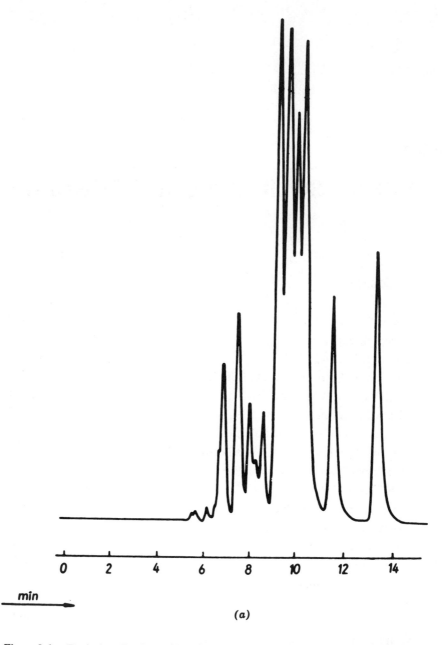

0   2   4   6   8   10   12   14

min

(a)

Figure 9.1*a* Typical applications of liquid-solid chromatography for nonpolar or moderately polar samples. (*a*) Separation of polychlorobiphenyl (PCB) distillate fraction. Column, 25 × 0.47 cm 5-μm silica (Pragosil SI 5); mobile phase, *n*-pentane; temp., ambient; 1.7 ml/hr; UV, 254 nm. Reprinted from (*4*) with permission.

9.1 INTRODUCTION

Liquid-solid or *adsorption* chromatography (LSC) is the oldest of the various LC methods. In its classical form, so-called *open-column* chromatography as in Figure 1.1, LSC was conceived and developed by Tswett at the turn of the century. *Thin-layer chromatography* (TLC), the open-bed version of adsorption chromatography (see Figure 1.1), was introduced in the early 1950s by Kirchner and later popularized by Stahl. Modern LSC, featuring automated operation plus fast, high-resolution separation, dates from 1967 (*1*). Because of its versatility, plus the earlier widespread use of TLC, adsorption chromatography has remained one of the more popular of the modern LC methods (although reverse-phase BPC has displaced it in some of its applications).

Despite the advantages of modern LSC with respect to improved separation and analysis, as well as convenience, both open-column and thin-layer chromatography are still widely practiced. It is sometimes convenient to carry out preparative separations with open-column chromatography in the case of easily separated samples. Because of its simplicity and flexibility, TLC can be used to great advantage in conjunction with modern LSC. It is well suited for exploring the best combination of mobile phase and stationary phase for a given sample

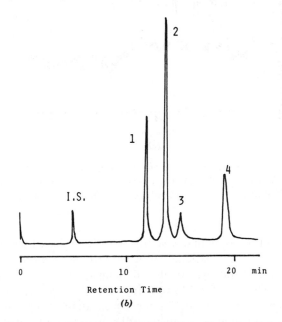

Figure 9.1*b* Separation of *cis-trans*-retinal isomers (from irradiated all-*trans*-retinal). Column, 25 × 0.79 cm Zorbax-Sil (6-μm silica); mobile phase, 12%v ethyl ether/hexane; temp., ambient; 2.0 ml/min (500 psi); UV, 254 nm; 1-μl (10 ng) sample. Reprinted from (*5*) with permission.

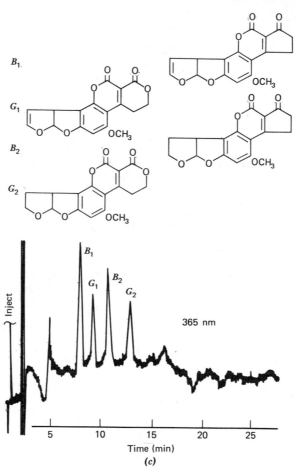

Figure 9.1c Separation of aflatoxins in peanut butter extract. Column, 25 × 0.21 cm Zorbax-Sil; mobile phase, 60%v $CH_2Cl_2/CHCl_3$ plus 0.1%v methanol; temp., ambient; 0.7 ml/min (1500 psi); UV, 365 nm; 15 ng each compound. Reprinted from (6) with permission.

(Section 9.3). Once the optimum separation conditions have been thus identified, the same mobile phase and packing can be used for modern LSC.

There are further advantages to the preliminary use of TLC in scouting unknown samples. The use of corrosive spray reagents plus charring [e.g., see (2)] is a sensitive, universal technique for the visualization of separated TLC bands, revealing many compounds that might not be picked up by a single LC detector. Also, every band is observable in the final TLC chromatogram, whereas strongly retained bands often escape detection in isocratic elution from an LC column. Thus, TLC serves to monitor compounds that might otherwise be missed. The literature of TLC (2, 3) includes many applications that can serve as guides for similar separations by modern LSC, providing that something is known about the composition of the sample.

LSC gives best results when applied to certain kinds of samples or where a certain kind of selectivity is required. Organic-soluble samples are handled best, whereas very polar and/or ionic samples often yield disappointing separations. Some typical applications of LSC are shown in Figure 9.1, where the samples range from the relatively nonpolar polychlorobiphenyl fraction in (a), to the slightly polar irradiated-retinal sample in (b), and the moderately polar aflatoxins in (c). Very polar, multifunctional compounds often require careful adjustment of chromatographic conditions in LSC to avoid tailing bands and low column N values. However, there are many reported examples of such samples being well resolved by LSC. One example (Figure 9.2) is that of the separation of tri- through dodecagalloyl glucose derivatives ($C_{27}H_{24}O_{18}$-$C_{90}H_{60}O_{54}$) by gradient elution from a pellicular silica column. Ionic compounds, on the other hand, are seldom well separated by LSC; typically, such samples give tailing bands that have low N values, particularly at higher k' values. An example of this problem is provided in Figure 9.3a, for the separation of various cationic

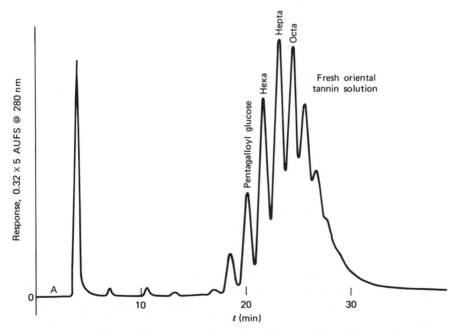

Figure 9.2 Separation of gallotannin extract from Chinese gall nuts by liquid-solid gradient elution chromatography. Column, 50 × 0.2 cm Corasil-II (pellicular silica); mobile phase, gradient from A (4% B in hexane) to B (25/74/1%v tetrahydrofuran/methanol/acetic acid); temp., ambient; 2 ml/min; UV, 280 nm; 5-μl sample. Reprinted from (12) with permission.

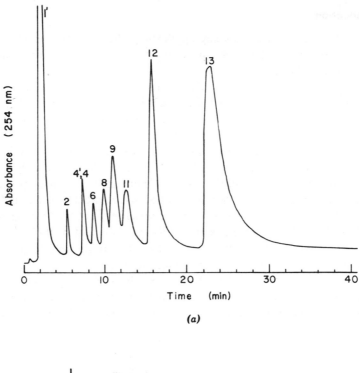

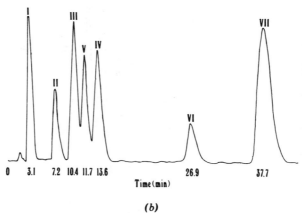

Figure 9.3 Separation of ionic samples by liquid-solid chromatography. (*a*) Degraded methylene blue dye sample (typical result for LSC separation of ionics). Column, 25 × 0.21 cm Zorbax-Sil; mobile phase, 80/19.5/0.5%v methanol/water/formic acid; temp., 50°C; 0.35 ml/min (2200 psi); UV, 254 nm; 10-μm sample. Reprinted from (*13*) with permission. (*b*) Optical brighteners of formula I. Column, 25 × 0.22 cm, MicroPak Si-5 (5-μm silica); mobile phase, benzene/dioxane/methanol/ammonia (30%) 40/50/20/10%v; ambient; 0.4 ml/min; UV. Reprinted from (*14*) with permission.

methylene-blue analogs on silica. On the other hand, careful attention to the composition of the mobile phase (see Section 19.2) allows anionic compounds to be well separated by LSC in favorable instances. An example is that of the various optical brighteners (I) shown in Figure 9.3b.

(I)

Samples that dissolve only sparingly in water and/or water-miscible organic solvents (e.g., polyaromatic hydrocarbons, fats, oils, etc.) are often better separated by LSC than by reverse-phase LC. This is especially true when larger sample sizes are required for increased detection sensitivity or for recovery and characterization of separated sample bands. Compared to bonded-phase packings, LSC packings are also cheaper, can often be used with higher sample loadings, and show greater stability toward pH extremes. These factors make LSC the preferred method for most preparative separations (Section 15.3).

Concerning separation selectivity, compounds of different chemical type are easily separated on silica or alumina, as illustrated in Table 9.1. Table 9.1 also shows that LSC can provide easy resolution of compounds with differing numbers of functional groups. Finally, LSC is unique among the four LC methods in providing maximum differentiation among isomeric mixtures; a few examples are provided in Table 9.1. [See also Figure 9.1b and the discussion in (15)].

It is interesting to contrast the selectivity in LSC with that of reverse-phase LC; in principle, either method might be used for many samples. The major differences in selectivity concern the separation of homologs, benzologs, and isomers; a few relevant examples are shown in Table 9.2. In general, *homologs* are much better separated by reverse-phase or polar bonded-phase LC than by LSC. The examples of Table 9.2 are rather typical in this regard. Each additional $-CH_2-$ group in the alkyl benzoates of Table 9.2 decreases k' by 2-3% for separation by LSC; this is a very small change in k', and therefore separation of adjacent homologs is difficult. For the reverse-phase separation of the alkyl anthraquinones in Table 9.2, each added $-CH_2-$ group increases k' by about 1.7-fold, allowing easy separation of adjacent homologs as in Figure 2.24.

Benzologs can be well separated in both LSC and reverse-phase LC, but generally separation is markedly better by reverse phase. In the examples of Table 9.2, addition of a benzo group increases k' (or α) by a factor of 1.1-1.2 in LSC but by 1.4-1.8-fold in reverse-phase LC. The effect of adding an aromatic ring to the sample molecule is greater in LSC for less-polar compounds. Thus,

Table 9.1   Separation selectivity in liquid-solid chromatography.

	Value of k'
Compound type[a] (silica)	
2-Methoxynaphthalene	0.6
1-Nitronaphthalene	1.8
1-Cyanonaphthalene	2.7
1-Acetonaphthalene	5.5
Number of functional groups[a] (silica)	
2-Methoxynaphthalene	0.6
1,7-Dimethoxynaphthalene	1.4
1-Nitronaphthalene	1.8
1,5-Dinitronaphthalene	6.1
Acetophenone[b]	1.1
3-Nitroacetophenone[b]	1.6
Isomers (alumina)	
m-Dibromobenzene[c]	3.8
p-Dibromobenzene[c]	6.9
Quinoline[d]	5.4
Isoquinoline[d]	18.6
1,2,3,4-Dibenzanthracene $(C_{22}H_{14})$[e]	0.6
Picene $(C_{22}H_{14})$[e]	12.0

[a] 23% CH_2Cl_2/pentane (7).
[b] 60%v CH_2Cl_2/pentane (7).
[c] Pentane (8).
[d] Benzene (9).
[e] CH_2Cl_2 (8).

the polyaromatic hydrocarbons of increasing ring size are well resolved in LSC (e.g., Figure 1.2), and oligomeric polystyrenes can be nicely separated on silica, as shown in Figure 9.4.

Isomer separations are usually much more pronounced in LSC than by reverse-phase LC. One example, that of the separation of 5,6- from 7,8-benzoquinoline, is provided in Table 9.2. Here the separation is trivially easy by LSC but rather difficult by reverse-phase BPC.

The basis of selectivity in LSC can be understood in terms of the adsorption process [for details, see (15, 16)]. Retention of a solute *S* in LSC requires displacement of an equivalent number of adsorbed solvent molecules *E*, as illus-

Table 9.2 Comparison of separation selectivity between liquid-solid and reverse-phase chromatography

Homologs

Value of k'

R	(liquid-solid)[a]	(reverse-phase)[b]
$R = C_1$	4.8	3.3
C_2		6.5
C_4	4.1	17
C_{10}	3.6	

	Value of α^c	
	Liquid-solid	Reverse-phase

Benzologs

	Liquid-solid	Reverse-phase
	1.2	1.4
	1.1	1.8

Table 9.2 (Cont.)

Isomers	Value of α^c	
	Liquid-solid	Reverse-phase
	12.5	1.06

[a] 5%v ether/pentane, alumina (10).

[b] 40%v methanol/water, C_{18} bonded-phase (6).

[c] (LSC) 1%v isopropanol/hexane, μPorasil (silica); (reverse-phase) methanol/water gradient on μ-Bondapak-C_{18} (11).

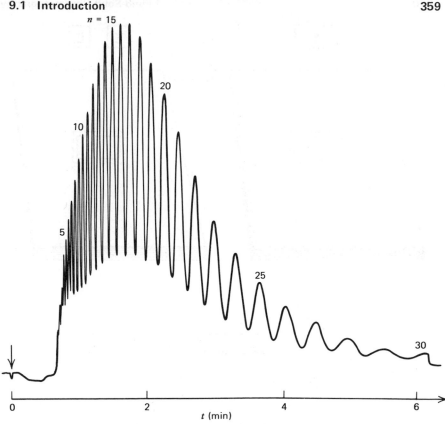

$n = 15$

20

10

5

25

30

0 2 4 6

t (min)

Figure 9.4 Separation of polystyrene oligomers $(\phi\text{-CH-CH}_2)_n$ in a 2100 molecular-weight polystyrene sample. Column, 25 × 0.4 cm LiChrosorb SI-60 (5-μm silica); mobile phase, 13%v tetrahydrofuran/pentane; 4.0 ml/min (2175 psi); UV, 254 nm; 10-μl (25 mg/ml) sample. Reprinted with permission of E. Merck, Darmstadt.

trated in Figure 9.5a. Because nonpolar hydrocarbon groups on a sample molecule are weakly attracted to the polar adsorbent surface, most solvents tend to displace such hydrocarbon groups from the surface, as in Figure 9.5b for n-butyl phenol. Therefore, hydrocarbon substituents contribute little to sample retention, and there is usually little difference in k' values among molecules differing only in their aliphatic substituents. Polar functional groups, on the other hand, are strongly attracted to the adsorbent surface, so that compounds with substituents of differing polarity (or a differing number of such groups) are readily separated.

Another feature of LSC is the presence of discrete adsorption sites on the surface of the adsorbent, as illustrated by the points A in Figure 9.5c. Optimum interaction between a solute molecule and the adsorbent surface occurs when

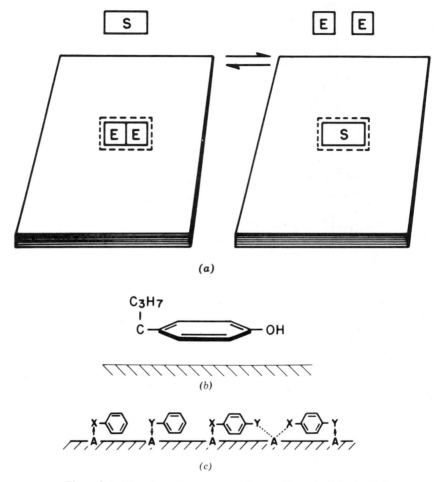

Figure 9.5   The adsorption process and separation selectivity in LSC.

solute functional groups exactly overlap these adsorption sites—as a hand fits a glove or a key its lock. This overlapping is readily possible for monofunctional solutes such as phenyl-X or phenyl-Y in Figure 9.5c, but not for polyfunctional solutes such as phenyl-X,Y. However, certain polyfunctional solutes will be better matched to the adsorbent surface than their isomeric counterparts, resulting in preferential retention of the compound that best fits the surface. In practice, it is rare that an isomeric pair of compounds cannot be separated by LSC, and sometimes the separation factors are surprisingly large, giving, for example, α values of 10 or more.

Apart from size-exclusion chromatography (Chapter 12), where the basis of separation is quite simple and relative retention is easily predicted for compounds of different molecular structure, retention in LSC is easier to understand and predict than for the other LC methods. Semiquantitative estimates of k' are often possible, knowing only the experimental conditions and the structure of the migrating compound (15). In some cases prediction of k' can be made within a few percent [e.g. see (7,17)] over a wide range of experimental conditions. This means that the control of separation in LSC (via changes in k' or α) can be relatively straightforward if we understand and utilize the basis of retention (see also Section 14.1).

It should be mentioned that the preceding simple picture of retention in LSC has recently been questioned (18, 19), at least for LSC with polar mobile phases. Final agreement over the details of retention in LSC has therefore not yet been reached. Fortunately, as pointed out by others (19, 20), because different theories of LSC retention lead to similar predictions of the effect of separation variables on retention, the practical significance of this continuing debate is relatively unimportant.

9.2 COLUMN PACKINGS

Packings for LSC (adsorbents) are described in terms of their chemical composition (silica, alumina, carbon), their surface area, and the geometry of the particle (diameter, porous vs. pellicular, and irregular vs. spherical). Chapter 5 provides a complete description of how particle geometry affects separation, which is the same for all the LC methods. Most LSC separations today make use of totally porous particles in the 5-10-μm range. For a detailed listing of the commercial packings available for LSC, see Table 5.5 for totally porous, small particles, Table 5.10 for pellicular packings, and Table 5.13 for packings used in preparative separation (large, porous particles).

Almost all present applications of modern LSC are limited to two adsorbent types: silica and alumina. Retention and separation on these two adsorbents are generally similar, with more polar sample components being preferentially held. The usual order of elution is

saturated hydrocarbons (small k') $<$ olefins $<$ aromatic hydrocarbons $\approx$ organic halides $<$ sulfides $<$ ethers $<$ nitro-compounds $<$ esters $\approx$ aldehydes $\approx$ ketones $<$ alcohols $\approx$ amines $<$ sulfones $<$ sulfoxides $<$ amides $<$ carboxylic acids (large k').

For a number of practical reasons most separations are today carried out on silica rather than alumina. Silica allows higher sample loadings, is less likely to promote unwanted reactions of the sample during separation, and is available in a wider range of chromatographically useful forms, and there is a wider literature covering previous applications of silica for different samples (both by column and TLC). When a change in separation selectivity is required, this is normally achieved via variation in the mobile phase (Section 9.3), rather than by a change in the column packing.

Occasionally, it is advantageous to consider the use of alumina in place of silica. The differences in selectivity between these two adsorbents have been reviewed in detail (15), and Table 9.3 summarizes some of these differences. Here the relative retention of different compound types on silica versus alumina is shown as the ratio of k' values. Most compound types show closely similar retention on each adsorbent, as seen in Table 9.3 for the various naphthalene derivatives. These compounds show variations in relative retention of no more than 30% at most, and usually much less (k' ratio of 0.9-1.2). Basic compounds, such as pyridines and amines, are usually more strongly held on silica. Table 9.3 illustrates this for two compounds of this type, with preferential retention on silica by a factor of 2 to 4. Acidic compounds are preferentially held on basic alumina (see below), frequently by large factors. This is illustrated in Table 9.3 for pyrroles, phenols, and thiophenols. Carboxylic acids are very strongly retained on alumina, sometimes irreversibly. Aromatic hydrocarbons, particularly those of larger ring size, are more strongly retained on alumina; for example, picene in Table 9.3. With optimum mobile phases and high-activity (i.e., low water content) adsorbent, alumina provides much better separations of aromatic hydrocarbon isomers than does silica (15, chapter 11).

Table 9.3 Selectivity on silica versus alumina.

Sample compound	(k' silica)/(k' alumina)
Substituted naphthalenes[a]	
$-OCH_3$, $-NO_2$, $-CO_2CH_3$,	
$-CN$, $-CHO$, $-COCH_3$	0.9-1.2
Basic compounds[a]	
2-Chloroquinoline	3.9
N-Methyl aniline	1.9
Acidic compounds[b]	
Pyrroles	0.06
Phenols	0.0002-0.02
Thiophenol	0.00001
Aromatic hydrocarbons[a]	
Picene ($C_{22}H_{14}$)	0.1

[a] 15%v benzene/pentane for silica; 25%v benzene/pentane for alumina (7).
[b] Approximate, average values from (21).

Another reason for preferring alumina over silica in certain applications is that silica is mildly acidic, whereas alumina (so-called basic alumina) is normally basic. Samples that are susceptible to acid or base-catalyzed reaction should therefore not be exposed to the adsorbent that promotes such unwanted alteration of sample components. Generally, alumina is most troublesome for base-sensitive samples, because basic alumina has a pH of about 12. Because the pH at the silica surface is only about 5, few acid-catalyzed reactions will take place on silica, and this adsorbent is properly regarded as the safest for general use in LSC. So-called neutral and acidic aluminas are also offered, but it is questionable whether these are to be preferred over silica.

The acidic nature of the silica surface also seems poorly adapted to the separation of cationic compounds (e.g., Figure 9.3a), or compounds possessing multiple basic groups such as $-N=$ or $-NR_2$. Presumably numerous, strong attachments are formed between the $\equiv Si-OH$ or $\equiv Si-O^-$ groups of the surface and corresponding basic or cationic groups in the adsorbing molecule. Cationic or basic compounds are usually better separated by ion-pair chromatography (Chapter 11). Recently a new adsorbent type has been reported: graphitized carbon-coated porous silica (21). Although this experimental packing is still relatively inefficient compared to the best present silicas and aluminas and is not commercially available, it offers some interesting future possibilities. Carbon is a nonpolar adsorbent, in contrast to silica and alumina, and retention on this packing is virtually the reverse of that for the polar adsorbents. In this respect carbon much resembles the nonpolar reverse-phase packings such as C_{18}-modified silica, although it is likely that carbon will provide increased selectivity for the separation of certain stereoisomers. At this point it is too early to know whether packings based on carbon will find significant use in modern LC.

A final factor that must be considered in characterizing an adsorbent such as silica is the variability of its surface, as a result of differences in surface area, average pore diameter, or water content. This is a somewhat complicated subject, but the essential features are fairly clear [e.g., see review in (15) and later work in (22-25)]. Adsorbent surface area in LSC is primarily important as a measure of the amount of stationary phase in the column. To a first approximation, sample k' values will be directly proportional to adsorbent surface area, other factors being equal. For this reason the surface area of an adsorbent must be held within narrow limits if its retention properties are to remain unchanged from batch to batch. Most porous silicas have surface areas of about $400 \, m^2/g$, and this is a practical optimum value for most applications. Higher surface areas can be achieved, but only at the expense of small pores (diameters less than 10 nm), which give poorer mass transfer and lower column N values. Lower surface areas mean decreased sample loadings and lower detection sensitivity, just as for pellicular packings versus totally porous packings; see, for example, the data of Figure 5.9, which show that Corasil-II ($25 \, m^2/g$) provides one-fifth the sample loading of a totally porous ($400 \, m^2/g$) silica. Because pellicular silicas have only

3-6% of the surface area of totally porous packings, k' values on pellicular packings will be about one-twentieth as large as on porous packings—other factors being equal (e.g., the same mobile phase).

Some workers have advocated the use of lower-surface-area silicas for the separation of more polar (and strongly retained) samples. It is not clear that this is good advice, since the solvent strength can always be changed to provide adequate k' values (Section 9.3). Although selectivity might be expected to differ somewhat between high and low surface-area adsorbents, and low-surface-area adsorbents are more readily regenerated after gradient elution or contamination with polar sample constituents, these possible advantages must be weighed against the lower sample capacity of the low-surface-area adsorbents. It is also inconvenient (and expensive) to keep several different kinds of silica columns on hand, when a single silica column (400 m^2/g) will be adequate for almost all separations.

The water adsorbed on the surface of the adsorbent also has an important effect on many of its properties. Surface water on silica exists in two forms: (a) chemically combined with the surface in the form of $\equiv$Si$-$OH groups, and (b) as molecular water which adsorbs onto strong adsorbent sites, as in Figure 9.5c. For silica, these strongly adsorbing sites are $\equiv$Si$-$OH groups of various kinds [e.g. see (15)]. Properly prepared silicas that have not been heated above 200-

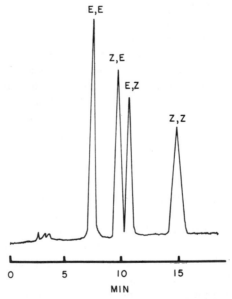

Figure 9.6 Separation of insect attractant isomers on silver-nitrate-coated silica. Column, 10 × 0.44 cm, 5% AgNO$_3$ on silica (LiChrosorb SI-60; 10-μm); mobile phase, benzene; ambient; refractive index detection. Reprinted from (26) with permission.

300°C have a maximum coverage of surface hydroxyl sites; all commercial silicas for LC are of this type. The molecular water adsorbed onto the silica (or alumina) surface plays an important role in the separation characteristics of the adsorbent. However, because the control of adsorbent water content is commonly achieved by varying the water content of the mobile phase, this aspect of the adsorbent is discussed separately in Section 9.3.

Silica coated with silver nitrate has found limited application for enhancing the separation of *cis/trans*-olefinic isomers. An example is shown in Figure 9.6: the separation of cis/cis (*Z, Z*), trans/cis (*E, Z*), and so on, isomers of a diunsaturated insect attractant. However, as seen in the example of Figure 9.1*b*, many olefinic isomers can be separated on silica directly (without addition of silver nitrate).

9.3 MOBILE PHASES

LSC separations are usually carried out with silica as adsorbent and at ambient temperature (Section 9.5). This means that variations in sample retention for optimum separation are achieved almost exclusively by changes in the mobile phase. Thus, the major decision required in the design of a satisfactory LSC separation is selection of the mobile phase. Fortunately, the mobile phase in LSC can provide enormous variation in sample retention and separation, and it is rare that two given compounds are not adequately separated in some LSC system. Furthermore, the basis of retention and separation in LSC is better understood than for most other LC methods. As a result, with a few simple rules we can utilize the enormous power of the mobile phase in LSC with minimum effort.

Solvent Strength

Solvent strength, which controls the k' values of all sample bands, is easily predicted in LSC. The solvent strength parameter $\epsilon°$ defines solvent strength quantitatively for a given adsorbent,* and Table 6.1 lists values of $\epsilon°$ (alumina as adsorbent) for a large number of different solvents. An increase in $\epsilon°$ means a stronger solvent and smaller k' values for all sample bands. Values of $\epsilon°$ for silica are generally similar to those for alumina, being about 0.8-fold smaller on average. Table 9.4 lists several useful solvents for LSC, in order of increasing $\epsilon°$ values for silica as adsorbent.

*The LSC solvent strength parameter $\epsilon°$ has been incorrectly referred to as the "Hildebrand solubility parameter". Actually, $\epsilon°$ is the adsorption energy per unit area of the solvent [see the discussion in (*15*)].

Table 9.4 Some useful solvents for use as LSC mobile phases.

Solvent	Solvent strength ϵ° Silica	Alumina	Selectivity group
3-M Fluorochemical FC-78	−0.2	−0.25	−
n-Hexane n-Heptane Isooctane	0.01	0.01	−
1-Chlorobutane	0.20	0.26	V
Chloroform	0.26	0.40	VIII
Methylene chloride	0.32	0.42	V
Isopropyl ether[a]	0.34	0.28	I
Ethyl acetate	0.38	0.58	VI
Tetrahydrofuran	0.44	0.57	III
Propyl amine	~0.5		I
Acetonitrile	0.50	0.65	VI
Methanol	~0.7	0.95	II

[a]Methyl t-butylether has recently been recommended (25a) as a preferred alternative to either isopropyl or ethyl ether in LSC and other LC procedures. Its advantages include much reduced susceptibility to peroxide formation (so much so that antioxidants need not be added during use), a solvent strength ϵ° similar to ethyl ether (and somewhat greater than for isopropyl ether), and less potential hazard.

The usual technique for determining the right solvent strength in LSC is similar to that followed in other LC methods (e.g., Section 7.3 for BPC and LLC). Two solvents whose ϵ° values are, respectively, too small (A) and too large (B) are blended together in various proportions to allow continuous variation in ϵ° from that of pure A to pure B. Some binary mixture of A and B will therefore provide just the right strength and sample k' values for a given separation. Because Table 9.4 provides solvents with an enormous range of solvent strength, some binary solvent A/B from Table 9.4 will allow adequate retention of any given sample. An increase in solvent ϵ° value by 0.05 units usually decreases all k' values by a factor of 3-4.. Thus, the solvents of Table 9.4 allow variation in sample k' values over a range of about 10^{10} (for silica).

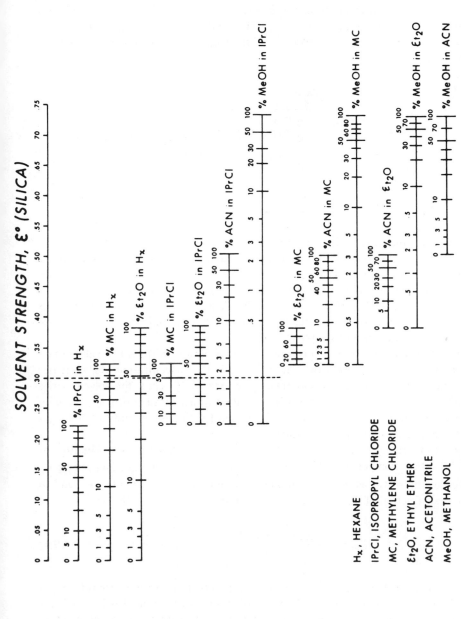

Figure 9.7 Solvent strength as a function of binary composition (silica as adsorbent). See text, Reprinted from (28) with permission.

367

Solvent strength in LSC does not vary linearly with the %v B in the binary mixture A/B. This is illustrated in the data of Figure 9.7, which plots ϵ° values (silica) for several binary solvent mixtures A/B. Each horizontal line in Figure 9.7 corresponds to compositions of a particular binary A/B; the first such line is for hexane (A) and isopropyl chloride (B). For this particular binary, the data of Figure 9.7 suggest the following ϵ° values (top of Fig. 9.7) as a function of composition.

%v B	ϵ°
0	0.00
20	0.08
40	0.14
60	0.17
80	0.20
100	0.22

In varying the mobile-phase composition for change in solvent strength, it is desirable initially to change ϵ° by some fixed amount each time, for example, by 0.05 units. For the preceding case of hexane/isopropyl chloride, this would correspond to the following compositions: 100%v B $(\epsilon^\circ = 0.22)$; 60% v B (0.17); 33%v B (0.12); 18%v B (0.07); 4%v B (0.02). The data of Figure 9.7 for these 13 solvent binaries allow the rational adjustment of binary composition in each case for the precise control of sample retention.

For binary mixtures of two solvents A and B of quite different ϵ° values (e.g., tetrahydrofuran/hexane), a twofold decrease in the concentration of B increases sample k' values by two to four times [e.g., see (16)]. Thus, the solvent series 100%v, 50%v, 25%v, 12%v, and 6%v tetrahydrofuran/hexane provides roughly equal spacing of ϵ° values and could be used for the initial study of the right solvent strength of a given separation.

We can generalize these observations on solvent strength as a function of mobile-phase composition. The value of ϵ° for a binary solvent (ϵ_{ab}) can be related to the ϵ° values of the components A and B of the solvent (ϵ_a, ϵ_b) and to the mole fraction of B (N_b) in the binary mixture:

$$\epsilon_{ab} = \epsilon_a + \frac{\log (N_b \, 10^{\alpha n_b(\epsilon_b - \epsilon_a)} + 1 - N_b}{\alpha n_b}. \tag{9.1}$$

Here α is a constant that varies with adsorbent activity or water content $(0.6 \leqslant \alpha$ 1.0) and n_b varies (from about 4 to 6) with the molecular size of B [see (15) for further discussion]. The dependence of solvent strength (ϵ_{ab}) on composition

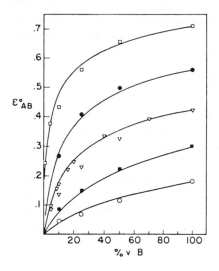

Figure 9.8 Variation of solvent strength in LSC as a function of binary composition. Binary solvent ϵ° values versus solvent composition for adsorption onto alumina; ○, pentane (A) -CCl_4, (B); ■, pentane (A); n-propyl chloride (B); ▲, pentane (A), CH_2Cl_2 (B); ●, pentane (A), acetone (B); □, pentane (A), pyridine (B). Solid lines calculated from Eq. 9.1. Reprinted from (15) with permission.

Table 9.5   Solvent strength in liquid-solid chromatography for carbon as adsorbent.

Solvent	ϵ° (carbon)
Methanol	0.00
Acetonitrile	0.04
Ethanol	0.05
Ethyl acetate	0.13
n-Hexane	0.10
n-Heptane	0.13
Methylene chloride	0.13
Butyl chloride	0.13
n-Octane	0.14
Tetrahydrofuran	0.14
n-Nonane	0.17
Chloroform	0.18
Benzene	0.20
Xylene	0.24

Source: From (27).

($\%v$ B) for several binaries and alumina as adsorbent is plotted in Figure 9.8. The solid curves are from Eq. 9.1 and the points are experimental values.

Colin et al. (27) have recently published a listing of solvent ϵ° values for carbon as adsorbent. As can be seen in Table 9.5, these are in the reverse order for the polar adsorbents; the more polar solvents in this case have lower ϵ° values.

Although the right solvent strength for a given LSC separation can be determined empirically as described earlier, other approaches are available that offer greater speed and convenience. First, for samples of known composition, a rough estimate of the optimum solvent ϵ° value is available from the procedure described by Saunders (28). This is summarized in Table 9.6 for different compound types and for solvents that are anhydrous or contain added water (see later discussion). A second approach to selecting the right solvent strength and composition is by means of TLC, as discussed in detail later.

Solvent Selectivity

Sample α values in LSC are varied by holding the solvent strength ϵ° constant, while exchanging one strong solvent B for another. In practice, this is done with the aid of solvent strength plots such as that of Figure 9.7. Typically a trial-and-error substitution of different solvents of equal ϵ° are tried. For example, if $50\%v$ ether/hexane provides the right solvent strength (e.g., $\epsilon^\circ = 0.30$, from Figure 9.7) but a change in solvent selectivity is desired, any of the solvents in Figure 9.7 falling on the vertical dashed line for $\epsilon^\circ = 0.30$ could be substituted for $50\%v$ ether/hexane as a possible alternative for improved solvent selectivity. Thus, looking from top to bottom in Figure 9.7, solvents of equal strength but different B would include $75\%v$ methylene chloride/hexane, $1.7\%v$ acetonitrile/isopropyl chloride, and $0.2\%v$ methanol/isopropyl chloride. One of these solvents would probably provide the selectivity desired. Following the selection of the best solvent in terms of selectivity, it is often required to fine-tune solvent strength again, by minor adjustment in the $\%v$ B of the final mobile phase chosen. The reason is that solvent strength scales such as that of Figure 9.7 are only approximate.

The trial-and-error approach to solvent selectivity as just described is tedious at best, and sometimes completely unsuccessful. When α must be increased, what we want is a set of simple rules that predict the variation of α with solvent composition. Fortunately, there are two general rules that can be used to advantage: the "B-concentration" rule and the "hydrogen-bonding" rule. We next discuss each in turn.

The "*B-concentration*" *rule* is illustrated in Table 9.7, which shows the separation of two substituted naphthalenes by different mobile phases on alumina as adsorbent [similar effects are observed for silica; see (7)]. Here the strong

Table 9.6 Correct solvent strength (silica as adsorbent) for compounds of different type.

Compound class	Optimum $\epsilon^{\circ a}$	
	Dry solvent	Water added[d]
Aromatic hydrocarbons	0.05-0.25	−0.2-0.25
Alkyl or phenyl halides R-X_n	0.0-0.3	−0.2-0.1
Mercaptans, disulfides	0.0	−0.2
Sulfides	0.1	−0.1
Ethers	0.1	0.0
Nitro-compounds[b]	0.2-0.3	0.1
Esters[b]	0.2	0.1
Nitriles[b]	0.2-0.3	0.1
Ketones[b]	0.3	0.1
Aldehydes[b]	0.2	0.1
Sulfones[b]	0.3-0.4	0.2
Alcohols[b]	0.3	0.2
Phenols[b]	0.3	0.2
Amines[c]	0.2-0.6	0.0-0.4
Acids[b]	0.4	0.3
Amides	0.4-0.6	0.3-0.5

[a]Weaker solvents are required if water is added to the mobile phase (see last column).
[b]For monofunctional compounds (e.g., nitrobenzene); polyfunctional compounds require larger ϵ° values.
[c]Tertiary amines require smaller ϵ° values, primary and secondary aliphatic amines require larger ϵ° values.
[d]50% water-saturated.

solvent B is changed for each binary mobile phase. Because solvent strength is held roughly constant, this requires a corresponding change in the concentration of B in each mixture. For the moderately weak solvent benzene as B, a concentration of 50%v B is required. For the very strong solvent dimethyl sulfoxide, only 0.05%v B in the weak solvent A (pentane) is required. As the strength of the pure solvent B is increased, and the concentration of B is therefore decreased (moving down Table 9.7), it is seen that the compound dinitronaphthalene is retained to a greater extent, relative to the compound acetyl naphthalene;

that is, solvent selectivity changes regularly and continuously. Thus, for the largest change in solvent selectivity and sample α values, a very dilute or a very concentrated solution of B in the weak solvent A should be used. As a corollary, maximum α values are generally found for either very small or very large concentrations of the solvent B. In the example of Table 9.7, maximum α values were found *both* at low and at high concentrations of B. In other cases, however, maximum α values might occur at only one end or the other of the %v B range, that is, either for low %v B or high %v B, but *not* for both.

Table 9.7   Solvent selectivity in LSC; the "B-concentration" effect.

| | | k' | | |
| | | COCH$_3$ structure | NO$_2$ structure | |
Solvent (pentane solution)			NO$_2$	α
50%	Benzene	5.1	2.5	0.5
25%	Ethyl ether	2.5	2.9	0.8
23%	CH$_2$Cl$_2$	5.5	5.8	0.95
4%	Ethyl acetate	2.9	5.4	1.8
5%	Pyridine	2.3	5.4	2.4
0.05%	Dimethylsulfoxide	1.0	3.5	3.5

Source: Data from (7).

Returning to our previous discussion of Figure 9.7, we assumed our initial separation was with 50%v ethyl ether/hexane. From Figure 9.7 we can select other solvents of the same strength: 75%v methylene chloride/hexane, 1.7%v acetonitrile/isopropyl chloride, or 0.2%v methanol/isopropyl chloride. Any of these particular solvents offers a large change in %v B, but the latter two solvents would probably give the largest change in α values, because the relative change in %v B is greater, that is (50/1.7) and (50/0.2) versus (75/50) for methylene chloride/hexane.

The *"hydrogen-bonding" rule* can be stated as follows: Any change in the mobile phase that results in a change in hydrogen bonding between sample and mobile-phase molecules generally results in large changes in α. An example is shown in Table 9.8, for the LSC separation of N-methyl aniline from 2-chloroquinoline on alumina, using different mobile phases (different strong solvents B)

of roughly equal strength. The different solvents B are classified according to their selectivity group (Section 6.5 and Table 6.1), group I corresponding to strong proton acceptors; group V, to neutral compounds (large dipole moment); and group VIII, to proton donors. For the two compounds being separated, both are bases, so their response to proton-donor or neutral solvents should be similar, (little difference in α). However, N-methyl aniline is a weak proton donor, and its interaction with basic solvents of group I results in a large change in α (Table 9.8).

Table 9.8 Solvent selectivity in LSC; the "hydrogen-bonding" effect.

| Mobile phase[a] | Solvent selectivity group | k' values[b] | | α |
		(1)	(2)	
5%v ethyl ether	I	6.5	4.0	1.6
5%v triethylamine	I	4.4	2.0	2.2
23%v CH_2Cl_2	V	2.2	2.3	1.05
15%v CH_2ClCH_2Cl	V	3.2	2.9	1.10
30%v $CHCl_3$	VIII	2.0	2.6	1.30

Source: Data from (7).
[a] %v B (pentane for A).
[b] (1) refers to N-methylaniline, (2) to 2-chloroquinoline.

The solvent selectivity group indicated in Tables 6.1 or 9.4 provides a ready classification of the hydrogen-bonding characteristics of the solvent. Maximum change in hydrogen bonding (and solvent selectivity) would be expected from one of the four extreme groups: I, II, V, or VIII. Thus (just as in BPC–Section 7.3), solvents from each of these four groups should be tried when attempting to maximize solvent selectivity. For example, from Table 9.4 we might select as possible B solvents for selectivity optimization (via hydrogen-bonding effects): n-propylamine or isopropyl ether (I), methanol (II), methylene chloride (V), and chloroform (VIII). Intermediate solvent selectivity can be obtained by blending ternary solvents. For example, some mixture of chloroform/hexane might result in overlapping of one pair of compounds, whereas another mixture of isopropyl ether/hexane might give poor resolution for another pair of compounds in the sample. Presumably, some mixture of chloroform/isopropyl ether/hexane (of the right strength) can be found that will provide intermediate retention and allow separation of all bands in the sample (see also Section 6.5).

Apart from the latter example, ternary solvent mixtures are rarely required in LSC for controlling selectivity. By means of the preceding two rules for maximizing solvent selectivity, most samples should be adequately resolved by simple binary mobile phases. Sometimes other solvent components are required, for reasons other than selectivity. We next discuss the addition of water or polar modifiers for adsorbent deactivation and general improvement in separation (apart from α values). In Section 9.5 we discuss the addition of other mobile-phase components to control band tailing.

Finally, it should be noted that the weak solvent component A normally plays little role in affecting the selectivity of the mobile phase. Therefore, the choice of A is relatively unimportant so far as separation per se is concerned, particularly as long as $\epsilon°$ for pure A is less than 0.2.

Addition of Water or Polar Modifiers to the Mobile Phase

In LSC it is customary to add small amounts of water or a polar "modifier" such as methanol to the mobile phase. These polar compounds are preferentially adsorbed onto strong adsorption sites (see discussion of Figure 19.2), leaving a more uniform population of weaker sites that then serve to retain the sample. This "deactivation" of the adsorbent leads to a number of improvements in subsequent separations, which we will shortly discuss. Unfortunately, the proper use of water or polar modifiers in LSC is not without difficulty. For this reason a somewhat detailed discussion of this procedure is given here. Once the various pitfalls are recognized, the actual use of adsorbent deactivation is relatively straightforward and amply justifies the extra effort involved.

Because of the nuisance involved in adsorbent deactivation procedures, there is increased interest in polar-phase BPC packings as an alternative to LSC with silica or alumina. These latter packings are an adequate substitute for LSC in some cases and do not require the addition of water to the mobile phase. However, the traditional adsorbents offer many features that have so far not been duplicated in BPC packings. The optimal use of LSC with adsorbent deactivation is therefore expected to find continued heavy use in the foreseeable future.

The Value of Adding Water. Among the beneficial effects of adding water to the mobile phase in LSC are the following:

- Less variation in sample retention from run to run.
- Substantially increased maximum sample loadings (greater sample capacity) for linear-isotherm separation.
- Higher column efficiencies and reduced band tailing in some cases.
- Partial compensation for lot-to-lot differences in adsorbents of the same type.
- Reduced tendency of the adsorbent to catalyze unwanted reactions of the sample during separation.

Some of the preceding effects are illustrated in Figure 9.9, for elution of a model solute (3-phenyl-1-propanol) from an 8-μm totally porous silica. Figure 9.9a

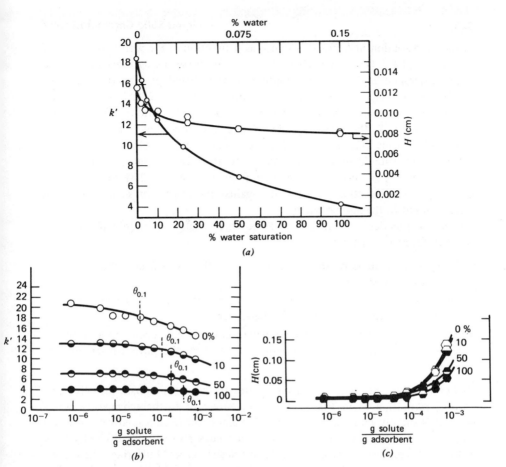

Figure 9.9 Effects of added water in mobile phase on separation of model compound (phenyl propanol) on porous silica. (*a*) Variation of k' and H values with % water saturation of mobile phase. (*b*) Variation of k' with sample size for different % water saturation of mobile phase. (*c*) Same as (*b*), except variation of H with sample size and % water saturation.

	% water saturation	% water
◯ ⬡	0	0.0007
◖ ⬢	10	0.015
◑ ⬢	50	0.075
● ⬢	100	0.15

Column, 25 × 0.32 cm, Zorbax-Sil (8-9 μm); mobile phase, methylene chloride with varying amounts of water; ambient; 2.0 ml/min; UV, 254 nm; 20-μg sample in (*a*). Reprinted from (*30*) with permission.

shows the variation in k' for this solute as a function of the water saturation of the mobile phase. Here the partial water saturation (see below) varies from 0 to 100%, corresponding to a water content of the solvent of from 0 to 0.15%w. The k' versus percent water-saturation curve of Figure 9.9a is seen to drop sharply with increasing water in the region 0-25% saturation, but for higher water contents of the mobile phase, k' changes more slowly with increasing water. This means that retention (k' values) is quite sensitive to small changes in mobile-phase water concentration when saturation is less than 25%, but much less so when water saturation is greater than 25%. Above 25% saturation the mobile phase is effectively "buffered" against the effects of small changes in water content of the mobile phase.

In actual practice it is difficult to avoid small changes in mobile-phase water content, for several reasons:

- The water contents of the starting (supposedly dry) solvents that comprise the mobile phase vary somewhat.
- Water is readily picked up from or lost to the atmosphere, depending on its humidity and the water content of the solvent.
- Further changes in mobile-phase water content can occur as a result of contact with the walls of intermediate containers, the solvent reservoir, and so on.

However, by adding some quantity of water to the mobile phase initially (e.g., 25-50% saturation as in Figure 9.9a), the effect of further changes in water content is sharply reduced, and sample k' values will remain relatively constant from run to run. In some LSC systems, α values also vary significantly with adsorbent water content [e.g., see (8)], and in such cases "buffering" the mobile phase with water minimizes changes in relative as well as absolute retention. For the same reason addition of more or less water to the adsorbent can be used to offset differences in retention for adsorbents from lot to lot.

Figure 9.9b shows the variation in k' as a function of sample size, for four different water-saturation values of the mobile phase (0, 10, 50, and 100%). The quantity $\theta_{0.1}$ in Figure 9.9b refers to the sample size that leads to a 10% reduction in k' from the value at small sample injections; it is arbitrarily defined as the *linear capacity* of the LSC system, that is, the maximum sample loading possible, assuming that decrease in k' by more than 10% is unacceptable. In Figure 9.9b, $\theta_{0.1}$ increases from 0.04 mg sample per gram adsorbent (0% saturation to 0.4 mg/g for 100% saturation. Thus, larger sample weights can generally be injected in LSC when water deactivation is used.

Figures 9.9a and 9.9c show the effect of water saturation of the mobile phase on column H values (i.e., column efficiency). Here a slight decrease in H with increasing water saturation is observed in the nonoverloaded separation of (a); the effect is larger when the column is overloaded, as in (c). Thus, column efficiency is often improved when water is added to the mobile phase.

A practical example of the beneficial effects of water addition in LSC is provided by Figure 9.10 for the separation of two steroids on silica. In (a) the separation with a dry mobile phase gives poor resolution and tailing bands. Addition of water to the mobile phase (b) improves peak shape and results in higher column efficiency.

Isohydric Solvents. Although the advantage of adding water to the mobile phase in LSC is relatively obvious, in actual practice certain problems arise. First of all, there is the question of how much water should be added to the mobile phase for optimum performance. In the example of Figure 9.9 it appears that the amount of added water should exceed 25% saturation, or about 0.04%w. Other studies [e.g., see (31)] have shown that the use of 100%-saturated solvents is undesirable, because such LSC systems are often unstable. Apparently under these conditions the pores of the adsorbent gradually fill with water, leading to changes in retention with time, and in some cases a change from LSC retention

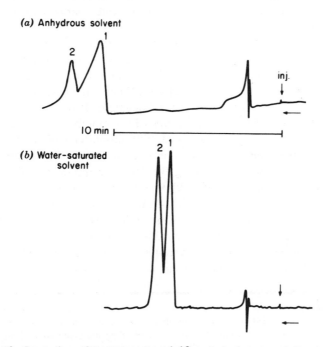

Figure 9.10 Separation of testosterone and 19-nortestosterone acetates on silica. (a) 0% water-saturated mobile phase; (b) 100% water-saturated mobile phase. Column, 25 cm, μ-Porasil (10-μm porous silica); mobile phase, 40%v ethyl ether/hexane; ambient; 2 ml/min; UV, 254 nm. Reprinted from (36) with permission.

to LLC retention (with water as stationary phase). Another study (*32*) finds that for water partial pressures above 50% (corresponding to the preceding 50% water saturation of the mobile phase), the amount of water taken up by the adsorbent (and sample k' values) varies with the history of the column (i.e., the order in which different mobile phases are used). For these and other reasons it seems desirable to maintain the water saturation of the mobile phase between certain limits, and *ideally at about 50% water saturation for the silicas* used in modern LSC. Other studies (*15, 37*) suggest a value of 25% water saturation for alumina.

A major problem in the use of water-containing mobile phases is that of slow-column equilibration when changing from one mobile phase to another—as required for the separation of different samples. This is particularly pronounced for totally porous adsorbents, but less so for pellicular particles. Slow column equilibration arises most commonly through the use of different mobile phases that are not *isohydric*. By definition (*33*), isohydric mobile phases can be used with a given column interchangeably, without changing the water content of the adsorbent. That is, when a second isohydric mobile phase is used, the adsorbent neither takes up nor loses water to the mobile phase. When the successive mobile phases are not isohydric, the adsorbent water content changes with change in the mobile phase.

Solvents with the same partial saturation by water (e.g., 50%) are roughly isohydric; they provide rapid column equilibration after a change in mobile phase, because the column is already roughly equilibrated with respect to water. For a change to a nonisohydric mobile phase, column equilibration can in some cases be very slow. This is illustrated in the example of Figure 9.11*a*, where an alumina column was initially in equilibrium with a mobile phase composed of 35%v CH_2Cl_2/pentane, 50% water saturated. With this initial mobile phase, injection of quinoline as sample gave a retention time of 13 min. The mobile phase was subsequently changed by increasing the water saturation to 75%. This resulted in a gradual accumulation of additional water onto the adsorbent, and a decrease in k' values (as in Figure 9.9*a*). While the second mobile phase was flowing through the column, injections of quinoline were made repeatedly, and the resulting t_R values were plotted versus the volume of second mobile phase that had passed through the column. As seen in Figure 9.11*a*, about 300 column volumes of mobile phase were required in this system before the retention of

Figure 9.11 Equilibration of adsorbent column with (*a*) water and (*b*) acetonitrile. (*a*) Retention time for quinoline as sample, alumina packing, and 35%v CH_2Cl_2/pentane as mobile phase; initial water saturation of solvent is 50%, then increased to 75% and sample injected for t_R measurement as a function of mobile-phase volume. Reprinted from (*37*) with permission. (*b*) Retention time for diphenoxybenzene as sample, silica as packing, and hexane as mobile phase; column was initially equilibrated with dry hexane, then 30 μl of acetonitrile was injected and elution continued with 0.05%v acetonitrile/hexane. Reprinted from (*34*) with permission.

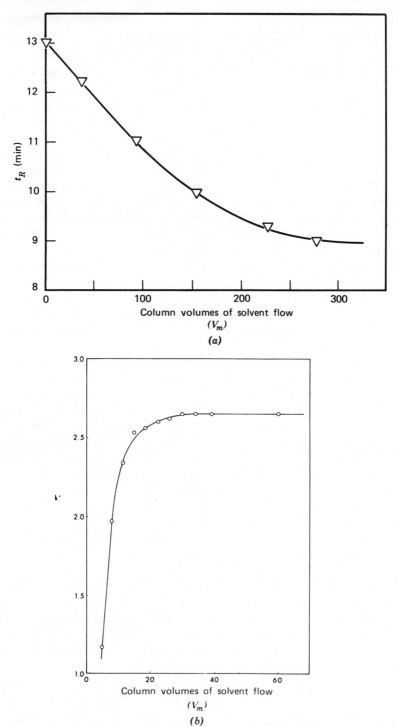

(a)

(b)

379

quinoline became constant (t_R = 9 min). At this point column equilibration was achieved, and separations with the new mobile phase could be carried out.

The slow equilibration of the column of Figure 9.11a, following a change in % water saturation of the mobile phase, is due to the large capacity of the adsorbent for water and the low solubility of water in this mobile phase. Typically, LSC adsorbents can adsorb 10% or more of water, whereas less than 1% of water can be dissolved in typical mobile phases (e.g., only 0.15%w water for 100%-water-saturated CH_2Cl_2 in Figure 9.9). If, following a change in mobile phase as in Figure 9.11a, samples are injected prior to equilibration of the column (i.e., before 300 column volumes in this particular case), generally unsatisfactory results are obtained: smaller N values, variable k' values, and poor resolution. Therefore, it is *always* advisable following *any* change in mobile phase to check for column equilibration by repeated injection of some retained sample (as in Figure 9.11a), until constant k' values are obtained.

The problem of slow column equilibration toward water can be reduced by using isohydric solvents where possible (e.g., 50%-saturated ethyl ether, 50%-saturated CH_2Cl_2, etc.). Therefore, if, for example, 50%-saturated mobile phases are used, each of these 50%-saturated solvents can be interchanged as mobile phase (for different samples), and column equilibration times will be minimal—usually within 20-30 column volumes of the new mobile phase.

The use of isohydric mobile phases solves the problem of long equilibration times for a column that is in use, and therefore already roughly equilibrated with respect to water. However, there is still the problem of columns that are not yet equilibrated with water: newly packed columns, regenerated columns (Section 9.5), and so on. For these cases it is desirable first to equilibrate the column with a mobile phase that has a high water solubility, for example, 50%-saturated ethyl ether (0.65%w water) or 50%-saturated ethyl acetate (4.9%w water). See Table 6.1 for solubility of water in other organic solvents. The high water content of these mobile phases in turn allows rapid equilibration of the column with water (within 20-30 column volumes). Following this initial column equilibration, any other isohydric mobile phase can be substituted for the latter solvent, with equilibration normally occurring within 30 column volumes. Use of the mobile phase of interest directly on an unequilibrated column can result in very much longer equilibration times, as in Figure 9.11a.

Isohydric solvents avoid still another problem, that of maintaining an optimum water content of the adsorbent. With such mobile phases the adsorbent is never "dry" (0% saturation) nor water saturated (100% saturation), thus avoiding all the problems associated with these extremes. Finally, the exact (%w) water content of a series of isohydric mobile phases need not be separately determined, because each is, for example, 50% water saturated.

Preparation of Isohydric Solvents The preparation of 50%-water-saturated mobile phases deserves some comment. The dry mobile phase should first be blended if required (e.g., 35%v CH_2Cl_2/pentane, as in Figure 9.11a), then divided into two equal volumes. One volume should be saturated with water, to give 100%-saturated solvent. Finally, the two volumes (dry and wet) should be combined for the final mobile phase. Note that it is *not* allowed to prepare 50%-saturated CH_2Cl_2 and 50%-saturated pentane, and then combine these to give 35%v CH_2Cl_2/pentane. The resulting solvent would not be isohydric, and water could actually come out of solution, because the solubility of water in organic mixtures is not a linear function of the %v composition.

In principle, dry solvents to be used as described here should be prepared by storing commercial solvents over a 4-Å molecular sieve (Linde). In practice, it is more convenient and generally adequate simply to assume that freshly opened bottles of LC-grade solvents are approximately "dry." The saturation of the mobile phase with water (to give 100%-water-saturated solvent) is more difficult than might be assumed. Simply shaking water and most water-immiscible solvents often requires several hours for complete equilibration. A more satisfactory alternative is as follows. First, prepare a quantity of 30% water/silica by mixing together water and Davison Code 62 grade silica (W. C. Grace & Co., Baltimore) in a closed bottle. Shake until the mixture is free-flowing. The mobile phase to be saturated with water (e.g., 500 ml) is added to a vessel equipped with a magnetic stirrer, and 10-20 g of the 30% water/silica are added. The mixture is gently stirred for 30 min, the silica is allowed to settle, and the water-saturated solution is decanted off and stored in a stoppered bottle. The resulting 100%-saturated solvent can then be blended 1:1 with dry solvent to give 50%-saturated mobile phase.

When the mobile phase can dissolve 0.5%w of water or more at saturation, an alternative procedure is preferable. Small aliquots (0.1-0.2%w) of water are added to the solvent with subsequent vigorous shaking in a suitable container. When the last aliquot fails to go into solution, the saturated mobile phase is separated from the residual water and stored as previously indicated.

Mobile phases that are miscible with water or contain water-miscible components must be handled differently. Here it is common simply to add 0.1-1%w water to the mobile phase. Although such solvents are no longer strictly isohydric with respect to 50%-water-saturated solvents, the effect of water in such mobile phases is less important; the water-miscible, polar mobile-phase components take the place of water (see the following discussion).

Use of Organic Modifiers Instead of Water. Addition of water to the mobile phase is not always the best approach to adsorbent deactivation. We recommend that 50%-water-saturated mobile phases be used for isocratic separations with water-immiscible solvents, whenever the mobile phase $\epsilon°$ value is greater than about 0.1. Addition of water to weaker solvents is not advisable for the following reasons:

- Solvents with $\epsilon° < 0.1$ are relatively nonpolar, and 50%-water-saturated mobile phases generally contain less than 0.02% water.

- It is difficult to prepare and handle solvents containing only traces of water, without changing their water content in the process (34).

- Limited data (33) suggest that 50%-water-saturated solvents such as isooctane and cyclohexane are no longer isohydric with more polar solvents (e.g., CH_2Cl_2) that are also 50% water saturated; 60-70% water saturation of these saturated hydrocarbon solvents is required.

In the case of gradient elution, use of 50%-water-saturated solvents A and B to form the gradient leads to the same problem noted earlier in the blending of isohydric solvents A and B. The admixture of 50%-saturated A and B during a gradient leads to intermediate compositions of A/B that are nonisohydric, with possible precipitation of water within the column. This adversely affects separation and leads to separation irreproducibility and long column-regeneration times.

For either weak solvents ($\epsilon° \leqslant 0.1$) or gradient elution, a preferred alternative to water deactivation of the adsorbent is the use of polar organic modifiers that mimic the effects of water. Such modifiers include methanol, acetonitrile, and isopropanol. Acetonitrile added to nonpolar mobile phases can largely duplicate the effects of adding water [i.e., effect on sample capacity, H values, k' and α values, etc.; see (34). In such cases it is generally necessary to add about four times as much acetonitrile as water. For less-polar solvents in LSC, the following acetonitrile concentrations are suggested:

Mobile phase	%w acetonitrile modifier
Hexane, heptane, isooctane	0.02
Benzene	0.12
CH_2Cl_2	0.34
$CHCl_3$	0.14
Isopropyl ether	1.2

Thus, for 5% v CH_2Cl_2/hexane ($\epsilon° = 0.08$ in Figure 9.7), one might add (0.02 × 0.95) + (0.34 × 0.05) = 0.036%w acetonitrile to the mobile phase in place of water. Similar calculations for other solvents can be made using the preceding data.

For water-miscible mobile phases or in gradient elution, it is common to add 0.05-0.2%w acetonitrile or methanol as a water substitute. A further advantage of using these organic modifiers in place of water is that the final mobile phases can be more simply prepared, by measuring in the requisite amount of modifier. Column equilibration with polar modifiers can be speeded up greatly by injecting an excess of the modifier onto the column, prior to column equilibration with the new mobile phase containing the modifier. This technique is illustrated in Figure 9.11b for equilibration of an initially activated column of silica (containing no water or modifier) with 0.05%w acetonitrile/hexane as mobile phase.

Equilibration is achieved within about 30 column volumes, compared to 300 column volumes in Figure 9.11*a*.

The use of organic modifiers in place of water is not itself without problems. Methanol and propanol can produce distorted bands for lower concentrations of the modifier (*30*). Also, in some systems organic modifiers do not give α values equivalent to those for addition of water to the mobile phase.

For a further discussion of the use of water and polar modifiers in LSC, see (*30, 33, 34, 37*) and the review in (*35*).

Thin-Layer Chromatography to Pilot LSC Separation

Because the selection of a mobile phase of the right strength and selectivity is the major task in designing an LSC separation for a given sample, anything that simplifies this task will prove useful. The search for the right solvent can be carried out on the LC unit directly, but this is often a tedious, time-consuming approach. A much easier and faster procedure is to use TLC in the initial scouting for an appropriate mobile phase. Under similar conditions separation by LSC and TLC proceed by the same retention mechanism, and R_f values in TLC can be used to predict k' values in LSC (same adsorbent and mobile phase) by means of the relationship.

$$k' = \frac{(1 - R_f)}{R_f}.\tag{9.2}$$

Table 9.9 Dependence of k' values in LSC on R_f values in TLC.

R_f	k'	R_f	k'
0.01	99	0.25	3.0
0.02	49	0.30	2.3
0.03	32	0.35	1.9
0.04	24	0.40	1.5
0.05	19	0.45	1.2
0.06	15	0.50	1.0
0.07	13	0.55	0.8
0.08	11	0.60	0.7
0.09	10	0.65	0.5
0.10	9.0	0.70	0.4
0.12	7.3	0.75	0.3
0.14	6.1	0.80	
0.16	5.3	0.85	≈ 0.0
0.18	4.6	0.90	
0.20	4.0	0.95	

Values of k' versus R_f from Eq. 9.2 are listed in Table 9.9. For reasons we consider shortly, Eq. 9.2 is seldom obeyed exactly, and TLC data should generally be used as *rough* estimates of retention in corresponding LSC systems.

The first step in the TLC selection of an appropriate mobile phase is to find the solvent strength that is correct for the sample to be separated. This is conveniently accomplished with the technique illustrated in Figure 9.12. A standard 5 × 20-cm TLC plate coated with silica is laid horizontally on the lab bench (this approach assumes that the LSC separation will be carried out with silica also). Two- to 10-μl aliquots of sample are now spotted equidistantly along the right-hand side of the plate. A series of seven solvents with $\epsilon°$ values equal to 0.1, 0.2, and so on, is now used to develop each of these spots. The solvent series shown in Figure 9.12 was chosen from the data of Figure 9.7. The development of each spot proceeΦ as follows. A fine-tipped medicine dropper is used to transfer the first solvent (25%v isopropyl chloride/hexane) to the first sample spot. Dropwise addition (2 or 3 drops) is continued until the spot diameter (solvent front) has increased to about 2 cm. This procedure is then repeated immediately to the left of the sample spot to provide a blank development for comparison. Following development of the first sample spot, the remaining spots are in turn developed with solvents 2, 3, and so on. At the conclusion of sample development, the plate is allowed to dry in air; then it is placed into a covered

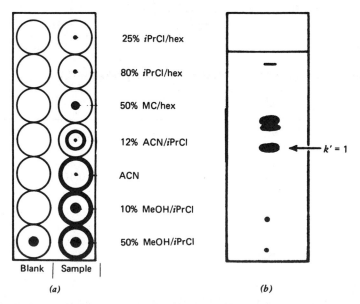

Figure 9.12 Use of TLC to scout mobile phases for LSC. (*a*) Radial TLC of a hypothetical sample, using seven solvents of increasing strength. (*b*) Normal TLC separation of same sample, using solvent (12% ACN) selected from (*a*). *i*PrCl, isopropyl chloride; hex, hexane, MC, methylene chloride; ACN, acetonitrile; MeOH, methanol.

jar containing a few crystals of iodine. The iodine vapor results in the visualization of the plate in 10-15 min, as illustrated in Figure 9.12a. Typically, the spots for the low-solvent-strength mobile phases appear as dark points at the center of a large ring that corresponds with the solvent front. As solvent strength increases along the plate (downward in Figure 9.12a), the sample spot begins to move at some point, and for very strong solvents (e.g., 50%v methanol/isopropyl chloride in Figure 9.12a) all sample bands will have moved to the outer solvent-front ring. For a solvent of some intermediate strength, (12%v acetonitrile/isopropyl chloride in Figure 9.12a), one or more sample bands will be seen to occupy an intermediate position within the developed spot, and this defines the approximate solvent strength that should be used in the LSC separation.

Normal TLC is used next to define the best solvent strength more accurately and to anticipate the possible need for change in solvent selectivity. The sample is spotted in the usual fashion on a 5×20-cm plate, developed with the solvent selected previously (12%v acetonitrile/isopropyl chloride), and visualized again in the iodine chamber. Figure 9.12b shows a possible result of this second separation. Three major bands are seen in the R_f range 0.5-0.7. Referring to Table 9.9, we see that the LSC k' values for this same solvent would be about 0.4-1.0. In this case the solvent is too strong to yield desired k' values in the range of $2 \leqslant k' \leqslant 5$. A decrease in $\epsilon°$ by 0.05 units, corresponding to 4%v acetonitrile/isopropyl chloride (Figure 9.7) should increase k' values by two to four times, which then places the sample k' values in about the right range. Therefore, at this point we would conclude that a mobile phase of 4%v acetonitrile/isopropyl chloride should be used for the LSC separation. Further fine tuning of solvent strength may be required, but this is best done on the LC unit by minor variation of the %v acetonitrile in the mobile phase.

The TLC chromatogram of Figure 9.12b might have revealed additional problems at this point:

- Tailing bands.
- Overlapping bands that require a change in selectivity.
- Additional strongly retained bands, at or near the origin.

If tailing bands are encountered in the TLC separation, they will probably also be a problem in the LSC separation. Therefore, changes in mobile-phase composition should be explored by TLC in an effort to eliminate tailing. A detailed discussion of band tailing in LSC (and its elimination) is provided in Sections 9.5 and 19.2. Once a mobile phase has been found that eliminates peak tailing in the TLC separation (and has the right solvent strength), it can be tried for LSC.

After solvent strength has been optimized, the TLC separation may show several bands with similar or identical R_f values. Alternatively, apparently single bands from the TLC separation may show up as partially resolved bands in the LSC separation with the same mobile phase. In any case a change in solvent selectivity is then required. This is best explored by further TLC separations, using the general rules in Section 9.3 for maximizing changes in α by proper choice of the polar mobile-phase component B. When a promising mobile phase has been thus identified, it can be tried in the LSC separation.

When the initial TLC separation shows strongly retained bands ($R_f \approx 0.00$) in addition to bands near the center of the plate, isocratic elution may not provide adequate separation because of the "general elution problem." In these cases gradient elution or a related technique is required (Chapter 16).

In carrying out the initial TLC separations as in Figure 9.12a, difficulty may be experienced in the dropwise addition of these volatile solvents. Because of the warming of the medicine dropper from handling, and the high vapor pressure of these solvents, the solvents tend to drip or even spurt without any pressure on the rubber bulb. This problem can be circumvented by simply squeezing the bulb several times, so as to suck up and discharge each solvent before dropwise addition of solvent to the plate. This technique cools down the dropper, eliminating the tendency for the solvent to drip out by itself.

Several studies [e.g., see (38-41)] have shown that LSC k' values correlate well with TLC R_f values, when such comparisons are carried out in an appropriate manner. However, in actual practice many workers report that TLC data often fail to correlate well with LSC data. There are several reasons for this:

- Solvent demixing.
- Variation in the adsorbent.
- Variation in adsorbent water content.
- Solvent concentration gradients in TLC.

Whenever the concentration of the stronger solvent B in a binary mobile phase is less than 10%v, *solvent demixing* may occur during the TLC separation, resulting in R_f values that cannot be related to LSC k' values. Solvent demixing refers to the preferential adsorption of the solvent B from the binary mobile-phase mixture A/B as it moves up the TLC plate. As a result, B is removed from the mobile phase until the adsorbent surface is saturated, and typically a "front" develops at some point along the plate, where the composition of the mobile phase changes abruptly from A/B to pure A [see discussion of (15), Section 8.2].

An example of solvent demixing in TLC is illustrated in Figure 9.13. In Figure 9.13a the separation of a two-component sample with 20%v benzene/pentane is shown. Here there is no problem and the resolution of the two compounds from their mixture (1 + 2) is apparent. In (b) the separation was repeated with the mobile phase: 1% acetonitrile/pentane. In this case the great difference in polarities or solvent strengths of the two solvent components (acetonitrile B vs. pentane A) resulted in solvent demixing and the formation of a front at about $R_f = 0.2$. As a result the various samples (1, 2, and 1 + 2) move at the demixing

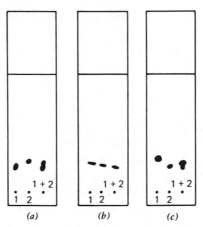

Figure 9.13 TLC separations of 1.7-dimethoxynaphthalene (1) and 1-nitronaphthalene (2) on silica plates; (*a*) 20%v benzene/pentane solvent, (*b*) 1%v acetonitrile/pentane solvent (no presaturation), (*c*) 1%v acetonitrile/pentane solvent (10-min presaturation). Reprinted from (*7*) with permission.

front, with severe band compression and the characteristic band shapes shown in Figure 9.13*b*. One might conclude from a superficial examination of this TLC separation that 1%v acetonitrile/pentane is incapable of separating the two compounds 1 and 2. However, because of solvent demixing, this conclusion is incorrect.

Solvent demixing in TLC can be avoided by preequilibration of the plate with the mobile-phase vapor, to adsorb the polar solvent component *B*, and more precisely mimic the corresponding LSC separation where column equilibration normally takes place before any separations are attempted. In the example of Figure 9.13*c*, preequilibration of the TLC plate was effected as follows. The various samples were first spotted onto the plate in the usual way, the solvent chamber for TLC was supplied with mobile phase [1%v acetonitrile/pentane, as in (*b*)], and a small metal block was placed in the chamber so as to allow positioning of the plate in the chamber without its touching the liquid at the bottom of the chamber. The plate was left in the closed chamber, out of direct contact with liquid, for 10 min. Following this preequilibration of the plate, the metal block was removed from the chamber, and development of the plate then proceeded in the normal manner. The resulting separation is shown in Figure 9.13*c*, which should be contrasted with the separation in (*b*). Observe that now the two sample components are resolved, with component 1 migrating faster. A subsequent LSC separation of the sample with the same 1% acetonitrile mobile phase gave the same result: faster migration of compound 1 and resolution of the two bands. Thus, preequilibration of the TLC plate allows a closer duplication of the LSC separation and gives retention data that can be correlated more closely with LSC.

Another reason for differences in TLC versus LSC retention data is that the adsorbents may differ in their separation characteristics, even though the TLC and LSC adsorbents are nominally similar (e.g., totally porous silica of similar surface areas). Some commercial suppliers offer TLC and LSC adsorbents that are certified as equivalent (e.g., Merck), and that may therefore circumvent this problem. Other than this, there is no simple answer in avoiding such differences in adsorbents. Fortunately, such effects are not too common and occur more often with polar samples and mobile phases.

Variations in adsorbent deactivation or water content can also lead to differences in retention on the TLC plate versus the LSC column. If the LSC separation is carried out with a 50%-water-saturated mobile phase, then the TLC separation will be equivalent if carried out in a room where the humidity is also 50%. If the room is dryer, the TLC separation will show smaller R_f values and larger k' values than the LSC separation. This means that a weaker mobile phase would be required in the LSC separation, to give an equivalent strength. For 25% room humidity, the k' values estimated from the TLC separation will be two to three-times too large, and the optimum 50%-water-saturated LSC mobile phase should therefore be made weaker by about 0.05 ϵ° units.

A final cause for difference between TLC R_f values and LSC k' values is the existence of *solvent concentration gradients* in TLC. These gradients occur because a thicker layer of mobile phase is adsorbed onto the bottom of a TLC plate than onto the top of the plate, following separation of the sample. This effect has been discussed by several workers, who have compared TLC R_f values with LSC (column) k' values for the same adsorbent, mobile phase, and samples [e.g., see (39-41) and Appendix I of (15)]. As a result of these solvent gradients in TLC, R_f values are reduced 1.2-1.5-fold versus the values calculated from Eq. 9.2 and a k' value from LSC in columns (same system). When solvent preequilibration is used in TLC, the latter reduction in R_f is closer to 1.5-fold. Thus, if an observed R_f value is 0.5, the "true" value should be 1.5 times greater, or 0.75. The corresponding LSC (column) k' value would then be 0.3, rather than 1.0 (Table 9.9).

If we combine the effects of a 25% room humidity (vs. 50% water saturation of the LSC mobile phase) plus solvent concentration gradients as earlier, the k' values calculated from Table 9.9 should be multiplied by about 4 to obtain the expected LSC values. Alternatively, this means that the optimum-strength LSC solvent will often have a ϵ° value 0.05-0.1 units less than predicted by the raw TLC data (for $0.2 \leqslant R_f \leqslant 0.4$).

Various commercial devices are available for improving or simplifying the use of TLC as a guide for LSC. The Vario KS chamber (Camag, about $500) allows control of humidity and mobile-phase preequilibration prior to TLC separation.

In this way LSC conditions can be duplicated more closely. A simpler and less expensive unit (Selectasol solvent-selector system, Schleicher & Schull) allows the simultaneous radial development of a sample with 16 separate mobile phases, similar to the seven mobile phases used in Figure 9.12a. This unit allows easy solvent preequilibration and better separations than can be obtained by the crude dropwise addition of mobile phase as in Figure 9.12a. Finally, the Camag U-chamber for HP-TLC separations ($3500) allows complete humidity and mobile-phase preequilibration, fast separations (1-2 min), and separation efficiencies (N values) that approach those of LSC.

TLC can also be used to scout solvents for reverse-phase BPC [e.g., see (42)], using silanized silica TLC plates. However, it is questionable whether this approach to solvent selection is really worthwhile. The choice of mobile phases in reverse-phase LC is much more restricted than in LSC, and most separations can be carried out with a single, two-component mobile phase such as methanol/ water. If the LC system is equipped with a gradient former, different methanol/ water mixtures can be prepared by dialing in the desired composition and run directly by LC (as in Fig. 2.24). Alternatively, a gradient elution run can be used to estimate the optimum mobile-phase composition for isocratic LC (Section 16.2). Thus, little time is saved in reverse-phase LC by using TLC as an initial guide.

The purification of some solvents used in LSC has been reviewed in Section 6.6. A further discussion of this subject is given in (43).

9.4 OTHER SEPARATION VARIABLES

Sample-size effects have been discussed in general terms for the various LC methods in previous sections (see also Figures 2.6, 5.9, and 9.8 for LSC data on separation performance vs. sample size). In general, the volume of the sample should be less than one-third the volume of the first band of interest in the chromatogram (Section 7.4), unless the sample is dissolved in a solvent that is weaker than the mobile phase. Under favorable circumstances (e.g., water deactivation of totally porous silica adsorbents as in Section 9.3), samples as large as 1 mg/g of adsorbent can be charged without adversely affecting separation. Pellicular silica packings (e.g., Figure 2.6) and alumina adsorbents generally have linear capacities of 0.2 mg/g or less. The maximum sample loading of very efficient (i.e., $\leqslant 10$-μm) LSC columns is also less if column efficiency is not to be degraded. This is illustrated by the data for a 10-μm silica plotted in Figure 9.14. In this case column H values rise appreciably for column loadings greater than 0.01 mg/g, particularly for sample bands with higher k' values. The effect of sample size on separation performance should be determined in every case where maximum sample loadings are desired.

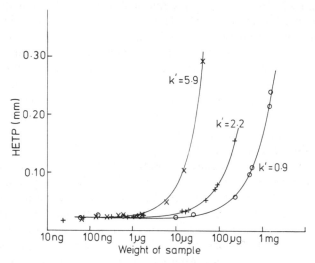

Figure 9.14 Variation in column H values with sample size for a very efficient LSC system. Column, 14.5 × 0.5 cm, Partisil-10 (10-μm silica); mobile phase, 8%v methylene chloride/hexane (50%-saturated); ambient; 0.25 cm/sec; UV, 254; samples are phenetole (k' = 0.9), nitrobenzene (k' = 2.2) and methyl benzoate (k' = 5.9). Reprinted from (44) with permission.

Sample size in LSC may also be limited by low solubility of the sample in the mobile phase, particularly for preparative separations. The use of other solvents for sample dissolution, of similar solvent strength as the mobile phase, but with a greater solubility for the sample, is discussed in Chapter 15 (See Figure 15.11).

Temperature is normally not used as a separation variable in LSC, and most applications are carried out at ambient temperature. Figure 9.15a illustrates the typical effect of an increase in temperature on separation; retention for all bands is decreased, but any changes in selectivity with temperature are slight. Sample k' values normally decrease by 2% for each 1°C increase in column temperature. For this reason, columns in LSC should be thermostatted when retention must be controlled within narrow limits, for example, for more precise identification of unknowns, based on t_R values, or for greater precision in quantitation by peak-height measurements.

Occasionally, significant changes in selectivity or band sequence occur in LSC upon a large change in column temperature. Such changes seem to occur more frequently when dilute solutions of a very polar solvent B are used as mobile phase, probably because under these conditions the polar modifier tends to desorb from the adsorbent at higher temperatures. An example of such a selectivity change with temperature is shown in Figure 9.15b.

9.5 SPECIAL PROBLEMS

Inability to reproduce adsorbents and packed LSC columns from batch to batch is a frequent complaint of users. We have already noted the similar problem of differences in the adsorbent used in TLC scouting experiments versus the final LSC column separation. Suppliers of LSC packings are attempting better control of the uniformity of adsorbents of a given type, and in some cases reproducible

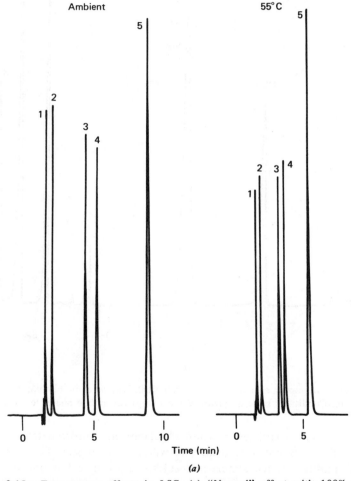

Figure 9.15a Temperature effects in LSC. (*a*) "Normal" effect with 100% methylene chloride as mobile phase. Samples are benzene (1), nitrobenzene (2), methyl benzoate (3), benzaldehyde (4), acetophenone (5), α-methyl benzyl alcohol (MBA), benzyl alcohol (BA), and 3-phenyl-l-propanol 3PP). Column, 25 × 0.46 cm, 5-μm silica; mobile phases, as indicated; temp., as indicated; UV, 254 nm. Reprinted from unpublished data of W. A. MacLean and A. F. Poile, Perkin-Elmer Corp., with permission.

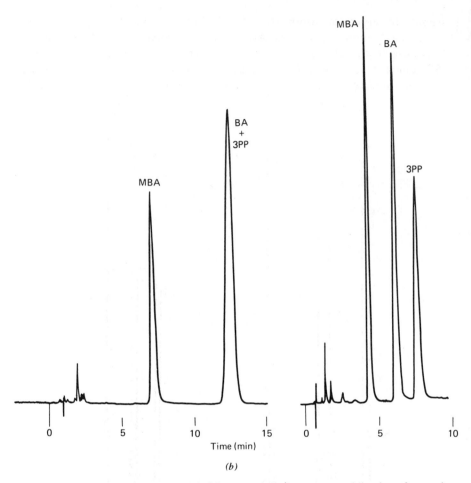

Figure 9.15*b* Selectivity change with 2%v acetonitrile/hexane as mobile phase for varying temperature. Conditions same as Figure 9.15*a* except for mobile phase, temperature, and sample.

retention for a standard test mixture is guaranteed. Unfortunately, adsorbent irreproducibility is not a problem that will soon disappear. The retention of complex, multifunctional sample molecules is a sensitive function of the configuration of the adsorbent surface, which apparently varies with minor changes in the process used to synthesize the adsorbent. The inclusion of traces of acid, base, or transition metals in the final adsorbent can further effect the retention of acids or bases, or of neutral compounds that can undergo complexation reactions.

Under these circumstances, what advice can be offered to the chromatographer? First, note that the problem of adsorbent variability is worse for separations of very polar, multifunctional molecules. If retention constancy is particularly important for a given application, one should consider the use of a more reproducible separation method, for example, reverse-phase or ion-pair BPC. Second, if only LSC will suffice, and repeatable separations are required over a period of time, then a quantity of the adsorbent from the same batch should be stockpiled for the preparation of identical columns at some later time. Alternatively, several packed columns should be purchased with the request that these be from the same batch of adsorbent. Finally, variation of the water content of the mobile phase may allow some control over adsorbent retention characteristics, such that differences in adsorbents from lot to lot can be partially offset by a change in adsorbent water content. For the separation of acidic or basic samples, differences in adsorbent basicity or acidity can be minimized by adding an acid or base to the mobile phase (see following discussion of band tailing).

Just as *addition* of water to the mobile phase solves some problems, any *variation* of the mobile-phase water content leads to still other problems. From the discussion of Section 9.3 it is apparent that such a variation in water content from run to run will lead to unacceptable variation in the t_R values of different sample components. This can be caused by exposure of the mobile phase to air of different humidity than the mobile phase, or to contact of the mobile phase with containers where loss or pickup of water can occur. Such effects are most important for less-polar solvents, and it is suggested that when they are being used methanol or acetonitrile be substituted for the water in the mobile phase (see Section 9.3). Another cause of water pickup by the mobile phase is contamination by moist air; when transferring the inlet pump line from one solvent container to another (as when changing mobile phases), it is important not to leave the inlet line filter out of the solvent for more than a few seconds. Otherwise the continuing action of the pump pulls air into the filter, the resulting evaporation of solvent cools the filter, and water from the air then condenses on the filter—to be mixed with incoming solvent when the filter is inserted into the next mobile phase to be used.

Column deactivation is a general problem in LSC, more so than for other LC methods. The highly active surface can tenaciously hold very polar compounds present as traces in injected samples or as low-level impurities in the mobile phase. Guard columns (Section 5.6) can be used to reduce this problem.

Typically, with continued use of an LSC column, sample k' values eventually begin to decrease, bands become unsymmetrical, the plate number N declines, and resolution gradually worsens with time. What is needed then is *column regeneration*, to flush out all the chemical "garbage" that has accumulated on the adsorbent over time. One such column regeneration procedure is as follows (*43*):

Wash the column in sequence with the following number of column volumes of each solvent: methylene chloride (10), methanol (20), and water (20). If it is known that either basic or acidic samples have been used to a great extent in prior LSC separations on the column, then pyridine or acetic acid, respectively, can be added to the methanol and water steps. Following the purging of the column, it must be reconditioned or reequilibrated to remove any excess water or polar modifiers. One such reconditioning scheme, when either dry mobile phases or mobile phases plus organic modifiers are to be used, is as follows (43): methanol (20), methylene chloride (20), and hexane (20). Alternatively, for use with 50%-water-saturated mobile phases, the following sequence is satisfactory: methanol (20), 50%-saturated ethyl ether (20), and final (50%-saturated) mobile phase (20). In all cases the final equilibration of the column must be checked by allowing the last mobile phase to flow through the column, with injection of a standard sample ($k' > 1$) every 5 to 10 column volumes. Only when the retention of the standard sample becomes constant should the column be used for LC separations.

Column deactivation by chemical garbage can also be avoided by cleaning up the sample prior to injection (Section 17.2), and by advance consideration of the sample and its possible reaction with the adsorbent. Thus, acidic samples in general tend to be irreversibly held on alumina. In one study (45), injection of amine hydrochlorides onto an alumina column gradually led to irreversible retention of further samples. In this case addition of acetic acid to the mobile phase solved the problem. However, a better solution is the avoidance of alumina columns for the separation of strongly acidic samples.

Tailing bands in LSC are common when very polar and/or acidic or basic samples are to be separated. Section 19.2 discusses this problem in more detail. Briefly, band tailing can be counteracted by using water deactivation (i.e., 50%-water-saturated mobile phases) or by adding triethylamine or ammonia to the mobile phase when separating basic samples, and acetic or formic acid when separating acidic samples (0.1-1%).

The undesired alteration or loss of sample components during separation by LSC is not a common problem. However, sample components that are labile toward oxygen, acid or base are potentially susceptible to reaction on the column. Adsorbents appear to increase the tendency of sample oxidation in the presence of molecular oxygen; alumina is normally strongly basic, and silica is mildly acidic. Oxidation of the sample is generally controllable by thoroughly degassing or sparging the mobile phase with an inert gas such as helium, so that all oxygen is purged from the column. In extreme cases an antioxidant such as butylated hydroxytoluene (BHT) can be added to the mobile phase if UV detection below 290 nm is not used. Addition of 0.05% BHT is effective and will not alter the strength or selectivity of the mobile phase.

Acid-sensitive samples can be separated on silica if a small amount of amine (e.g., triethylamine) is added to the mobile phase. Similarly, base-sensitive samples can be protected by adding acetic acid to the mobile phase and by avoiding alumina. In one study (46), rearrangement of labile terpenes during separation on silica was traced to contamination of the silica by heavy metals. Here the problem was eliminated by an initial acid wash of the silica, followed by rinsing and drying, plus final add-back of 5% water to deactivate the silica. Addition of chelating agents such as EDTA to the mobile phase would probably also help when separating compounds susceptible to heavy-metal catalysis.

Solvent demixing can occur in LSC as well as in TLC. Whenever a new mobile phase is used, a certain volume of that mobile phase (e.g., 10-20 column volumes) is required to flush out the old solvent and equilibrate the adsorbent with the new solvent. For mobile phases that consist of a small concentration of a very polar solvent B in a relatively nonpolar solvent A, this equilibration process may require more than 20 column volumes of the new solvent. In any

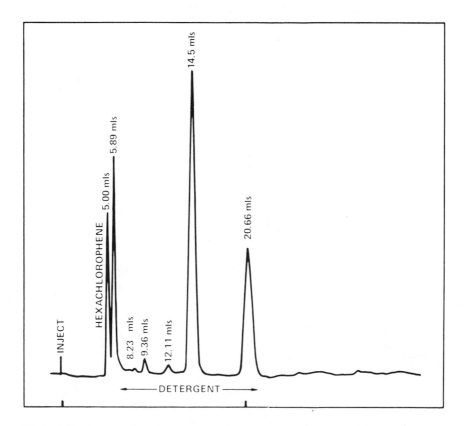

Figure 9.16 Assay for hexachlorophene in germicidal soap. Column, 123 × 0.2 cm, Corasil-II (pellicular silica); mobile phase, 75%v chloroform/isooctane; temp., ambient; 1.0 ml/min; refractive index detection. Reprinted by permission of Waters Associates.

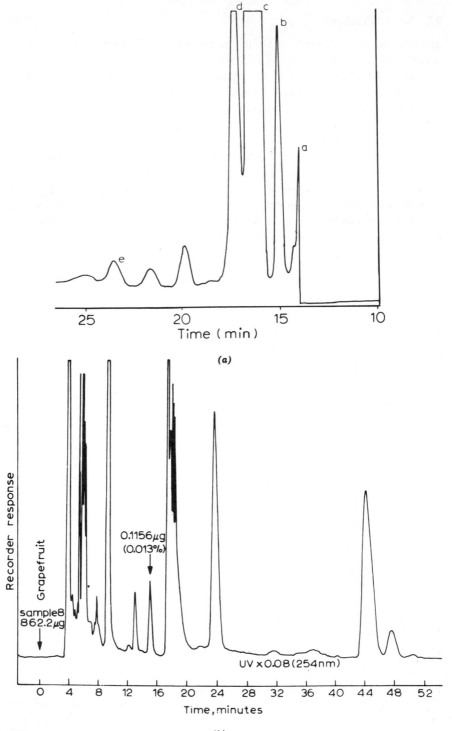

(a)

(b)

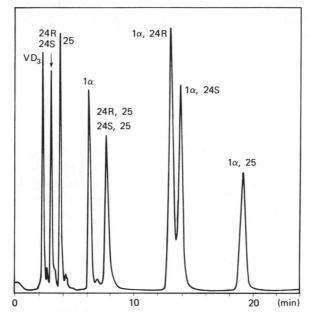

Figure 9.18 Separation of vitamin D₃ and various dihydroxy metabolites by LSC. Column, 25 × 0.21 cm, Zorbax-Sil; mobile phase, 2%v methanol/methylene chloride; temp., ambient; 1300 psi; UV. Reprinted from (*52*) with permission.

case, equilibration can be verified by repeated injection of some standard sample, with column flushing continued until retention times for the sample become constant. Solvent demixing in LSC is a particular problem in gradient elution, as discussed in Section 16.1.

A question that is often asked concerning LSC separation is whether large proportions of methanol or water can be used in the mobile phase. Apparently, some suppliers of silica columns have warned against these solvents as detrimental to column life. It is possible that some poorly prepared silica columns will be adversely affected by such conditions, but well-made columns are apparently more durable. One study (*47*) reported the assay of morphine in urine with a small-particle silica column and 30%v methanol/water (with 2 *N* NH₄⁺

Figure 9.17 Assay of natural oils by LSC. (*a*) Flavor constituents in cassia oil; a, benzaldehyde; b, eugenol; c, cinnamaldehyde; d, cinnamyl acetate; e, cinnamyl alcohol. (*b*) Determination of bergapten (0.013%) in grapefruit oil. Conditions for (*a*): Column, 92 × 0.22 cm, Corasil-II; mobile phase, gradient elution from 0-40%v ethyl acetate/cyclohexane; 1.0 ml/min; UV, 260 nm; 10-μl sample (10%v). Reprinted from (*49*) with permission. Conditions for (*b*): Column, 30 × 0.4 cm, μ-Porasil (10-μm silica); mobile phase, ethyl acetate/isopropanol/isooctane 1/1/80; ambient; 1.0 ml/min; UV, 254 nm. Reprinted from (*51*) with permission.

buffer) as mobile phase; the column was in use "without noticable deterioration" for several months.

9.6 APPLICATIONS

Almost every possible type of sample has by now been separated by TLC [see examples given in (2)] or by modern LSC. However, as indicated at the beginning of this chapter, certain kinds of samples are better handled by LSC than by other LC methods. Although gas chromatography competes with LSC for the analysis of nonpolar samples that are thermally stable and volatile, many workers have noted that LSC has certain advantages even for such samples. Similarly, for more polar, water-soluble, and/or ionic samples, reverse-phase LC is being used more and more frequently, but even among this class of compounds there are striking examples of the successful application of LSC.

The assay of various less-polar additives in a wide range of commercial products has been reported by LSC. Figure 9.16 shows one such example, the analysis of hexachlorophene (hexachlorocyclohexane) in soap, using a pellicular silica column. Both fat-soluble and water-soluble vitamins have been determined in a variety of food products (48). Concerning the separation of Figure 9.16, it is *always* inadvisable to have the compound of interest elute at t_o—as is done here. There is no way of ensuring that interfering compounds are not coeluting with the hexachlorophene peak in Figure 9.16. In this case it would probably be better to carry out the separation with gradient elution (LSC) or to switch to reverse-phase LC to provide later elution of the band of interest. A similar problem can be noted for the internal standard (I.S.) of Figure 9.1b.

The less-polar plant pigments, such as carotenoids and porphyrins, have been successfully separated by LSC for over 40 years. A recent application of modern LSC to this area (50), for the fingerprinting of these compounds in recent sediments, noted that LSC gave many more resolved peaks than did reverse-phase BPC.

Natural oils and flavor extracts are readily assayed by LSC. Figure 9.17a shows the separation of the flavor components of cassia oil by gradient elution with a pellicular silica. Figure 9.17b portrays the determination of the natural phototoxin bergapten (t_R = 15 min) in grapefruit oil, using a 10-μm porous-silica column. This assay allows the setting of standards to limit the maximum content of this toxic substance in commercial citrus oils.

The assay of various pharmaceutical products by LSC continues to receive heavy emphasis. The separation of vitamin D_3 and seven of its metabolites (isomeric diols) is shown in Figure 9.18, using a 7-μm porous-silica column. The ability of LSC to separate structurally similar isomers is well illustrated in this example. The decomposition of the active ingredient(s) of a drug product under

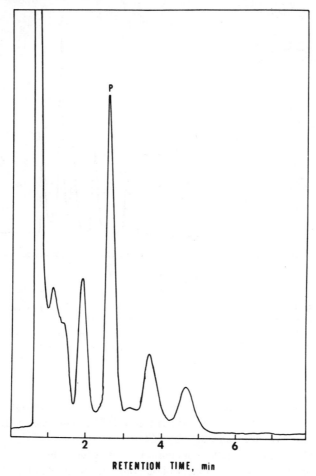

Figure 9.19 Separation of heat-stressed sample of propoxyphene hydrochloride by LSC. Column, 25 × 0.32 cm, Merck 10-μ silica; mobile phase, 0.65%v basic isopropanol/hexane; temp., ambient; 1.5 ml/min; UV, 254 nm; 10-μl sample. Reprinted from (*53*) with permission.

thermal stress can be followed by LSC in many cases. Figure 9.19 shows the separation of a thermally stressed sample of propoxyphene hydrochloride, using a 10-μm porous silica. Separation of this sample on alumina was less successful (*53*), as could have been predicted in view of the problems associated with injecting strongly acidic samples onto this adsorbent. The LSC analysis of a cough syrup that contains several alkaloids is shown in Figure 9.20. In this case the sample was first derivatized (dansylated) to render these amines detectable

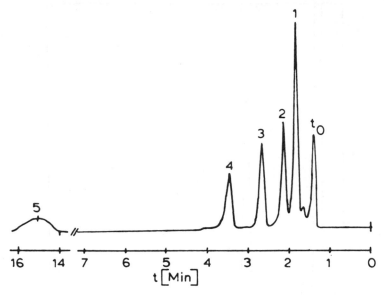

Figure 9.20 Separation of alkaloids in cough syrup after dansylation of sample. (1) Dns-ephedrine; (2) narcotine, Dns-cephaeline; (4) Dns-emetine; (5) codeine. Column, 25 × 0.28 cm, Merck SI 100 (10-μm silica); mobile phase, isopropanol/ammonia/isopropyl ether 2/0.3/48; temp., ambient; UV, 254 nm; 10-μl samples. Reprinted from (*54*) with permission.

by fluorescence. Derivatization of basic compounds before separation on silica also tends to improve their separation characteristics. Many steroids of pharmaceutical interest have been separated by LSC.

Liquid-solid chromatography is increasingly being used by synthetic organic chemists to assay reaction products, isolate compounds for characterization, and purify final products. Figure 9.21 shows a typical application of this kind, the separation of two pyrazoline isomers. The power of LSC in the separation of isomers is again confirmed in this example. Section 15.5 provides several further examples of the preparative separation of synthetic organic mixtures by LSC.

Many of the drugs of abuse have been assayed by LSC. Figure 9.22 shows the detection of LSD in a tablet sold on the street. Such assays allow the confirmation of a given drug and can further distinguish drug samples from different sources ("fingerprint" technique, involving comparison of other bands in the chromatogram). The separation of all the PTH-amino acids by gradient elution LSC is shown in Figure 9.23*a* (with the exception of arginine and histidine). The application of this assay in the Edman sequencing of a protein with terminal sequence ser-tyr-ser-met-glu- is shown in Figure 9.23*b*, *c*. The tyrosine band is

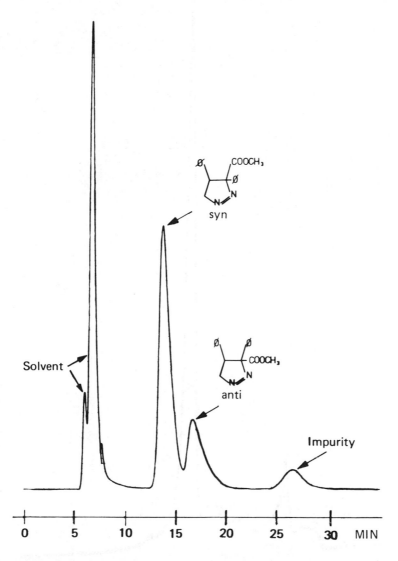

Figure 9.21 Separation of *syn-* and *anti-*pyrazoline isomers on silica. Column, 100 × 0.2 cm Perisorb A (pellicular silica); mobile phase, 50%v methylene chloride/isooctane; temp., ambient; 0.25 ml/min (300 psi); UV, 254 nm; 5-μl sample. Reprinted with permission of Hewlett-Packard Co.,

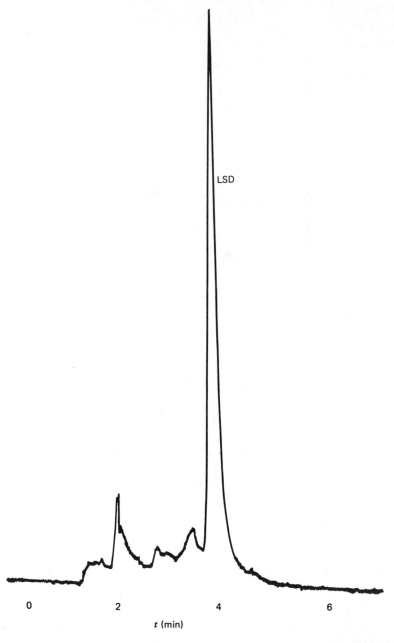

LSD

0   2   4   6

t (min)

Figure 9.22 Analysis of LSD tablet. Column, 25 × 0.21 cm, Zorbax-Sil; methanol/acetic acid/methylene chloride 30/0.1/70; temp., ambient; 0.5 ml/min (1200 psi); fluorescence, 334-nm excitation, 408-nm emission; 5-μl sample. Reprinted by permission of the DuPont Instrument Products Division.

seen in (*b*), and the glutamic acid band in (*c*), corresponding to the second and fifth Edman cycles, respectively.

In addition to the preceding more or less typical applications of LSC, we have seen earlier applications of this LC method to both water-soluble and ionic compounds (Figures 9.2, 9.3). LSC has even been applied to the trace detection of metals as their colored complexes. Figure 9.24 shows the separation of three metals as their diacetylbisthiobenzhydrazones.

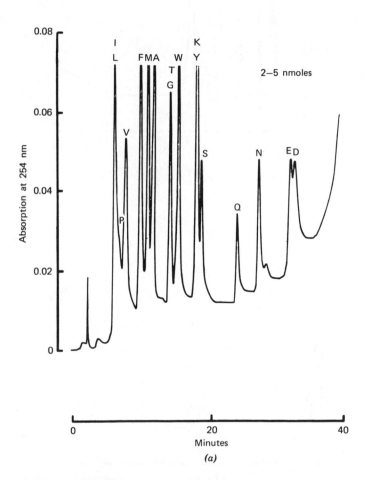

Figure 9.23a Separation of PTH amino acids by LSC. (*a*) Separation of 17 amino acids (except arginine and histidine). Column, 25 × 0.21 cm, Zorbax-Sil; mobile phase, gradient elution from 9/11/3980 methanol/propanol/hexane to methanol/propanol 9/11; ambient; UV, 254 nm. Reprinted from (*55*) with permission.

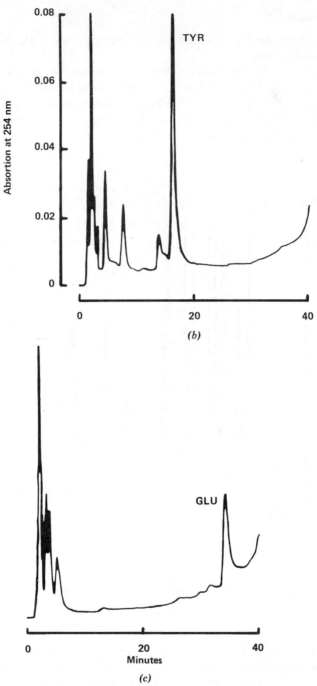

Figure 9.23b, c Separation of PTH amino acids. (b) analysis of second cycle from Edman degradation of protein. (c) analysis of fifth cycle from Edman degradation of same protein. Conditions as in Figure 9.23a.

404

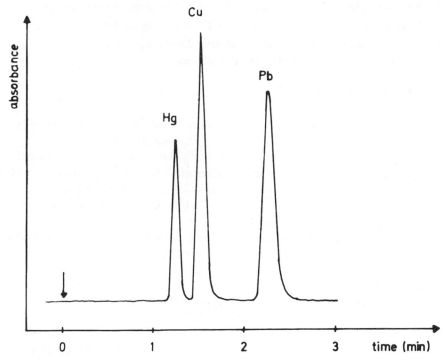

Figure 9.24 Determination of trace metals by LSC. Column, 30 × 0.23 cm, Nucleosil 100-10 (10-μm silica); mobile phase, benzene; temp., 23°C; 1.5 ml/min; UV, 360 nm. Reprinted from (*56*) with permission.

9.7   THE DESIGN OF AN LSC SEPARATION

Selecting the Column

Today most LSC separations are carried out on one of the small-particle, totally porous silicas (e.g., μ-Porasil, Zorbax-Sil, Partisil-5 or -10, etc.; see Table 5.5). For routine applications in which the ruggedness and reliability of the LSC system are important, as for control of a plant process, and where the separation is not too difficult (or analysis time is less important), a pellicular silica column may be preferred (e.g., Corasil-II, Vydac, etc.; see Table 5.10). Columns of pellicular packings are easier to prepare and less subject to plugging; they are also easier to equilibrate and regenerate. For preparative separations of large samples, 50-μm particles of porous silica are usually used (Chapter 15).

An alumina column is occasionally preferable to:

- Alter separation selectivity, after different mobile phases have been tried.
- Eliminate band tailing, after changes in the mobile phase have been tried.
- Avoid reaction and/or loss of the sample during separation; acids are generally better separated on silica, bases on alumina.

Selecting the Mobile Phase

The effect of the mobile phase on separation should be determined by initial TLC runs. A series of solvents of different strength should be tried first, as in Figure 9.12. A normal TLC separation using the solvent of about the right strength should then be carried out. Values of k' for the same solvent in LSC can be estimated from Table 9.9, keeping in mind that k' values in LSC are often two to four times smaller than in TLC, so that the LSC solvent should be about 0.05 ϵ^0 units weaker than the best TLC solvent.

If solvent selectivity is marginal when using the solvent of the right strength (i.e., there are overlapping bands in the preceding TLC separation), other solvents of similar strength should be tried. The *"B*-concentration" rule should be used first (e.g., Table 9.7), trying different binary solvents of either very low (0.5-2%v) or high (50-100%v) concentrations of the strong solvent component B in the mobile-phase binary. For some possible examples of such solvents, see Figure 9.7. For further changes in solvent selectivity, attempt to vary hydrogen bonding between sample and solvent molecules. This can be achieved by using strong solvents B of differing hydrogen-bonding character, for example, a solvent from selectivity group I, II, V, or VIII (see examples of Table 9.4). These further studies of solvent selectivity should also be carried out by TLC.

If the initial TLC separation shows several compounds with both small (0.00-0.05) and large (0.7-1.0) R_f values, then isocratic elution will probably not be successful. Gradient elution or other techniques for solving the general elution problem should be considered (Chapter 16).

When a solvent of the right strength and selectivity has been selected, it should then be tried for the LSC separation of the sample. For water-immiscible solvents with $\epsilon^0 > 0.1$, use 50%-water-saturated mobile phase. For weaker mobile phases, add 0.02-0.1%v acetonitrile in place of water. For water-miscible mobile phases, add 0.05-0.2%v water.

If tailing bands are observed in the initial TLC separations, try adding either acid or base to the mobile phase: 0.1-0.2%v acetic or formic acid for acidic samples, 0.1-0.2%v ammonia or triethylamine for basic samples. When the final mobile phase is tried for LSC, allow 20-30 column volumes of solvent to flow through the column before injecting any samples; then repeatedly inject some standard sample, to ensure that column equilibration has occurred before attempting the separation of "real" samples.

Temperature

The temperature of the column should be at ambient, or thermostatted near ambient if retention times must be held within tight limits. Sample k' values will change by about 2%/°C change in column temperature.

Sample Size

Up to 1 mg sample per gram of adsorbent can be injected in favorable cases (totally porous silica particles). However, for very efficient columns, N values may be reduced when sample size exceeds 0.05-0.1 mg/g. The effect of sample size on separation should be checked when injecting samples larger than 50 μg/g.

The volume of injected sample can usually be 25-50 μl for very efficient columns of small volume (e.g., 15 × 0.3 cm of 5-μm particles), and 50-400 μl for longer, wider columns. Larger volumes can be handled if the sample is dissolved in a solvent that is weaker than the mobile phase. In some cases larger weights of sample can be injected if the sample is made up as a larger volume of more dilute sample.

Other Considerations

When further increase in sample resolution is required, column efficiency can be increased as in the discussion on controlling resolution in Section 2.5. Special problems and their solution are further discussed in Section 9.5 and Chapter 19.

REFERENCES

1. L. R. Snyder, *Anal. Chem.*, **39**, 698, 705 (1967).
2. E. Stahl, *Thin-Layer Chromatography: A Laboratory Handbook*, 2nd ed., Academic, New York, 1969.
3. J. G. Kirchner, *Thin-Layer Chromatography*, Wiley-Interscience, New York, 1967.
4. J. Krupcik, J. Kriz, D. Prusova, P. Suchanek, and Z. Cervenka, *J. Chromatogr.*, **142**, 797 (1977).
5. K. Tsukida, A. Kodama, and M. Ito, *ibid.*, **134**, 331 (1977).
6. Du Pont Instrument Products Division.
7. L. R. Snyder, *J. Chromatogr.*, **63**, 15 (1971).
8. *Ibid.*, **20**, 463 (1965).
9. *Ibid.*, **8**, 319 (1962).
10. *Ibid.*, **15**, 344 (1964).
11. M. Dong and D. C. Locke, *J. Chromatogr. Sci.*, **15**, 32 (1977).
12. T. H. Beasley, H. W. Ziegler, and A. D. Bell, *Anal. Chem.*, **49**, 238 (1977).

13. W. W. Dean, G. J. Lubrano, H. G. Heinsohn, and M. Stastny, *J. Chromatogr.*, **124**, 287 (1976).

14. D. Kirkpatrick, *ibid.*, **139**, 168 (1977).

15. L. R. Snyder, *Principles of Adsorption Chromatography*, Dekker, New York, 1968.

16. L. R. Snyder, *Anal. Chem.*, **46**, 1384 (1974).

17. L. R. Snyder, *Adv. Chromatogr.*, **4**, 3 (1967).

18. R. P. W. Scott and P. Kucera, *J. Chromatogr.*, **112**, 425 (1975); *ibid.*, **149**, 93 (1978).

19. E. H. Slaats, J. C. Kraak, W. J. T. Brugman and H. Poppe, *ibid.*, **149**, 255 (1978).

20. E. Soczewinsky, *J. Chromatogr.*, **130**, 23 (1977).

21. H. Colin, N. Ward, and G. Guiochon, *ibid.*, **149**, 169 (1978) (and previously referenced papers by this group).

22. R. P. W. Scott and P. Kucera, *J. Chromatogr. Sci.*, **13**, 337 (1975).

23. R. P. W. Scott and P. Kucera, *ibid.*, **12**, 473 (1974).

24. J. M. Bather and R. A. C. Gray, *J. Chromatogr.*, **122**, 159 (1976).

25. Z. E. Rassi, C. Gonnet, and J. L. Rocca, *ibid.*, **125**, 179 (1976).

25a. C. J. Little, A. D. Dale, J. A. Whatley and J. A. Wickings, *J. Chromatogr.*, **169**, 381 (1979).

26. R. R. Heath, J. H. Tumlinson, and R. E. Doolittle, *J. Chromatogr. Sci.*, **15**, 10 (1977).

27. H. Colin, C. Eon, and G. Guiochon, *J. Chromatogr.*, **122**, 223 (1976).

28. D. L. Saunders, *Anal. Chem.*, **46**, 470 (1974).

29. J. J. C. Scheffer, A. Koedem, and A. B. Svendsen, *Chromatographia*, **9**, 425 (1977).

30. J. J. Kirkland, *J. Chromatogr.*, **83**, 149 (1973).

31. J. H. M. van den Berg, J. Milley, N Vonk, and R. S. Deelder, *ibid.*, **132**, 421 (1977).

32. K. Chmel, *ibid.*, **97**, 131 (1974).

33. J. -P. Thomas, A. Brun, and J. -P. Bounine, *ibid.*, **139**, 21 (1977).

34. D. L. Saunders, *ibid.*, **125**, 163 (1976).

35. H. Engelhardt, *J. Chromatogr. Sci.*, **15**, 380 (1977).

36. R. D. Burnett, In *High Pressure Liquid Chromatography in Clinical Chemistry*, P. F. Dixon et al., eds., Academic, New York, 1976, p. 45.

37. L. R. Snyder, *J. Chromatogr. Sci.*, **7**, 595 (1969).

38. H. Schlitt and F. Geiss, *J. Chromatogr.*, **67**, 261 (1972).

39. H. Soczewinski and W. Golkiewics, *ibid.*, **118**, 91 (1976).

40. A. M. Siouffi, A. Guillemonat, and G. Guiochon, *ibid.*, **137**, 35 (1977).

41. S. Hara, *ibid.*, p. 41.

42. R. K. Gilpin and W. R. Sisco, *ibid.*, **124**, 257 (1976).

43. D. L. Saunders, *J. Chromatogr. Sci.*, **15**, 372 (1977).

44. J. N. Done, *J. Chromatogr.*, **125**, 43 (1976).

45. J. H. Knox and J. Jurand, *ibid.*, **103**, 311 (1975).

45a. J. F. Lawrence and R. Leduc, *Anal. Chem.*, **50**, 1161 (1978).

46. J. J. C. Schoeffer, A. Koedem, A. Baerheim, *Chromatographia*, **9**, 425 (1976).

47. I. Jane and J. F. Taylor, *J. Chromatogr.*, **109**, 37 (1975).

48. T. van de Weerdhof, M. L. Siersum, and H. Reissenweber, *ibid.*, **83**, 455 (1973).

49. M. S. F. Ross, *ibid.*, **118**, 273 (1976).

50. S. K. Hajibrahim, P. J. C. Tibbetts, C. D. Watts, J. R. Maxwell, G. Eglinton, H. Colin, and G. Guiochon, *Anal. Chem.*, **50**, 549 (1978).

51. C. K. Shu, J. P. Walradt, and W. I. Taylor, *J. Chromatogr.*, **106**, 271 (1975).

52. N. Ikekawa and N. Koizumi, *ibid.*, **119**, 227 (1976).

53. R. K. Gilpin, J. A. Korpi, and C. A. Janicki, *ibid.*, **107**, 115 (1975).

54. R. W. Frei, W. Santi, and M. Thomas, *ibid.*, **116**, 365 (1976).

55. E. W. Matthews, P. G. H. Byfield, and I. MacIntyre, *ibid.*, **110**, 369 (1975).

56. P. Heizmann and K. Ballschmiter, *ibid.*, **137**, 153 (1977).

BIBLIOGRAPHY

Engelhardt, H., *J. Chromatogr. Sci.*, 15, 380 (1977) (a review of the use of moderators, including water, in LSC).

Geiss, F., *Die Parameter der Dünnschichtchromatographie*, Vieweg, Braunschweig, 1972 (very thorough and theoretical treatment of TLC).

Saunders, D. L., *J. Chromatogr. Sci.*, 15, 372 (1977) (a review of practical aspects of LSC; various hints used by this particular worker).

Snyder, L. R., *Principles of Adsorption Chromatography*, Dekker, New York, 1968 (a detailed review of the basis of retention in LSC).

Snyder, L. R., *Anal. Chem.*, 46, 1384 (1974) (a more recent review of solvent effects in LSC).

Snyder, L. R., and D. L. Saunders, in *Chromatography in Petroleum Analysis*, K. Altgelt and H. Gouw, eds., Dekker, 1979, Chap. 10 (a review of LSC applications in the petroleum industry, including classical LC procedures).

Stahl, E., *Thin-Layer Chromatography: A Laboratory Handbook*, 2nd ed., Academic, New York, 1969 (an excellent summary of previous TLC separations for use as models for the LSC separation of similar compounds).

TEN

ION-EXCHANGE CHROMATOGRAPHY

10.1	Introduction	410
10.2	Column Packings	414
10.3	Mobile Phases	419
10.4	Other Separation Variables	426
10.5	Special Problems	427
10.6	Applications	429
	Ion Chromatography	438
10.7	The Design of an Ion-Exchange Separation	445
	Selecting the Column	445
	Selecting the Mobile Phase	447
	Temperature	450
	Sample Size	450
	Other Considerations	450
References		450
Bibliography		452

10.1 INTRODUCTION

Ion-exchange chromatography (IEC) was the first of the various LC methods to be used widely under modern LC conditions. Automated, high-resolution IEC dates from the early 1960s, with the introduction of routine amino acid analysis. Basically the same technique was later extended to the analysis of literally hundreds of different compounds in physiologic fluids such as urine and serum, although these analyses typically took 10-70 hours to complete. During this same period, IEC methods for the separation and analysis of protein mixtures were assuming major importance in the biochemical area. However, until recently these protein separations by IEC have been carried out by classical, low-pressure LC procedures.

The renaissance in LC that began in the late 1960s featured IEC applications from the very start (1) [e.g., the separation and analysis of nucleic acid bases, nucleosides, and nucleotides on pellicular ion exchangers, as reviewed in (1a)]. For various practical reasons, however, IEC has remained less popular than the other LC methods, and more recently the new technique of ion-pair chromatography (Chapter 11) has begun to displace IEC in some of its traditional areas of application. Nevertheless, for certain separation problems IEC is still uniquely useful. The recent development of new column packings for the separation of high-molecular-weight, ionic samples promises to give IEC a new lease on life.

Ion-exchange chromatography is carried out with packings that possess charge-bearing functional groups. The most common retention mechanism is *simple ion exchange* of sample ions X and mobile phase ions Y with the charged groups R of the stationary phase:

$$X^- + R^+Y^- \rightleftharpoons Y^- + R^+X^- \quad \text{(anion exchange)}, \tag{10.1a}$$

$$X^+ + R^-Y^+ \rightleftharpoons Y^+ + R^-X^+ \quad \text{(cation exchange)}. \tag{10.1b}$$

For anion-exchange chromatography, as shown in Eq. 10.1a, the sample ion X^- is in competition with the mobile phase ion Y^- for the ionic sites R^+ of the ion exchanger. Similarly, in cation-exchange chromatography (eq. 10.1b), sample cations X^+ compete with the mobile-phase ions Y^+ for the ionic sites R^- of the ion exchanger. Sample ions that interact weakly with the ion exchanger (in the presence of competing mobile phase ions) will be weakly retained on the column and elute early in the chromatogram. Sample ions that interact strongly with the ion exchanger will be retained more strongly and elute later.

Although IEC is sometimes applied to the separation of ionic compounds (i.e., salts), more commonly it is used for separating organic acids (HA) and bases (B) that can exist as ions under suitable conditions of pH:

$$HA \rightleftharpoons H^+ + A^-, \tag{10.2a}$$

$$B + H^+ \rightleftharpoons BH^+. \tag{10.2b}$$

The extent of ionization of acids and bases in an IEC system can be controlled by varying the pH of the mobile phase. Increasing pH leads to increased ionization of acids and lesser ionization of bases, and vice versa for decrease in pH. An increase in ionization in each case leads to increased retention of the sample.

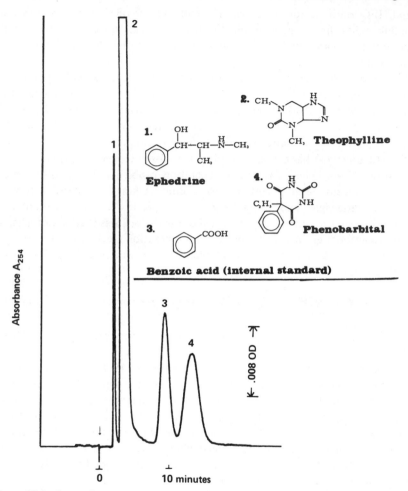

Figure 10.1 Separation of an asthma tablet by anion exchange chromatography. Column, 150 × 0.2 cm, Zipax-SAX (pellicular anion exchanger) mobile phase 0.01 M NaNO$_3$ (pH 5.7) in water; 37°C; 1.0 ml/min, ΔP = 1800 psi; UV, 254 nm, 0.06 AUFS. Reprinted by permission of Spectra-Physics, Inc.

Figure 10.1 illustrates a typical application of IEC for the analysis of the three components of an asthma tablet, plus the added internal standard, benzoic acid. The neutral ephedrine is unretained and elutes at t_0, whereas the acidic compounds theophylline, benzoic acid, and phenobarbital are selectively retained.

   Many ion exchangers for LC consist of a polymeric stationary phase with ionic functional groups [e.g., $-SO_3^-$ groups for cation exchangers and $-N(CH_3)_3^+$ groups for anion exchangers]. Other IEC packings have the ionic functional groups grafted onto a silica surface via organic connecting groups. In either case, retention by the ion exchanger can be regarded as the result of two sequential processes: (a) distribution of sample compounds between the mobile phase (usually water or an aqueous mixture) and an organic stationary phase, much like that occurring in reverse-phase BPC or LLC (Chapters 7 and 8) and (b) reaction with ionic sites within the stationary phase (i.e., ion exchange). Sample retention depends on both processes, and the control of separation can be understood in terms of the experimental factors that affect each process separately. Because a distribution or partitioning of sample between the aqueous mobile phase and an organic stationary phase is involved in IEC, even nonionic compounds are in some cases retained and can be separated by IEC columns. For example, mixtures of sugars were first separated by LC in this fashion.

Ion-exchange chromatography is quite flexible, in that various other mechanisms of separation can be invoked, although these are used less frequently in modern LC. *Ionic complexes of neutral molecules* can be formed and then separated—for example, the IEC separation of sugars as borate complexes:

$$\text{Sugar} + \text{BO}_3^= \ \rightleftharpoons \ \text{sugar} \cdot \text{BO}_3^=. \tag{10.3}$$

Ligand complexes of ions can be formed that alter their relative retention or that even change the sign of the product ion completely: for example,

$$\text{Fe}^{+3} + 4\text{Cl}^- \rightleftharpoons \text{FeCl}_4^-. \tag{10.4}$$

In the latter case, ferric ions can now be separated on an anion-exchange column. Ion exchangers in the acid or base form can also be used to carry out separations via *neutralization reactions*:

$$\begin{aligned} \text{H}A + R^+\text{OH}^- &\rightleftharpoons R^+A^- + \text{H}_2\text{O}, \\ B + R^-\text{H}^+ &\rightleftharpoons R^-B\text{H}^+. \end{aligned} \tag{10.5}$$

The latter technique is useful for separating very weak acids (HA) or bases (B) as a group, using nonaqueous mobile phases such as methanol or benzene (e.g., 2). In these nonaqueous solvents, such weak acids as phenols and even pyrroles are retained, as are the weakly basic amides and sulfoxides.

Finally, ion-exchange columns can also function for *ligand-exchange reactions*:

$$RM-L + X \rightleftharpoons RM-X + L. \tag{10.6}$$

Here RM refers to a metal/ion-exchanger ion-pair, L is a mobile-phase ligand that can form a complex with the metal M, and X is a sample ligand. Ligand-exchange chromatography has been used with cation exchangers in the nickel or copper form for the separation of amino acids and other bases (*3*). More recently, Karger et al. (*4*) have shown excellent separations of ligand-like molecules by a form of *soap chromatography* (Section 11.3), involving the use of reverse-phase BPC packings plus strongly retained alkyl chelators. Various metals (e.g., Cd) can then be added to the stationary phase, providing the basis for ligand-exchange chromatography of compounds such as the sulfonamides.

A technique related to ligand exchange (Eq. 10.6) that is used occasionally in modern LC is also based on metal-ion-loaded cation exchangers (*RM*), with silver ion being used most frequently. Olefinic or amine samples form complexes with silver ion and are thereby retained by the ion exchanger. Mercuric ion-loaded cation exchangers have also been used for the separation of sulfides and mercaptans as a group, although these metal-ligand complexes are quite strong and are of limited value for modern LC separations. However, metals such as zinc that form weaker complexes may prove useful.

10.2 COLUMN PACKINGS

Many different ion-exchange packings have been used in LC, ranging from soft gels suitable for the separation of macromolecules such as proteins, to very rigid gels useful only for the separation of inorganic ions and small molecules. Applications of high-performance IEC have used one of the following kinds of packing:

- Polymeric porous particles (conventional ion-exchange resins).
- Bonded-phase pellicular particles.
- Bonded-phase porous particles.

Historically, the polymeric porous packings were used widely for the separation of such samples as amino acids, peptides, and carbohydrates. These packings consist of microporous spheres, typically with a diameter of about 10 μm, and are made from a polystyrene-divinylbenzene copolymer that is subsequently substituted with ionic functional groups. The structure of the polymer is illustrated in Figure 10.2 for the most commonly used cation and anion exchangers. Typically, sulfonate groups serve for cation exchange, whereas trialkylammonium groups provide anion-exchange properties. Although these porous polystyrene resins are still used for the separation of small molecules by high-pressure LC, the resulting columns are less efficient than corresponding ion-exchange columns of other types. The reason is very slow diffusion of sample molecules in the micropores within the polymer matrix. Consequently, most applications of the porous polymer packings are in areas that were developed prior to the advent of modern LC.

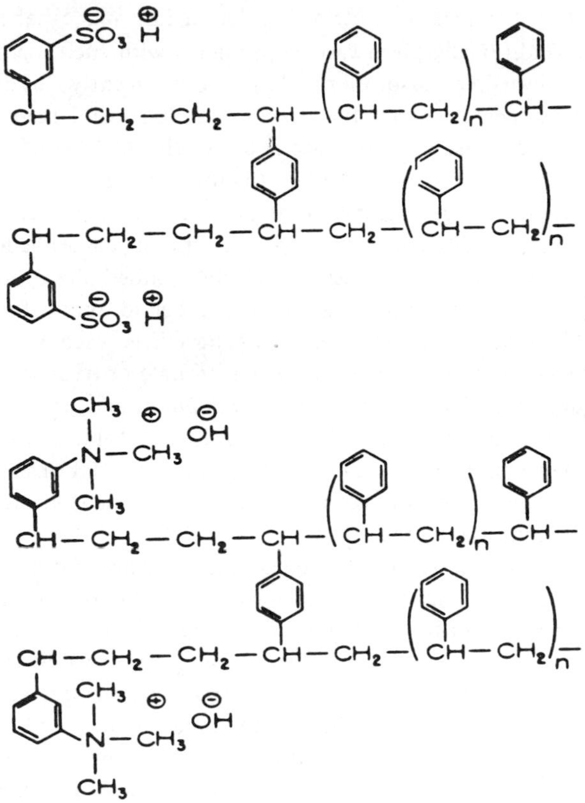

Figure 10.2 Structure of typical cross-linked polystrene ion exchangers: (a) cation exchanger and (b) anion exchanger. Reprinted by permission of Varian Aerograph.

The bonded-phase particles begin with a silica matrix, which is then coated with a polymeric covering similar to that shown in Figure 10.2. Other polymers besides polystyrene have been used (e.g., silicone, fluorocarbon, etc). In the case of pellicular particles, either a glass bead or a superficially porous particle such as Zipax can serve as the matrix. Porous microparticles of silica can also be coated. In either case, the slow diffusion within a totally polymeric particle (i.e., conventional ion-exchange resin) is now replaced by much faster diffusion within a thin layer of polymer. The result is a considerable improvement in column efficiency of these bonded-phase particles, compared to totally polymeric particles.

Until recently, all ion-exchange packings have been based on a stationary phase that is hydrophobic in the absence of ionic functional groups (e.g., polystyrene, silicone, etc.). Samples of proteins, nucleic acids, and other large, ionic molecules have traditionally given various problems with such ion exchangers (denaturation, irreversible adsorption, etc.). More recently, hydrophilic polymer coatings have been developed that, in combination with large-pore silica particles (so-called Glycophase/CPG packings), allow the high-speed separation of the largest molecules by ion exchange [see (5) for a review].

Modern ion-exchange packings come in a variety of particle sizes, shapes, and functionalities, as discussed in Chapter 5 and listed in Tables 5.8 and 5.12. Highest column efficiencies result with the bonded-phase porous particles of $5-10-\mu m$ diameter, but other packings have found application in one area or another. Many different ionic substituents have been used in ion exchangers, but most applications of modern LC involve sulfonate or trialkylammonium groups, as in Figure 10.2. These *strong cation exchangers* and *strong anion exchangers* often carry the label *SCX* or *SAX*, respectively (cf. Table 5.12). Weak cation and anion exchangers, based on either carboxylate ($-COO^-$) or amine (e.g., $-NH_3^+$) groups, have been used occasionally. Other substituent groups have so far not found significant use in modern LC.

The totally polymeric ion exchangers imbibe water and swell in the presence of aqueous mobile phases, and the relative swelling of the particle determines both its porosity and its ability to function as an ion exchanger for larger molecules. The greater the swelling, the larger the pores, so that larger sample-molecules can then enter the particle and undergo ion exchange. The relative swelling of the ion exchanger is controlled by the amount of divinylbenzene added for cross-linking, with higher concentrations giving a more rigid particle that swells less. At the same time, the strength of the particle increases with increasing divinylbenzene content, and in practice a compromise between increased swelling and particle strength is sought. Conventional polymeric ion exchangers contain 4-12% divinylbenzene, with 8% cross-linked particles being most popular.

Ion-exchange separations can also be carried out with *liquid ion exchangers* that have been physically coated onto porous or pellicular supports. These liquid ion exchangers are typically organic liquids, immiscible with water, that carry ionic or ionizable groups. They function in essentially the same way that ionic organic coatings do in the bonded-phase ion exchangers. Examples of liquid ion exchangers for cation exchange include dialkyl esters of phosphoric acid and alkyl esters of alkane phosphonic acids. Liquid anion exchangers are usually high-molecular-weight amines (e.g., tri-*n*-octyl amine) or quaternary ammonium compounds (e.g., trioctylmethylammonium). For more discussion of liquid ion exchangers and the related process of ion-pair chromatography, see Chapter 11 and (6). For a discussion of the special problems involved in the use of these mechanically held stationary phases, see the related discussion of Sections 8.3 and 8.6.

Relative sample retention and maximum sample size in IEC increase with the *ion-exchange capacity* of the column packing. This is equal to the concentration of measurable (titratable) ionic groups $R^{\pm}$ within the particle, but also varies with pH. Thus, at low pH, cation exchangers are neutralized by proton addition:

$$R^- + H^+ \rightleftharpoons RH^+. \tag{10.7a}$$

At high pH, anion exchangers are likewise neutralized by base:

$$R^+ + OH^- \rightleftharpoons R^+OH^-. \tag{10.7b}$$

The effect of pH on the exchange capacity of both strong and weak cation and anion exchangers is illustrated in Figure 10.3. Because the ion-exchange capacity of strong cation exchangers drops to zero at low pH, these packings cannot be used much below pH 1. Weak cation exchangers must be used above pH 6. Similarly, strong anion exchangers can only be employed below pH 11, whereas

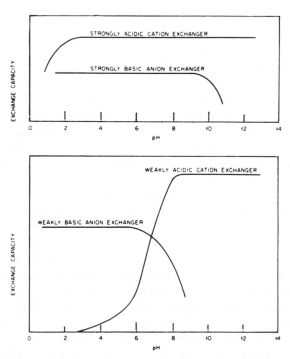

Figure 10.3 Effect of pH on capacity of ion exchangers. C. D. Scott, in *Modern Practice of Liquid Chromatography*, J. J. Kirkland, ed., Wiley-Interscience, New York, 1971.

weak anion exchangers are limited to pH 8 and below. We can regard the effects illustrated by Figure 10.3 as the result of increasing competition at pH extremes of either H^+ or OH^- ions for ion-exchange sites, with a resulting exclusion of sample ions from these same sites.

Figure 10.3 shows that strong ion exchangers can be used over a wider range in pH than can weak exchangers, and this accounts for the greater popularity of the strong exchangers. Thus, a wider range of compound types can be separated on strong exchangers, and more compounds can be separated in a single pH-gradient elution separation. However, strongly retained compounds that do not tolerate extremes in pH are more readily recovered from weak ion-exchange resins [e.g., at pH's below 4 (cation exchange) or above 10 (anion exchange)]. The weak ion exchangers, particularly anion exchangers with diethylaminoethyl (DEAE) substituents, have been favored mainly for peptide and protein separations, where elution must often take place under mild conditions.

The exchange capacity of the various ion exchangers also depends on the type of particle used. Typical values are

> bonded-phase pellicular particles   5-50 μeq/g,
> bonded-phase porous particles   100-1000 μeq/g,
> polymeric porous particles 10^3-10^4 μeq/g.

Thus, pellicular packings give lower k' values and require smaller sample sizes, other factors being equal.

The choice of a particular ion-exchange packing may also depend on the selectivity desired. In general, differences in separation sequence or selectivity can be expected for packings based on different polymeric phases and for different substituent groups (e.g., weak vs. strong exchangers). For amphoteric compounds such as amino acids, proteins, and so on, the sample can be separated by anion exchange at one pH or by cation exchange at another pH. Considerable change in separation selectivity would be expected in a cation exchange system, compared to separation of the same sample by anion exchange.

The packing of ion-exchange columns was discussed in general terms in Section 5.4. The convenience of dry-packing pellicular particles suggests that these materials should be considered first for the routine analysis of relatively simple mixtures. Bonded-phase porous particles in the 5-10-μm size range are required for more difficult separations and for applications where larger samples must be injected (e.g., trace analysis). Simple separations where column stability is a paramount consideration are often best handled by conventional ion-exchange resins (i.e., polymeric porous particles) of ~10 μm diameter.

The important characteristics of the three general types of ion-exchange packings are summarized in Table 10.1.

Table 10.1 Comparison of ion-exchange packings.

Property	Pellicular	Silica	PS-DVB resins
Typical d_p, μm	30-40	5-10	7-10
Typical ion exchange capacity, meq/g	0.01-0.1	0.5-2	3-5
Rigidity to pressure deformation	Excellent	Very good	Fair to poor[b]
Shape	Spherical	Spherical or irregular	Spherical
Pressure drop	Low	High	Highest
Efficiency	Moderate	High	Low
Packing technique	Dry	Slurry	Slurry
Cost, bulk	$2-3/g	$10-20/g	$15-25/g
pH Range	2-9 (coated) 2-7.5[a] (bonded)	2-7.5[a]	0-12 (anion) 0-14 (cation)
Regeneration rates	Fast	Moderate	Slow

Source: From (7) with permission.
[a] Some manufacturers claim upper limit of 9.
[b] Depending on degree of cross-linking.

10.3 MOBILE PHASES

The mobile phase in IEC is chosen to (a) provide adequate solubility for the various salts and buffers necessary for ion exchange, (b) control sample retention by use of the right solvent strength, and (c) provide separation selectivity as required by a particular application. IEC separations are usually carried out with aqueous salt solutions as mobile phase. The mobile phase is usually buffered, and moderate amounts of water-miscible organics (e.g., methanol, acetonitrile) are sometimes added. Solvent strength and selectivity depend on several variables: the type and concentrations of buffer ions and other salts, the pH, and the type and concentrations of added organic solvents. Sample retention in IEC is less well understood than in the other LC methods, and exceptions can be found for any general rule that is proposed. Nevertheless, useful predictions can be made on the basis that retention is the result of two processes: *sample distribution* between an aqueous mobile phase and an organic stationary phase, followed by *ion-pair formation* according to Eq. 10.1. The factors that affect

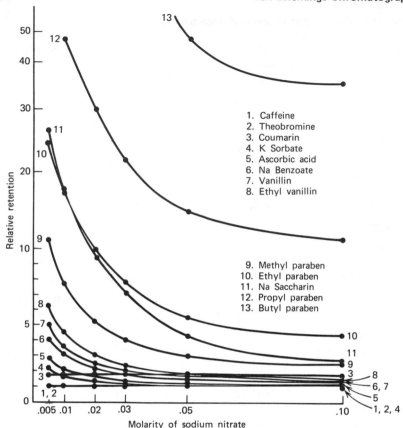

Figure 10.4 Effect of ionic strength on retention of various food additives in anion-exchange chromatography. Column, 100 × 0.21 cm, Zipax coated with 1% quaternary ammonium liquid-anion-exchanger; mobile phase, 0.01 M sodium borate (pH 9.2) with added NaNO$_3$ in water, 24°C; 0.85 ml/min, ΔP = 1000 psi; UV, 254 nm. Reprinted from (19) with permission.

ion-pair formation normally dominate, with sample distribution being of lesser importance.

Consider first the effect on sample retention of a change in *ionic strength of the mobile phase.* Either the concentration of the buffer can be varied (holding pH constant) or additional salt can be added when it is undesirable to increase buffer concentration. In either case, solvent strength normally increases as ionic strength is increased, as illustrated by the data of Figure 10.4. As the concentration of sodium nitrate is increased in this anion exchange system, sample retention decreases regularly. The reason is that an increase of mobile-phase ions leads to stronger competition of these ions for a place within the ion exchanger; that is, in Eq. 10.1 an increase in concentration of either Y^- or Y^+ drives the reaction to the left, resulting in smaller sample k' values.

For monovalent sample (X^-, X^+) and mobile phase (Y^-, Y^+) ions, simple theory (3) predicts that k' will be proportional to $1/[Y^-]$ in anion exchange and to $1/[Y^+]$ in cation exchange. For IEC systems that contain only one mobile-phase salt (including buffer), this is usually approximately the case. For divalent sample ions (X^{-2}, X^{+2}), k' varies as either $1/[Y^-]^2$ or $1/Y^+]^2$. As a result, a change in ionic strength normally changes the k' values of all sample ions by the same factor, providing that the various sample ions have the same charge (monovalent, divalent, etc.). Thus, a change in ionic strength normally provides a change in solvent strength, with minimal change in solvent selectivity, as illustrated by the data of Figure 10.4. The major exception is the case of sample ions of mixed charge (e.g., a mixture of naphthalene mono- and disulfonates).

Retention on various ion exchangers of the same chemical type increases with increasing ion-exchange capacity of the particle. Generally, sample k' values are proportional to ion-exchange capacity. The greater k' values for porous versus pellicular particles can be compensated for by the use of lower-ionic-strength solvents in the case of pellicular particles. For example, if an initial separation is carried out on a porous polymeric ion exchanger with a $NaNO_3$ concentration in the mobile phase of $0.5\,M$, it can be estimated that roughly the same k' values would be obtained on a pellicular packing with a $NaNO_3$ concentration of 0.002 M (assuming monovalent sample ions). This 250-fold reduction in ionic strength is proportional to the 250-fold reduction in ion-exchange capacity for pellicular versus porous-polymer packings.

Occasionally, the *partitioning* of sample ions (or neutral molecules) into the polymeric stationary phase is the major retention process (without ion exchange). Then an increase in ionic strength of the mobile phase may *salt out* such sample molecules (i.e., reduce their solubility in the mobile phase). In these cases it is possible for sample k' values to *increase* with increase in ionic strength, although this is not commonly observed. Solvent strength in IEC is also a function of the *type* of mobile-phase counter-ion. Thus, for equal concentrations of different mobile-phase ions Y^- in anion exchange, the retention of a given sample will generally increase in the sequence:

$$\text{citrate} < SO_4^{-2} < \text{oxalate} < I^- < NO_3^- < CrO_4^{-2} < Br^- <$$

$$SCN^- < Cl^- < \text{formate} < \text{acetate} < OH^- < F^-. \qquad (10.8)$$

(Note that halide ions should generally *not* be used in the mobile phase, because these ions attack stainless steel.) This solvent strength sequence varies somewhat

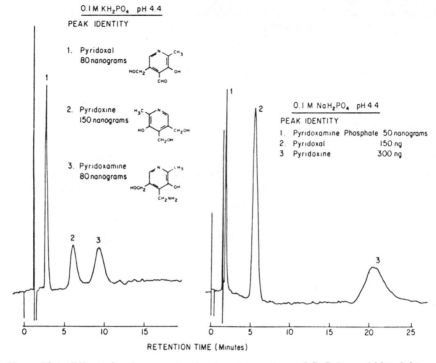

Figure 10.5 Effect of cation on retention in cation-exchange LC. Column, 100×0.21 cm, Zipax-SCX (pellicular cation exchanger); mobile phase, 0.1 M KH$_2$PO$_4$ (a) and 0.1 M NaH$_2$PO$_4$ (b) (pH 4.4); ambient; 1200 psi; UV detection at 254 nm (0.01 AUFS). Reprinted from ($19a$) with permission.

from one anion exchanger to another, as well as with experimental conditions, and should therefore be considered approximate.

Similarly, in cation exchange, retention of sample ions increases for the following sequence of mobile-phase counter-ions:

$$Ba^{+2} < Pb^{+2} < Sr^{+2} < Ca^{+2} < Ni^{+2} < Cd^{+2} < Cu^{+2} < Co^{+2} < Zn^{+2} <$$

$$Mg^{+2} < UO_2^{+2} < Tl^+ < Ag^+ < Cs^+ < Rb^+ < K^+ < NH_4^+ < Na^+ < H^+ < Li^+.$$
$$(10.8a)$$

The preceding cation series shows somewhat less variation from one IEC system to another than for the preceding anion series. Thus, solutions of fluoride ions are weak solvents in anion exchange, whereas solutions of lithium are weak solvents in cation exchange. An example of the effect of replacing one mobile-phase cation (K$^+$) by another (Na$^+$) is shown in Figure 10.5 for the separation of three pyridoxal-related compounds. As predicted by the preceding cation-strength sequence, Na^+ is a weaker-binding cation compared to K^+, leading to generally larger k' values in Figure 10.5.

A change in the mobile-phase counter-ion can lead to useful change in separation selectivity in some cases, although the dominant effect is a simple change in solvent strength. Thus, in Figure 10.5 the separation sequence does not change, but the α values for compounds 1 and 2 change from 3.2 (K^+) to 4.8 (Na^+). Similarly, in the routine analysis of physiologic fluids for amino acids and peptides, a change in mobile-phase ion from Na^+ to Li^+ allows the separation of the previously unresolved amino acids asparagine and glutamine (8).

Sample retention in IEC can also be controlled through variation of the *mobile-phase pH*. Sample retention in anion exchange LC generally increases with increase in pH, whereas retention in cation exchange decreases with increase in pH. The primary effect of pH is to control the acid/base equilibria and ionization of sample acids and bases, according to Eq. 10.2a,b. For example, consider the retention of an aliphatic amine B with pK_a equal to 10.0 in a cation exchange system. Assume that ionized B, i.e., BH^+, strongly favors the stationary phase (its concentration there is much larger than in the mobile phase). The concentrations of un-ionized base $[B]$ in the mobile phase and of ionized base $[BH^+]$ in the stationary phase are related by a variation of the well-known Henderson-Hasselbalch equation (9):

$$\log ([B]/[BH^+]) = pH - pK_a + C. \qquad (10.10)$$

Here C is a constant for a given sample ion and IEC system, which accounts for differences in solubilities of B and BH^+ in the mobile and stationary phases; C can be ignored in the case where only pH is being varied. Equation 10.10 predicts (with $C = 0$) for our aliphatic amine example the following dependence of $[B]/[BH^+]$ on pH:

pH	$[B]/[BH^+]$	Relative k'
8	0.01	0.99
9	0.10	0.91
10	1.00	0.50
11	10.00	0.09
12	100.00	0.01

At low pH values (below 9), the amine is largely in the ionized form and k' does not vary with pH. As the mobile-phase pH approaches the sample pK_a value (10 in this case), an increasing fraction of the sample is in the un-ionized form; when pH = pK_a, the sample is exactly half-ionized. Thus, in this same pH range (9-11) bracketing the sample pK_a value, the retention of the sample changes sharply with change in pH. Finally, as the pH exceeds a value of 11 in the preceding example, the ionization of the sample is largely suppressed, and it is retained only slightly (if at all).

For anion exchange, we can write a similar expression corresponding to Eq. 10.10, simply by substituting $[A^-]/[HA]$ for $[B]/[BH^+]$. $[A^-]$ and $[HA]$ refer to the concentrations of ionized and un-ionized acid HA as a function of pH.

It should be apparent from the preceding example of the aliphatic amine that continued increase in pH eventually results in all sample bases becoming un-ionized, and therefore unretained. Thus, an increase in pH in cation exchange is equivalent to an increase in solvent strength. Likewise, a decrease in pH in anion exchange is equivalent to an increase in solvent strength. However, variations in pH have their largest effect on sample retention when pH is close to the pK_a value of the sample molecule (see data of Table 10.2). Thus, in the preceding example of the cation-exchange separation of an aliphatic amine, change in pH from 1 to 8 causes little change in the k' value of this compound.

Although a change in pH can be used as a means of varying solvent strength in IEC, as in gradient elution separations, it should be apparent that large changes in selectivity can occur as a result of a change in pH. This can be illustrated by the separation of two amines, one aliphatic with pK_a equal to 10 and the other aromatic with pK_a equal to 6, and $k_1 = k_2$ at low pH. Again, Eq. 10.10 can be used to calculate relative k' values.

Relative k' values

pH	Aliphatic amine	Aromatic amine	α
4	1.00	0.99	1.01
5	1.00	0.91	1.09
6	1.00	0.50	2.00
7	1.00	0.09	11

Figure 2.28 shows an actual example of the usefulness of pH variation in anion-exchange LC to enhance separation selectivity.

When matching pH to a particular separation problem, it is useful to have some idea of the pK_a values of the sample components. The pK_a value of a compound is largely determined by the functional groups in that compound. Table 10.2 provides a brief summary of the pK_a values of some important compound types. For additional information on compound pK_a value, see (*10, 11*). Table 10.3 lists some common buffers for controlling mobile-phase pH.

If the major process in an IEC system is not ion exchange but a simple partitioning of sample between aqueous and organic phases, then a change in pH has

Table 10.2 pK$_a$ values of representative compound types.

Compound type	Acids	Bases
	pK$_a$	
Amides		~2-1
Pyrroles		0.3
Aliphatic sulfoxides		1.5
Thiazoles		1-3
Aromatic amines		4-7
Amino acids	2-4	9-12
Carboxylic acids	4-5	
Aromatic thiols	6.5	
Alkyl thiols	10.5	
Phenols	10-12	
Aliphatic amines		9-11
Carbazoles	12	

Source: Data from (*7,8*).

Table 10.3 Buffers for ion-exchange chromatography.

Buffer[a]	pK$_a$	Buffer range
Phosphate[b]		
pK$_1$	2.1	1.1-3.1
pK$_2$	7.2	6.2-8.2
pK$_3$	12.3	11.3-13.3
Citrate[b]		
pK$_1$	3.1	2.1-4.1
pK$_2$	4.7	3.7-5.7
pK$_3$	5.4	4.4-6.4
Formate[b]	3.8	2.8-4.8
Acetate[b]	4.8	3.8-5.8
Tris (hydroxymethyl) aminomethane	8.3	7.3-9.3
Ammonia[b]	9.2	8.2-10.2
Borate[b]	9.2	8.2-10.2
Diethylamine	10.5	9.5-11.5

[a] For multiply ionizable buffers, the various pK$_a$ values (pK$_1$, pK$_2$, etc.) are indicated; buffers for ion exchange normally involve pK$_1$ only, because of the strong retention of polyvalent buffer ions.
[b] Commonly used in high performance IEC.

a quite different effect on retention. In this case, as the ionization of the sample is decreased, the un-ionized compound will be preferentially retained by the organic phase. This leads to increasing retention of bases with increase in pH and to decreasing retention of acids [e.g., see examples in (*12*)].

The effect of *adding an organic solvent* to the mobile phase in IEC separations is similar to the same change in mobile-phase composition in reverse-phase LC [see data of (*12-14*)], for addition of up to 35%v organic solvent. Thus, sample retention is generally reduced with increase in the amount of organic solvent added, and the effect is greater for less-polar solvents. Changes in selectivity can be achieved by adding different organic solvents. For example, the difficult separation of the amino acids threonine and serine in conventional cation exchange LC is much improved by adding up to 10% of various organics to the mobile phase (*8, 15*). Organic solvents that have been used in IEC include methanol, ethanol, acetonitrile, and dioxane.

On the basis of a large number of modern-LC ion-exchange separations summarized elsewhere [e.g., see (*16-18*)], it is possible to set down some general conditions or rules for selecting the mobile phase in IEC. These are reviewed in Section 10.7.

10.4   OTHER SEPARATION VARIABLES

IEC separations are carried out at temperatures between ambient and 60°C. The reduction in mobile-phase viscosity that results at higher temperatures usually provides much more efficient separations than at lower temperatures, because of a lowering of D_m (Eq. 5.16). Therefore, if sample and column stability allow, separation temperatures of 50-60°C are usually advantageous. Most bonded-phase ion exchangers can be used up to 60°C, except at high pH values. The polymeric cation exchangers are generally stable to 80°C. Biochemical separations by ion exchange (e.g., enzymes) have often been run at 4°C, using classical LC techniques. However, modern LC allows such separations to be achieved in short times (e.g., 10-20 min), which decreases the risk of sample degradation during separation. Consequently, there is little need for the use of subambient temperatures.

An increase in separation temperature usually results in a lowering of all sample k' values, but this effect can be counterbalanced by decrease in the ionic strength of the mobile phase. Small changes in temperature often result in dramatic changes in separation selectivity for IEC, particularly for compound pairs of different structure or differing charge. Therefore, selectivity changes via temperature adjustments can be tried, particularly if changes in mobile-phase pH are ineffective.

The weight of sample that can be separated on a particular IEC column depends on the ion-exchange capacity of the column (Section 10.2). Normally, sample loads less than 5% of the total ion-exchange capacity are required if column efficiency is not to be seriously degraded. For typical sample components, with molecular weights of 200-500, this means that 0.05-1 mg of sample can be charged per gram of pellicular packing, and up to 50 mg of sample per gram of polymeric packing.

In one study (*21*) involving the separation of inorganic ions on a porous, bonded-phase cation exchanger, it was found that sample sizes in excess of 1% of the column capacity resulted in serious additional band broadening. Although this may have been the result of the very efficient columns used (H = 0.02 mm), maximum sample sizes lower than the 5% column capacity cited previously should be expected in some IEC systems.

The volume of the sample solution charged should normally not be greater than one-third the volume of the first eluted peak of interest (cf. Section 7.4). However, in ion exchange it is easy to use a solvent for the sample that is quite weak (e.g., very low ionic strength, a pH that favors sample ionization, etc.). Under these conditions it is possible to inject quite large sample volumes, particularly for gradient elution separations.

10.5 SPECIAL PROBLEMS

The IEC column packings developed so far for modern LC have generally not been totally satisfactory. Current packings tend to be (a) less efficient than comparable packings for the other LC methods; (b) less stable, particularly for anion exchange; and (c) less reproducible. There is little that the user can do about this situation, except to be aware that these limitations exist, so that steps can be taken to minimize the consequences. Thus, if higher column efficiencies are needed, consideration should be given to an alternative LC method; for example, ion-pair chromatography (Chapter 11) or reverse-phase partition LC with ion suppression (Section 7.5). However, the efficiency of IEC columns can usually be improved by operation at 60°C, and some studies have shown that a change in pH or addition of a few percent organic solvent (e.g., ethanol) to the mobile phase can yield a dramatic improvement of both column efficiency and band symmetry (tailing) [e.g., see (*13, 22*)].

Because present IEC columns (particularly anion exchangers) tend to be less stable, more attention should be paid to sample cleanup (Section 17.1) and to the avoidance of certain experimental extremes. Column stability generally decreases (a) with elevated temperature operation, (b) with addition of organics to the mobile phase, and (c) at high pH. The bonded-phase ion exchangers, and particularly the pellicular packings, tend to be less stable, as are anion exchangers in general. Although it may be advantageous for other reasons to use conditions that adversely affect column stability, at least the compromises involved should be understood. Certainly if it is found that the column is degrading rapidly, the experimental conditions and procedure should be reevaluated. As with all LC procedures, the recommendations of the manufacturer for a particular IEC packing should be heeded.

Because of the special problems associated with stationary phase and column reproducibility in IEC, caution is indicated in setting up routine assays based on this LC method. The preferred approach is to use columns that can be packed by the user; then stockpile enough packing from a single lot to last for all the columns that will ever have to be prepared. Where a method is to be developed for an entire industry or where the method will be the basis of regulatory enforcement by a governmental agency, thought should be given to the use of another LC method such as ion-pair chromatography. The eventual development of stable, reproducible packings for IEC will make this LC technique more attractive for routine assays that are to be used in different laboratories.

IEC columns are subject to the same deactivation processes as those in adsorption and partition chromatography. That is, strongly adsorbed components that are present either as contaminants in the mobile phase or as constituents in the sample can accumulate on the column. This leads in time to a decrease in sample k' values and lowering of column efficiency. The solution to this problem is the use of either precolumns or guard columns (Section 5.6), proper sample pretreatment (Section 17.1), or column regeneration. Column regeneration in IEC can be achieved by washing the column with a very strong solvent (see

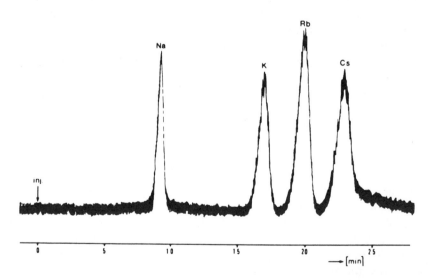

Figure 10.6 Separation of alkali metals by cation-exchange LC: $^{24}Na^+$, $^{42}K^+$, $^{86}Rb^+$, $^{137}Cs^+$. Column, 45 × 0.5 cm, Aminex Q-1505 (polymeric 20-35-μm cation exchanger); mobile phase, 1.6 M HCl in water. Reprinted from (24) with permission.

Section 10.3 for guidelines), followed by extended washing with the final mobile phase to be used. For example, superficially porous cation exchangers have been regenerated by treatment with 1 M nitric acid (23).

10.6 APPLICATIONS

Modern IEC spans a wide range of applications, ranging from simple mixtures of the elements, on one hand, to the resolution of closely similar macromolecules, on the other. Between these extremes, the technique is being used for the analysis of samples of biochemical or biomedical interest, for the measurement of additives such as vitamins and preservatives in foods and beverages, for various active ingredients in medicinal formulations, for drugs and their metabolites in serum and urine, for residue analysis in food raw materials, and for many other separation problems. Despite the limitations of IEC described previously and in

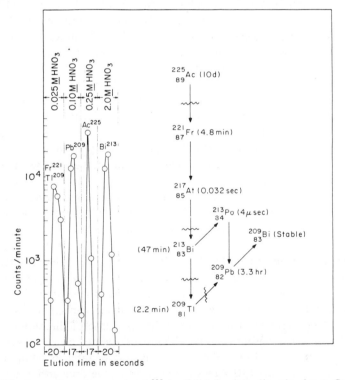

Figure 10.7 Separation of a sample of $^{225}_{89}$Ac, following its radioactive decay. Column, 1.0 × 0.2 cm, Zorbax-Sil (5-μm) coated with di-(2-ethylhexyl) phosphoric acid in dodecane; mobile phase, 0.025-2.0 M HNO$_3$ in water; 50°C; u = 17 cm/min; detection by measurement of radioactivity. Reprinted from (25) with permission.

the following chapter, for many applications it is still a uniquely useful method. As the quality and variety of IEC packings continue to improve, and with increased understanding of how separation in IEC can be controlled, the present disadvantages of this LC method will become less important.

Classical applications of IEC concentrated heavily on the separation of inorganics, particularly closely related elements in the lanthanide and actinide series, as well as other radioisotopes of interest in the atomic energy industry [e.g., see (6)]. Modern LC techniques can be adapted to these same separations, with important advantages in terms of automation, increased analysis speed, and improved assay precision. Figure 10.6 shows the separation of several radioactive alkali metals on a porous polymeric cation-exchanger. The elution of the separated ions was followed in this case by radiometric detection. Of particular significance is the possibility of adapting high-speed IEC to the detection of unstable radioisotopes with half-lives of a few minutes; Figure 10.7 shows the IEC separation of the radioactive-decay products from $^{255}_{89}Ac$, in about 1 min. In this case, a very short column (10 X 2 mm) plus gradient elution allows the rapid separation of very small sample-volumes with minimal dilution of the separated fractions. Collection of separated bands plus off-line counting was used in this analysis. High-efficiency IEC also allows the separation of closely similar radio-isotopes in very high purity (decontamination factors of 10^6 or greater), when particular attention is paid to minimizing band tailing. Horwitz and coworkers [e.g., see (26)] have made important contributions in this area. The separation of Figure 10.7 and earlier high-purity fractionations of radioactive mixtures (see Figure 19.6) have relied mainly on pellicular particles with a mechanically held liquid-ion-exchanger [e.g., di-(2-ethylhexyl) phosphoric acid in Figure 10.7]. This approach follows the classical emphasis on liquid ion exchangers in so-called extraction chromatography (6).

Modern IEC has played an important role in the pharmaceutical industry in the analysis of mixtures of water-soluble drugs. Figure 10.8 shows one such application, using a bonded-phase, 10-μm porous-particle cation exchanger. The quality control of pharmaceutical products represents one of the more important applications, as illustrated by the analysis of an asthma tablet in Figure 10.1 and an analgesic in Figure 10.9. In each case the only sample preparation involved was simple dissolution of the tablet. Because the mobile phase in the separation of Figure 10.9 is distilled water, the retention mechanism in this case must be reverse-phase partition, without actual ion exchange of the sample. Modern LC has also proved useful in studying the fate of ingested drugs, by assaying for the drug and its metabolites in the urine or plasma of subjects taking the drug. As an example, Figure 10.10 shows the analysis of the drug antipyrine plus its major metabolites in a urine sample. For this separation, cation exchange on a bonded-phase pellicular packing was used.

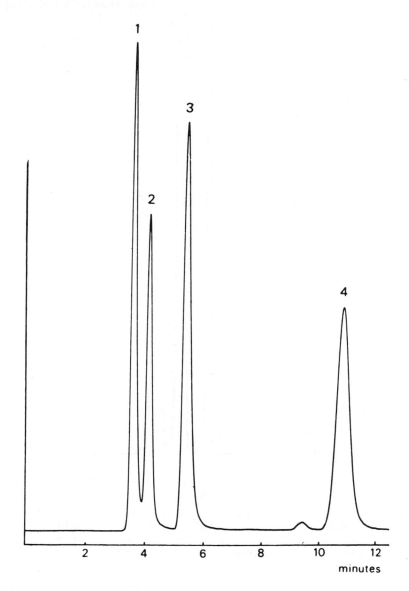

Figure 10.8 Separation of benzodiazepins: (1) oxazepam, (2) nitrazepam, (3) diazepam, and (4) chlordiazepoxide. Column, 25 × 0.46 cm Partisil-10-SCX (bonded-phase 10-μm porous particles); mobile phase, 0.05 *M* ammonium phosphate (pH 3.0) in 60%v methanol/water; 22°C; 1.0 ml/min, $\Delta P \approx 350$ psi; UV, 254 nm; 2-μl sample. Reprinted from (*12*) with permission.

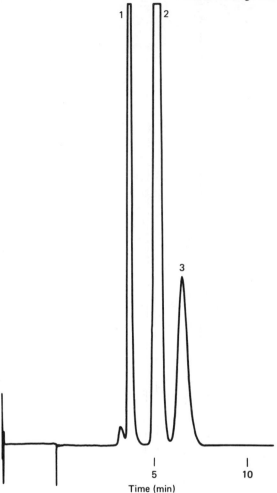

Figure 10.9 Analysis of Bromoseltzer tablet for (1) Caffeine, (2) phenacetin, and (3) acetanilide. Column, 100 × 0.21 cm Zipax-SAX (pellicular anion exchanger); mobile phase, water; ambient; 0.6 ml/min, ΔP = 600 psi; UV, 254 nm, 0.16 AUFS; 5-μl sample. Reprinted by permission of R. C. Williams, DuPont Instrument Products Division.

The IEC analysis of intentional or accidental additives in foodstuffs at low or trace levels is fairly common, because these compounds are often ionic and/or water-soluble. Figure 10.11 shows the assay of both riboflavin and niacin in a cake extract, using cation exchange on a bonded-phase pellicular packing. Figure 10.12 is the separation on a similar column of trace levels of contaminants in milk: the three metabolites of the fungicide benomyl (23).

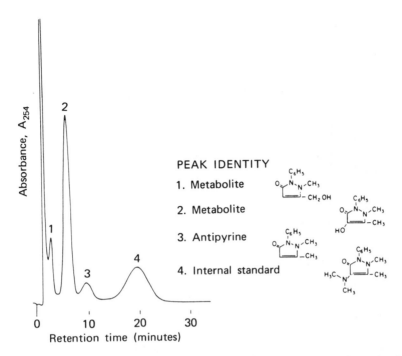

Figure 10.10 Cation exchange separation of a urinary extract containing antipyrine and its metabolites. Column, 100 × 0.21 cm, Zipax SCX; mobile phase, borate buffer, pH 9; flow-rate, 1.0 ml/min; temp., ambient; pressure, 1000 psi; detector, UV, 254 nm, 0.04 AUFS. Reprinted by permission of D. R. Baker, DuPont Instrument Products Division.

Modern IEC has been applied to classical biochemical separations of macromolecules such as proteins and nucleic acids. In the past, such separations have consumed hours or days, and overall resolution has been unimpressive by today's modern LC standards. Figure 10.13 shows the separation of a crude alkaline phosphatase mixture on a 37-74-μ DEAE Glycophase/CPG column. This is a porous, wide-pore silica with a bonded-phase coating of a hydrophilic (glycerol-based) polymer, subsequently substituted with DEAE anion-exchange groups $[-C_2H_4N(C_2H_5)_2^+]$. The separation in Figure 10.13a is monitored at 280 nm with a UV detector, whereas in Figure 10.13b a reaction detector (see Section 17.4) is used, based on the formation of p-nitrophenol from p-nitrophenylphosphate, catalyzed by the enzyme alkaline phosphatase. Thus, detection in (b) is highly specific for the enzyme and provides confirmation that peak 3 in (a) is the compound alkaline phosphatase.

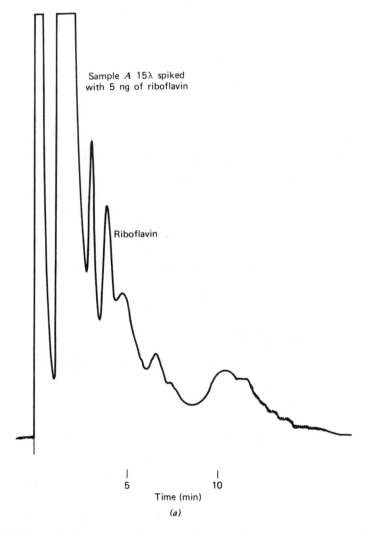

Sample *A* 15λ spiked
with 5 ng of riboflavin

Riboflavin

5 10

Time (min)

(*a*)

Figure 10.11 Analysis of water-soluble vitamins in a cake extract. (*a*) Column, 100 × 0.21 cm, Zipax-SCX (pellicular cation exchanger); mobile phase, 0.003 M H_3PO_4 in water; ambient; 1.0 ml/min, ΔP = 1500 psi; UV, 254 nm, 0.04 AUFS; 30 μl sample. (*b*) Conditions same, except mobile phase is 0.01 M KH_2PO_4 plus 0.009 M H_3PO_4. Reprinted by permission of R. C. Williams, DuPont Instrument Products Division.

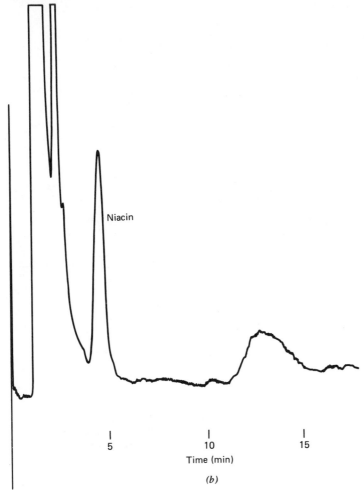

Niacin

5   10   15

Time (min)

(b)

Fig. 10.11 Continued.

The analysis of biochemical mixtures of smaller molecules by modern IEC is now an established field. Figure 10.14 shows a separation of a protein hydrolysate for its constituent amino acids, using a 10-μ polystyrene cation exchanger and a reaction detector based on the ninhydrin reaction. Figure 10.15 shows the separation of nucleotides in a rat liver extract, using a bonded-phase pellicular anion exchanger.

Clinical and biomedical applications of modern IEC are growing steadily. Figure 10.16 shows the rapid analysis of the four polyamines—putrescine, cadaverine, spermidine, and spermine—in both urine and whole blood samples. These natural metabolites have been linked to the management of various can-

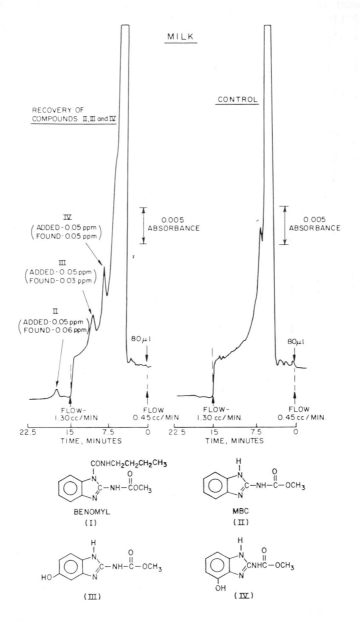

Figure 10.12 Recovery of benomyl metabolites in cow milk. Column, 100 × 0.21 cm; Zipax-SCX; mobile phase, 0.1 *M* acetic acid plus 0.1 *M* sodium acetate in water; 60°C; ml/min; UV, 254 nm. 0.05 AUFS. Reprinted from (*23*) with permission.

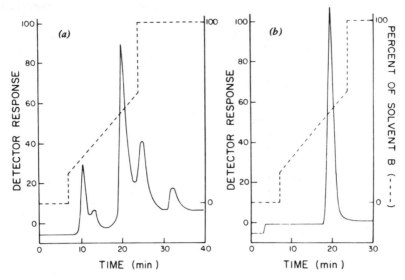

Figure 10.13 Separation of a crude alkaline phosphatase enzyme sample. Column, 50 × 0.5 cm, DEAE Glycophase (37-74-μm bonded-phase porous particles); mobile phase, Tris/NaCl gradient (pH 8.0); temp., ambient. (*a*) UV, 280 nm; (*b*) reaction detector with measurement of *p*-nitrophenol product at 410 nm. Reprinted from (*5*) with permission.

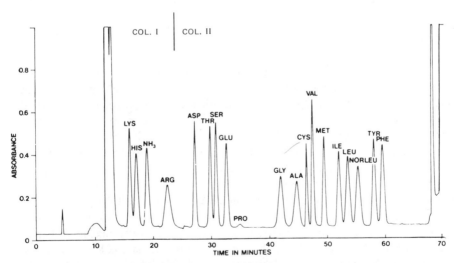

Figure 10.14 Synthetic mixture of protein-hydrolysate amino acids, separated by cation-exchange gradient elution, with ninhydrin reaction detection. Column, (I) 4 × 0.4 cm and (II) 22 × 0.5 cm of 10-μm strong cation resin (as in Figure 10.2*a*); mobile phase, gradient elution with citrate buffers; 60°C; 0.5 ml/min, ΔP = 150-600 psi; photometric detection at 570 nm after ninhydrin reaction; 0.03 μmoles each amino acid. Reprinted from (*27*) with permission.

437

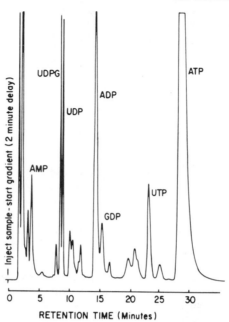

Figure 10.15 Separation of nucleotides in rat-liver extract. Column, 50 × 0.21 cm, Permaphase-AAX (bonded-phase pellicular anion exchanger); mobile phase, linear gradient from 0.003 M to 0.5 M KH$_2$PO$_4$ (pH 3.3) in water; temp., 40°C; 1.0 ml/min, ΔP = 1000 psi; UV, 254 nm, 0.04 AUFS; 5-μl sample. Reprinted with permission of DuPont.

cers, an increase in tumor mass being associated with elevated polyamine levels. In this example a simple sample pretreatment is followed by IEC separation and reaction detection with ninhydrin. Various other disease states can be accurately assessed from the relative concentrations of certain groups of closely similar natural molecules (e.g., hemoglobin variants and so-called isoenzymes). The rapid separation and analysis of these biochemical markers has posed a serious problem for the clinical laboratory, because the time-consuming classical separations are in conflict with the need for high sample throughput with minimum special handling. The development of the DEAE Glycophase/CPG packings now permits the analysis of this kind of sample within a few minutes. Figure 10.17 shows the separation by 5-10-μm particles of (Figure 10.17a) a mixture of hemoglobin variants, (Figure 10.17b), the creatine phosphokinase (CK) isoenzymes, and (Figure 10.17c) the lactate dehydrogenase (LDH) isoenzymes, all by gradient elution.

Ion Chromatography

A novel form of ion-exchange chromatography using conductimetric detection

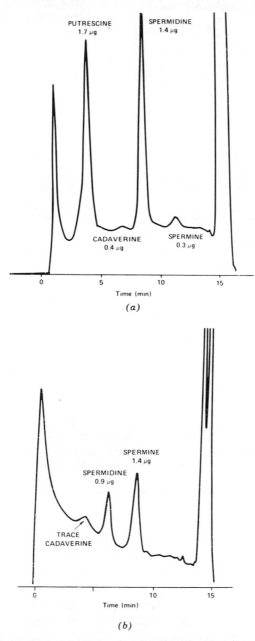

Figure 10.16 Separation of hydrolyzed polyamines in (*a*) urine and (*b*) whole blood. Column, 4.5 × 0.4 cm of same porous polymer cation exchanger as in Figure 10.14; mobile phase, gradient elution with citrate buffer; temp., 80°C; 0.9 ml/min; photometric detection at 570 nm after ninhydrin reaction; 2-ml sample with preseparation cleanup on precolumn. Reprinted from (*22*) with permission.

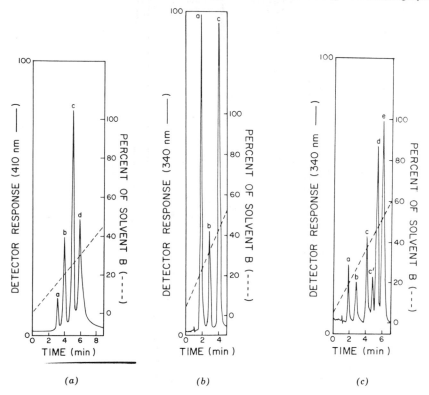

Figure 10.17 Separation of (*a*) hemoglobin variants, (*b*) CK isoenzymes, and (*c*) LDH isoenzymes. Column, 25 × 0.4 cm, DEAE-Glycophase/CPG (macroporous 5-10-μm particles of bonded-phase anion exchanger); mobile phase, linear ionic strength gradient of Tris/NaCl at pH 7.5-8.0; temp., 25°C; 2.5-3.0 ml/min, ΔP = 2200-2500 psi. (*a*) Photometric detection at 410 nm; (*b* and *c*) UV detection at 340 nm after reaction of enzymes with substrate in reaction detector. Reprinted from (*28*) with permission.

has emerged as a powerful method for measuring a wide variety of cations and anions, in both content and trace analyses. Previously the analysis of non-UV-absorbing ionic species by IEC had been limited by the lack of a suitable detector. Conductivity has often been proposed for monitoring ionic species in column effluents, because this parameter is a universal property of ionic species in solution and shows a simple dependence on the concentration of ionic species. In an ion-exchange separation, however, conductivity response is generally overwhelmed by the mobile-phase electrolyte. Small et al. (*29*) solved this detection problem by using a combination of ion-exchange resins to remove the background electrolyte, leaving only the ionic solute of interest as the major conducting species in the column effluent. Thus, a simple conductivity cell can be used

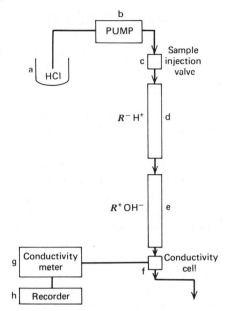

Figure 10.18 System for cation analysis by ion chromatography. Reprinted from (*29*) with permission.

as a general detector for all ionic solutes in this new analytical approach, called *ion chromatography* (IC). The general technique used in IC is illustrated in Figure 10.18, in this case for cation analysis. The system uses (*a*) an eluent reservoir, (*b*) a pump, (*c*) a sample injection device, (*d*) the separation column (where solutes are separated by conventional elution chromatography), followed by the eluent stripper column ("stripper") (*e*), where the eluent from the separating column is neutralized. The ionic solutes of interest elute from the stripper in a background of deionized water, to be monitored by the conductivity cell (*f*), conductivity meter (*g*), and recorder (*h*) (or integrator) combination. Consider, for example, the analysis of the cation mixture Li^+, Na^+, and K^+ (as chlorides). The eluent, in this case dilute HCl, is pumped to the two columns in series (Figure 10.18): a cation-exchange column d, and a strong-base resin column e in the OH^- form. If a sample of Li^+, Na^+, and K^+ chlorides is injected into the separating column d, these cations will be resolved and elute from column d at various times in a background of HCl mobile phase (eluent). On entering the stripper column, two reactions take place: (a) HCl is removed by the strong basic resin,

$$HCl + resin\text{-}OH^- \rightarrow resin\text{-}Cl^- + H_2O;$$

(6) the alkali metal chlorides (and other anions) are converted to the hydroxides

$$M^+Cl^- + \text{resin-OH}^- \rightarrow M^+OH^- + \text{resin-Cl}^-$$

and pass directly through the stripper and into the conductivity cell, where they are monitored in a background of deionized water. The peaks so formed can be quantitatively evaluated by conventional height or area measurements (Chapter 13).

Anion analysis is carried out in an analogous scheme; e.g., with sodium hydroxide as eluant, an anion exchanger as the separating column, and a strong-acid (cation) exchanger in the H^+ form for the stripper.

IC practice differs from conventional ion-exchange chromatography in that the stripper column must periodically be regenerated. To keep regeneration at a minimum and at the same time maintain overall chromatographic efficiency, the relative volumes and specific ion-exchange capacities of columns d and e in Figure 10.18 must be properly adjusted. To preserve high chromatographic efficiency (minimize band broadening), the volume ratio of column e to column d is kept as low as possible, a ratio of unity being desirable, but a value of $\leqslant 10$ acceptable. Regeneration requirements are minimized by using low-capacity pellicular particles or surface-modified resins in the separating column d. Thus, low ionic strength mobile phases can be used with small sample sizes, and the conventional, high capacity, porous ion-exchange resin in the stripper columns e can then be used for many analyses before regeneration is required. The mobile-phase (eluting) ion must obviously be removable by the stripper.

A number of separating schemes that have been proposed for IC are shown in Table 10.4. These systems are not inclusive, and others are being developed as more experience in gained with the technique. The apparatus for IC, distributed exclusively by Dionex Corporation (under license by the Dow Chemical Co.), is relatively conventional. Several models are available, ranging from rather simple instruments to those capable of fully automatic operation, including regeneration of the stripping column.

Applications of IC are quite diverse and many of its uses have been directed toward the determination of trace ions in various water samples, as shown in Figure 10.19. Here we see the anion chromatogram of "blow-down" water, to assist in monitoring the concentration of treatment chemicals that are used in boilers. Trace cations can also be measured in aqueous systems, as shown in Figure 10.20 for the alkali metals in fruit juice. In this case, small amounts of Na^+ and NH_4^+ can be easily determined in the presence of a large excess of K^+. IC also appears to hold promise for assaying hetero-atoms in elemental analyses; in Figure 10.21 a multiple-halogen standard was combusted with sulfamic acid to yield good agreement (by IC) with theoretical values for F^-, Cl^-, Br^-, and SO_4^{-2}. (The I^- peak was too broad at the eluent strength used to be accurately measured.) Both organic anions and cations have been analyzed by IC at trace and assay levels. Figure 10.22 shows the cation chromatogram of a sample of methylamines plus ammonium and sodium ions. Precision in the assay of these

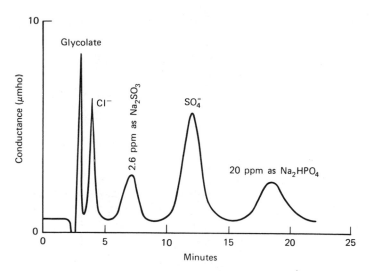

Figure 10.19 Typical boiler blowdown water sample. Analytical column 100×0.28 cm, Chromex anion exchanger (50-μm surface-layer-active type); stripper column, 30×0.28 cm, Dowex 50-2X-16; mobile phase, 0.005 M sodium carbonate, 0.004 M sodium hydroxide in deionized water; flowrate, 1.75 ml/min; temp., ambient; sample volumn 500 μl; detector sensitivity, 10 μmho full scale; stripper regenerant, 1 M H_2SO_4 in deionized water. Reprinted from (*34*) with permission.

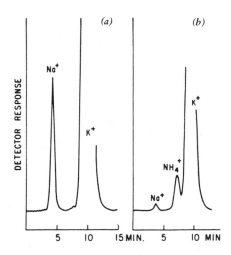

Figure 10.20 Analysis of (*A*) orange juice and (*B*) grape juice. Analytical column, 25×0.9 cm glass surface sulfonated styrene/divinylbenzene, 44-88 μm; stripper column, 27×0.9 cm, Dowex 1-X80H⁻, 37-74 μm; mobile phase, 0.01 NHCl; temp., ambient; flowrate, 2.67 ml/min; sample, 100-μl undiluted juices; detector, conductivity. Reprinted from (*29*) with permission.

443

Table 10.4 Schemes for ion chromatography.

Separating column	Eluant	Stripper column	Stripping reaction
Cation analysis			
$Resin-H^+$	HCl	$Resin-OH^-$	$Resin-OH^- + HCl \rightarrow resin-Cl^- + H_2O$
$Resin-Ag^+$	$AgNO_3$	$Resin-Cl^-$	$Resin-Cl^- + AgNO_3 \rightarrow resin-NO_3^- + AgCl$
$Resin-Cu^{2+}$	$Cu(NO_3)_2$	Resin-amine	$Resin-amine + Cu(NO_3)_2 \rightarrow resin-amine \cdot Cu(CO_3)_2$
Resin-anilinium	Aniline hydrochloride	$Resin-OH^-$	$Resin-OH^- + aniline \cdot HCl \rightarrow aniline + Resin-Cl^- + H_2O$
$Resin-Ag^+$	$AgNO_3$-HNO_3	1st $resin-Cl^-$	$Resin-Cl^- + AgNC_3 \rightarrow resin-NO_3^- + AgCl$
		2nd $resin-OH^-$	$Resin-OH^- + HNO_3 \rightarrow resin-NO_3^- + H_2O$
Anion analysis			
$Resin-OH^-$	NaOH	$Resin-H^+$	$Resin-H^+ + NaOH \rightarrow resin-Na^+ + H_2O$
Resin-phenate	Na-phenate	$Resin-H^+$	$Resin-H^+ + PhO^-Na^+ \rightarrow resin-Na^+ + PhOH$

Source: From (29).

444

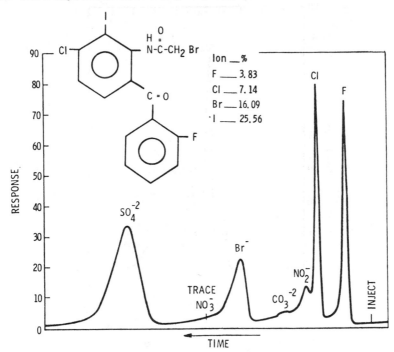

Figure 10.21 Analysis of combusted multiple-halogen standard mixed with sulfonic acid. Sample, 6.362 mg of standard plus 2.100 mg of sulfonic acid; 0.1% N_2H_4 scrubber; Dionex Corp. instrument. Reprinted from (*36*) with permission.

synthetic mixtures was excellent (CV's $< 3\%$); the cation bases were converted to chloride salts by use of a short column in the Cl^- form between the stripper and the conductivity detector. Table 10.5 illustrates the wide range of IC applications that have been reported. It should be noted that IC calibration plots are often nonlinear.

10.7 THE DESIGN OF AN ION-EXCHANGE SEPARATION

Selecting the Column

For the analytical separation of mixtures of small molecules (MW < 2000), the best initial choice of column packings is a bonded-phase ion exchanger with 5-10-μm particles (e.g., Partisil-10 SCX). It must first be determined whether an anion or cation exchanger is appropriate for the sample of interest, usually on the basis of a knowledge of the structure and acid/base properties of individual

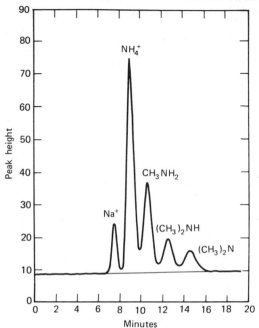

Figure 10.22 Separation of methylamine mixture. Conditions as in Figure 10.20, except stripper column 25 × 0.9 cm AG-1X10 anion-exchange resin, 37-74 μm, hydroxide form; chloride-form column; 12 × 0.28 cm AG-X10 anion-exchange resin 37-74 μm, chloride form; flowrate, 3.8 ml/min; sample, 50 μl of standards, 10-600 ppm. Reprinted from (*35*) with permission.

sample compounds. For the routine separation of simple mixtures by IEC, pellicular packings (e.g., Zipax SAX) have certain advantages. A large enough batch of column packing should be secured initially so that identical columns with reproducible properties can be prepared as the need arises.

For the preparative separation of easily resolved mixtures, porous polymeric particles (e.g., Durrum DA-X8) offer better reproducibility, larger ion-exchange capacity, and the possibility of larger sample sizes. However, the efficiency of these latter packings is much lower than that of the bonded-phase porous particles.

For the separation of large molecules (MW > 2000) there is a more limited choice of possible column packings (e.g., CPG-Glycophase DEAE). Here only porous packings are so far available, with 5-10-μm particles used for analytical separations and 37-74-μm particles available for preparative applications. Packings for the IEC separation of high-molecular-weight samples have not yet been fully developed to the point where reliable materials are routinely available. Potential problems that the user should be aware of include incomplete elution (and recovery) of sample components due to denaturation and/or adsorption, as well as instability of these packings at pH's (e.g., $6 < pH < 9$) required for many separations.

Tables 5.8 and 5.12 list a number of acceptable packings of each type previously discussed.

Selecting the Mobile Phase

The mobile phase is normally water plus one or more of the following additives: buffer, neutral salt, or organic solvent.

The *buffer* is selected for its ability to control or maintain pH at the value selected. Table 10.3 lists several common buffer systems, with the pH range over which the buffer is usable. For a listing of additional buffers of special interest in biochemical separations, see (*37,38*). Many reported IEC separations use phosphate buffers, particularly at low pH. It should be kept in mind that multiply charged ions are generally held on ion exchangers more strongly, which means that phosphate buffers at high pH (where PO_4^{-3} is a major ion) will compete strongly for ion-exchange sites in anion-exchange systems. Although this effect can be offset by lowering the concentration of the buffer, the buffer is then less able to maintain the pH of the system constant. Similar considerations apply to the use of other buffer systems that involve multiply charged buffer ions.

Many IEC separations have been reported in which the buffer is omitted. For example, cation-exchange separations at low pH often use either phosphoric acid or perchloric acid solutions. In such systems all moderately strong bases are fully ionized, regardless of moderate fluctuation in pH, and a buffer may then not be needed. Although the similar use of ammonia or NaOH solutions in anion exchange could be used in principle, many ion exchangers (e.g., bonded-phase silicas) are not stable at pH values above 9.0.

The *molarity* of the buffer, acid, or neutral salt in the mobile phase should range from about 0.001-0.5 M for most separations. Within this range of concentrations, sample k' values can be varied over wide limits by a variation of buffer or salt concentration.

Neutral salts are often added to the buffer, for the further control of solvent strength. Of the various salts used in this fashion, $NaNO_3$ seems by far the most popular, presumably because it is innocuous and has an intermediate retention strength (see retention sequence of Eq. 10.8). Halide ions should be avoided, because they attack stainless steel.

Organic solvents such as methanol, acetonitrile, and *n*-propanol can be added to the mobile phase for further control of selectivity, to improve column efficiency and/or peak shape, or as another way to adjust solvent strength and k' values. Organic solvent is sometimes added in amounts of up to 40%, but more often 3-10%. Care should be taken that the particular column packing used is

Table 10.5 Illustrative application of ion chromatography.

Material analyzed	For	Concentration range	Remarks	Ref.
Human urine	Na^+, NH_4^+, K^+	N.A.		(29)
Dog-blood serum	Na^+, K^+	N.A.		(29)
Fruit juice	Na^+, NH_4^+, K^+	N.A.		(29)
Organic amines	Methylamines	N.A.	Synthetic mixture	(29)
Organic amines	Ethylamines	N.A.	Synthetic mixture	(29)
Quaternary ammonium compounds	TEA, TBA	N.A.	Synthetic mixture	(29)
Divalent metal ions	Ca^{+2}, Mg^{+2}	N.A.	Synthetic mixture	(29)
Organic acids	Chloroacetic acids	N.A.	Synthetic mixture	(29)
Ambient aerosols	SO_4^{-2}, NO_3^-	ppm		(30)
Hatchery water	NO_3^-, PO_4^{-3}	ppb		(31)
Paper mill effluent	Cl^-, SO_4^{-2}, oxalate	ppm		(31)
Pond water	NO_3^-, PO_4^{-3}	ppb		(31)
Pulping liquors	SO_4^{-2}, S^{-2}	ppm		(31)
Soil extracts	Cl^-, NO_3^-, SO_3^{-2}, SO_4^{-2}	N.A.		(31)
Engine exhausts	Cl^-, NO_3^-, SO_3^{-2}, SO_4^{-2}, Br^-, NH_4^+, amines	N.A.		(31)
Ambient aerosols	Cl^-, NO_2^-, PO_4^{-3}, Br^-, NO_3^-, SO_3^{-2}, SO_4^{-2}, NH_4^+	N.A.		(31)

Sample	Species	Level	Notes	Ref.
Upper atmosphere	F^-, Cl^-, PO_4^{-3}, NO_3^-, SO_4^{-2}	N.A.		(31)
Elemental analysis	S, P, Cl	N.A.		(31)
Electronic devices	Cl^-, Br^-, SO_4^{-2}, Na^+, K^+, NH_4^+, amines	ppm		(31)
Pulping liquid (green)	S^{-2}, SO_4^{-2}	High		(31)
Plating bath water	F^-, Cl^-, SO_4^{-2} CrO_4^{-2}	ppm		(32)
Well water	Cl^-, PO_4^{-3}, Br^-, SO_4^{-2}	ppm		(32)
Brine solution	SO_4^{-2}	ppm		(32)
Rainwater	Cl^-, PO_4^{-3}, SO_4^{-2}	ppm		(32)
Caustic solution (50%)	Cl^- and ClO^-, ClO_3^-, SO_4^{-2}	ppm		(32)
Corrosion leaking water	F^-, Cl^-	ppm		(32)
Surfactant	Glycolate, chloroacetate, Cl^-	N.A.		(32)
Tapwater	F^-, CO_3^{-2}, SO_4^{-2}	ppm		(33)
Organic acids	Lactic, pyruvic acids	N.A.	Synthetic mixture	(33)
Broiler blow-down water	Glycolate, Cl^-, SO_3^{-2}, SO_4^{-2}, PO_4^{-3}	ppm		(34)
Organic amines, ammonia	Methylamines, NH_3	ppm	Synthetic mixture, assay	(35)
Schoniger oxygen flask combustion	S, Cl, P, N	Low	Assay	(36)
Parr oxygen bomb	S, Cl, P, N	Low	Assay	(36)

compatible with addition of the amount and type of organic solvent desired. Typically a twofold reduction in k' will result from adding 7% methanol, 3% acetonitrile, or 1% propanol to the aqueous mobile phase [e.g., see (13)]. Addition of more than 35% organic solvent can lead to an *increase* in k' (14).

Temperature

Temperatures of 50-80°C are normally recommended for maximum column efficiency. Bonded-phase ion exchangers are less stable at higher temperatures, and manufacturer recommendations should be consulted on this point. A temperature of 60°C is a good, commonly used compromise. The main reason for lower temperatures is to avoid reaction of sample components, and for this reason the separation of enzymes, coenzymes, and other biologicals is commonly carried out at ambient temperature.

Sample Size

The quantity of sample injected should not exceed 1-5% of the total column capacity (in μeq/g). For bonded-phase porous particles, this means maximum sample sizes of 5-10 μeq/g of packing. Pellicular packings allow maximum sample injections of only one-tenth as much—for example, 0.5-1 μeq/g. The volume of sample injected should normally not exceed one-third of the volume of the first band of interest (cf. Section 7.4). However, larger volumes can be tolerated if the sample ionic strength is reduced below that of the mobile phase or if the pH is changed to give larger k' values for all sample components.

Other Considerations

Where further increase in sample resolution is required, column efficiency can be increased as described in the discussion on controlling resolution in Section 2.5c. Special problems and their solution are described in Section 10.5 and Chapter 19.

REFERENCES

1. C. G. Horvath, B. A. Preiss and S. R. Lipsky, *Anal. Chem.*, **39**, 1422 (1967).

1a. P. R. Brown, *High Pressure Liquid Chromatography: Biochemical and Biomedical Applications*, Academic, New York, 1973.

2. L. R. Snyder and B. E. Buell, *Anal. Chem.*, **40**, 1295 (1968).

3. W. Reiman, III, and H. F. Walton, *Ion Exchange in Analytical Chemistry*, Pergamon, New York, 1970, p. 170.

4. N. H. C. Cooke, R. L. Viavattene, R. Eksteen, W. S. Wong, G. Davies, and B. L. Karger, *J. Chromatogr.*, **149**, 391 (1978).

5. F. E. Regnier, K. M. Gooding, and S. -H. Chang, in *Contemporary Topics in Analytical and Clinical Chemistry*, Vol. 1, D. Hercules et al., eds., Plenum, New York, 1977, p. 1.

6. T. Braun and C. Ghersini, eds., *Extraction Chromatography*, Elsevier, New York, 1975.

7. R. E. Majors, *J. Chromatogr. Sci.*, **15**, 334 (1977).

8. J. V. Benson, M. J. Gordon, and J. A. Patterson, *Anal. Biochem.*, **18**, 228 (1967).

9. A. L. Lehninger, *Biochemistry*, Worth, New York, 1970, p. 48.

10. A. Albert and E. P. Serjeant, *Ionization Constants of Acids and Bases*, Wiley, New York, 1962.

11. L. R. Snyder and B. E. Buell, *J. Chem. Eng. Data*, **11**, 545 (1966).

12. P. J. Twitchett, A. E. P. Gorvir, and A. C. Moffat, *J. Chromatogr.*, **120**, 359 (1976).

13. J. H. Knox and J. Jurand, *J. Chromatogr.*, **87**, 95 (1973).

14. R. Eksteen, J. C. Kraak, and P. Linssen, *ibid.*, **148**, 413 (1978).

15. G. Ertingshausen and H. A. Adler, *Am. J. Clin. Pathol.*, **53**, 680 (1970).

16. J. N. Done, J. H. Knox, and J. Loheac, *Applications of High-Speed Liquid Chromatography*, Wiley-Interscience, New York, 1974.

17. *The HPLC Applications Book*, Vol. 1, Hewlett-Packard, Avondale, Pa.

18. J. J. Kirkland and L. R. Snyder, *Solving Problems with Modern Liquid Chromatography*, *Am. Chem. Soc.* Short Course Manual, Washington, D.C., 1974.

19. J. J. Nelson, *J. Chromatogr., Sci.*, **11**, 28 (1973).

20. R. C. Williams, D. R. Baker, and J. A. Schmidt, *ibid.*, **11**, 618 (1973).

21. E. P. Horwitz, C. A. A. Bloomquist, and W. H. Delphin, *J. Chromatogr. Sci.*, **15**, 41 (1977).

22. H. Adler, M. Margoshes, L. Snyder, and C. Spitzer, *J. Chromatogr. Biomed. Appl.*, **143**, 125 (1977).

23. J. J. Kirkland, *J. Agric. Food Chem.*, **21**, 171 (1973).

24. J. F. K. Huber and A. M. Van urk-Schoen, *Anal. Chim. Acta*, **58**, 395 (1972).

25. E. P. Horwitz, W. H. Delphin, C. A. A. Bloomquist, and G. F. Vandegrift, *J. Chromatogr.*, **125**, 203 (1976).

26. E. P. Horwitz and C. A. A. Bloomquist, *J. Chromatogr. Sci.*, **12**, 200 (1974).

27. G. Ertingshausen, H. A. Adler, and A. S. Reichler, *J. Chromatogr.*, **42**, 355 (1969).

28. S. H. Chang, K. M. Gooding, and F. E. Regnier, *ibid.*, **125**, 103 (1976).

29. H. Small, T. S. Stevens, and W. C. Bauman, *Anal. Chem.*, **47**, 1801 (1975).

30. J. Mulik, R. Puckett, D. Williams, and E. Sawicki, *Anal. Lett.*, **9**, 653 (1976).

31. "Applications of Ion Chromatography," Dionex Corporation, Sunnyvale, Calif., October, 1976.

32. R. C. Chang, "Ion-Chromatography: A New Technique in Water Analysis," Dionex Corporation, 1976.

33. Dionex Corporation, Bulletin 10-76-3M, 1976.

34. T. S. Stevens, V. T. Turkelson, and W. R. Albe, *Anal. Chem.*, **49**, 1176 (1977).

35. S. A. Bouyoucous, *Anal. Chem.*, **49**, 401 (1977).

36. F. C. Smith, Jr., A. McMurtrie, and H. Galbraith, *Microchem. J.*, **22**, 45 (1977).

37. N. E. Good, G. D. Winget, W. Winter, T. N. Connolly, S. Izawa, and R. M. Singh, *Biochemistry*, **5**, 467 (1966).

38. D. E. Gueffroy, ed., *A Guide for the Preparation and Use of Buffers in Biological Systems, Calbiochem.*, La Jolla, Calif., 1975.

BIBLIOGRAPHY

Braun, T., and C. Ghersini, eds., *Extraction Chromatography*, Elsevier, New York, 1975 (review of classical applications of inorganic separations, using liquid ion exchangers).

Brown, P. R., *High Pressure Liquid Chromatography: Biochemical and Biomedical Applications*, Academic, New York, 1973 (early review of modern LC applications of pellicular ion exchangers to biochemical area).

Eksteen, R., J. C. Kraak, and P. Linssen, *J. Chromatogr.*, **148**, 413 (1978) (excellent and detailed discussion of mobile phase in IEC with polymeric stationary phases; useful for cases where full optimization of separation in IEC is required).

Helfferich, F., *Ion Exchange*, McGraw-Hill, New York, 1972 (general discussion of ion exchange, particularly for inorganic samples and classical exchangers).

Horvath, C., in *Ion Exchange and Solvent Extraction*, J. Marinsky and Y. Marcus, eds., Vol. 5, Dekker, New York, 1973, p. 207 (review of pellicular IEC packings).

Ion Chromatography Dateline, Dionex Corp. (a continuing series of technical reports on IC, beginning Jan., 1979).

McDonald, J. C., *Amer. Lab.*, Jan. 1979, p. 45 (good review of ion-chromatography).

Perrin, D. D., and B. Dempsey, *Buffers for pH and Metal Ion Control*, Halsted Press (Wiley), New York, 1974 (description of buffer systems with specific experimental details, background information, special buffers, etc.).

Regnier, F. E., K. M. Gooding, and S. -H. Chang, in *Contemporary Topics in Analytical and Clinical Chemistry*, D. Hercules et al., eds., Plenum, New York, 1977, p. 1 (a thorough review of the CPG-Glycophase ion exchangers, along with many applications to biochemical and clinical separations).

Rich, W. E., *Anal. Instrum.*, **15**, 113 (1977) (a review of ion chromatography).

Sawicki, E., J. D. Mulik, and E. Wittgenstein, eds., *Ion Chromatography Analysis of Environmental Pollutants*, Ann Arbor Science Publishers, Ann Arbor, Mich., 1978.

Scott, C. D., in *Separation and Purification Methods*, Vol. 3, E. S. Perry, C. J. van Oss, and E. Grushka, eds., Dekker, New York, 1974, p. 263 (review of high-resolution IEC analysis of physiological fluids by Oak Ridge National Laboratory group).

ELEVEN

ION-PAIR CHROMATOGRAPHY

11.1	Introduction	454
11.2	Column Packings	457
11.3	Partitioning Phases	458
	General Considerations	458
	Role of the Counter-ion	459
	Solvent Polarity	464
	Secondary Ions	467
	pH Effects	467
	Solvent Selectivity	469
11.4	Other Separation Variables	470
11.5	Special Problems	471
11.6	Applications	473
11.7	The Design of an Ion-Pair Separation	477
	Selecting the Column	477
	Selecting the Mobile Phase	478
	Temperature	480
	Sample Size	481
	Other Considerations	481
References		481
Bibliography		482

453

11.1 INTRODUCTION

Ion-pair chromatography as adapted to modern LC is of comparatively recent origin, being first applied in the mid-1970s. However, its use in classical LC and liquid-liquid extraction is considerably older. The rapid acceptance of ion-pair chromatography (IPC) as a new HPLC method owes much to the work of Schill and coworkers [e.g. see (1)] and to its unique advantages. At various times IPC has also been called extraction chromatography, chromatography with a liquid ion-exchanger, soap chromatography and paired-ion chromatography (PIC^{TM}). Each of these techniques is basically similar, and we will use the term *ion-pair chromatography* for all of them.

The current popularity of ion-pair chromatography arises mainly from the limitations of ion-exchange chromatography and from the difficulty in handling certain samples by the other LC methods (e.g., compounds that are very polar, multiply ionized, and/or strongly basic). Present ion-exchange packings suffer from the following disadvantages.

- Columns tend to be less efficient than other LC columns.
- Columns are often irreproducible from lot to lot.
- Columns tend to be less stable.
- Limited choice among packings, resulting in little ability to vary selectivity by changing the packing.

As we will see, IPC can overcome all these objections.

IPC can be carried out in either normal-phase or reverse-phase modes, each of which has its own advantages (Table 11.1). Here we describe IPC in terms of the more popular reverse-phase mode, although the basic process is similar for either technique. The stationary phase in reverse-phase IPC can consist either of a silanized silica packing such as that normally used in bonded-phase LC (e.g., C_8 or C_{18} bonded phase; see Section 7.2) or of a similar packing with a mechanically held, water-immiscible organic phase such as 1-pentanol. For the moment we assume that the stationary phase consists of a C_8 or C_{18} BPC packing.

The mobile phase consists of an aqueous buffer (plus an added organic co-solvent such as methanol or acetonitrile for bonded-phase separations) and an added *counter-ion* of opposite charge to the sample molecule. For example, assume the separation of a group of carboxylic acids, using a mobile phase buffered at pH 7.0 so that all sample compounds are in the form RCOO⁻ (represented hereafter as R^-). The counter-ion in this case might be tetrabutyl ammonium ion Bu_4N^+ (or TBA^+). In the simplest case of ion-pair chromatography, it can be assumed that the sample and counter-ions are soluble only in the aqueous mobile phase, and the ion pair formed from these ions is soluble only in the

Table 11.1 Advantages of normal-phase versus reverse-phase operation in ion-pair chromatography.

Normal-phase operation
1. Can use absorbing or fluorescing ions as counter-ions for easy detection of nonabsorbing sample compounds
2. Easy to vary selectivity by varying organic phase composition

Reverse-phase operation
1. Does not require mechanically held (less-stable) stationary phase
2. Aqueous samples (most common form in IPC separations) can be injected directly.
3. Gradient elution is possible, by varying (a) counter-ion concentration in mobile phase or (b) solvent polarity with bonded-phase columns
4. Counter-ion bleed from stationary phase is eliminated
5. Biological samples require less cleanup prior to injection, because sample constituents of little interest are usually hydrophylic and elute at t_o
6. Solvent strength easily varied by change in counter-ion concentration.
7. Separation selectivity easily varied by changing pH of mobile phase
8. Bonded-phase columns (no organic liquid added as stationary phase) can be quickly converted for reverse-phase partition LC, and vice versa; this increases the utility of a single column by allowing rapid interchange between two LC methods.

---

organic stationary phase. In this case we can write Eq. 11.1 for distribution of sample R^- between the two phases:

$$R^-_{aq} + TBA^+_{aq} \rightleftharpoons R^-TBA^+_{org} \tag{11.1}$$

ion pair

The subscripts *aq* and *org* refer to aqueous and organic phases, respectively. The extraction constant E is next defined in terms of the concentrations ( ) of the preceding species:

$$E = \frac{(R^-TBA^+)_{org}}{(R^-)_{aq}(TBA^+)_{aq}} \tag{11.2}$$

where E is a constant for a particular IPC system but varies with mobile-phase pH and ionic strength, concentration and kind of organic co-solvents in the mobile phase (e.g., acetonitrile or methanol), and temperature. The capacity factor k' is related to E as follows:

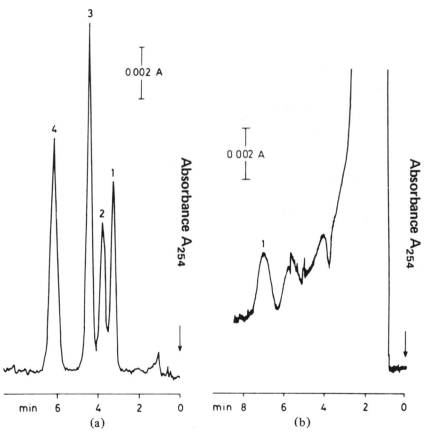

Figure 11.1 Reverse-phase IPC separation of pyridine derivatives with 1-pentanol as stationary phase. (*a*) 1, nicotinic acid; 2, isonicotinic acid; 3, 5-fluoro-3-hydroxymethyl pyridine; 4, 5-fluoropyridine-3-carboxylic acid. (*b*) Assay for nicotinic acid in serum. Column, 20 × 0.32 cm, silizanized LiChrosorb SI-60 (10 μm) with 1-pentanol as stationary phase; 25°C; 0.8 ml/min, ΔP = 1000 psi; UV, 254 nm. (*a*) Mobile phase, 0.03 *M* tetrabutylammonium NaSO$_4$ (counter-ion) plus 0.04 *M* phosphate (pH 7.4) in water; 90-220 ng each compound injected. (*b*) Mobile phase, same except counter-ion concentration is 4 times larger (0.12 *M*); 55-μl serum sample plus 50-ng nicotinic acid. Reprinted from (*11*) with permission.

$$k' = \frac{V_s}{V_m} \frac{(R^-TBA^+)_{org}}{(R^-)_{aq}}$$

$$= \frac{V_s}{V_m} E(TBA^+)_{aq}. \qquad (11.2a)$$

Thus, the k' values of all sample compounds (of unit negative charge, -1) are predicted to be proportional to the concentration of the counter-ion TBA^+. Note that the distribution coefficient K (eq. 2.5) is related to E as $E = K/(TBA^+)$.

The variation of $(TBA^+)_{aq}$ provides a way of controlling solvent strength while holding selectivity more or less constant. Figure 11.1a shows a typical application of reverse-phase IPC for the separation of several acids of physiological interest. The measurement of one of these compounds in serum is shown in Figure 11.1b, using a higher concentration (vs. Figure 11.1a) of the counter-ion TBA^+ to increase k' of compound 1 (nicotinic acid) and separate it from early-eluting sample components.

The application of IPC and the control of separation in this LC method are readily understandable in terms of preceding discussion in Chapters 6, 7, and 10, as reviewed in Section 11.3 of this chapter. The present chapter describes the use of IPC in sufficient detail for the average reader. For additional information the reader is directed to (1-17) and the Bibliography.

More recently, the question has arisen of whether the equilibrium implied by Eq. 11.1 really describes IPC under normal circumstances. Some workers have argued that the counter-ion is in fact adsorbed onto the organic stationary phase, with subsequent retention of charged sample ions by simple ion exchange. Experimental studies summarized in (18, 19, 19a) indicate, for normal IPC with reasonably small counter-ions, that the counter-ion is not adsorbed and Eq. 11.1 applies. However, for much larger counter-ions as in "soap chromatography" (see the following discussion), retention of the counter-ion occurs and the resulting process for sample ions is very much like ion-exchange chromatography.

11.2   COLUMN PACKINGS

Most applications of modern IPC have been carried out with small, porous particles (5-10 μm). Reverse-phase separations are usually carried out with BPC packings such as Zorbax ODS or LiChrosorb RP-2. Alternatively, a mechanically held organic liquid such as 1-pentanol can be used as stationary phase, with silanized silica particles. For normal-phase separations, a porous silica packing such as that used in LSC can be employed. In this case the stationary phase consists of a mechanically held aqueous solution of buffer plus counter-ion.

Columns for IPC are prepared in the same way as columns for BPC (Chapter 7) or LLC (Chapter 8). The packing is first slurry-packed into the column. For mechanically held stationary phases, one of two techniques is generally used to add the stationary phase to the packed column (see Section 8.3): (a) *in situ* injection [e.g., see (*13*)], where the stationary phase is injected into the column and excess liquid is washed off by continued flushing with mobile phase and (b) equilibrium coating [e.g., see (*11*)], where mobile phase saturated with the stationary phase is washed through the column until an equilibrium layer of stationary phase has accumulated on the packing.

Apart from the use of bonded-phase packings in IPC, supports for mechanically held phases should be selected for the same characteristics as in partition LC with mechanically held stationary phases (Section 8.3). It has been noted [e.g., see (7)] that wide-pore supports can more readily lose stationary phase as a result of mechanical shear forces within the column, suggesting the use of narrower-pore supports. IPC columns with mechanically held stationary phase must be treated like their partition-LC counterparts (Section 8.3): pre-equilibration of mobile phase with stationary phase, use of precolumns, thermostatting of the unit, and so on.

11.3 PARTITIONING PHASES

General Considerations

Tables 11.2 and 11.3 summarize several IPC systems that have been reported in the literature; the mobile and stationary phases, the counter-ion, and the type of sample are listed for each application. At first glance, it is apparent that there are wide variations in both technique and experimental conditions. In addition to reverse-phase and normal-phase options, reverse-phase separations can be further subdivided into the four groups shown in Table 11.3: (a) stationary phase consisting only of the hydrocarbon groups bonded to the packing; (b) soap chromatography, which is similar to (a) but features the use of detergents as counter-ions; (c) stationary phase consisting of an added, mechanically held organic liquid; and (d) stationary phase consisting of or containing a water-immiscible liquid ion exchanger, which also doubles as the counter-ion. This diversity of IPC techniques makes it possible to tailor the method to individual samples for optimum results. However, the large number of options is also likely to confuse beginning and experienced chromatographers alike.

The comparison of normal-phase versus reverse-phase IPC in Table 11.1 shows that reverse-phase operation will generally be preferable, and for this reason it is suggested that this approach be tried first. In choosing among the various reverse-phase options of Table 11.3, we recall from Chapter 8 that the use of bonded-phase partition systems is generally preferred over mechanically held stationary phases, and this gives the advantage to techniques (a) and (b) of Table 11.3. At this point there seems little to choose among the latter two techniques, each of which probably has its preferred areas of application. However, a good general rule at this point for IPC seems to be that if one type of system gives tailing bands and/or low column plate numbers, one should switch to another of the various operational modes for IPC.

The general principles for controlling separation in IPC are similar for any of the various techniques of Tables 11.2 and 11.3 and can be readily understood from the discussion of Chapters 7 and 10. The *first requirement* is that the system be physically stable, which means that any organic-rich phase used must be immiscible with the water-rich phase. This, in turn, requires that the mobile phase be preequilibrated with stationary phase in the case of mechanically held liquids and that the temperature of the LC system be held within narrow limits by thermostatting. For normal-phase systems, where the counter-ion is loaded into the stationary phase and where sample ions leave the column as ion-pairs, loss of counter-ions from the stationary phase must be prevented. This can be achieved either by forming ion pairs of the sample before injection or by using a counter-ion that is somewhat soluble in the mobile phase (and then adding the counter-ion to the mobile phase).

A *second requirement* is that tailing bands be avoided. Tailing bands are not uncommon in IPC, because IPC is normally selected for samples that are prone to tailing in other LC systems. Although further discussion of this problem is given in Sections 11.5 and 19.2, the best general rule is to try another IPC system (different counter-ion, stationary phase, and/or mobile phase) when tailing bands are observed in the first attempt at separation.

A *third requirement* is that sample k' values as defined by Eq. 11.2a not change with sample concentration, because this can also result in tailing bands. Therefore, the IPC aqueous phase must be adequately buffered with respect to both pH and concentration of the counter-ion. Conventional buffers such as citrate and phosphate have been used, and in some cases the counter-ion itself is an adequate buffer. For separations at low pH, 0.1-0.2 M solutions of a strong acid provide adequate buffering.

Role of the Counter-ion

After the preceding basic requirements of the IPC system are met, solvent strength and possibly selectivity must be adjusted for a given application. In reverse-phase systems, solvent strength is readily varied by changing the counter-ion or its concentration. For reverse-phase systems, Eq. 11.2a can be generalized to give

Table 11.2 Normal-phase ion-pair chromatography systems.

Stationary phase[a]	Mobile phase	Counter-ion	Samples	Ref.
pH 9.0	Cyclohexane/$CHCl_3$/1-pentanol	N,N-dimethyl-protriptyline[b]	Carboxylic acids	(1)
0.1 M $HClO_4$	Various mixtures of tributyl phosphate, ethyl acetate, butanol, CH_2Cl_2, and/or hexane	ClO_4^-	Amines	(3)
HPO_4^{-2}/PO_4^{-3} buffer	Butanol/CH_2Cl_2/hexane	$(Butyl)_4N^+$	Carboxylic acids	(3)
0.1 M $HClO_4$	CH_2Cl_2, $CHCl_3$, butanol and/or pentanol	ClO_4^-	Amines	(4)

pH 5-6	CH_2Cl_2, and/or $CHCl_3$	Picrate[b]	Amines	(7)
pH 6-8.5	Butanol/heptane	$(Butyl)_4N^+$	Sulfonamides	(8)
0.2-0.25 M HClO$_4$	Butanol/CH_2Cl_2/hexane	ClO_4^-	Amines and quaternary ammonium compounds	(13)
0.1 M methanesulfonic acid	Butanol/CH_2Cl_2/hexane	$CH_3SO_3^-$	Amines	(13)
pH 8.3	Butanol/CH_2Cl_2/hexane	$(Butyl)_4N^+$	Carboxylic acids	(13)
pH 7.4	CH_2Cl_2/$CHCl_3$/butanol and/or pentanol	$(Ethyl)_4N^+$ $(Butyl)_4N^+$ $(Pentyl)_4N^+$	Glucuronide and sulfate conjugates	(15)

[a]Water buffered by various salts or counter-ion.
[b]Colored counter-ions for detection of nonabsorbing compounds.

Table 11.3 Reverse-phase ion-pair chromatography systems.

Stationary phase	Mobile phase[a]	Counter-ion	Sample	Ref.
(a) Bonded phases				
ODS-silica[b]	0.1 M HClO$_4$/water/acetonitrile	ClO$_4^-$	Amines	(5)
LiChrosorb RP-2[b]	pH 7.4	(Butyl)$_4$N$^+$	Carboxylic acids	(15)
μ-Bondapak C$_{18}$[b]	Methanol/water; pH 2-4	(Butyl)$_4$N$^+$	Dyes	(20)
(b) "Soap chromatography"				
SAS-silica[b]	Water/propanol and/or CH$_2$Cl$_2$	(Hexadecyl)(CH$_3$)$_3$N$^+$	Sulfonic acids	(9)
ODS-silica[b]	Water/methanol/H$_2$SO$_4$	(Dodecyl)SO$_3^-$	Amines	(13)
(c) Organic stationary phase				
1-Pentanol	pH 7.4	(Butyl)$_4$N$^+$	Carboxylic acids, sulfonates	(6,11,15)
(d) Liquid ion exchangers				
Bis-(2-ethylhexyl) phosphoric acid/CHCl$_3$	pH 3.8	Bis-(2-ethyl-hexyl) phosphate	Phenols	(1)
Tri-n-octylamine	0.05 M HClO$_4$	(Octyl)$_3$NH$^+$	Carboxylic acids, sulfonates	(2)

[a] Water plus buffer, unless noted otherwise.
[b] Reverse-phase partition packings, used without added organic stationary phase.

$$\text{(anionic samples)} \quad k' = \frac{V_s}{V_m} E (C^+). \qquad (11.3)$$

$$\text{(cationic samples)} \quad k' = \frac{V_s}{V_m} E (C^-). \qquad (11.3a)$$

Here (C^+) and (C^-) refer to the concentration of a cationic or an anionic counter-ion, respectively, and E is constant if other conditions are unchanged. Thus, increasing the concentration of the counter-ion in the mobile phase causes an increase in k' for reverse-phase IPC (and a decrease in normal-phase IPC). Equations 11.3 and 11.3a apply for the case of singly ionized sample ions. For bivalent or trivalent sample ions, k' changes as $(C^+)^2$ or $(C^+)^3$, respectively.

In normal-phase IPC, k' can be varied by changing the concentration of counter-ion in the stationary phase. However, this is less convenient, because it means changing the stationary phase (i.e., reloading the column). In either normal-phase or reverse-phase operation, k' can also be varied by changing the type of counter-ion (e.g., substituting pentane sulfonate for heptane sulfonate). The addition of one $-CH_2-$ group to the counter-ion molecule results in a change of all k' values by a factor of up to 2.5 (the effect is larger at low concentrations of the counter-ion); larger counter-ion molecules give larger k' values in reverse-phase IPC and smaller k' values in normal-phase IPC. Thus, in one study (15) the counter-ion was varied from tetraethylammonium ion to tetrapentyl-ammonium, a change that by itself allowed the variation of k' values over a range of $(2.5)^{12}$, or about 10^5 times.

The relative efficacy of different counter-ions in promoting ion-pair formation and extraction into chloroform from water is illustrated in Table 11.4. Here a difference in k' values of about 10^{10} can be obtained in going from Cl^- as counter-ion to dipicrylamine. Perchlorate ion is seen to be a surprisingly good extractant (for an inorganic counter-ion); for this reason it has been widely used in IPC for separating cations.

When all sample ions are fully ionized, a change in solvent strength via a change in counter-ion concentration leads to minimal changes in separation selectivity. This is illustrated in Figure 11.2a, where log k' for several sulfonamides is plotted against the log of inverse counter-ion concentration. For compounds 6, 9, and 12, which are fully ionized in this system, no change in selectivity occurs as counter-ion concentration is varied. Compound 13 is partially ionized and compound 14 is un-ionized; for these compounds selectivity does vary with change in counter-ion concentration.

A novel application of IPC for the separation of peptides and proteins has recently been reported (15a), using a hydrophylic counter-ion rather than the usual hydrophobic counter-ion. Higher-molecular-weight peptides and proteins tend to be retained irreversibly on reverse-phase BPC packings because of the numerous hydrophobic groups within these molecules. With the addition of

Table 11.4   Extraction power of different
counter-ions; chloroform/water with tetrabutylammonium ion as sample.

Counter-ion	$\log E^a$
Cl^-	-0.11
Benzoate	0.39
Br^-	1.29
Toluene-4-sulfonate	2.33
Salicylate	2.42
Naphthalene-2-sulfonate	3.45
ClO_4^-	3.48
Picrate	5.91
Dipicrylamine	9.6

Source: From (*1*).
^{a}Equation 11.2.

hydrophylic counter-ions such as PO_4^{-3}, ion pairing between counter-ion and $-NH_3^+$ groups in the protein leads to a net increase in charge on the protein molecule and a decrease in retention. This approach provides a means for controlling the retention of both high-molecular-weight and very hydrophobic molecules in reverse-phase systems.

The anionic counter-ion trifluoracetic acid has been recommended for pre-parative IPC (*15b*), since it is volatile and can be removed from recovered sample fractions by lyophilization.

Solvent Polarity

Solvent strength in either normal-phase or reverse-phase IPC can also be varied by changing the polarity of the mobile phase. When this is done by varying the relative concentrations of a given two-component mixture (e.g., butanol/hexane), changes in selectivity are usually minor. Again, this is illustrated in Figure 11.2*b* for several sulfonamide samples, where now $\log k'$ is plotted versus the concentration of butanol in the butanol/hexane mobile phase. The one exception, compound 14, is nonionized under these conditions and again behaves atypically.

For reverse-phase systems without added organic stationary phase, mixtures of water with either methanol or acetonitrile are generally used as mobile phase. As the percent of water is decreased, the solvent becomes stronger and sample k' values decrease. Mixtures of butanol or pentanol diluted with CH_2Cl_2, $CHCl_3$, and/or hexane are common mobile phases in normal-phase IPC systems. (Here more-polar solvents are stronger and give smaller k' values.) Solvent polarity in

IPC does not follow the solvent polarity parameter P' (see Table 6.1) exactly, as shown by the sequences of increasing solvent strength for normal-phase IPC in Table 11.5. The reason is that solvent strength in IPC is a function of the ability of the solvent to dissolve or stabilize *ions*, and *ion pairs*, whereas P' is mainly a function of the solvent's ability to dissolve polar *nonionic* compounds. We would, therefore, expect that solvent strength in IPC should be a function of both P' and the dielectric constant ϵ of the solvent, and this is seen to be the case in Table 11.5. The solvent function $(P' + 0.25\epsilon)$ seems to predict relative solvent strength for this series of solvents and can be used to estimate approxi-

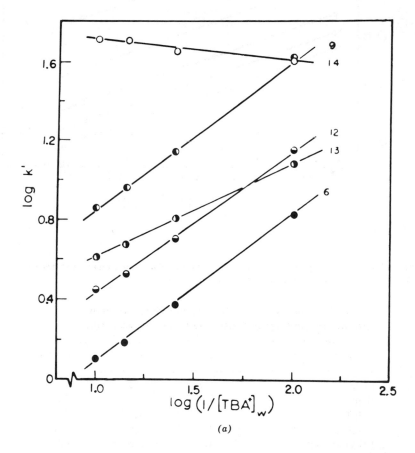

(a)

Figure 11.2a Selectivity changes as the concentration of counter-ion is varied to change k' values. Sulfonamide separations by normal-phase IPC. Column, 25 × 0.32 cm LiChrospher SI-100 (10 μm) coated with 1.5-5.0%w 0.1-0.25 M borate (pH 8.5) plus counter-ion (TBA⁺) and Na₂SO₄ in water; mobile phase, 25-50%v butanol/heptane mixtures; 27°C; UV, 254 nm. Reprinted from (8) with permission.

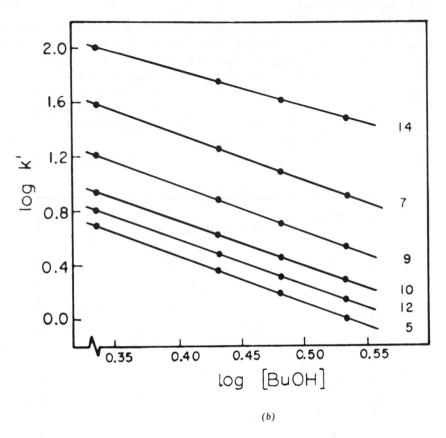

(b)

Figure 11.2*b* Selectivity changes as the polar mobile phase component is varied to change k' values. Other conditions as in Figure 11.2a.

Table 11.5 Solvent strength in normal-phase ion-pair chromatography; solvents listed in increasing strength.

Solvent	P'	ϵ	$P' + 0.25\epsilon$
CCl_4	1.7	2.2	2.3
Benzene	3.0	2.3	3.6
$CHCl_3$	4.4	4.8	5.6
CH_2Cl_2	3.4	8.9	5.6
Ethyl acetate	4.3	6.0	5.8
Methylisobutyl ketone	3.5	13.1	6.8
1-Pentanol	3.6	14.7	7.3
1-Butanol	3.9	17.5	8.3

Source: Data from (*1,4*).

mate solvent strength for other IPC solvents not listed in Table 11.5 (see Table 6.1, last column).

Secondary Ions

Apart from an increase in the counter-ion concentration, an increase in ionic strength of the aqueous phase generally reduces the formation of ion-pairs (2,8), as a result of competition of secondary ions in forming ion pairs with the counter-ion. Thus, an increase in ionic strength decreases k' values in reverse-phase IPC and increases k' in normal-phase IPC. One study (8) showed a two- to three-fold change in k' for each doubling of ionic strength. Those secondary ions whose charge is of the same sign as that of sample ions (i.e., cationic or anionic) have the largest effect on sample k' values. In one study (2), involving the IPC separation of sample anions, the effect of secondary ions on k' increased in the sequence $NO_3^- < Br^- < Cl^- < SO_4^{-2}$.

pH Effects

Sample k' values are much affected in IPC by a change in pH or by a change from one solvent to another (e.g., from ethyl acetate as mobile phase to butanol). However, these modes of changing k' are usually accompanied by large changes in separation selectivity and normally are not used for simple adjustments in solvent strength. The effect of pH variation on separation in IPC is illustrated in the theoretical plots of Figure 11.3 and is quite similar to the case of varying pH in ion-exchange LC (cf. discussions of Chapter 10, especially Figure 10.3 and Eq. 10.8). Similar plots as in Figure 11.3 are observed for either anion or cation separations, but we will illustrate the case of anionic samples here. A reverse-phase system is assumed for the plots of Figure 11.3a, and a normal-phase system is assumed in Figure 11.3b.

In reverse-phase IPC (Figure 11.3a) maximum k' values are obtained at intermediate values of pH, where the sample compounds are completely ionized and ion-pair formation is at a maximum. As the pH of the mobile phase is lowered, sample anions X^- begin to form the un-ionized acids HX, leading to a smaller number of sample ion pairs in the stationary phase. Therefore, the value of k' for each anion begins to decrease as pH is lowered. Sample compounds such as C in Figure 11.3, with a pK_a value of about 6, begin to elute more quickly as soon as the pH is lowered below 7, and they are only weakly retained at pH values less than 5. Stronger acids, such as A in Figure 11.3 with a pK_a value near 2, are strongly retained until the pH drops below 2. If all three anions, A, B, and C, are retained equally at pH 8, it is clear from Figure 11.3a that each will be well separated from the other at a pH of 4. Thus, a change in pH offers a powerful tool in IPC for changing separation selectivity. If we

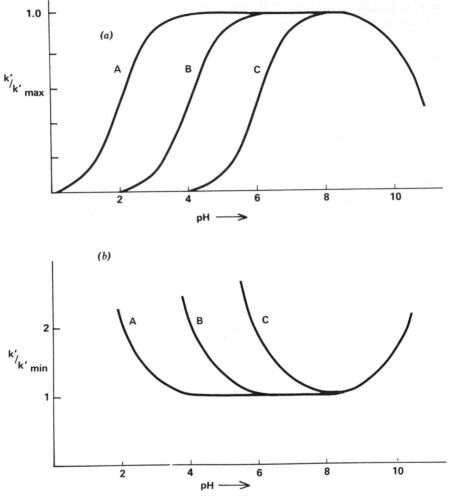

Figure 11.3 Variation of relative k' values with pH in IPC of anions. (a) Reverse-phase and (b) normal-phase. k'_{max} refers to maximum (constant) value of k' for fully ionized sample molecules. Curves A, B, and C are for compounds with pK_a values equal to 2, 4, and 6, respectively.

know the pK_a values of the acids or bases we are separating (see Table 10.1), we can predict how the relative retention of each compound will vary with pH [e.g., examples of (8)].

The drop in k' values at higher pH in Figure 11.3a is similar to the decrease in exchange capacity in Figure 10.3. In each case, mobile-phase OH⁻ ions are beginning to tie up the counter-ion and compete with sample anions in the formation of ion pairs. If the counter-ion were itself a relatively weak base, this drop in k' in Figure 11.3a would occur at a lower pH, thus restricting the pH range for useful IPC separation. For this reason the use of weak acids or bases as counter-ions in IPC is generally avoided. For normal-phase IPC, as in Figure 11.3b, a similar—but inverted—dependence of k' on pH results. Sample compounds are now more strongly retained at low and at high pH values. The examples of Figure 11.3 assume that un-ionized sample ions are not retained by the aqueous phase. When this is not the case, the resulting curves are somewhat more complex, as discussed in (2). Finally, it should be noted that silica particles are unstable at pH values above 8.5-9, so that IPC is normally carried out at lower pH.

Solvent Selectivity

Selectivity changes via a change in solvent proceed as in the case of partition LC, and the rules of Sections 6.5 and 7.3 apply equally. Thus, mixtures of different solvents can each be adjusted for the right solvent strength, and the use of polar solvents from different selectivity groups should provide maximum changes in selectivity. An example [data from (3)] is shown in Figure 11.4, where k' values for four catechol amines are shown for three different mobile phases (normal phase IPC system): butanol/hexane (II), ethyl acetate/hexane

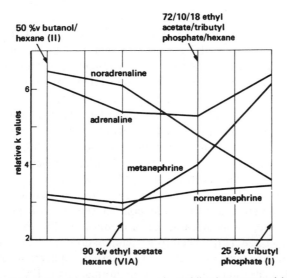

Figure 11.4 Selectivity changes with change in mobile-phase composition for normal-phase IPC. Separation of catechol amines with 0.1 M HClO$_4$/0.9 M NaClO$_4$ in water as stationary phase. Data of (3).

(VIA), and tributyl phosphate/hexane (I). Quite different selectivities are observed in each system, with a mixture of ethyl acetate and tributyl phosphate in hexane providing the best separation.

11.4 OTHER SEPARATION VARIABLES

Sample size effects for IPC and the other LC methods are generally similar. The volume of injected sample should not be so large as to significantly broaden bands of interest (Section 7.4), and the quantity of injected sample should be small enough so that k' values are not affected and column efficiency is not significantly degraded. In most IPC separations it is possible to increase the maximum possible sample size by increasing the concentration of counter-ion in the aqueous phase, because the latter is often a limiting factor.

If counter-ion concentration is increased for greater sample capacity, this will simultaneously change sample k' values (i.e., Eqs. 11.3, 11.3a). However, the effect on k' can be minimized by simultaneously adding excess neutral salt to the aqueous phase, which tends to tie up the counter-ion as a separate ion pair, as discussed in the preceding section. In effect, this approach buffers the counter-ion concentration in much the same way that a pH buffer works (i.e., the *total* quantity of counter-ion is available to react with sample, but its effect on k' is reduced by forming an intermediate ion-pair with the secondary ion from the added neutral salt). The overall result is an IPC system with a high capacity for sample yet with control over k' values by varying the concentration of neutral salt in the aqueous phase [for further details see (6)]. The maximum sample size in reverse-phase IPC can also be increased by adding the sample as the preformed ion pair (6); in this case, a roughly stoichiometric quantity of counter-ion is mixed with sample before injection, followed by adjustment of sample pH to favor formation of ion pairs [e.g, see (7)].

Temperature effects in IPC are more important than in some other LC methods. *First*, in the case of mechanically held stationary phases, the column must be adequately thermostatted. *Second*, because the mobile phases used in IPC are generally rather viscous, there is an advantage to elevated-temperature separations. However, with mechanically held phases, increase in temperature often means greater difficulty in holding the stationary phase intact during separation, and therefore a decrease in column stability. *Finally*, changes in selectivity with temperature appear more pronounced in IPC than for the other LC methods [e.g., see (2)]. Therefore, temperature may be an important variable for optimizing separation selectivity in certain applications of IPC.

The use of UV-absorbing counter-ions in normal-phase IPC allows sample bands to leave the column as detectable ion pairs. This approach permits the analysis of non-UV-absorbing compounds, without derivitization as a separate step outside the column (e.g., as in Section 17.3). An example is shown in Figure 11.5 for the separation of several drugs as the picrate ion pairs. The applicability of this technique requires counter-ions that are relatively insoluble in the organic phase; otherwise the baseline absorbance will be too high for detection of low concentrations of absorbing ion pairs in the column effluent. Apart from the use of picrate ion (1) and 2-naphthyl sulfonate (10) in the separation and detection of amines, N,N-dimethylprotriptyline has been used in similar fashion for the analysis of anions (1).

11.5 SPECIAL PROBLEMS

Column instability is a general problem in LC, and one would expect additional problems in IPC because of the "dirty" samples that are commonly encountered, as well as from the occasional use of mechanically held stationary phases. One study (7) found rather short column-lifetimes for normal-phase IPC, particularly for larger-particle packings and higher mobile-phase flowrates. This instability was attributed to mechanical stripping of the stationary phase by the shear forces generated within the column. However, another study (11), with reverse-phase IPC and 1-pentanol as stationary phase, found that thousands of serum samples could be run before the column had to be recoated. Possibly key features of this procedure were (a) the use of the equilibrium-coating technique, which would seem to preclude any possibility of stationary phase loss via shear forces and (b) the use of preequilibrated phases plus careful column thermostatting. Interestingly, this same study (11) found that reverse-phase columns without added organic stationary phase were less stable in use. However, this may be due to improperly manufactured packings (e.g., inadequate coverage by alkyl groups), which are known to be less stable in reverse-phase (non-IPC) use. Another study (12) involving the routine analysis of dye mixtures by reverse-phase IPC (no added organic stationary phase) found columns stable and reproducible after injection of 800 samples.

Normal-phase columns will gradually lose counter-ion from the stationary phase, because sample compounds leave the column as the ion pairs. However, this effect can be eliminated by preforming the ion pairs of the sample.

The possibility of band tailing in IPC has already been noted. In reverse-phase separations it seems important to have maximum coverage of the silica surface by trialkylsilyl (or other) groups [e.g., see (5)]. Band tailing can also be caused by dissociation of ion pairs in the organic phase, which in some cases can be overcome by increasing the counter-ion concentration. Inadequate buffering of the aqueous phase is another source of band tailing in IPC. In some cases, band tailing and low plate numbers have dramatically improved on changing from a conventional ion-pair system to "soap chromatography" [e.g., see (9)]

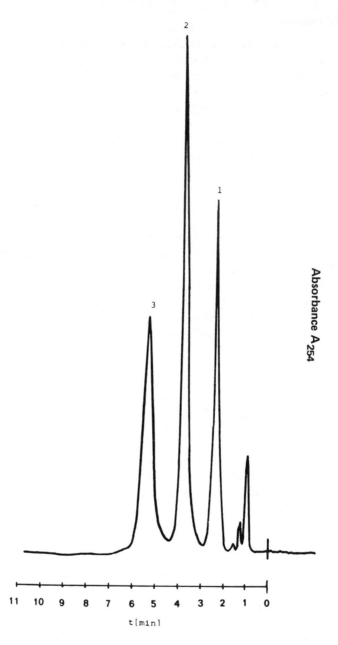

Figure 11.5 Normal-phase IPC separation of basic drugs with detection of picrate ion pairs: hyoscyamine (1), ergotamine (2), and scopolamine (3). Column, 10 × 0.3 cm, Merckosorb SI-100 (5 μm), coated with 0.6 M picric acid plus citrate buffer (pH 6.0) in water; mobile phase, chloroform; temp., ambient; 0.6 ml/min; UV, 254 nm; 1-3 μg each compound injected. Reprinted from (7) with permission.

472

which seems particularly well suited for the separation of very polar, multiply charged molecules such as the sulfonated dyes. In some cases reduced band tailing is found for long-chain phases (e.g., octadecyl); in other cases better results are obtained with short-chain coatings.

Unlike the case of ion-exchange LC, column reproducibility in IPC has been found to be generally satisfactory. It should be noted that the performance of reverse-phase columns with *mechanically* held organic stationary phase can vary with the nature of the silanization of the packing surface [e.g., see (*15*)].

The stationary phase 1-pentanol has been used in reverse-phase IPC. The aqueous mobile phase must then be saturated with 1-pentanol, in which case any increase in temperature of the mobile phase results in precipitation of 1-pentanol. For this reason, the detector is normally held at a temperature lower than that of the column [e.g., see (*15*)].

In several cases [e.g., see (*5, 13*)] slow equilibration of IPC columns with the mobile phase was noted immediately after coating the column with stationary phase or changing the mobile-phase composition. These observations seem generally associated with either reverse-phase columns without added organic stationary phase or normal-phase columns where change in the counter-ion concentration is attempted. On the other hand, one study of columns with 1-pentanol as stationary phase remarked on the very rapid equilibration even of new columns; only 30 column volumes of mobile phase sufficed to give complete equilibrium and subsequent stable column operation.

11.6 APPLICATIONS

In general, ion-pair chromatography has been applied to polar samples such as bases, ionics, and compounds with multiple, ionizable groups. Aqueous samples, including various physiological fluids, have received considerable attention. A few examples will illustrate the range of applicability of IPC and pinpoint specific cases where hard-to-handle samples gave good separations. The tetracycline derivatives, an example of which is provided by tetracycline (I),

(I)

are very polar bases that give tailing and low column efficiencies when their separation is attempted by adsorption, reverse-phase partition, or ion-exchange LC (5). However, good separations were obtained using reverse-phase IPC without added organic stationary phase, as shown in Figure 11.6. The resolution

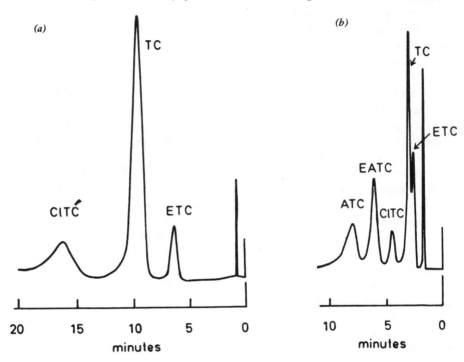

Figure 11.6 Reverse-phase IPC separation of tetracyclines on (a) ODS-silica and (b) short-chain alkyl-silica; no added organic stationary phase. Column, 12.5 × 0.5 cm, silylated Partisil (18 μm). (a) Mobile phase, 0.1 M HClO₄/0.3 M NaClO₄/0.002 M citric acid in 15%v acetonitrile/water; 0.21 cm/sec. (b) Mobile phase, 0.1 M HClO₄ in 25%v acetonitrile/water; 0.13 cm/sec. Reprinted from (5) with permission.

obtained in this study was adequate for the characterization of crude tetracycline samples for the various impurities present.

Various amino and/or hydroxy sulfonates of naphthalene are important as dye intermediates or final dyes. These highly ionic, very polar compounds have traditionally resisted attempts at their separation by other LC methods, including ion exchange. As seen in Figure 11.7, however, soap chromatography with the counter-ion cetrimide provides excellent resolution and good band shape for mixtures of these compounds. Some authors have referred to the ability of IPC to carry out compound-type separations, but it is apparent in Figure 11.7 that closely related isomers can also be resolved by IPC. In fact,

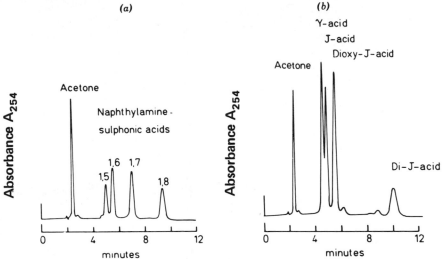

Figure 11.7 Separation of dye intermediates by soap chromatography. Column, 12 × 0.5 cm, short-chain silylated Partisil 10 (7 μm); mobile phase, 1%w certimide in 30%v propanol/water; temp., ambient; UV, 254 nm; 1-5 μl of sample containing 1-5 μg each compound. Reprinted from (9) with permission.

IPC has exhibited a general ability to effect good separations of closely related isomers, as illustrated by several additional examples in (16).

Figure 11.8 shows another typical application of IPC: the separation of an ionic antiarrythmia drug, QX-572, (II), from urine by normal-phase IPC.

$$\text{C}_6\text{H}_5\text{—NH—CO—CH}_2\text{—N}^+\text{—CH}_2\text{—CO—NH—C}_6\text{H}_5 \quad \text{Br}^-$$

with CH$_3$ groups above and below the N$^+$

(II)

Figure 11.9 shows the separation of several polar, ionic steriod-conjugates by normal-phase IPC.

Paired-ion chromatography, or PIC^{TM}, as introduced by Waters Associates, is simply reverse-phase IPC without an added organic stationary phase. In this case an already convenient technique has been made even easier to use by pre-packaging the counter-ion plus buffer for direct addition to the mobile phase (e.g., a methanol/water mixture). The counter-ions available from Waters Associates include

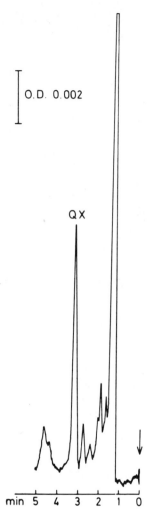

Figure 11.8 Separation of quaternary ammonium drug QX-572 from urine extract by normal-phase IPC. Column 15 × 0.45 cm, Partisil 10 (10 μm) coated with 0.2 M HClO$_4$/0.8 M NaClO$_4$ in water; mobile phase, 10/10/80%v 1-butanol/CH$_2$Cl$_2$/hexane; ambient; UV, wavelength not specified; 100-μl sample containing 10 ng QX-572. Reprinted from (*13*) with permission.

PIC Reagent A: 0.005 M tetrabutylammonium phosphate (pH 7.5).
PIC Reagent B-5:   0.005 M pentane sulfonic acid (pH 3.5).
PIC Reagent B-7:   0.005 M heptane sulfonic acid (pH 3.5).

These buffer/counter-ion combinations are suitable for the separation of most samples for which IPC is applicable. A representative separation with PIC

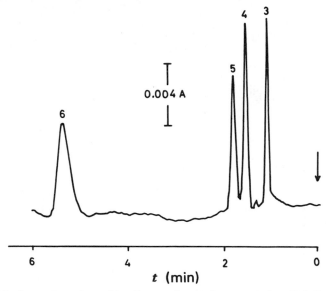

Figure 11.9 Separation of steroid sulfate conjugates by normal-phase IPC. Compounds are sulfates of 3-estradiol (3), 17-estradiol (4), 17-estriol (5), and 3-estriol (6). Column, 15 × 0.45 cm, LiChrospher SI-100 (10 μm) coated with 0.6 ml/g of 0.1 M tetraethylammonium phosphate (pH 7.4); mobile phase, 10%v 1-pentanol/CH_2Cl_2; 25°C; 0.3 cm/sec, 450 psi; UV, 254 nm. Reprinted from (*15*) with permission.

Reagent A is shown in Figure 11.10 for impurities in Red Dye 40. For further details, see (*15*). A similar separation of several dyes and dye intermediates on a reverse-phase bonded-phase IPC column, using tetrabutyl ammonium ion as counter-ion, is shown in Figure 11.11. Here peaks 3 and 5 are added internal standards (3-chlorobenzoic acid and 3-nitrosalicylic acid, respectively).

11.7 THE DESIGN OF AN ION-PAIR SEPARATION

Selecting the Column

Unless there is a specific reason to choose a normal-phase system (see Table 11.1), IPC should be carried out in the reverse-phase mode. In this case, the preferred column will be packed with 5-10-μm particles of an alkyl-bonded-phase; [e.g., C_{18} bonded phase (see Table 5.6)]. IPC has so far not been used much for preparative separations, possibly because counter-ions and buffers complicate the recovery of separated components from the mobile phase.

The starting column packing for IPC with added organic stationary phase is the same as that for reverse-phase IPC (i.e., an alkyl-bonded phase). Normal-phase

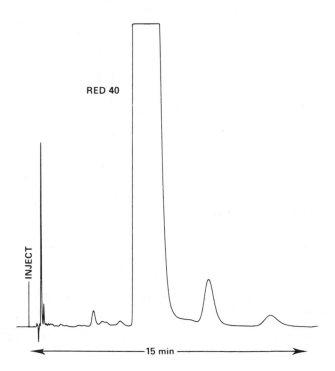

Figure 11.10 Separation of impurities in Red Dye No. 40 by reverse-phase IPC. Column, 30 × 0.4 cm, μ-Bondapak-C$_{18}$ (10 μm); mobile phase, 50%v methanol/water plus PIC reagent A: ambient; 2.0 ml/min; UV, 254, 0.1 AUFS. Reprinted from (20) with permission of Waters Associates.

IPC should be carried out with 5-10-μm porous silica packings of the type used for conventional liquid-solid LC (e.g., LiChrosorb SI 100).

Selecting the Mobile Phase

For reverse-phase separations without added organic stationary phase, water/organic solvent mixtures are used with addition of counter-ion, buffer, and (sometimes) neutral salts. The best counter-ion and pH depend on the kind of sample to be separated (i.e., acids, bases, anions, cations). Sample anions and cations that are ionized regardless of pH do not require a buffer. Acids are usually separated at a pH of 7-9, whereas bases are separated at a pH of 1-6. The pH of the mobile phase can be varied to adjust separation selectivity and improve individual α values.

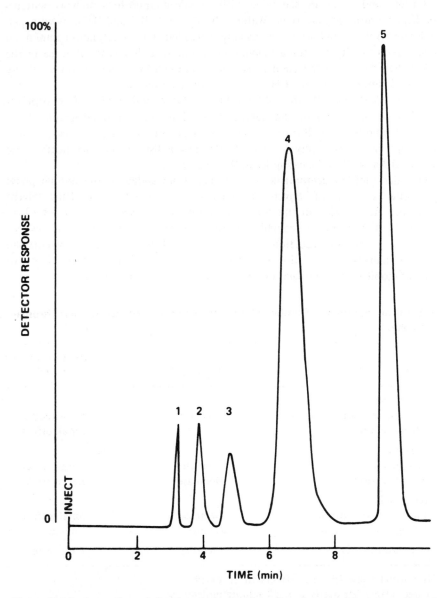

Figure 11.11 Separation of the dye, tartrazine (no. 4) and some of its intermediates by reverse-phase IPC. Column, 30 × 0.4 cm, μ-Bondapak-C$_{18}$ (10 μm); mobile phase, 0.003 M tetrabutylammonium hydroxide/0.00006 M tridecylamine/0.125%v formic acid in 50%v methanol/water; ambient; UV, 280 nm. Reprinted from (*12*) with permission.

Cations and bases are separated with anionic counter-ions such as pentane-or heptane-sulfonic acid (e.g., Waters PIC Reagents B-5 and B-7), or with perchlorate. Anions and acids are usually separated with a tetraalkylammonium counter-ion, such as tetrabutylammonium ion. In either case, increase in the alkyl chain length of the counter-ion increases retention in reverse-phase IPC by up to 2.5 times per added $-CH_2-$ group in the counter-ion.

The concentration of the counter-ion is usually 0.005-0.05 M, except for perchlorate (0.5-1 M) or the detergents used in soap chromatography (e.g., 1%w of counter-ion). Buffer concentrations are similar to those used in ion-exchange chromatography (0.001-0.5 M). For a list of possible buffers, see Table 10.2 as well as the examples in this chapter.

Organics such as acetonitrile and methanol are added to the mobile phase primarily as a way of conveniently changing k' values (i.e., adjusting solvent strength). An increase in the concentration of organic solvent produces a lowering of all k' values by roughly the same degree as in conventional reverse-phase LC separations (Section 7.3). The use of different organic solvents can lead to changes in separation selectivity. Table 11.6 summarizes the effects of various mobile-phase variables on both solvent strength and selectivity.

Table 11.6   Summary of mobile-phase effects in reverse-phase ion-pair chromatography.

Variable	Effect on k'	Effect on solvent selectivity
Counter-ion		
Increase concentration	Increase[a]	No effect[b]
Change counter-ion	Change	Small effect
Increase pH		
Sample anions	Increase[a]	Large effect
Sample cations	Decrease[a]	Large effect
Organic solvent		
Increase concentration	Decrease[a]	Small effect[b]
Change organic solvent	Change	Change
Temperature	Increase[a]	Large effect

[a] In normal-phase IPC, the effects are reversed.
[b] Larger effect for partly ionized sample molecules.

Temperature

Most IPC separations have been carried out at ambient temperatures. However, reverse-phase applications (without added organic stationary phase) should benefit from increased temperatures (e.g., 50-70°). Separation selectivity in IPC can often be changed substantially by a relatively small change in separation temperature.

Sample Size

No systematic studies have been made on possible sample sizes in IPC. Generally these should be comparable to those for LLC (Section 7.4), particularly if the counter-ion concentration in the mobile phase is sufficiently high. Addition of sample as the preformed ion pair also increases the allowable maximum sample size.

Other Considerations

See the discussion of controlling resolution in Section 2.5 for increasing resolution via changes in column efficiency. For special problems in IPC, see Section 11.5.

REFERENCES

1. S. Eksborg, P. -O. Lagerström, R. Modin, and G. Schill, *J. Chromatogr.*, **83**, 99 (1973).

2. J. C. Kraak and J. F. K. Huber, *ibid.*, **102**, 333 (1974).

3. B. -A. Persson and B. L. Karger, *J. Chromatogr. Sci.*, **12**, 521 (1974).

4. J. H. Knox and J. J. Jurand. *J. Chromatogr.*, **103**, 311 (1975).

5. J. H. Knox and J. J. Jurand, *ibid.*, **110**, 103 (1975).

6. K. -G. Wahlund, *ibid.*, **115**, 411 (1975).

7. W. Santi, J. M. Huen, and R. W. Frei, *ibid.*, **115**, 423 (1975).

8. S. C. Su, A. V. Hartkopf, and B. L. Karger, *ibid.*, **199**, 523 (1976).

9. J. H. Knox and G. R. Laird, *ibid.*, **122**, 17 (1976).

10. J. Crommen, B. Fransson, and G. Schill, *ibid.*, **142**, 283 (1977).

11. K. -G. Wahlund and U. Lund, *ibid.*, **122**, 269 (1976).

12. D. P. Wittmer, N. O. Nuessle, and W. G. Haney, Jr., *Anal. Chem.*, **47**, 1422 (1975).

13. B. -A. Persson and P. -O. Lagerström, *J. Chromatogr.*, **122**, 305 (1976).

14. J. H. Knox and J. Jurand, *ibid.*, **125**, 89 (1976).

15. B. Fransson, K. -G. Wahlund, and G. Schill, *ibid.*, **125**, 327 (1976).

15a. W. S. Hancock, C. A. Bishop, R. L. Prestidge, D. R. K. Harding, and M. T. W. Hearn, *ibid.*, **153**, 391 (1978).

15b. G. E. Dunlap III, S. Gentleman, and L. I. Lowney, *ibid.*, **160**, 191 (1978).

16. G. Schill, R. Modin, K. O. Borg, and B. -A. Persson, in *Drug Fate and Metabolism*, Vol. 1, E. R. Garrett and J. L. Hirtz, eds., Dekker. New York, 1977, Chap. 4.

17.   S. Eksborg and G. Schill, *Anal. Chem.*, **45**, 2092 (1973).

18.   C. Horvath, W. Melander, I. Molnar, and P. Molnar, *ibid.*, **49**, 2295 (1977).

19.   R. P. W. Scott and P. Kucera, *J. Chromatogr.*, **142**, 213 (1977).

19a. R. S. Deelder and H. J. M. Linssen, *ibid.*, in press.

20.   Waters Associates, *Paired-ion Chromatography. An Alternative to Ion Exchange.*

BIBLIOGRAPHY

Kraak, J. C., K. M. Jonker, and J. F. K. Huber, *J. Chromatogr.*, **142**, 671 (1977) (good general discussion of selectivity effects on soap chromatography, with specific application to amino acid separations).

Schill, G., in *Ion Exchange and Solvent Extraction*, Vol. 6, J. A. Marinsky and Y. Marcus, eds., Dekker, New York, 1974, Chap. 1 (very detailed review of ion-pair formation as applied to extraction).

Schill, G., R. Modin, K. O. Borg, and B. -A. Persson, in *Drug Fate and Metabolism*, E. R. Garrett and J. L. Hirtz, eds., Dekker, New York, 1977, Chap. 4 (review of ion-pair extraction and chromatography; detailed discussion of solvent effects).

Terweij-Groen, C. P., and J. C. Kraak, *J. Chromatogr.*, **138**, 245 (1977) (excellent theoretical and experimental discussion of factors that affect separation in IPC with liquid ion exchangers as stationary phase).

Tomlinson, E., T. M. Jeffries and C. M. Riley, *J. Chromatogr.*, **159**, 315 (1978). (A 44-page review of ion-pair chromatography; thorough discussion plus 124 references).

TWELVE

SIZE-EXCLUSION CHROMATOGRAPHY

12.1	Introduction	484
12.2	Column Packings	487
	Available Types	487
	Semirigid Organic Gels	488
	Rigid Packings	489
	Selection of Packing	489
	Bimodal Pore-Size Configuration	494
	Other Considerations	499
12.3	Mobile Phases	500
12.4	Other Separation Variables	503
12.5	Molecular-Weight Calibration	509
12.6	Recycle Chromatography	519
12.7	Problems	522
12.8	Applications	525
12.9	The Design of an SEC Separation	534
	Selecting the Column	534
	Selecting the Mobile Phase	537
	Operating Variables	538
	References	538
	Bibliography	540

12.1 INTRODUCTION

The newest of the four LC methods, size-exclusion liquid chromatography, is also referred to as gel chromatography, gel filtration, and gel permeation chromatography (GPC), as well as other names. Size-exclusion chromatography (SEC) is the easiest of the LC methods to understand and use, and the most predictable. Despite its simplicity, this powerful technique has been applied to a broad variety of sample types to solve widely different separation problems.

Consider the types of samples and separation problems for which size exclusion is a logical first choice. First, SEC is the preferred method for separating higher-molecular-weight components (MW > 2000), particularly those that are nonionic. The method is most widely used to obtain the molecular-weight distribution of synthetic polymers (e.g., Figure 1.4). However, individual macromolecules such as proteins and nucleic acids also are best separated by SEC. Second, SEC can be used to separate simple mixtures conveniently and rapidly, when the components of the mixture have sufficient difference in molecular size. Third, SEC is a powerful method for the initial exploratory separation of unknown samples, quickly providing an overall picture of sample composition with a minimum of separation development. This approach promptly indicates whether the sample is a simple or complex mixture and gives the approximate molecular weight of the sample components. Preliminary separations by SEC often define which LC method, or combination of methods, will be best for a given sample. An initial SEC separation is often an essential first step in the resolution of a complex sample by use of more than one LC method, as described in Section 18.2.

SEC separates molecules according to their effective size in solution, using a column packing with pores of a particular average size. Figure 12.1 illustrates

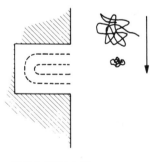

SINGLE PORE SIZE SEPARATES
MOLECULES = $10^{1.5}$ Δ MW

Figure 12.1 Size-exclusion fractionation within a single pore. Reprinted from (9) with permission.

the reason for the relative retention of different sample molecules by a porous packing particle. If some of the sample molecules are too large to enter the pores, they are totally excluded from the particle. These large sample molecules move directly through the column and appear first in the chromatogram. Molecules that are so small that they can permeate almost the entire particle (dotted line closest to pore wall, Figure 12.1) are most retained and move through the column most slowly; these solutes appear last in the chromatogram. Solutes of intermediate size are sterically less able to approach the walls of the pores (dotted line in the center of the pore wall in Figure 12.1) and, on the average, spend less time in the pores; these solutes move through the column at speeds dependent on their relative size. Because solvent molecules are usually quite small, they are eluted last at the normal t_0. Therefore, in SEC the entire sample elutes *before* t_0 (i.e., $k' < 0$), and this is a fundamental difference from the other LC methods.

As noted in Figure 12.1, a single-pore-size packing is theoretically capable of separating molecules falling within a 1.5 decade of molecular weight. The usual pore-size distribution in commercial packings actually permits separations over molecular-weight ranges of 2-2.5 decades for a single pore size.

Separation occurs in SEC strictly on the basis of molecular size. Retention of solutes by other mechanisms (e.g., adsorption) is undesirable and usually does not occur with the proper combination of substrate and mobile phases. The characteristic elution behavior of SEC leads to several distinct advantages in its application:

Narrow bands make possible easy detection. Because all solutes leave the column relatively quickly (before t_0), resultant retention volumes V_R are small. All peaks in SEC give good signals, even with detectors that are relatively insensitive (e.g., the refractometer), as predicted by Eq. 13.2. For this reason the "universal" refractometer-detector is widely used in SEC with consistently good results.

Short separation times without gradient elution is another advantage of SEC. Because all components are eluted quickly, there is no general elution problem, as is often the case with the other LC methods.

Predictable separation times are characteristic of SEC because the beginning and the end of each chromatogram are known and constant for a particular set of operating conditions. All compounds elute at retention volumes between total exclusion (i.e., very large molecules) and total permeation (t_0). Thus, it is possible to inject a series of different samples over a predetermined time series without danger of overlapping one chromatogram with another, permitting the use of automatic (unattended) sample injections. The predictable separation times in SEC also mean less time wasted in waiting for the last components in an unknown sample to elute (as in the case for the other LC methods).

Predictable elution according to molecular size offers a substantial advantage in the identification of unknown components, because retention in SEC is determined strictly by molecular size, and can be predicted for a solute of known molecular structure. With the other LC methods, retention times give little information about molecular structure, because many different aspects of molecular structure can contribute to V_R.

Freedom from sample loss or reaction during separation is typical of SEC. Such effects are not often found in any of the other modern LC methods, but they do occur occasionally. However, SEC is one of the gentlest separation techniques, because of the normal absence of higher-energy sorptive forces characteristic of the other LC methods.

An absence of column deactivation problems is a strong advantage of SEC, for the column packing does not tend to accumulate strongly retained compounds, either from impure solvents or dirty samples. Thus, with proper handling, SEC columns do not degrade with time because of the buildup of such materials on the packing. Precolumns and guard columns, which are often used with the other LC methods to protect the main column, are unnecessary in SEC. One danger to an SEC column is its clogging by samples that contain particulates. Such samples should be filtered before injection, as described in Section 5.6.

The unique characteristics of SEC also present some disadvantages:

Limited peak capacity is the most serious disadvantage of SEC. Only a few separated bands can be accommodated within the separation, because the chromatogram is quite short (all compounds should elute before t_0). With the other LC methods, hundreds of compounds can be resolved in a single separation (Figure 1.3). However, in SEC more than about a dozen distinct bands are seldom seen in a single chromatogram. Because of the relatively low resolution of SEC (due to very short retention), this method is usually incapable of completely resolving a complex, multicomponent sample; further separation by other LC methods generally is required. Fortunately, polymers need not be separated completely for determination of molecular-weight distribution.

Inapplicability to some samples, particularly those whose components are of similar size, is a second disadvantage of SEC. Generally, this method cannot resolve compounds of similar size (e.g., with <10% difference in molecular weights). Thus, many important separations, such as the resolution of isomers, are generally not possible with SEC.

The different principles of separation in SEC require somewhat different separation optimization, compared to other LC methods; the concepts of separation factor α and capacity factor k' are not normally used. As far as resolution is concerned, the types and composition of the mobile phase are also relatively unimportant in SEC, which is unlike the other LC methods. These differences in the separation principles of SEC are not a real disadvantage; however, a different approach is required in establishing a separation.

It is useful to classify the various applications of SEC, as these lead to different experimental approaches and utility of the final results. First, SEC is commonly divided into the techniques of gel filtration chromatography (GFC), using aqueous solvents, and gel permeation chromatography (GPC), using organic solvents, for application to water-soluble or organic-soluble samples, respectively. Although the mechanism of separation for these two methods is identical, the column packings and techniques used can be quite different. Second, SEC can be carried out with either rigid or nonrigid column packings. Rigid packings are required for high-pressure (modern) LC, but some samples are better separated on the nonrigid gels. Finally, SEC can be used either to resolve individual compounds or to partially separate polymers for calculating a molecular-weight distribution (e.g., see Figure 1.4).

In this chapter we emphasize the SEC separation of samples using columns of rigid particles. Separations on nonrigid packings are important, but often fall outside the scope of modern LC, because high-pressure operation is precluded; such separations also are adequately reviewed elsewhere (*1-5*). The determination of polymer molecular-weight distributions by SEC also is a broad, detailed subject, demanding more attention than it can be given here. A recent book on modern size-exclusion chromatography treats this subject in detail, including the newer aspects of gel filtration chromatography (*6*). Therefore, in this chapter on SEC we only briefly describe the determination of molecular-weight distributions and primarily focus attention on the separation of individual compounds.

12.2 COLUMN PACKINGS

Available Types

The choice of an appropriate column is a major decision when setting up an SEC system, because the separation is controlled largely by the packing material used. As discussed in Sections 2.1 and 5.7, column plate height varies essentially with the square of the particle diameter d_p. The effect of particle size is especially important for macromolecules, which have small diffusion coefficients; here the use of columns with very small, totally porous particles is particularly favored in SEC. Therefore, our discussion in this chapter highlights the use of high-efficiency small-particle SEC columns for analysis. Columns with larger particle sizes are used mainly for preparative applications, where relatively large amounts of sample must be isolated for subsequent identification.

Traditionally, SEC analyses of synthetic organic polymers have been made with cross-linked, semirigid polystyrene gel packings. Small, rigid, inorganic particles (e.g., silica), however, have recently shown several significant advantages over organic gels:

- Rigid particles are relatively easily packed into homogeneous columns that are mechanically stable over long periods.
- A much wider range of mobile phases can be used, for greater versatility and increased convenience.
- Rapid equilibration occurs with new solvents, allowing rapid solvent change-over.
- Rigid packings are stable at high temperatures, and with certain solvents required for characterizing particular synthetic macromolecules.
- Rigid particles can also be used in aqueous systems for separating high-molecular-weight water-soluble solutes by GFC.

A potential disadvantage of the rigid, inorganic particles is retention of solutes by adsorption, or degradation of certain materials (e.g., denaturing of proteins). However, siliceous particles can be surface-modified with appropriate organic functional groups to eliminate this possible disadvantage for most applications.

Column packings for high-performance, analytical SEC can be arbitrarily defined as those having particle sizes of $\lesssim 30$ μm. Columns of these materials can be used at relatively fast flowrates and high input pressures to produce superior separations within short times.

Semirigid Organic Gels. Table 5.9 lists some of the semirigid organic gels useful for HPSEC. The most popular of these materials are cross-linked styrene-divinyl-benzene copolymers with d_p less than 10 μm. These packings are available in pore sizes that will fractionate a molecular-weight range of about $50\text{-}10^7$. Columns of packings with small pores (e.g., μ-Styragel 100 Å) allow the separation of relatively small molecules, those with molecular weights up to about 1000; larger pore sizes will fractionate higher-molecular-weight samples.

Porous vinylacetate-copolymer gels are similar to the polystyrene gels, but cover a somewhat smaller molecular-weight range. Neutral, hydrophilic polyacryl-amide gels (e.g., Biogel P-2) are useful in aqueous systems for gel filtration chromatography of water-soluble materials. Although several polyacrylamide gels are available with exclusion limits ranging up to about 400,000 MW, only the small pore size (Biogel P-2) has sufficient mechanical strength to allow use at pressures up to 150-200 psi (see Table 5.9).

Beads of polydextrans also have utility in modern SEC, and several products (e.g., Sephadex) are available with molecular-weight exclusion limits ranging up to 10^8. Although most of these materials are relatively soft, some (e.g., Sephadex G-25) have sufficient strength for efficient GFC separations in aqueous systems. A hydroxy-propylated derivative, Sephadex LH-20 is a similar gel that can be used for SEC with polar organic solvents. With both G-25 and LH-20, column input pressure must be limited to 150-200 psi, which restricts their application in modern SEC. These latter semirigid packings are generally only available in particle sizes >30 μm; therefore, column performance is relatively poor, compared to the ≤10 μm polystyrene gels and porous silica packings.

New columns of small-particle (8-12-μm) hydrophilic gels for SEC separation in aqueous mobile phase are also now available (TSK-Gels, Toyo Soda Manufacturing Co.). The composition of these gels has not yet been described, but apparently columns can be used at relatively high pressures (up to at least 100 atm) without problems (*6a*).

Rigid Packings. Silica-based packings for modern SEC have been of increasing interest and utility, and Table 5.9 lists some materials that are commercially available. Some of these packings can be obtained only in packed columns; others can be purchased in bulk. Some are untreated; others have been modified with certain organic substituents to reduce adsorption. For example, μ-Bondagel is substituted with aliphatic ether groups; Glycophase-G/GPC is modified with glyceryl-groups (I) to reduce potential retention by adsorption (*7,8*). Further

$$Si-O-\underset{\underset{O}{|}}{\overset{\overset{O}{|}}{Si}}-(CH_2)_3-O-CH_2-\underset{\underset{}{\overset{OH}{|}}}{CH}-CH_2OH$$

(I)

work is required to define all the problems associated with adsorption in SEC and to develop packings that can overcome adsorption for all samples.

Column packings that are primarily useful for preparative-scale SEC are given in Table 5.13. In general, these packings consist of larger particles of the same materials used for high-performance SEC. Although columns of the preparative packings in Table 5.13 also can be used for analysis, longer separation times are required—other factors being equal. Columns of Styragel, a cross-linked styrene-divinylbenzene copolymer, have long been used for GPC analyses and are still valuable for applications in which large sample sizes are required. These packings are compatible with most organic solvents, but not with acetone, alcohols, or water.

Selection of Packing

The ability of an SEC packing to separate the constituents of a sample according to molecular weight is defined by *calibration plots* for the various packing materials. A calibration plot for a particular packing or column is a graph of sample molecular size versus relative retention or retention volume. Figure 12.2 is a hypothetical example, where sample molecular weight is plotted against

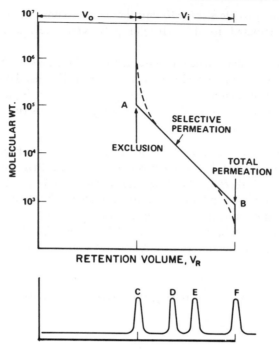

Figure 12.2 Hypothetical calibration plot for a given size-exclusion packing or column.

retention volume V_R. Other measures of molecular size and retention are used, and will shortly be discussed. In Figure 12.2 there is an *exclusion limit* (point A) for the packing, corresponding in this case to a molecular weight of 10^5. Molecules larger than the exclusion limit are too large to enter the pores of the packing and are totally excluded; they co-elute as a single band C, with a retention volume V_o. The *interparticle volume* V_o is equal to the volume of solvent within the column but outside the packing particles.

Also shown in Figure 12.2 is a total *permeation limit* (point B, for molecular weight of 1000 in this case), such that all compounds with molecular weights smaller than B are able to permeate the packing particles totally. In practice, these compounds elute as a single band F, with retention volume V_t that corresponds to retention time t_0 and is equal to the total volume of solvent within the column. The quantity V_i, equal to $(V_t - V_o)$, is the volume of solvent held within the packing particles, or the *intraparticle volume*.

Compounds with molecular weights that decrease from A to B progressively spend more time within the pores of the column packing, as a result of their increasingly smaller size. Such compounds (e.g., D, E) can be separated when their molecular sizes are different. This intermediate region ($A <$ mol. wt. $< B$) is referred to as the *fractionation range* of the packing.

Solute size in size-exclusion chromatography is actually determined by *molecular hydrodynamic radius*, or the radius of gyration of the solute in the mobile-phase solvent. Thus, the effective size of an eluting solute is governed by its geometrical shape (i.e., hard sphere, random coil, or rigid rod), the association between solute molecules, and the association of the solute with the mobile-phase solvent. Retention in SEC is measured in terms of the *distribution coefficient*:

$$K_0 = \frac{V_R - V_0}{V_i}. \tag{12.1}$$

Values of K_0 range between zero (total exclusion) and unity (total permeation). K_0 values normalize the size-exclusion retention in the various packings, regardless of the volume of the column or the internal volume of the particles; K_0 is

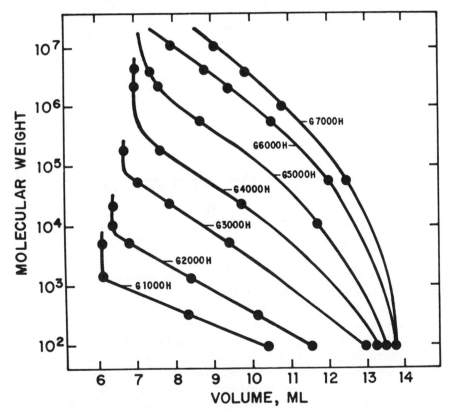

Figure 12.3 Calibration curves for microparticulate organic gels. Varian MicroPak polystyrene gels; mobile phase, tetrahydrofuran; solutes, polystyrenes. Reprinted with permission of Varian Associates.

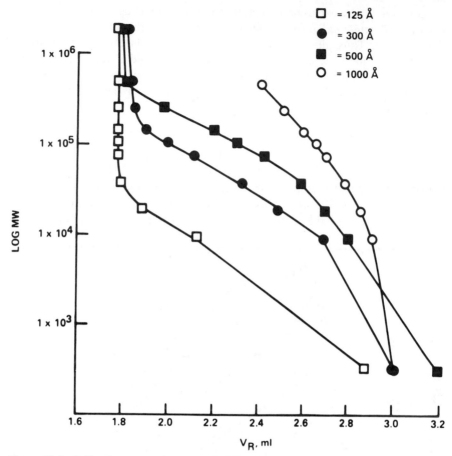

Figure 12.4 Calibration curves for oganic-modified silica packings. Columns, μ-Bondagel E; mobile phase, water; samples, dextrans. Reprinted with permission of Waters Associates.

the preferred retention parameter for comparing data between different packings and under different experimental conditions.

Approximate calibration plots for a given packing are normally provided by the manufacturer. Figure 12.3 shows calibration plots of microparticulate, polystyrene gels (Varian) for gel permeation chromatography. Similarly, Figure 12.4 shows calibration plots of ether-modified, porous-silica μ-Bondagel packings (Waters) now being used for both high-performance gel filtration and gel permeation chromatography. For actual use, *exact* calibration curves should be determined for every SEC column, using macromolecule standards available from the suppliers of SEC packings. Molecular-weight calibration procedures are discussed more fully in Section 12.5 and in (6).

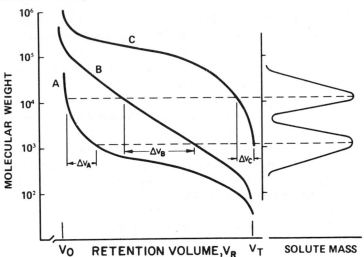

Figure 12.5 Selection of the proper packing for maximum resolution of two adjacent bands.

Control of separation in size-exclusion chromatography is simple and predictable. Band widths and resolution vary with column plate number N, as described in Section 2.3. Separation is also measured by the spacing of band centers; the further apart the two bands, the better is their resolution. Two compounds of differing molecular size are best separated when they elute near the middle of the SEC fractionation range, as illustrated in the hypothetical example of Figure 12.5. Here are shown calibration plots for columns of three different porous packings (A, B, C), where the separation of two compounds (compound 1, molecular weight, 10^3; compound 2, molecular weight, 10^4) is of interest. With gel A the two compounds elute near the exclusion limit, so that the spacing between band centers (ΔV_A) is small and separation is therefore poor. In this case a packing with larger pores and a higher exclusion limit should be used. Column C has a higher exclusion limit, but now the two compounds elute near the total permeation limit. Thus, the peak spacing (ΔV_C) is again small, and separation is still poor. With column B, however, the molecular sizes of compounds 1 and 2 fall near the center of the fractionation range. With this column, the spacing of band centers (ΔV_B) is large, and separation is much improved.

Choice of pore size normally is the most significant parameter to be considered in the selection of the column, because pore size dictates the range of molecular-weight separation. Columns of the smallest pore size (e.g., $\leqslant 60$ Å) are desirable for separating mixtures of small molecules (<5000 molecular weight). For separating samples with a relatively narrow molecular-weight range (e.g., <2.5 decades) columns of a single pore size normally are used. For instance, proteins in the 10^3-10^5 molecular weight range can be separated with a single pore size (e.g., ~300 Å). Once the optimum pore size is selected, resolution can be increased by coupling two or more columns of the same pore size to increase N. In this case the range of molecular-weight separation remains the same, but the volume in which the separation is made (V_R axis) is increased by a factor equal to the column length increase.

A different sort of separation problem is encountered with broad range samples, that is, samples containing several species covering a wide molecular-weight range. In this case column fractionation must span a wide range in molecular size. Maximum overall separation range is achieved with packings having nonoverlapping but very close or adjacent separation ranges, as illustrated in Figure 12.6. The calibration plots for four SEC packings of different pore sizes, A, B, C, and D, are given in Figure 12.6a. Retention volumes for a series of connected columns are additive, so that the calibration plot for a connected column set can be determined from the calibration data for these individual columns. Maximum molecular-weight range and calibration linearity is accomplished by connecting columns A and D, which have close but nonoverlapping molecular-weight separation ranges. The calculated curve for two columns of A and D combined is shown in Figure 12.6b (solid line). Note that the molecular-weight fractionation range for this four-column set now spans about 4 decades; samples with components of widely different molecular weight can now be separated, as illustrated for the hypothetical sample in Figure 12.6c (where four separate bands appear to be resolved).

Although such a column set ($2A + 2D$) can separate compounds with molecular weights varying by at least 4 decades, it should be emphasized that only a moderate separation of any two adjacent bands of similar size is achieved with such a combination. If compounds that are close in size must be resolved, columns of a single pore size with a narrow fractionation range that overlaps the two compounds of interest should be used. For example, if we suspect that band J, Figure 12.6c, is really two bands that are unresolved, we should reseparate the sample on a four-column set of column C (Figure 12.6b), because band J falls near the center of the fractionation range for this column. The resulting separation is shown in Figure 12.6d, where now band J is split into two bands. At the same time, however, bands K, L, and M have been merged into a single band that elutes near the total permeation limit of the column C.

It should be emphasized that different packings (e.g., A and D) should *not* be mixed into the same column. First, this precludes the use of several (connected) columns in different combinations (as in the examples of Figure 12.6c vs. 12.6d). Second, mixing different LC packings into the same column can result in particle sizing and a lowering of column efficiency.

Bimodal Pore-Size Configuration. (This section is somewhat advanced and of particular interest for molecular-weight (MW) distribution assays. The reader

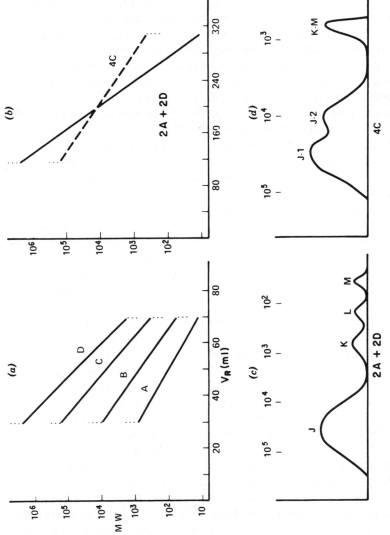

Figure 12.6 Combining size-exclusion columns for separating broad-range mixtures or two compounds of similar size.

may wish to skip to the following section: "Other Considerations".) When determining the MW distribution of polymeric samples, values of V_R from the chromatogram must be converted to values of MW—using the calibration curve. The most accurate determination of MW results if this calibration curve is a straight line over the entire range of V_R in the chromatogram (6). A wide, linear MW calibration is best accomplished by combining columns that exhibit an optimum *bimodal pore-size distribution.* This is achieved by coupling SEC columns containing only *two* discrete pore sizes (bimodal), having about a *1-decade difference in pore size, moderately narrow pore-size distributions, and approximately equal pore volumes for the two pore sizes* so that the linear portions of the individual calibration plots are essentially parallel (9).

The bimodal pore-size configuration previously defined is superior in MW range and MW calibration linearity (therefore, MW accuracy) to a single, wide-range pore size represented by column sets with several overlapping MW ranges—the traditional method of obtaining wide MW separating ranges in SEC (9). The latter approach produces relatively large deviation from a linear fit of calibration points, with a linear range generally less than 3 decades of molecular weight. On the other hand, properly constructed bimodal-column sets exhibit at least 4 decades of MW range, with a close linear fit of the calibration points; up to 5 decades of molecular separation with good calibration linearity have been reported (9). Thus, for the preliminary separation of unknown samples, columns with a bimodal pore-size distribution are best for a versatile, general-purpose column set producing moderate resolution in a very broad MW separation range. Conversely, connected columns of the *same* packing are used when the separation of two (or more) compounds of similar size must be optimized (as in Figure 12.6d). Bimodal column sets designed for optimum performance for one polymer-solvent system also function equally well for other systems.

It should be stressed that in setting up bimodal column systems, there is some leeway in the pore-size separation of the two pores and the relative pore volumes, because of the relatively moderate influence of these parameters in the optimum range (9). Thus, useful bimodal column sets can still be developed even though characteristics of the packings may vary somewhat around the optimum (i.e., about a decade of pore-size separation and equal pore volumes for the two sizes).

Porous silica packings may be more convenient for the bimodal approach than organic gel-type packings because (a) pore size and pore-size distribution can be measured experimentally for silica packings to predict the column arrangement without having to construct MW calibration curves (see later discussion); (b) certain silica packings have more consistent pore volumes and pore-size distributions for convenient bimodal optimization. An actual example of the bimodal arrangement with rigid silica particles is illustrated in Figures 12.7 and 12.8. Inspection of Figure 12.7 shows that the 60-Å porous-silica microsphere column has a linear fractionation range of about 10^2 to 10^4 MW, whereas the 750-Å

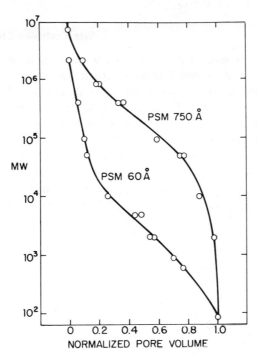

Figure 12.7 Comparison of molecular-weight fractionation range for pore sizes differing by about 1 decade. Columns, 10 × 0.78 cm each, porous-silica microspheres (10); mobile phase, tetrahydrofuran; temp., 22°C; flowrate, 2.5 ml/min; UV detector, 254 nm; sample, 25-μl solution of polystyrene standards. Reprinted from (9) with permission.

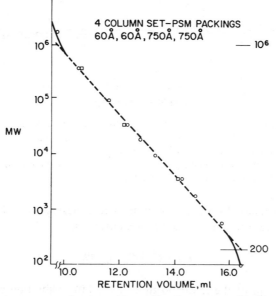

Figure 12.8 Polystyrene molecular-weight calibration curve with bimodal column set. Conditions same as for Figure 12.7. Reprinted from (9) with permission.

column separates from about 10^4 to 10^6 MW. Because the slopes of the calibration curves (the so-called *column separation capacity*) are nearly identical within these respective MW limits, the combination of these two columns gives the linear calibration curve shown in Figure 12.8, with about 4 decades of linear MW range (200-10^6 MW shown at right) and an excellent fit of the plot to the actual calibration data. Resolution of a linear column set such as that in Figure 12.8 can be increased by using additional columns of the same pore sizes (but always in the same proportion) to provide equivalent pore volume for each pore size.

If the calibration curve for each packing pore size is known, the arrangement of a bimodal pore-size column set can be predicted. In Figure 12.9 are the linear

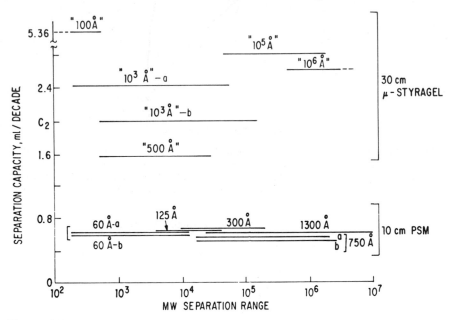

Figure 12.9 Separation capacity versus calibration range for μ-Styragel and porous-silica microsphere columns. Data on PSM columns same as for Figure 12.7; μ-Styragel data obtained using: mobile phase, tetrahydrofuran; temp., $23°$C; flowrate, 1.0 ml/min; detector, RI; sample, 100 μl of polystyrene standard solutions (a) and (b) represent data on different packing lots. Reprinted from (9) with permission.

MW separation ranges for various columns plotted against the column separation C_2 for the linear portion of the calibration curve, where

$$V_R = C_1 - C_2 \log \text{MW}, \qquad (12.2)$$

C_1 is the intercept of the linear portion of the MW calibration curve, and C_2 is the slope of the linear portion in ml per decade of molecular weight. Suppose, for example, that the "500-Å" and the "10^5-Å" μ-Styragel columns in Figure 12.9 were connected. The MW separation range of the "500-Å" column does not overlap that of the "10^5-Å" column, but is sufficiently close to provide continuity in separating molecular weights from about 500 to 2,000,000 MW. However, the calibration curve will show a sharp change of slope nearer 20,000 MW because of the differences in separation capacities (C_2) of the two columns. Because separation capacity for each pore size must be equal to obtain a linear calibration plot, the pore volume associated with the "500-Å" μ-Styragel then must be increased about two fold. Thus, in this example a linear μ-Styragel set could be obtained with 2-30-cm "500-Å" μ-Styragel columns coupled to a single "10^5-Å" μ-Styragel column. With this approach column sets having bimodal pore-size distributions can usually be obtained with commercial columns.

Direct substitution of a new column for another in a bimodal set without knowing the calibration curve for the new column can cause problems. For example, exchanging the two "10^3-Å" μ-Styragel columns (Figure 12.9) differing significantly in pore size (different MW separation ranges) and internal volume (different separation capacities C_2) would lead to changes in the shape of the calibration curve of the set and alteration of the calibration linear range and fit. Thus, lot-to-lot variations can limit the versatility of organic gel packings for wide-range linear column set arrangement. On the other hand, some porous silicas such as those given in Figure 12.9 are particularly suited for bimodal arrangement, because they have virtually identical and reproducible separation capacities and MW range resulting from the consistent pore structure and porosity defined by the method of synthesis (*10*).

In constructing columns for wide MW fractionation range, it is feasible to use individual columns of different pore size (e.g., Zorbax-PSM-100 Å and 1000 Å) or to mix the two packings in a single column. As discussed earlier, however, the former approach is preferable. Single particles with a bimodal pore-size distribution have also been reported (*9*).

Other Considerations

Column-packing techniques for size-exclusion chromatography were discussed in Chapter 5. Columns of porous silicas and semirigid organic gels are packed by the high-pressure wet-fill method. Prepacked columns of many of these packings can be purchased. Because of the care needed in preparing high-efficiency size-exclusion columns, and the greater importance of maximum column efficiency in SEC, beginners are advised to purchase prepacked columns.

Column efficiency in size exclusion is based on the same factors as for the other LC methods (see, e.g., Chapters 2 and 5). Columns of small-particle silicas often give the highest resolution for the same polymers—other factors being equal (*11*). It should be stressed that column efficiency is more important in

SEC than for the other LC methods, because of the low peak capacity of SEC. Thus, marginal columns in SEC are not useful and should be discarded. Typically, minimum plate counts of 4000 and 5000 are specified by manufacturers for 25-30-cm lengths of <10-μm organic gel and silica-based particles, respectively. Very large plate counts are needed to separate mixtures of small molecules having modest differences in molecular weight, and this condition is obtained by connecting several columns of the smallest pore-size packings, as illustrated in Figure 12.10.

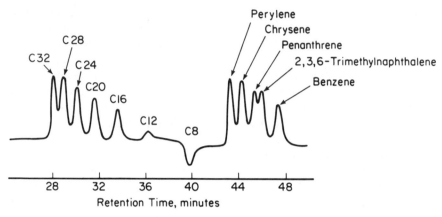

Figure 12.10 High-resolution size-exclusion separation of small molecules. Column, 150 × 0.8 cm (five 30 cm columns), μ-Styragel 100 Å; mobile phase, tetrahydrofuran; flowrate, 1 ml/min; detector, RI. Reprinted from (*12*) with permission.

12.3 MOBILE PHASES

Unlike the other LC methods, in SEC the mobile phase is *not* varied to control resolution. Rather, the mobile phase is chosen for its ability to dissolve the sample and for low viscosity at the temperature of separation (i.e., for higher N values). A low-viscosity mobile phase has a boiling point that is about 25-50°C higher than the temperature of the column. For samples that are difficult to dissolve, the solvent most often is selected to provide adequate sample solubility. The solvent is also chosen to conform to detector requirements (Chapter 4).

The mobile phase also must be selected for compatibility with the column packing. For example, very polar solvents such as acetone, alcohols, dimethylsulfoxide, and water generally are not used with polystyrene packings. Also, with nonrigid gel packings for gel filtration chromatography, salt should be added to the mobile phase to maintain constant ionic strength (*12a*). A wide range of solvents can be used with the silica-based packings, but in aqueous systems the mobile phase must be maintained within the range pH ~2-8.5;

otherwise, degradation of the packing can occur. Dissolution of these silica packings is accelerated by high pH and high ionic strength.

The mobile phase is often chosen to eliminate or reduce interaction of the sample with the packing surface (e.g., retention by adsorption, in addition to the desired SEC retention). Unwanted solute-retention effects have been widely recognized for organic gels (13). For example, Sephadex and Bio-Gel P gels contain a small number of carboxylic acid groups that can exhibit cationic exchange properties in distilled water or low-ionic-strength eluents. Generally, an ionic strength of 0.05-0.1 in the mobile phase removes these undesirable effects. In some cases, ionic interaction between the solute and the carboxylic acid groups on the matrix is so strong that suppression of the carboxylic acid ionization is necessary (e.g., by using 0.01 N HCl as the mobile phase).

Retention of solutes by effects other than the desired size exclusion can be a particular problem with silica-based packings. With unmodified silicas, a good rule of thumb is to use a mobile phase that is much more strongly adsorbed than the solute. The solute then is less able to compete for active sites on the packing. For example, N,N-dimethylformamide is preferred over tetrahydrofuran for separating polyurethane samples. Adsorption often can be recognized by the elution of solutes after the total permeation volume. Peak tailing also results from adsorption effects, but is not a good criterion, because many polymeric samples have asymmetric MW profiles that result in tailing peaks.

Because of acidic SiOH groups, an unmodified siliceous packing can behave as a weakly acidic ion exchanger at pH $>$ 4-5, resulting in retention other than size exclusion. Therefore, a mobile phase for separating solutes that might be retarded by ion exchange (e.g., organic bases) must be appropriately adjusted so that only the desired size-exclusion process occurs. The ion-exchange behavior of unmodified silica is demonstrated in Figure 12.11. Here we see that Na^+, NH_4^+, K^+, and trimethylammonium (TMA^+) ions from various salts are increasingly adsorbed in this order (pH 5, column of unmodified silica particles, aqueous sodium acetate mobile phase). At pH $>$ 4 the TMA^+ cation can only be desorbed at high sodium concentrations. At pH $<$ 4, where silica is neutral or even positively charged, TMA^+ is still retained, unlike Na^+, K^+, and NH_4^+. These findings suggest that of Na^+, NH_4^+, K^+, and TMA^+, the last cation is the most effective physical modifier in preventing adsorption on silica for a number of neutral and cationic species. As the TMA^+ cation renders the silica surface hydrophobic, a reverse-phase effect may occur. However, by adding organic solvent to the mobile phase (e.g., methanol), this reverse-phase effect can also be eliminated. For example, polyvinylalcohol samples tend to adsorb onto untreated silica from aqueous mobile phases in SEC separations. In one study, use of a mobile phase of 0.025 M tetramethylammonium phosphate (pH 3) in water/methanol (1:1) prevented adsorption of the sample during its analysis for MW distribution by SEC (15). The low pH prevented dissociation of the silica and was a key factor in eliminating adsorption.

The type of anion used in the aqueous mobile phase also has a significant influence on unwanted adsorption of solutes to silica surfaces during SEC. This effect has not been carefully studied, but seems to involve counter-ions at the silica surface. Most effective in reducing adsorption are the polyvalent anions; sulfate and phosphate appear to be more effective than monovalent anions such as chloride, perchlorate, and acetate, as expected from other ion-exchange data (Section 10.3).

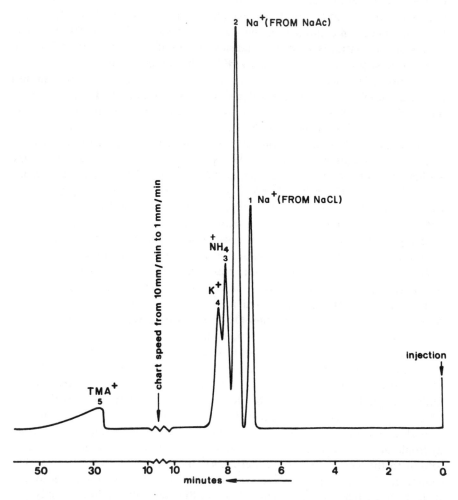

Figure 12.11 Adsorption effects on unmodified porous silica. Columns, 30 × 0.46 cm, LiChrospher SI-100; mobile phase, 0.5 *M* aqueous sodium acetate (pH 5); flowrate, 0.6 ml/ min; sample, 20 μl of 3 mg/ml ammonium acetate, 3 mg/ml potassium acetate, and 6 mg/ml of tetramethylammonium chloride; (1) NaCl, (2) NaAc, (3) NH$_4$Ac, (4) KAc, (5) TMA-Ac. Reprinted from (*14*) with permission.

To eliminate adsorption on silica substrates in aqueous SEC, ionic strength should be greater than 0.05 M, with 0.1-0.5 M often being satisfactory. However, the effect of mobile-phase ionic strength on retention is unpredictable for some solutes. For example, certain proteins elute at very low ionic strengths (0.001 M) but will adsorb when the ionic strength is greater than about 0.05 M. Therefore, the ionic strength of aqueous mobile phases must be experimentally optimized for each system.

Adsorption of solutes to porous-silica packings sometimes can be reduced or eliminated by coating the surface of the substrate with highly polar macromolecules such as Carbowax 6000 and Carbowax 20-M (16), ethylene glycol, or aliphatic alcohols. In general, the approach is to use a modifier that is more tightly bonded to the support than the solute, so that the solute is less able to compete for active sites.

Although porous silica packings can be surface-modified with organic functional groups to reduce adsorption of solutes, careful selection of the mobile phase still is important for two reasons. First, surface coverage can be imperfect; for example, a controlled-pore glass modified with glyceryl groups (Glycophase G/GPC) exhibited a slight anionic character that affected the SEC of certain enzymes at ionic strengths below 0.1 (17). However, at ionic strengths greater than 0.1 this residual ionic character had no effect on protein elution. Thus, even with surface-modified supports, it appears desirable to utilize a buffered mobile phase with an ionic strength of at least 0.1. Addition of ethylene glycol or Carbowax can also help with adsorption problems. Second, with polar-modified substrates, the mobile phase sometimes must contain organic solvents to eliminate hydrophobic interactions with solutes. For example, some proteins are retained on ether-modified μ-Bondagel E columns beyond the total permeation volume (18). In this case, increasing salt concentration in the mobile phase sometimes leads to increased retention volumes, which suggests reverse-phase retention; also, changing pH has little effect. This problem was solved by adding 0.2-1% sodium dodecylsulfate to the mobile phase, which gave a quantitative recovery of proteins. Figure 12.12 shows a human plasma profile, with possible peak identifications, using this separation approach.

The manufacturer's literature should always be consulted to ensure compatibility of the packing with the various potential mobile phases. Several commonly used solvents for SEC are listed in Table 12.1 along with various properties of interest.

12.4 OTHER SEPARATION VARIABLES

Temperature in SEC can be increased for difficulty soluble samples or to lower solvent viscosity for improved column performance. For convenience, most SEC separations are carried out at room temperature. However, some high MW poly-

Table 12.1 Properties of solvents commonly used in size-exclusion chromatography.

Solvents	Melting point (°C)	Boiling point (°C)	Density @ 20°C	UV cutoff (nm)	Viscosity @ 20°C cP	Refractive index @ 20°C	Flash point (°C)	Oral LD-50* (mg/Kg)(rat)	TLV^R (rat, ppm)	Irritant to skin & eye	Toxicity
Tetrahydrofuran[a]	-65	66	0.8892	220	0.55	1.4072	14	3000	200	Mild	Slight
1,2,4-Trichlorobenzene (TCB)[b]	17	213	1.4634	307	1.89(25°C)	1.5717	99	756	5	Moderate	Slight
o-Dichlorobenzene[b]	-19	180	1.3048	294	1.26	1.5515	66	500	50	Moderate	Moderate
Toluene	-95	110.6	0.8669	285	0.59	1.4969	4	5000	100	Moderate	Slight
N,N-Dimethylformamide (DMF)[c]	-61	153	0.9445	275	0.90	1.4294	58	3500	10	Moderate	Slight
Methylene chloride (dichloromethane)	-97	40.1	1.3266		0.44	1.4237	None	2136	200[g]	Severe	Slight
Ethylene dichloride (dichloroethane)	-36	84	1.235		0.84	1.4443	13	680	50	Slight	Slight
N-Methyl pyrrolidone[c]	-24	202	1.027	262	1.65	1.47	95.4	7000	Not set	Moderate	Very low
m-Cresol	12	202.8	1.034	302	20.8	1.544	94	242	5	Severe	Moderate
Benzene	5.5	80.1	0.8790	280	0.652	1.5011	27	3800	1[e]	Mild	Slight
Dimethylsulfoxide (DMSO)	18	189	1.014	260	2.24	1.4770	95.0	20,000	Not set	None	Very low
Perchloroethylene	-19	121	1.622	290		1.505	None	5,000	100[f]	Severe	Very low
o-Chlorophenol	7	175.6	1.241		4.11	1.5473[40]	None	670	Not set	Severe	Slight
Carbon tetrachloride	-23	76.8	1.589	265	0.969	1.4630	None	1770	10[f]	Moderate	Slight

Water	0	100.0	1.00	1.00	2.00°C	1.33	None	—	—	—	Moderate
Trifluorethanol		73.6	1.382	190		1.2910	40.6	240	Not set	Mild	Slight
Chloroform[d]	−64	61.7	1.483	245	0.58	1.4457	None	2000	10[g]	Moderate	Slight
Hexafluoroisopropanol	−3.4	58.2	1.59	190	1.02(25°C)	1.2752	None	1040	Not set	Very severe	Slight

Source: From (6) with permission.

[a] Generally contains butylated hydroxy toluene at a few hundreths of a percent as stabilizer.

[b] These solvents are usually used at 135°C. The use of an antioxidant is recommended: Ionol (Shell) 5 g/gal or Santonox R (Monsanto) 1.5 g/gal.

[c] Quite hydroscopic. Relatively large amounts of water (several percent) may drastically affect fractionation.

[d] Ordinarily contains 0.75% ethanol as stabilizer.

[e] Recently proposed level because of carcinogenicity.

[f] Value may be lowered because of possibility of carcinogenicity.

[g] Proposed new level.

The toxicity data were collected and evaluated by Dr. Clifford Dickinson of the Haskell Laboratory for Toxicology and Industrial Medicine, Central Research & Development Department, E. I. Du Pont De Nemours, Inc., Newark, Delaware.

The LD_{50} data refer to the dosage (mg chemical/kg body weight) which kills 50% of rats tested in time. A somewhat arbitrary correspondence with toxicity used in the table is:

<50 mg/kg	=	highly toxic
50–500 mg/kg	=	moderately toxic
500–5000 mg/kg	=	slightly toxic
5000 mg/kg	=	very low toxicity

The TLV[R] is the *threshold limit value* for airborne contaminants set by the American Conference of Governmental and Industrial Hygienists.

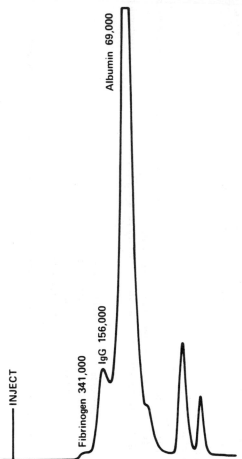

Figure 12.12 Separation of human plasma proteins. Column, μ-Bondagel E, 2-1000 Å, 500 Å, 125 Å; mobile phase, 0.05 Trizma buffer, pH 7.4, 1% sodium dodecylsulfate; flowrate, 0.3 ml/min; detector, UV, 280 nm. Reprinted from (*18*) with permission.

olefins and polyamides require temperatures of 100-135°C, because these samples are not sufficiently soluble at lower temperatures. If the samples are stable, then gel filtration chromatographic separation of biologically important macromolecules also can be carried out at above-ambient temperatures (e.g., 50°C), and the resulting lowering of the viscosity of the aqueous mobile phase provides improved column efficiency and better resolution.

Sample size in SEC is limited by the usual requirement that the total sample volume be sufficiently small to eliminate band broadening due to the sample (Section 7.4). When separating macromolecules, sample size also may be limited

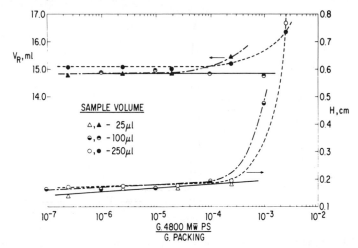

Figure 12.13 Influence of sample weight and volume on SEC column performance. Column, 10 × 0.62 cm; porous-silica microspheres, 40 Å (silanized); mobile phase, tetrahydrofuran; flowrate, 1.5 ml/min; temp., 22°C; detector, UV, 254 nm; solute, 4800 MW polystyrene. Reprinted from (*19*) with permission.

by sample viscosity. A rough guide is that the sample solution injected onto the column should have a viscosity no greater than twice that of the mobile phase.

Table 12.2 shows typical sample concentrations for μ-Styragel columns as a function of molecular weight. These values also appear to be typical for columns of the other types of smaller than 10-μm packings. Because viscosity decreases sharply with temperature, more concentrated samples can be charged at higher column temperatures. Figure 12.13 shows that no more than 0.5-1 mg of solute per gram of packing can be injected into a high-performance SEC column before significant changes in retention volume and plate height occur. Typically, for a

Table 12.2 Typical molecular weight versus sample concentrations for μ-Styragel columns.

Molecular-weight range	Sample concentration (%w)
Up to 20,000	0.25
34,000 to 200,000	0.10
400,000 to 2,000,000	0.05
2,000,000 plus	0.01

Source: From Waters Associates with permission.

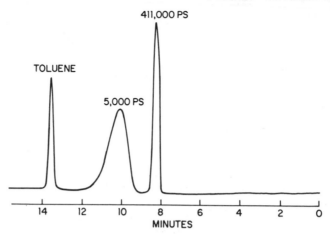

Figure 12.14 Band widths for solutes of different molecular weight. Columns, 10 × 0.62 cm, porous-silica microspheres 50 Å (silanized); mobile phase, tetrahydrofuran; flowrate, 1.6 ml/min; temp., 22°C; detector, UV, 0.2 AUFS at 254 nm; sample, 25 μl-2 mg/ml, 411,000 MW polystyrene; 4 mg/ml, 5000 MW polystyrene; 2 mg/ml toluene. Reprinted from (*19*) with permission.

25 × 0.8-cm column, about 15 mg of sample can be charged (somewhat more for lower MW samples, less in the case of higher MW). However, much larger samples can be charged in preparative separations when the compounds are easily resolved; see Table 15.5. This loading is roughly comparable to that for silica liquid-solid columns (Section 9.3), even though different mechanisms control the overloading of SEC and LSC columns. In SEC columns, injection of macromolecules at higher concentrations can result in slight changes in conformation (decrease in size) of the molecule, causing increased retention volume.

Band broadening in SEC with polymers differs in some respects from that in the other LC methods. In lower-efficiency SEC systems, band widths can remain approximately constant throughout the chromatogram, because molecular size decreases sharply with V_R in size exclusion, and plate heights H increase with molecular size. In this case these two effects combine to cancel roughly the normal increase in band broadening for more strongly retained bands. However, in high-efficiency SEC, peaks can have varying widths. For example, in Figure 12.14 we see that the 411,000 MW polystyrene peak is relatively narrow because all components are totally excluded from the packing and the different molecular weights are "squeezed" together. The toluene monomer peak also is narrow because of its large diffusion coefficient. On the other hand, the 5000 MW polystyrene peak is relatively broad because the high-efficiency column used partially fractionates this polymer standard.

Band broadening in size-exclusion chromatography is given by

$$\sigma_t^2 = \sigma_c^2 + \sigma_d^2 \tag{12.3}$$

where σ_t^2 is the observed variance of the peak, σ_c^2 is the variance contribution due to the normal peak broadening within the column, and σ_d^2 is the contribution of the peak variance caused by the molecular size fractionation of the polymer (σ_d is proportional to the MW range of a polymer band).

12.5 MOLECULAR-WEIGHT CALIBRATION

The key to the measurement of accurate molecular weights or molecular-weight distribution by SEC is the availability of an exact calibration plot. Obtaining such a plot is accomplished in various ways. When macromolecules involving a single molecular weight (e.g., proteins) or a narrow-distribution of molecular weights are involved, the simple *peak-position calibration method* can be utilized. In this case a series of standards of known molecular weight is chromatographed using column packings whose pore structures are optimized for the molecular weight of interest (Section 12.2). If the standards are sufficiently close to monodisperse ($\bar{M}_w/\bar{M}_n < 1.1$), then all the "average" molecular weights are nearly equal, and

$$M_{\text{peak}} = \bar{M}_w = \bar{M}_n, \tag{12.4}$$

where M_{peak} is the molecular weight corresponding to the band in the chromatogram, $\bar{M}_w$ is the weight-averaged molecular weight, and $\bar{M}_n$ is the number-averaged molecular weight. By plotting log MW of the standard ($\bar{M}_w$) versus peak retention volume, the calibration curve is then established as previously illustrated in Figure 12.7 for single columns and Figure 12.8 for a column set. Unknown samples are then chromatographed in the same manner, and the elution volumes of sample peaks are compared to the calibration curve for estimation of molecular weight. This is a reliable experimental procedure, as no assumptions are made regarding the MW distribution of the calibration standards (except that they are almost monodisperse and that there is a constant molecular conformation between the calibration standards and the unknown solutes). In Figure 12.15 we see the log MW versus retention volume V_R calibration plot for a series of protein-sodium dodecyl sulfate complexes. Here molecular weight is plotted versus retention calculated as the distribution coefficient. The molecular weights of unknown proteins can be accurately determined with this calibration, *providing the unknown also is in the same conformation.*

Unfortunately, the peak-position method is limited by the availability of appropriate standards of narrow MW distribution. Large errors in molecular weight can result when calibrations prepared with narrow standards of one

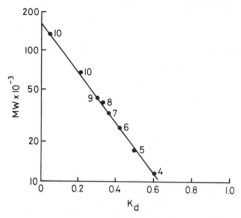

Figure 12.15 Peak-position calibration for protein-sodium dodecyl sulfate complexes. Column, 175 × 0.9 cm, controlled-pore glass GPC, 255 Å; mobile phase, 6 M urea, 0.5% SDS, 0.05 M phosphate buffer, pH 7; flowrate, 0.66 ml/min; sample, 200 μl; solutes: (4) cytochrome C; (5) apomyoglobin; (6) α-chymotripsinogen A; (7) rabbit troposyosin; (8) aldolase; (9) ovalbumin; (10) bovine serum albumin (small amount of dimer also present). Reprinted from (20) with permission.

polymer are used to characterize polymers of other types. Accurate molecular weights are obtained *only* if the unknown polymer and the separating system are identical for the standard and the unknown. However, it sometimes is convenient to use a peak-position calibration curve prepared with readily available standards (e.g., polystyrenes) and to utilize the MW determinations of unknown samples made with this calibration for comparison only. Molecular weights obtained in this fashion are based on an *equivalent* molecular weight. It should be recognized, however, that actual molecular weights obtained by this method often can be in error by a factor of 2 to 3, and by as much as an order of magnitude under the most unfavorable conditions.

The simple peak-position calibration method cannot be used when there are no narrow MW distribution standards. If a computer is available and the $\overline{M}_w$ and $\overline{M}_n$ values for a single broad MW polymer can be determined (i.e., by light scattering and osmometry), then the *single-broad-standard* calibration method provides accurate MW measurements. However, this method is too involved to discuss here, and (6) should be consulted for details. If a computer is not available, it is still possible to establish approximate calibrations with broad standards that can provide useful MW information. Let us now examine how this approach might be used on a real sample to obtain MW estimations.

In Figure 12.16 we see the elution pattern for an unknown chitosan: a carbohydrate polymer of different MW ranges that is manufactured from the chitin of shrimp and crab wastes. Also shown are the composite elution curves for a

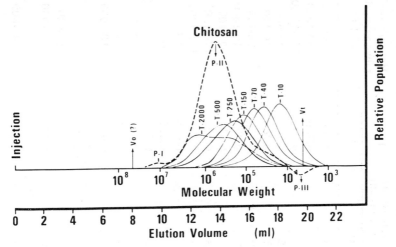

Figure 12.16 Molecular-weight determination of chitosan. Column, 30 × 0.23 cm lengths in the ratio (2-4-6-2-2-2) of 2500, 1500, 550, 250, 140 Å, respectively (total, 18 ft), 37-74 μm, Glycophase G-CPG; mobile phase, 2% acetic acid; flowrate, 1 ml/min; temp., ambient; detector, RI; sample, 50 μl, 5 mg/ml in mobile phase. Reprinted from (21) with permission.

series of broad MW-distribution dextrans. In this approach at least two (or more of these polymers are selected as "standards," (using samples that are very different in molecular weight). (If the weight-average molecular-weight $\bar{M}_w$ values for these two widely variant polymers are unknown, they are determined by light scattering or ultracentrifugation.) A calibration curve then is constructed by plotting the $\bar{M}_w$ versus the SEC peak position for these broad standards. $\bar{M}_w$ values for the unknown sample then can be estimated by comparing the observed elution volume for the unknown with the calibration curve.

It should be stressed that the method just discussed produces only *approximate* molecular weights and should only be used when more accurate methods (e.g., the single broad standard method or the universal calibration method, discussed later) are not feasible. The MW accuracy of this method is dependent on the unknown polymer having the *same* structure and MW distribution as the standards. In the case shown in Figure 12.16, chitosan standards are unavailable, so readily available dextran standards of similar structure were employed; therefore, errors in the estimated molecular weight due to polymer structural differences could be anticipated. The obvious difference in the MW distribution profile for the standards and sample in Figure 12.16 would also result in MW error. Better MW accuracy would have been obtained if standards with a MW distribution similar to the unknown could have been used.

MW determinations for known polymers often can be conveniently carried out by the *universal calibration* method. This method is based on the fact that if

Table 12.3 Mark-Houwink constants for molecular-weight determination by the universal calibration method.

Polymer	Solvent	Temperature (°C)	Mark-Houwink constants		Molecular-weight range applicable × 10^{-3}
			$K \times 10^4$	a	
Polystyrene	THF	25	1.60	0.706	>3
Polystyrene	THF	23	68.0	0.766	50-1000
Polystyrene (comb)	THF	23	2.2	0.56	150-11,200
Polystyrene (star)	THF	23	0.35	0.74	150-600
Polyvinylchloride	THF	23	1.63	0.766	20-170
	–	–	1.50	0.77	20-105
	–	–	1.60	0.77	10-1000
	–	–	7.2	0.61	27-100
Polymethylmethacrylate	THF	23	0.93	0.72	170-1300
	–	–	1.28	0.69	150-1200
	–	–	21.1	0.406	<31
	–	–	1.04	0.697	>31
Polycarbonate	THF	25	3.99	0.77	–
Polycarbonate	THF	25	4.9	0.67	7-77
Polydioxalane	THF	25	0.937	0.874	–
Polyvinyl acetate	THF	25	3.5	0.63	10-1000
Polyvinyl bromide	THF	20	1.59	0.64	–
Polyvinyl ferrocene	THF	30	0.72	0.72	–
Polyisoprene	THF	25	1.77	0.735	40-500
Natural rubber (poly-*cis*-isoprene)	THF	25	1.09	0.79	10-1000
Butyl rubber (isobutene coisoprene)	THF	25	0.85	0.75	4-4000
Poly-1,2-butadiene	THF	20	$Mn(PB) = 0.167$	$Mn(PS)$	9-25

Polymer	Solvent	T			
Poly-1,4-butadiene	THF	40	5.78	0.67	10-100
Poly-1,4-butadiene	THF	25	76.0	0.44	270-550
Poly-1,4-butadiene (cis-trans ≅ 0.8)	THF	25	—	—	—
8% Vinyl	THF	25	4.57	0.693	80-1100
28% Vinyl	THF	25	4.51	0.693	20-200
52% Vinyl	THF	25	4.28	0.693	20-200
73% Vinyl	THF	25	4.03	0.693	20-200
Polybutadiene 20% cis, 20% vinyl	THF	25	2.36	0.75	3-6
SBR (25% styrene)	THF	40	3.18	0.70	70-1000
SBR (25% styrene)	THF	25	4.1	0.693	24-40
SBR 1507	THF	30	3.0	0.70	10-1000
SBR 1808	THF	30	5.4	0.65	10-1000
Cellulose nitrate	THF	25	25.0	1.0	95-2300
Cellulose trinitrate	THF	25	3.21	0.83	60-6000
Amylose acetate	THF	25	108.0	0.70	20-500
Amylose butyrate	THF	25	111.0	0.70	20-500
Amylose proprionate	THF	25	248.0	0.61	20-500
Polystyrene	ODCB	135	1.38	0.70	2-900
Polyethylene	ODCB	135	4.77	0.70	6-700
Polyethylene	ODCB	135	5.046	0.693	10-1000
Polyethylene	ODCB	138	5.06	0.70	0.2-200
Polybutadiene (hydrogenated)	ODCB	135	2.7	0.746	10-500
Polypropylene	ODCB	135	1.30	0.78	28-460
Polydimethyl siloxane	ODCB	138	3.83	0.57	25-300
Polydimethyl siloxane	ODCB	87	8.19	0.50	20-800
Polystyrene	m-Cresol	135	2.02	0.65	4-2000

Table 12.3 (Cont.).

Polymer	Solvent	Temperature (°C)	Mark-Houwink constants $K \times 10^4$	a	Molecular-weight range applicable $\times 10^{-3}$
Polyethylene terephthalate	m-Cresol	135	1.75	0.81	2.7-32
Polyethylene terephthalate	m-Cresol	135	2.0	0.90	0.45-0.80
Polyethylene terephthalate	m-Cresol	25	0.077	0.95	–
Nylon 66	m-Cresol	130	0.40	1.00	8-24
Nylon 6	m-Cresol	25	32	0.62	0.5-5
Nylon 610	m-Cresol	25	1.35	0.96	8-24
Nylon 66	m-Cresol	25	$0.15 + 3.53 \times 10^{-4}\,M$	0.79	0.15-50
Nylon 6	OCP	90	6.2	0.64	10-1000
Polyethylene terephthalate	OCP	–	3.0	0.77	1-300

Source: In part from (6) with permission.
Note: THF = tetrahydrofuran; ODCB = o-dichlorobenzene; OCP = o-cresylphosphate.

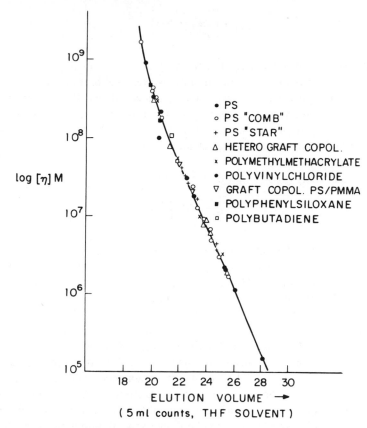

Figure 12.17   Universal SEC calibration curve. Reprinted from (23) with permission.

only the solute hydrodynamic volume controls the size-exclusion process, a plot of log intrinsic viscosity times molecular weight (log $[\eta]M$) versus retention volume V_R provides a calibration curve that is approximately valid for all polymers. [M here is more nearly $\bar{M}_n$ according to (22).] This approach produces a universal calibration curve (Figure 12.17) that is the same for virtually all polymers, whether random coil, rigid rod, or spherical. Let us now describe how this approach can be used to calculate the molecular weight of unknown polymers.

First, a peak-position calibration is prepared for the range of MW interest, using readily available, narrow MW standards (e.g., polystyrene). Next, using this peak-position calibration the molecular weight of the unknown is calculated according to

$$M_1 = \left[\frac{k_2}{k_1} (M_2{}^{\alpha_2}) \right]^{1/\alpha_1} , \tag{12.5}$$

where M_1 is the molecular weight of the unknown polymer; M_2 is the molecular weight obtained from the peak-position polystyrene calibration curve; k_1 is a coefficient for the polymer to be analyzed; k_2 is a coefficient for the MW standard (in this case, polystyrene); α_1 is a second coefficient for the polymer to be analyzed; and α_2 is the second coefficient for the MW standard (in this case, polystyrene). Values of k and α are calculated by

$$k = 6.19 \cdot 10^9 (K)^{1/3} \tag{12.6}$$

$$\alpha = \left[\frac{(1+a)}{3} \right] \tag{12.7}$$

where K and a are *Mark-Houwink constants* such as those given in Table 12.3. More extensive lists of Mark-Houwink constants are given in (6) and (24). These constants account for the MW dependency of the intrinsic viscosity for the different polymers involved, and as illustrated in Table 12.3, the values are solvent and temperature dependent.

The universal calibration approach is broadly applicable as long as Mark-Houwink constants are available for the polymer of interest and size exclusion is the only retention phenomena. Serious errors can result if other effects (e.g., adsorption) cause retention. The validity of the universal calibration also can be markedly affected by the choice of solvent. Accurate GPC measurements with the universal calibration approach are only obtained when the polymer is dissolved in "good" solvents, in which polymer-solvent interactions are favored over polymer-polymer interactions. The universal calibration method is conceptually sound, but its use is still rather limited, and the accuracy and precision of the method have not yet been fully evaluated.

Size exclusion can also be used to obtain valuable molecular weight or size data on small molecules. As with macromolecules, the effective size of an eluting small molecule determines its elution by SEC. Several MW or size relationships have been found useful in the SEC of small molecules. First, correlations between elution volume and *molecular length* have been noted (25, 26). The sizes of molecules are calculated from atomic radii and bond angles for various functional groups. n-Hydrocarbons are used as standards, for which molecular lengths L_m are calculated as follows:

$$L_m = 2.5 + 1.25n \quad \text{(in Å)}. \tag{12.8}$$

Here n is the number of carbon atoms in the molecule. For example, L_m for n-pentane is $2.5 + 5(1.25) = 8.75$ Å. The molecular lengths of other compounds can also be calculated as described in (*25*). With this approach the retention volume V_R for the standards (*n*-alkanes) are determined by chromatography, and these values are plotted versus L_m to obtain a calibration curve. The observed L_m values obtained on other solutes by referral of their V_R values to this calibration curve are termed *effective molecular length*. It has been observed that many molecules do not elute according to the calculated L_m values predicted by Eq. 12.8. Thus, experimentally derived corrections have been established to account for these differences. In addition, effective molecular lengths have been shown to be a function of the eluting solvent; therefore, correction factors have also been introduced to allow for solute-solvent associations. Molecular length values can be useful for predicting or confirming the structure of an unknown compound, or for predicting whether or not an SEC separation is feasible for compounds of different structures but the same molecular weight (see Figure 12.10).

Molar volume has also been used as a size parameter in the SEC of small molecules (*27*), and values for compounds are calculated from their densities at 25°C

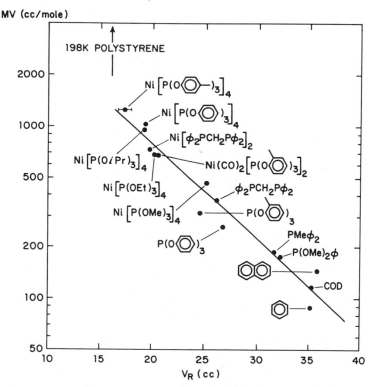

Figure 12.18 Molar volume calibration for nickel complexes. Column, 120 × 0.8 cm, μ-Stryagel 100 A; mobile phase, *dry* tetrahydrofuran; flowrate, 3.4 ml/min; detector, UV and RI; sample, 25 μl, ~2% solution in THI. Reprinted from (*30*) with permission.

and expressed as ml/mol. Again, *n*-alkanes are used as standards for the calibration curve, which is obtained by plotting observed V_R versus molar volume. When compared to this calibration plot, experimentally observed values for other solutes then are termed *effective molar volumes*. Because the observed elution characteristics of various solutes (other than aklanes, branched alkanes, and alkenes) can be somewhat different from that calculated from their densities, effective molar volumes must be corrected according to experimentally derived guidelines (*28, 29*). Nevertheless, calculated molar volumes for various types of compounds often are relatively consistent even when there are gross structural differences. In Figure 12.18 we see that the data for a series of nickel complexes and other solutes containing no nickel do not deviate greatly from the calibration curve for polystyrene standards (solid).

Log molecular weight versus V_R data for small molecules also can show acceptable correlation. In Figure 12.19 is shown a MW versus V_R calibration plot for a series of drugs having grossly dissimilar structures. As with the MW versus V_R calibration methods just discussed, this approach also provides only reasonable estimates, but often can be quite useful in predicting the molecular weight (or identity) of an unknown in a sample.

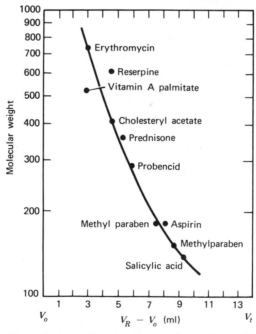

Figure 12.19 Size-exclusion calibration for a series of drugs. Column, 90 × 0.8 cm, μ-Styragel 100 Å; mobile phase, tetrahydrofuran; flowrate, 2 ml/min; detector, UV, 0.32 AUFS, 254 nm. Reprinted with permission of Waters Associates.

12.6 RECYCLE CHROMATOGRAPHY

Resolution in SEC is restricted by a limited peak capacity n, which can be estimated by (31):

$$n \simeq (1 + 0.2N^{1/2}). \tag{12.9}$$

For example, a set of columns with a plate count of 15,000 would have a peak capacity of about 25, which means that a *maximum* of 25 monomer peaks with peak resolutions of unity can be placed within the total-exclusion and total-permeation volume of this column set. For real samples this situation is very unlikely, and it is rare that more than a dozen or so distinctive peaks are seen in SEC separations (see Figure 12.10). The limited peak capacity of SEC is further aggravated in columns of organic gels, because long columns and high pressures cannot be used to obtain maximum plate count per unit of time. This problem can be solved by using *recycle chromatography*. One approach for carrying out recycle chromatography is the *closed-loop* technique illustrated in Figure 12.20. This technique allows partially resolved bands to be sent back to the same column for further separation. Operation is begun in a normal mode, with mobile phase flowing from the solvent reservoir through the pump into the column and through the detector to the fraction collector. When a partially separated band passes through the detector, valve (V_2) is switched to allow these bands to return to the low-volume reciprocating pump and reenter the column

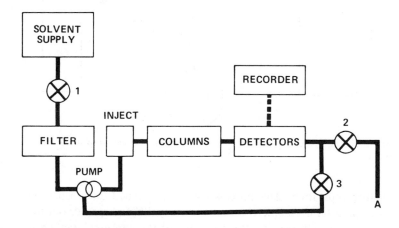

For RECYCLE — Valve 1 & 3 Open, Valve 2 Closed
For DRAWOFF — Valve 1 & 2 Open, Valve 3 Closed

Figure 12.20 Equipment schematic for closed-loop recycle chromatography. Reprinted with permission of Waters Associates.

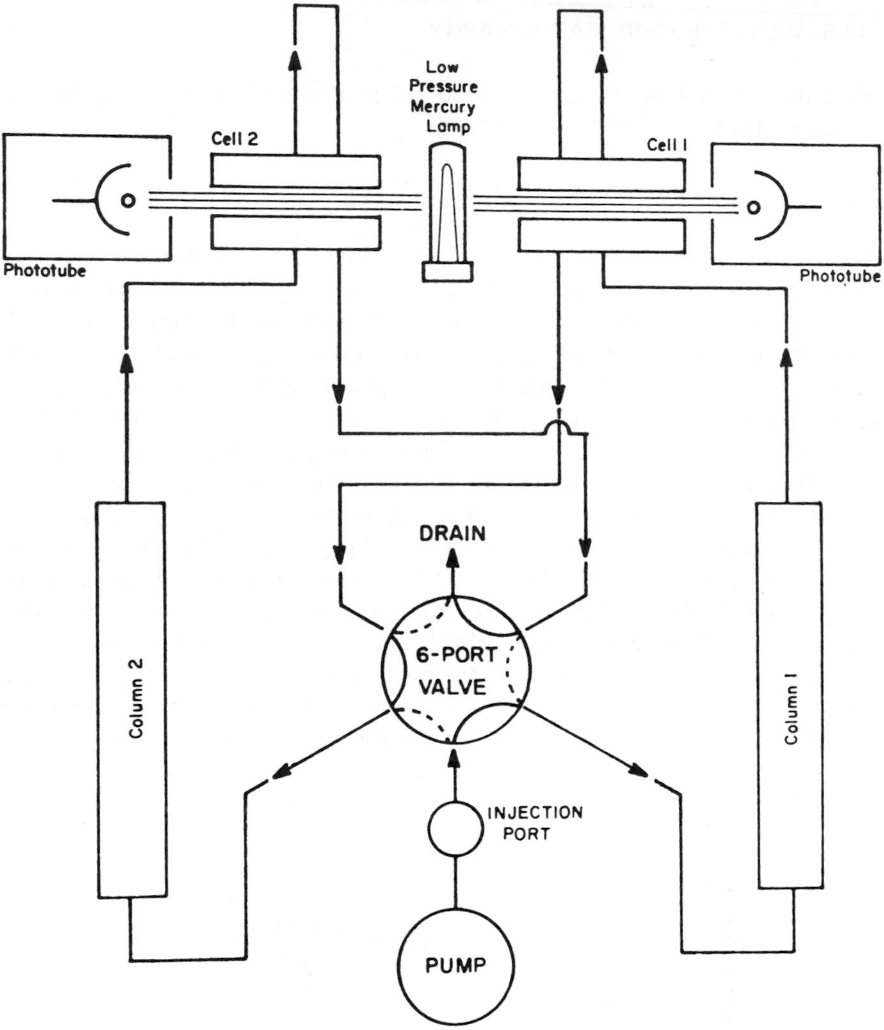

Figure 12.21 Equipment schematic for dual-column recycle. Reprinted from (*32*) with permission.

for further separation. The process can be continued until separation of the bands in question is complete, or until the recycle band spreads out to cover the entire column length (so that the two ends of the chromatogram begin to over-lap).

An alternative to the closed-loop approach for recycling in SEC is the *dual-column* recycle system illustrated in Figure 12.21. A sample containing the peaks to be resolved by recycling is introduced into column 1; effluent from

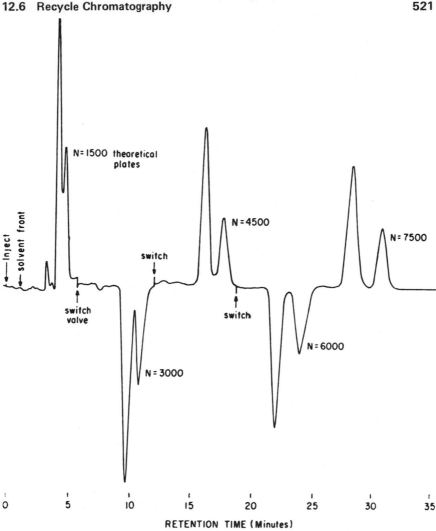

Figure 12.22 Separation of endo-exo isomers with dual-column recycle. Column, 2-50 ×
0.8 cm, Permaphase-ODS (alternating in series); mobile phase, 65% methanol, 35% water;
flowrate, 10 ml/min; temp., ambient; detector, UV, 0.64 AUFS, 254 nm. Reprinted from
(*32*) with permission.

column 1 is monitored by cell 1 of the dual-cell UV photometer, before it passes
into column 2. In this valve position, cell 1 is at high pressure and cell 2 is at
ambient. While the sample is in column 2, the valve is switched to divert the flow
back into column 1. In this valve position, cell 2 becomes the high-pressure cell.
As the sample emerges from column 2, cell 2 provides a record to indicate

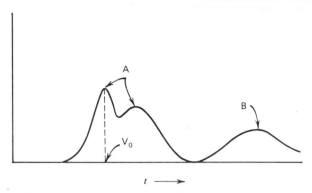

Figure 12.23 False band at the exclusion limit in size-exclusion chromatography.

whether or not adequate resolution has been obtained. A sample can be recycled through the column system in this manner until the broadened peaks completely occupy one column volume or until adequate resolution has been obtained. The advantage of this system is that the sample does not have to pass through a pump chamber, so that the peaks are not broadened by extracolumn effects due to the pump volume. It is not essential to monitor the peaks eluting from both columns if they are matched in performance (e.g., same elution volume). In this case a detector cell is not required for the second column because its switching cycle can be calculated, based on data displayed by the detector on the first column.

Recycle chromatography can also be used with the other LC methods for separating single pairs of compounds (but complex mixtures cannot be handled since overlap of peaks from different cycles is likely during the recycle operation). Figure 12.22 shows the dual-column alternate-pumping recycle separation of methyl-1,1-(spirocyclopropyl)-indene isomers by reverse-phase chromatography. Note that in this system, each pass through 50-cm pellicular columns added about 1500 plates to the system. After five passes, 7500 plates were generated and baseline resolution of the two isomers was achieved. Recycle chromatography also is finding increasing use in the other LC methods for very difficult preparative separations (Chapter 15). This technique effectively yields longer column lengths, without an accompanying increase in column pressure. When combined with automatic equipment for continuous, unattended operation, recycle chromatography can yield extremely large plate counts ($>10^5$) in a relatively short time.

12.7 PROBLEMS

Changing solvents with columns of microparticulate gels can lead to collapse of the particles, resulting in severe degradation of performance. Such solvent effects are most prominent with the small-pore microparticulate polystyrene gels (e.g., μ-Styragel 100 Å), and manufacturer recommendations on solvents should be followed carefully. Table 12.4 lists solvents that are and are not acceptable for μ-Styragel 100 Å packings. Because of possible gel collapse even when manufacturers' recommendations are followed, some workers recommend that the smallest-pore microparticulate polystyrene gels be utilized with a single solvent

Table 12.4   Solvent compatibility with μ-Styragel.

Satisfactory	Marginal	Damaging
Benzene	Cyclopentane	Acetone
Carbon tetrachloride	Cycloheptane	Benzyl alcohol
Chloroform	Diethylamine	Butyl acetate
p-Dioxane	Methylene chloride	Carbon disulfide
Diethylbenzene	Diethyl ether	m-Cresol
Divinylbenzene	Ethyl acetate	o-Chlorophenol
Pyridine	Methylethyl ketone	Dimethylformamide
Toluene	Triethylamine	Dimethylacetamide
Tetrahydrofuran	90% Tetrahydrofuran/	n-Dodecane
Trichlorobenzene	methanol	50% Tetrahydrofuran/
Tetrachloroethane		methanol
Tetrahydropyran		Acetic acid
Tetrahydrothiophene		Acetic anhydride
p-Xylene		Acetonitrile
		Cyclohexane
		Hexafluoroisopropanol
		Heptane
		Hexane
		Isooctane
		Trifluoroethanol
		Alcohols
		Water

Source: From Waters Associates with permission.

(e.g., tetrahydrofuran) for routinely analyzing samples of small molecules (*33*). If another solvent must be used for sample dissolution, then a duplicate column set is recommended for this different solvent. Columns of rigid siliceous particles can be used with virtually any solvent of pH < 8.5.

Columns of polystyrene gels can be damaged (i.e., N is greatly decreased) if air bubbles are allowed to enter. The connecting and disconnecting of this column type must be carried out with the manufacturer's recommendations, to ensure that air is excluded and column performance remains unchanged during handling.

Solvent degassing can be a potentially serious problem with polystyrene packings, because such columns can be ruined by the release of air bubbles within the column. This effect is magnified by the use of elevated separation temperatures. Therefore, it is essential that all solvents be thoroughly degassed before use (e.g., by boiling the solvents for 5-10 min).

False bands in SEC can arise when polymeric samples are separated. This is illustrated in Figure 12.23 for the hypothetical separation of a monomodal polymer fraction (A). When a significant part of such a polymer fraction consists of oligomers that are totally excluded (i.e., have a molecular weight greater than the exclusion limit), these elute together as an *apparent* second peak at V_0. Thus, the presence of a separate polymer band at V_0 should always be regarded with suspicion. In such cases, the sample should be reseparated on a column with a larger exclusion limit (larger pores). If the band at V_0 (from the first separation) disappears, then a false exclusion band was present. If the initial band is still seen in the second separation, a true bimodal MW distribution exists for the sample.

Self-association of sample molecules in SEC is an occasional problem. For example, ionic surfactants form micelles in some solvents. The result is asymmetrical bands whose V_R values change with sample size. Sample association can usually be controlled by using a mobile phase that is a better solvent for the sample, for example, for ionic surfactants, a more-polar solvent. Sample association can occasionally be used to advantage. For example, carboxylic acids dimerize in nonbasic solvents such as benzene, but dissociate in more basic solvents such as tetrahydrofuran (i.e., remain monomeric). Therefore, the V_R value of a carboxylic acid varies with the solvent, because its apparent molecular size is doubled in nonbasic solvents. In such cases the solvent can be used to control separation by adjusting the V_R values of carboxylic acids independently of other compounds in the sample. Association of solvent and sample molecules (e.g., alcohols plus ether solvents) can be used in a similar fashion to control separation, but the changes in V_R usually are small. Occasionally it may be advantageous to increase molecular size artificially through association, by adding a large associating compound to the solvent (e.g., an acid to complex a basic sample).

As noted in Section 12.3, adsorption effects in size-exclusion chromatography can be an important problem with porous-glass or silica gels. Aside from the use of special deactivated packings (e.g., μ-Bondagel E), various materials can be added to the mobile phase (e.g., 0.05% Carbowax 20-M) to suppress sample adsorption by competing for active adsorption sites on the particle surface. Adsorption is normally not observed for samples that do not permeate the porous particle. In gel permeation chromatography, modification of siliceous particles with chlorotrimethylsilane can reduce adsorption effects without significantly affecting pore volume or pore size. For details on this treatment, see (6).

Retention by effects other than size exclusion also has been reported with organic gel packings. For example, aromatic compounds can be retarded by Sephadex, but this can be eliminated by the choice of solvent. Although in SEC the elution of solutes is predictable, nonelution (or very large retention volumes) sometimes occurs both with silica-type and organic microparticulate gels. To determine whether a sample is totally eluting, a material balance should be carried out. The weight of recovered material eluting up to the total permeation volume (after solvent evaporation) should equal the weight of injected sample.

12.8 APPLICATIONS

A wide range of individual SEC separations have been carried out, and many of these are described in the bibliography for this chapter. Let us now look at some illustrative samples that show the unique capabilities of modern SEC.

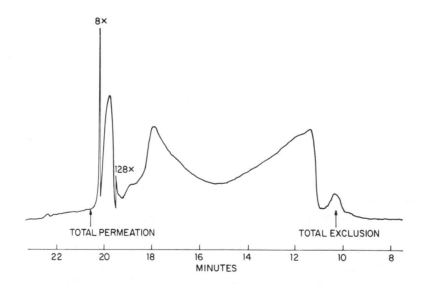

Figure 12.24 First analysis of polymer film acetone extract. Column, 100 × 0.62 cm, porous-silica microspheres, 50 Å (silanized); mobile phase, acetone; flowrate, 1.23 ml/min; detector, RI, 8×; temp., 22°C; sample, 50 μl of acetone extract. Reprinted from (19) with permission.

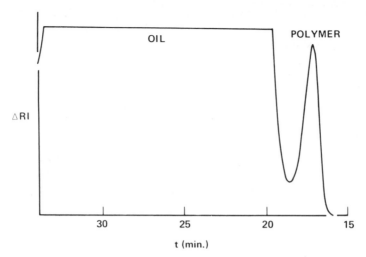

Figure 12.25 Determination of polymer additives in lubricating oil. Columns, 120 × 0.8 cm, 2-Styragel 100 Å, 1-Styragel 10^3 Å, 1-Styragel 10^4 Å; mobile phase, tetrahydrofuran; temp., ambient.

Size exclusion is quite useful in providing initial information about unknown samples, particularly those containing polymeric constituents. This approach is often a preferred "first-analysis" method that can provide much useful information in a relatively short time (15-20 min) and sometimes solve the problem without additional work. The first-analysis capability of size exclusion on a column of small, rigid siliceous particles is illustrated in Figure 12.24. Here we see the initial chromatogram of an acetone extract of a photographic film. This chromatogram showed a small amount of material eluting at the total exclusion volume, suggesting polymer greater than about 5000 MW (presumably the principal film polymer). This peak was followed by two different groups of broad, lower MW materials, representing other types of polymer in the film coating. A peak of smaller molecules near total permeation contained dyes and other nonpolymeric additives in the film.

The concentration of high-molecular-weight additives or reaction products in a lower-molecular-weight matrix can be readily determined by SEC. Figure 12.25 illustrates the determination of a polymeric additive in lubricating oil, where the polymer band at V_R = 17 ml is well separated from the oil. Despite the marginal sensitivity of refractometer detectors, as little as a few hundredths of a percent of the polymer additive could be accurately determined in this system, because it was possible to charge relatively large amounts of sample (note that the large lubricating oil band is off-scale). Similarly, small amounts of low-molecular-weight compounds can be determined in a polymer matrix by

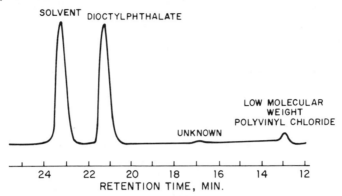

Figure 12.26 Separation of additives in commercial polyvinylchloride. Column, 2-MicroPak BKG-2000H; mobile phase, tetrahydrofuran; detector, UV. Reprinted with permission of Varian Associates.

proper selection of the column pore size. An illustration of this is shown in Figure 12.26 for additives in a commercial polyvinylchloride sample. Di-*n*-octylphthalate (DOP) was easily determined and an unknown was detected at about 17 min, as well as very low-molecular-weight materials at the total permeation volume (about 23 min). In this case the main polyvinylchloride peak of no interest was barely sensed by the UV detector.

SEC is widely used to show qualitative changes in polymeric systems due to processing, aging, and so on. In Figure 12.27 resin degradation during regrinding can be seen by the changes at the higher molecular weight, when compared to fresh (virgin) material. Also shown is an acceptable blend of fresh and "regrind" resins. By comparison with this acceptable blend, optimum use may be made of regrind, without excessively compromising polymer physical properties. This "fingerprinting" approach is further illustrated in Figure 12.28, which shows SEC chromatograms of a "standard" glue used by a plywood plant for Douglas fir stock and a resin glue from an alternate source (dashed curve). Two things are worth noting in this latter curve. First, there is a hump on the high-molecular-weight portion due to the presence of material added by the manufacturer of the alternate resin to increase the tackiness desirable for plywood production. Second, the resin in the alternate glue showed less high-molecular-weight material than the standard glue. To make it usable, the alternate resin glue was further cross-linked with more formaldehyde to make the MW distribution similar to that of the standard resin and produce the properties required for end use.

Oligomers often can be best determined by high-performance SEC, as shown in Figure 12.29 for a phenolic resin. Using a long (180-cm) column, the resin eluted at the total exclusion volume (about 50 ml), followed by a series of

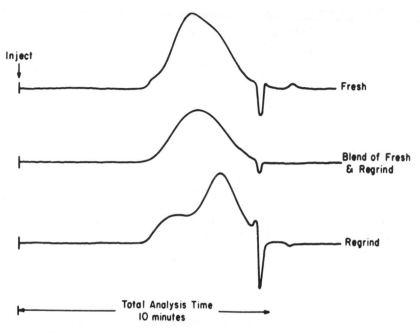

Figure 12.27 Characterization of fresh and regrind resin. Column, 25 × 0.62 cm ea., DuPont SE, 1-1000 Å, 1-500 Å, 1-100 Å; mobile phase, tetrahydrofuran; flowrate, 1.5 ml/min; detector, UV, 254 nm. Reprinted from (*34*) with permission.

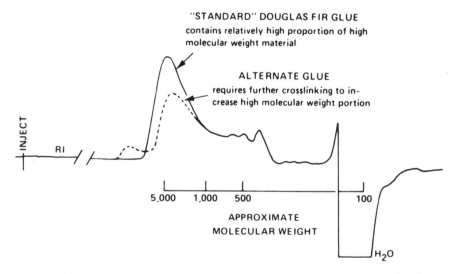

Figure 12.28 Comparison of wood glues by SEC. Reprinted from (*35*) with permission.

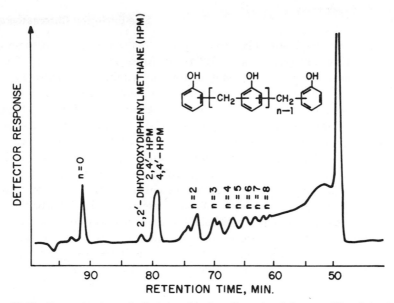

Figure 12.29 Determination of oligomers in phenolic resin. Column, μ-MicroPak BKG-2000H; mobile phase, tetrahydrofuran; flowrate, 0.74 ml/min; detector, UV. Reprinted with permission of Varian Associates.

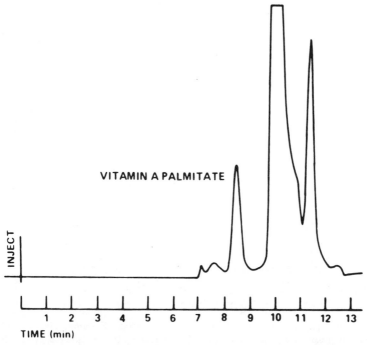

Figure 12.30 Analysis of vitamin A in vitamin drops. Column, 90 × 0.8 cm, μ-Styragel, 100 Å; mobile phase, tetrahydrofuran; flowrate, 2 ml/min; detector, UV, 254 nm. Reprinted with permission of Waters Associates.

529

oligomers for which reasonable resolution was obtained to $n = 6$. Also resolved were dihydroxydiphenylmethane (HPM) additives in the polymer. Such a chromatogram is useful in accounting for the number of reactive sites per molecule, and using this approach a reactivity index can be calculated to provide a basis for selecting the proper curing parameters for various resins. Such oligomer separations can be quite useful in controlling the manufacture of plastics. For example, epoxy and other thermosetting resins are cured to produce hard molded parts, and in many operations, control of curing is critical. Resin reactivity, which must be controlled, varies from batch to batch, and depends on the number of reactive sites present in a given weight of resin. By determining the amount of each oligomer present, and accounting for the number of reactive sites per molecule for each oligomer peak, a "reactivity index" for each batch of resin can be calculated to provide a basis for proper curing. Additives in polymer systems also can be monitored for quality control.

In recent years there has been a rapid increase in the use of SEC for separating small molecules. Many applications dealing with the characterization of fossil fuels, refined products, by-products, fats and oils, additives in plastics, and various others have been summarized (*36, 36a*). SEC separation of small molecules has been utilized for qualitative comparisons and to isolate fractions for subsequent characterization by supplemental techniques. Also, for appropriate samples, quantitative analysis of small molecules by high-performance SEC can be carried out with reproducibilities of about 1% relative (*37*). Figure 12.30

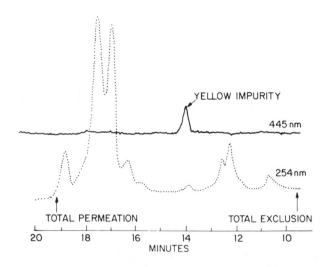

Figure 12.31 Characterization of yellowed polymer extract. Column, 100 × 0.62 cm, porous-silica microspheres, 50 Å (silanized); mobile phase, methanol; flowrate, 1.25 ml/min; detectors: UV photometer, 0.02 AUFS, 254 nm; UV spectrophotometer, 0.02 AUFS, 445 nm; temp., 22°C; sample, 50 μl of methanol extract. Reprinted from (*19*) with permission.

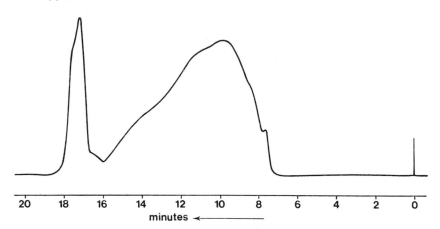

Figure 12.32 Molecular-weight distribution of a carboxymethylcellulose (CMC) sample. Columns, 30 × 0.46 cm, 1-LiChrospher SI-100, and 1-LiChrospher SI-500; mobile phase, 0.5 M aqueous sodium acetate, pH 6; flowrate, 0.5 ml/min; sample, 100 μl of 0.5% solution; detector, RI. Reprinted from (*14*) with permission.

shows the SEC analysis of vitamin A palmitate in vitamin drops, carried out by injecting a solution of the sample without pretreatment. The analysis is rapidly performed without interference from higher- and lower-molecular-weight excipients.

Extremely labile substances often can be separated on microparticulate gel columns when all other separating systems fail; these columns probably represent the gentlest separation technique available. For example, Figure 12.18 shows the log molar volume calibration for a series of oxygen-sensitive and easily hydrolyzed nickel complexes. Separation on a set of microparticulate gel columns using carefully dried solvents and total exclusion of air allowed a series of these compounds to be analyzed and other similar compounds to be tentatively identified by their molecular volume elution characteristics (*30*).

High-performance SEC in combination with selective detection sometimes permits the rapid solution of a difficult problem. Figure 12.31 is a chromatogram of a methanol extract of a yellowed polymer where the problem was to determine the origin of the undesirable yellow color. The yellow impurity is clearly apparent as a single peak in the upper chromatogram in Figure 12.31, using a spectrophotometric detector operating at 455 nm. In the lower chromatogram of Figure 12.31, the recording with a UV detector at 254 nm is shown. The band width found for the yellow impurity peak was similar to that exhibited for single compounds by this column set, suggesting that the yellow material was not polymeric. Comparison of the elution volume with a polystyrene-hydrocarbon calibration curve indicated a molecular weight of about

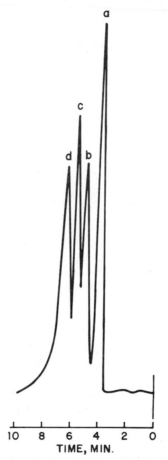

Figure 12.33 Separation of proteins on "diol"-modified porous silica. Column, 30 × 0.4 cm LiChrospher SI-100–diol, 10 μm; mobile phase, 0.05 *M* sodium phosphate buffer, 0.1 *M* sodium chloride, pH 7.5; flowrate, 2.85 ml/min; detector, UV, 210 nm; sample, 10^{-4} *M* or lower in buffer: (*a*) adolase (rabbit); (*b*) chymotripsinogen A (beef); (*c*) lysozyme (human); (*d*) reduced glutathione. Reprinted with permission of K. K. Unger.

1000. This yellow peak was trapped and characterized by mass spectrometry that confirmed the material was a single compound with molecular weight of about 900. The use of porous silica in this problem was particularly important; it was possible to use methanol as mobile phase and to inject the crude extract directly after filtration, without any experimentation on the mobile-phase-support combination.

Columns of silica-based SEC supports can be conveniently used with aqueous mobile phases to characterize a wide variety of water-soluble solutes, as illus-

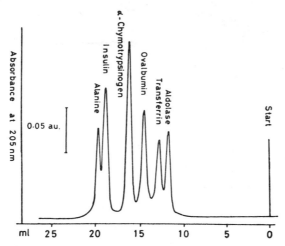

Figure 12.34 Separation of proteins on small-particle gel column. Column, 60 × 0.75 cm, TSK-Gel 2000 SW; mobile phase, 0.01 M phosphate buffer (pH 6.5) with 0.2 M sodium sulfate; flowrate, 0.3 ml/min; detector, UV, 205 nm; sample. Reprinted with permission from (*39*).

trated in Figure 12.32 for a carboxymethylcellulose (CMC) sample. Determination of the MW characteristics of CMC and other soluble cellulose ethers and esters has been greatly simplified, because the silica columns can be used at elevated temperatures to reduce the high viscosity of these polymer solutions (*14*). Modified silica substrates are also quite versatile for separating a wide variety of macromolecules by gel filtration chromatography. For example, proteins can be determined by high-performance GFC on polar bonded-phase silica. As illustrated in Figure 12.33, a mixture of enzymes is separated on a single 30-cm column of bonded-"diol" silica in less than 10 min. Separations of certain proteins also can be carried out on unmodified porous silica packings that are deactivated with polyglycols (e.g., Carbowax 6000 or Carbowax 20-M added to the mobile phase; Section 12.3).

Small-particle hydrophilic gel columns have recently become available (Toya Soda) for the SEC of water-soluble solutes. Figure 12.34 shows the separation of a complex mixture of protein standards using a column of 8-12-μm particles.

SEC is a very useful prefractionation or clean-up step prior to analysis by other high-resolution methods. As discussed in Section 16.2, this operation can be carried out manually, or automatically with switching valves, as required. Figure 12.35 shows the clean-up separation of an extract of plant tissue to be tested for biological activity. The crude 1:1 isopropanol/water plant extract was injected directly into the column after filtration and chromatographed with the same solvent. The components of interest could only be observed with a RI

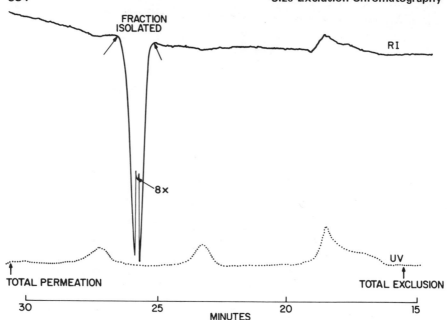

Figure 12.35 SEC cleanup of plant extract. Column 100 × 0.62 cm; porous-silica micro-spheres, 50 Å (silanized); mobile phase, 1:1 isopropanol/water; flowrate, 0.78 ml/min; temp., 22°C; detectors, UV, 0.04 AUFS, 254 nm; RI, 4×; sample, 50 μl of extract residue at 1.4 mg/ml in mobile phase. Reprinted from (*19*) with permission.

detector, and a significant amount of UV-absorbing material was eliminated in the isolation.

A new and unusual application of SEC is the characterization of colloidal particles. Figure 12.36 shows a particle-diameter calibration for a series of polymer latex standards and silica sols. This technique has been used to investigate the growth of polymer particles in the emulsion polymerization of styrene and of vinylacetate, in which particles in the size range 200-6500 Å were characterized (*38*). Particulates of ⩽ 1 μm can be fractionated by SEC, and particle size distribution determinations can be carried out with packings of the appropriate pore size.

12.9 THE DESIGN OF AN SEC SEPARATION

Selecting the Column

Proper selection of the column set is the key to optimum results in SEC separa-

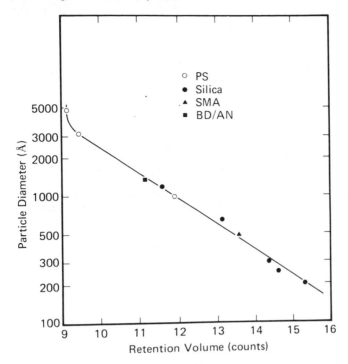

Figure 12.36 Particle-diameter SEC calibration. Column, 120 × 0.8 cm, 1-CPG 2500, 1-CPG 1500 (38-74 μm); mobile phase, 1 g/l KNO₃ and 1 g/1 Aerosol OT in water; flowrate, 7.5 ml/min; detector, UV, 254 nm; PS = polystyrene, silica = silica sol, SMA = styrene-methacrylic acid copolymer, BD/AN = butadiene-acrylonitrile copolymer. Reprinted from (*38*) with permission.

tions. The basic guide is to obtain maximum resolving power in the MW range of interest, with minimum column band spreading. For completely unknown samples, the range of molecular weight involved can be quickly determined experimentally by using a wide MW range column set utilizing a bimodal-pore-size configuration (e.g., two columns each of 1000 and 100 Å). This column set, capable of separation over about 4 decades of molecular weight, is best for a wide-range analysis such as that shown in Figure 12.37a. If with this set the sample elutes nearer the total exclusion volume as in Figure 12.37b, a column set with a single large pore size (e.g., 1000 Å) should be utilized. If the sample elutes in a relatively narrow MW range in the middle of the fractionation range as shown in Figure 12.37c, then a column set of a single intermediate pore size (e.g., 250 Å) should be employed. Finally, if the sample mostly elutes at the total permeation limit for the bimodal column set as in Figure 12.37d, then a column set using only small pores (e.g., 60 Å) should be employed for maximum

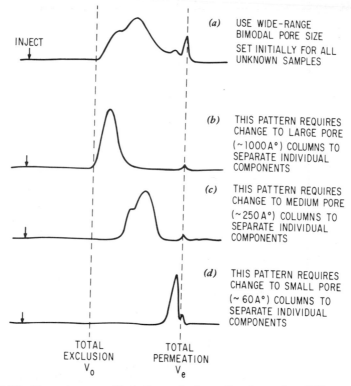

Figure 12.37 Chromatograms illustrating selection of column for SEC separations.

resolution of these low-molecular-weight constituents. For mixtures with a relatively narrow molecular-weight range, only columns with a single pore size are needed, because they have fractionation ranges of 2-2.5 decades of molecular weight (see Figure 12.7). A single column length (e.g., 25 cm of a 5-10-μm packing) does not provide the resolution required in most SEC analyses. Therefore, a series of columns (e.g., 60-120 cm for 10-μm packings; 200-400 cm for 50-μm packings) is usually needed.

Choosing between organic-gel or porous-silica SEC packings often is dependent on the solvent that must be used to dissolve the sample. There are many systems that can be analyzed equally well with columns of either the porous silicas or the microparticulate gels. However, in high-performance SEC (i.e., <10-μm particles), porous-silica packings often are preferred for wide molecular-weight samples because of greater versatility and generally higher performance.

For separating small molecules (i.e., <1000 MW), columns of the smallest pore, microparticulate polystyrene gels (e.g., μ-Styragel 100 Å and MicroPak

BKG 1000H) usually provide the highest resolution. However, a greater useful separation range (up to $\sim 10^4$ MW) can be obtained with columns of 60-Å porous silica (e.g., Zorbax PSM-60), and this substrate can be utilized with a much wider range of solvents, including water. For good resolution of molecules up to about 10,000 MW, a 100-120-cm set of 60-Å porous-silica columns is generally preferred as a single, versatile separating system.

Deactivated silica substrates (e.g., μ-Bondagel E or diol) should be employed when sensitive solutes (e.g., proteins) are involved or when adsorption problems on untreated porous-silica columns cannot be overcome by proper adjustment of the mobile phase (Section 12.3).

Selecting the Mobile Phase

In SEC the mobile phase is normally selected to ensure good sample solubility. Tetrahydrofuran is a preferred solvent because of its good solubility properties and excellent chromatographic characteristics. This solvent also is a good choice with small-pore microparticulate polystyrene gel columns because of its desirable properties in swelling the gel. The sample itself probably will dictate the solvent required for the separation. Highly polar solvents such as hexafluoroisopropanol, methanol, acetone, dimethylsulfoxide, and water all can be used with the modern <10-μm silica-based column packings.

SEC separations involving aqueous mobile phases are required for many macromolecules of biological origin, and some synthetic polymers. Whether using unmodified or modified silica-based particles or hydrophilic organic gels (e.g., Bio-Beads P-2), the mobile phase in all cases should contain some salt to eliminate adsorption or matrix effects with the substrate. Ionic strengths of about 0.1-0.5 are usually adequate. Sodium, ammonium, potassium, and tetramethylammonium cations are increasingly adsorbed to unmodified silica in this order, and, therefore, are increasingly more effective in eliminating cationic adsorptive effects. Di- or trivalent anions such as sulfate and phosphate, respectively, are often more effective than monovalent anions in eliminating adsorption. Particularly effective in eliminating adsorption with unmodified porous silica packings is 0.025-0.05 M tetramethylammonium-phosphate adjusted to pH 3. However, the net effect may result in some reverse-phase adsorption, which can be eliminated by the addition of methanol (e.g., 1:1). Methanol also is added to the mobile phase to suppress reverse-phase effects sometimes found with hydrophilic-modified silica packings (e.g., μ-Bondagel E).

Adsorption of anionic polyelectrolytes on unmodified silica using aqueous mobile phases generally is not observed.

Adjustment of pH also can be effective in eliminating adsorption in some gel filtration separations. Mobile phases with pH < 4 suppress the ionization of free-COOH or SiOH on the packing paticles, making retention by ion exchange less likely. Sample stability may dictate the pH of the mobile phase, and for each separation the optimum working pH must be determined.

In gel filtration separations of especially sensitive solutes such as proteins, addition of other modifiers such as Carbowax 6000, Carbowax 20-M, or ethylene glycol to the mobile phase may be required to eliminate adsorption. For characterizing polyelectrolytes on both chemically modified silica and unmodified silica, it is sometimes necessary to add other modifying agents such as sodium dodecylsulfonate (e.g., 0.5%) to the mobile phase.

Wherever possible, the viscosity of the mobile phase should be kept low to maintain high column efficiency. Detector compatability may also influence the selection of the mobile phase, particularly when the differential refractometer is required.

Operating Variables

SEC separations normally are performed at ambient temperatures, but the temperature should be controlled ($\pm 1°C$) for maximum MW accuracy. Elevated temperatures must be employed for certain polymers (e.g., polyethylene, certain polyamides) because of the solubility requirements. Higher temperatures may be restricted for gel filtration separations of many biologically active macromolecules, and separations near ambient temperature are common. Less than ambient temperatures are required only for very labile solutes.

Samples should generally be dissolved in the solvent that is used as the mobile phase. However, in special cases the sample can be dissolved and injected in a solvent different from the mobile phase, as long as it is completely miscible.

Optimum mobile-phase flowrate in SEC depends on packing particle size, column internal diameter, operating temperature, and mobile phase. However, convenient flowrates are about 1-2 ml/min for 0.4-0.8-cm i.d. columns of 5-10-μm porous particles.

Sample weight permitted in SEC generally is a function of the solute and its solubility in the mobile phase. Typically, a maximum solute concentration of 0.25% is used, with 0.1% solutions often preferred. Injected sample volume generally should not exceed one-third of the baseline volume of a totally permeating monomer peak. However, larger volumes are tolerated if some resolution can be sacrificed.

REFERENCES

1. H. Determann, *Gel Chromatography: Gel Filtration, Gel Permeation, Molecular Sieves,* Springer-Verlag, New York, 1968.
2. I. Fisher, *An Introduction to Gel Chromatography,* North-Holland, Amsterdam, 1969.

3. K. H. Altgelt, *Theory and Mechanism of Gel Permeation Chromatography*, *Adv. Chromatogr.*, **7**, 3 (1968).

4. H. Determann and J. E. Brewer, "Gel Chromatography," in *Chromatography*, E. Heftmann, ed., 3rd ed., Van Nostrand-Reinhold, New York, 1975.

5. K. J. Bombaugh, in *Modern Practice of Liquid Chromatography*, J. J. Kirkland, ed., Wiley-Interscience, New York, 1971, Chap. 7.

6. W. W. Yau, J. J. Kirkland, and D. D. Bly, *Modern Size-Exclusion Chromatography*, Wiley-Interscience, New York, 1979.

7. F. E. Regnier and R. Noel, *J. Chromatogr. Sci.*, **14**, 316 (1976).

8. F. E. Regnier, U. S. Patent 3,983,299, September **28**, 1976.

9. W. W. Yau, C. R. Ginnard, and J. J. Kirkland, *J. Chromatogr.*, **149**, 465 (1978).

10. J. J. Kirkland, *J. Chromatogr.*, **125**, 231 (1976).

11. W. W. Yau, J. J. Kirkland, D. D. Bly, and H. J. Stoklosa, *J. Chromatogr.*, **125**, 219 (1976).

12. A. P. Graffeo, Association of Official Analytical Chemists Meeting, Washington, D.C., October 19, 1977.

12a. A. R. Cooper and D. P. Matzinger, in *Chromatography of Synthetic and Biological Polymers*, Vol. 1, R. Epton, Ed., Ellis Horwood, Ltd., Chichester, Eng., 1978, p. 354.

13. K. W. Williams, *Lab. Pract.*, **21**, 667 (1972).

14. F. A. Buytenhuys and F. P. B. van der Maeden, *J. Chromatogr.*, **149**, 489 (1978).

15. F. P. B. van der Maeden, P. T. van Rens, and F. A. Buytenhuys, *ibid.*, **142**, 715 (1977).

16. I. Schechter, *Anal. Biochem,*, **58**, 30 (1974).

17. H. D. Crone and R. M. Dawson, *J. Chromatogr.*, **129**, 91 (1976).

18. R. Vivilecchia, B. G. Lightbody, N. Z. Thimot, and H. M. Quinn, *J. Chromatogr. Sci.*, **15**, 424 (1977).

19. J. J. Kirkland and P. E. Antle, *ibid.*, **15**, 137 (1977).

20. M. J. Frenkel and R. J. Blagrove, *J. Chromatogr.*, **111**, 397 (1975).

21. A. C. M. Wu, W. A. Bough, E. C. Conrad and K. E. Alden, Jr., *ibid.*, **128**, 87 (1976).

22. A. E. Hamielec and A. C. Ouano, *J. Liq. Chromatogr.*, **1**, 111 (1978).

23. Z. Grubisic, P. Rempp, and H. Benoit, *J. Polym. Sci.*, Polym. Lett. Ed., 753 (1967).

24. J. Brandrup and E. H. Immergut, ed., *Polymer Handbook*, 2nd Ed., Wiley-Interscience, New York, 1975, Chap. IV.

25. J. G. Hendrickson and J. C. Moore, *J. Polym. Sci.*, Polym. Chem. Ed., **4**, 167 (1966).

26. J. G. Hendrickson, *Anal. Chem.*, **40**, 49 (1968).

27. W. B. Smith and A. Kollmansberger, *J. Phys. Chem.*, **69**, 4157 (1965).

28. A. Lambert, *J. Appl. Chem.*, **20**, 305 (1970).

29. A. Lambert, *Anal. Chim. Acta.*, **53**, 63 (1971).

30. C. A. Tolman and P. E. Antle, *J. Organomet., Chem.*, **159**, C5 (1978).

31. J. C. Giddings, *Anal. Chem.*, **39**, 1027 (1967).

32. R. A. Henry, S. H. Byrne, and D. R. Hudson, *J. Chromatogr. Sci.*, **12**, 197 (1974).

33. P. E. Antle, E. I. du Pont de Nemours & Co., private communication, 1976.

34. *Du Pont Instruments Liquid Chromatography Review*, "Size-Exclusion Chromatography," Bulletin E-14063, 1977.

35. J. Cazes and N. Martin, "Solution of Materials Problems in Forest Products," in *Liquid*

Chromatography of Polymers and Related Materials, J. Cazes, ed., Dekker, New York, 1977, p. 121.

36. V. F. Gaylor and H. L. James, *Anal. Chem.*, **48**, 44R (1976).

36a. R. E. Majors and E. L. Johnson, *J. Chromatogr.*, in press.

37. A. Krisher and R. G. Tucker, *Anal. Chem.*, **49**, 898 (1977).

38. S. Singh and A. E. Hamielec, *J. Liq. Chromatogr.*, **1**, 187 (1978).

39. S. Rokushika, T. Ohkawa, and H. Hatano, Joint U.S.-Japan Seminar on Modern Techniques of Liquid Chromatography, Boulder, Colo., June 28-July 1, 1978.

BIBLIOGRAPHY

Ackers, G. K., *Adv. Protein, Chem.*, **24**, 343 (1970) (analytical aspects of traditional GFC).

Anderson, D. M. W., "Gel Permeation Chromatography," in *Practical High Performance Liquid Chromatography*, C. F. Simpson, ed., Heyden, London, 1976 (recent review review of SEC practice, including some high-performance).

Andrews, P., *J. Biochem.*, **96**, 595 (1965) (review of classical gel chromatography procedures for determining protein molecular weights).

Billingham, N. C., "Characterization of High Polymers by Gel Permeation Chromatography," in *Practical High Performance Liquid Chromatography*, C. F. Simpson, ed., Heyden, London, 1976 (recent comprehensive review of molecular-weight determination by GPC).

Bombaugh, K. J. "The Practice of Gel Permeation Chromatography," in *Modern Practice of Liquid Chromatography*, J. J. Kirkland, ed., Wiley-Interscience, New York, 1971 (discussion primarily of the chromatographic aspects of classical GPC).

Determann, H., *Gel Chromatography*, 2nd ed., Springer-Verlag, New York, 1969 (laboratory handbook of conventional gel chromatography, with emphasis on GFC).

Epton, R., ed., *Chromatography of Synthetic and Biological Polymers,* Vol. 1, Ellis Horwood Ltd., Chichester, England, 1978 (reviews and research papers-many on modern SEC).

Hjerten, S., "Molecular Sieve Chromatography of Proteins," in *New Techniques in Amino Acid, Peptide and Protein Analysis,* A. Niederweiser and G. Pataki, ed., Ann Arbor-Humphrey Science Publishers, Ann Arbor, Mich., 1973 (conventional GFC of proteins).

Quano, A. C., and J. F. Johnson, "Gel Permeation Chromatography" in *Polymer Molecular Weights,* Part II, P. E. Slade, Jr., ed., Dekker, New York, 1975 (recent review of the more classical aspects of GPC and excellent list of applications).

Yau, W. W., J. J. Kirkland, and D. D. Bly, *Modern Size-Exclusion Chromatography,* Wiley-Interscience, New York, 1979 (comprehensive book on modern SEC).

THIRTEEN

QUANTITATIVE AND TRACE ANALYSIS

13.1   Introduction   542
   Sources of Error   542
     Sampling   543
     Chromatographic Separation   543
     Detection   544
     Peak Measurement   545
13.2   Peak-Size Measurement   545
   Peak Height   545
   Peak Area   546
13.3   Calibration Methods   549
   External Standard   549
   Internal Standard   552
13.4   Selection of Calibration Method   556
   Peak Height or Peak Area?   556
   Isocratic versus Gradient Elution   559
   Conclusions   560
13.5   Trace Analysis   560
   Introduction   560
   Factors Affecting Trace Analysis   561
     Column Resolution   561
     Sample Injection   564
     Detection   566
     Calibration   570

          Sample Pretreatment   572
          Preferred Conditions for Trace Analysis   572
References   573

13.1 INTRODUCTION

The ability to analyze quantitatively a wide variety of materials is probably the most important aspect of modern LC. Prior to 1968, column chromatography generally was difficult to use for precise quantitative analysis. However, modern equipment has made LC a highly accurate and precise technique. The precision of results now obtainable by LC is comparable or superior to that provided by GC and exceeds that obtainable by other high-resolution separation methods (e.g., TLC, electrophoresis). Modern LC can be used both for the high-precision assay of major components in simple samples and for the analysis of minor or trace constituents in complex mixtures. Under carefully controlled, optimum conditions, assays can be performed with standard deviations (1σ) of $\pm0.25\%$, relative (*1*), and parts per billion (ppb) trace analyses are now commonplace. A particular advantage of quantitative analysis by LC is that the extent of sample separation (sample cleanup and derivative formation) required prior to performing the final measurement is usually less than that for analogous GC methods.

Sources of Error

In quantitative LC, several possible sources of error exist:

- Sampling.
- Choice of chromatographic parameters.
- Detection.
- Measurement and calibration.

The *accuracy* of results is dependent on calibrating the system with reliable standards and minimizing overlap of component bands. The ability to define and control instrumental operating conditions for a particular analysis determines the *precision* of results; that is, how reproducible the results will be at different times, between different analysts, and at different locations. Precise and accurate quantitative analyses can only be obtained when careful attention is given to all phases of the analysis, from initial sampling to final calculation. Let us now examine each source of error, so that LC procedures can be designed to obtain the desired results.

Sampling. Good LC analyses depend first of all on the collection of the sample. For example, any portion of a single liquid phase may be considered as representative, provided the sample has been thoroughly mixed before taking an aliquot. With bulk materials (e.g., foods) that are not homogeneous, it is more difficult to ensure a representative sample. In these cases, suitable procedures involve collecting the largest practical sample, thoroughly mixing, and "quartering" according to well-recognized procedures (2). With this approach a small sample that is representative of the total can be obtained. When collecting samples, it is recommended where possible to take several subsamples from each source. These can then be analyzed individually and averaged; marked deviations of results from the average indicate a possible sampling problem.

Proper sampling also involves the accurate preparation of solutions for analysis. Samples introduced into the LC instrument must be totally dissolved and homogeneous; techniques for filtering samples are discussed in Section 5.6. Centrifugation sometimes is an acceptable alternative for removing particulates from a suspension. Solutions must be prepared to ±0.1% accuracy to meet the requirements of precise LC analyses, which means that accurately calibrated volumetric flasks and pipettes of no less than 25-ml and 10-ml, respectively, should be used with appropriate technique and at a controlled temperature.

In quantitative analysis, the solvent used to dissolve the sample should be, if possible, the initial mobile phase. This minimizes changes in band shape and k' values that can adversely affect precision, and it eliminates any possibility of sample precipitation on injection. When it is not possible to dissolve the sample in the mobile phase, the solvent used should be miscible with the mobile phase, and only small sample volumes (e.g., <25 μl) are permitted; in such cases, somewhat poorer quantitative results are likely.

Valves with sampling loops are always preferable to syringe injection because of better precision, less operator dependency, and greater convenience. Modern high-pressure microsampling valves (Section 3.6) can yield overall precisions of <±0.1% under optimum conditions. For routine analyses, automated sample-injection equipment should be considered (Section 3.6), because highest analysis precision is obtained with this approach. The relationships discussed in Section 7.4 predict the maximum sample volume allowed without significantly affecting column resolution. Rarely is <25 μl of sample injected for precise quantitative analysis, and 50-200-μl volumes are typical. Larger sample volumes increase the precision of the sampling and often reduce the need for operating at higher detector sensitivity, where noise and drift make quantitative measurements less precise.

Chromatographic Separation. The LC separation itself can lead to errors in quantitation. Although sample decomposition is rare in modern LC, problems should be suspected if a linear increase in solute peak size is not found for a linear increase in concentration with a series of standards of known concentrations.

Potential errors in quantitation due to tailing peaks are common and are best avoided by selecting a better column or separating system that produces symmetrical peaks (see Section 19.3 for discussion of tailing peaks). Odd-shaped or spurious peaks of any type should be avoided for quantitation. It is particularly important *not* to use peaks at (or near) t_0 for quantitation, because insufficient resolution can result in large errors. Generally, quantitative analyses are carried out only with peaks of $k' \geqslant 0.5$. Errors due to overlapped peaks can be further minimized by ensuring adequate separation, as mentioned in the following discussion.

The type of LC column can significantly influence analytical results. As discussed in Section 5.1, small-particle columns produce higher resolution and sharper peaks and are thus favored for complex mixtures and trace analysis. However, columns of $\sim$30-μm pellicular packings are often well suited for routine analyses of less complex systems, as discussed in Section 5.2.

Best quantitative results are obtained when the sample load is well below the capacity limit for the column, so that peak retention times and column plate count do not significantly change with variation in sample size. Highest column efficiency and peak-height reproducibility occur at about 10^{-5}-10^{-6} g of solute per gram of most types of packing (*3*). If adequate detection is a problem, larger samples (10^{-4}-10^{-3} g/g) can be used with some loss in plate count, as long as resolution is adequate. However, with larger sample sizes, peak-height response may not be linear and peak-area measurements may be required.

Detection. Analysis errors can arise from variations in detector output. As indicated in Section 4.2, detector sensitivity, baseline stability, and linearity are important specifications for quantitative analysis, and these parameters must be maintained stable for best analytical precision. Detector sensitivity is especially dependent on flowrate, as discussed in Section 13.4. Detector response should be linear throughout the concentration range to be used; otherwise, significant errors can result. Too high a sample concentration can produce nonlinear peak heights and areas.

Detector selectivity is desirable for the quantitative measurement of a component in the presence of other materials of no interest, and this is particularly important in trace analysis (see Section 13.5). Detector response generally varies for different compounds; for example, responsive solutes of similar functionality can vary as much as fivefold for electrochemical detectors and up to 10^4 with UV detectors. Therefore, detectors must be calibrated for *each* compound to be measured. It is important to utilize a detector with sufficient sensitivity so that the size of the sample needed for the quantitative measurement will not overload the column.

Peak Measurement. Errors in quantitative analysis can arise during handling of data from the detector. For example, significant errors can result if a recorder is not in correct adjustment. The method of quantitating the measurement also influences the accuracy and precision of the analysis. Two approaches are used for measuring peak size; the first is simply to measure the *height* of the peak; the second involves the measurement of *peak area* with one of several methods. For details on errors associated with band-size measurements in chromatography, consult (*3a-3c*).

13.2 PEAK-SIZE MEASUREMENT

Peak Height

The simplest method involves peak height h' values, measured as the distance from baseline to peak maximum, as illustrated by peak 2 in Figure 13.1. Peak heights are easily determined with a ruler or by counting chart divisions. Some electronic integrators and computers also report peak heights. Baseline drift can be corrected for by interpolating the baseline between the start and finish of the peak; this is shown for peaks 1 and 3 in Figure 13.1. However, a flat, stable baseline is essential for maximum precision, and this should be a first requirement in developing a quantitative LC procedure.

In the discussion of R_s in Section 2.5, peak-height measurements are more accurate than peak areas, because peak heights are less subject to interference by

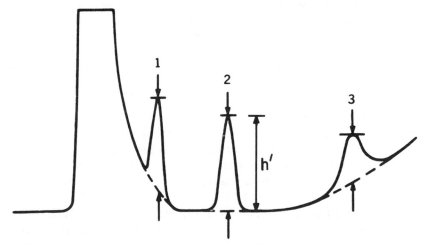

Figure 13.1 Peak-height measurement. Reprinted from (*4*) with permission.

adjacent, overlapping peaks. For $R_s = 1$ and an accuracy of at least ±3%, it is possible to vary the relative heights of adjacent bands by about a factor of 10^3. On the other hand, for the same accuracy, relative peak areas cannot be varied more than ninefold. This means that for equivalent accuracy, less resolution is required for quantitating by peak height than by peak area. Peak heights are almost always used for trace analysis, as discussed later in this chapter.

Peak Area

Normally the precision of methods using peak-area values (A) is less influenced by changes in instrumental and chromatographic parameters (see Section 13.4). In addition, peak-area measurements often are preferred with asymmetrical peaks (as in Figure 19.1a). However, asymmetrical bands should only be used when symmetrical bands cannot be attained. Severely tailing peaks (as in Figure 19.5a) cause special problems with area measurement, and in this case peak heights may yield better precision. A disadvantage of peak-area methods is that they are more affected by neighboring peaks. A rough rule of thumb is that peak area should be used when maximum precision is more important, and peak height when accuracy (or possible interferences) is of major concern.

Various techniques for measuring the areas of chromatographic peaks are illustrated in Figure 13.2. A simple but effective manual technique that measures a quantity proportional to peak area is to determine the product of the peak height times width at one-half the peak height (Figure 13.2a). This is a reasonable approach with peaks that closely approximate a Gaussian distribution but is less satisfactory for unsymmetrical or very broad, low peaks. The precision of the height-width method is much improved if recorder chart speed is increased to the point where peak width can be measured with precision. Measuring peaks on the recorder trace from the inside edge of one side to the outside edge of the other will avoid precision errors due to the recorder pen width, which can be significant in the case of narrow peaks.

The triangulation technique (Figure 13.2b) calculates a quantity proportional to area, using the product of the peak height and width of the peak base; the latter is determined by tangents constructed from the sides of the peaks so that they interesect the baseline. This technique is particularly demanding of operator technique, and therefore is less precise; it exhibits no advantage over method (a).

Areas may be determined by planimeter (Figure 13.2c), a device that mechanically integrates peak area by tracing the perimeter of the peak. The precision and accuracy of this rather tedious procedure is somewhat dependent on operator skill. Although improved precision can be obtained by making repetitive tracings, this makes an already time-consuming technique even more tedious.

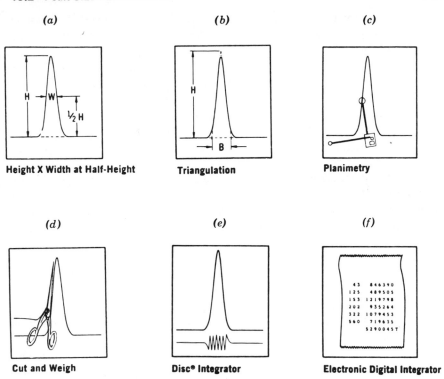

Figure 13.2 Techniques for measuring peak areas. Reprinted with permission of Varian Associates.

On the other hand, this method can be used with peaks that are not symmetrical and it is especially used for manual molecular weight calculations in gel permeation chromatography (5).

Another approach (Figure 13.2*d*) is to cut the peak out of the recorder trace and weigh it. This approach is inconvenient and destroys the original recorder trace (of course, it may be copied), but with method (*c*) it represents another approach to manually determining peak areas for unsymmetrical bands. The precision of this method depends on the weight constancy of the chart paper and the care used in cutting out the peak. Accuracy of the cutting process can be improved by keeping the ratio of peak height to the width at half-height in the range of 1-10.

Some recorders are equipped with mechanical (e.g., ball-and-disc) integrators (Figure 13.2*e*) that automatically produce tracings indicating peak area. This method provides a degree of automation, but the accuracy is limited by the performance of the recorder. The ball-and-disc integrator often can be used with good accuracy for irregularly shaped peaks. However, adjustment for baseline drift and incompletely resolved peaks decreases accuracy and increases the the analysis time.

Digital electronic integrators (Figure 13.2*f*) now are widely used in LC for measuring peak areas. These devices automatically sense peaks and print out the areas in numerical form. The exception is negative peaks (e.g., from refractive index detectors), which generally cannot be handled.

Computing integrators (including "smart" recorders) are even more sophisticated and offer a number of features in addition to basic digital integration. Because these devices have both memory and computing capabilities, they can be programmed for a particular analysis. Thus, it is possible to set the device for a specific analysis by calling up a preset program. Computing integrators generally are more sophisticated in the sensing of a peak, including capabilities to upgrade integrating parameters to maintain accuracy as the separation progresses and eluting peaks become broader. These devices also are better equipped to measure partially resolved peaks and separations in which significant baseline shift occurs. A particular advantage of the computing integrator is its ability to apply detector-response factors to raw peak areas to compensate for differences in detector sensitivity. Such devices are of considerable value for routine quantitative analysis, providing savings in analysis time and improved analytical accuracy.

Peak areas in LC also can be conveniently measured with the many different computer systems that are used in gas chromatography. These are the most sophisticated data-handling systems available. In general, most computer systems are capable of serving gas and liquid chromatographic systems simultaneously. Many of these devices print out a complete report, including names of the compounds, retention times, peak areas, area correction factors, and the corrected weight percent of the various sample components. A detailed discussion of computer-handling of chromatographic data is beyond the scope of this chapter. A recent review of this subject is recommended for further reading (*6*).

Typical precisions that one can expect from LC measurements using the various integration methods are given in Table 13.1. If separation and area-measuring techniques are carried out with extreme care, the values of Table 13.1 can be improved as much as twofold.

In manual methods for measuring peak areas, precision degrades significantly as the ratio of peak height h' to width half-height $W_{1/2}$ becomes extreme. For example, problems occur with very sharp peaks or very flat peaks having $h'/W_{1/2}$ values of greater than 10 and less than 0.5, respectively. Errors in manual area measurements can be kept to a minimum by optimizing the $h'/W_{1/2}$ values in the range of 1-10 via control of detector attenuation and adjustment of chart speed. In all cases, the larger the peak area, the better the precision of the measurement. Changing chart speed does not affect peak-height measurements.

Table 13.1 Precision of peak-area measurement techniques.

Method	Relative precision, 1σ (%)
Planimeter	3
Triangulation	3
Cut and weigh	2
$H \times W_{1/2}$	2
Ball and disc integration	1
Electronic digital integrator	0.5
Computer	0.25

To summarize the preceding discussion:

● For manual data reduction, measurement of peak height is preferred because of its simplicity, speed, and greater accuracy. However, variation in separation conditions may affect peak-height values more than peak areas (see Section 13.4).

● Electronic or computer integration is the fastest and preferred approach for measuring peak areas. These more sophisticated devices allow more accurate measurement of fused peaks and better handling of drifting baselines.

● The technique using height times the width at half-height is the preferred manual area method when symmetrical peaks are involved. The ball-and-disc integrator, planimeter, and cut and weigh methods, in decreasing order of preference, are applicable for asymmetrical peaks.

13.3 CALIBRATION METHODS

External Standard

A calibration plot is first constructed, based on samples (standards) that contain known concentrations (or weights) of the compounds of interest. A fixed volume of each standard solution is then injected and processed as in the assay procedure; peak size (height or area) is plotted versus concentration (or weight) for each compound. An example of such a calibration is shown in Figure 13.3. If the calibration plot is linear and intercepts the origin, the *calibration factor S* (peak size/concentration or alternatively, peak size/weight) is determined as the slope of the calibration plot through the origin. Thus in Figure 13.3, the calibration factor S in terms of peak height (mm) for a constant sample volume would be

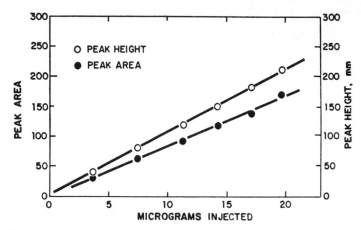

Figure 13.3 Peak-height and peak-area calibrations with external standard. Column, 100 × 0.21 cm, Corasil II, 50-75 μm; mobile phase, 1% isopropanol in hexane; flowrate, 0.95 ml/min; sample, 10.6 μl of antioxidant CAO-14 in mobile phase. Reprinted from (7) with permission.

$$S = \frac{218}{20} = 10.9 \text{ mm}/\mu g.$$

For an unknown sample, the amount of the component of interest Y would be

$$\mu g \text{ of } Y = \frac{h_y(\text{mm})}{S}$$

$$= \frac{h_y}{10.9}$$

where h_y is the peak height for the component of interest. The concentration of Y in the original sample would then be

$$\text{wt } \% \ Y = \frac{\text{wt of } Y \text{ found } (\mu g) \times \left[\dfrac{\text{total sample volume, ml}}{\text{sample volume injected, ml}}\right]}{\text{weight of original sample, g}} \times 10^{-4}$$

$$(13.1)$$

For example, if 10 μg of Y were found in an injected sample volume of 0.100 ml and if the sample solution resulted from dissolution of 0.25 g of original sample in 25 ml of solvent,

$$\text{wt } \% \ Y = \frac{10 \times (25/0.100) \times 10^{-4}}{0.2500} = 1.00.$$

In a properly designed system, calibration plots are linear and extrapolate through the origin. In those rare cases where calibration plots exhibit significant nonlinearity, then the concentration of an unknown can be read directly from the best curve through the calibration points. Interpolation between data points for nonlinear curves should not be attempted unless the calibration points are very precise and close together and the entire calibration curve is rechecked frequently. On the other hand, with linear calibrations only the calibration factor S value needs frequent checking, using single-point measurements.

In all trace analyses conducted by any calibration procedure it is very desirable to carry out analyses on *blank or control samples* (i.e., similar samples that contain none of the compound of interest). In this way, the level of expected interference for each matrix can be established. Analysis of a series of control samples from various sources is valuable in establishing the level of possible interference due to sample variability.

A principal source of error with the external-standard calibration procedure comes from sample injection. When using syringe injection, analysis precision is usually not much better than about ±5% relative, because the results depend greatly on the ability to reproducibly inject the sample into the instrument. However, with a sampling microvalve, overall analysis precision of about 1% can easily be achieved with the external-standard method. By combining valve injection with careful control of instrumental parameters (e.g., as provided by modern digitally controlled instruments), plus frequent calibration, analytical precisions of ±0.25% can be obtained with simple peak-height measurements by the external standardization method. This level of precision is readily achieved with automatic sampling systems (Section 3.6), where operator variability is eliminated and all analysis conditions are precisely reproduced.

The inherent simplicity and flexibility of the peak-height procedure makes this approach more attractive for every kind of LC analysis (e.g., "one-time" assays, trace determinations, and routine assays requiring the highest precision). However, peak-height analysis can in some cases yield nonlinear calibration curves, where the same analysis by peak area gives a linear calibration. In this case, peak-area analysis is usually preferred. Causes of such nonlinearity for peak-height response include tailing bands, column overload (too large a sample), detectors with large response times, and so on.

For less demanding applications, adequate precision is available by chromatographing standards once or twice daily and making appropriate adjustments to the calibration factor S to compensate for changes in instrumental parameters. Changes in the calibration factor S are considered valid when the factor calculated from the most recent run deviates from the previous calibrating factor by more than the 2σ precision of the method. At this point a new value of S should be determined as the average of several new standard runs. For highest precision, using automatic sampling devices (Section 3.6), standard solutions should be run between every pair of unknowns and the calibration factor recalculated for every analysis. With this approach, the average S value from a designated number (i.e., >6) of previous standard runs is used for analyzing unknowns; any single S value deviating by more than the 2σ precision of the previous standards is not used in calculating the average S value used for a particular analysis.

Calibrations by peak area are prepared and used in the same manner as those used for peak heights. With frequent calibration, peak-height methods have the same reproducibility as peak-area methods. However, because peak areas are generally less affected by changes in some separation parameters (see Section 13.4), the reproducibility of peak-area methods often is better than peak heights when infrequent calibration (e.g., once daily) is used. However, as previously noted, the accuracy of peak-area methods is inferior to that obtainable with peak heights when overlapping or trace-level peaks are involved, and for this reason, peak-height calculations are almost always preferred for trace analysis. The precision of both peak-height and peak-area analyses can be improved by running replicates, but with a sacrifice in overall analysis time.

The *peak-normalization* calibration method used widely in gas chromatography is not generally utilized in LC analysis. The method is based on measuring the area of every peak in the chromatogram and reporting the percent area for each component based on the sum of all peak areas. This approach should only be used in LC after detector-response correction factors are applied to allow for the differences in the S values of each component in the sample.

A variation on simple peak-height and peak-area calibration, the *method* of *standard additions*, is generally appropriate for analysis of certain minor and trace components and therefore is discussed later in this chapter.

Internal Standard

A widely used technique of quantitation using both peak heights and peak areas involves the addition of an *internal standard* to compensate for various analytical errors. With this approach a known compound at a fixed concentration is added to the unknown sample to give a separate peak in the chromatogram. This known compound is used as an internal marker to compensate for the effect of minor variations in separation parameters on peak size, including sample-size fluctuations. However, because the delivery of sample volumes is quite precise with microsampling valves, the main utility of the internal standard technique in LC is in assays that require sample pretreatment (and/or solute derivatization) where variable recoveries of compounds of interest may occur. It is not widely appreciated that use of an internal standard actually *increases* the analysis precision error by $\sqrt{2} = 1.4$ times, compared to the external calibration method with frequent calibration, because of the uncertainty of *two* peak-size measurements rather than one.

To compensate for losses of the compound of interest during sample workup, an internal standard that is structurally similar to the compound(s) of interest is added at a known concentration to the original unknown sample, the pretreatment is carried out, and the resulting sample is analyzed. In this approach any loss of the component of interest will (it is hoped) be accompanied by the loss of an equivalent fraction of internal standard. The accuracy of this approach is obviously dependent on the structural equivalence of the compound(s) of interest and the internal standard [i.e., for best results the internal standard and the compound(s) of interest should extract equally, react equally, etc.].

Internal-standard calibrations are constructed by chromatographing appropriate volumes of calibration mixtures containing the compound(s) of interest with a constant concentration of the internal-standard compound. Peak heights or peak areas of the compounds of interest are determined, and the compound to internal-standard peak-height or peak-area ratios are plotted versus the con-

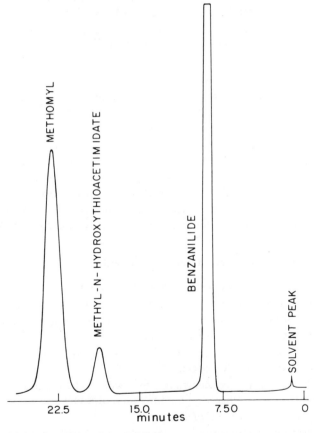

Figure 13.4 Separation of calibration mixture containing Lannate methomyl insecticide with internal standard. Column, 100 × 0.21 cm, 1% β,β′-oxydipropionitrile on Zipax, <37 μm; mobile phase, 7% chloroform in n-hexane; flowrate, 1.3 ml/min; detector, UV, 254 nm, 0.8 AUFS; sample, 20 μl. Reproduced from (8) with permission.

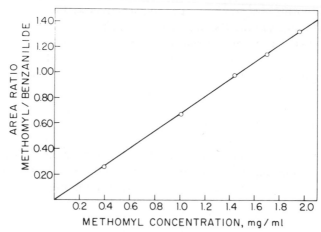

Figure 13.5 Peak-area ratio calibration with internal standard. Conditions: Same as in Figure 13.4; methomyl = *S*-methyl-*N*-[(methylcarbamoyl)oxy-thioacetimidate. Reprinted from (*8*) with permission.

centration of the compound of interest. If calibration mixtures have been prepared properly, this calibration plot is linear and intercepts the origin. Figure 13.4 is a chromatogram of an internal-standard calibration mixture for analyzing technical samples and formulated mixtures containing Lannate methomyl insecticide and a minor precursor. Benzanilide was used as an internal standard and peak-area ratio measurements were made by electronic integration. Figure 13.5 shows the peak-area ratio calibration curve used for this analysis.

The selection of the internal standard is critical for both peak-height and peak-area measurements, and the requirements are summarized in Table 13.2.

Table 13.2   Requirements for an internal standard.

- It must have a completely resolved peak; no interferences
- It must elute close to compound(s) of interest (similar k' values)
- It must behave equivalently to compound(s) of interest for analyses involving pretreatments, derivative formation, etc.
- More than one internal standard may be required for multicomponent mixtures to achieve highest precision.
- It must be added at a concentration that will produce a peak-area or peak-height ratio of about unity with compound(s) of interest
- It must not be present in original sample
- It must be stable; unreactive with sample components, column packing, or mobile phase
- It is desirable for it to be commercially available in high purity

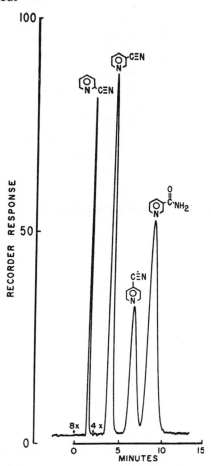

Figure 13.6 Analysis of substituted pyridine isomers with internal standard. Column, 100 × 0.21 cm; Zipax-SCX (strong cation ion exchanger), <37 μm; mobile phase, 0.1 N sodium nitrite/0.1 N phosphoric acid; flowrate, 1.7 ml/min; temp., 27°C; sample, 5 μl; detector, UV, 254 nm; attenuator settings, 8X and 4X = 0.08 and 0.04 AUFS, respectively; last peak, internal standard. Reprinted from (9) with permission.

A practical problem with the internal standard technique is that the standard must be located in a "vacant" region in the sample chromatogram. For simple mixtures such as that shown in Figure 13.4, this is usually not difficult. However, for complex samples the selection of an internal standard can be tedious, and in such cases, quantitation may be restricted to the external-standard method. It should also be kept in mind that unexpected interferences can adversely affect analysis accuracy by overlapping *either* the assay-peak *or* the internal standard band; this effectively doubles the chance of error from this source, when using an internal standard.

A satisfactory internal standard often is a compound that is structurally related to the component to be measured (e.g., an isomer or close homolog). The internal standard should have similar k', solubility, and detection response but be adequately separated from other sample components. In situations where such closely similar compounds are not available, selection (meeting the criteria in Table 13.2) should be made on the basis of similar solubility in the mobile phase and similar detector response. Figure 13.6 illustrates the use of the internal standard technique using the related compound nicotinamide for the analysis of the isomeric cyanopyridines by ion-exchange chromatography.

13.4 SELECTION OF CALIBRATION METHOD

Peak Height or Peak Area?

The choice of peak-height or peak-area measurements requires an understanding of the effect of chromatographic parameters on the precision of each approach. Detector response or the size of a chromatographic peak is proportional to the weight of the solute contributing to that peak. The detectors in common use in modern LC are all concentration-dependent devices; that is, they respond to sample concentration and are independent of sample flowrate (e.g., if the mobile-phase flowrate is stopped, detector signal remains essentially unaffected). Based on the inherent characteristics of concentration-dependent detectors, we can develop some guidelines for predicting the effect of changing various chromatographic parameters on the detector response and, consequently, on the resulting quantitative analysis. Table 13.3 summarizes the influence of certain chromatographic parameters on the precision of quantitative analyses in LC with the concentration-dependent detectors that are now widely employed. *These*

Table 13.3 Effect of chromatographic parameters on precision of quantitative analysis in LC using concentration-dependent detector.

Changing parameter	Approximate effect on quantitation method	
	Peak height, h'	Peak area, A
k'	$\dfrac{1}{1 + k'}$	No change
N	$N^{1/2}$	No change
u	$\leq u^{-0.2}$	$\dfrac{1}{u}$

Table 13.4 Preferred quantitation method for changing LC parameters.

Changing experimental condition	Possible cause	Parameter changed	Quantitation method preferred	
			Area	Peak height
Mobile phase/stationary phase	• Gradient elution; mobile-phase fractionation • Change in adsorbent activity • Loss of stationary phase	k'	x	
Velocity	• Pumping imprecision, flowrate change	u		x
Column efficiency	• Compression of column bed • Loading of column inlet with strongly retained components • Column packing degradation	N	x	
Temperature	• Column not thermostatted	T	x	
Peak shape	• Non-Gaussian peaks from chemical effects, slow detector response, poorly packed column, etc.	—	x	x[a]
Sample volume	• Irreproducible injection	V_s	—	—

[a]For badly tailing peaks.

relationships assume that the reduced velocity v is greater than approximately 20. Based on these relationships, we can then evaluate the effects of varying experimental conditions and from this determine which quantitative approach, peak height or peak area, is preferred. Table 13.4 indicates the preferred quantitative method when various LC separation parameters are subject to change. These results indicate that if control of flow is adequate and solvent composition cannot be maintained precisely (e.g., highly volatile solvents or in gradient elution), peak-area measurements are better because they are relatively independent of mobile-phase composition (conversely, changes in k' caused by variation in mobile-phase composition significantly affect peak heights). However, if detector response varies for any reason (e.g., change in UV absorption maximum with change in mobile-phase composition), the peak-area response can also change. On the other hand, if the composition can be maintained precisely (as in isocratic operations) but control of mobile-phase flowrate is poor, peak-height measurements generally yield more precise quantitative results, because peak heights are less dependent on flowrate than peak areas, as illustrated in Figure 13.7. Peak heights are particularly favored when working at the approximate column plate height minimum (reduced velocity of 3-5, see Figure 5.1), for in this range, peak heights are virtually unaffected by minor flowrate changes. In applications where frequent standardizations are not feasible or where column efficiency can be expected to change over a period of time (e.g., routine in-line process-analyzer), peak areas generally will yield more precise data (however, greater resolution is required to maintain accuracy, as discussed previously). When temperature is uncontrolled or peak shape is non-Gaussian, peak-area

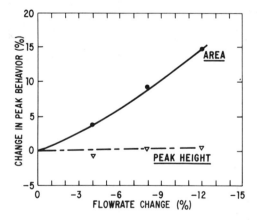

Figure 13.7 Effect of flowrate on peak height and peak area in isocratic analysis. Column, 25 × 0.31 cm, Spherisorb ODS, 10 μm; mobile phase, water/methanol (1:1); initial flowrate, 2.00 ml/min; temp., 40°C; detector, UV, 254 nm; sample, 10 μl, propyl *p*-hydroxybenzoate ($k' = 4.8$). Redrawn from (*11*) with permission.

measurements often are preferred. However, peak heights should be used for badly tailing peaks (but only when better peaks cannot be produced by appropriate action!), because peak areas are then difficult to measure accurately even by electronic integration. With unreproducible sample-volume injections, both methods suffer equivalent imprecision.

Isocratic versus Gradient Elution

For the most precise quantitative data, isocratic separations are generally preferred. Gradient elution only should be used in instances when other approaches are not adequate. Although fairly reproducible gradients are characteristic of modern LC instruments, use of gradient elution in quantitative analysis involves several practical disadvantages. First, gradient elution often results in baseline shifts that make the measurement of peak height or area more difficult. Second, spurious peaks can arise during gradient elution, because of solvent impurities (such as impure water; see Figure 7.19) or sharp peaks resulting from solvent demixing (Figure 16.8). Third, quantitative reproducibility is generally poorer because of difficulties in maintaining constant separation conditions compared to isocratic approaches. Finally, between successive chromatograms in gradient elution, additional time is required for the column to be reequilibrated with the initial mobile phase. This process often can double the time required for each analysis.

Just as in isocratic separations, in gradient elution peak area (and retention time) increases significantly with decrease in flowrate, whereas peak height varies much less. Also, as in isocratic elution, a change in mobile-phase composition in gradient elution leads to decrease in peak height with increase in retention, whereas peak area is unaffected by change in the mobile phase.

To summarize, in quantitation by gradient elution the fundamental need is to control carefully total flowrate when areas are measured and to maintain the mobile-phase gradient constant for peak heights. In instruments with high-pressure mixing systems, it is necessary to control the delivery precisely from both pumps so as to maintain constant flowrates and mobile-phase gradients. These parameters must be held constant for each separation in a series of gradient elution runs. This is difficult with certain types of pumping systems (*12*). Constancy of mobile-phase delivery, both in volume flowrate and solvent strength, is not easily maintained for high-pressure mixing gradient systems with two pumps; therefore, the most precise quantitation may not be possible with these devices. Recent studies suggest that certain low-pressure gradient elution systems with a single pump (Section 3.5) can produce more precise quantitative analyses than two-pump high-pressure mixing gradient systems (*13, 14*). For example, retention time, peak-height, and peak-area reproducibility for modern low-pressure mixing systems are almost equivalent to isocratic operation (indicating precise formation of solvent gradients); see Table 3.4. Where applicable, coupled columns (Section 16.2) with an isocratic system are clearly preferred over gradient elution for the routine analysis of mixtures containing a wide range of k' values. In situations where gradient elution must be used, simple step-gradients involving a sudden or step-change in solvent strength (simple or multiple) are sometimes suitable for routine analysis (*15*).

Conclusions

- Peak-height measurement with external standards is the recommended method of LC quantitation for most applications. In its simplest form this method produces fairly precise results with minimum effort, and it is particularly suited for trace analysis (see Section 13.5). With close instrumental control and frequent recalibration the method also produces highly precise analysis (≤0.25% relative is possible, ~1%, relative is typical).

- Peak-height measurements are more accurate than for peak areas but generally are somewhat more affected by variation in separation parameters (other than flowrate).

- Peak-area calibration may be desirable for analysis of simple, well-resolved mixtures, particularly with moderately asymmetrical peaks or if only simple LC equipment is available.

- Peak-height measurements require less resolution of components and are better suited for analysis of more complex mixtures and for trace analysis.

- Internal standard calibration should be considered if sample pretreatment (or derivitization) results in variable and less than complete recovery of the component(s) of interest.

- Quantitation by gradient elution usually should be utilized only for less precise or nonroutine, survey-type analysis of mixtures containing components of a wide k' range.

- Isocratic coupled-column techniques provide faster, more precise analyses of mixtures and are much better suited for routine analysis of samples with wide k' values.

13.5 TRACE ANALYSIS

Introduction

There are several reasons why LC is a powerful and widely used technique for the analysis of trace (≤0.01%) components. First, because of the wide variety of selective interactions in LC, the high resolving power needed for trace analysis in complex mixtures often is readily available. Second, some trace analyses are best carried out by modern LC because of the availability of a number of very selective devices (e.g., amperometric, fluorometric) that uniquely detect certain components at very low concentrations. Third, LC (unlike GC) results in sample concentration, rather than sample dilution, in many cases. Finally, there is often less sample cleanup required prior to LC analysis than in other methods (e.g., GC); in favorable cases, samples can be injected without prior treatment or sample derivitization.

Several experimental factors have a significant effect on the sensitivity, accuracy and reproducibility of LC trace analyses. These include column resolution, chromatographic sample injection technique, detection, analytical calibration procedure, and pretreatment of the sample. The influence of each of these factors on trace analyses by LC is examined next. For detailed treatments on the influence of separation parameters and sample volume on detection limits in LC, see also (*10*) and (*16*).

Factors Affecting Trace Analysis

Column Resolution. Frequently the analytical problem is to find a trace component in a complex mixture, sometimes composed of components with widely varying k' values. It is often feasible with reverse-phase separations to elute the bulk of the sample constituents at or near t_0, with the trace constituent being sufficiently retained for accurate measurement without interference. When two peaks are close together, measurement of the trace component can be most accurate when the peak elutes prior to the major constituent in the sample. Unfortunately, the trace constituent often elutes instead on the trailing edge of a principle peak, seriously interfering with the determination. In some cases the trace component may even be completely masked by early-eluting chemical "garbage," and resolution must be substantially increased to permit measurement.

The three terms of the resolution equation (Eq. 2.10) can be regarded as effectively independent, and each parameter can be varied and optimized separately. Thus, α, N, and k' can be adjusted to allow the trace analysis to be carried out with highest sensitivity and accuracy. Quantitative analysis in trace LC is usually carried out by peak-height measurements, because this approach permits highest accuracy (freedom from possible interfering components) and adequate precision (see Section 13.2). Analysis sensitivity can be improved by adjusting the chromatographic parameters α, k', and N to provide maximum peak heights for the trace components (i.e., the highest possible concentration in the column effluent) commensurate with adequate resolution from possible interferences.

During isocratic separations where the sample is injected in the mobile phase as solvent, the chromatographic column invariably dilutes the sample and makes the trace component more difficult to detect. The practical answer to this problem is to obtain sufficient separation of the trace component, but with a minimum of dilution for maximum sensitivity. The amount that the injected sample is diluted can be approximated by (16):

$$\frac{c_{max}}{c_0} = \frac{V_s(N)^{0.5}}{V_r(2\pi)^{0.5}},$$ (13.2)

where c_{max} is the concentration at the peak maximum, c_0 is the initial concentration in the sample, V_s is the injected volume, N is the column plate count and V_r is the retention volume of the trace component. However, Eq. 13.2 is inaccurate when c_{max}/c_0 values exceed about 0.8.

For highest sensitivity, c_{max}/c_0 (which is proportional to peak height, h') should be maximized. Extent of sample dilution can be easily estimated for a particular separating system. First, the solute retention volume and the column plate count is measured from a test injection using isocratic conditions. For example, if for a 15 × 0.4-cm-i.d. column packed with 5-μm particles the retention volume of the solute is 6.0 ml and the column plate count is 10,000 for an injected volume of 100 μl, then

$$\frac{c_{max}}{c_0} = \frac{0.1 \times (10,000)^{0.5}}{6 \times (2\pi)^{0.5}} = 0.66.$$

Thus, under these conditions the sample has been diluted by about one-third, a typical level of dilution for a trace analysis in which the peak is eluted at a low k' value (e.g., 0.5-1.5); greater dilution occurs at larger k' values. Thus, the minimum detectable concentration of the trace component can be calculated if the absolute detector response for a particular concentration and the detector noise level are known.

Equation 13.2 predicts that increased sensitivity for a trace component (decreased dilution) is obtained by (a) using a column with a high plate count, (b) increasing the injection volume, and (c) reducing the retention volume of the trace component. However, these variables are not independent. For example, the column plate number N will eventually decrease as the injection volume V_0 is increased (Section 7.4), and if N is increased by using a longer column, then V_R will also increase. Therefore, Eq. 13.2 should not be interpreted or applied too literally.

Equation 13.2 further indicates that c_{max}/c_0 (or h') is related to the square root of the column plate number. The most effective means of increasing peak height by increasing N is to use efficient columns of <10-μm particles. For trace analysis short columns (10-15 cm) with 5-6-μm particles should be utilized whenever possible. In this case, column volume and V_R values are minimized and the concentration of the eluting trace component is maximized, to increase detection sensitivity.

Because column plate count N depends on the velocity of the mobile phase u, it can be shown that when the reduced velocity v is 20 or greater (see Section 5.7), then (10):

$$h' \propto u^{-0.2} . \qquad (13.3)$$

However, in the reduced velocity range of about 3-10, which is commonly used with columns of <10-μm particles, peak height changes very little, and it is in this range that maximum peak-height reproducibility and optimum peak-height response occur. In Figure 13.8a is shown the effect of decreasing mobile-phase velocity to increase solute peak heights. From this we can conclude that lower mobile-phase velocities should be used for high sensitivity in trace analysis. If the analysis time is not critical, reduced velocities of about 5 produce highest sensitivity. This roughly corresponds to about 0.2 cm/sec and 0.08 cm/sec for columns of 5-μm and 10-μm particles, respectively.

The sensitivity of trace analysis by peak-height measurement is particularly affected by the retention V_R or the capacity factor k' of the solute peak. It can be shown that (10):

$$h' \propto (1 + k')^{-1} . \qquad (13.4)$$

Figure 13.8b shows the significant effect of increasing peak height by decreasing k' (e.g., by changing the strength of the mobile phase). Thus, although k' values

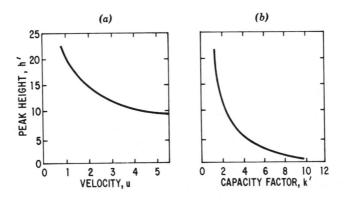

Figure 13.8 Effect of mobile-phase velocity and k' on peak height. (a) Liquid-solid chromatography, 25 cm × 0.32-cm-i.d. porous silica microsphere, 8-9 μm, 7.5-nm pores; mobile phase, dichloromethane (50% water saturated); test solute, benzhydrol, $k' = 2$. (b) Liquid-liquid chromatography, 100 cm × 0.21 cm i.d., 3% β,β'-oxydipropionitrile on <38 μm Zipax; mobile phase, hexane; solute, 2,6-dimethylphenol.

of about 1-10 are optimum in terms of resolution, Eq. 13.2 suggests that small k' and V_r values can be utilized to increase peak height and sensitivity. If possible, trace analyses using peak-height measurements should be carried out with solute of interest having lower k' values, preferably in a range of 0.5-1.5. However, resolution still must be adequate from "garbage" eluting at or just after t_0 to achieve accurate analyses. In reverse-phase LC it is often possible to arrange the system so that unwanted components elute at t_0 and the desired component elute without interference in a h' range of 2-5, with few components eluting afterward. This approach is particularly desirable, because long periods of purging strongly retained "garbage" from the column are not required.

In summary, to maximize peak height and sensitivity, k' values should be kept as small as possible (but with adequate component resolution) and N should be maximized and V_R minimized by using high plate count, short, small-particle columns (e.g., 10-15 cm of 5-μm particles).

An increase in the separation factor α value is a powerful technique for improving the accuracy of a trace LC analysis. Serious overlap of a trace component peak with that of a major component often can only be overcome by changing the selectivity of the system. This is accomplished by varying the mobile-phase and/or stationary-phase combination and useful guidelines for improving α in separations are given in Chapters 7-11 for each of the LC methods.

If the increase in α and resulting increase in resolution are sufficient, analysis time can then be decreased by reducing the column length (usually better for trace analysis) or by increasing mobile-phase velocity (e.g., p. 64). It can be further shown that for a particular resolution value (16):

$$\frac{c_{max}}{c_0} \propto \left(\frac{\alpha-1}{\alpha}\right),\qquad(13.6)$$

assuming constant H and k'. In difficult separations (α approaches unity), the selectivity effect can significantly affect peak dilution. For example, if α is improved only from 1.01 to 1.025, peak dilution can be reduced 2.5 times (constant k') by appropriately shortening the column to maintain constant R_s.

Sample Injection. For maximum sensitivity, as large a sample volume as possible should be injected. Ideally, sample injection should be as a plug or sharp band that displaces the mobile phase at the column inlet. Poor design of the injection system or improper injection technique can lead to unwanted mixing of the sample solution and the mobile phase and can reduce peak heights. Sample valves with large-volume loops are particularly convenient for trace analysis.

Smaller S values (decreased peak height per concentration unit) will be obtained if too large a sample volume is introduced into the column, because of band broadening. Generally, maximum sample size can be predicted by Eq. 7.2, although larger sample volumes can be used if resolution is still adequate. A good rule of thumb is to start with a sample volume that is one-fifth of the volume of the peak of interest (measured with a very small sample volume). The sample volume then is increased until the expected increase in peak height per concentration unit (or resolution from neighboring components) becomes limiting. If too large a sample mass is injected, the column inlet can become mass-overloaded by major components, leading to similar effects as too large a sample volume. Typically, $100\text{-}500\text{-}\mu l$ samples can be used with highly efficient 0.3-0.4-cm-i.d. analytical columns in isocratic systems. Larger volumes of more dilute samples often cause less band broadening than smaller volumes of higher concentrations. The relatively large sample volumes that can be used in LC largely compensate for the lower sensitivity of some LC detectors compared with GC detectors, the overall result being a similar practical sensitivity for the two techniques.

Equation 13.2 predicts that if the size of the sample is limited, smaller internal-diameter columns (e.g., 0.2-0.3-cm i.d.) should be used to maintain as small a V_R and as low a sample dilution as possible. However, if sufficient sample is available, larger-diameter columns ($\geqslant0.4$ cm) are preferred, because the greater dilution can be appropriately compensated by a larger injection volume; also, the larger-i.d. columns are somewhat more efficient (Section 5.4), and extra-column effects are minimized.

Very large samples often can be charged to the LC column if the initial mobile phase is a weak solvent. In this instance the sample accumulates at the column inlet, because the k' values of all sample components are large. This *on-column trace enrichment* represents a powerful technique for greatly enhancing the detection sensitivity of certain sample components. For example, less-polar organics (e.g., polycyclic aromatic hydrocarbons) present in waste-water samples at trace levels can be concentrated during injection by use of a reverse-phase column (e.g., C_{18}). For maximum enrichment, the starting solvent used as mobile phase should be as weak as possible (e.g., 90-95% water), and then solvent strength should be sharply increased, using either gradient elution or a step-gradient change in the mobile phase. Figure 13.9 is an example of this approach, showing the assay of cyclosporine (a peptide of molecular weight 1202) in urine at the 360 ppb level (Figure 13.9b); the separation in Figure 13.9a is of a urine blank. In this case 1.8 ml of urine could be charged to the column; the column was then washed with a relatively weak mobile phase (water, followed by 32% acetonitrile/water) to purge initially eluting "garbage." A step-gradient change to 50% acetonitrile/water then eluted the cyclosporine as a sharp, easily detected peak. Finally, the more strongly retained sample components were stripped from the column with another step-gradient change to 70% acetonitrile/water, and the column was regenerated with a final water flush.

The on-column enrichment technique allows the injection of a liter or more of sample in favorable cases, with the final volume of the peak(s) of interest being less than 50 μl in favorable cases (a 20,000-fold concentration of the compound). In another example of this technique, Figure 16.16 shows the results of injecting 200 ml of waste-water as sample onto a 30 X 0.4-cm reverse-phase column.

As described in Section 17.2, it is usually possible to enhance the sensitivity of a trace analysis by first concentrating the solute (e.g., by solvent evaporation). This is particularly important if the trace component is at very low concentrations and detection is a problem. Unfortunately, preconcentration of the sample may not be useful if the preconcentration step results in a loss or alteration of the trace component or if the solubility of the major components in the sample is low. Another problem that can arise is that the concentration of the major components in the preconcentrated sample may increase to a level where the column is overloaded with the major components, making resolution of the trace component more difficult.

Specialized sampling techniques can be utilized for trace analysis in LC if simple preconcentration steps are not feasible. These techniques are not unlike those that have been widely used in gas chromatographic application, and they involve the use of devices such as prefractionating columns, coupled columns, and back-flushing techniques, as discussed in Section 16.2 and in (18-20).

The determination of trace components in extracts of complex naturally occurring systems can be particularly difficult. Even if adequate resolution and detector sensitivity for the early-eluting trace component of interest can be achieved, there are often many later-eluting components in the extract that continue to disturb the detector baseline. Subsequent samples cannot be analyzed until these strongly retained components are cleared from the column and a stable detector baseline is reestablished. As discussed previously, more strongly retained compounds can be eluted by increasing the strength of the mobile phase after the trace component of interest has been eluted. An alternative solution to this rather general problem in trace analysis is to use column backflushing or column-switching techniques as described in Section 16.2. There is one important caution with column backflushing operations—this approach only can be used with analytical columns that have been sufficiently well packed so that the bed does not shift during the backflushing operation.

Detection. The selectivity of the detector is generally important in developing sensitive and accurate trace analysis methods, because sample components that can overlap or mask the trace component of interest are ignored by certain detectors or detector settings. At present there are LC detectors available for many compound types that permit their selective measurement at very high sensitivity [see especially (16a)]. Ultimate detector sensitivity is a function of

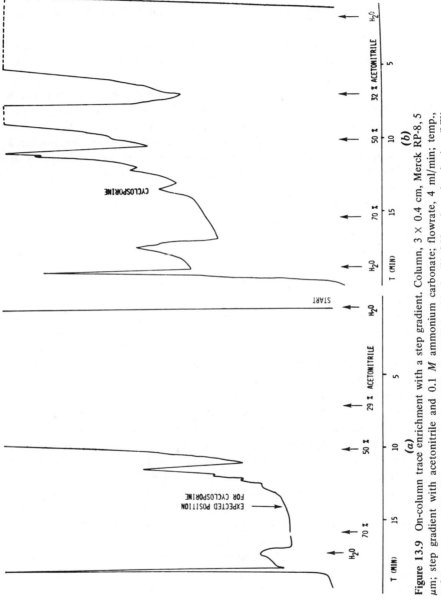

Figure 13.9 On-column trace enrichment with a step gradient. Column, 3 × 0.4 cm, Merck RP-8, 5 μm; step gradient with acetonitrile and 0.1 M ammonium carbonate; flowrate, 4 ml/min; temp., 70°C; detector, UV, 215 nm; sample, 1.78 ml. (*a*) Urine blank; (*b*) 360 ppb cyclosphorine (MW 1202) in urine. Reprinted from (*17*) with permission.

both the signal-to-noise ratio of the device and the ability to discriminate for the trace component in the total sample. Before accurate trace analyses can be performed, it is necessary to identify positively the expected trace component in the sample. See the discussion of how large R_s should be in Section 2.5 and Chapter 14.

The LC pump can limit the sensitivity of a trace analysis, because pulsations and flow changes generate baseline noise with various detectors. Trace analyses generally are best carried out with pumps delivering a nonpulsating flow, such as the positive displacement (syringe), constant-pressure gas displacement, or the pneumatic- and hydraulic-amplifier types. When using reciprocating pumps, efficient pulse-dampening systems (e.g., electronic feedback; see Section 3.4) may be required before the analysis can be carried out at highest detector sensitivity. Detector cell design also has a great influence on the sensitivity of the detector to flow noise, and cells with special flowpaths and thermal lagging are best in this regard (see Section 4.3).

Several commercially available detectors that are preferred for trace analysis in LC are listed in Table 13.5, along with typical specifications and the influence of various operating parameters.

Ultraviolet photometers and spectrophotometers are the most widely used detectors for trace analysis in modern LC. UV photometers with 1% noise at less than 0.002 absorbance units (AU) full scale now are commercially available. With this high sensitivity, solutes with relatively low absorptivities can be monitored, and it is possible to detect a few nanograms of a solute having only moderate ultraviolet absorption. For maximum sensitivity, it is desirable to work at the peak-absorption maximum. However, multiple wavelength photometric and spectrophotometric detectors also allow the analyst to select the wavelength that produces the highest response for the trace component of interest, relative to that for possible interfering substances (e.g., see Figure 4.7).

Fluorometers and spectrofluorometers are particularly useful for the selective detection of compounds that fluoresce or can be made to fluoresce by suitable derivatization. As suggested in Table 13.5, compared with UV absorbance, fluorescence detection can provide an increase of up to 100 times in sensitivity, allowing measurement of a few picograms in favorable cases (e.g., Figure 4.21).

Trace analysis for compounds having little or no response to the selective detectors in Table 13.5 can sometimes be carried out by forming suitable derivatives prior to or after the chromatographic separation (Sections 17.3, 17.4). A trace analysis accomplished by forming a fluorogenic derivative prior to separation is shown in Figure 17.4. Because of the high selectivity of *postcolumn derivative reactions*, trace analyses can often be accomplished with a minimum of sample cleanup. Figure 13.10 shows a chromatogram for a mixture of methomyl and *N*-nitrosomethomyl, yielding two peaks with the UV detector and only one peak by the selective reaction detector for the nitroso-compound.

Table 13.5 Useful detectors for trace analysis by LC.

Parameters	Detector			
	UV photometer, spectrophotometer	Fluorimeter, spectrofluorimeter	Reaction detector (postcolumn)	Electrochemical detector
Approximate sensitivity at 1% noise, full scale	0.002 AU	a	0.005 AU	10^{-10} amps
Useful with gradients	Yes	Yes	Yes	b
Possible sensitivity variability for responsive compounds	Very large	Very large	Small	2- to 5-fold
Sensitivity to pump pulsations, flowrate changes	Low	Low	Moderate	Moderate to high
Approximate sensitivity to favorable sample, g/l	2×10^{-7}	1×10^{-9}	5×10^{-7}	$\leqslant 10^{-9}$

Source: In part from (*10*).
[a] Appropriate specifications not available.
[b] Not available, but probably restricted.

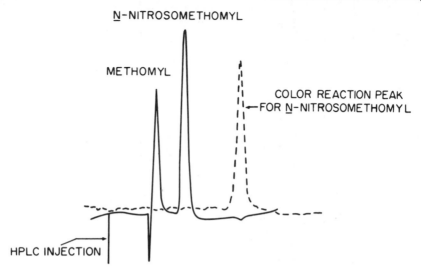

Figure 13.10 Reaction detector for nitroso-compounds. Column, 30 × 0.4 cm, μ-Bondapak-C$_{18}$; mobile phase, 20% 1-propanol/water; flowrate, 1.5 ml/min.; detectors: solid line, UV, 254 nm; dashed line, nitroso-compound detector. Reprinted from (*21*) with permission.

This detector is based on reaction with the Griess reagent, which is specific for compounds that hydrolyze to nitrite in dilute acid. The limit of sensitivity for this device is about 35 ng of sample injected (calculated as sodium nitrite).

As discussed in Section 4.7 electro-reducible and electro-oxydizable compounds can be detected in column effluents at very low concentrations by electrochemical detectors. For certain trace analyses, the amperometric detector is the detector of choice, because it has the highest sensitivity of any of the devices known. In addition, the amperometric detector can be made selective by adjusting the potential of the electrochemical process (see Table 4.5) or by using pulsed-potential operation (*22*).

Calibration. In trace analysis, accuracy is most important; hence, bands free of interfering components are sought. High precision usually is not as important in trace analysis, and reproducibilities of about 10% relative are often adequate. As a result, peak-height methods should be used for trace analysis, as discussed earlier in this section. Fortunately, peak-height calibrations also are the most convenient, and satisfactory trace analysis methods can usually be developed with modest effort.

Calibration plots in trace LC analysis generally are linear and pass through the origin. This can be contrasted with corresponding GC plots, which are often nonlinear and/or do not extrapolate through the origin, because of sample degradation, irreversible retention, and so on. As discussed in Section 13.2, peak-height-ratio (internal standard) calibration is sometimes used in trace analysis to compensate partially for losses that can occur during workup procedures. The internal standard is added to the sample before workup at a known concentration and is then carried through the entire procedure. Any losses that occur during the workup (by incomplete extraction, volatization, etc.) are thus compensated for by relating the results to the known concentration of the similar internal standard.

The method of *standard addition* (or "spiking") can be useful in certain cases involving complex samples, where sample-matrix effects may be involved and where a sample-blank cannot be conveniently obtained for constructing a calibration plot. Thus, the presence of other compounds in the sample may affect the retention and/or the peak-height of the compound of interest. The approach of standard addition is then as follows: the sample(s) is first submitted to LC analysis, and the height h_y of the band of interest is measured. A known weight w_y of the compound (Y) to be determined is then added to the sample. Now the sample is reassayed by LC and the new height of the band Y determined: h_y'. A calibration factor S_y for compound Y can now be determined:

$$S_y = \frac{h_y' - h_y}{w_y}.$$

The weight of Y in the original sample (before addition of the quantity w_y of Y) is then given as h_y/S_y. If concern exists over the linearity of the calibration plot for Y, values of S_y can be determined for several different values of w_y. It should be stressed that the method of standard addition does *not* correct for baseline variation, other sample interferences, and so on. These must be handled in the usual way *before* the standard-addition procedure is applied. In effect, this approach is a form of *in situ* calibration. An alternative to the standard-addition procedure is simply to obtain a typical sample that is known to be free of the compound Y and to use this sample as a matrix for preparing standards for normal calibration (i.e., add in various amounts of Y to the sample blank, and prepare a calibration plot).

Figure 13.11 shows a standard-addition calibration curve resulting from calibrating with four successive standard additions of 5-hydroxy-indole acetic acid (5HIAA) in human cerebrospinal fluid. The concentration of 5HIAA in the original sample was determined by extrapolating the standard-additions calibration plot to zero concentration to determine the peak height of the 5HIAA in the original sample (1.5 mV/5 here). The concentration of 5HIAA in the original sample then corresponds to that concentration that, when added,

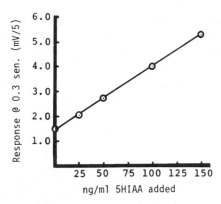

Figure 13.11 Standard-additions calibration. Column, 30 × 0.4 cm, μ-Bondapak-C₁₈; mobile phase, 89% 0.01 *M* sodium acetate, pH 4.0 (with acetate acid)/11% acetonitrile; flowrate, 1.9 ml/min; detector, fluorometer; sample, 10 μl, 5-hydroxy-indole acetic acid (5HIAA). Reprinted from (*23*) with permission.

produces a 1.5 mV/5 peak-height increase (i.e., 3.0 mV/5 on the calibration plot) or 50 ng/ml of 5HIAA.

Sample Pretreatment. Trace analyses often require that the concentration of the component of interest be enhanced to attain the desired sensitivity. Section 17.2 describes a variety of sample enrichment or clean-up procedures that can be employed.

LC is readily adapted to automation, and automatic sample pretreatment and sample injection equipment are now available (e.g., Du Pont Instruments Products Division and Technicon Instruments Corp.) to improve the convenience and precision of trace analyses. This apparatus is capable of automatically performing the usual steps in sample workup (e.g., grinding, extraction, filtration, etc.) for a wide variety of sample types, both solid and liquid. The combined technology of automatic sample treatment, followed by automatic LC separations, and finally automatic data handling and analysis printout, greatly facilitates the use of LC for the routine trace analysis of clinical, drug, food, pesticide, and environmental samples.

Preferred Conditions for Trace Analysis

Based on the discussion in this chapter, preferred conditions can be suggested for trace analysis by LC:

• Use a selective detector with a large signal-to-noise ratio for compound(s) of interest.

- Use chromatographic systems that exhibit large separation factors for the trace compound of interest. Separations with $\alpha > 1.2$ provide resolution that greatly simplify trace analysis.

- Set up separations with smaller k' values (e.g., $k' = 0.5$-1.5, if resolution permits); reverse-phase separations with the "garbage" eluting at t_0 are fast and convenient, where possible.

- Use a short, efficient column, preferably with more than 5000 theoretical plates. Columns with <10-μm particles are particularly advantageous, and the shortest column possible commensurate with adequate resolution should be used (but watch for extracolumn band broadening).

- Use column diameters $\geqslant 0.4$ cm if sample volume is not limited; if sample size is limited, 0.2-0.3-cm-i.d. columns should be used.

- Use reduced mobile-phase velocities $\nu = 5$-10, or ~ 0.2 cm/sec. for 5-μm particles and ~ 0.08 cm/sec. for 10-μm particles.

- Use large sample volumes. Start with sample that is one-fifth the volume of the peak of interest, increase until peak-height per concentration unit or resolution from neighboring components become limiting.

- Use a pulseless pump capable of precise mobile-phase delivery.

- Use peak-height measurements with isocratic separations for highest accuracy and adequate reproducibility.

- Preconcentrate sample for highest sensitivity; if required, use a clean-up procedure for complex systems to simplify final measurement.

REFERENCES

1. R. P. W. Scott and C. E. Reese, *J. Chromatogr.*, **138**, 283 (1977).

2. J. I. Hoffman, "Principles and Methods of Sampling," in *Treatise on Analytical Chemistry*, Vol. I, Interscience Encyclopedia, Inc., New York, 1959.

3. J. N. Done, *J. Chromatogr.*, **125**, 43 (1976).

3a. D. L. Ball, W. E. Harris, and H. W. Habgood, *Sep. Sci.*, **2**, 81 (1967).

3b. D. L. Ball, W. E. Harris, and H. W. Habgood, *J. Gas Chromatogr.*, **5**, 613 (1967).

3c. D. L. Ball, W. E. Harris, and H. W. Habgood, *Anal. Chem.*, **40**, 1113 (1968).

4. N. Haddon et al., *Basic Liquid Chromatography*, Varian Aerograph, Walnut Creek, Calif., 1971.

5. ASTM Method D-3536-76.

6. H. L. Pierson and D. J. Steible, Jr., in *Modern Practice of Gas Chromatography*, R. L. Grob, ed., Wiley-Interscience, New York, 1977, Chap. 8.

7. R. E. Majors, *J. Chromatogr. Sci.*, **8**, 338 (1970).

8. R. E. Leitch, *ibid.*, **9**, 531 (1971).

9. C. P. Talley, *Anal. Chem.*, **43**, 1512 (1971).

10. J. J. Kirkland, *Analyst (London)*, **99**, 859 (1974).

11. S. R. Bakalyar and R. A. Henry, *J. Chromatogr.*, **126**, 327 (1976).

12. M. Martin, G. Blu, C. Eon, and G. Guiochon, *Ibid.*, **112**, 339 (1975).

13. S. Mori, K. Mochizuki, M. Watanabe, and M. Saito, *Am. Lab.*, **9**, 21 (1977).

14. F. W. Karasek, *Res. Dev.*, **28**, 32 (1977).

15. F. Erni, R. W. Frei, and W. Lindner, *J. Chromatogr.*, **125**, 265 (1975).

16. B. L. Karger, M. Martin, and G. Guiochon, *Anal. Chem.*, **46**, 1640 (1974).

16a. P. T. Kissinger, L. J. Felice, D. J. Miner, C. R. Preddy, and R. E. Shoup, in *Contemporary Topics in Analytical and Clinical Chemistry*, Vol. 2, D. H. Hercules, ed., Plenum, New York, 1978, p. 55.

17. P. Schauwecker, R. W. Frei, and F. Erni, *J. Chromatogr.*, **136**, 63 (1977).

18. J. Svojanowvsky, M. K. Tasorik, and J. Janak, *J. Chromatogr. Rev.*, **8**, 9 (1966).

19. S. Dal Nogare and R. S. Juvet, Jr., *Gas-Liquid Chromatography*, Interscience, New York, 1962, p. 307.

20. Product Bulletin on GC Sampling and Switching Valves, Carle Instruments, Inc., Anaheim, Calif., 1973.

21. G. M. Singer, S. S. Singer, and D. G. Schmidt, *J. Chromatogr.*, **133**, 59 (1977).

22. D. G. Swartzfager, *Anal. Chem.*, **48**, 2189 (1976).

23. G. M. Anderson and W. C. Purdy, *Anal. Lett.*, **10**, 493 (1977).

FOURTEEN

QUALITATIVE ANALYSIS

14.1	Retention Data for Sample Characterization	576
	Direct Comparison of t_R Values	576
	Prediction of t_R Values	578
	General Considerations	579
	Size-Exclusion Chromatography	585
	Liquid-Liquid Chromatography	585
	Liquid-Solid Chromatography	585
	Bonded-Phase Chromatography	587
	Ion-Exchange Chromatography	588
	Ion-Pair Chromatography	589
14.2	Qualitative Analysis of Sample Bands from Analytical-Scale LC Separations	589
	Chromatographic Cross-Check	591
	Relative Detector Response	592
	Identification by Supplementary (Ancillary) Techniques	592
	Peak Isolation	593
	Identification Methods—General	595
	Mass Spectrometry	595
	Infrared Spectroscopy	597
	Nuclear Magnetic Resonance Spectroscopy	600
	Color Reaction (Spot Analysis)	602
14.3	On-Line Spectroscopic Analysis of LC Peaks	603
	Mass Spectrometry	603
	Infrared Spectroscopy	607
	References	612
	Bibliography	614

The identification of a separated sample band in LC can be accomplished in various ways. Thus, we have seen that the retention time t_R of a compound in a given LC system is characteristic of that compound. Therefore, if the t_R value of a known compound matches that of an unknown band, this suggests that the two compounds are the same. Alternatively, an initial analytical separation can be scaled up to the point where a large quantity of pure compound is isolated; the resulting material can then be characterized by conventional chemical or spectroscopic analysis. In a variation on the latter approach, a band can be isolated in small quantity (e.g., micrograms or nanograms) from an analytical-scale separation, and with suitable manipulation, this minute quantity of sample can be identified by spectroscopic or colorimetric means. Finally, in some cases it is convenient to interface a spectroscopic analyzer directly to the outlet of an LC column. In this way, on-line analysis of separated bands becomes possible.

The recovery of milligram or gram quantities of sample for conventional characterization is discussed in Chapter 15. No more will be said here of this approach. Following sections of this chapter address the remaining three approaches to LC qualitative analysis: (a) use of retention data, (b) adaptation of spectroscopic or chemical analysis for use with LC bands from analytical-scale separations, and (c) direct interfacing of spectroscopic analyzers to an LC column.

14.1 RETENTION DATA FOR SAMPLE CHARACTERIZATION

The use of the t_R value of an unknown band for its preliminary characterization can follow one of two routes. First, compounds suspected to be present in the sample can be separated in the same LC system, and their t_R values can be measured. Comparison of these t_R values with that of the unknown band may then lead to the finding that the unknown band has the same retention as some known compound and is therefore probably the same compound. Second, following such a comparison of t_R values, no match may be found for the unknown band. In this case, it may be possible to estimate the t_R value of different known compounds that are unavailable for direct measurement of their retention. Each of these two possibilities will be considered in turn.

Direct Comparison of t_R Values

Little needs to be said about this approach, because it is very simple. The unknown sample is separated in a given LC system; then various known compounds are injected in turn. The retention times of the latter are compared with various bands in the starting sample. The likelihood of a successful match in t_R values depends on our prior knowledge of the sample and, therefore, our ability to anticipate the presence of specific compounds in the sample. Successful matching of retention times also requires the availability of likely reference compounds.

For simple mixtures, an accurate match between sample bands and known reference compounds is usually obvious: either two t_R values are the same or they are not. In the case of more complex samples, several possible compounds may elute with similar t_R values, and actual bands within the sample chromatogram are more likely to overlap—with resultant uncertainty in the assignment of accurate t_R values to various sample bands. In this situation the reliability of a tentative characterization based on matched t_R values becomes greater, the more precisely we can show that the two t_R values are identical.

Increased precision of t_R measurements is facilitated by increased resolution (e.g., Figure 2.18a and related discussion), by close control over separation conditions, and by repetitive measurement of a t_R value plus averaging of individual data points. Those separation conditions that are most critical include pump flowrate F, mobile-phase composition, and temperature. If t_R values are determined in replicate and averaged, the reference sample and unknown should be run in an alternating sequence to average out long-term drift, and standard deviations for each t_R value should be determined. This allows statistical evaluation of the goodness of fit between known and unknown bands.

An additional complication for complex samples is that sample-matrix effects may slightly alter the t_R value of a given compound. This is particularly likely for early-eluting and/or overlapped bands. In this situation, the method of *standard addition* (Section 13.5) can be employed to verify the t_R value of the compound in question in the actual sample matrix. The t_R value of the original sample band should not change after addition of the compound in question if the two compounds are the same.

As additional retention data are published for new LC separations, an everincreasing number of t_R values for different sample compounds and different LC systems are to be found in the literature. It would be highly convenient if these values were usable for purposes of comparison and qualitative analysis, but at present such data are of limited value. Previous chapters on the individual LC methods have commented on the relatively poor reproducibility of some LC columns at present, even for columns from the same lot. As a result, a t_R value from the literature will generally offer only a rough approximation to t_R for the same compound in a nominally equivalent LC system. However, as discussed later in connection with prediction of LSC retention data, there are techniques for relating literature t_R values to newly measured data for the same compounds. Possibly these techniques can be improved and generalized for broad application.

A second problem in using literature t_R values for comparison with data for unknown bands is that of locating literature data on a given compound. The

Journal of Chromatography bibliography reviews (Table 1.1) provide a running index of compounds by structure, for ready access to such information.

Prediction of t_R Values

Before discussing how we go about estimating retention data for a given compound in different LC systems, it should be noted that data of this type have many uses. Aside from the value of predicted t_R values for qualitative analysis (for comparison with t_R for an unknown band), the following possibilities exist:

- Predicting the right solvent strength for eluting a particular compound at a convenient k' value (e.g., Table 9.6).

- Predicting the possibility of separating two compounds in a given LC system, based on estimating the difference in their relative retention; in this way the most promising LC system for a given separation can be selected in advance of actual experimental study.

- For complex separations requiring the use of several individual methods to resolve the sample fully, predicted retention data can allow the optimum selection of the minimum number of individual LC methods required [e.g., see (1) for the separation and analysis of petroleum hetero-compounds].

- Predicting the structure of a possible internal standard whose t_R value must be sufficiently different from that of other sample bands to avoid overlapping those bands (Section 13.3).

We do not present here a detailed discussion of all that is known concerning the prediction of retention in LC. Rather, we will provide a general review with reference to other, more detailed treatments and we emphasize approaches that are particularly suited for qualitative analysis.

The ability to predict accurate retention data varies from one LC system to another. In size-exclusion chromatography reliable estimates of t_R as a function of molecular structure are usually possible (Section 12.5). Accurate predictions of relative retention are sometimes achievable in LSC and LLC, and to a lesser extent in BPC. Only rough predictions of relative retention can presently be made in ion-exchange and ion-pair chromatography, and only for compounds of known pK_a values.

It is always easier to predict relative retention or α values, as opposed to absolute k' values. Furthermore, relative t_R values are easier to estimate for related compounds or derivatives (e.g., naphthalene versus 2-nitronaphthalene or acetic versus propionic acid). However, these are just the situations that are often of interest to the chromatographer, because typical samples are mixtures of related compounds. In favorable cases, the accuracy of such α-value predictions can approach ±2-5%, but poorer agreement is more often the rule.

If we are to attempt the accurate prediction of t_R values for a wide range of possible sample compounds, as a function of different experimental conditions, then a fairly detailed and accurate understanding of the mechanism of retention is required for an LC method of interest. At the moment there is a lack of agreement among different workers as to the precise retention mechanism for any of the LC methods, with the possible exception of SEC. Enough is known, however, to provide the more venturesome chromatographer with a basis for predicting retention in selected cases.

General Considerations.   Values of t_R usually vary in a regular and predictable fashion with repeated substitution of some group i into a sample molecule (e.g., as in series of homologs, benzologs, or oligomers). Often some function of t_R or k' will be linear with n, the number of repeating groups i within the sample

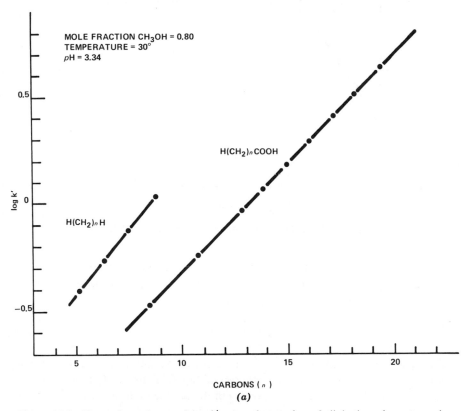

Figure 14.1   Linear dependence of log k' on n, the number of aliphatic carbon atoms, in reverse-phase BPC. (*a*) Homologous carboxylic acids, 89%v methanol/water as mobile phase. Reprinted from (*2*) with permission.

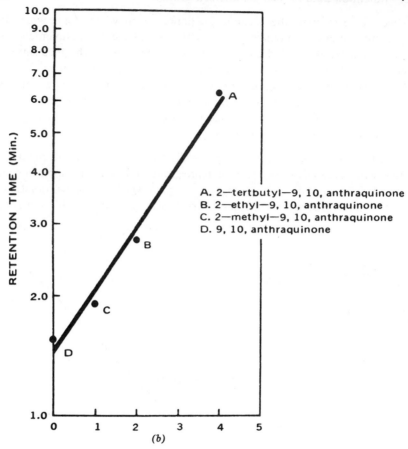

Figure 14.1b   Alkyl anthraquinones (from Figure 2.26).

molecule (e.g., $-CH_2$-groups for a homologous series). For isocratic elution, a general relationship that is often obeyed in such cases is the Martin rule:

$$\log k' = A + Bn. \tag{14.1}$$

Here A and B are constants for a given sample series, and a specific LC system (same column, mobile phase, and other conditions). Thus, $\log k'$ is predicted to vary linearly with the number of groups i in the sample molecule. An example of the validity of Eq. 14.1 is shown in Figure 14.1a, where $\log k'$ values for a homologous series of carboxylic acids $[CH_3-(CH_2)_n-COOH]$ and n-alkanes (C_nH_{2n+2})

are plotted versus n, in the separation of these compounds by reverse-phase BPC. Figure 14.1b shows a similar plot, derived from the data of Figure 2.26 (reverse-phase BPC separation of alkyl anthraquinones). In the latter plot n is taken as the total number of alkyl carbons, rather than the number of $-CH_2-$groups in the molecule. Because the repeating unit is variously $-CH_3$, $-CH_2-$, and $=C=$, the resulting plot is only approximately linear.

Equation 14.1 is obeyed more often in LLC or BPC systems. However, the linearity of log k' with n is often observed approximately for LSC systems as well. Figure 14.2a shows a plot of log k' versus the number n of aromatic double bonds or $-C=C-$ groups, for various aromatic hydrocarbons separated on silica.

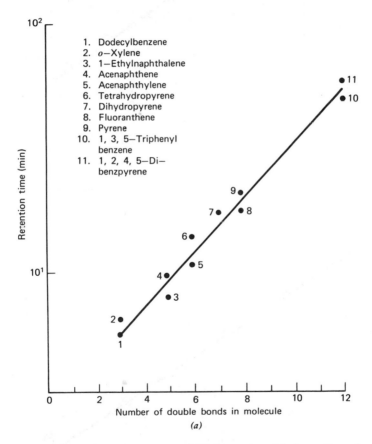

1. Dodecylbenzene
2. o—Xylene
3. 1—Ethylnaphthalene
4. Acenaphthene
5. Acenaphthylene
6. Tetrahydropyrene
7. Dihydropyrene
8. Fluoranthene
9. Pyrene
10. 1, 3, 5—Triphenyl benzene
11. 1, 2, 4, 5—Di— benzpyrene

Number of double bonds in molecule

Retention time (min)

(a)

Figure 14.2 Applicability of Eq. 14.1 to LSC Separation. (a) Separation of aromatic hydrocarbons on silica, n equals number of $-C=C-$ groups. Reprinted from (3) with permission.

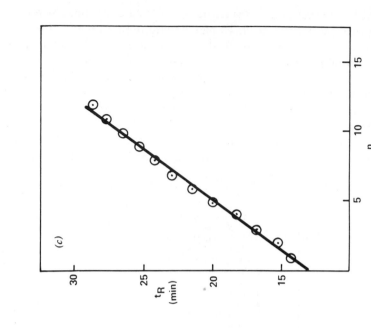

Figure 14.1c Separation of gallotannins on silica by gradient elution. n equals number of gallic acid units (data of Figure

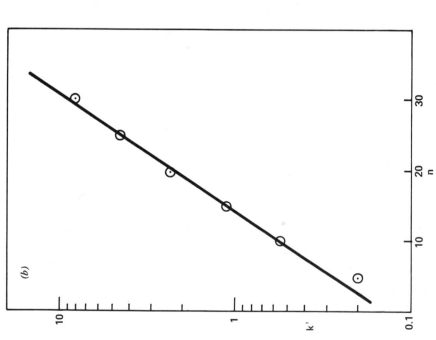

Figure 14.1b Separation of polystyrene oligomers on silica, n equals number of styrene units (data of Figure 9.4).

In LSC the effect of aliphatic carbons in the molecule plays a lesser role in retention, as can be seen in Figure 14.2*a* (where these groups are ignored in calculating *n*). Another demonstration of Eq. 14.1 for an LSC separation is shown in Figure 14.2*b*, using the separation shown in Figure 9.4. The sample in this case is an oligomeric series resulting from the polymerization of styrene, so the repeating unit *i* is (I).

$$\phi\diagdown$$

$$\diagup \overset{\phi\diagdown}{\diagup}CH-CH_2-$$

(I)

Equation 14.1 applies only to isocratic separations. For separations by gradient elution (with LSS gradients; see Section 16.1), it can be shown that Eq. 14.1 reduces to

$$t_R = A' + B'n. \tag{14.1a}$$

That is, values of t_R are now linear in *n*. This is illustrated in the plot of Figure 14.2*c*, based on the gradient elution LSC separation of Figure 9.2. The sample series for this latter example consists of various gallotannins, where the repeating unit *i* is the galloyl group (II).

$$HO\diagdown$$

$$-O-\langle\ \rangle-COO-$$

$$HO\diagup$$

(II)

As a corollary to Eq. 14.1, the introduction of a given functional group *i* into any sample molecule will change *k'* for that compound by some constant factor α_i, in a given LC system (fixed column, mobile phase, etc.). These group constants α_i for different substituents *i* generally depend on the polarity of *i*; α_i increases with polarity for all but reverse-phase systems and decreases with the polarity of *i* in reverse-phase systems. Values of α_i for some selected LC systems are listed in Table 14.1 and are further discussed later. Because these α_i values of Table 14.1 vary with the mobile phase, the data shown in Table 14.1 are only roughly applicable to any given LC system. However, measurement of a value of α_i in a particular LLC or BPC system allows fairly accurate estimates of t_R for other compounds containing the group *i*.

For example, in a reverse-phase BPC separation with 60%v acetonitrile/water as mobile phase, k' values were determined for toluene (3.68) and p-cresol (0.78). The α_i value for an $-OH$ substituent was then calculated: $\alpha_i = 0.78/3.68 = 0.21$ (versus the value of 0.3 given in Table 14.1 for a different mobile phase). That is, addition of an $-OH$ group to a sample molecule in reverse-phase BPC decreases k' and t_R. It was then desired to predict k' for the compound p-hydroxy butylbenzoate in this same LC system, using the related compound butylbenzoate as starting point. The value of k' for the latter compound in this LC system was found equal to 7.76, so that the k' for p-hydroxybutylbenzoate could be estimated equal to this value times α_i for an $-OH$ substituent: $k' = 7.76 \times 0.21 = 1.63$. The latter value of k' would be expected to be fairly accurate, because an experimental value of α_i was used.

For further discussion of the Martin rule, see (6a).

Table 14.1 Group constants α_i for various substituents i in different LC systems. Values are for substitution onto an aromatic ring.

Substituent i	α_i		
	LSC on silica, $\epsilon^\circ = 0.3^a$	LLC, reverse phase: water/ n-octanol[b]	BPC, reverse-phase; 50%v methanol/water[c]
$-Br$	0.5	7.2	
$-Cl$	0.5	5.1	2.0
$-F$	0.7	1.4	
$-CH_2-$	0.8	3.6	1.9
$-C=$ (aromatic)	1.2		1.3
$-S-CH_3$	1.7	4.1	
$-O-CH_3$	2.1	1.0	0.9
$-NO_2$	2.3	0.5	0.7
$-C\equiv N$	3.7	0.3	0.5
$-CHO$	4.9		0.5
$-CO_2CH_3$	5.2	1.0	1.1
$-COCH_3$	22	0.3	0.6
$-OH$	22	0.2	0.3
$-NH_2$	55	0.06	0.3
$-COOH$	200	0.5	0.1^d
$-CONH_2$	450		

[a] Calculated from data of (4), for an intermediate solvent strength.
[b] Hydrophobic substituent constants of Hansch (5).
[c] Experimental data of (6).
[d] Probably for ionized group $-COO^-$.

Size-Exclusion Chromatography. Values of t_R or V_R for a compound of known molecular structure are readily estimated in SEC separations if a calibration of molecular length or weight versus retention volume (or time) is available. Examples of this approach are provided in Figures 12.17-19 and related discussion, as well as the discussion of (7). In general, the addition of a group i to a sample molecule decreases t_R because of the resulting increase in molecular size. Equation 14.1 is inapplicable to SEC separations.

Liquid-Liquid Chromatography. A good deal is known about the factors that influence the distribution of sample molecules between two liquid phases, primarily as the result of equilibrium extraction studies. A good general reference is (5), which describes in detail one approach to the prediction of extraction coefficients (relative k' values) as a function of solute structure and the nature of the two immiscible liquid phases. Extrapolation of these data to LLC then assumes only that the mechanism of retention in the LLC system is the simple equilibrium distribution of sample molecules between the two liquid phases (stationary and mobile).

Very reliable predictions of retention in LLC are also possible through the semi-empirical approach of Huber (5a), providing that certain experimental data on related solutes and LC systems are available.

Liquid-Solid Chromatography. Apart from SEC, retention in LSC is perhaps the best understood of any LC method, and reasonably reliable predictions of retention can be made for a wide range of compounds under a variety of separation conditions (different adsorbents, mobile phases, etc.). A detailed review of retention in LSC along with a general approach for predicting t_R values is given in (4), with additional material provided in (8-9). Equation 14.1 is, by itself, of more limited value for LSC systems, because of the pronounced isomer effects that are encountered in LSC. Thus, Eq. 14.1 and the α_i values of Table 14.1 assume that all isomers have the same value of k', whereas this is usually not true in LSC (see discussion of Section 9.1). Some of the factors that can have a large effect on α_i values in LSC (for a given group i) include

- Electronic or steric interaction of i with adjacent groups in the sample molecule.
- Localization effects, as illustrated in Figure 9.5c and discussed in Section 9.1.
- Relative planarity of the sample molecule.

These are further discussed in Chapters 10 and 11 of (4). Reliable prediction of the relative retention of different isomers of a given compound in LSC is usually possible [see (4)], which can often facilitate qualitative analysis by LSC.

Differences in the adsorbent or column packing in LSC can be corrected for, when comparing k' values between two different columns or between a TLC system and an LSC system (same mobile phase and adsorbent type for both

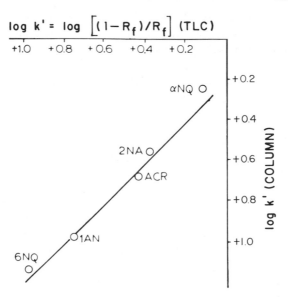

Figure 14.3 Correlation of k' values between two equivalent LSC systems (TLC vs. column LCS.) System is silica (adsorbent) with 10%v methylethylketone/cyclohexane (mobile phase). Compounds are 6-nitroquinoline (6NQ), 1-amino-naphthalene (1AN), acridine (ACR), 2-nitroaniline (2NA) and α-naphthoquinoline (αNQ). Reprinted from (9a) with permission.

systems). Thus, if log k' values for one column are plotted against log k' values from the second column (or TLC system, see Table 9.9), a linear plot is generally observed. This is illustrated in Figure 14.3, for the separation of several nitrogen compounds on silica. Here column data were compared with TLC data, using 10%v methyl ethylketone/cyclohexane as mobile phase.

Plots such as that of Figure 14.3 are one way to improve the reliability of literature t_R values for extrapolation to an LSC system of immediate interest. For example, assume we have separated a sample into six bands, four of which we have been able to identify by comparing t_R values with known reference compounds. Further assume that someone has reported retention data in the same LSC system for the four reference compounds for which we have t_R values (in our system), plus several other compounds that might be present in our sample but are not available to us for direct comparison. In this situation we can plot log k' values for our four reference compounds versus log k' for these same compounds from the literature study. Now the resulting linear plot allows us to estimate k' values in our system from k' values for the remaining compounds reported in the literature. This general approach of using double-log k' plots is probably useful for improving the transferability of k' values from the literature for other LC systems as well (e.g., BPC, IEC), but it has not been tested to any extent.

The mechanism of retention in LSC appears to be well described by the displacement model discussed in Section 9.1 and (4). Although some workers have suggested a different mechanism for separation (see Section 9.1) in some LSC systems, the displacement model does an excellent job in predicting actual k' values [e.g., see (4) (8), and (10)].

Bonded-Phase Chromatography. The mechanism of BPC retention is still a matter of debate [e.g., see review of (11)]. Most of the practical interest and theoretical attention in this area has been focused on reverse-phase systems. Regardless of mechanism, accurate predictions of t_R are often possible, based on relationships such as that in Eq. 14.1. For example, one such study (12) describes the prediction of retention for various aromatic hydrocarbons in a typical reverse-phase BPC system. Other workers have suggested that BPC separation is essentially similar to corresponding LLC separations (with analogous phases), and, therefore, data from equilibrium extraction systems can be used to

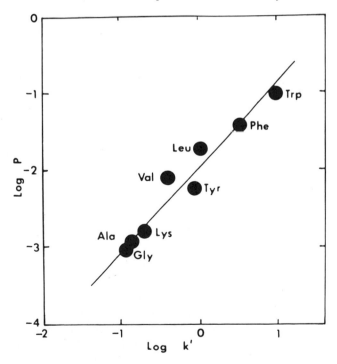

Figure 14.4 Correlation of partition coefficients in octanol/water (*P*) values) with k' values in reverse-phase BPC for amino acid samples. Column, LiChrosorb RP-8 (5 μm); mobile phase, 0.1 *M* phosphate (pH 6.7); temp., 70°C; 2 ml/min (2200 psi). Reprinted from (13) with permission.

predict retention in BPC. That this is indeed the case is suggested by the similarities between LLC and BPC as regards the variation in retention with experimental conditions. An example of this is provided by Figure 14.4, which shows log k' values for several amino acids (reverse-phase BPC) plotted against log P values (extraction coefficients) from an analogous equilibrium distribution (LLC) system. Because the latter P values can be calculated fairly precisely for a wide range of compounds (5), the probable generality of the relationship of Figure 14.4 provides a powerful means of predicting t_R values in reverse-phase BPC systems. For other correlations of k' values in reverse-phase BPC systems with solute distribution constants (K values) for related liquid-liquid partition systems, see (12a-c). For yet another example (alkyl phenols) of the similarity of retention in LLC and related BPC systems, see the study of (12d) where several LC systems are cross-compared, including cyano-bonded-phase versus various nitrile-substituted liquid phases.

Ion-Exchange Chromatography. The mechanism of retention for IEC was discussed in Section 10.3. Three main factors affect the retention of a given compound: (a) its degree of ionization (for acids and bases), (b) the net charge on the ionized molecule, and (c) the tendency of the molecule to distribute into an organic phase from the customary aqueous mobile phase used in IEC. According to the first factor (a), stronger acids or bases are more strongly retained in, respectively, anion and cation exchange separations. Or, within a series of acids separated by anion exchange, retention generally increases with decrease in the pK_a value of individual sample acids. For separation of bases by cation exchange, retention generally increases with increase in pK_a value of the sample compound. An example (14) is provided by the ion-exchange separation of various weakly basic petroleum compounds, with pK_a values indicated in parentheses:

indoles (-2.4) < sulfoxides (1.4) ≈ pyridones (0.7-1.9) < pyridines (3-7).

In this case, factors (b) and (c) discussed below were unimportant, because the preceding petroleum compounds were of similar molecular weight and carried a single ionizable functional group. Therefore, the pK_a value of each compound type was the predominant factor in determining retention.

The second factor (b) in determining retention in IEC is the net charge on the molecule (e.g., in cation exchange, retention increases as the molecular charge goes from 0 to +1, +2, and so on). Thus, in cation exchange separations of the amino acids (Figure 10.14), the so-called acidic amino acids (two $-COO^-$ groups, one $-NH_3^+$ group), which include aspartic and glutamic acid, elute early from the column, whereas the basic amino acids (one $-COO^-$ group, two $-NH_3^+$ groups; lysine, arginine, histidine) elute late from the column.

The final factor (c) that affects sample retention in IEC is the preference of the molecule for the organic stationary phase versus the aqueous mobile phase, very much as in the case of LLC or BPC. The principal factor here is the molecular weight or hydrophobicity of the sample molecule, which is well illustrated by the "neutral" amino acids (those with only one $-COO^-$ and $-NH_3^+$ group). Thus, in the IEC amino acid separation of Figure 10.14, these compounds are seen to be retained in increasing order of molecular weight of the hydrocarbon substituent R:

$$\text{glycine } (R = -H) < \text{alanine } (-CH_3) < \text{valine } (-i\text{-propyl})$$

$$< \text{leucine } (-i\text{-butyl}) < \text{phenylalanine } (-\text{benzyl}).$$

As seen in Figure 14.4, this is the same order of retention as that in reverse-phase LLC and BPC systems. Similarly, the insertion of an $-OH$ group into an amino acid significantly reduces its t_R value, versus the parent compound (cf. $\alpha_i = 0.3$ in Table 14.1); for example, from Figure 10.14, serine $<$ alanine, and tyrosine $<$ phenyl alanine.

Despite the obvious regularities in IEC retention as the sample molecule is altered in various ways, accurate prediction of t_R values in IEC systems is not yet possible.

Ion-Pair Chromatography. Retention in IPC is governed by very similar principles as those for IEC separation. As an example, the effect of sample pK_a value on the k' values of several related sulfa drugs in a typical IPC system is shown in Figure 14.5. There is an obvious trend to higher k' values with increase in pK_a or decrease in acid-strength for these compounds. In this case, a normal-phase system was employed, so the usual relationships between retention and acid or base strength of sample compounds in reverse-phase IPC are inverted.

Considerable scatter in the plot of Figure 14.5 is also noted, which emphasizes again our inability to make precise estimates of t_R in either IEC or IPC at the present time. The precise mechanism of retention in IPC is the subject of continuing controversy, with some workers favoring retention of the free counter-ion in reverse-phase BPC systems and others disagreeing [see discussion of (16) and (17)].

14.2 QUALITATIVE ANALYSIS OF SAMPLE BANDS FROM ANALYTICAL-SCALE LC SEPARATIONS

The preceding section describes the use of chromatographic retention data as a means of characterizing unknown bands from an LC separation. In many cases

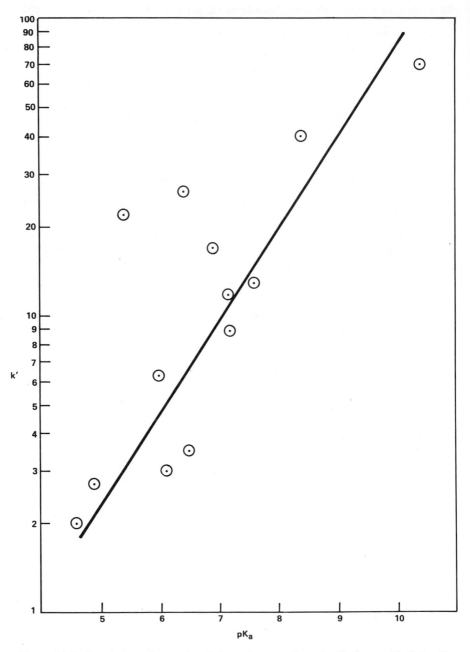

Figure 14.5 Correlation of retention in ion-pair separation of sulfa drugs with their pK$_a$ values. Data calculated from Fig. 5 of (*15*).

this approach is simply not applicable, and in most cases only a tentative con-firmation of structure is possible, although the confidence in a final structure can be increased by obtaining t_R values in more than one LC system and by noting relative response factors by more than one detector. More commonly, however, final confirmation requires use of an ancillary method (e.g., mass spectrometry, infrared spectroscopy).

Chromatographic Cross-Check

As discussed in Section 14.1, elution of a component of suspected structure at a particular retention time is only partial proof of its qualitative identity; that is, other compounds can elute identically. Although obtaining the expected reten-tion time for an unknown component by an alternative separation is not an absolute proof of identity, this may be the only feasible technique if the amount of the solute of concern is insufficient for confirming tests by an ancillary method. Use of capacity factor k' values often is preferred when assembling retention data for qualitative comparisons, because this term (unlike t_R) is not influenced by mobile-phase flowrate or column geometry. However, the reliabi-lity of such an identification can be greatly enhanced by using another LC method or another chromatographic technique (e.g., GC, TLC) to verify that the unknown elutes at the retention time (or k', R_f) for the suspected component. When using the retention-time cross-check technique, it is important that the two or more separation methods involve different solute/stationary-phase inter-actions. For example, a tentative identification by LSC might be best confirmed using reverse-phase BPC (not TLC). This is illustrated in Table 14.2, showing that the capacity factors of several substituted-urea herbicides are decidedly different for reverse-phase and normal-phase systems with a polar bonded-phase column. In addition to the use of different LC methods for confirming the qualitative identity of the component, use of the same column with different mobile phases also can be of value, but is a less severe test of structure.

Table 14.2   Effect of mobile phase on solute retention.

Mobile phase	Capacity factor of compound, k'^a		
	Diuron	Monuron	Neburon
Methanol/water (35:65)	4.3	1.0	25.6
Dioxane/hexane (1:99)	1.5	1.1	0.3

Source: From (*17a*).
[a]Column, 100 $\times$ 0.21 cm. Permaphase ETH; temperature, 50°C for methanol/water, 27°C for dioxane/hexane; detector, UV, 254 nm.

One of the difficulties in using LC retention for sample analysis is uncertainty that the entire sample has eluted from the column, even for the case of gradient elution (Section 16.2). Thus the possibility exists of missing a major component of the sample, and (worse) confusing the structure of a minor eluting band with that of the main sample constituent. One approach is to inject a known mass of sample, collect the *total* column effluent, remove the solvent, and weigh the residue to ensure elution of the principal sample components. Although this is an acceptable approach in preparative LC, it is experimentally inconvenient with an analytical column, since usually no more than a milligram of sample can be injected. Also, blanks must be run to ensure that contaminants in the solvents or from the column packing do not bias the results. Of particular aid in qualitative LSC are independent studies by thin-layer chromatography, provided the mobile phase is the same, and the coating on the TLC plate and the column packing are comparable. Materials not eluted or difficult to elute from the column are suggested by spots that remain at or near the origin on the TLC plate.

Relative Detector Response

Another technique for confirming solute identity is to use the relative response of two (or more) detectors. This approach is conveniently carried out by placing two different detectors in series following the column (e.g., UV and fluorescence). Identity of a solute can then be verified by determining the response ratio for the unknown and comparing this ratio with that for an authentic sample of the expected component under identical separation conditions. A modification of this technique is to record the response of the solute of interest at two wavelengths with a spectrophotometric detector; solute identity is again confirmed if the response ratio is identical to that for a standard of the anticipated compound. For LC instruments that permit stop-flow operation, confirmatory identification can be obtained with a variable wavelength UV detector by determining the absorption spectrum to characerize the compound of interest. Absorbance ratios are independent of the concentration of compounds in the detector flow cell (*18*).

Identification by Supplementary (Ancillary) Techniques

In many cases, positive identification of an unknown can only be accomplished by isolating the peak during a chromatographic run and subsequently analyzing it by a supplemental method. A number of techniques can be used to obtain qualitative information on peaks that have been isolated from LC separations. Table 14.3 lists some instrumental and chemical techniques for identifying LC isolates, in approximate order of overall usefulness and convenience. Included in this table is the type of information that can be obtained by using each technique, as well as the approximate minimal sample required for a test. Even though some of these techniques provide good evidence for the identity of an unknown, it is often desirable to use two or more tests when an unequivocal structure proof is required. For example, a combination of mass spectrometry and infrared spectroscopy (or nuclear magnetic resonance spectroscopy) is usually conclusive for the structure of an unknown.

Table 14.3 Some techniques for identifying LC isolates.

Technique	Approximate minimum sample required (μg)	Information obtained
MS	0.005	Molecular weight and empirical formula; structural information; confirmation of structure
IR	5(0.05)[a] (solid) 50(0.05)[a] (soln.)	Functional groups, possible structure; confirmation
NMR	1000 (5)[a]	Structures containing ^{1}H, ^{11}B, ^{13}C, ^{11}F, ^{31}P; confirmation
Elemental analysis	200-1000	Percent of elements, element ratios
Color reactions	0.1	Functional groups
UV	0.1	Possible structure of conjugated system; partial confirmation
Polarography	0.01	Redox potential, number of electrons transferred; partial confirmation
Spectrofluorimetry	0.05	Partial structure confirmation

[a] Using Fourier transform methods.

Peak Isolation. For identification, the unknown peak must be isolated in a form compatible with the supplementary technique. Table 14.4 lists some important characteristics of LC phases in this connection. First, the mobile and stationary phases must not react with the solute of interest, during either the LC separation or subsequent removal of the solvent. Most important, these phases must be highly purified to minimize interfering residues when the solvent is removed; redistilled or distilled-in-glass solvents generally are required. If possible, mobile phases should be volatile for easy removal by evaporation, and stationary phases should be either volatile or "nonbleeding," so as not to leave any residue in the isolated sample. Finally, if there is a residue from either the mobile or stationary phase, it should be compatible with the supplementary identification technique.

Table 14.4   Desirable characteristics of chromatographic phases for LC identification.

The chromatographic phases should be:
   Highly purified to minimize "background" (very important)
   Nonreactive with solutes
   Compatible with the chromatographic system for separating and sensing peak
   "Nonbleeding" or volatile for easy removal by evaporation
   Compatible with supplementary identification technique

The trapping of peaks in modern LC is quite convenient; most modern LC instruments provide a collection outlet or port. Trapping is usually accomplished manually, because peaks elute rapidly; fraction collectors are required only for very long separations. To maintain the resolution (and purity) of the peak to be isolated, the volume of tubing between the detector and trapping port should be at a minimum. Also, the volume between the detector and the trapping port should be measured so as to identify the "lag time" between the sensing of a peak and its actual arrival at the trapping port. This volume can be calculated if the dimensions of the detector cell and connecting tubing are known. Alternatively, a highly colored dye can be injected onto the column, and the transit time of the resulting band between detector and point of collection can be determined (by combining the detector signal with visual detection of the band at the point of collection).

It is convenient to collect isolated fractions in small screw-capped vials, preferably fitted with Teflon liners to reduce contamination. These vials must be carefully cleaned and dried before use, because "as-received" vials often contain impurities that significantly contaminate isolated peaks.

Isolated fractions should be analyzed as soon as possible after collection, to reduce the possibility of chemical alteration. If a delay in analysis is anticipated, it is desirable to sparge the trapped fraction with pure nitrogen or helium, wrap the vial with aluminum foil to exclude light, and store in a freezer until analysis is possible.

In some cases the pure isolate will be required for subsequent identification. Volatile organic solvents (e.g., dichloromethane, hexane) are conveniently evaporated from the trapping vial, using a slow stream of pure nitrogen while warming with an infrared lamp. Water or water-organic solvent mixtures also can be removed by evaporation; however, in some cases freeze-drying (lyophilization) is more convenient (see Section 15.4). Volatile buffers (e.g., ammonium formate) sometimes can be eliminated at elevated temperatures, in the case of nonvolatile solutes from ion-exchange separations. Volatile buffers for ion exchange also include the use of carbonate solutions. In this case, fractions containing bicarbonate in the mobile phase are first passed through a cation exchanger (acid form) that converts the bicarbonate to CO_2 (19). Finally, solutes can usually be isolated from mobile phases containing nonvolatile buffers, by extraction or gel filtration chromatography.

Identification Methods—General. Although each of the methods in Table 14.3 can provide useful qualitative information for certain samples, only a few are routinely employed in LC. As might be anticipated, the most-used techniques are those that generally provide definitive information. We now briefly discuss the more useful supplemental identification techniques for qualitative LC studies.

Mass Spectrometry. Mass spectrometry (MS) is probably the most powerful technique available for structural characterization, and it is often the method of choice for the rapid identification of compounds separated by LC. Among the advantages of MS analysis are (a) ability to provide the elemental analysis and molecular weight of the compound; (b) ability to identify various fragments of the molecule; (c) requirement of very small sample size. It is beyond the scope of this treatment to discuss the details of MS for characterizing unknown compounds; see (20) and (21) for information of this type. Although ionic solutes generally cannot be handled by the widely used electron-impact MS method, a surprising variety of relatively nonvolatile solutes have been characterized by this approach (see Table 14.7). For less volatile or ionic compounds, chemical ionization (21a), field desorption (21b), and other special MS sampling techniques can usually be employed (21).

The manual sampling approach in MS is very simple and can be carried out without the complex interfacing system described in Section 14.3. However, manual sampling can be tedious if a very large number of peaks from a separation must be identified. Another potential (but less common) limitation of the manual method is the loss of sample during handling and workup.

Transfer of an isolated LC fraction into the mass spectrometer inlet usually is readily accomplished by first evaporating the trapped fraction to one or two drops (25-50 μl). This small volume is then transferred with a microsyringe and slowly evaporated on the MS insertion probe for subsequent characterization.

The sensitivity of MS analysis depends on the instrumentation and the nature of the sample, but under particularly favorable circumstances less than 1.0 ng of a component can be characterized. Typically, a minimum sample size of 5-10 ng is preferred for MS characterization. Often the limitation of a characterization is

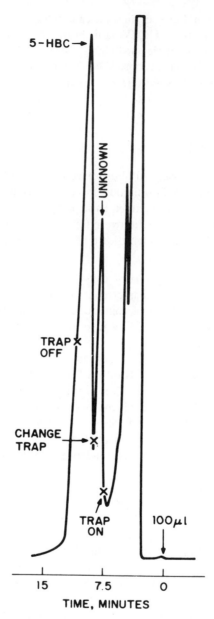

Figure 14.6 Isolation of pesticide metabolite from cow urine extract. Column, 100 × 0.21 cm, Zipax SCX (strong cation resin); mobile phase, 0.1 *N* acetic acid-0.1 *N* sodium acetate, pH 4.5; flowrate, 0.45 ml/min; temp., 60°C; detector, UV, 254 nm; sample. 100 μl of extract. 5-HBC = 5-hydroxy-2-benzimidazole-carbamate. Reprinted from (*22*) with permission.

not the inherent sensitivity of the mass spectrometer, but rather the level of background impurities that are in the isolate. It is not always appreciated that very slowly eluting solutes from previous separations can create substantial MS background peaks that obscure component identification. Thus, it is imperative that a column be carefully purged before undertaking the isolation and characterization of trace components. It is also important to obtain a control or "blank" (trapped) sample to ensure that the background is adequately free of possible interferences to the desired characterization, whether it is done by mass spectrometry or some other technique. This is conveniently accomplished by collecting a volume of eluent from a "clean" column prior to the injection of the sample containing the peak(s) of interest. The volume of this blank or control fraction should be the same as the fraction to be characterized. This blank isolate is then evaporated and analyzed in the same fashion as the unknown.

The identification of unknown components by MS can be illustrated by an actual example involving metabolites of the fungicide benomyl (*22, 23*). In the LC analysis of milk from cows fed with benomyl-fortified feeds, an unknown component was observed eluting adjacent to the anticipated 5-hydroxy metabolite 5-HBC (Figure 14.6). This unknown peak was found in the milk only at the highest feeding rate (50 ppm benomyl in feed), but was at levels too low to permit convenient identification. However, at the highest feeding rate this unknown component was found at sufficient levels in the cow urine (about 2 ppm) to attempt the desired characterization. The cow urine was extracted and separated by the same LC method. Five 100-μl sample aliquots were chromatographed, and the unknown peak (IV in Figure 14.6) was isolated. The combined isolated fractions were adjusted to pH 7, extracted in ethyl acetate, back-extracted with water, filtered through anhydrous sodium sulfate to eliminate water, and evaporated to 25-50 μl for MS (electron-impact) analysis. Figure 14.7*A* shows the mass spectrum of the unknown, which suggested methyl-4-hydroxy-2-benzimidazolecarbamate (by comparison with a synthesized sample of this material, Figure 14.7*B*). The structure was further confirmed by infrared spectrophotometric analysis, using the KBr technique described below for comparison to the authentic sample. The MS pattern shown in Figure 14.7*A* was obtained on approximately 100 ng of metabolite; smaller amounts could have been identified if the situation had demanded.

Infrared Spectroscopy. Specific information on functional groups, possible structure assignment, and confirmation of postulated structure are the particular strengths of infrared spectroscopic (IR) analysis. IR is probably the second most common technique for identifying LC isolates and is particularly useful in supplementing mass spectrometry information (e.g., for defining isomeric structures). In the example of Figure 14.7, IR was important in establishing that the positional isomer was in fact the 4-hydroxy structure, rather than another isomer. Positive assignment of this configuration was not possible by MS alone.

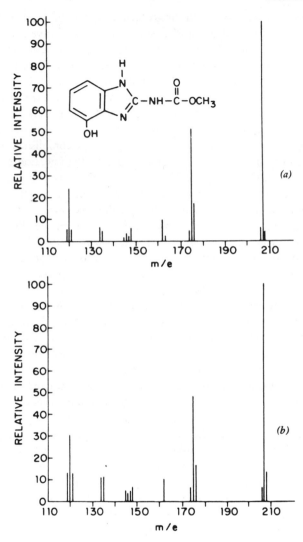

Figure 14.7 Mass-spectrometric identification of pesticide metabolite. Mass spectra: (A) isolated cow urine metabolite; (B) synthetic 4-hydroxy-2-benzimidazolecarbamate. Reprinted from (22) with permission.

There are several IR sampling techniques that are convenient for the characterization of peaks from modern LC separations [see (24) for a general discussion]. One of the more useful IR sampling methods for identifying LC isolates is the KBr-disk technique (25, 26). With this approach the isolated sample fraction is first evaporated to 1-2 drops. This concentrated solution is carefully and slowly deposited from a microsyringe onto 5-10 mg of infrared-quality, potassium bromide powder; each microdrop of the solvent is evaporated onto the salt with a gentle stream of dry nitrogen. Alternatively, sample fractions may be concentrated onto KBr powder using a special vial with a convex bottom (Figure 14.8), which helps to localize the concentrating solute onto the salt during solvent evaporation. With this approach the entire isolated fraction may be placed without preevaporation into the vial containing KBr powder.

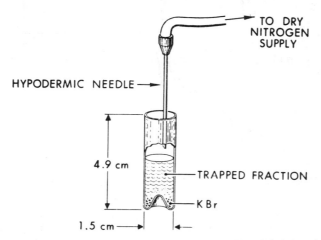

Figure 14.8 Sample concentration technique for micro KBr-disk infrared spectrophotometry. Reprinted from (*26a*) with permission.

After completely removing the solvent by either technique, the potassium bromide powder is mixed, pressed into a micropellet (e.g., <1.5-mm o.d.), and scanned in an infrared spectrometer (with a 5-7 X beam condensor, if highest sensitivity is required). With this procedure good spectra can be obtained using as little as 2 μg of a solute. However, because of losses involved in trapping and sampling, 5-20 μg of solute often is the minimum amount needed to obtain good-quality spectra on a routine basis.

In some cases the potassium bromide disk containing the material to be identified can also be subsequently transferred to the direct insertion probe of a mass spectrometer, where, on heating the probe, the sample is vaporized from the disk into the spectrometer inlet to obtain the mass spectrum (*26a*).

If a Fourier-transform infrared (FTIR) spectrometer is available, the amount of required sample is greatly reduced. In Figure 14.9*a* is the FTIR spectrum of 2-6-dimethoxyphenol obtained by the micro KBr disk technique (0.5-mm i.d.) on 0.5 μg of sample using 300 scans with no scale expansion. Figure 14.9*b* shows 50 ng of the same compound in a 0.2-mm-i.d. KBr disk using 400 scans with 7X scale expansion by the FTIR technique.

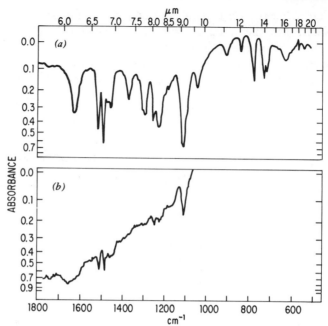

Figure 14.9 Identification by Fourier transform infrared spectrophotometry using Micro KBr-disk technique. IR spectra of 2,6-dimethyloxyphenol in 0.5 mm KBr disks; (*a*) sample in 0.5 μl of 0.1% CS_2 solution (0.5 μg sample) transferred to 0.2-mg KBr, 300 scans, no-scale expansion; (*b*) sample in 0.5 μl of 0.01% CS_2 solution (50-ng sample) transferred to 0.2-mg KBr, 400 scans, 7× scale expansion. Reprinted from (*25*) with permission.

Infrared identification of unknown peaks also can be carried out in solution, using microsampling techniques. The general approach is to load the concentrated isolate into a microcell and scan using solvent compensation. The FTIR method provides the highest sensitivity for obtaining useful spectrum in solution; with care, 300-500 ng of material is sufficient to provide a spectra suitable for qualitative identification or confirmation. However, control or background runs should be made to correct for possible impurities.

Another IR sampling technique that is sometimes useful for identifying LC peaks involves attenuated total reflectance (ATR). With this method the solvent containing the isolated peak is carefully evaporated onto an ATR plate, in a narrow band to just fill the infrared sampling beam. The spectrum is then obtained with normal ATR techniques. The amount of sample required for a satisfactory spectrum normally is several times that of the KBr technique, although ATR/FTIR can be used to decrease sample requirements.

Nuclear Magnetic Resonance Spectroscopy. One very powerful technique for qualitative identification is nuclear magnetic resonance (NMR) spectroscopy;

unfortunately, relatively large amounts of solute are required for satisfactory application. Thus, to fulfill the requirements of NMR examination, repetitive analytical runs or preparative separations often must be carried out to obtain sufficient unknown for characterization. Fourier-transform NMR techniques can greatly increase sensitivity, because repetitive scans over long time periods can be utilized to improve the signal-to-noise ratio. Under favorable conditions, useful spectra on as little as 5 μg can be obtained. With the NMR technique, the LC solvent usually must be completely removed from the isolate and the sample dissolved in an appropriate NMR solvent.

Table 14.5 Illustrative color reactions (spot analysis) for classifying functional groups.

Type of compound	Reagent	Positive reaction	Det. limit (μg)	Compounds tested
Alcohols	$K_2Cr_2O_7-HNO_3$	Blue color	20	C_1-C_8
	Ceric nitrate	Amber color	100	C_1-C_8
Aldehydes	2,4-DNP	Yellow ppt	20	C_1-C_6
	Schiff's	Pink color	50	C_1-C_6
Ketones	2,4-DNP	Yellow ppt	20	C_3-C_8 (methyl ketones)
Esters	Ferric hydroxamate	Red color	40	C_1-C_5 (acetates)
Mercaptans	Sodium nitroprusside	Red color	50	C_1-C_9
	Isatin	Green color	100	C_1-C_9
	Pb (OAc)$_2$	Yellow ppt	100	C_1-C_9
Sulfides	Sodium nitroprusside	Red color	50	C_2-C_{12}
Disulfides	Sodium nitroprusside	Red color	50	C_2-C_6
	Isatin	Green color	100	C_2-C_6
Amines	Hinsberg	Orange color	100	C_1-C_4
	Sodium nitroprusside	Red color, $1°$	50	C_1-C_4
		Blue color, $2°$		Diethyl and diamyl
Nitriles	Ferric hydroxamate-propylene glycol	Red color	40	C_2-C_5
Aromatics	HCHO-H_2SO_4	Red-wine color	20	$\phi H-\phi C_4$
Aliphatic, unsaturated	HCHO-H_2SO_4	Red-wine color	40	C_2-C_8
Alkyl halide	Alc. AgNO$_3$	White ppt	20	C_1-C_5

Source: from (*18a*) with permission.

Color Reaction (Spot Analysis). As illustrated in Table 14.5, selective color reactions for a wide range of functional groups and compound types can be utilized to obtain qualitative information on an isolated unknown component. The sensitivity of this approach is generally much less than that of the MS or IR methods discussed earlier. However, color reactions are simple and can be carried out without expensive equipment. Often the expected retention time plus an appropriate color reaction will be adequate to confirm the structure of a suspected component.

In many cases it is not necessary to eliminate the chromatographic solvent completely; the desired color tests can be made by concentrating the isolated fraction in a small tube for reaction. However, for some reactions the chromatographic solvent must be completely eliminated and a small amount of a different solvent used.

Functional-group analysis as discussed earlier can also be used for qualitative analysis in a somewhat different fashion. Assume that the original sample has been separated in some LC system, and that it is desired to determine the presence (or absence) of a particular group (e.g., $-NH_2$) in one or more separated sample bands. The sample can be subjected to reaction conditions that will lead to the conversion of the functional group of interest; for example, reaction of the $-NH_2$ group with 2,4-dinitro-1-fluorobenzene (DNFB). Reseparation of the reacted sample now results in a decrease in the concentration of all sample bands that contain the amino group, as well as the reappearance of bands that are the products of reaction with various amino derivatives. In certain cases it may be possible to infer additional information on the molecular structure of a compound of interest from the number of derivative bands observed in the chromatogram. For example, a single monofunctional amine will normally give a single derivative with DNFB; a diamino compound could yield two or three derivatives, and so on.

The advantages of the preceding approach to qualitative analysis with respect to color reactions or "spot analysis" include the following:

- Spot analysis is relatively insensitive (see Table 14.5), and many compounds react to only a limited extent with a given derivatizing reagent; the preceding technique requires only a measurable decrease in the concentration of a particular compound of interest.

- Spot analysis is limited to derivatizing reactions that yield colored reaction products, the preceding technique works with any chemical reaction, regardless of whether the derivative formed is colored.

- The reaction technique described allows the use of chemical logic to infer additional information on the sample compound, as in the case of polyfunctional compounds.

A number of reactions specific for various functional groups are listed in Table 14.5 and in Section 17.3 in the discussion of sample derivitization. These reactions are generally limited to those that yield colored products, but of course many additional chemical reactions are available. Specific reactions that provide positive identification of a given sample band are provided by enzymatic analysis. A given biological compound often serves as a substrate for a highly specific enzyme. In such cases analysis of both the original sample and the product of the reaction of the sample in the presence of such an enxyme shows the disappearance of a single band from the chromatogram of the original sample. Then, the identity of the reacted band can be inferred with a high degree of confidence.

The general technique of sample characterization via reaction with group-specific or compound-specific reagents has been employed previously in separations by thin-layer chromatography [e.g., see the discussion in (28) and references cited there; additional examples of possible characterizing reactions are listed in these references].

14.3 ON-LINE SPECTROSCOPIC ANALYSIS OF LC PEAKS

Coupled, on-line instrumental systems have been developed to allow the qualitative identification of an unknown LC peak. Although these on-line systems eliminate some of the disadvantages of the manual methods discussed earlier, they have unique limitations of their own, and may not be applicable in some cases. In addition, coupled, on-line equipment is expensive and requires considerable operator training for optimum results.

Mass Spectrometry

Several methods have been proposed for the direct coupling of a liquid chromatograph to a mass spectrometer:

- Use of atmospheric pressure ionization (API) (29).
- Enrichment of reverse-phase eluents using a membrane interface (30).
- Chemical ionization (CI) using a small fraction of the mobile-phase solvent as a reagent gas (31).
- Transport of solute through differential vacuum locks on a wire or ribbon of metal or plastic (32).
- Reduction of solute to a hydrocarbon and subsequent MS analysis of the reductant (33).
- Formation of a molecular beam by laser vaporization of mobile phase (34).

It is not clear that any one of these methods is superior, because LC/MS is in an early stage of development; each of the cited interface systems has its particular advantages and disadvantages. Just as in GC/MS, there is no single LC/MS interface for all samples and all LC conditions. Rather, one approach may be preferred for a specific sample, whereas another method may be best for another sample.

Table 14.6   Desirable characteristics of an LC/MS interface.

The interface should:
  Allow efficient sample transfer, $\geqslant 30\%$
  Provide reasonably precise sample transfer
  Permit free choice of LC method
  Permit free choice of MS operating mode
  Vaporize sample without decomposition (at least as well as solid probe)
  Retain chromatographic peak integrity
  Permit speed, convenience, reliability, minimum operator skill

---

Source: From (*35*).

Table 14.6 lists what a satisfactory LC/MS interface should do. First, the system should provide for an efficient transfer of the solute peak to maintain high sensitivity; usually $\geqslant 30\%$ recovery is desired. Sample transfer also should be reasonably reproducible. The LC/MS interface should permit a free choice of both the LC method and the MS operating mode for most efficient operation. The sample must be vaporized without decomposition, and in this respect should be at least as good as a solid insertion probe in the manual mode. The LC/MS interface also should retain the integrity of the chromatographic separation; bands should not be significantly broadened by the interface. Finally, the interface should permit fast, convenient, and reliable analysis, with a minimum of operator involvement and skill.

The two most popular and potentially useful LC/MS interface systems are chemical ionization with mobile phase as the reagent gas, and a belt-transport interface. The former has the advantage of simplicity and economy; the latter allows relatively free selection of the LC or MS method and gives a relatively high yield of solute, but is much more expensive.

The belt-interface system used by Finnigan Instruments is shown schematically in Figure 14.10. The LC eluent is introduced directly onto a belt (stainless steel or Kapton) and carried under an infrared evaporator, which aids in solvent evaporation. The solute residue continues along with the belt through two differential vacuum chambers and into the mass spectrometer inlet. There the belt passes through a heated vaporizer, where the sample is flash-vaporized into the

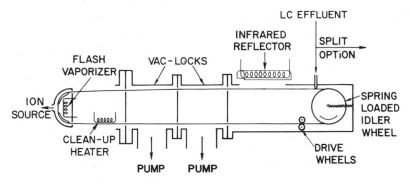

Figure 14.10 Schematic of belt interface for LC/MS. Reprinted with permission of Finnigan Instruments.

electron-impact ion chamber. Residual samples that might cause recycle peaks are removed by a clean-up heater.

The range of chromatographic effluent flow that can be accepted by the belt interface depends on the solvent. Volatile nonpolar solvents (e.g., hexane) can be used at a flow of up to 2.0 ml/min; solvents of intermediate polarity can be used in the range of 0.6-1.2 ml/min. However, aqueous solution must be kept below a flow of 0.2-0.3 ml/min. If it is necessary to exceed these flows to interface a desired chromatographic method, a zero-dead-volume splitter is used.

Although the belt LC/MS interface has been in use a relatively short time, there are a number of examples that illustrate its versatility and flexibility. Table 14.7 lists some of the applications that have been carried out by this approach. Ideally, LC/MS is desirable for any sample that cannot be handled by GC/MS.

Table 14.7   Illustrative LC/MS applications.

Type of sample	Component examples
Mycotoxin	Aflatoxin, zeralenone
Triglycerides	Coconut oil, porcine fat
Prostaglandins	
Bile acids	Cholic acid, cholanic acid
Nucleosides	Cytidine, uridine
Disaccharides	Lactose, trehalose
Porphyrins	Octaethyl porphyrin, meso-porphyrin

Source: From (*35*).

Perhaps surprisingly, there are many types of compounds that are readily amenable to LC/MS analysis, because their vapor pressure and stability are generally sufficient for fast vaporization without excessive decomposition. An illustration of this is shown in Figure 14.11 in the LC/MS analysis of aflatoxins. As seen in the UV chromatogram and the total-ion chromatogram obtained with the mass spectrometer operating in a chemical ionization mode with methane, the four aflatoxins are not completely separated and B_2 and G_1 overlap. However, by utilizing the high specificity of the mass spectrometer, the individual components can be assayed independently, as shown by the ion chromatograms for masses 313, 315, 329, and 331. Although in this case detection limits of 3-30 ng were obtained, a detection limit of 1 ng or less could be anticipated if the mass spectrometer had been operated in a selected ion mode (vs. <1 pg by laser fluorescence; see Section 4.5).

On-line MS identification also can be carried out using the direct chemicalionization technique. About 50 μl/min of the chromatographic effluent is fed to

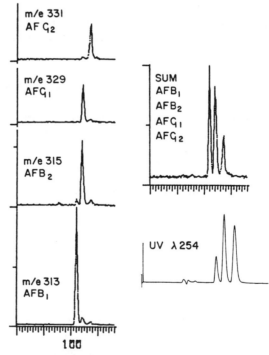

Figure 14.11 LC/MS analysis of aflatoxins. Column, 30 × 0.4 cm, μ-Porasil; mobile phase, $CHCl_3/CH_3OH(75/25)$; flowrate, 0.8 ml/min; UV detector, 254 nm; Finnigan Model 3200 mass spectrometer; chemical ionization mode with CH_4. Reprinted from (*35*) with permission.

the inlet of the mass spectrometer, where this solvent is vaporized and used as the reagent gas for the solute-ionization process (36). A splitter is normally used to select a small portion of the column effluent to be supplied to the mass spectrometer inlet. The CI technique allows relatively simple spectra (e.g., characteristic molecular ions). This is in contrast to electron-impact ionization, where similar molecules are more completely fragmented, leading to a spectrum of lower masses. The CI technique also is capable of detecting nanogram amounts of solute, but overall sensitivity is limited by the fact that only a small portion of the chromatographic effluent is normally utilized.

Atmospheric pressure ionization (API) also has been used as an interface for LC/MS. With this approach a quadrapole mass spectrometer is fitted with a ^{63}Ni radioactive ionization source on the inlet and is operated essentially at atmospheric pressure (29). The ionization source is situated at the end of a heated capillary from which the column effluent is vaporized into a stream of nitrogen immediately adjacent to a small orifice leading into the mass spectrometer. The sensitivity of the API interface for LC is similar to the CI approach, but further development is required to define its potential applicability and limitations.

Infrared Spectroscopy

Fourier-transform infrared (FTIR) spectroscopic techniques applied to LC provide a useful detector for the on-line characterization of unknown peaks. Infrared chromatograms on the various LC peaks allow the usual functional group analysis. In addition, complete absorption spectra can be retrieved at any point on the LC chromatogram (except for regions of solvent opacity) because of the high speed, large signal-to-noise ratio (high sensitivity), and computational power of FTIR systems.

A "moving-band" solvent removal method has been described for the on-the-fly detection of chromatographic effluents with an interferometric infrared spectrometer (37). This detector uses the same principle as the transport detector for LC/MS described earlier, and the LC flame-ionization detector mentioned in Table 4.7. Unfortunately, this detector is mechanically complex, and no such device is available commercially. Recently the Nicholet Instrument Corporation introduced an LC accessory for their FTIR spectrometer that offers several advantages as an LC detector (38). First, several IR bands may be simultaneously monitored in real time. Second, complete spectra corresponding to each point on the chromatogram are available. Finally, the computation equipment required for Fourier transformation is readily available for reduction and display of the large amounts of data generated.

With the FTIR approach, a 3-mm fixed cell having 21 μl of internal volume is utilized with the LC instrument to collect the desired infrared spectrograms "on-the-fly." The equipment permits the monitoring of up to five IR bands simul-

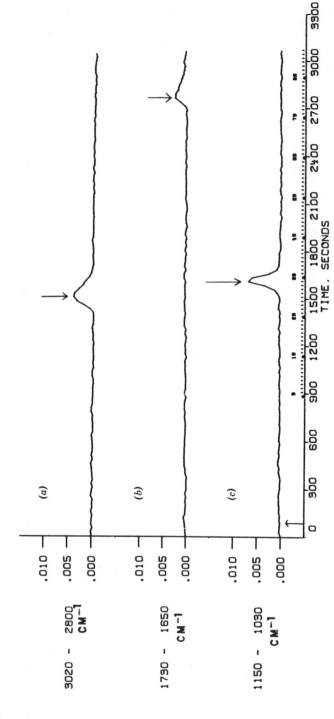

Figure 14.12 LC/IR characterization of SEC separation. Columns, 2-30 × 0.8 cm, μ-Styragel 100 Å and 1000 Å; mobile phase, CCl$_4$; flowrate, 0.5 ml/min; Nicolet LC/IR system, 0.2-mm KBr flowcell; 20 μl sample; (a) 99-ng "Nujol" parafin oil, (b) 178-ng phenyl siloxane oil DC704, and (c) 61-ng acetophenone in CCl$_4$. Reprinted from (38) with permission.

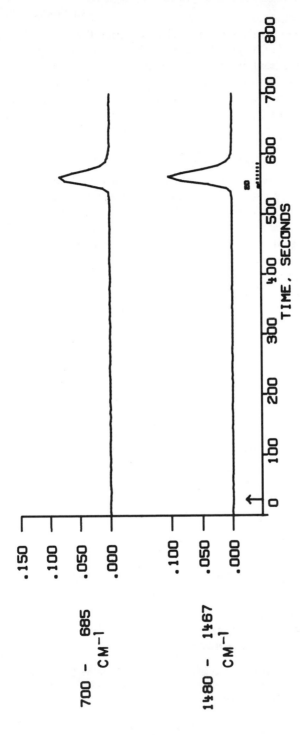

Figure 14.13 LC/IR "chemigram," from SEC of commercial xylenes. Column, 30×9 0.8 cm, μ-Styragel 100 Å; rest as for Figure 14.12: absorbance plots: 700–685-cm window for *m*-xylene, 1480–1467 cm⁻¹ window for *o*-xylene. Reprinted from (38) with permission.

609

taneously, to be plotted continuously in real time during the run. In favorable situations about 100 ng of a typical solute can be detected by this approach, using the C-H stretching band at 3.5 μm and a data acquisition time of 2-4 sec. Figure 14.12 shows the absorption versus retention time plots in various wavelength regions for detection of 99 ng of Nujol (curve *A*), 61 ng of acetophenone (curve *B*), and 178 ng of CD 704 silicone oil (curve C), by monitoring at the C—H, C=O, and C—O stretching absorption bands, respectively, for each solute separated by size exclusion.

The large number of absorption bands present in an IR spectrum offer the simultaneous possibility of both universal and chemically specific detection. In using LC/FTIR, compounds with certain functional groups can usually be identified by simple inspection of the chromatograms at appropriate wavelengths. It is often desirable to plot several chromatograms at once, each one specific for a particular IR band. Because of the multiplex character of FTIR, a complete IR spectrum is available for any point of interest on the chromatogram; therefore, chemical identity of chromatographic peaks can be confirmed by comparison with library spectra.

As discussed in Section 4.6, some LC solvents are too opaque for use in conventional infrared detection with typical cells of about 3-mm path length. However, because many common LC solvents can be utilized with FTIR (e.g., tetrahydrofuran), LC separations requiring the use of relatively opaque solvents often can yield useful spectral information. The polar solvents used in most reverse-phase LC have large opaque infrared regions. However, with thinner flow cells (e.g., 0.1 mm), important portions of the spectrum remain usable when FTIR is employed.

Because the IR detector is insensitive to flowrate, flow programming requires no special adaptation to LC/FTIR systems. Gradient elution normally is a diffi-

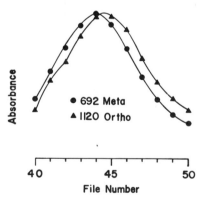

Figure 14.14 Partial SEC resolution of xylene isomers confirmed by LC/IR. Same conditions as Figure 14.13 except plots at frequencies noted. File number also represents time base. Reprinted from (*38*) with permission.

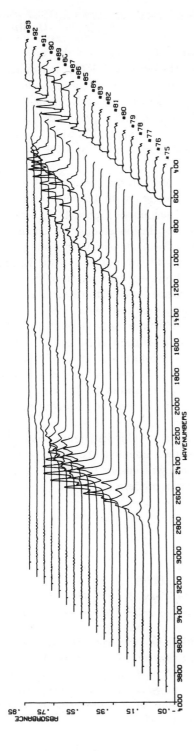

Figure 14.15 Three-dimensional plot (time, frequency, absorbance) of SEC separation by LC/IR. Columns 3-30 × 0.8 cm, μ-Styragel 100 Å, 1000 Å, 10000 Å; rest same as for Figure 14.12, except: sample, 0.5 ml containing 0.8% DC701 silicone oil, 1.3% mineral oil, and 0.3% benzene. Each spectrum slightly offset to represent a Z axis in time. The 680-660-cm⁻¹ window sensitive to phenyl-silicone DC701 and benzene; 1100-1040-cm⁻¹ window sensitive only to silicone 701; 2900-2850-cm⁻¹ window sensitive to paraffin oil only. Reprinted from (38) with permission.

cult technique to accomplish with LC/IR, because the solvent background spectrum is continuously changing. However, with FTIR, individual background spectra for each point of the chromatogram can be acquired either by synthesizing background spectra from the spectra of the two solvents used in the gradient or by acquiring solvent spectra from an identical preliminary chromatographic run without sample injection. The second method is widely applicable, and is usually the method of choice for peak identification by LC/FTIR when gradient elution is used.

The sophisticated data system and software programs available with the FTIR system permit a wide range of application with LC. For instance, chromatographic peak homogeneity can be determined. Figure 14.13 is an IR "chemigram" from a 100-Å μ-Styragel size-exclusion column, of commercial xylenes containing mostly the meta-isomer, with some para-isomer and a small amount of ortho-isomer. The spectra of early and late portions of the chromatographic peak were not obviously different; however, subtraction of the early from the late part of the peak resulted in a spectrum of ortho-xylene. Plotting peak absorbance values for frequencies characteristic of meta- (692 cm^{-1}) and ortho-xylene (986 cm^{-1}) confirmed that the isomers spearate with R_s ~0.1, as shown in Figure 14.14. Such an incomplete resolution of isomers is undetectable and useless for LC analyses, but adequate resolution is available by using the FTIR approach. LC/FTIR intrinsically produces three-dimensional information (time, frequency, and abosrbance). Figure 14.15 shows the three-dimensional plot of the separation of DC704 silicone oil, mineral oil and benzene by size-exclusion chromatography.

As with LC/MS, the LC/IR technique is still in its infancy as an integrated tool for the qualitative analysis of unknown components. It can be predicted with confidence that future developments will ensure that both techniques play a prominent part as real-time interfaces in modern LC.

REFERENCES

1. L. R. Snyder, *Acc. Chem. Res.*, 3, 290 (1970).

2. N. Tanaka and E. R. Thornton, *J. Am. Chem. Soc.*, 99, 7300 (1977).

3. N. Hadden et al., *Basic Liquid Chromatography,* Varian Assoc., Walnut Creek, Calif., 1971, Chap. 7.

4. L. R. Snyder, *Principles of Adsorption Chromatography,* Dekker, New York, 1968.

5. R. F. Rekker, *The Hydrophobic Fragmental Constant,* Elsevier, Amsterdam, 1977.

5.a J. F. K. Huber, C. A. Meijers, and J. A. R. J. Hulsman, *Anal. Chem.,* 44, 111 (1972).

6. S. R. Bakalyar, R. McIlwrick, and E. Roggendorf, *J. Chromatogr.,* 142, 353 (1977).

6a. B. L. Karger, L. R. Snyder and C. Horvath, *An Introduction to Separation Science,* Wiley-Interscience, New York, 1973, pp. 55-57.

7. J. G. Hendrickson and J. C. Moore, *J. Polym. Sci.,* Part A-1, **4**, 167 (1966).

8. L. R. Snyder, *J. Chromatogr.,* **63**, 15 (1971).

8a. L. R. Snyder and B. E. Buell, *J. Chem. Eng. Data,* **11**, 545 (1966).

9. L. R. Snyder, *Anal. Chem.,* **46**, 1384 (1974).

9a. E. Soczewinski and W. Golkiewicz, *J. Chromatogr.,* **118**, 91 (1976).

10. L. R. Snyder, *Adv. Chromatogr.,* **4**, 3 (1967).

11. H. Colin and G. Guiochon, *J. Chromatogr.,* **141**, 289 (1977).

12. J. F. Schabron, R. J. Hurtubise, and H. F. Silver, *Anal. Chem.,* **49**, 2253 (1977).

12a. J. K. Baker, R. E. Skelton and C. -Y. Ma, *J. Chromatogr.,* **168**, 417 (1979).

12b. M. Kuchar, V. Rejholec, M. Jelinkova, V. Rabek and O. Nemecek, *ibid.,* **162**, 197 (1979).

12c. S. H. Unger, J. R. Cook and J. S. Hollenberg, *J. Pharm. Sci.,* **67**, 1364 (1978).

12d. K. Callmer, L.-E. Edholm, and B. E. F. Smith, *J. Chromatogr.,* **136**, 45 (1977).

13. I. Molnar and C. Horvath, *J. Chromatogr.,* **142**, 623 (1977).

14. L. R. Snyder and B. E. Buell, *Anal. Chem.,* **40**, 1295 (1968).

15. S. C. Su, A. V. Hartkopf, and B. L. Karger, *J. Chromatogr.,* **119**, 523 (1976).

16. C. Horvath, W. Melander, I. Molnar, and P. Molnar, *Anal. Chem.,* **49**, 2295 (1977).

17. R. P. W. Scott and P. Kucera, *J. Chromatogr.,* **142**, 213 (1977).

17a. J. J. Kirkland, *Anal. Chem.,* **43**, 43A (1971).

18. R. Yost, W. MacLean, and J. Stoveken, *Chromatogr. Newsl.,* **4**, 1 (1976).

18a. J. T. Walsh and C. Merritt, Jr., *Anal. Chem.,* **32**, 1378 (1960).

19. W. Loesche, R. Bublitz, A. Horn, W. Koehler, H. Petermann, and U. Till, *J. Chromatogr.,* **92**, 166 (1974).

20. M. C. Hamming and N. G. Foster, *Interpretation of Mass Spectra of Organic Compounds,* Academic, New York, 1972.

21. M. L. Gross, ed., *High Performance Mass Spectrometry: Chemical Applications,* ACS Symposium Series 70, American Chemical Society, Washington, D.C., 1978.

21a. M. A. Baldwin and F. W. McLafferty, *Org. Mass Spectrom.,* **7**, 111 (1973).

21b. N. Evans, D. E. Garnes, A. H. Jackson, and S. A. Matlin, *J. Chromatogr.,* **115**, 325 (1975).

22. J. A. Gardiner, J. J. Kirkland, H. L. Klopping and H. Sherman, *J. Agric. Food Chem.,* **22**, 419 (1974).

23. J. J. Kirkland, *J. Agric. Food Chem.,* **21**, 171 (1973).

24. R. G. J. Miller and B. C. Stace, eds., *Laboratory Methods in Infrared Spectroscopy,* 2nd ed., Heyden, London, 1972.

25. J. J. Kirkland, *Anal. Chem.,* **29**, 1127 (1957).

26. R. W. Hannah and S. C. Pottacini, Pamphlet no. IRAS-8-PEP57110, Perkin-Elmer Corp., Norwalk, Conn., 1971.

26a. A. A. Juhasz, J. O. Doali, and J. J. Rocchio, *Am. Lab.,* **6**, 23 (1974).

27. S. S. T. King, *J. Agric. Food Chem.,* **21**, 526 (1973).

28. L. R. Snyder, *Principles of Adsorption Chromatography,* Dekker, New York, 1968, p. 351.

29. D. I. Carroll, I. Dzidic, R. N. Stillwell, K. D. Haegele, and E. C. Horning, *Anal. Chem.,* **47**, 2369 (1975).

30. P. R. Jones and S. K. Yang, *Anal. Chem.,* **47**, 1000 (1975).

31. F. W. McLafferty, R. Knutti, R. Venkataraghavan, P. J. Arpino, and B. G. Dawkins, *Anal. Chem.,* **47**, 1503 (1975).

32. W. H. McFadden, H. L. Schwartz, and S. Evans, *J. Chromatogr.,* **122**, 389 (1976).

33. W. L. Erdahl and O. S. Privett, *Lipids,* **12**, 797 (1977).

34. C. R. Blakley and M. L. Vestal, "Application of Crossed-Beam LC/MS to Involatile Biological Samples," Paper D7, 25th Annual Conference on Mass Spectrometry, Washington, D.C., May, 1977.

35. W. H. McFadden, D. C. Bradford, D. E. Games, and J. L. Gauer, *Am. Lab.,* **9**, 55 (1977).

36. P. J. Arpino, H. Colin, and G. Guiochon, "On-line Liquid Chromatography-Mass Spectrometry," Paper A8, 25th Annual Conference on Mass Spectrometry, Washington, D. C., May, 1977.

37. P. R. Griffiths and D. Kuehl, *Anal. Chem.,* **50**, 418 (1978).

38. D. W. Vidrine, in *Fourier Transtorm Infrared Spectroscopy,* J. Ferraro and L. Basile, eds., Academic, New York, in press.

BIBLIOGRAPHY

Budzikiewicz, H., C. Djerassi, and D. H. Williams, *Interpretation of Mass Spectra of Organic Compounds,* Holden-Day, San Francisco, 1964 (excellent coverage on interpreting mass spectra).

Burlingame, A. L., C. H. L. Shackleton, I. Howe, and O. S. Chizhov, *Anal. Chem.,* **50**, 346R (1978) (review of MS literature since 1974).

Hamming, M. C., and N. G. Foster, *Interpretation of Mass Spectra of Organic Compounds,* Academic, New York, 1972 (general aspects of mass spectrometry).

McEwen, C. N., and C. Merritt, Jr., eds., *Practical Mass Spectroscopy,* Practical Spectroscopy, Vol. 3, Dekker, New York, 1978 (state-of-the-art articles on latest MS techniques).

Miller, R. G. J., and B. C. Stace, eds., *Laboratory Methods in Infrared Spectroscopy,* 2nd ed., Heyden, London, 1972 (comprehensive treatment of IR laboratory techniques).

Snyder, L. R., *Principles of Adsorption Chromatography,* Dekker, New York, 1968 (theory of correlating LSC retention with solute structure).

Vidrine, D. W., in, *Fourier Transform Infrared Spectroscopy,* J. Ferraro and L. Basile, eds., Academic, New York, in press (state-of-the-art treatment of LC/IR technology).

FIFTEEN

PREPARATIVE LIQUID CHROMATOGRAPHY

15.1	Introduction	615
15.2	Separation Strategy	617
15.3	Experimental Conditions	623
	Columns	624
	Column Packings	626
	Solvents	634
	Equipment	636
15.4	Operating Variables	641
	Sample Volume	641
	Sample Mass	642
	Optimizing Preparative Throughput	643
	Recovery and Purity Determination	646
15.5	Applications	647
15.6	A Preparative Separation Example	654
	Summary of Preparative LC Conditions	660
	References	661

15.1 INTRODUCTION

Previous chapters in this book have emphasized the analytical aspects of LC, in which the emphasis is on acquiring qualitative and/or quantitative information on the sample. In this chapter we discuss the preparative aspects of LC, as an effective and convenient technique for isolating relatively large amounts of purified components. These purified materials may be needed as standards, synthesis intermediates, testing materials, and so on.

The phrase, *preparative liquid chromatography* often is used to describe the isolation process, rather than defining the quantity of material isolated. However, in this chapter we define preparative LC as a technique for isolating significant amounts (>0.1 g) of pure compounds using large-diameter columns. As indicated in Table 15.1, the amount of purified sample to be isolated by LC depends on the objective. Generally, sufficient quantities for identification by instrumental methods can be obtained with analytical systems. More elaborate confirmations may require greater amounts of sample, which can be isolated by scale-up or semipreparative methods. Large amounts of purified materials usually require operating conditions different than those for analysis, and it is to this area that the discussion in this chapter is mainly directed.

Table 15.1   Requirements for purified samples.

Objective	Sample weight required (mg)
Tentative identification by instrumental methods	<1
Positive identification by instrumental methods; confirmation of structure by chemical reactions	1-100
Use as analytical standard, in synthesis, or for testing	>100

As discussed under controlling resolution in Section 2.5, the three main goals of any chromatographic separation (resolution, separation speed, and sample loading or capacity) are interrelated. If we want to emphasize any one of these three goals, the other goals must usually be compromised. Thus, in analytical LC, information on samples is required; speed and resolution are emphasized, and sample size is of necessity reduced. However, in preparative LC we seek purified sample; to achieve it we emphasize capacity, and usually we compromise on speed and/or resolution.

Because sample capacity is a prime goal of preparative LC, technique and equipment are often different from those used for analysis. One of these differences can be illustrated in Figure 15.1. These plots show that analytical separations by LSC typically are performed with sample weights of less than 1 mg/g of adsorbent, which roughly represents the column linear capacity limit, $\theta_{0.1}$ (cf. Figure 2.6). In this range of sample weights (<1 mg/g), peak retention (k' values) and column efficiency change little with sample size, and sample component isolations can be carried out with larger diameter columns in much the same manner as analytical separations. This so-called *scale-up chromatography* performs separations with resolutions equivalent to analytical separations. As

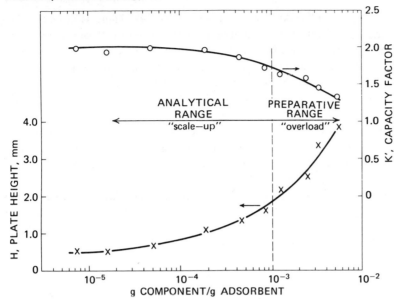

Figure 15.1 Effective sample weight in analytical and preparative liquid chromatography. Upper curve, solute capacity factor, k'; lower curve, column plate height, H; column, 50 × 10.9 cm i.d. with 10%(w/w) water on 35-75-μm Porasil-A; mobile phase, chloroform (50% water saturated); mobile-phase velocity, 0.25 cm/sec; sample, diethylketone in chloroform. Reprinted from (*1*) with permission.

discussed later in this chapter, scale-up often is used with columns of <10-μm particles for very difficult separations, to prepare modest amounts of highly purified components from complex mixtures or other samples requiring high resolution. On the other hand, to enhance *preparative throughput* (i.e., yield of component of a certain purity per unit time), preparative separations are performed with wide-diameter columns using large sample loads (greater than 1 mg/g of packing). We can define *column overload* in preparative chromatography as the condition in which the column linear capacity $\vartheta_{0.1}$ values are exceeded (i.e., k' values are lowered by 10% or more from the values for analytical separations). We should also note that column efficiency decreases greatly in overload, as indicated in Figure 15.1.

15.2 SEPARATION STRATEGY

Optimization of conditions for all forms of preparative LC proceeds similarly to that for analytical separations (Chap. 2). It is necessary to use the proper solvent strength (k'), adequate column efficiency (N) (but less important in column overload), and good separation selectivity (α). There is not yet agreement on the best k' values in preparative LC for all cases (2-4); optimum k' may depend on the amount of sample injected. When the column is operated in an overload condition, column efficiency can be increased more profitably by increasing column length (because of packing volume increase), rather than by reducing column pressure (or flowrate). However, other factors discussed in the next section should be considered. Good separation selectivity (or favorable α values) are more important in preparative LC than in analytical separations (cf. Figure 2.27). Because overall separation time can be greater for preparative applications, it is usually worthwhile to spend more time initially exploring how selectivity varies with experimental conditions.

The effect of chromatographic parameters on an analytical separation can be predicted by the resolution (R_s) relationship (Section 2.4):

$$R_s = \frac{1}{4}(\alpha - 1) \sqrt{N} \left(\frac{k'}{1 + k'} \right) \tag{5.1}$$

With a nonoverloaded column, resolution is relatively predictable; α and k' are essentially independent of mobile-phase velocity and other kinetic effects. However, when a column is overloaded, the commonly accepted relationships involving chromatographic resolution no longer apply quantitatively. Large sample weights grossly affect column equilibrium, and all three terms of the resolution expression, N, k', and α, are altered as the degree of overload varies. As exempli-

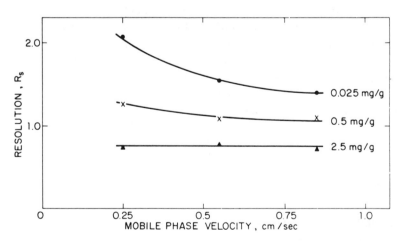

Figure 15.2 Effects of mobile-phase velocity and solute weight on resolution. Mobile phase and column as shown in Figure 15.1; sample, diethylketone in chloroform; plots show milligram of total sample injected per gram of packing. Reprinted from (1) with permission.

fied in the data of Figure 15.2, the resolution of a column in a nonoverload condition (e.g., <0.5 mg of solute per gram of adsorbent) is significantly dependent on mobile-phase velocity u; both α and k' remain essentially constant during the separation, and only the plate number N is affected by u. However, when the column is overloaded, as is often the case in preparative LC (e.g., greater than 1 mg of solute per gram of adsorbent in LSC), both α and k' are drastically decreased with increasing sample load, and the effect of mobile-phase velocity on N (and resolution) is now greatly reduced. One result of this effect is that maximum sample throughput per unit of time is achieved by deliberately overloading the column and using the highest possible mobile-phase velocity (which still permits adequate resolution). In a sample-overload condition, column plate number is not so important; rather, it is the quantity of packing required to accommodate a band that largely determines the final overall resolution.

To summarize, there are *two* different approaches in preparative isolation, depending on the separation goal. First, in scale-up separations, efficient large-diameter columns are used in essentially an analytical (nonoverload) mode to prepare modest amounts of materials that may be very difficult to separate. Resolution is optimized by adjusting the separation parameters (Eq. 5.1) in the

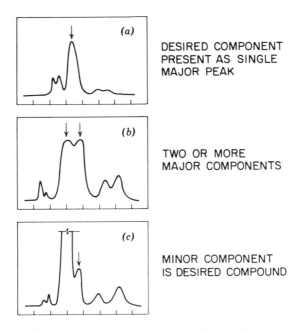

Figure 15.3 Typical situations encountered in preparative liquid chromatography. (*a*) Desired component present as single major peak. (*b*) Two or more major components. (*c*) Minor component is desired compound. Reprinted from (5) with permission.

normal fashion (Section 2.4). In this preparative approach, small-particle (~10-μm) columns often are used for highest resolution and highest product purity, and for fast separation times. Longer columns of larger particles can accomplish the same separation, but with significantly increased separation times, as discussed later. In the second preparative approach, large-diameter columns of large particles (~50-μm) are operated in an overload fashion to prepare relatively large amounts of purified material, particularly where α values can be made fairly large (e.g., >1.2). Separation time can be relatively short in column overload, because relatively high flowrates can be used without seriously degrading resolution.

The most commonly encountered situations in preparative LC are illustrated in Figure 15.3. In the situation of Figure 15.3a, the preparative isolation of a single major component is best handled by using the scheme in Figure 15.4. Starting with the analytical separation (Figure 15.4a), resolution first is enhanced (Figure 15.4b) by optimizing k', α, and N. Next, the weight of sample introduced into the column is increased until the peaks begin to overlap (Figure 15.4c). If only modest amounts of sample components are required or if maximum resolution is needed for separating closely eluting components, purified materials can be isolated at this sample loading in amounts that will be adequate for some purposes. In this *loading-limit* condition, the column is operated very much as for analytical separations, but the diameter is scaled up to allow maximum sample size. Repetitive separations can be carried out as in Figure 15.4c to increase the total yield of purified material to desired levels.

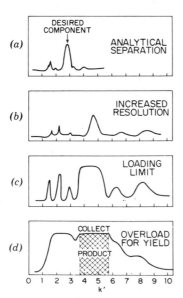

Figure 15.4 Maximizing preparative yield. Reprinted from (5) with permission.

If much larger amounts of purified material are needed, the column should be overloaded (Figure 15.4*d*). Even though peaks are overlapping, a *heart-cut* (Figure 15.4*d*, cross-hatched portion) will produce a highly purified component with a greater throughput per unit of time than is possible by separation at the loading limit (Figure 15.4*c*). It is often desirable to collect smaller fractions through the heart-cut range (Figure 15.4*d*), then assay these by TLC or analytical LC for purity. Sufficiently pure fractions can then be blended into the final purified fraction.

The heart-cut technique can be used advantageously as a prefractionation step to expedite the purification of a closely eluting pair of components (as in Figure 15.3*b*), or of a single component surrounded by closely eluting impurities. This approach is illustrated by the hypothetical separation in Figure 15.5. In Figure 15.5*a* we see the analytical chromatogram of a mixture in which

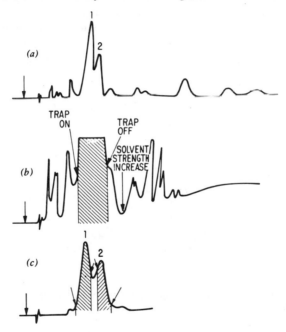

Figure 15.5 Optimizing isolation of closely eluting solute pair. (*a*) Isocratic analytical separation. (*b*) Isolation of combined solutes of interest using gross column overload with solvent strength increase to purge column. (*c*) Isolation of purified individual components.

large amounts of closely similar components 1 and 2 each are desired in high purity. Because there are many additional components eluting well after the peaks of interest, normal isocratic isolation would require a considerable time before these undesired impurity peaks cleared the column and a second separation could be made. A useful approach for this situation is illustrated in Figure 15.5*b*, where the preparative column is first heavily overloaded with the sample. Resolution is not critical here, so very large sample weights can be injected. Using an isocratic spearation mode, a heart-cut is taken to collect a fraction containing peaks 1 and 2 (cross-hatched portion in Figure 15.5*b*). Immediately after collecting this fraction, the strength of the mobile phase is sharply increased to elute quickly the components of no interest from the column. (Additional impurities not seen in an isocratic separation sometimes will elute under this condition). After reequilibrating the column to the initial isocratic state, peaks 1 and 2 may be rechromatographed at higher resolution, as shown in Figure 15.5*c* (e.g., by decreasing sample size with repetitive sample-injection, or by recycle, as discussed later).

Manual or automatic recycle (Section 12.6) can be used to effect the resolution of closely eluting components. As illustrated in Figure 15.6, if an abundance of sample is available for injection (such as that prepared by the heart-cut technique just described), direct collection of components *A* and *B* in the cross-hatched front and back "wing" portions of the overlapping peaks will produce

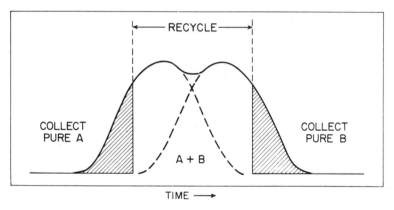

Figure 15.6 Recovery of two incompletely resolved components. Reprinted from (5) by permission.

highly purified materials. Again, repetitive separations may be required to obtain very large amounts of purified materials. If the starting sample is in short supply, the overlapping middle portion *A* + *B* can be recycled for further collection of the purified component "wings."

Because the bulk of the original impurities in a sample are eliminated by using the heart-cut technique illustrated in Figure 15.5*b*, repetitive isolations can be carried out rapidly, or recycle separation of the desired compounds can be more readily accomplished. If significant impurities do not elute after the peaks of interest, increasing solvent strength is not required (e.g., as in Figure 15.5*b*) and the column can be allowed to clear in the normal isocratic mode. This eliminates the necessity of reequilibrating the column after a mobile-phase change.

When the desired material is a minor or trace component, as in Figure 15.3*c*, a slightly different preparative procedure is used, as illustrated in Figure 15.7. The trace component seen at the loading limit of the column (*a*) is enriched by overloading the column after optimizing resolution as discussed earlier. As in (*b*),

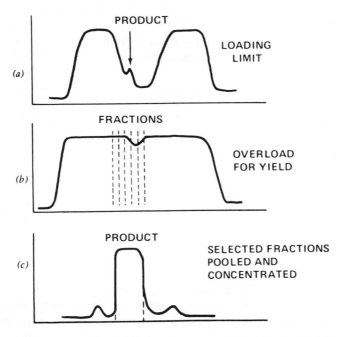

Figure 15.7 Isolation of minor or trace components. Reprinted from (*5*) with permission.

fractions are collected in the elution region expected for the desired component. The isolated fractions then are analyzed by an appropriate technique and those containing higher concentrations of the trace component are pooled and concentrated. This enriched sample is then rechromatographed for final purification (Figure 15.7*c*) using, for example, the approach described in Figure 15.3*a*.

15.3 EXPERIMENTAL CONDITIONS

Optimizing the preparative LC system for sample capacity involves a different set of operating conditions than are normally used for analytical separations, as exemplified by comparing typical separation parameters listed in Table 15.2. In the following sections we discuss in more detail the unique experimental conditions that should be used for preparative LC.

Table 15.2 Comparison of typical operating parameters.

Parameter	Analytical	Preparative
Column diameter	2-5-mm i.d.	>10-mm i.d.
Column packing	10-μm totally porous or 30-μm pellicular	10-μm or (better) 50-μm totally porous
Mobile-phase flowrate	~1 ml/min	>10 ml/min
Mobile-phase velocity	0.1-1 cm/sec	0.01-0.5 cm/sec
Sample volume	<200 μl	>1000 μl
Detector sensitivity required	High	Low
Weight of injected sample	<0.5 mg	>100 mg

Columns

In preparative LC, sample capacity usually is increased by using columns of larger internal diameter. As discussed in Section 5.4 for analytical columns, scaling up the diameter of a preparative column does *not* reduce chromatographic efficiency. For example, for nonoverloaded columns, the plate count of dry-packed ~50-μm particles can increase by more than twofold with increase in column internal diameter (6). The same relationship has been noted for columns with particles as small as 5 μm (7) and may involve the infinite-diameter column effect (Section 5.6), a complicated interaction between particle size, column internal diameter, column length, and packing technique. Consequently, for nonoverloaded systems, separations with large-diameter columns frequently are superior to those obtained with narrow-bore columns, provided the same ratio of sample weight to cross-sectional areas is maintained and the remaining chromatomatographic parameters are held constant. There have been no detailed studies on the effect of column internal-diameter on efficiency or resolution for overloaded columns. However, it is reasonable to assume that overload separations are no less efficient as column diameter increases. A practical upper limit on column diameter in the overload mode has not yet been established, but commercial apparatus with 8-cm i.d. is used effectively. Columns with even larger internal diameters for large-scale plant processing appear feasible for producing very large amounts (kilograms) of highly purified products.

Stainless steel tubing that can withstand pressures of up to 3000 psi is used most commonly for making preparative columns. However, for pressures up to 150-200 psi, large-diameter glass columns can be used.

As with analytical columns (Section 5.4), preparative LC columns are prepared in straight sections that can be connected when longer columns are needed. One commercial supplier (Varian Associates) furnishes semipreparative, 0.8-cm i.d. columns of 10-μm particles with tapered inlets, and another (Whatman) supplies columns with tapered outlets. Both approaches are reputed to improve the efficiency of scale-up separations. However, additional studies are needed to prove the practical value of these approaches, because the column blanks are considerably more expensive and it is not clear whether the claimed advantage of the tapered configuration is maintained with columns connected in series.

Techniques for packing preparative LC columns are similar to those used for analytical units (Section 5.4). However, for < 20-μm particles, slurry-packing procedures have to accommodate the lower pressure limits of the larger-diameter column blanks. Also, larger slurry reservoirs are needed to deliver the volume of packing material into the columns. For example, 1-l reservoirs are desirable for filling 25 $\times$ 2-cm i.d. column blanks. The tap-fill dry-packing procedure (Section 5.4) is quite satisfactory for >30-μm particles. However, particular care should be used to tap the columns relatively lightly during the filling, to ensure that sizing of the particles is minimized.

Packed glass columns of about 50-μm silica adsorbent are available in 25 $\times$ 1.0-cm, 31 $\times$ 2.5-cm, and 44 $\times$ 3.7-cm sizes for preparative separations at about 100 psi ("Lobar," E. Merck; EM Laboratories). These columns are packed with the same silica used for Merck TLC (i.e., same selectivity), in a standard, activated state. In addition, the columns are prepared with a uniform efficiency (more than 1200 plates for the smaller sizes) for reproducible separations. Because of their relatively low cost, these columns can be used for a single preparative separation and discarded if desired, or they can be reequilibrated and reused in the usual manner. The former is particularly attractive when the sample contains very strongly held components (as indicated by TLC separation, see Section 15.3), whose elution would require significant time and solvent before the column could be cleaned and reequilibrated for another separation.

To improve the convenience of commercial preparative columns, unique methods for packing the stationary phase into large-diameter columns have been developed. For example, *radial compression* technology is used in preparative LC instrumentation by Waters Associates to produce Prep Pak/500 columns (8). This approach automatically transforms a flexible-wall (plastic) cartridge prefilled with chromatographic packing (e.g., ~75-μm silica or reverse-phase packing) into a preparative LC column by radially compressing the cartridge with nitrogen within a special metal chamber (see Figure 15.14). A differential pressure of at least 200 psi above system back-pressure must be maintained on the external column wall. Each 30×5.7-cm prepacked cartridge can be connected in series to increase effective column length in the usual fashion, or a cartridge can be used for recycle (Section 12.6). These radially compressed columns can only be used in conjunction with the associated commercial apparatus (see below); they have the advantage of relatively low cost plus repeatable column performance (about 400 plates), along with the ability to handle 10-50 g of sample per run.

Longitudinal loading and compression is used in commercial 8-cm-i.d. columns (Chromatospec, Jobin Yvon, J. Y. Optical Systems), as illustrated in Figure 15.8 (*9*). After removing the column head, the packing is loaded into the empty column. The column is then closed and pressure (100-150 psi) is applied to a piston at the bottom of the column. The pressure from this piston tightly packs the stationary-phase particles. The sample and the mobile phase are introduced while the piston is held in the upstroke position. The piston also makes it possible to remove the stationary phase from the cylinder, so that the separation can be carried out with incomplete elution. For example, the separation can be stopped as soon as the first component has eluted, the column head removed, and the adsorbent extruded by pushing the piston up. A UV lamp can detect the different fractions on the column for subsequent elution and recovery. The packing material can be reused for another separation if this is desired.

Column Packings

Totally porous particles should be used in preparative LC, because a large sample capacity is normally the separation goal. Superficially porous or pellicular supports rarely are utilized for preparing large quantities of sample, because these materials have a sample capacity less than one-fifth that of totally porous particles of the same type. Although milligrams of purified components have been isolated with columns of pellicular packing, it is inconvenient and costly to isolate larger amounts of materials using this approach. Table 15.3 lists representative low-cost, high-capacity, totally porous packings useful for preparative separations by the four HPLC methods. Other packings can be purchased in prepacked columns, both for preparative and semipreparative applications (e.g., larger particles: "Lobar," Merck; radially compressed columns, Waters; small particles: "Magnum," Whatman; "Silarex," Prep A, Lion Technology; Chrompack columns, Chrompack, Nederland, B. V., etc.). Additional preparative packings and packed columns are given in Section 5.3.

Liquid-solid (adsorption) chromatography should be used to carry out preparative separations whenever possible. This LC method provides the greatest flexibility and convenience for preparative isolations, with high resolution, good

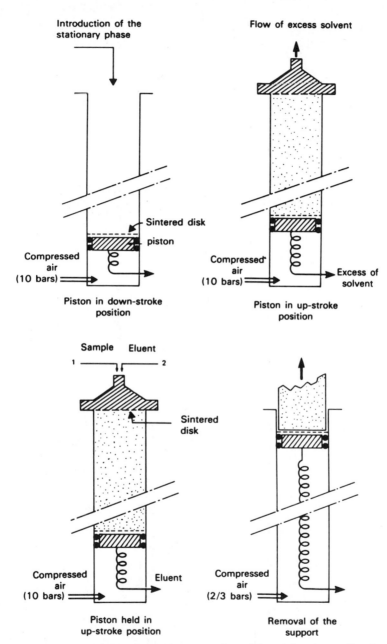

Figure 15.8 Longitudinal loading and compression of 8-cm-i.d. columns. Reprinted with permission of J. Y. Optical Systems.

Table 15.3 Illustrative bulk packings for different preparative LC methods.

Method	Name	Supplier	Type	Size (μm)	Approximate surface area (m²/g)[d]
SEC(GPC-nonaqueous)[a]	Styragel (various pore sizes)	Waters	Cross-linked poly-styrene	<37	N.A.[d]
	Spherosil[a,c] (various pore sizes—also useful for GFC)	Supelco, others	Porous-silica beads	<40	10-400
SEC(GFC-aqueous)[b]	Bio-Gel P-2	Bio-Rad	Polyacrylamide	37-75	N.A.
LSC	Spherosil—XOA—400[c]	Supelco, others	Porous silica	<37	400
	LiChroprep	Merck	Irregular porous silica	10; 15-25; 25-40	500
	Woelm W-200	ICN Pharmaceuticals	Alumina	18-30	200
	Zorbax-Sil	DuPont	Porous silica beads	7	400

LLC	Spherosil –XOBO30	Supelco, others	Porous-silica beads	<40	50
	Dia-Chrom	Applied Science	Diatomaceous earth	37-44	1-3
BPC	Bondapak–C$_{18}$	Waters	Porous-silica beads	37-75	175
	Porasil–B		with bonded C$_{18}$		
			(for reverse-phase)		
	LiChroprep	Merck	Irregular silica	15-25 or 25-40	400
			with C$_2$, C$_8$, or C$_{18}$		
Cation exchange	Rexyn 101 (Na)	Fisher Scientific	Cross-linked polystyrene, sulfonic acid	37-75	N.A.
Anion exchange	Rexyn 201 (Cl)	Fisher Scientific	Cross-linked polystyrene, amine quarternary	37-75	N.A.

[a] For high & low MW

[b] MW = 100-1800

[c] Also various grades of 37-75 μm Porasil from Waters Associates.

[d] N.A. = not applicable.

sample throughput, and moderate cost. LSC also can be used for a broad range of compound types by appropriate selection of the mobile phase [e.g., see (*10*) for the great variety of samples that have been separated with adsorbents by TLC]. Another advantage is that LSC permits the use of mobile phases that have a higher solubility for many compounds, thus allowing higher sample loading and preparative throughput. Furthermore, solvents generally used for LSC are relatively volatile and are easily removed from the isolated fraction containing the desired components. Although most of the information developed has been found with the LSC approach, preparative separations also have been made with the other methods, (e.g., liquid-liquid chromatography). Reverse-phase preparative separations with bonded-phase columns are convenient as a follow-up to analytical separations in this mode. Unfortunately, this method often suffers from relatively low sample capacity because of the poor solubility of many constituents in aqueous mobile phases. Although reverse-phase preparative separations have not yet reached great popularity, this approach has considerable merit for some samples.

No systematic study has been made of the effect of particle size for preparative HPLC with totally porous particles. However, depending on the preparative goals, both small particles (5-10 μm) for high-resolution scale-up separations and coarse particles (30-60 μm) for high-capacity overload separations have been used. Sample capacity increases proportionally with column cross-sectional area, regardless of particle size. For high sample loading, large-diameter columns filled with coarse, uniform particles should be used (*7*). The experimental data in Figure 15.9 show that columns with increasing lengths and larger particles have equivalent separation performances for small samples (0.001-0.01 g in Figure 15.9). However, for equivalent resolution, separation times are longer, relative to columns of small particles. On the other hand, for a given pressure drop, long columns filled with coarse packing materials operate more efficiently at higher sample loads (>0.1 mg for 30-μm particles [○] in Figure 15.9), primarily

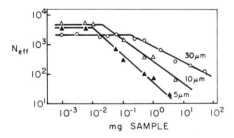

Figure 15.9 Effect of particle size and column length on sample capacity. Columns, ▲, LiChrosorb SI-100, 5 μm, 10 × 0.3 cm; △ LiChrosorb SI-100, 10 μm; 50 × 0.30 cm; ○, LiChrosorb SI-100, 30 μm, 200 × 0.3 cm; mobile phase, hexane, (saturated 50% with water); velocity, 0.8 cm/sec; temp., 22°C; sample, phenetol (k' = 3). N_{eff} = effective plates. Reprinted from (*7*) with permission.

because of the larger weight of packing available for the sample in the longer column. Of course, separation time is usually longer for longer columns of large particles, relative to shorter columns of small particles.

The question often arises: In preparative LC, should columns of small (5-10-μm) particles or larger (30-50-μm) particles be used? The answer to this question depends on the specific preparative goal and experimental capabilities of the laboratory. In difficult preparative separations where α approaches unity, larger N values are required. In these cases small-particle columns used with relatively small sample sizes in a nonoverload (essentially analytical) mode are favored. However, when α is much greater than unity, columns of large particles operated at high sample loads (i.e., overloaded) are advantageous. In Table 15.4 we see a summary of the effects of particle size and column length parameters, for analytical, semipreparative, and preparative columns with equivalent resolution operated in a nonoverloaded state (i.e., <1 mg of sample per gram of packing). Direct comparisons cannot be made in column overload, because the concept of resolution is no longer valid (Section 15.1). In small-particle columns that are mainly used for the preparative separation of compounds with small α values, column overload is normally avoided, because a loss of resolution often cannot be tolerated.

Column A in Table 15.4 shows typical data for a 25 X 0.46-cm i.d. column of 10-μm porous LSC particles used for analysis. When operated with a mobile-phase velocity of about 0.5 cm/sec, this column exhibits a plate count of about 4000. For a typical solute ($k' = 2$-5), an average separation time is about 10 min with maximum weight of solute of about 3 mg before sample overloading causes a 10% decrease in k'. Column B shows that a 2.3-cm i.d. preparative column of 10-μm porous particles similarly operated would exhibit a plate number (and resolution) at least equivalent to the analytical column in the same separating time. However, because of the increased weight of adsorbent in this larger i.d. column, a much larger quantity (about 78 mg) of a single component could be isolated before column overload. This configuration is especially useful for scale-up preparative isolations of modest samples (e.g., 50-200 mg) of very difficultly separable components (4).

Intermediate i.d. (0.8-1.0-cm) "semipreparative" columns of 10-μm particles such as that shown in column C of Table 15.4 represent a practical alternative for certain preparative situations. Such columns are particularly useful to rapidly isolate milligrams or fractional-gram quantities (14 mg for the case in Table 15.3) of difficulty separable components in the scale-up mode. To isolate purified components in amounts equivalent to that from a single separation with a larger (2.3-cm i.d.) column, repetitive isolations are required with the semipreparative column.

Longer columns of larger, less expensive porous particles often represent a convenient alternative for effective preparative separations. Large-diameter, long columns filled with coarse, uniform packing materials can be operated at a high

Table 15.4 Analytical and preparative parameters for equivalent resolution (nonoverloaded columns).

Experimental parameter	(A) Analytical (0.46-cm i.d.)	(B) Preparative (2.3-cm i.d.)	(C) Semipreparative (1.0-cm i.d.)	(D) Preparative (2.3-cm i.d.)
Porous particle size, μm	10	10	10	30
Column length, cm	25	25	25	100
Mobile-phase velocity, cm/sec	0.5	0.5	0.5	~0.08
Theoretical plates	4000	4000	4000	4000
Approximate weight of silica in column, g	3	78	14	250
Maximum sample per separation, mg[a]	3	78	14	250
Typical separation time, min	5-10	5-10	5-10	50-200
Throughput of product, mg/min (same purity)	0.6	16	3	5
Cost of column	Low	High	Modest	Modest

[a]Silica adsorbent. Assumes ~1 mg of sample per gram of packing. Greater sample loadings often can be used in column-overload preparative applications at higher mobile-phase velocities, for increased throughput and shorter separation times; see text.

throughput with a relatively low pressure drop, and have the advantage that they can be conveniently dry-packed (Section 5.4). As indicated in column D of Table 15.4, a 100 × 2.3-cm-i.d. column of 30-μm particles also exhibits a plate count equivalent to the shorter columns of smaller particles. However, for resolution equivalent to that for small-particle columns, a much lower mobile-phase velocity (e.g., 0.08 cm/sec) must be used; as a result, separation times are much longer. Because of the greater weight of adsorbent in this longer column, up to about 250 mg of a component in this particular example can be isolated in a single, nonoverloaded separation.

A characteristic of longer, larger-volume columns is that they tolerate a large sample volume without decrease in column efficiency, relative to shorter columns. This feature is important in separations involving samples with limited solubility that must be injected at low concentrations (Section 15.4).

The practical use of larger particles in preparative LC is illustrated in Figure 15.10. Chromatogram (a) is the analytical separation of an impure aromatic-

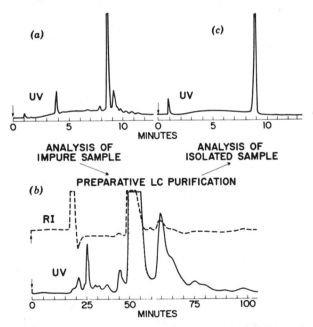

Figure 15.10 High-resolution preparative isolation with larger particles. Analytical chromatograms (a) (impure) and (c) (purified): column, 25 × 0.48-cm i.d., 6-μm porous-silica microsphere adsorbent, 300 m^2/g; mobile phase, linear gradient elution, 10% dioxane in isooctane to 50% dioxane in isooctane in 8 min; flowrate, 1.0 ml/min; sample, 50 μl of 40 mg/ml in 1:1 isooctane-dioxane; detector, UV, 254 nm, 0.32 AUFS. Preparative chromatogram (b): column, 88 × 2.3-cm i.d., <38-μm, Spherosil-XOA-400; mobile phase, 1:1 pentane/dioxane with 1.5% water; flowrate, 14 ml/min; sample, 5 ml of 200 mg/ml of impure aromatic-substituted, acetylated polysaccharide; detector: UV, 254 nm, 0.08 AUFS; RI, 128 × 10^{-5} RIUFS. Reprinted from (1) with permission.

substituted, acetylated polysaccharide by LSC with gradient elution on an analytical column of 6-μm porous silica. Chromatogram (*b*) is a preparative isolation with a longer, wider column of 30-μm porous silica operated isocratically with equivalent resolution. In this separation about 0.5 g of the main component (cross-hatched portion) was isolated in a single run. Note that resolution equivalent to (or better than) the analytical separation was obtained in this preparative run, but the separation took about 10 times longer. The final purity of the isolate is indicated in analytical chromatogram (*c*). Large-diameter columns of 30-60-μm LSC particles operated at low mobile-phase velocities under sample-overload conditions are an effective and convenient approach for the high-resolution isolation of gram quantities of a wide variety of compounds.

Solvents

The selection of the mobile phase in LC has already been discussed in detail in Chapter 6 and in chapters on the individual LC methods (Chapters 7-12). As discussed in Section 9.3, thin-layer chromatographic separations can be used to optimize the mobile phase initially (k' and α values) for LSC separations. After further optimizing the mobile phase on an analytical column of the same packing as for the preparative separation, the system usually can be used with the preparative column directly or with only minor changes. TLC is particularly useful in alerting the operator to strongly retained compounds of no interest ("garbage"), which might slowly elute from the column to contaminate subsequent isolated fractions.

As with analytical LSC applications, care must be taken with a new separation to ensure that the preparative column is equilibrated with the mobile phase. This equilibration may require relatively large volumes of the new mobile phase before injection of a sample in preparative applications. Careful equilibration is especially important if substantial changes in adsorbent activity can occur as a result of the loss or gain of water or other very polar solvent modifiers. One advantage of constant-activity, prepacked preparative columns (e.g., Lobar-Merck, PrepPak-Waters) is that equilibration with the mobile phase often is not required for the initial separation on a fresh column. However, a subsequent repeat run may differ from the original separation because of a change in adsorbent activity, and the separation selectivity may differ from that of a mobile-phase-equilibrated column. These factors suggest that gradient elution is less convenient in preparative LSC than for analytical studies, because of the large volume of solvent required for column reequilibration.

As in analysis, mobile phases of relatively low viscosity are favored in preparative LC (to maintain high column efficiency), and the solvent must be compatible with the detector used to monitor the separation. Higher column temperatures can be used to decrease mobile-phase viscosity or increase the concentration of thermally stable samples having poor solubility; however, this approach can be experimentally inconvenient.

For convenient removal from isolated fractions, the mobile phase should be relatively volatile. Volatile solvents normally are used anyway in LSC, but less volatile additives (e.g., acetic acid, morpholine) used in the mobile phase to reduce peak tailing can be bothersome. Of special importance is that solvents must be highly purified. Nonvolatile impurities are concentrated when the mobile phase is removed from an isolate, and significant contamination can then result. This problem can be minimized by using freshly distilled or specially purified commercial LC solvents (e.g., "distilled-in-glass").

Choice of the proper solvent for dissolving the sample is a special problem in preparative LC, because the goal often is to inject a relatively large weight of sample into the column for high preparative throughput. Fortunately, LSC generally allows a relatively high concentration of sample in the mobile phase, because moderately polar solvents often are used for proper elution with this method. With any LC method a good technique is to dissolve the sample in the mobile phase to be used for the separation. Injecting a solution of the sample in a solvent stronger than the initial mobile phase often leads to severe degradation of column resolution. It is always preferred, however, to inject larger volumes of sample contained in a solvent that is weaker than the initial mobile phase, if the required sample solubility is obtained.

Sometimes sample solubility in the mobile phase (of a weaker solvent) is inadequate, so that the largest practical sample volume does not allow the desired sample weight to be injected onto the LSC column. In these cases one or more sample components are usually highly polar, and thus their solubility in the mobile phase is limited by a low solvent P' value (Chapter 6). Simply increasing the polarity of the solvent used to dissolve the sample is not adequate because ϵ^0 generally increases with P', resulting in a sample solution that is dissolved in too strong a solvent (ϵ^0 for the sample solvent should never be greater than ϵ^0 for the mobile phase). However, there is a way out of this dilemma, because the dependence of ϵ^0 on P' is variable for binary solvents. This is illustrated in Figure 15.11, which shows the general relationship between P' and ϵ^0, for several common binary solvents. This dependence of P' on ϵ^0 is the result of the linear increase in P' with change in volume fraction ϕ_b (Eq. 6.2), versus the convex change in ϵ^0 with ϕ_b (as in Figure 9.7a). The practical consequence of these relationships and Figure 15.11 is as follows. For a solvent mixture of given strength ϵ^0 (e.g., 0.2 in Figure 15.11), a dilute solution of a more polar solvent B (e.g., ethyl ether) will have a smaller P' value than a more concentrated solution of a less polar solvent B (e.g., chloroform). As a result, for increased P' and higher solubility of sample in the solvent, it is better to dissolve the sample in 50%v chloroform/hexane ($\epsilon^0 = 0.2$, $P' = 2.3$) than in 5% ethyl ether/hexane ($\epsilon^0 = 0.2$, $P' = 0.2$). Thus, alternative solvents of the same solvent strength ϵ^0

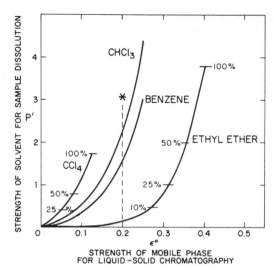

Figure 15.11 Ability of solvent to dissolve sample (P') versus strength of that solvent in LSC ($\epsilon°$). Plots show various solvents in binary mixtures with hexane. Reprinted from (*11*) with permission.

value but higher solubility (P' values) can be used to inject the sample into the column without adverse effects. Extensive tables of ϵ^0 and P' values are given in Chapter 6 to assist in the selection of appropriate alternative solvents for injection.

Equipment

Equipment for preparative LC is not as critical as for analysis, and less-sophisticated, lower-cost systems can be used satisfactorily. However, to optimize preparative LC separations it often is necessary to use different pumps, sampling systems, and detectors than those normally required for analytical LC.

Pumping systems should have a solvent delivery of up to 100 ml/min for large-bore (e.g., 2-cm-i.d.) columns. Pumps with very-high-pressure capabilities are not required in preparative LC, and pressure limits of about 3000 psi (~200 bar) generally are adequate. Furthermore, the precision and accuracy specifications of the pump are not critical, because analytical information from the chromatogram usually is not the goal. Reciprocating and pneumatic-amplifier pumps generally are preferred, because of their capability for higher pumping rates and continuous solvent output. "Mini" pneumatic-amplifier pumps are particularly convenient and are relatively inexpensive. Commercial analytical LC instruments containing reciprocating pumps often deliver no more than 10 ml/min of solvent, which represents the lower end of the desired preparative flowrate range. This flowrate limitation does not in itself preclude the use of analytical equipment for preparative applications, but much longer separation times can be expected. Pulsations from certain pumps (e.g., reciprocating) that affect detector baselines are generally not a disadvantage in preparative applications.

Because of the large volumes of mobile phase used in preparative LC, relatively large solvent reservoirs ($>$ 2 l) are required. Particular emphasis on safety should be given, because of the possibility of fire and the large volumes of highly flammable solvents that are used. Adequate laboratory ventilation is important in preparative LC, because of the large volumes of potentially toxic solvents that may be used.

Sensitive detectors are not required in preparative LC, because solute concentrations are relatively high. Actually, high sample concentrations can cause problems, because it is then difficult to determine whether the overlapping peaks seen on the recorder are due to an overloaded column or a nonlinear detector response. For example, in Figure 15.12 the UV detector output (nonlinear) suggests incomplete resolution of the two components of interest. On the other hand, the less sensitive refractive index (RI) detector shows excellent separation; much more sample could be injected before column resolution would be seriously degraded. UV photometers with short path length (1 mm) cells are commercially available for preparative use (e.g., Du Pont).

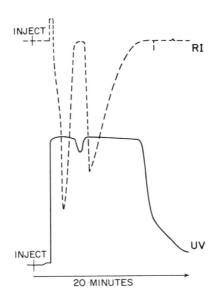

Figure 15.12 Detector-overload effects. Reprinted from (5) with permission.

The RI detector appears to be generally suitable for preparative separations, although a combination of both RI and UV detectors may be required for the accurate monitoring of all peaks in a particular system. Use of only a UV detector can cause occasional errors, as illustrated in Figure 15.13. In this case chromatogram (*a*) was obtained by using a UV detector on an impure sample known to contain a pair of isomeric compounds. However, preparative isolation

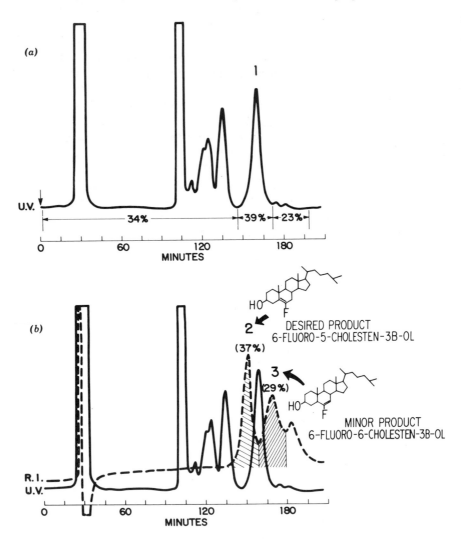

Figure 15.13 Detector anomoly in preparative isolation. Conditions same as in Figure 15.10 (*b*) except: mobile phase, 1.5% isopropanol in pentane; flowrate, 10.7 ml/min; sample, 15 ml of 17 mg/ml mixture in mobile phase; RI detector, 64×10^{-5} RIUFS. Reprinted from (*6*) with permission.

of the apparent major peak (1), believed to be one of the desired components, gave only a 39% recovery of the injected sample weight. In addition, material representing approximately 23% by weight eluted after this peak, even though only very minor components were indicated in the chromatogram by the UV detector. Rerunning this sample with both the RI and UV detector produced chromatogram (*b*). Collection of the major components indicated by RI showed that the cross-hatched peaks (2) and (3) were the desired isomeric compounds, representing a total of about 66% by weight of the injected sample. Thus, although the UV detector showed peak (1) as a major component, in reality it was only a strongly absorbing minor impurity with a retention time similar to that of the desired constituents. Use of the general RI detector rather than the selective UV device increases the probability that peaks of interest (and possible interfering contaminats) will be properly monitored. An effective technique for many preparative applications is to use both the UV and RI detectors in series.

UV spectrophotometric detectors are quite useful for preparative separations. These units are more versatile than single-wavelength UV photometers, and have the capability to operate in the 190-220-nm region, which allows the detection of many compounds not sensed at longer wavelengths. Operation at much higher wavelengths (i.e., >300 nm) is useful in isolating desired colored components. A particular advantage of spectrophotometric detectors is in "detuning" the wavelength from the solute adsorption maximum, to decrease the sensitivity of the detector and thus reduce the potential for detector overload in preparative work.

Extra-column effects are not a problem with large-diameter columns used in preparative LC, because of the relatively large internal volumes of the columns. Consequently, low dead-volume tubings and fittings are not as critical as for analysis. It is of particular importance that the detector *not* be made with very-narrow-bore tubing, which can severely limit the flow of mobile phase and cause considerable back pressure at the flowrates needed for wide-diameter columns. Both RI and UV detectors with larger bore tubing (e.g., 0.75-mm i.d.) are commercially available (e.g., Waters, Du Pont). Alternatively, a stream splitter on the exit of the large-diameter column can supply a small fraction of the effluent to an analytical detector.

The large sample aliquots used in preparative LC can be conveniently introduced with a valve, without interrupting the flow of the mobile phase during the separation. Sample loops up to about 10 ml (e..g, 167 cm of 0.28-cm-i.d. tubing) can be used. If a syringe is to be used, injection is normally made by the stop-flow technique, whereby the sample is introduced at atmospheric pressure; the pump then is restarted. Very large volumes can be delivered with a syringe by septum injection (e.g., 100-200 ml), but experimental problems are common (e.g., pieces of the septum obstructing the column inlet).

Large sample volumes also can be conveniently injected with a syringe through a high-pressure valve in the stop-flow mode. The approach is to attach the syringe to the sample loop position in a normal high-pressure sampling valve. With the pump turned off, and valve in the "inject" position (Figure 3.17, "inject"), the left-hand loop connection is broken and the sample is fed into the column through this valve port, using a large syringe. The valve is then rotated to "load" position, and the pump is restarted. For very large samples (e.g., >100 ml) the sample can be metered onto the column by means of a low-volume pump, using the valve-injection stop-flow technique just described. Automatic sample-metering systems can be advantageous in repetitive isolations, and such devices are commercially available for semipreparative systems (e.g., Siemans, Hewlett-Packard).

Manual fraction collection usually is adequate for isolating one or a few separated components. However, automatic fraction collectors are convenient when a large number of components must be isolated in a single run or when repetitive runs are used to increase the weight of one or more components. The detector signal can actuate either fraction collection or recycling of mobile phase to the pump. More elaborate systems have been devised to repurify the mobile phase automatically after use and return it to the solvent reservoir in a "closed-loop" operation (*12, 13*). Computer-controlled systems have also been used to effect automatic collection of chromatographic peaks (*14*).

An example of an integrated commercial instrument for preparative LC is shown in Figure 15.14. This Prep LC/System 500 instrument (Waters Associates) has a pumping system (*A*) that delivers flowrates up to 500 ml/min at 500 psi pressure, from a dual solvent inlet (*B*). The sample is injected through a port (*C*) into (in this case) the serially connected columns (*D*), which have been radially

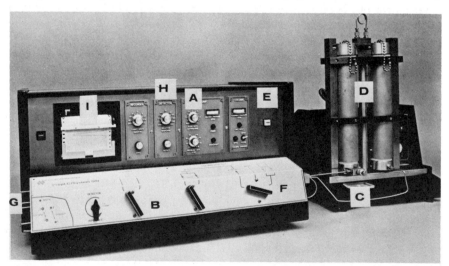

Figure 15.14   Commercial preparative LC instrument. Reprinted with permission of Waters Associates.

compressed by means of electronic controls (E). A multifunction valve (F) is used to divert column effluent to waste, recycle samples, or direct flows to ports (G) for peak collection. Separations are monitored with a differential refractometer (controls at H), and the output signal is presented on a recorder (I).

15.4 OPERATING VARIABLES

Sample Volume

Sample injection in preparative LC is an important but imperfectly understood operation. Loading the sample across the entire column cross section is generally preferred in preparative LC, because this permits better use of the total column packing, with reduced column overload at the inlet. Compared to center-injection techniques, increased sample capacities of at least 20% have been found (7) by loading the sample across the entire column cross section, using a sample inlet distributing head such as that shown in Figure 15.15. When possible,

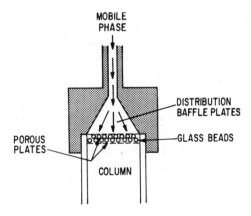

Figure 15.15 Inlet for preparative LC column. Reprinted from (7) with permission.

samples should be injected as relatively large volumes of lower concentration, rather than as smaller volumes of more concentrated solutions. This approach reduces the effect of overloading the column-bed inlet, thus increasing column loadability and performance.

The volume of sample that can be introduced depends on the column internal diameter and length, sample solubility, the mobile-phase/stationary-phase combination, and the resolution goal. For maximum resolution (e.g., in a "scale-up," nonoverloaded column), this volume generally should not exceed one-third the

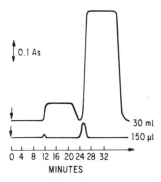

Figure 15.16 Sample-volume overload in LC. Column, 25 × 0.46 cm, LiChrosorb RP-8, 10 μm; mobile phase, water/acetonitrile/triethylamine (870:124:6); flowrate, 2.5 ml/min at 22°C; feed volume: lower, analytical chromatogram, 150 μl; upper, preparative chromatogram, 30 ml. (Reprinted from (*15*) with permission.

volume of the earliest peak of interest (Section 7.4). However, if resolution permits (e.g., for large α values), much larger sample volumes are tolerated. Thus, in Figure 15.16 we see that a 30-ml sample was injected into this 25 × 0.46-cm i.d. reverse-phase column for the upper preparative chromatogram, compared to the typical 150-μl volume for the lower analytical chromatogram. The large α value for this separation allowed this very large sample volume, which would correspond to *600 ml* for a comparable 2.3-cm-i.d. column. In this case, adequate resolution of the desired components was obtained, even though the resultant peaks were rectangular and greatly broadened by the large injection volume. Use of very large feed volumes sometimes can be used to overcome sample solubility limitations, and this approach is particularly effective in reverse-phase operations where sample solubility problems are common. For difficult separations (i.e., α < 1.2), 20-100-ml sample volumes typically are used with 100 × 2.3-cm-i.d. preparative LSC columns. However, as suggested in Figure 15.16, the sample volume can be further increased in separations having large α values, without seriously degrading resolution. Generally, the best approach is to use the largest feed volume that still permits adequate resolution of the peaks of interest.

Sample Mass

The weight or "loading" of sample that can be injected into a preparative LC column is dependent on the chromatographic system and the resolution required. In LLC the quantity of stationary phase determines the capacity; in LSC, sample capacity is determined by packing surface area; in ion-exchange chromatography it is the number of ion-exchange groups. Sample loading limit in SEC is dependent on the molecular weight of the solute and may also be a function of packing type. Table 15.5 indicates typical weights of sample that can be injected into LSC columns of varying internal diameters. For very difficult separations

Table 15.5 Approximate sample weight for preparative LSC separations (50 cm of totally porous column packings).

Column internal diameter (mm)	Difficult Separation ($\alpha < 1.2$) (mg)	Easy Separation ($\alpha > 1.2$) (mg)
2	0.2-2	5-25
8	10-50	100-500
25	100-500	1000-5000

($\alpha < 1.2$) the column generally should be operated in a scale-up or nonoverload condition, to obtain maximum resolution (as in analytical separation). However, for easier separations ($\alpha > 1.2$), much larger sample weights can be injected, because the column can be heavily overloaded while still producing the desired resolution. Resolution appears to be governed by the injected weight of individual solutes, rather than by the total weight of the sample mixture—when the column is operated in a nonoverload state. However, this sampling effect has not been studied under overload conditions.

The sample loading capacity of a nonoverloaded column increases linearly with the column cross-sectional area (see Eq. 15.1). (Note that increasing column radius does not increase the retention time of the solute, providing the same mobile-phase velocity is maintained by proportionate increase of column flow-rate.) It also follows that when the column radius is increased, more mobile phase is used for each separation. However, the amount of solvent used per unit mass of solute separated is independent of column radius if the same mobile-phase linear velocity is used.

Optimizing Preparative Throughput

Present knowledge concerning nonoverloaded columns indicates that high throughput is largely a function of column operation. Thus, optimization of the usual separation variables (α, N, k') will assist in achieving a high sample throughput with the desired sample purity. Information on sample throughput for mass-overloaded columns is sparse. Because the effect of resolution parameters (α, k', N) is different in overload, optimized separation variables are different and are largely determined empirically.

Conditions for maximum sample throughput can be summarized as follows:

- Throughput is increased with higher column resolution. This is primarily obtained by optimizing selectivity α with the proper choice of mobile phase. Increasing column plate count N can provide somewhat higher throughput per unit time, but only with increased pressure.
- Throughput is enhanced at lower k' values for nonoverloaded columns; effect of k' on overloaded columns is poorly defined.
- Throughput is increased with the use of totally porous, rather than pellicular particles.
- For very difficult separations with small α values, small-particle columns (5-10 μm) are favored in preparative separations. For easier separations, larger-particle columns give higher throughput per unit time.
- Column loading capacity and throughput increase with increase in column cross section.
- Throughput is increased with increased column length, coupled with increased mobile-phase flowrate, particularly in overload.

The sample throughput data for various systems in Table 15.6 verify some of the preceding guidelines for improving preparative throughput. The sample throughput increase between methods (a) and (b) corresponds to the increase in column cross section. In both cases the columns are operated in a nonoverload condition but at the loading limit. The throughput increase between methods

Table 15.6 Sample throughput for various column types.

	Particle size; Sample throughput (mg/min)		
Method	Pellicular	Totally porous particles	Gel particles (for SEC)
(a) 50 × 0.2-cm analytical column	30 μm; 0.01-0.02	10 μm; 0.1-0.2	10 μm; 1-2
(b) 50 × 2-cm scale-up at loading limit	30 μm; 1-2	30 μm; 5-20	50 μm; 100-200
(c) 50 × 2-cm preparative, overloaded	-	30 μm; 50-200	50 μm; 1000
(d) 50 × 5-cm preparative, overloaded	-	70 μm; 200-1000	50 μm; 2500

Source: In part from (7).

(*b*) and (*c*) is achieved by mass-overloading the column. To achieve the indicated fivefold improvement, resolution is actually reduced to about 50%, and yield losses may amount to 10-25%, depending on the trapping technique used to overcome the overlapping of components. The increase in throughput in (*d*) relative to (*c*) is achieved by further increase in column internal diameter. In this illustration, lower-cost larger particles are used in (*d*), somewhat increasing band overlap and decreasing yield relative to that obtained with smaller particles (i.e., 30 μm). These data indicate that with column overload there can be an increase in the yield per unit time of separated material, without a loss in purity, even though resolution may be decreased siginificantly.

As suggested in Figure 15.2 and the preceding guidelines, preparative throughput can also be increased by increasing mobile-phase flowrate, because efficiency is affected very little by velocity in column overload. Actually, a larger reduction in resolution often occurs with column overloading than with increased flow, and a twofold sample throughput improvement through increased flow has been demonstrated compared to column overloading at lower flowrates (*7*). Flow programming also can reduce the time needed for a preparative separation if peaks of interest are widely separated. As illustrated in Figure 15.17, an early-

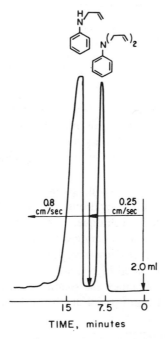

Figure 15.17 Flow-programmed preparative separation of substituted anilines. Column as in Figure 15.1; mobile phase, 0.5% methanol (v/v) in cyclopentane (50% water saturated); sample, 2 ml, 50 mg/ml. Reprinted from (*16*) with permission.

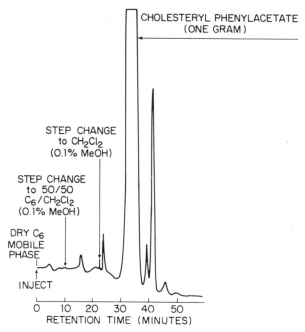

Figure 15.18 Gradient elution in preparative LC. Column, Spherosil XOA-400, two 50 X 2.3-cm i.d. columns in series; mobile phase, step gradient (two changes) dry hexane to dichloromethane (0.1% methanol); flowrate, 30 ml/min; pressure, 1000 psi; temp., 23°C; detector, UV, 254 nm, 0.32 AUFS. Reprinted from (*17*) with permission.

eluting aromatic amine was preparatively isolated with the column first operated at a relatively low flowrate for higher resolution. Following the collection of this first peak, the flowrate was significantly increased to collect a later-eluting second peak with adequate resolution, but greatly increased speed and preparative throughput.

Although requiring more manipulation, gradient elution can sometimes be used effectively in preparative LC separations. Figure 15.18 shows the isolation of a steroid using a step-gradient technique. This sample, which was soluble at effective concentrations only in a moderately strong solvent, was injected into the column initially operated with a low-strength mobile phase. The strength of the mobile phase then was increased in two steps so that 1 g of the main component eluted in an optimum k' range for isolation with the required purity. Of course, the gradient technique is not practical with an RI detector.

Recovery and Purity Determination

The solvent from isolated fractions usually can be eliminated by evaporation under a stream of dry nitrogen with warming (e.g., with an infrared lamp).

Techniques that do not condense water in the isolate should be used. Large volumes of mobile phase can be conveniently removed with a vacuum rotary evaporator. Freeze-drying, a very mild process where applicable, can be effectively used to eliminate particular solvents (e.g., water, dioxane, and benzene). Special care should be taken to minimize contamination and degradation of the carefully purified sample, whatever the concentration process used.

Following the final collection of the fraction of interest, purity should be measured by high-efficiency analytical LC, TLC, or another appropriate technique. If the isolated component is not sufficiently pure, it can be further purified by LC or crystallization. Prediction of the purity and recovery of a particular isolate from overlapping bands is sometimes possible with the standard peak resolution system described in Figures 2.11-2.16, which assumes that the two overlapping bands have approximately equal detection sensitivities (and are Gaussian).

It is desirable to conduct a material balance from the preparative separation, to ensure that all sample components have eluted from the column. This is accomplished by trapping and weighing each fraction after evaporating the solvent, then comparing the total weight of the trapped fractions with the weight of the injected sample. Yield of purified components also can be determined by using this procedure.

Establishing the purity of the final isolated fraction by LC analysis can involve significant error if only peak-area response is used, because LC detectors often have different responses for different compounds. One approach to verifying the purity of a preparative fraction is to isolate a very narrow heart-cut of the peak of interest, in the region of expected highest purity. This isolate is then used as an analytical standard for subsequent LC analysis of the preparative isolate. In critical cases, the purity of the heart-cut "standard" can be further verified by alternative analytical methods or by another LC method (e.g., reverse-phase rather than LSC).

15.5 APPLICATIONS

Although LSC often is the preferred method for preparative LC, size exclusion can be used to advantage for certain studies, for example, to prepare narrow-molecular-weight polymer standards. High-efficiency SEC sometimes is useful for isolating individual compounds, as illustrated with a μ-Styragel column in Figure 15.19. Size exclusion also is important in clean-up operations (e.g., in pesticide and drug-metabolite analysis). Figure 15.20 shows size-exclusion separations of the lipid material in fish extracts from ^{14}C-labeled pesticide residue, using a polystyrene-divinylbenzene-copolymer gel packing and organic solvents. These chromatograms show the separation of a mixture containing up to 500 mg of lipids and trace amounts of pesticides to baseline resolution in

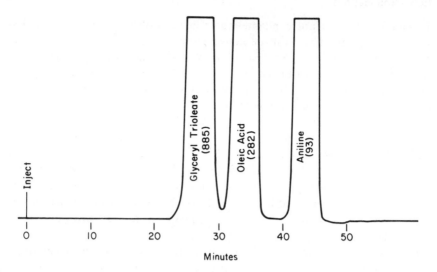

Figure 15.19 Preparative isolation with high-efficiency size-exclusion columns. Column, 150 × 0.8 cm, μ-Styragel 100 Å; mobile phase, tetrahydrofuran; flowrate, 1.0 ml/min; detector, RI; sample, 150 mg each compound. Reprinted from (*23*) with permission.

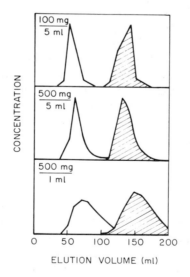

Figure 15.20 Isolation of pesticides from lipid extracts by preparative-scale SEC. Column, 60 × 2.5 cm, Biobeads SX-2; mobile phase, cyclohexane; white peak: lipids; cross-hatched peak: ^{14}C-DDT. Reprinted from (*19*) by permission.

648

approximately 35 min. Note in this example that a larger volume (500 mg/5 ml) of the same quantity of sample gave better resolution than a more concentrated solution (500 mg/1 ml). The fractions associated with this material were externally monitored by scintillation counting.

Liquid-liquid chromatography can be used advantageously for certain samples. Figure 15.21 shows the separation of pyrethrins using a stationary phase of nitromethane supported on diatomaceous earth particles, with a mobile phase of

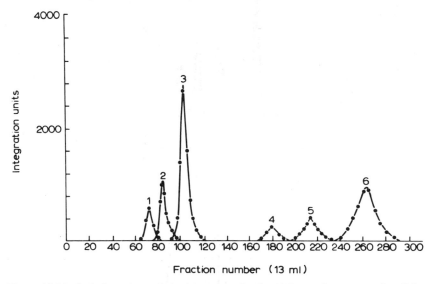

Figure 15.21 Isolation of pyrethrins by preparative liquid-liquid chromatography. Column, 100 × 3.8 cm, nitromethane on diatomaceous earth, <40 μm; mobile phase, diethyl ether/hexane (1:3); flowrate, ~4 ml/min; pressure, 25 psi; sample, 0.3 g in hexane. Reprinted from (20) by permission.

1:3 ether/hexane. Fractions were collected and monitored by gas chromatography to identify component peaks. While this separation was in progress, the mobile phase was changed to carbon tetrachloride/hexane (1:3) saturated with nitromethane, so a precolumn was required to maintain equilibration between the mobile and stationary phases, to prevent stationary phase bleed from the column (Section 8.6). Because volatile solvents were used in both the mobile and stationary phases, highly purified fractions could be obtained by simple evaporation of the solvents from the isolated fractions.

As indicated previously, preparative LC separations are usually best carried out by LSC. Comparison of an analytical LSC separation with semipreparative and large-diameter column separations is shown in Figure 15.22 for 1- and 4-hydroxy-1,2,3,4-tetrahydrophenanthrene (80:20 mixture). Chromatogram (a) is

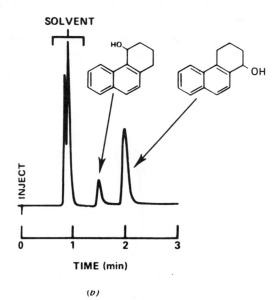

(a)

SOLVENT

INJECT

0   1   2   3

TIME (min)

(b)

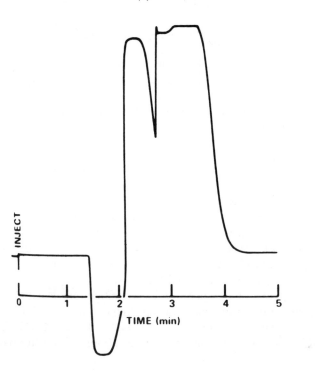

INJECT

0   1   2   3   4   5

TIME (min)

650

the analytical separation in a nonoverload condition using a 100-μg sample. Chromatogram (*b*) shows the same analytical system in a semipreparative mode using a 54-mg sample in a column-overload condition. Chromatogram (*c*) is for a 10-g sample on a large-diameter radially compressed LSC column operated in overload with a relatively high mobile-phase velocity (for high throughput). Figure 15.23*a* shows the purity of the 4-hydroxy isomer (1.7 g) in fraction 2 (Figure 15.22*c*), and Figure 15.23*b* shows that of the 1-hydroxy isomer from combined fractions 4 and 5 (total 6.3 g). Both fractions were more than 99% pure.

Isomers are particularly amenable to separation by preparative LSC, as further illustrated in Figure 15.24. Here all-*trans*-(peak 1) and 9-*cis*-(peak 2)-C_{18}-ketone intermediates in the synthesis of vitamin A are shown separated for collection into relatively pure fractions (cross-hatched areas) in less than 15 min. This

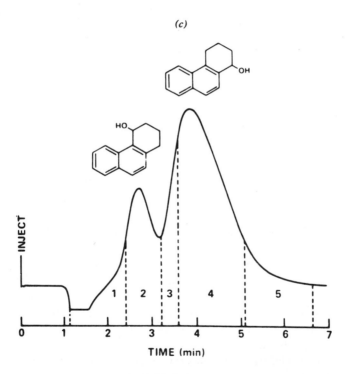

Figure 15.22 Purification of 1- and 4-hydroxy-, 1,2,3,4-tetrahydorphenanthrene (80/20 mixture. (*a*) Analytical separation: column, 30 × 0.42 cm, μ-Porasil; mobile phase, methylene chloride/ethyl acetate (95:5); flowrate, 4.0 ml/min; detector, RI, 16×; sample, 100 μg in 10 μl. (*b*) Semipreparative separation: mobile phase and column as in (*a*); flowrate, 2.5 ml/min; detector, RI, 128×; sample, 54 mg in 1.75 ml. (*c*) Preparative separation: column, 30 × 5.7 cm, 50-μm silica gel; mobile phase as in (*a*); flowrate, 300 ml/min; detector, RI; sample, 10 g in 45 ml. Reprinted from (8) with permission.

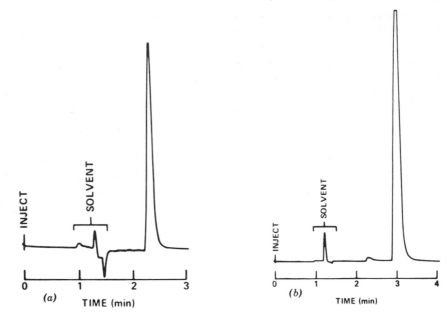

Figure 15.23 Analysis of isolated fractions from Figure 15.22c. (a) Analysis of fraction 2; solvent and column as in 15.22a; flowrate, 2.5 ml/min; detector, RI, 8×; sample, 20 μl of fraction 2. (b) Analysis of fractions 4 plus 5. Conditions as in (a); sample, 20 μl of fractions 4 plus 5. Reprinted from (8) with permission.

separation was unobtainable by classical, gravity-fed LSC methods, demonstrating the unique power of preparative LC to solve purification problems rapidly in organic syntheses.

Preparative LSC also can be used to isolate relatively large amounts of rather polar solutes. Figure 15.25 shows the isolation of a protected synthetic peptide (21). In this case, 0.47 g of sample in a chloroform/methanol/acetic acid solvent was loaded onto a commercially available silica gel column, and elution was carried out with a convex gradient using the solvents indicated. In other similar separations, loading of smaller volumes of less-soluble, protected peptides dissolved in dimethylformamide and acetic acid was feasible, without significant loss in column resolution.

An interesting separation of some rather exotic compounds is illustrated in Figure 15.26. Chromatogram (a) is the analytical separation of intermediates in a vitamin B-12 synthesis with a column of approximately 50-μm pellicular silica adsorbent. Chromatogram (b) is the preparative separation on a 2.3-cm-i.d. column of approximately 60-μm porous silica with a similar mobile phase. A very low mobile-phase velocity was used to isolate gram quantities of the main components in a single run with this long column of relatively large particles, result-

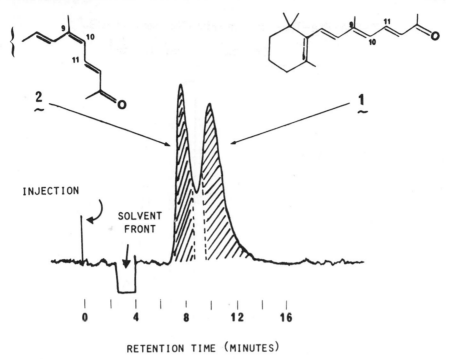

Figure 15.24 Preparative separation of isomers by LSC. Column, 30 × 4.7 cm, PrePak/ 500 silica column; mobile phase, 11% diethylether in hexane; flowrate, 250 ml/min; detector, RI; sample, 820 mg of mixture. Reprinted from (*24*) with permission.

ing in a relatively long separation time. Much faster separation of these components with equivalent resolution could be expected with a column of smaller particles (e.g., 30 μm).

Even larger sample quantities can be isolated with columns of larger internal diameters, as illustrated in Figure 15.27. Here a mixture of steroids was separated into purified components on a longitudinally compressed 100 × 8-cm-i.d. column of silica gel. Although this is not a "pretty" chromatogram, very large amounts of highly purified material (e.g., 33 g of androstenedione) were isolated in a single run. Columns of this internal diameter can be used to prepare kilogram quantities of purified materials within a reasonable time, if desired. Plant processing by LC is feasible, if end use of the isolated product has sufficient economic value.

Automatic preparative LC can be used to isolate relatively large amounts of highly purified components. Figure 15.28 shows the repetitive separation of diastereomeric carbamates on a column of acidic alumina (about 3 hours required per individual run). This system cleanly separates the diastereoisomers (α =

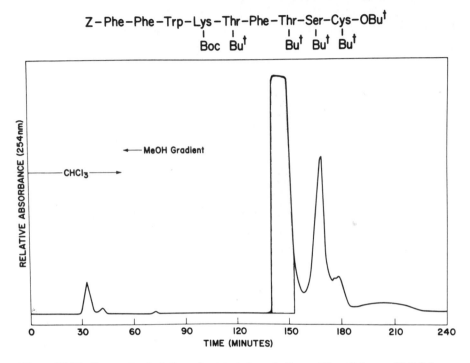

Figure 15.25 Preparative isolation of protected synthetic peptides. Column, 43 × 3.8 cm, silica gel 60 ("Lobar"); mobile phase, chloroform initial, convex gradient with methanol; flowrate, 7.5 ml/min; detector, UV, 254 nm; sample, 474 mg in 8 ml of $CHCl_3$: MeOH: AcOH (88:10:2); Boc = $N\text{-}t$-butyloxycarbonyl, Bu^t = s-t-butyl. Reprinted from (*21*) with permission

1.37) as determined by UV absorbance. (Because of the saturation of the 280-nm detector, peak overlap appears to be greater than is actually the case). This is an example of using a diastereomeric isocyanate derivative $[(R)\text{-}(-)\text{-}1]$ to provide resolution of an enantiomeric fluoroalcohol. With this automatic, repetitive separation system, 6-10 g of the diastereomeric mixture were separated in 24 hours.

15.6   A PREPARATIVE SEPARATION EXAMPLE

Let us now review the history of an actual example, to illustrate the various steps required in setting up a preparative LC separation. The particular problem to be discussed as the iolation of synthetic diastereomeric 8-azaprostaglandins from a small-scale crude reaction product (*22*).

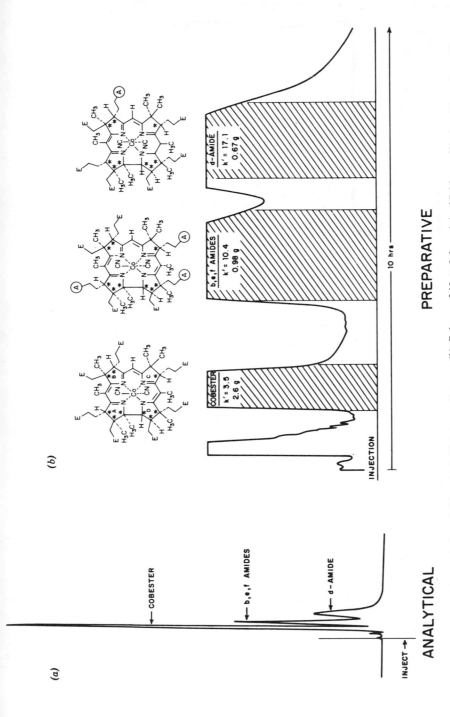

Figure 15.26 Preparative isolation of Vitamin B-12 intermediates. (a) Column, 180 × 0.2-cm i.d., Corasil II, 37-50 µm; mobile phase, hexane/isopropanol/methanol. (b) Column, 240 × 2.3-cm i.d., 37-80 µm; Woelm silica gel; mobile phase, hexane/isopropanol/methanol (5:2:1); flowrate, 34 ml/min; injected sample, 5 g. Reprinted from (25) with permission.

655

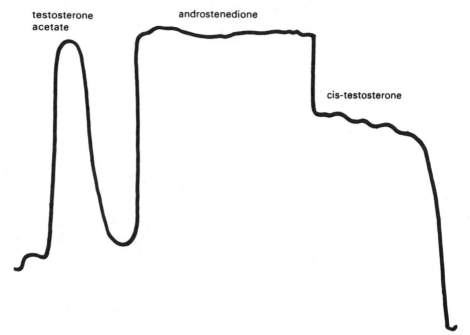

Figure 15.27 Large-scale separation of steroids. Column, 100 × 8 cm, silica gel, Merck H 60 (40 μm); mobile phase, chloroform/methanol (94:4); detector, UV, 250 nm; sample, 3 g testosterone acetate, 16 g *cis*-testosterone. 33 g androstenedone. Reprinted with permission of J. Y. Optical Systems.

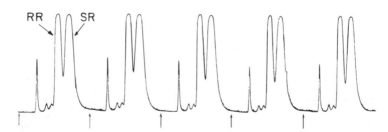

Figure 15.28 Automatic repetitive preparative separation of diastereomeric carbamates. Column, 122 × 6.4 cm, acidic alumina; mobile phase, benzene; sample, 1 g.; detector UV, 280 nm. Reprinted from (*13*) with permission.

656

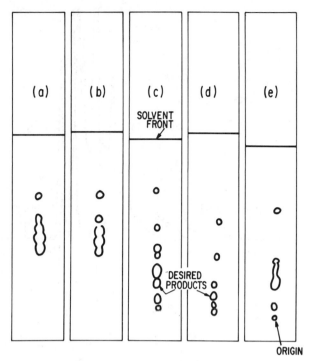

Figure 15.29 Scouting of solvent systems for separating 8-azaprostaglandin diastereomers by thin-layer chromatography. TLC plates, silica gel 60 F-254, 250 microns (Merck); visualization, potassium dichromate/sulfuric acid; sample size, 2.5 μg crude reaction product; conditions, 20°C/20% relative humidity; mobile phase: (a) tetrahydrofuran; (b) 0.5% methanol in tetrahydrofuran; (c) 10% methanol in diethylether; (d) 2.5% methanol in chloroform; (e) 5% methanol in chloroform. Reprinted from (22) with permission.

The first step in developing the preparative isolation was to select candidate mobile-phase systems for the separation. Using a crude reaction product, solvents were tested by the "spot" TLC technique described in Section 9.3, to screen for useful mobile-phase solvents. As a result of these rapid tests, several potential solvents were selected for conventional TLC separations, the results of which are shown in Figure 15.29. From these TLC separations it was decided that the best overall mobile-phase system to produce desirable k' and α values would be 2.5% methanol in chloroform (Fig. 15.29d). A mobile phase of 10% methanol in diethylether (Figure 15.29c) produced an equally promising TLC separation, but the use of the highly flammable and potentially oxidizable diethylether was considered to be less desirable for laboratory operation. In each of the better solvent systems (i.e., 2.5% methanol/chloroform and 10% methanol/diethylether) the diastereomers were not well separated by TLC. However, the desired

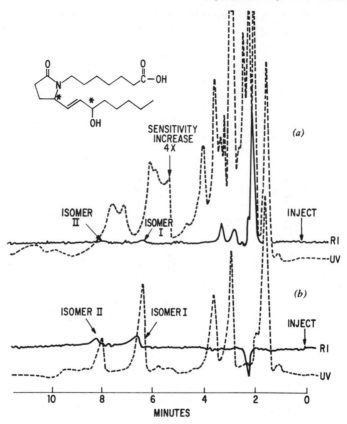

Figure 15.30 LC analysis of 8-azaprostaglandin diasteromer samples. Column, 25 × 0.32 cm porous-silica microspheres, 60 Å, 300 m²/g; mobile phase, 2.5% methanol in chloroform, 22°C; flowrate, 1.25 ml/min; samples: (*a*) Crude reaction product, 10 μl, 50 mg/ml; (*b*) Partially purified product, 50 μl, 6.5 mg/ml; UV detector, 0.04 AUFS; RI detector 1×. Reprinted from (*22*) with permission.

materials did elute at R_f values that should be equivalent to optimum k' values in the LC column.

Using the solvent developed from the TLC screening, crude reaction product and a small sample of material that had been partially purified by recrystallization were run on a small-particle analytical LSC column, as shown in Figure 15.30. It was obvious that a large number of UV-absorbing impurities were present in both the crude reaction product (Fig. 15.30*a*) and the partially purified sample (Fig. 15.30*b*). Nevertheless, these analytical chromatograms clearly showed that for the partially purified product (*b*), the required diastereomers

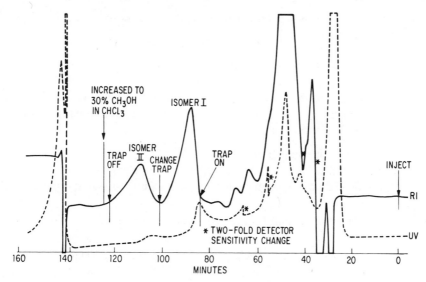

Figure 15.31 Preparative isolation of 8-azaprostaglandin diastereomers. Column, 100 × 2.3 cm Spherosil-XOA-400, <38 μm; mobile phase, 2.5% methanol in chloroform, 22°C; flow-rate, 12.5 ml/min; pressure, 2000 psi; detector: UV, 0.32 AUFS, 254 nm; RI, 1/16 full-sensitivity; sample, 10 ml, 1.25 g in mobile phase. Reprinted from (22) with permission.

(I and II) were resolved and well separated from other sample impurities with this solvent system.

Using the analytical LC system as a model, the separation was then carried out on a 100 cm × 2.3-cm-i.d. column of approximately 30-μm porous silica beads, as shown in Figure 15.31. In this case, 10 ml of mobile phase containing all (1.25 g) of the partially purified sample was injected into the preparative column via a valve. With the RI detector as a guide, the diastereomers (I and II) were individually isolated in this separation. Following collection of II, the strength of the mobile phase was increased to 30% methanol in chloroform, to elute strong-ly retained materials known from TLC to be present in the sample (see Figure 15.29). After purging with this solvent for about 20 min, the column was then clean and ready for another separation (after equilibration with the appropriate mobile phase).

Collected isomers I and II (Figure 15.31) amounted to 12% and 9%, respective-ly, of the sample injected into the column. Purity of the isolate was better than 99.5% in both cases, based on analytical LC. Had more crude reaction product been available, much larger sample weights could have been injected in a single separation, to obtain larger amounts of purified substances with little or no degradation of product purity. However, in this case the amounts isolated were sufficient for the purposes intended, and the preparative LC problem was successfully completed.

Summary of Preparative LC Conditions

Existing knowledge of preparative techniques suggests that approaches different from those normally used for analysis are sometimes needed to obtain significant amounts of purified components:

- Use liquid-solid chromatography, if possible. TLC and analytical-scale separations are important as "pilot" runs. Optimize α values for compound(s) of interest, for highest preparative throughput of purified material per unit of time.
- Use wide-diameter columns for increased sample throughput. Columns of up to at least 8-cm i.d. provide excellent resolution.
- Use totally porous, larger-particle packings and higher stationary phase loadings to optimize sample capacity. Particles of 30-50 μm provide a good compromise between column efficiency, operating pressure, and cost.
- For higher resolution with large porous particles, use lower mobile-phase velocities. For highest sample throughput per unit time, use the highest velocity permitting adequate resolution.
- Overload columns to increase sample throughput (but with decreased resolution). Isolation of heart-cut portions of peaks improves purity but decreases yield.
- Use recycle if difficult separations (after optimizing α) require larger plate count than permitted by increasing column length.
- Use wide columns of <10-μm particles in scale-up mode (analytical conditions) for most difficult separations requiring highest resolution in least time.
- If feasible, adjust the retention of the compound to be isolated to $1 < k' < 5$ to optimize sample throughput.
- Use larger sample volumes of more dilute solutions to minimize column overload at the inlet.
- Use high-capacity, continuous-pumping systems (e.g., reciprocating or pneumatic-amplifier pumps); pressures above 3000 psi rarely are required.
- Sensitive detectors usually are not necessary; dual UV and RI detectors generally are satisfactory.
- Samples are more conveniently introduced with a valve, but syringe injection can be used.
- Use purified, low-viscosity mobile phases and higher temperatures if sample permits.

- For difficult separations ($\alpha \approx 1$), use smaller d_p and avoid column overload.
- Prefractionate samples containing significant amounts of strongly retained components of no interest (e.g., use simple gravity column chromatography or low-cost "throw-away" pressurized columns).

REFERENCES

1. J. J. DeStefano and J. J. Kirkland, *Anal. Chem.*, 47, 1103A (1975); 47, 1193A (1975).
2. K. J. Bombaugh and P. W. Almquist, *Chromatographia*, 8, 109 (1975).
3. R. P. W. Scott and P. Kucera, *J. Chromatogr.*, 119, 467 (1976).
4. J. Krusche, P. Roumetiotis, and K. K. Unger, Third International Symposium on Column Chromatography, Salzburg, Austria, September 1977.
5. G. J. Fallick, *Am. Lab.*, 5, 19 (1973).
6. J. J. DeStefano and H. C. Beachell, *J. Chromatogr. Sci.*, 10, 654 (1972).
7. A. Wehrli, *Z. Anal. Chem.*, 277, 289 (1975).
8. J. N. Little, R. L. Cotter, J. A. Prendergast, and P. D. McDonald, *J. Chromatogr.*, 126, 439 (1976).
9. F. W. Karasek, *Res. Dev.*, 28, 32 (1977).
10. E. Stahl, ed., *Thin-Layer Chromatography*, 2nd ed., Springer-Verlag, New York, 1969.
11. J. J. Kirkland and L. R. Snyder, *Solving Problems with Modern Liquid Chromatography*, American Chemical Society, Washington, D. C., 1974.
12. W. H. Pirkle and R. W. Anderson, *J. Org. Chem.*, 39, 3901 (1974).
13. W. H. Pirkle and M. S. Hoekstra, *ibid.*, 39, 3904 (1974).
14. P. A. Bristow, *J. Chromatogr.*, 122, 277 (1976).
15. A. Wehrli, U. Hermann, and J. F. K. Huber, *ibid.*, 125, 59 (1976).
16. J. J. DeStefano in *Introduction to Modern Liquid Chromatography*, by L. R. Snyder and J. J. Kirkland, Wiley-Interscience, New York, 1974, Chap. 12.
17. DuPont Instrument Products Division, Liquid Chromatography Applications Report, 73-03, 1973.
18. L. R. Snyder, *J. Chromatogr. Sci.*, 10, 200 (1972).
19. A. R. Cooper, A. J. Huges, and J. F. Johson, *Chromatographia*, 8, 136 (1975).
20. F. E. Rickett, *J. Chromatogr.*, 66, 356 (1972).
21. T. F. Gabriel, M. H. Jimenez, A. M. Felix, J. Michalewsky, and J. Meienhofer, *Int. J. Pept. Protein Res.*, 9, 129 (1977).
22. P. E. Antle, E. I. du Pont de Nemours & Co., private communication, January, 1977.
23. A. P. Graffeo, Association of Official Analytical Chemists Symposium, Washington, D. C., October, 1977.
24. M. J. Pettei, F. G. Pilkiewicz, and K. Nakaniski, *Tetrahedron Lett.*, (24), 2083 (1977).
25. Waters Associates, Milford, Mass., Technical Bulletin An 72-120, 1972.

SIXTEEN

GRADIENT ELUTION AND RELATED PROCEDURES

16.1	Gradient Elution	663
	Introduction	663
	Design of Optimum Solvent Gradients	668
	The Two Solvents A and B	670
	Gradient Shape	676
	Gradient Steepness	680
	Other Separation Variables	686
	Varying N	686
	Varying α	688
	Other Aspects of Gradient Elution	691
	Column Regeneration	691
	Applications	693
16.2	Column Switching and Stationary-Phase Programming	694
	Column Combinations: General Considerations	696
	Column Switching	698
	Stationary-Phase Programming	700
	Selectivity Switching	706
	Coupled-Column Operation	707
	Column Backflushing	709
	Experimental Considerations	710
16.3	Flow Programming	712
16.4	Temperature Programming	715
16.5	Practical Comparison of Various Programming and Column-Switching Procedures	715
	References	717
	Bibliography	718

For the most part our discussion in previous chapters has assumed normal or *isocratic* elution: A sample is injected onto a given column and the mobile phase is unchanged throughout the time required for the sample to elute from the column. In most cases, isocratic separation is well adapted to individual applications or samples, and no more complex form of separation need be considered. In some cases, however, complex samples are encountered that have a wide range of compound retention times or k' values. Then we are faced with the so-called *general elution problem*, as was previously illustrated in Figures 2.25 and 3.11. The isocratic separation of such samples typically exhibits poor resolution of early-eluting bands, difficult detection of late-eluting bands, and unnecessarily long separation times. The solution to the "general elution problem" is a change in conditions *during* separation, so as to allow separate optimization of the k' values of individual bands. This can be achieved by means of gradient elution, which was discussed briefly in Chapters 2 and 3, or by certain other techniques that we review in this chapter. In following sections we discuss each of these methods, including gradient elution, and describe how these techniques can be best applied to individual separation problems.

Various nonisocratic techniques can also be applied to certain other problems:

- Enhancement of detection sensitivity in trace analysis.
- Improvement of resolution of retained bands from large concentrations of unretained components.
- Increase of peak capacity and/or resolving power for very complex samples that contain a large number of individual components.
- Decrease of column deterioration due to accumulation of strongly retained chemical "garbage" from certain samples.

The preceding list is by no means complete, and it should be stressed that the procedures described in following sections enormously increase the ability of LC to handle difficult samples and to deal with unusual situations or requirements. Although these techniques can entail additional complexity, for the right problem this extra effort will be fully repaid.

16.1 GRADIENT ELUTION

Introduction

The theory of gradient elution (GE) has been discussed in considerable detail [see (1)-(8a)], as befits a technique that is both uniquely useful and inherently complex. Here we summarize current theory in terms of simple pictures of what takes place in gradient elution and offer a number of general rules for optimizing the performance of practical GE separations. Equipment for carrying out GE was discussed in Chapter 3. GE devices provide for the admixture of two (or more) solvents, a weak initial solvent A and a strong solvent B, so that the concentration of B in the mobile phase entering the column increases throughout the separation. As a result, the mobile phase provides large k' values for the sample of interest at the beginning of separation and small k' values at the end of separation. That is, the solvent is initially weak and becomes progressively stronger as separation proceeds.

It is instructive to look more closely at the change in sample k' values during a GE separation and the corresponding migration of these bands through the column. Figure 16.1 illustrates this interrelation of solvent strength and band migration. Here three sample components (X, Y, and Z) are considered, with X least strongly retained and Z most strongly retained. The solid curves for each component mark the position of the band within the column as a function of time; that is, these curves plot band position as the fractional distance r between column inlet and outlet. The migration of each band follows a similar pattern, illustrated here for compound X. At some time t_x, the k' value of X will decrease to the point where X begins to migrate along the column. As the k' value of X continues to decrease with time (dashed curve A in Figure 16.1), X moves more rapidly along the column, yielding the characteristic concave curve of r versus separation time t for X in Figure 16.1. Finally, when $r = 1$ for compound X, the band leaves the column at its retention time t_X. Similarly, compound Y begins its migration along the column at the time t_y, and elutes from the column at time t_Y; and likewise for compound Z.

Note first that the instantaneous values of k' (k_t) during migration of X (or Y, or Z) fall within the range of (roughly) $1 \leqslant k' \leqslant 5$. This means that the resolution of X from adjacent bands will be favored; that is, the *average* or effective k' value *during* its migration and separation is optimum (see discussion of Eq. 2.10, pp. 51-52).

A second feature that emerges from Figure 16.1 is the fact that k' for each band is fairly small (≈ 1) at the time the band leaves the column. Small k' values at the time of elution mean narrow bands and increased sensitivity, as discussed in Section 13.5. This results in a potential twofold or greater increase in sensitivity for gradient elution versus isocratic elution, even for simple samples that do not require gradient elution. Thus, in gradient elution it is possible to obtain both maximum resolution and sensitivity *for every band in the sample*.

A third point worth noting in Figure 16.1 is that compounds such as Z with very large k' values at the beginning of the separation are nevertheless eluted fairly rapidly, because of the exponential decrease in all k' values with time. Finally, the gradient in k' shown in Figure 16.1 as a function of time (dashed curves A, B, C) implies a corresponding gradient along the column at any given time. That is, k' for a given band decreases along the column, so that the tail of the band moves continually in a region of smaller k' than the band front. This

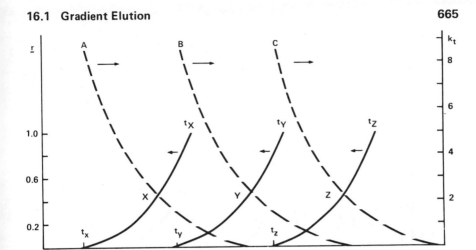

Figure 16.1 Illustration of band migration in gradient elution. Solid line shows fractional migration r of band along column, as function of time; dashed line shows band k' value as a function of time (instantaneous value k_t).

feature of gradient elution is useful in the case of bands that would tail in isocratic elution, as in Figure 16.2*a*. Superimposed onto Figure 16.2*a* is shown the k' gradient that exists in gradient elution; the reduction in k' for the band tail causes its faster migration relative to the band center and a correction of tailing as illustrated in Figure 16.2*b*. Because of this, band tailing is generally less of a problem in gradient elution.

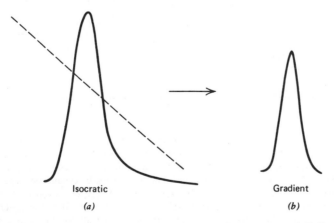

Isocratic   Gradient

(*a*)   (*b*)

Figure 16.2 The improvement of tailing bands in gradient elution. (*a*) Tailing band in isocratic separation, with k' gradient (for gradient elution) superimposed. (*b*) Same band in gradient elution (2× attenuated).

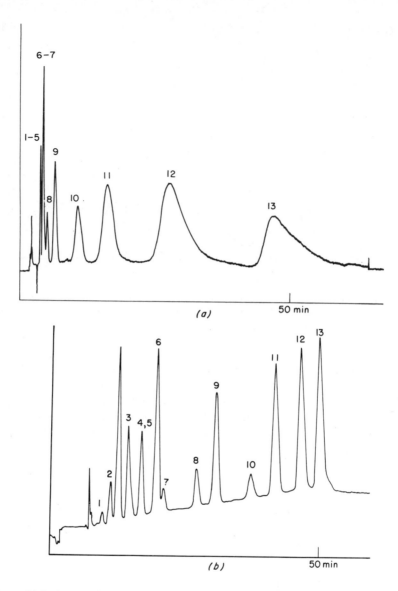

Figure 16.3 Separation of carboxylic acid mixture by (a) isocratic ion-exchange separation, (b) gradient elution, and (c) GE with steeper gradient. Bands are following acids: 1, o-toluic; 2, benzoic; 3, maleic; 4, phthalic; 5, fumaric; 6, terephthalic; 7, isophthalic; 8, 1,2,3-tricarboxybenzene; 9, 1,2,4-tricarboxybenzene; 10, 1,3,5-tricarboxybenzene; 11, 1,2,4,5-tetracarboxybenzene; 12, pentacarboxybenzene; 13, hexacarboxybenzene. Column, DuPont SAX Permaphase; mobile phase, 0.01 M borate buffer at pH 9.7 in water; temp., ambient; UV, 254 nm; 5-μl sample, 1% each compound. (a) Mobile phase also contains 0.055 M NaNO$_3$. (b) Mobile phase gradient from 0.01-0.1 M NaNO$_3$ at 2%/min. (c) Same at 5%/min. Reprinted from (9) with permission.

Figure 16.3 provides a classic illustration of the value of gradient elution, in this case for the ion-exchange separation of various carboxylic acids. In the isocratic separation of Figure 16.3a, seven unresolved bands are bunched together into two initial peaks, whereas the last two bands show reduced sensitivity and moderate to severe tailing. In Figure 16.3b the gradient elution separation of this same mixture (with varying ionic strength of the mobile phase) results in (a) separation of all but two compounds (peaks 4 and 5), (b) the resolution of additional (unknown) bands eluting before peak 1, (c) good detection sensitivity for the last eluting bands, and (d) elimination of band tailing for all compounds (the apparent small tail on peak 13 is actually a partially resolved band, which is more obvious in Figure 16.3c). Figure 16.3c, which involves GE separation of the same mixture with a steeper solvent gradient, is further examined in the following discussion.

Band migration, resolution, and peak width in GE separations are largely controlled by the form of the solvent gradient or *mobile phase program* [i.e., mobile-phase composition (%B) as a function of separation time t]. Other variables such

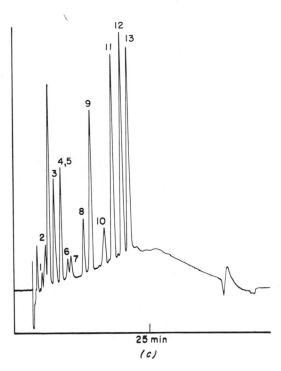

Figure 16.3 Continued

as column length L and flowrate F also play a role in gradient elution, but this role is usually subordinate to that of the solvent gradient. Therefore, the selecttion and further adjustment of the solvent program is normally the first step in the design of a successful GE separation.

Design of Optimum Solvent Gradients

Previous papers (*1-5*) have developed a general rule for choosing the right solvent gradient for any GE separation: the use of so-called *linear-solvent-strength* (LSS) gradients. Most such gradients consist of binary mixtures of some weaker solvent A with a stronger solvent B. Later we consider more complex gradients, where we begin with some solvent A, carry out an initial gradient from A to solvent B, and subsequently extend the gradient to mixtures of B with some third solvent C (ternary gradient). For the moment we assume that the gradient is formed from only two solvents A and B, where either A and/or B can consist of pure solvent or some solvent mixture [e.g., a reverse-phase gradient from 30%v methanol/water (A) to 80%v methanol/water (B)] .

An LSS gradient for a given sample and LC method is of the form

$$\log k_x = \log k_0 - b \, [t/t_0].\tag{16.1}$$

Here k_x refers (approximately) to the value of k' for a given band X as a function of time t after the start of the gradient; k_x, of course, decreases as the strength of the mobile phase increases with time. The parameter b should be roughly constant for all sample bands, with an optimum value of 0.2-0.3. The quantity k_0 is equal to the value of k_x at the beginning of separation (for mobile phase equal to pure A) and should be reasonably large for all sample components ($k_0 \gg 1$). This description of optimum GE gradients may appear unnecessarily abstract, but the application of this concept to practical GE separations will be illustrated with reasonably obvious examples.

Referring back to Figure 16.1, an LSS gradient will provide identical migration curves (e.g., solid curves of Figure 16.1) for all sample components, beginning at characteristic times t_x and terminating in the retention time t_X. The time t_x for each compound increases according to its k_0 value in a given GE system. As a direct consequence of this migration behavior (in LSS GE systems), the average or effective value of k' during migration is the same for every band, and the terminal k' value at the time of elution from the column will also be identical for each band. This in turn results in equivalent resolution of every band (so far as k' and Eq. 2.10 are concerned) and equal bandwidths for all components (which should mean roughly equal detection sensitivity). Thus, LSS gradients are directly related to the primary goals of gradient elution separation.

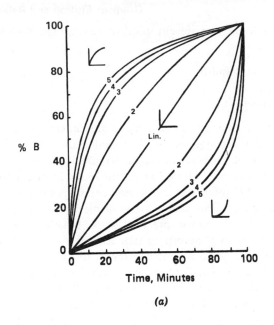

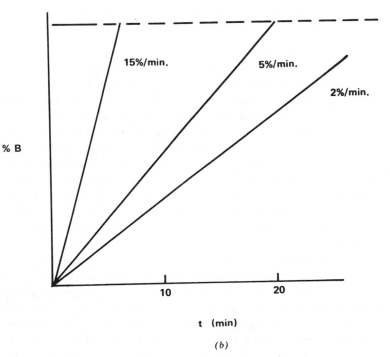

Figure 16.4 (a) Gradient shape, (b) steepness.

In more direct terms, the solvent gradient can be described by three different characteristics:

- The two solvents A and B.
- Gradient shape.
- Gradient steepness.

The strengths of the two solvents (or solvent mixtures) at the beginning and end of the gradient play a major role in determining the adequacy of the solvent gradient. The shape of the gradient can be described as linear, or varying degrees of convex or concave, as illustrated in Figure 16.4a. Most commercial gradient devices for use in LC offer a series of gradient shapes similar to those shown here. More complex gradient shapes are available with some commercial units, but these are rarely required in practical work. Finally, the steepness of the gradient is determined by the arbitrary time scale at the bottom of Figure 16.4a. Thus, if the time to carry out a gradient (from A to B) is doubled, the gradient is said to be half as steep. Gradient steepness can be expressed variously as the %/min change in the concentration of B entering the column or, for nonlinear gradients, as the *average* %/min change in the concentration of B. The latter would be simply 100%, divided by the time during which the gradient is operated (or B continues to change). Figure 16.4b illustrates several linear gradients of varying steepness.

The Two Solvents A and B. Consider first the choice of the two solvents A and B that form the gradient. With the right solvent strength for B, all bands of interest will be eluted during the gradient, or soon after the completion of the gradient with continued elution by pure B. That is, B will be strong enough to elute all sample components with k' values no greater than about 5-10. Similarly, with the right solvent strength for A, no partially resolved or unresolved bands will be eluted near t_0. Furthermore, the final chromatogram should contain no blank spaces near the beginning or end of the gradient, as these represent wasted time in gradient elution. That is, the strength of A should not be too weak, nor that of B too strong. A few examples will further illustrate some of these points.

Figure 16.5a illustrates an adequately designed gradient for reverse-phase LC, as applied to the separation of a mixture of aromatic hydrocarbons. All bands are reasonably well resolved, and there are no large gaps in the chromatogram where peaks are absent. In this case, solvent A is 50%v methanol/water and B is pure methanol. In figure 16.5b, solvent A is replaced with the weaker solvent: 20%v methanol/water. Now the initial part of the chromatogram is only sparsely occupied with sample bands, and the resulting separation requires about 15 min longer. However, there is no benefit in terms of sample resolution, and the detection sensitivities of compounds 1 and 2 have been adversely affected. Had we started with the separation of Figure 16.5b, the obvious improvement would have been an increase in the strength of solvent A.

Figure 16.6 shows the reverse-phase GE separation of a vitamin extract, with solvent *A* chosen as 10%v methanol/water and *B* as pure methanol. In this case the gradient is completed during the first 15 min, but sample components continue to elute without any apparent end. Later-eluting bands have broadened to the point where their detection is becoming difficult. Here the problem is that

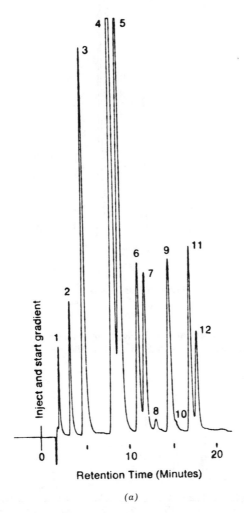

(a)

Figure 16.5 Effect of strength of initial solvent *A* on gradient separation. (*a*) 50%v methanol/water. Samples are 1 (benzene) through 11, 12 (benzpyrenes). Column, 100 × 0.21 cm, ODS-Permaphase (pellicular reverse-phase); mobile phase, gradient at 2%v/min beginning with 50%v methanol/water; temp., 50°C; 1.0 ml/min, ΔP = 1000 psi; UV, 254 nm. Reprinted from (*10, 11*) with permission.

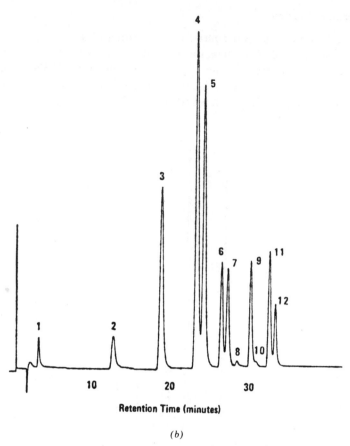

Retention Time (minutes)

(b)

Figure 16.5*b* Effect of strength of initial solvent on gradient separation. Conditions as in Figure 16.5*a*, except gradient starts with 20%v methanol/water.

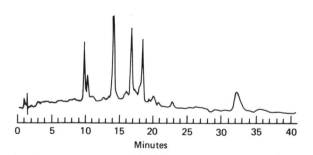

Minutes

Figure 16.6 Effect of strength of final solvent *B* on gradient separation. Analysis of a vitamin extract. Column, 25 × 0.3 cm, Spherisorb ODS (10 μm); mobile phase, 10-100%v methanol/water gradient in 15 min; UV, 254 nm; 10-μl sample. Reprinted by permission of Spectra-Physics Corp.

solvent B is not sufficiently strong, so that many compounds elute with final k' values that are too high. The solution would be the use of a stronger solvent B; for example (Section 7.3), propanol, dioxane, or THF.

Figure 16.7 illustrates a GE separation where solvent A is too strong and solvent B is too weak. The result is marginal resolution of early-eluting bands, and excessive broadening plus tailing of later-eluting bands, just as in isocratic separation. In this liquid-solid separation on silica, ethyl ether or ethyl acetate might have been a better choice for solvent B (stronger solvent). Similarly, solvent A could be further weakened by adding 10-20%v 3M Fluorochemical (Table 6.1) as a weaker solvent to the hexane. A gradient such as that of Figure 16.7 is said to have too narrow a *range*. Similarly, a gradient with too wide a range would show all bands in the center of the chromatogram, with empty space at the beginning and end (e.g., see Fig. 16.15a). In the latter case A would be too weak and B too strong. Chapters 6-11 contain information on solvent strength for the various LC methods. Table 16.1 provides a summary of the solvent property that should be changed to yield a stronger solvent B for a given gradient.

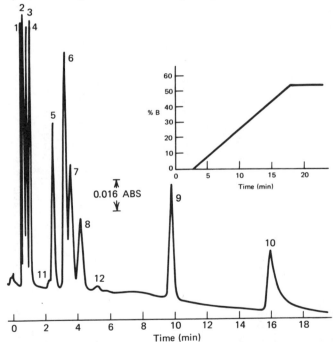

Figure 16.7 A solvent gradient with a range that is too narrow. Separation by gradient elution on silica of various dyes. Column, 15 × 0.21 cm, MicroPak Si-10 (10-μm silica); mobile phase, gradient from 0.2%v isopropanol/hexane to 0.2%v isopropanol/methylene chloride as shown in figure; ambient; 1.0 ml/min; UV, 254 nm; 4-μl sample with 0.3 mg/ml each compound (0.1 mg/ml 9, 10). Reprinted with permission of Varian Aerograph.

Table 16.1 Summary of gradient characteristics for optimized LSS gradient elution[a].

LC method	Solvent strength increases with	Preferred gradient shape
Liquid-liquid or bonded-phase		
Reverse-phase	Increasing %v organic; decrease in polarity or P'	Linear %v B
Normal-phase	Increasing polarity or P'	Linear %v B
Liquid-solid	Increasing polarity or $\epsilon°$	Concave %v B
Ion-exchange	Increasing salt concentration; increased pH (cation separations); decreased pH (anions)	Concave salt concentration; linear pH
Ion-pair		
Reverse-phase	Decreasing counter-ion concentration; other variables same as reverse-phase or ion-exchange	Concave in counter-ion concentration; other variables, see reverse-phase LLC or ion-exchange
Normal-phase	Reverse of above	Same as above

[a]For optimum gradient steepness, see Table 16.3.

Occasionally a GE separation on silica or alumina requires a very wide-range gradient, or solvent strengths for A and B that are quite different. This would be the case for samples whose components cover an unusually wide range in k' values (e.g., $>10^4$) in isocratic separation. If a GE separation is attempted with two solvents A and B of widely different strength, *solvent demixing* can occur during separation, just as in the case of separation by TLC (Section 9.3 and Fig. 9.13b). An example is provided in Figure 16.8 for the separation of several compounds on silica, using a gradient of hexane (A) to isopropanol (B). In this case the two solvents are too different in strength and solvent demixing occurs. The symptoms are easily recognized; thus, in Figure 16.8, there is a sudden narrowing of sample bands at one point in the middle of the chromatogram (noted by an asterisk), with bands being wider on each side of the peak in question. For more complex samples, several compounds might be merged together into a poorly resolved cluster, just as sometimes occurs at t_0 in isocratic elution. In the example of Figure 16.8, the very polar solvent isopropanol (B) is adsorbed onto the column during the inital stages of the GE separation. At some later time the column becomes saturated with B, at which point the composition of mobile phase leaving the column changes abruptly from pure A to some mixture of A and B. The sudden change in solvent strength results in rapid elution of compounds of intermediate retention strength, and generally poor resolution at

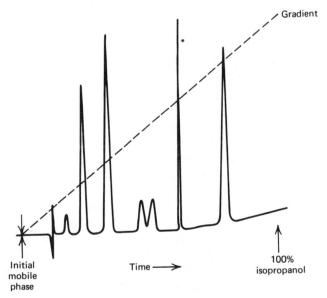

Figure 16.8 Solvent demixing in gradient elution. Separation of unknown compounds on silica. Column, 25 × 0.32 cm, Zorbax-Sil; mobile phase, gradient from 0.1%v isopropanol/hexane to isopropanol; temp., ambient. Reprinted courtesy of J. J. Kirkland.

this point in the separation. Solvent demixing may also result in a gradual widening of sample bands just before the breakthrough of solvent B.

Where a wide-range gradient is required, as for mixtures of very polar and nonpolar species, it is necessary to use multisolvent gradients: e.g., A to B, then B to C. The strengths of solvents A, B, and C increase in this order. Examples of the use of such multisolvent gradients have been discussed in (2,4,12), for application to liquid-solid separations. Multisolvent gradients are only rarely required in liquid-solid LC, and not at all with other LC methods.

Gradient Shape. We have already noted that commercial gradient formers can produce gradients of any shape, varying from markedly concave to markedly convex (Figure 16.4). The appropriate gradient shape varies from one LC method to another, and is determined by Eq. 16.1 (for LSS gradients) and the dependence of k' on solvent composition in a given case. For a detailed discussion of these relationships, see for liquid-liquid LC (5, 13-15), for liquid-solid LC (1,4,16); for ion-exchange LC (1,14); and for ion-pair LC, Chapter 11. Suffice it to say that the optimum gradient shape for a given situation can be predicted in advance, as summarized in Table 16.1. Several representative solvent gradients for different LC methods are summarized in the following section.

What happens if the gradient shape selected differs from the recommendations of Table 16.1? In most cases, no serious loss in resolution will result, because gradient shape is not a very critical parameter in GE separation. In a few cases, departure from the gradient shapes suggested in Table 16.1 can actually improve separation, for reasons we will examine in a moment. Sometimes, however, the wrong gradient shape yields an obviously inferior separation.

Consider first, however, the changes in a GE chromatogram that can be expected as the gradient shape is changed. In Figure 16.9 we illustrate this with some hypothetical separations based on the accompanying gradients: convex (a), linear (b), and concave (c). We assume that a linear gradient happens to be optimum for the example of Figure 16.9. As a result, the chromatogram for this gradient (Figure 16.9b) shows bands of constant width from beginning to end, with good resolution throughout the chromatogram. As the gradient becomes more convex, as in Figure 16.9a, the faster increase in %v B at the beginning of the separation leads to elution of these bands with lower average values of k' and lower final k' values (see the following section). As a result, initially eluting bands are sharpened and pushed together, with a loss in resolution. At the same time, during the latter part of the separation, the %v increase in B (per unit time) is decreased, leading to the opposite effects: wider bands and increased resolution. The opposite overall changes are seen in Figure 16.9c versus Figure 16.9b: wider, better-resolved bands initially and sharper, less-resolved bands finally.

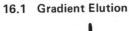

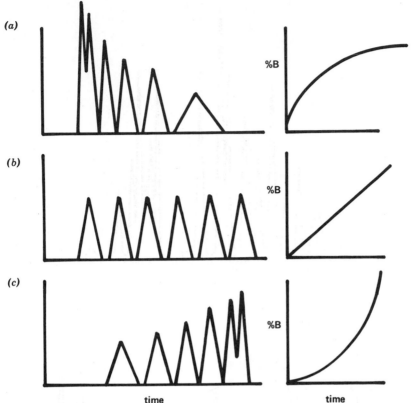

Figure 16.9 Effect of gradient shape on the resulting chromatogram. (*a*) Convex gradient, (*b*) linear gradient, (*c*) concave gradient.

The trends observed in Figure 16.9 can be generalized beyond the case of gradients that should be linear. Whenever a GE chromatogram resembles that of Figure 16.9*a*, it is necessary to make the gradient less convex, or more concave; if the gradient is already linear, then it should be made concave. If the gradient is already concave, it should be made more concave. For GE chromatograms resembling those of Figure 16.9*c*, more convex gradients are required.

When the correct gradient shape has been selected (e.g., the linear gradient for the reverse-phase separation of Figure 16.5), the bands should be of equal width across the chromatogram. With a slightly nonoptimum choice of gradient shape, as in the linear (instead of concave) salt gradient in the ion-exchange separation of Figure 16.3*b*, band width is seen to increase for later-eluting bands (cf. Figure 16.9). However, the separation is otherwise fully acceptable, and few workers

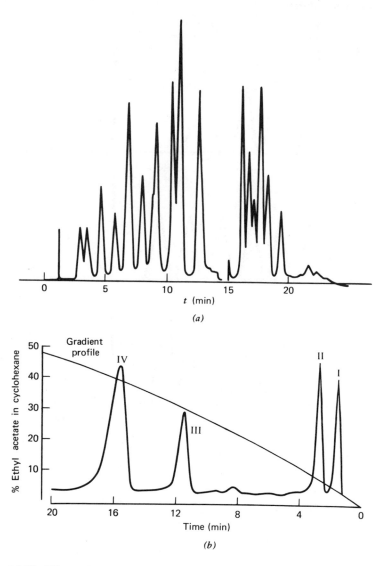

Figure 16.10 Effect of gradient shape in separations by liquid-solid LC. (*a*) Concave gradient, (*b*) convex gradient. (*a*) Column, 130 × 0.7 cm 20-μm silica; mobile phase, gradient from pentane to ethyl ether; temp., ambient; 26 ml/min. Substituted aromatic compounds as sample. Reprinted from (*2*) with permission. (*b*) Column, 185 × 0.32 cm, Corasil-II (pellicular silica); mobile phase, cyclohexane to ethyl acetate; temp., ambient; UV, 280 nm; 10-μl sample. Natural anthraquinone derivatives as sample. Reprinted from (*17*) by permission.

678

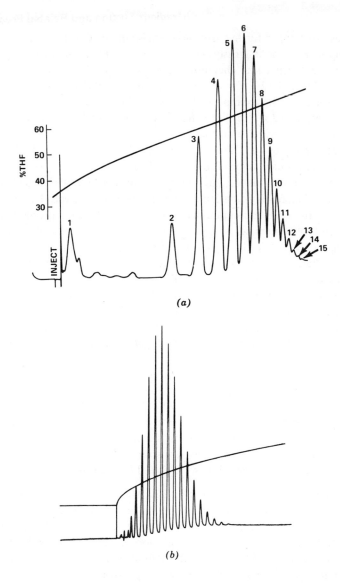

Figure 16.11 Need for convex gradients in separations of oligomeric mixtures by reverse-phase LC. (a) Polystyrene (600 MW), (b) fabric-finishing mixture. (a) Column, BondaPak-C_{18}/Corasil (reverse-phase pellicular, 122 × 0.23 cm; mobile phase, water/THF gradient: ambient, 1.0 ml/min; UV, 256 nm. Reprinted by permission of Waters Associates. (b) Column, HC-ODS/Sil-X; mobile phase, 20-30%v acetonitrile/water gradient, temp., ambient. Reprinted with permission of Perkin-Elmer Corp.

would choose to tamper further with gradient shape in this example. However, when the gradient shape is grossly wrong and where the gradient range is fairly wide, obvious deficiencies in the resulting separation will be apparent. This is illustrated in Figure 16.10 for separations by liquid-solid LC. With the correct (concave) gradient in the example of Figure 16.10a, all bands are observed to be of approximately equal width. With the convex gradient in Figure 16.10b, later bands are much wider and exhibit tailing.

For certain samples a more convex gradient is required than that recommended previously or in Table 16.1. These are samples composed of a regular series of compounds (e.g., polymer oligomers, aromatic hydrocarbons, or carbohydrates of increasing ring number) that yield successively larger b values (Eq. 16.1) for later-eluting compounds. Most commonly, such problem samples are mixtures of oligomers of widely varying molecular weight. An example is shown in Figure 16.11a for the separation of polystyrene oligomers varying from $1 \leqslant n \leqslant 12$, where n refers to the number of monomers in the oligomer. For the oligomers with $n \geqslant 2$, the water (A)/THF(B) gradient in this reverse-phase separation is roughly linear. However, the bunching together of later-eluting bands suggests that the gradient should actually be convex (see also example of Figure 7.35). In a related separation of an oligomeric mixture in a similar system (Figure 16.11b), a convex gradient gives excellent spacing and band widths for the various sample components.

Finally, in some separations the weakest solvent A available may still result in poor resolution at the front end of the chromatogram. Here the only improvement possible is the use of so-called *gradient delay*. By starting the gradient *after* elution of bands near t_0, some improvement in resolution is often possible. An example is shown in Figure 16.7, where it is seen that the gradient was delayed for about 3 min, during which time the four closely bunched peaks at the beginning of the chromatogram (1-4) were eluted isocratically.

Gradient Steepness. The parameter b in Eq. 16.1 is directly proportional to gradient steepness. The larger the change in %B/min, the larger is b. This steepness parameter b is also related (5) to the average value of k' during separation and to the final value of k' at the time of elution. In fact, it is roughly correct to equate the average value of k' to $1/b$ and the final value of k' to $1/2b$. Therefore, just as an increase in k' for a band in isocratic separation increases resolution and decreases sensitivity, so in gradient elution resolution decreases and sensitivity increases with increasing gradient steepness b. Table 16.2 provides a more quantitative summary of these relationships.

The effect of gradient steepness (or b) on resolution and peak sensitivity is illustrated in Figure 16.3c versus Figure 16.3b. The gradient steepness in (c) is increased by a factor of 2.5 (5%/min versus 2%/min). As a result, the bands are bunched more tightly together in (c), the bands are narrower and higher in (c), and separation time in (c) is shortened by a factor of about 2. These are exactly the effects that would result in isocratic elution if a stronger solvent were used in

Table 16.2 Relationship of resolution and peak sensitivity in gradient elution to the gradient steepness b.

b	Relative R_s	Relative sensitivity[c]
0.05	0.94	0.11
0.1	0.79	0.21
0.2[a]	0.63	0.37
0.3[b]	0.54	0.51
0.5	0.39	0.72
1.0	0.20	1.03
2.5	0.06	1.35

Source: From (5).
[a] Optimum value when column length L is held constant (for maximum R_s).
[b] Optimum value when column pressure P is held constant.
[c] Relative to an isocratic band at t_0.

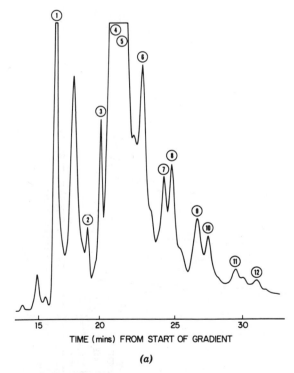

TIME (mins) FROM START OF GRADIENT

(a)

Figure 16.12 Separation of polycyclic aromatic hydrocarbons by reverse-phase gradient elution. (a) 3%/min increase in methanol. Column, (a, b) 50 × 0.26 cm, (c) 150 × 0.26 cm, Sil-X-II RP; mobile phase, water to 90%v methanol/water convex gradient; temp., 60°C; 1.0 ml/min; UV, 254 nm. Reprinted by permission of Perkin-Elmer Corp.

681

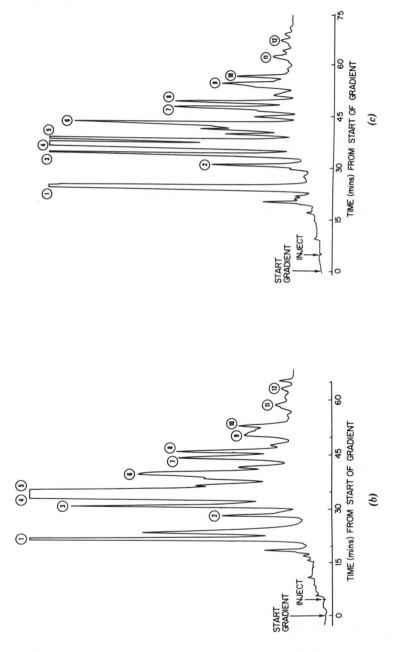

Figure 16.12b, c Separation of polycyclic aromatic hydrocarbons by reverse-phase gradient elution. Conditions as in Figure 16.1*a*, except: (*b*), 1%/min increase in methanol; (*c*), same as (*b*), with 3-fold increase in column length.

(*c*) versus (*b*), pointing up again the similarity of gradient steepness *b* and solvent strength (or the reciprocal of k') in isocratic separation. A similar example is provided in the reverse-phase GE separation of a polyaromatic hydrocarbon sample in Figures 16.12*a* and 16.12*b*. Here the same water-to-90%-methanol gradient was used in both separations, but 30 min were allowed for the separation in (*a*) versus 90 min in (*b*). The less steep gradient in (*b*) has resulted in an obvious improvement in resolution.

Just as there exists an optimum value of k' for resolution in isocratic separations, in GE there exists an optimum value of *b* or optimum gradient steepness that will provide maximum resolution for a given separation time. This optimum in *b* is equal to 0.1-0.3, corresponding to optimum k' in isocratic elution of $3 \leqslant k' \leqslant 5$. In further discussions of gradients that are optimized in terms of steepness, we assume that the optimum *b* = 0.2. From Eq. 16.1 and the dependence of k' on solvent composition (i.e., %v *B*), an optimum value of gradient steepness $\triangle$ %v *B*/min can be determined, corresponding to *b* = 0.2. Because of the form of Eq. 16.1, this optimum gradient steepness varies with t_0:

$$\angle\text{\%v } B/\text{min} = \text{constant} / t_0. \qquad (16.2)$$

Based on the preceding discussion and relevant experimental data, it is possible to offer several specific examples of optimized gradient steepness for various LC methods, as summarized in Table 16.3 (see also Table 16.1).

As an example, consider how Table 16.3 and Eq. 16.2 are used to estimate the gradient for a reverse-phase separation. If acetonitrile/water is chosen for the gradient system, from Table 16.3 we can obtain the optimum gradient steepness: equal to 5%v B/t_0. The constant in Eq. 16.2 is therefore equal to 5%v. If t_0 is 1.5 min for the particular column and flowrate, from Eq. 16.2: $\triangle$%v B/min = 5/1.5 = 3.3%v B/min. Alternatively, if the total gradient to be covered in the separation is 20-100%v acetonitrile, the elapsed time for the gradient should be set equal to $(100 - 20)/(\triangle\text{\%v } B/\text{min}) = 80/3.3 = 24$ min. Many commercial gradients specify this elapsed time for the gradient, rather than %v B/min.

For normal-phase BPC systems, Table 16.3 specifies that an optimum gradient is one that provides a change in P' for the mobile phase of 0.5 units per t_0. For the case of pure solvents *A* and *B*, with P' values equal to P_a and P_b, this is equivalent to an optimum gradient

$$\triangle \text{ \%v } B/\text{min} = \frac{50}{(P_b - P_a) t_0}.$$

Thus, for $CHCl_3$ (*B*) and hexane (*A*), with P_b = 4.1 and P_a = 0.1 (Table 6.1), and for a t_0 value of 0.5 min, the optimum gradient would be 50/(4.1 - 0.1) x

Table 16.3 Some optimized gradients for LC separation.

LC Method	B/A
Reverse-phase BPC[a]	
Vary organic content of mobile phase *linearly*	Methanol/water, 7%v B/t_o
	Acetonitrile/water, 7% B/t_o
	Dioxane/water, 6%v B/t_o
	Isopropanol/water, 5%v B/t_o
	Tetrahydrofuran/water, 4.5%v B/t_o
Normal-phase BPC[a,b]	
Vary composition of binary mobile phase *linearly*, to give change in polarity equal to $0.5P'/t_o$	CH_2Cl_2/hexane, 15%v B/t_o
	Tetrahydrofuran/hexane, 12%v B/t_o
	$CHCl_3$/hexane, 12%v B/t_o
	Propanol/hexane, 12%v B/t_o
Liquid-solid LC[c]	
Vary composition of binary mobile phase in concave fashion, to give change in solvent strength equal to 0.02 ϵ°/t_o	1-Chlorobutane/hexane, 10 t_o for 0-100%
	$CHCl_3$/hexane, 13 t_o for 0-100%
	CH_2Cl_2/hexane, 16 t_o for 0-100%
	Isopropyl ether/hexane, 17 t_o for 0-100%
	Tetrahydrofuran/hexane, 22 t_o for 0-100%
	Acetonitrile/chlorobutane, 15 t_o for 0-100%
	Methanol/isopropyl ether, 18 t_o for 0-100%

Ion-exchange LC

Increase salt concentration by 1.6-fold per t_o in a *concave* gradient — 0.005-0.5 N NaNO$_3$ in 10 t_o

Increase pH by 0.2 units/t_o (cation exchange) — pH varied from 2 to 6 in 20 t_o

Decrease pH by 0.2 units/t_o (anion exchange) — pH varied from 8 to 2 in 30 t_o

Ion-pair LC (reverse-phase, bonded-phase)

Vary organic content of mobile phase *linearly* (most common) — See above for reverse-phase BPC

Decrease counter-ion concentration by 1.6-fold per t_o in a *concave* gradient — 0.1-0.01 M tetrabutylammonium ion in 5 t_o

Increase pH by 0.2 units/t_o (anionic counter-ion)

Decrease pH by 0.2 units/t_o (cationic counter-ion)

---

[a] Same rules apply for LLC, but gradient elution usually not applicable; see Chap. 8.

[b] Little experimental data for these systems, so less reliable prediction can be made; rate of change of mobile-phase composition = 100%v/t_o divided by twice the difference in P' values for the two solvents A and B.

[c] The t_o interval for gradient equal to difference in $\epsilon^°$ values for solvents A and B, divided by 0.02.

$0.5 = 25\%v \, B/$min. The optimum gradient steepness in liquid-solid LC is determined similarly, as outlined in Table 16.3, but with ϵ° values substituted for P' values.

Ion-exchange systems most commonly use salt gradients, and a 1.6-fold increase in salt concentration per t_0 is required for optimum resolution. Thus, if a separation time equal to $10t_0$ is planned, the ratio of salt concentrations for B versus A should be $(1.6)^{10}$, or about a 100-fold difference (e.g., a gradient from 0.005 to 0.5 M salt). Ion-pair separations by gradient elution normally use the reverse-phase mode with a linear change in solvent composition (i.e., exactly as for reverse-phase BPC).

Other Separation Variables

In isocratic elution, the selection of the correct solvent strength is commonly the most important step, and many separations are acceptable once sample k' values have been properly adjusted. Likewise, in gradient elution the selection of the right solvent gradient often provides satisfactory separation of a given sample. However, further adjustments in column N values or α values are occasionally required to achieve complete separation of a particular sample. Our approach in gradient elution for these cases is quite similar to that in isocratic elution. The solvent gradient, or its steepness as measured by b, is maintained constant (and optimum) while we systematically vary N or α.

Varying N. The same options are available for increasing N in gradient elution as for isocratic elution. The two major options are a decrease in flowrate F (holding column length L constant) or increase in L with a proportionate decrease in F (holding pressure P constant). For a required increase in R_s in gradient elution via a change in N, proceed exactly as outlined in the discussion on controlling resolution in Section 2.5 for isocratic elution. The only requirement during this change in L and/or F by some factor X is that the gradient steepness be held constant, in terms of the change in $\triangle\%v \, B/t_0$. A summary of the necessary changes in the gradient steepness accompanying these two options for increasing N and R_s is given in Table 16.4, along with the resulting change in other separation variables. According to Table 16.4, gradient steepness b can be maintained constant if the gradient time (for change from A to B during the separation) is changed to equal the new separation time predicted in Section 2.5 (Tables 2.2, 4-7) for increase in R_s.

It is important to note that if the gradient steepness (as $\triangle\%v \, B/$min) is left unchanged when flowrate is decreased and/or column length is increased, the true steepness measured as b increases, because t_0 is increased in each case. This in turn means a decrease in the effective value of k' during separation, and a loss in resolution. Generally this effect more than compensates any increase in N associated with these changes in separation conditions. Resolution in the re-

Table 16.4 Increasing resolution and N in gradient elution.

Comment	Variable	Column length constant	Column length varied
For increased R_s, change conditions as noted:	Flowrate F	Decrease by factor x^a	Decrease by factor x^a
	Column length L	No change	Increase by factor x^a
Then gradient steepness (%v B/min) must be decreased as follows:		By factor x	By factor x^2
And other variables are affected as follows:	Separation time t	Increased by x	Increased by x^2
	Column pressure P	Decreased by x	No change

[a]The factor x can be estimated from Tables 2.2-2.7, just as for isocratic elution.

sulting separation will often be worse, not better, if flowrate is decreased or column length increased, and the gradient steepness ($\triangle$%v B/min) is not appropriately decreased as in Table 16.4.

Figure 16.12 provides an illustration of how resolution is best optimized in gradient elution. The reverse-phase separation in (a) is carried out with a methanol/water gradient of about 6%v B/t_0, which from Table 16.3 is roughly optimum. However, as in isocratic elution, some improvement in R_s can usually be achieved by increasing the average k' value beyond the optimum range of 3-5. Thus, in Figure 16.12b, slowing down the gradient to 2%v B/min does result in a significant further increase in resolution, although with a doubling of separation time. However, it could have been predicted in this case that a better alternative would have been retention of the optimum gradient ($\sim$6%v B/t_0), but with an increase in column length L. Indeed, when this was done (Figure 16.12c), significantly better resolution was found for most bands in the sample, and with no further increase in separation time. Had the initial gradient not been optimum for the separation of Figure 16.12 (e.g., %v B/t_0 = 15%), then a simple decrease in the gradient steepness would have been more effective than an increase in column length.

Before leaving this discussion of column efficiency and N in gradient elution, it should be noted that the narrower bands in gradient elution (vs. isocratic separation) do *not* reflect higher N values, but smaller k' values at the time of elution. Because of this change in k' during separation for each band, a column N value cannot be determined by the usual calculation based on retention time and band width (Eq. 2.6 or 2.6b). For a more detailed discussion, see (*5*).

Varying α. As in isocratic separation, α values in gradient elution normally do not vary much as gradient steepness or N are varied. To change α values deliberately in gradient elution, it is necessary to change either the mobile phase or stationary phase, while holding k' values (and b) constant. Normally the mobile-phase composition will be varied, in one of two ways. First, another solvent B can be selected and the gradient reoptimized for this new solvent. For example, if our initial gradient were 7%v B/t_0 methanol/water for reverse-phase separation (cf. Table 16.3), 4.5%v B/t_0 tetrahydrofuran/water can be substituted. It is hoped that this change in mobile phase will provide some change in α values but leave average k' and N values at their original optimum levels. In the second approach for change in α values, a third solvent C can be added to both solvents A and B [e.g., see (*18*)].

An illustration of the first approach is shown in Figures 16.13a and 16.13b. Here the reverse-phase separation of a six-component mixture containing benzene was attempted, using conventional solvent systems: methanol/water in (a) and acetonitrile/water in (b). Although the α values for benzene are much changed between these two systems, overlap occurs onto band 3 in the first case

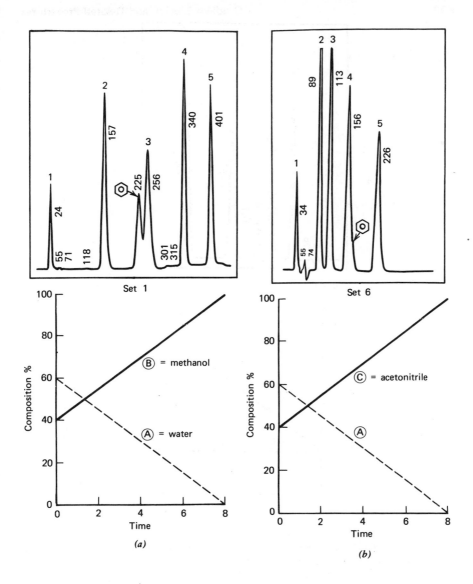

Figure 16.13 Separation of synthetic mixture including benzene by reverse-phase LC. Varying solvent selectivity by altering the gradient composition. Column, 25 cm 10-μm Spherisorb ODS; mobile phase, (*a*) 40%v methanol/water to methanol gradient; (*b*) 40%v acetonitrile/water to acetonitrile gradient; temp., 50°C; 1.5 ml/min; UV. Reprinted by (*18*) with permission.

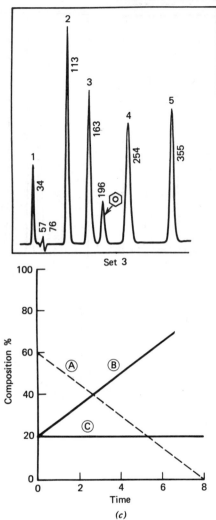

Figure 16.13c Separation as in Figures 16.13a, b, except 20%v methanol/20%v acetonitrile to 80%v methanol/20%v acetonitrile gradient.

and onto band 4 in the second case. Obviously, an intermediate gradient of methanol/acetonitrile/water should be more successful. This was accomplished according to the second approach, as illustrated in Figure 16.13c. Here the concentration of acetonitrile was held constant at 20%v, replacing 20%v of the methanol in solvent B. Now the benzene peak is moved between bands 3 and 4 and is nicely resolved. This technique does not require the use of a three-solvent gradient device as in (18). Simply diluting the water (A) and methanol (B) in the starting gradient (Figure 16.13a) with 20%v acetonitrile would duplicate this separation (Figure16.13c) and require only a conventional two-solvent gradient device.

Minor changes in the α values of adjacent bands sometimes result when the gradient steepness ($\triangle$%v B/t_0) is changed. These changes in α are analogous to those occurring in isocratic separation when solvent strength is varied. See (5) for a further discussion.

A general approach to designing a successful gradient elution separation is summarized in Table 16.5. This scheme should be particularly useful to those unfamiliar with gradient elution, but it can also serve as a checklist for more experienced workers.

Table 16.5 Designing a gradient elution separation.

---

1. Pick the two solvents A and B on the basis of their strengths, as discussed in Chapters 6-11.

2. Estimate the optimum gradient shape from Table 16.1.

3. Estimate optimum gradient steepness from Table 16.3.

4. Carry out initial separation.

5. Decrease strength of starting solvent A if bands bunched together near t_0; increase strength of A if bands are far apart in initial part of chromatogram; increase strength of final solvent B if bands continue to elute for a considerable time after completion of the gradient; reduce strength of B if all bands eluted before end of gradient.

6. Correct gradient steepness if necessary; increase for added sensitivity, decrease for increased resolution.

7. Adjust column N value (Table 16.4) as necessary for increased resolution; see Tables 2.2-2.7.

8. Vary solvent B for improved selectivity, if necessary.

9. Solve any special problems: solvent demixing, use of gradient delay for poor front-end resolution when weaker solvent A is not available, etc.

Other Aspects of Gradient Elution

Column Regeneration. At the conclusion of a gradient elution separation, the column must be reequilibrated with solvent A to remove all traces of solvent B before the next sample is injected. In principle, this can be done simply by washing the column with enough solvent A. In fact, column regeneration in this way can require a rather long time, particularly for liquid-solid LC columns and porous polymeric ion-exchangers. Pellicular packings, on the other hand, regenerate quickly.

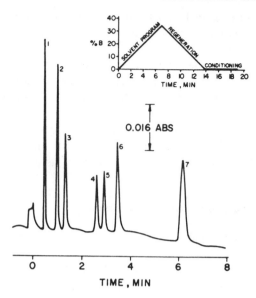

Figure 16.14 Faster column regeneration after gradient elution using a reverse gradient. Separation of antioxidants by liquid-solid LC on silica. Column, 15 × 0.24 cm 5-μm Li-Chrosorb; mobile phase, hexane (*A*), methylene chloride (*B*) gradient; ambient; 1.0 ml/min; 1-μl sample, 2 mg/ml each compound (1, BHT; 2, CAO-14; 3, triphenyl phosphate; 4, antioxidant 754; 5, BHA; 6, Goodrite; 7, Santowhite). Reprinted from (*19*) with permission.

It has been found (*19,20*) that column regeneration in LSC can be accelerated by using a *reverse gradient*, following completion of the gradient separation. This is illustrated in Figure 16.14 for the GE separation of a mixture of antioxidants on an alumina column. Following the 7-min separation, the gradient was simply reversed and the column finally purged with solvent *A*. In this way the column was fully conditioned (restored to its initial state) in 12 min. Using solvent *A* as purge immediately following the separation, a much longer period (30-45 min) was required. For reverse-phase BPC separations (*20a*), it has recently been noted that a simple change from 100% *B* to 100% *A* (step-gradient) at the end of separation leads to faster column regeneration, at least when the mixing volume between pump and column is small.

Column regeneration following GE separation is dependent on the total liquid *volume* passing through the column, not the *time* of column conditioning. Therefore, it is generally recommended to regenerate columns after gradient elution by use of the highest flowrate possible (for fastest column regeneration). Regeneration of a column after gradient elution can be checked by repeated injection of a standard sample-mixture, as discussed in Section 9.3 (see discussion of Fig. 9.11).

Applications. Gradient elution (or similar technique) is required for those samples that cannot be handled isocratically. Previous figures in this chapter provide many examples of this situation, which represents the commonest case where gradient elution is applied. Gradient elution is also often used for "scouting" unknown samples. Use of a reasonably broad-range gradient (e.g., 5-95%v methanol/water* in reverse-phase BPC) is likely to resolve most samples at least partially and will almost always suggest what further action is required, if separation is incomplete in the initial try. This approach to LC method development is further discussed in Chapter 18.

Following an initial "scouting" separation by gradient elution, one will sometimes wish to repeat the separation under isocratic conditions, to maximize the resolution of a given compound, or simply to determine that compound more conveniently and/or more precisely. In this case, a general rule is needed for selecting the proper isocratic solvent strength on the basis of the initial GE separation. As discussed in (5), if the GE retention time of the compound of interest is t_X, then the solvent entering the column at time $(t_X - 2.5t_0)$ should be used in the isocratic separation. This is illustrated with the GE separation of Figure 16.15a, for the separation of a hydrogenated-quinoline mixture on an alumina column. The compound V-b in Figure 16.15a was only partly resolved in this inital GE separation, and it was desired to repeat the separation isocratically with solvent strength optimized for the resolution of compound V-b. In this case t_X for V-b is about 20 min, t_0 is 1.7 min, and therefore $t_X - 2.5t_0 = 15.8$ min. At 15.8 min the composition of the mobile phase entering the column in the GE separation of Figure 16.15a was about 35%v CH_2Cl_2/pentane, and this composition was selected for a repeat isocratic separation of a similar sample. The resulting separation shown in Figure 16.15b yielded good resolution of the V-b band, in a shorter overall separation time than in Figure 16.15a.

Gradient elution can be a very powerful technique for trace analysis. Large sample volumes (many times the volume of the column) can be charged if the sample is much more weakly retained than the component of interest, and gradient elution can then be used to elute the peak of interest as a very narrow, easily detected band. One very practical example is provided by the measurement of trace levels of nonpolar substances in water, using reverse-phase GE-LC. Figure 16.16 illustrates such an application. Here 200 ml of river water (sample) were charged to a 30 x 0.4-cm reverse-phase BPC column, followed by gradient elution. A sharply convex gradient was used, which provides extreme sharpening of the last eluted bands. In this case the last bands off the column were concentrated about 100-fold relative to the initial sample, whereas chromatography normally leads to a broadening (and dilution) of the initial sample during the course of separation (see also Figure 13.9).

*Note that 0-100%v reverse-phase gradients generally give problems, because of degassing of the water/organic mobile phase after mixing, nonwetting of the packing by pure water, and slow regeneration; 5-95%v gradients are preferable.

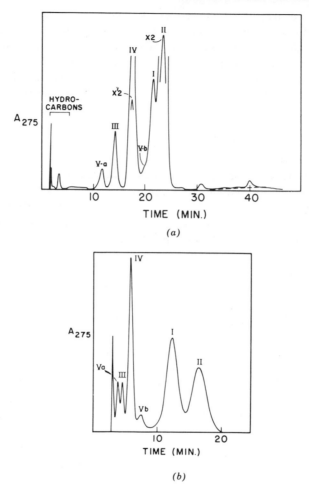

Figure 16.15 Use of gradient elution (*a*) to select correct solvent strength for subsequent isocratic separation (*b*). Separation of hydrogenated-quinoline sample by liquid-solid LC on alumina. Column, 130 × 0.7 cm 88-μm alumina; mobile phase, pentane, methylene chloride, acetonitrile gradient (*a*), 35%v CH$_2$Cl$_2$/pentane (*b*); ambient; 7.8 ml/min (*a*), 10 ml/min (*b*); UV, 275 nm. Reprinted from (*21*) with permission.

16.2 COLUMN SWITCHING AND STATIONARY-PHASE PROGRAMMING

The present section describes several special-purpose LC techniques that can have a variety of different goals and that are carried out in a number of different ways. The common feature of this group of procedures is the use of mobile-phase switching valves in conjunction with two or more separation

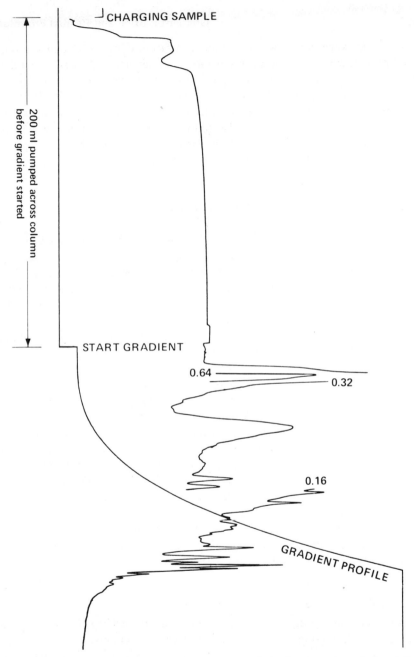

Figure 16.16 Trace enrichment of nonpolar compounds in river water sample by gradient elution reverse-phase LC. Column, 30 × 0.4 cm μ-BondaPak-C$_{18}$ (10-μm reverse-phase); mobile phase, 5-95% acetonitrile/water gradient; flowrate, 2.0 ml/min; ambient; UV, 200-ml sample. Compounds are presumed to be polycyclic aromatic hydrocarbons. Reprinted from (22) with permission.

columns. In some cases there is further utilization of two different stationary phases (or packings) for the same separation. In general, switching allows diversion of different sample components to different columns for maximum resolution with minimum time, during the separation of a single sample. These procedures have only recently found significant use in modern LC, but their popularity is increasing rapidly (see the following discussion). Column-switching procedures have been applied widely in gas chromatography during the past 20 years [(e.g., see (23)], and their use in LC seems destined to enjoy a similar success.

Column Combinations: General Considerations

Consider first the simple combination of different columns, without benefit of switching valves. Columns of the same type are commonly joined in series, to achieve increased column length (Section 5.6). In a few cases it is advantageous to join columns with different packings, although this is the exception, rather than the rule. For example, columns of different pore sizes, are commonly joined in SEC separations, to allow the separation of samples with a broad molecular-weight range (Chapter 12). In ion-exchange chromatography, columns of cation and anion exchange resins have been combined to allow simultaneous separation of all ionic species in one step [(e.g., see (24)] or to carry out ion chromatography (Section 10.7). Where "guard" columns are used to protect the main column (Section 5.6), it is advantageous to use pellicular packings for the guard column and porous packings for the main column. In other cases, however, it is often inadvisable to combine columns packed with different particles, except where switching valves are used to achieve a specific objective. The reason is that the N value obtained for a set of dissimilar columns in series can be significantly less than the N value calculated by adding the individual column N values. That is, there can be a serious overall loss in column efficiency unless certain factors are considered.

The loss in column efficiency as a result of combining columns with different packings has been discussed in detail (25). To summarize, column efficiency in such cases is maximized by attention to the following rules:

- Use columns with similar H values (e.g., same d_p).
- Use columns of the same internal diameter.
- Where columns of equal length are combined, try to keep k' values for compounds of interest in the optimum range for both columns.

The first two rules are important; the last rule is desirable, when it is compatible with the overall separation goals. Finally, maximum separation efficiency is also promoted by minimizing dead volumes in the connections between columns, and in any switching valves (see Section 3.2).

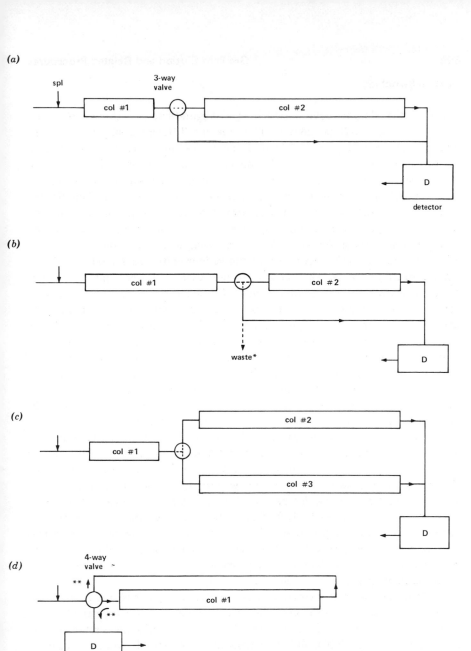

Figure 16.17 Different experimental configurations for column-switching separations.

Column Switching

Some of the simpler column-switching configurations are shown in Figure 16.17, and these will be used to illustrate the different applications of this general technique. Figure 16.17a is designed for use with the same column packing material in each column. In this case column 1 is one-half to one-fifth shorter than column 2. This design can be used as a solution to the general elution problem, when the range in sample isocratic k' values is relatively limited (e.g., $1 < k' < 40$). For this case, separation time can be significantly decreased, and the sensitivity for later-eluting bands can be somewhat increased. An example is provided in Figure 16.18, for the analysis of pesticides in milk (26). Figure 16.18a shows the separation of the various pesticides in milk on a 20-cm column of 10-μm silica. The total separation time is 30 min, and the dieldrin peak is considerably broadened. Figure 16.18b shows the separation of the same sample on a 5-cm column of the same type, with separation time reduced to about 10 min. The dieldrin peak is here considerably narrower (Eq. 2.6) and more readily detectable. However, early bands are insufficiently resolved. Using a column-switching scheme similar to Figure 16.17a, with a 5-cm column of silica for column 1 and a 15-cm column for column 2, all bands from HCB to α-BHC were allowed first to elute from column 1 onto column 2. The switching valve was then changed to allow elution of the remaining compounds (γ-BHC through dieldrin) directly from column 1 to the detector. Figure 16.18c shows the resulting separation, with initial elution of the latter compounds. Following the elution of dieldrin, the valve is again switched to allow elution of the various compounds from column 2; these bands then appear in the latter part of the chromatogram of Figure 16.18c. In this way total separation time is reduced by about one-third and detection sensitivity for dieldrin is increased about two-fold. However, the resolution of β-BHC from γ-BHC is somewhat reduced. Overall, there is a moderate improvement in the separation of Figure 16.18c versus Figure 16.18a. This general technique can be regarded as a possible solution to the general elution problem for fairly simple samples but has rather limited potential compared to gradient elution.

An alternative usage of the column configuration of Figure 16.17a (same packing each column) is for *precolumn venting*. In many LC assays, particularly by reverse-phase separation, the compound(s) of interest will be moderately strongly retained ($2 < k' < 10$), and the remainder of the sample will elute near t_0. Particularly in trace assays, the initial band at t_0 may tail to the point of overlapping later bands of higher k' value. This is illustrated in Figure 16.19a, for the determination of nicotinic acid (band 1) in serum by means of reverse-phase ion-pair chromatography. Band 1 sits on top of the tail from the t_0 band, making the quantitation of nicotinic acid difficult. With a system similar to that of Figure 16.17b, this separation was repeated. Column 1 was initially eluted to

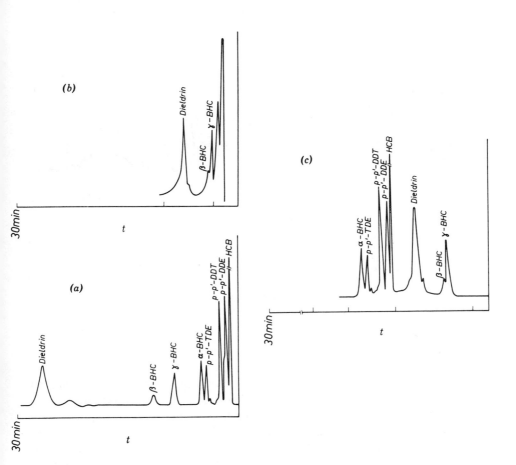

Figure 16.18 Separation of pesticides in milk by means of column switching with porous-silica columns. (*a*) Separation on 15-cm column, (*b*) separation on 5-cm column, (*c*) separation as in Figure 16.17*a* with 5-cm (1) and 15-cm (2) columns. Columns, 5 or 15 × 0.21 cm of Partisil (col. 1, 5 cm; col. 2, 10 cm); mobile phase, *n*-heptane; ambient; 1.0 ml/min; electron capture detection. Reprinted from (*26*) with permission.

waste (not to the detector; see Figure 16.17*b*), until most of the t_0 band had eluted from column 1. When band 1 was about to elute from column 1, the valve was switched to allow elution of band 1 from column 1 onto column 2. In this way much of the t_0-band was eliminated prior to the final separation of

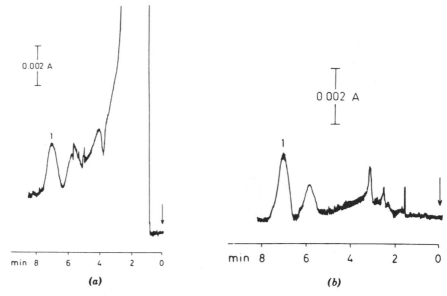

Figure 16.19 Separation of nicotinic acid (1) from serum (*a*) without and (*b*) with precolumn venting. Column configuration as in Figure 16.17*a*; ion-pair chromatography. Columns, 5 (1) and 15 (2) × 0.32 cm, silanized LiChrosorb SI-60, plus 1-pentanol stationary phase; mobile phase, 0.12 M tetrabutylammonium. pH 7.4, in water; temp., 25°C; 0.80 ml/min, 1000 psi; UV, 254 nm; 55-μl serum. Reprinted from (*27*) with permission.

the sample on column 2. The result, as shown in Figure 16.19*b*, is a much-improved chromatogram for the assay of band 1.

The efficacy of precolumn venting for removal of t_0-components that tail can be considerably improved if the precolumn (1 in Figure 16.17*b*) is eluted with a weaker solvent for some period before beginning elution with the solvent used to separate bands of interest. This is illustrated in Figure 16.20*a* for the ion-exchange separation and assay of polyamines in urine. In Figure 16.20*a* the sample was allowed to elute through both columns 1 and 2, without venting of early-eluting components. The result is a drastic overlapping of the compounds of interest by the tail of the t_0-band. In Figure 16.20*b* precolumn venting was employed with initial elution by a weaker solvent. Following this initial elution (and venting), the valve was switched and elution through columns 1 and 2 was continued with a stronger mobile phase. Almost total elimination of the t_0-tail was achieved in this way.

Stationary-Phase Programming. The ability of column switching to provide a solution to the general elution problem is considerably enhanced if the phase ratio of stationary phase in the two columns is allowed to vary. In this case, the configuration of Figure 16.17*b* is generally used, with columns 1 and 2 of equal

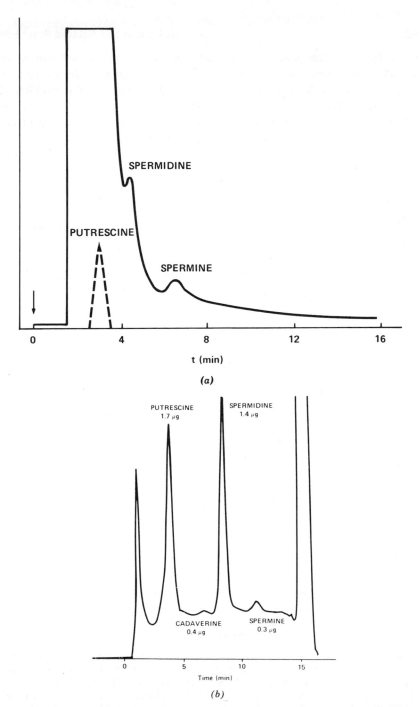

Figure 16.20 Separation of polyamines in urine, using precolumn venting. (*a*) Without precolumn venting, (*b*) with venting. See (*30a*) for details. Conditions same as for Figure 10.16 (gradient elution ion-exchange chromatography, using porous-polymer 10-μm particles).

length. Typically, the phase ratio of stationary phase in column 1 will be substantially lower than that in column 2. This can be achieved, for example, by using a pellicular packing in column 1 and a totally porous packing in column 2. Even better is the use of a low-surface-area packing in column 1 and a high-surface-area packing in column 2, because in this case the efficiency of each column can be maximized with small-particle packings. In either case, it is

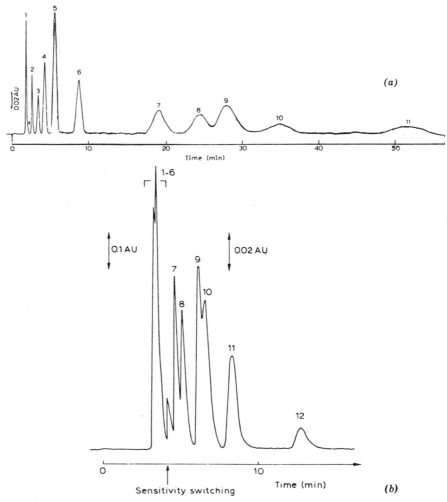

Figure 16.21 Column switching with stationary-phase programming for separation of steroid mixture. Column configuration as in Figure 16.17b. (a) Elution of sample through both columns 1 and 2 (no switching); (b) elution through column 1 only; (c) elution through both columns, with switching. Columns, 25 × 0.27-cm kieselguhr, 2 m²/g (1) and 15 m²/g (2); mobile phase, water/ethanol/isooctane; temp., ambient; UV, 236 nm. Reprinted from (*31*) with permission.

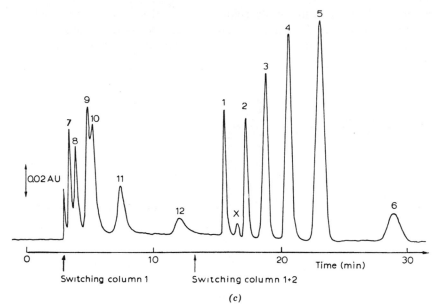

Figure 16.21 Continued.

essential that the same type of stationary phase be used in each column (e.g., silica, reverse-bonded-phase, ion-exchange, etc).

An example of this application is shown in the steroid separation of Figure 16.21a, using normal-phase liquid-liquid chromatography. Column 1 was packed with a 2-m^2/g support, whereas column 2 contained 15-m^2/g support. The amount of liquid stationary phase in column 2 was therefore about seven times greater than that in column 1, and sample k' values were increased in the same proportion. Figure 16.21a shows the separation of the sample by elution through both columns, with resulting slow separation and difficult detection of bands 7-12. Figure 16.21b shows the separation of the sample on the first column, with poor resolution of bands 1-6 (but good separation of bands 7-12). In the column-switching mode, bands 1-6 were first allowed to pass from column 1 onto column 2, followed by switching of the valve for elution of further bands from column 1 directly to the detector. Following elution of bands 7-12 from column 1, the valve is again switched, for continued elution through columns 1 and 2. In this way, bands 1-6 are now eluted from column 2. The composite chromatogram is shown in Figure 16.21c, for the column-switching separation. A substantial improvement in separation time and detection sensitivity of bands 7-12 is observed in this separation versus that of Figure 16.21a. The reason is that bands 7-12 are separated on the low-capacity column 1, where the lower k' values result in fast elution and narrow bands. Similarly, bands 1-6 are separated on the high-capacity column 2, with the higher k' values required for good resolution of these less strongly held components. The overall separation can be referred to, therefore, as *stationary-phase programming,* analogous to solvent programming in gradient elution.

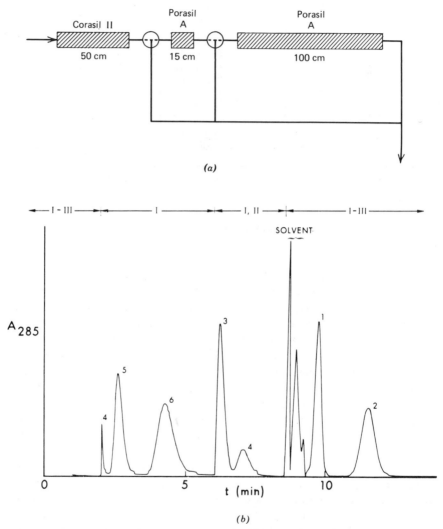

(a)

(b)

Figure 16.22 Column switching with stationary-phase programming for separation of antioxidants. (*a*) Column configuration; (*b*) separation of antioxidant mixture, ranging from 1 nonpolar to 6 polar; (*c*) separation of polymer extract. Reprinted from (*30*) with permission. Columns, as shown, 0.28-cm i.d.; 17%v CH_2Cl_2/pentane as mobile phase; ambient; UV, 285 nm.

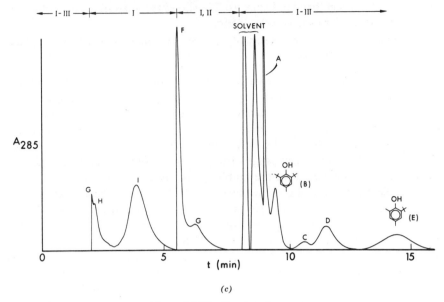

(c)

Figure 16.22 Continued.

The selection of the best time for initial switching of the valve (to allow retention of only bands 1-6 on column 2 in Figure 16.21c) is usually done empirically. Generally, the optimum switching point will be at about $1.5\text{-}2t_0$ for column 1 (i.e., at a time after sample injection equal to $1.5\text{-}2t_0$, when only column 1 is connected to the detector). Automatic selection of the switching-time under microprocessor control has been discussed by Willmott et al. (*28a*).

Column-switching as above was first used for low-pressure LC separation (*29*) and later for high-pressure LC (*30*), using three columns rather than the two shown in Figure 16.17b. The latter application for the separation of a mixture of polar and nonpolar antioxidants, and for analysis of antioxidants in a polymer extract, is shown in Figure 16.22. The column arrangement in Figure 16.22a provides for initial elution of sample through columns 1-3. Before the elution of initial bands from column 3, the first three-way valve is switched to allow elution of strongly retained bands from column 1 directly to the detector. At the completion of elution of these components from column 1, the two valves are switched to allow elution to the detector of columns 1 and 2 in series. In this way, bands of intermediate retention are rapidly eluted from column 2. When all bands on column 2 have been eluted, the two valves are again switched to allow serial elution through columns 1-3. Elution is continued until all bands are eluted. The resulting separation of polar plus nonpolar antioxidants in this fashion is shown in Figure 16.22b: compound polarity and retention increase from band 1 through band 6. Figure 16.22c shows the analysis in this fashion of a polymer extract for various antioxidants.

For some more recent examples of column switching with stationary-phase programming, see (*31, 32*).

Selectivity Switching. An alternative objective of the column arrangement of Figure 16.17b is for use with very complex samples (i.e., containing so many components that band overlap cannot be avoided even for large *N* values and with gradient elution). In this case, two columns are selected with completely different stationary phases. The only requirement of the two columns is that they give k' values in the right range with the same solvent or that fractions from column 1 are compatible with the mobile phase used for column 2. The objective of selectivity switching is the same in LC as in gas chromatography (*33*). The problem involved is illustrated in Figure 16.23. Here it is assumed that so many sample components are present that there is no space in the chromatogram from any single column to allow room for all the separated bands. Figure 16.23a shows a part of the chromatogram from column 1, with several overlapping bands indicated. Now on some other column 2, with different stationary phase, the resolution of the doublets of Figure 16.23a might be achievable, as shown in Figures 16.23b-d. The problem however, is that separation of the total sample on column 2, as shown in Figure 16.23e, results in remixing of several band pairs that were resolvable on column 1.

One solution to the preceding problem is selectivity switching, using the column configuration of Figure 16.17b. Referring now to Figure 16.23, as bands *a* + *b* leave column 1 in unresolved form, they are diverted to column 2, where their complete separation occurs, as in Figure 16.23b. Following diversion of bands *a* + *b* to column 2, the valve is again switched to avoid elution of bands c-f onto column 2, where they would overlap the separated bands *a* and *b*. Selectivity switching normally does not allow separation of all sample components but is intended instead for the resolution of a few selected components from a very complex mixture. For recent applications and further discussion of this technique, see (*34, 35*).

A modification of the arrangement of Figure 16.17b allows transfer of several fractions from the first column onto the second column, for further separation (e.g., *35, 36*). If a slow separation in the first column is combined with rapid separation in the second column, the result can be the complete analysis of very complex samples in a single operation. For example, Erni and Frei (*36*) used gel filtration in column 1 in combination with reverse-phase separation in column 2. The separation time in the first column was about 10 hr, versus 30-min separations in the second column. Successive 30-min fractions from the first column were automatically injected onto the second column, with a great increase in the resolving power of the total system. In this case, the mobile phase for the first

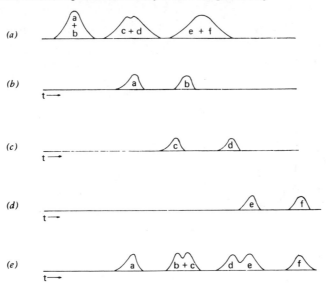

Figure 16.23 Selectivity switching of hypothetical sample with configuration of Figure 16.17b. See text. Reprinted from (33) with permission.

separation (water) was completely compatible with the second separation, because water is a very weak solvent in reverse-phase LC; see Figure 18.14.

Coupled-Column Operation

A more efficient form of column switching with stationary-phase programming was first introduced in 1970 (3), using the column configuration of Figure 16.17c. A short column 1 is packed with a low-capacity packing (e.g., pellicular, low surface area, lightly loaded, etc.), a longer column 2 is packed with the same material, and a final column 3 is packed with a high-capacity packing that will give larger k' values for all sample components. The principle of operation is illustrated in Figure 16.24. Columns 1 and 3 are first connected via the three-way valve, sample is injected, and elution is continued for a time equal to about $2t_0$ (for column 1). The passage of bands from column 1 is illustrated in Figure 16.24a. In this fashion bands 1-4 are passed onto column 3. The valve is then switched, with elution of bands 5-9 through columns 1 and 2. These low-capacity columns (1,2) provide smaller k' values for later-eluting bands; the result is their rapid separation as narrow, easily detected bands (see Figure 16.24b). Following elution of the sample from columns 1 and 2, the valve is switched to allow elution through columns 1 and 3. Bands 1-4 are now eluted

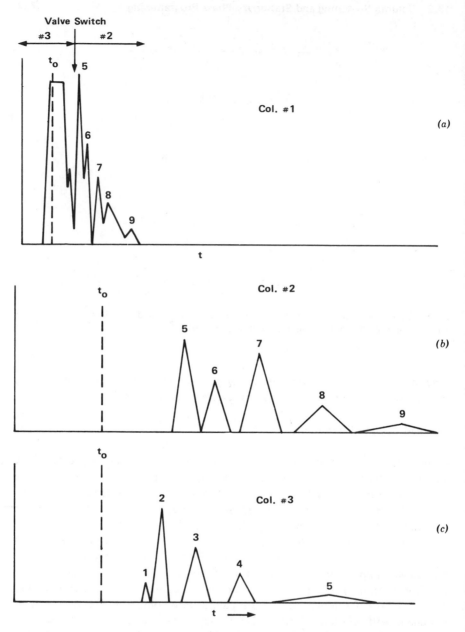

Figure 16.24 Coupled-column operation for configuration of Figure 16.17c with hypothetical sample. See text.

from the high-capacity column 3, providing adequate resolution and otherwise satisfactory separation, as in Figure 16.24c.

Coupled-column operation can be optimized to provide good resolution for moderately complex samples, using up to four total columns [see discussion of (3)]. It is somewhat more complex than simple column switching with stationary-phase programming, but in principle represents a more effective utilization of the total LC system. It should be considered particularly for the routine assay of samples with a very wide range in k' values.

Column Backflushing

Column backflushing has been widely used in gas chromatography, and the simplest column configuration is shown in Figure 16.17d. The four-way valve between the sample injector and column allows for flow of mobile phase through the column in either direction. Following elution of bands of interest, the valve is switched to allow reverse flow through the column. This results in rapid purging of strongly retained sample components from the column. In some cases, the total quantity of strongly retained material may be of interest, whereas in other cases it is desired only to clean out the column prior to injecting the next sample. An application of backflushing in LC is illustrated in Figure 16.25, for the analysis of gasoline in terms of its various compound types. In Figure 16.25a the elution of a total gasoline sample from a silica column is shown, without backflushing. While the saturates and olefins bands are readily eluted and easily quantifiable, the aromatics fraction is spread out over a considerable time and is difficult to measure. By using backflushing plus valve-switching immediately after elution of olefins, the chromatogram of 16.25b was obtained. Here the total aromatics are eluted as a sharp, easily quantitated band.

Some columns cannot be subjected to a reversal of the usual direction-of-flow of mobile phase (i.e., the inlet and outlet of the column cannot be reversed). Such columns exhibit shifts in the column packing under these conditions, with development of voids within the column, and a serious loss in column N value. Before using any column in a backflushing mode, it should be confirmed that no adverse effect on column efficiency will result. When in doubt, consult the supplier who has furnished the column.

Backflushing or other column-purging techniques are even more effectively used with column arrangements such as Figure 16.17a. Here the presence of a short precolumn (1) allows for faster purging of strongly retained compounds from this column. Indeed, following elution of major sample components, continued flushing of the precolumn may allow complete regeneration of column 1 without a change in flow direction (i.e., without backflushing). However, backflushing of the precolumn can be much faster.

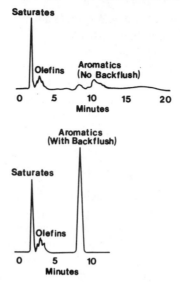

Figure 16.25 Separation of gasoline sample into compound-type fractions by liquid-solid LC and column backflushing. Configuration as in Figure 16.17d used, (a) without backflushing and (b) with backflushing. Column, 30 × 0.4 cm, μ-Porasil; mobile phase, 3-M Fluorochemical FC-78; ambient; 3.5 ml/min; refractive index detection; 3-μl sample. Reprinted from (37) with permission.

Experimental Considerations

Although column packings and switching valves suitable for the techniques described in this section have only recently become widely available, there is now no lack of such materials for any conceivable application of column switching. The various valves required are essentially similar to the injection valves described in Section 3.6, and suitable packing materials are listed in Chapter 5. Table 16.6 lists several column-packing combinations that can be used in the stationary-phase programming procedures previously described; see also (37a). The optimum combination of packings, valves, and column dimensions can be inferred from the preceding discussion, as well as the preceding references.

Although these column-switching operations can be carried out in a manual mode, normally such separations would be effected with automatic switching valves interfaced to a timer and actuated by initial sample injection. The selection of valve-switching times (relative to sample injection) can be determined empirically, by trial and error (i.e., various times can be selected, and the resulting separations evaluated). Some estimates of preferred times have been given previously for individual procedures.

Table 16.6 Possible column packings for column-switching with stationary phase programming (Figure 16.17b or c).

LC mode	Low-capacity packing[a]	High-capacity packing
Reverse-phase BPC	Pellicular C_{18} (30-μm) Porous C_2 (5- or 10-μm) Porous C_8 or C_{18} with surface area less than 50 m^2/g (5- or 10-μm)	Porous C_{18} (5- or 10-μm), with surface area $>$ 200 m^2/g
Liquid-solid (silica)	Pellicular silica (30-μm) Porous silica with surface area less than 50 m^2/g	Porous silica (5- or 10-μm) with 400 m$_2$/g surface
Ion-exchange Anionic	Pellicular anion exchanger (30-μm)	Porous-polymer anion exchanger (10-μm) BPC anion exchanger (5- or 10-μm)
Cationic	Similar, but cationic	Similar, but cationic
Other phases	Similar approach	Similar approach

[a]Column #1.

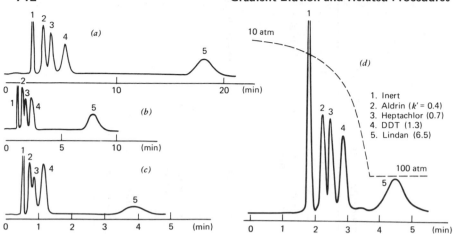

Figure 16.26 Flow (or pressure) programming separation of pesticide mixture by liquid-solid chromatography. (*a*) 140-psi column pressure; (*b*) 350-psi pressure; (*c*) 700-psi pressure; (*d*) pressure programmed from 140 to 1400 psi. Column, 50 × 0.2 cm, Merckogel SI-200 (28-μm); mobile phase, *n*-heptane; temp., ambient; mobile-phase velocity (*a*) 0.35 cm/sec, (*b*) 0.88, (*c*) 1.75. Reprinted from (*28*) with permission.

16.3 FLOW PROGRAMMING

Flow programming was originally suggested as another solution to the general elution problem, but it is in fact quite limited and is rarely used. Its one virtue is that it can be carried out in a manual mode without any additional equipment, unlike gradient elution and column switching. The principle of operation of flow programming is illustrated in Figure 16.26, for the separation of several pesticides on silica. In Figures 16.26*a-c* the pressure (and flow) is continually increased, with resulting faster separation—but poorer front-end resolution (bands 2, 3). This suggests that separation can be improved by beginning at a low pressure, to provide adequate resolution of initially eluting bands, then increasing the pressure to allow fast separation of strongly retained compounds. Figure 16.26*d* shows such a separation, and both resolution and separation speed are improved over any of the constant-pressure separations of Figures 16.26*a-c*.

Flow programming is actually a technique for increasing the front-end resolution of a sample at the expense of back-end resolution. For chromatograms such as that of Figure 16.26*c*, this approach can significantly improve the overall separation. However, it should be noted that flow programming does not increase the sensitivity of later-eluted bands; instead, peak heights are somewhat reduced because band width in volume units is increased at higher flowrates

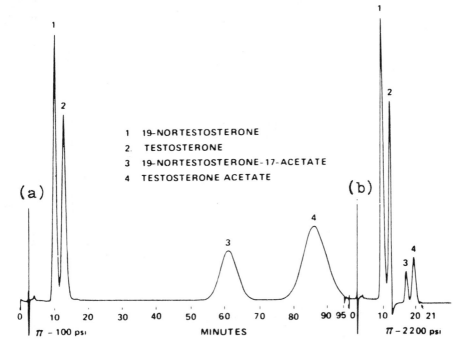

Figure 16.27 Flow programming separation of steroid mixture by reverse-phase BPC. (*a*) Without programming, (*b*) with flow programming. Column, 60 × 0.23 cm BondaPak C₁₈/ Corasil; mobile phase, acetonitrile/water (30/70); detector, UV, 254 nm. Reprinted with permission of Waters Associates.

(and lower resulting *N* values); for example, see Figure 16.27*b* and 16.27*a*.

A near ideal application of flow programming for faster separation is shown in the example of Figure 16.27. The initial separation (Figure 16.27 (*a*) required about 1.5 hr, at a pressure of 100 psi. By increasing the column pressure (and flowrate) following elution of bands 1 and 2 (Figure 16.27*b*) separation time could be reduced to about 20 min. See also the example of Figure 15.17, where the last band is rapidly eluted by flow programming.

It is interesting to compare the separation of Figure 16.18 (by column switching) with the alternative of flow programming. The separation of Figure 16.18*a* certainly suggests that flow programming would be beneficial in this instance. However, although flow programming would probably result in a separation resembling that of Figure 16.18*c* (with dieldrin and preceding BHC peaks eluted last), the peak heights of these latter bands would be significantly reduced relative to the column-switching separation of Figure 16.18*c*. For a further discussion of the theory of flow programming, see (*3*).

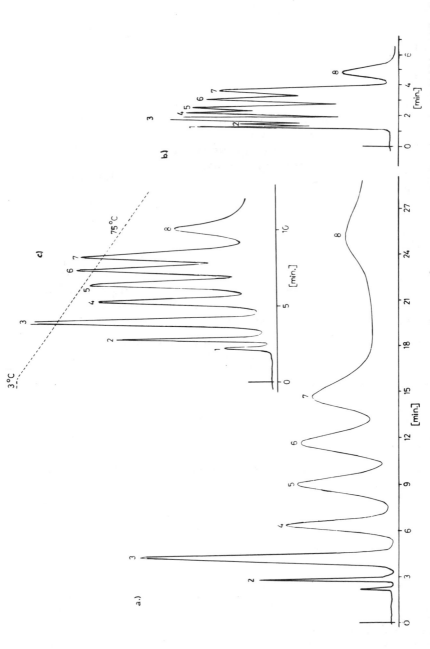

Figure 16.28 Temperature programming for separation on silica of mixture of polyaromatic hydrocarbons (benzene, 2, to picene 8). (*a*) Isothermal at 22°C, (*b*) isothermal at 70°C, (*c*) temperature program from 3–75°C. Column, 50 × 0.2 cm, Merckogel SI-200; mobile phase, *n*-heptane; temperature as stated earlier; mobile phase velocity 0.5 cm/sec. Reprinted from (*28*) with permission.

16.4 TEMPERATURE PROGRAMMING

An increase in separation temperature in LC generally results in lowered sample retention, just as in gas chromatography. Therefore, temperature programming—a continued increase in column temperature during separation—in principle offers still another solution to the general elution problem. An example of temperature programming in LC is shown in Figure 16.28. The isothermal runs at 22°C (*a*) and 70°C (*b*) exhibit various features of the isocratic separation of a sample with an overly wide range of k' values. Temperature programming from 3-75°C in (*c*) nicely overcomes these difficulties and resembles the result that would have been expected from gradient elution.

Although temperature programming has occasionally been advocated (28, 37) as a substitute for gradient elution or column switching in LC, it is an even more limited technique than flow programming, as discussed in (3, 5). Thus, only a limited control of k' values is possible, between the temperature limits set by mobile-phase boiling point on the upper end and excessive solvent viscosity at lower temperatures. Furthermore, column efficiency decreases with increasing mobile-phase viscosity (Section 5.7), so that in temperature programming the column efficiency is adversely affected over much of the initial part of the temperature program. Because these limitations of temperature programming are not offset by other considerations, few workers use this technique today.

16.5 PRACTICAL COMPARISON OF VARIOUS PROGRAMMING AND COLUMN-SWITCHING PROCEDURES

Which of the various techniques should be used for solving the general elution problem? Several considerations are involved in evaluating and comparing the relative merits of these different approaches. Table 16.7 summarizes the major factors that determine the relative value of each procedure. Because the primary goal of most of these procedures is dealing with the general elution problem, we will for the moment ignore such procedures as precolumn venting and selectivity switching. With respect to the remaining procedures, resolution and separation speed are optimal for gradient elution, and poor for flow and temperature programming or column switching with the same packing in columns 1 and 2. Reliability, simplicity, and stability of operation are less favored for gradient elution versus the various column switching techniques. For very complex samples with a wide range in k' values, the favored techniques are gradient elution when the entire sample is of interest and column backflushing when only a few compounds are to be determined. For an analysis of resolution/speed versus sample k' range, see (*3*). In terms of cost of the related equipment, gradient elution is the most expensive and flow programming is the cheapest.

Table 16.7 Comparison of different programming and column-switching procedures.[a]

Procedure	Resolution/ speed	Reliability/ simplicity	Sample k' range	Cost	Detector restraints	Column regener- ation
Gradient elution	++	-	++	-	-	-
Column switching, same packing (Figure 16.17a)	-	+	-	+	++	++
Precolumn venting (Figure 16.17b)	++	+	+	+	++	++
Stationary-phase programming (Figure 16.17b)	+	+	+	+	++	++
Selectivity switching (Figure 16.17b)	++	+	-	+	++	++
Coupled-columns (Figure 16.17c)	+	+	+	+	++	++
Backflushing (Figure 16.17d)	++	++	++	+	++	++
Flow programming	-	++	-	++	+	++
Temperature programming	-	+	-	+	+	-

[a] ++, very favorable; +, adequate; - , unfavorable.

The other techniques are moderately priced. Considering detector constraints, the various column-switching procedures are preferred, because the same mobile phase always passes through the detector. Gradient elution cannot be used with refractometer detectors, and many solvents give shifting or drifting baselines in gradient elution for other detectors as well. Column regeneration is not required in the column-switching procedures but is necessary in gradient elution and temperature programming.

On balance, which technique represents the best compromise? For scouting separations of unknown samples and for other one-time separations, the flexibility and power of gradient elution will continue to make it a popular technique for solving the general elution problem. For many routine, quantitative assays of complex samples, however, the stationary-phase programming procedures (including coupled columns) are the most valuable. They are preeminently reliable and cheap, and overall they will often provide the fastest separations possible, particularly when the need for column regeneration in gradient elution is considered.

For more specialized problems, precolumn venting, selectivity switching, and backflushing should find limited application.

REFERENCES

1. L. R. Snyder, *Chromatogr. Rev.*, 7, 1 (1965).

2. L. R. Snyder and D. L. Saunders, *J. Chromatogr. Sci.*, 7, 195 (1969).

3. L. R. Snyder, *ibid.*, 8, 692 (1970).

4. L. R. Snyder, *Anal. Chem.*, 46, 1384 (1974).

5. L. R. Snyder, J. W. Dolan, and J. R. Gant, *J. Chromatogr.*, 165, 3 (1979).

6. C. Liteanu and S. Gocan, *Gradient Liquid Chromatography*, Halsted Press (Wiley), New York, 1974.

7. P. Jandera, M. Janderova, and J. Churacek, *J. Chromatogr.*, 115, 9 (1975), and preceding articles in this series.

8. P. J. Schoenmakers, H. A. H. Billiet, and R. Tijssen, *ibid.*, 149, 519 (1978).

8a. L. R. Snyder, in *High Performance Liquid Chromatography — Advances and Perspectives*, C. Horvath, ed., Vol. 1, Academic, New York, 1980, Chap. 4.

9. J. Aurenge, *J. Chromatogr.*, 84, 285 (1973).

10. J. A. Schmidt, R. A. Henry, R. C. Williams, and J. F. Dieckman, *J. Chromatogr. Sci.*, 9, 645 (1971).

11. Du Pont Instruments Product Division, Product Bulletin 820PB4, August, 1971.

12. R. P. W. Scott and P. Kucera, *Anal. Chem.*, 45, 749 (1973).

13. G. Vigh and J. Inczedy, *J. Chromatogr.*, 129, 81 (1976).

14. P. Jandera and J. Churacek, *ibid.*, 91, 207 (1974).

15. L. R. Snyder and J. J. Kirkland, *Introduction to Modern Liquid Chromatography*, Wiley-Interscience, New York, 1974, p. 469.

16. L. R. Snyder, *Principles of Adsorption Chromatography,* Dekker, New York, 1968, Chap. 8.

17. P. P. Rai, T. D. Turner, and S. A. Matlin, *J. Chromatogr.,* **110**, 401 (1975).

18. L. B. Sybrandt and E. F. Montoya, *Am. Lab.,* 79 (August, 1977).

19. R. E. Majors, *Anal. Chem.,* **45**, 755 (1973).

20. H. Engelhardt, *Z. Anal. Chem.,* **277**, 267 (1975).

20a. G. Hewett, *Altex Chromatogr.,* **1** (4), 3, (September, 1978).

21. L. R. Snyder, *J. Chromatogr.,* **7**, 595 (1969).

22. J. N. Little and G. J. Fallick, *ibid.,* **112**, 389 (1975).

23. O. E. Schupp III, *Gas Chromatography,* Wiley-Interscience, New York, 1968, pp. 253-258.

24. C. D. Scott, D. D. Chilcote, S. Katz, and W. W. Pitt, Jr., *J. Chromatogr. Sci.,* **11**, 96 (1973).

25. J. Kwok, L. R. Snyder, and J. C. Sternberg, *Anal. Chem.,* **40**, 118 (1968).

26. R. J. Dolphin, L. P. J. Hoogeveen, and F. W. Willmott, *J. Chromatogr.,* **122**, 259 (1976).

27. K.-G. Wahlund and U. Lund, *ibid.,* **122**, 269 (1976).

28. H. Engelhardt, *Z. Anal. Chem.,* **277**, 267 (1975).

28a. F. W. Willmott, I. Mackenzie, and R. J. Dolphin, *J. Chromatogr.,* **in press.**

29. L. R. Snyder, *Anal. Chem.,* **36**, 774 (1964); **37**, 713 (1965).

30. L. R. Snyder, in *Modern Practice of Liquid Chromatography,* J. J. Kirkland, ed., Wiley-Interscience, New York, 1971, pp. 232-236.

30a. H. Adler, M. Margoshes, L. R. Snyder, and C. Spitzer, *J. Chromatogr.,* **143**, 125 (1977).

31. J. F. K. Huber and R. v. d. Linden, *ibid.,* **83**, 267 (1973).

32. J. F. K. Huber, *ibid.,* **149**, 127 (1978).

33. M. C. Simmons and L. R. Snyder, *Anal. Chem.,* **30**, 32 (1958).

34. J. F. K. Huber, in press.

35. E. L. Johnson, R. Gloor, and R. E. Majors, *J. Chromatogr.,* **149**, 571 (1978).

36. J. Erni and R. W. Frei, *ibid.,* p. 561.

37. J. C. Suatoni, H. R. Garber, and B. E. Davis, *J. Chromatogr. Sci.,* **13**, 367 (1975).

37a. J. F. K. Huber, *J. Chromatogr.,* **149**, 127 (1978).

BIBLIOGRAPHY

Engelhardt, H., (Z. Anal. Chem., **277**, 267 (1975) (brief review of gradient elution, column switching, temperature programming, and flow programming, with examples).

Janders, P., M. Janderova, and J. Churacek, *J. Chromatogr.*, **115**, 9 (1975) (this paper and preceding articles in same series deal with theory and practice of gradient elution).

Liteanu, C., and S. Gocan, *Gradient Liquid Chromatography,* Halsted Press (Wiley), New York, 1974 (detailed review of gradient elution in LC and TLC).

Snyder, L. R., *Chromatogr. Rev.,* **7**, 1 (1965) (an early review of gradient elution, emphasizing theory for various LC methods).

Snyder, L. R., *J. Chromatogr. Sci.*, 8, 692 (1970) (theory of gradient elution, coupled-columns, flow programming, and temperature programming for modern LC).

Snyder, L. R., in *High Performance Liquid Chromatography — Advances and Perspectives*. C. Horvath, ed., Vol. 1, Academic, New York, 1980, Chap. 4.

(detailed discussion of all aspects of gradient elution; especially LSS separations).

Snyder, L. R., and D. L. Saunders, *J. Chromatogr. Sci.*, 7, 195 (1969) (theory and practice of gradient elution for modern liquid-solid chromatography).

Snyder, L. R., J. W. Dolan and J. R. Gant, *J. Chromatogr.*, 165, 3, 31 (1979) (detailed discussion of theory and practice of gradient elution for modern reverse-phase LC).

Schoenmakers, P. J., H. A. H. Billiet, and R. Tijssen, *J. Chromatogr.*, 149, 519 (1978) (theory and practice of gradient elution in reverse-phase LC).

SEVENTEEN

SAMPLE PRETREATMENT AND REACTION DETECTORS

17.1	Introduction	720
17.2	Sample Cleanup	722
17.3	Sample Derivatization	731
17.4	Reaction Detectors	740
17.5	Automation of Sample Pretreatment	746
References		748
Bibliography		750

17.1 INTRODUCTION

Other chapters of this book deal with LC separation per se: the conditions of separation, the equipment required, and the interpretation of results. However, many samples require (or benefit from) additional treatment prior to their injection onto the column. Among the reasons why some samples cannot or should not be injected directly onto an LC column are the following:

- The initial sample should be a homogeneous solution in a solvent that is no stronger than the initial mobile phase for the separation; solid or particulate–containing samples, two-phase liquids, and samples dissolved in strong solvents usually give poor separation, and in some cases may damage the column.

- For adequate sensitivity, trace assays may require a preconcentration of the sample.

- Very complex samples containing many components may require a pre-separation into simpler fractions to avoid overlapping bands in the final LC separation.

- Some samples contain components that could harm the column; their removal prior to injection of the sample can greatly increase average column life.

Other samples may present a problem in terms of detector sensitivity. For example, the absorptivity of key compounds in the sample may be too low to measure with a photometric detector. Or the compounds of interest may be overlapped in the final chromatogram by various interfering compounds. In these cases, *sample derivatization* is often helpful. The sample is first reacted with an appropriate reagent to give derivatives of the compounds of interest. These derivatives are selected to give a good detector response (usually in a photometric or fluorescent detector) relative to other sample components. The derivatized sample is then injected for final LC separation.

In some cases, the derivatization of a sample for improved detection of bands of interest is best done on-line with a so-called *reaction detector,* immediately following the elution of sample bands from the column. Because of the similarity of the chemistry involved and because the goals of derivatization are generally the same, pre-LC derivatization and post-LC reaction detectors are discussed together in this chapter.

Sample pretreatment is usually carried out manually. However, when the pretreatment procedure is fairly involved and time-consuming and when large numbers of samples must be processed for the same analysis, one should consider the possibility of automating the pretreatment step. As we will see, virtually all the common chemical operations and manipulations can be readily automated: chopping or high-speed blending, extraction and phase separation, evaporation to dryness, chemical reaction, dilution and concentration, replacement of one solvent by another, filtration, and so on, The present chapter concludes with a brief description of how sample pretreatment can be automated in LC analysis.

The subject of this chapter extends well beyond the usual practice of LC and, in fact, embraces much of the general field of chemistry and biochemistry. Therefore, despite the frequent importance of these ancillary considerations in developing an overall LC procedure, our present discussion and coverage are of necessity brief. Our main goal in this chapter is to orient the chromatographer to the various ways chemical knowledge plus specific information about the sample can be applied to a given LC problem. We should always remember that the chemistry of the sample can be as important in designing a good separation or analysis as the actual chromatography.

In addition to this chapter dealing with sample pretreatment, the reader should be aware of books that deal specifically with this subject (*1, 2*) plus innumerable articles that provide specific examples of cleanup of samples for LC. In addition, one can find literature describing sample preparation for gas chromatography that may require only a few changes to be applicable to liquid chromatography.

17.2 SAMPLE CLEANUP
Contributed by John W. Dolan,
Technicon Instruments Corporation

Consider first the problem of *insoluble samples* such as cross-linked synthetic polymers, dried paints or inks, mixtures containing water-insoluble inorganics, and plant or animal tissue. Before a sample can be separated by LC, it must first be converted into soluble form. There are almost no exceptions to this requirement. Many of the preceding samples cannot be readily solubilized, but in most cases (fortunately) we are not interested in the total sample. Rather, we are more often concerned with soluble compounds that are held within an insoluble sample matrix. In these cases, it is possible to carry out a simple extraction of the sample, so as to separate extractible compounds of interest from an insoluble residue.

Extraction can be carried out in various ways. Finely divided sample plus solvent can be measured into a suitable container (bottle or flask), swirled occasionally, and the supernatant eventually sampled for LC analysis. This procedure is effective when near complete extraction of compounds of interest is easily achieved. More difficult samples can be extracted by a vigorous shaking of the sample/solvent mixture or by ultrasonic agitation. Frequently the processes of sample subdivision plus vigorous extraction are combined, by adding sample and solvent to a high-speed blender. More exhaustive extraction can be achieved by multiple, sequential extraction of the sample, either with successive aliquots of the same solvent or with a series of different solvents. Finally, continuous extraction with hot solvent (e.g. Sohxlet extraction) can be reserved for relatively intractable cases.

Many detailed examples of the extraction of plant and animal tissue, including foodstuffs and clinical specimens, are given in (*1, 2*). The general considerations pertaining to selection of the extraction solvent and other experimental conditions can be inferred from common chemical experience and the discussion of Chapter 6; we attempt only a brief summary here. Maximum or complete extraction of compounds of interest is generally desired, and this can be determined by several factors: the polarity of the extracting solvent (i.e., its P' value, Table 6.1), the pH of the sample/solvent mixture, the relative volumes of solvent and sample, and the method of contacting sample and solvent (see earlier). Maximum extraction is generally favored with solvents whose P' values are similar to the P' values of the compounds being separated. Thus, for extraction of nonpolar or fatty compounds, solvents such as hexane ($P' = 0.1$), toluene ($P' = 2.4$), and ethyl ether ($P' = 2.8$) are commonly used. For the extraction of more-polar compounds, on the other hand, polar solvents are required: acetone ($P' = 5.1$), acetonitrile ($P' = 5.8$), methanol ($P' = 5.1$), or mixtures of these solvents with water ($P' = 10.2$). Chloroform ($P' = 4.1$) has an intermediate polarity; for this and other reasons, it is often used as an extraction solvent. Samples of plant or animal tissue generally contain water, and for these samples more polar solvents (e.g., ethyl acetate) generally give more complete extraction of all sample components, as long as the latter are soluble in the extraction solvent.

The pH of the sample/solvent system can be varied to enhance the extraction of acids or bases. The pH can be adjusted to give either un-ionized sample components (low pH for acids, high pH for bases) or ionized compounds. In the first case (un-ionized sample components), a relatively nonpolar extraction solvent is normally used (e.g., the chloroform extraction of basic alkaloids from plant tissue, following addition of base to the sample). For extraction of compounds in the ionic form, the extraction solvent is usually water or alcohol/water—plus added acid or base, as in the extraction of alkaloids from plant tissue as their salts, via sulfuric acid/water/ethanol mixtures.

Where extraction gives low recovery of desired sample components despite optimum adjustment of solvent polarity and pH, recovery can be increased by using larger volumes of solvent for the same sample quantity, and especially by using successive or continuous extraction. In some cases, extraction may be poor, because the extraction process is slow. In these cases, the following expedients should be considered: (a) greater subdivision of sample for better contacting with solvent, as in a high-speed blender; (b) vigorous mixing of the two phases, as in a blender or by means of ultrasonification; (c) use of a solvent that better wets the sample (e.g., water-miscible solvents for wet samples, nonpolar solvents for oily samples, etc.); (d) heating of sample and solvent, plus continuous extraction as in Sohxlet extraction.

Compounds that are very volatile or unstable at ambient or higher temperatures are adversely affected by the heat developed during normal high-speed blending. In these cases, the sample can first be subdivided in a blender with addition of dry ice or liquid nitrogen. Following evaporation of the coolant, the cold sample can then be extracted in a separatory funnel at ambient or lower temperature.

In many cases the extraction procedure results in an emulsion that is not readily separated. The most common method for breaking an emulsion is the addition of salt (e.g., NaCl) either as a solid or a saturated aqueous solution. Settling of the lower phase can be accelerated in a separatory funnel by the application of heat, as with an infrared lamp. More difficult emulsions may require high-speed centrifugation to separate the two phases. Another technique that is finding increasing favor is the use of low-melting-point organic solvents such as ether for the extraction of aqueous samples. Emulsions in this case are first separated by centrifugation (so far as possible), then the mixture is frozen in dry ice or liquid nitrogen. Finally, the organic phase is simply poured off, leaving the frozen aqueous phase behind. Separation of organic and aqueous phases can also be achieved using hydrophobic or phase-separating filter paper. These filter papers come with a hydrophobic coating, so that only the organic phase will pass through, leaving the aqueous phase in the funnel.

Following extraction of the sample, the solvent phase is normally filtered and may be further treated prior to injection: dried over anhydrous sodium sulfate, concentrated by evaporation, subjected to further clean-up procedures, and so on. In some cases, sample compounds may be chemically bound to an insoluble matrix; then a chemical or enzymatic treatment is required for solubilization and extraction of fragments of interest. The usual chemical degradation reactions—acid or base hydrolysis, thermal decomposition, oxidative or reductive splitting, and so on—are available for this purpose, but obviously must be used with care in order to preserve the identity of final reaction products. With biochemical or clinical specimens, compounds may be bound to carrier proteins (e.g., drugs or hormones in serum). Here a simple denaturation of the protein is generally effective in releasing the compound. Such procedures as addition of strong acid or base, phosphotungstic acid, organic solvents and so on—with or without heating—generally suffice.

Further sample pretreatment may be required in the case of *very complex samples,* even when these are homogeneous solutions in a suitable solvent. For example, initial extracts of plant or animal tissue usually consist of a large number of individual compounds—only a few of which are of interest to the analyst. Commonly, these few compounds of interest occur as low-level or trace components in the total sample—for example, environmental contaminants or pesticide residues in food, food additives (including vitamins), drugs or hormones in serum or urine, and so on. When such a sample in injected directly onto a column, the compound(s) of interest may be completely obscured by surrounding bands. An example is shown in Figure 17.1a, for the assay by cation-exchange LC of niacin in bread. Here injection of the total bread extract gives a broad, multicomponent band near t_0, which totally overlaps and obscures the niacin band (shown as dashed curve). Generally a preseparation of the sample, in addition to an initial extraction, can greatly reduce sample complexity and facilitate the separation and detection of compounds of interest in a final LC analysis. In the example of Figure 17.1a, it was found that batch adsorption of the bread extract onto magnesium silicate, followed by extraction of the adsorbent and injection of the extract, gave the separation shown in Figure 17.1b. Now the niacin band is cleanly resolved from other sample components, allowing its easy quantitation.

Preseparation of the sample as in Figure 17.1 can be carried out in various ways. The easiest technique is a simple liquid-liquid extraction between two solvents—as in a separatory funnel. For example, an initial hexane extract can be further cleaned up by shaking it with acetonitrile or water. The sample compound of interest will end up in one of the two phases, and (it is hoped)

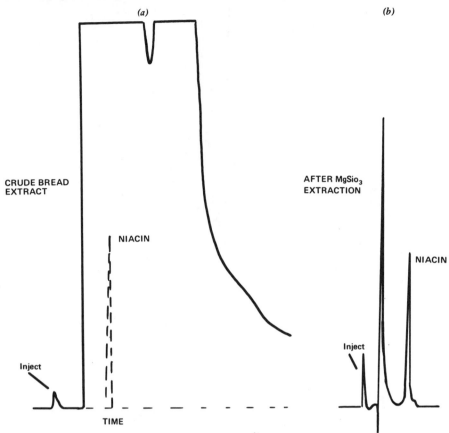

Figure 17.1 Determination of niacin in bread by cation exchange LC. (*a*) Crude bread extract, (*b*) magnesium silicate extract of (*a*). Column, 100×0.23 cm, Zipax-SCX (bonded-phase pellicular cation exchanger); mobile phase, $0.004 \ M \ NaH_2PO_4$ in water; temp., ambient; UV, 254 nm. Courtesy of V. A. De Stefannis, General Foods Corp.

most of the interfering sample components will end up in the other phase. An example of extraction to provide sample simplification is given in Figure 7.26.

Open-column or low-pressure LC can also be carried out on the sample, prior to a final high-performance LC (HPLC) analysis. For example, preseparation of a crude sample on columns of silica, Florisil, or charcoal can be used prior to reverse phase HPLC. Normal-phase or reverse-phase separations at low pressure by liquid-liquid LC can also be used for sample pretreatment by coating a silica column with a suitable stationary phase (e.g., acetonitrile/water, isooctane; etc.). Either low- or high-pressure separations by size-exclusion chromatography (SEC) are convenient and effective in simplifying many samples prior to their

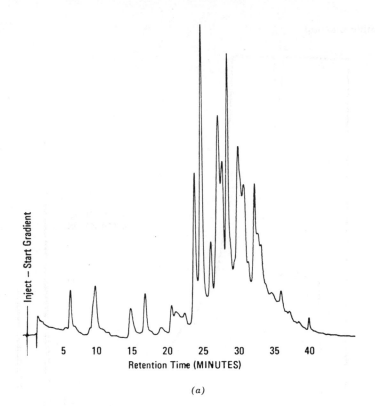

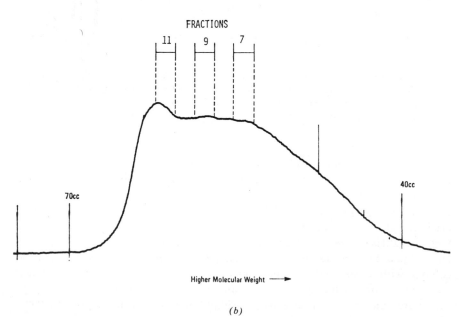

final LC analysis. Figure 17.2*a* shows the initial BPC separation of an auto-exhaust condensate, and it is apparent that there are too many compounds present in the sample for its resolution on this column. Figure 17.2*b* shows an SEC separation of the same sample, using a set of polystyrene gel columns. Fraction 7 from this SEC separation was isolated and reinjected onto the column of Figure 17.2*a*, with the final separation shown in Figure 17.2*c*. Now the chromatogram is much simpler, with less overlapping of adjacent bands compared to that of Figure 17.2*a*.

There are a large number of other separation methods that can be used for sample cleanup prior to LC [e.g., see discussion of (*4*)]. Low-pressure column separations are often used instead of batch methods, because column separations have better separation efficiency. Either adsorption or partition separations can be employed, with glass columns and large-particle packings being preferred (cf. Table 5.13). For routine application of low-pressure chromatography in the cleanup of large numbers of similar samples, several samples can be processed

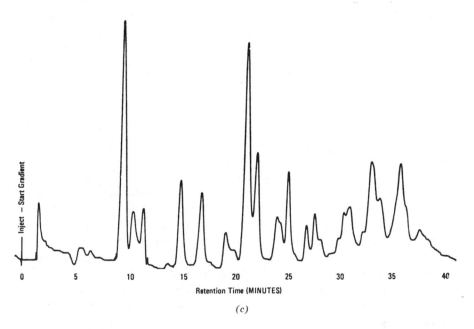

(c)

Figure 17.2 Simplification of a complex sample by size-exclusion chromatography prior to final separation by reverse-phase LC. (*a*) Separation of auto-exhaust condensate, (*b*) separation of same sample by SEC, (*c*) separation of fraction 7 from (*b*) by reverse-phase LC. (*a* and *c*), Column, 100 × 0.21 cm, ODS Permaphase (bonded-phase pellicular C_{18}); mobile phase, linear gradient 20-100%v methanol/water; temp., 60°C; 1.0 ml/min, P = 1000 psi; UV, 254 nm. (*b*) Column, 2-100 × 0.21 cm, SX-2 and SX-8 Biobeads (polystyrene SEC gel); mobile phase, CHCl$_3$; temp., ambient; 1.0 ml/min; UV, 254 nm. Reprinted from (*5*) with permission.

at the same time, as described in (6) for the isolation of aromatic hydrocarbons prior to their analysis by GC.

Disposable minicolumns ("Sep-Paks") containing silica or C_{18} bonded to silica are available for sample cleanup (Waters Associates). With the reverse-phase material, one loads the column with aqueous-phase sample using a syringe. All but the most polar material is strongly adsorbed onto the column with the polar mobile phase. The column is then flushed with a polar solvent to elute the unwanted polar material that is more polar than the sample of interest. Next, the sample of interest is released with an intermediate polarity solvent, leaving more strongly retained nonpolar material adsorbed on the column, which is then discarded. Although the clean-up principles for these small columns are no different from those for other open-column techniques, the convenience of prepackaged columns, use of a hypodermic syringe as a low-pressure pump, and the disposability of the columns can be cost effective in many cases.

Another option is the use of a disposable precolumn in the LC system [e.g., see (6a)]. Here untreated samples are injected directly onto the LC precolumn, which is followed by the analytical column. When the precolumn is degraded to the point where it is no longer useful, it is replaced, and analysis continues. The time and money saved in bypassing manual pretreatment may, in some cases, justify the expense of replacing the precolumn every 20-50 samples, especially if the precolumns are dry-packed in-house with larger particles (e.g., 40 μm), to reduce the expense.

Often more than one clean-up procedure is required prior to LC analysis of the final sample extract. Back extraction is a convenient example that is often employed. Thus, an initial pentane extract might be further simplified by back extraction with acetonitrile, with polar sample components ending up in the acetonitrile phase. Or back extraction of less polar compounds from an initial acetonitrile extract can be achieved by adding water plus chloroform to the acetonitrile solution. In this case, a water/acetonitrile phase results, with less-polar compounds preferring the chloroform phase. Similar extraction and back extraction of sample acids or bases can be achieved by intermediate pH adjustment. For example, barbiturates in urine can be extracted (as un-ionized acids) into chloroform, after acidification of the urine. Back extraction of the chloroform phase with aqueous sodium hydroxide then transfers the barbiturates into the water phase (as the ionized anions), leaving neutral nonpolars behind.

The choice of different methods for an overall clean-up procedure, and the sequence in which these methods are applied, is beyond the scope of the present chapter, but is reviewed in Chapter 19 of (4). However, several final points can be made:

- If properly chosen, each successive clean-up step in an overall pre-LC separation will result in further simplification of the sample and easier final LC separation (improved resolution and accuracy, greater column stability, etc.).

- Low-pressure column separations provide more efficient separation and better recovery of compounds of interest than corresponding batch separations; however, column separations are somewhat more tedious.

- When excessive sample interferences elute near t_0 in the final LC separation, these can usually be reduced by carrying out a pre-LC separation similar to the LC method; for example, a low-pressure silica separation preceding HPLC on a silica column.

- When sample interferences are excessive but elute close to the compound of interest (similar k' values, $k' \neq 0$), these can best be reduced by carrying out a pre-LC separation that differs from the LC separation (e.g., extraction followed by HPLC on silica, or low-pressure silica separation followed by reverse phase liquid-liquid HPLC).

So far, we have not discussed the possible need for concentrating the sample prior to injection or for replacing the solvent that contains the sample initially. These examples of sample pretreatment are interrelated, in that each aims at increasing the amount of sample that can be injected, thereby lowering detection limits. Ideally, the sample will be present as a concentrated solution in a solvent that is weak compared to the mobile phase that will be used for LC. When this is the case, relatively large amounts of sample can be injected without degrading the LC separation [see (6c) and (6d), and Section 7.4]. Alternatively, concentrated solutions of the sample in the final mobile phase are acceptable.

If the sample is in too large a volume of volatile solvent, such as dichloromethane or hexane, the excess solvent can be removed by evaporation. Usually the evaporation is preceded by drying over a suitable agent, such as anhydrous sodium sulfate, to remove traces of water. Large volumes of volatile solvent can be removed with a rotary evaporator using vacuum plus heat (if stability of the sample is no problem). When only a small volume of solvent remains, it is most effectively removed by evaporation under a gentle stream of dry nitrogen. One should be especially careful as the sample volume approaches dryness, because volatile samples can also evaporate. The concentrated sample may now be suitable for LC analysis, or one may wish to evaporate it to dryness and redissolve it in another solvent. For instance, a sample in hexane may be amenable for injection into a normal-phase LC system but should be transferred to an aqueous solvent (e.g., the mobile phase) for compatibility with reverse-phase LC.

Samples in aqueous systems can also be concentrated by evaporation, although this is usually more difficult than with volatile organic solvents. Solvent-solvent extraction may be possible if sample polarity permits, thus transferring the sample from an aqueous to a volatile organic phase. As mentioned in Section 13.5, large columns of aqueous samples can be charged into RP-LC systems

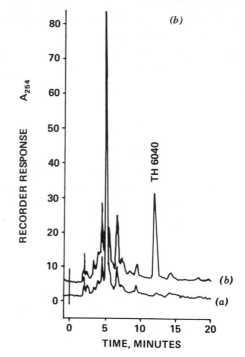

(a)

(i)   Manure plus acetonitrile combined
      in blender

(ii) Acetonitrile phase filtered and
      evaporated to dryness

(iii) Residue partitioned between
      acetonitrile and hexane,
      acetonitrile phase evaporated
      to dryness

(iv) Residue separated on Florisil,
      fraction containing regulator
      collected, evaporated to
      dryness, taken up in acetonitrile

(v)   Sample separated as in (B) by LC

Figure 17.3 Determination of insect growth-regulator residue in bovine manure by sample pretreatment plus LC. (*a*) Pretreatment scheme, (*b*) chromatograms of (*a*) manure extract and (*b*) fortified (500 ppb) extract. Column, 30 × 0.6 cm μ-Bondapak-C_{18} (10 μm); mobile phase, 57%v acetonitrile/water; ambient; UV, 254 nm. Reprinted from (*7*) with permission.

under suitable conditions, which provides sample concentration within the LC columns. Very dilute samples, such as pesticides and their metabolites in water, can also be concentrated before HPLC, using chromatographic techniques. In the case of pesticide metabolites, large volumes of water are often flushed through a polystyrene-divinyl benzene copolymer resin (XAD) on which the pesticides are held until released by an organic wash (*6b*). The resultant sample solution can be analyzed directly, further concentrated by evaporation, or evaporated to dryness and transferred to another solvent.

All sample solutions except those of pure standards should be filtered prior to LC analysis. Modern LC columns of 5- or 10-μm particles usually have a 2-μm frit or screen at the head of the column to contain the packing and provide some protection from particulates. Any particulate matter that is injected onto the column is usually stopped by the frit, but once the frit is plugged, the column is useless until the frit is cleaned or replaced. Small microfiltration units containing filters of 0.2-0.5 μm pore size are used in conjunction with hypodermic syringes to provide rapid filtration of low-volume samples (Millipore Corp.). The extra effort required for filtering each sample is certainly worthwhile when one considers the cost of the maintaining or replacing the chromatographic column.

A practical example of pretreatment plus LC is shown in Figure 17.3 for the analysis of bovine-manure samples. Here residues of the insect growth-regulator "Thompson-Hayward 6040" are to be determined at the ppb level. Figure 17.3a shows the pretreatment scheme, beginning with extraction of the manure in a blender, followed by liquid-liquid partition cleanup and Florisil (low-pressure) chromatography. The final pretreated sample is then analyzed by reverse-phase LC, as shown in Figure 17.3b.

In another application of sample pretreatment prior to LC (8), the use of a large starting sample (100 g of fish) followed by concentration of extract solutions allowed determination of polyaromatic hydrocarbons at the low parts-per-trillion level.

17.3 SAMPLE DERIVATIZATION

Derivative formation of sample components prior to their LC separation can be useful for several reasons: (a) to improve the initial extraction or preseparation of compounds of interest from the sample matrix, (b) to improve the subsequent LC separation by reducing sample polarity or changing the α values of selected compound pairs, (c) to change the relative detector response of different compounds in the sample, so that compounds of interest can be detected and quantitated in the presence of overlapping bands of no interest, and (d) simply to increase the detector response for certain compounds in the sample. In most cases, the latter two goals are of major interest in LC, and this area receives primary emphasis in the following discussion.

Derivatization for the purpose of improving the initial extraction or pre-separation of compounds of interest from a sample is usually confined to pH adjustment of the sample or the use of ion-pair extraction. The use of pH adjustment in extraction was discussed in the preceding section. Ion-pair extraction proceeds in much the same way as ion-pair chromatography (see Chapter 11). In each case the derivative formed involves a reversible reaction, so that the original compounds can be recovered easily. Other derivatizations to improve sample preseparation are less common but are employed occasionally. A related use of sample derivatization is to protect very labile compounds that might otherwise be destroyed during sample handling and LC separation. An example is the conversion of arylhydroxylamines to the corresponding ureas (reaction with methyl isocyanate) prior to sample workup and LC analysis (8a).

Derivatization for improved separation in LC is rarely needed, because of the diverse LC systems that are available for the resolution of almost any sample type. An important exception is the separation of optical isomers, where the usual LC approach is through the diastereomeric derivatives [e.g., separation of amino acid enantiomers in (8b)].

The general aspects of sample derivatization in LC for increased detectability have been covered in detail elsewhere [(9-10a) are especially recommended]. Therefore, such topics as the fundamental origin of molecular absorption and fluorescence, the role of solute structure in determining relative detection sensitivity, and the special requirements of the detector will not be treated further here. We will concentrate instead on a description of common reactions and derivatizing agents that have been used in modern LC, thereby providing a basis for selecting the best derivatization procedure for a particular application.

An obvious need for sample derivatization arises in the case of samples that do not absorb or fluoresce strongly in the UV or visible, because most LC detectors now in use are either photometers or fluorometers. The photometric detection of 1 μg of a sample component as a separated band typically requires a molar absorptivity ϵ for that component of at least 10; usually even lower levels of detection are desired. Therefore, most derivatizing agents used in LC have been selected to achieve ϵ values in the final derivatized compound of at least 1000. Some commonly used LC derivatizing agents are summarized in Table 17.1. Also shown here are the compound types that will react with the specific derivatizing agent, and the absorption characteristics (λ_{max}, ϵ) of the resulting derivative. In most cases, these derivatives are readily detectable at 254 nm. Some of the derivatives of Table 17.1 are intended primarily for fluorescence detection, and in these cases the wavelength maxima for excitation (primary, λ_1) and emission (secondary, λ_2) are given. Table 17.1 also indicates whether the reaction product following derivatization can be injected directly onto the column, or whether a further workup of the reaction mixture is required. Finally, Table 17.1 provides specific references to procedures for each of these reactions, including commercial suppliers of reagent kits for LC derivatization. Table 17.2 shows the general reactions involved for each of the derivatizing agents of Table 17.1.

Several points should be made with respect to the data of Table 17.1. *First,* all the reactions listed in Table 17.1 have been applied to high-performance LC separations, so their adaptation to other LC problems should be relatively straightforward. Because the yield of derivatized product does vary with the specific compounds involved, the recovery of derivatized sample must be checked in each case. Where the derivatization yield is unsatisfactory for a particular derivatizing agent, other reactions can be selected from Table 17.1.

Second, the list of derivatizing agents listed in Table 17.1 is by no means complete. Many reactions cited in (9) have been omitted, because these reactions duplicate other reactions listed in Table 17.1 and because the omitted

Table. 17.1 Derivatizing agents for enhancing photometric or fluorometric detection via pre-column reaction of the sample.

Derivatizing agent[a]	Reactive sample compound types	Reaction workup[b]	Derivative characteristics			References
			λ_1[c]	λ_2	$\epsilon \times 10^{-4}$[d]	
p-Bromophenacyl bromide (I)[e]	Carboxylic acids	Yes	260		1.8 (254)	(9-11)[f,g]
O-p-Nitrobenzyl-N,N'-diisopropylisourea (II)	Carboxylic acids	?	265		0.62 (254)	(10)[f,h]
1-(p-Nitro) benzyl-3-p-tolyltriazine (III)	Carboxylic acids	No	265		0.62 (254)	(9)
3,5-Dinitrobenzoyl chloride (IV)	Alcohols, amines, phenols	No			<1.0 (254)	(9,10)[f]
Pyruvoyl chloride (2,6-nitrophenyl) hydrazone (V)	Phenols, amines, alcohols, mercaptans	No				(12)
p-Iodobenzenesulfonyl-chloride (VI)	Phenols, alcohols	Yes				(12)
Benzoyl chloride	Alcohols	Yes	230		1.3(230)	(9,13)
p-Nitrobenzoyl chloride	Alcohols	Yes	254		1.0	(16)
p-Methoxybenzoyl chloride	Amines	Yes	254		1.5	(40c)
p-Nitrobenzyloxyamine·HCl (VIII)	Aldehydes, ketones	?	265		0.62(254)	(10,14)[f]
2,4-Dinitrophenylhydrazine	Aldehydes, ketones	Yes	430		2.0 (430)	(9,15)
N-Succinimidyl-p-nitrophenyl acetate (IX)	Amines[i]	?	265		0.62 (254)	(10)

Table 17.1 (Continued)

Derivatizing agent[a]	Reactive sample compound types	Reaction workup[b]	Derivative characteristics			References
			λ_1[c]	λ_2	$\epsilon \times 10^{-4}$ [d]	
2,4-Dinitro-1-fluorobenzene (X)[k]	Amines[i]	Yes	360			(9)
Dansyl chloride (5-N,N-dimethylaminonaphthalene-1-sulfonyl chloride) (XI)	Amines[i], amino acids, peptides, phenols	Yes	360	510	i	(9)
Pyridoxal (XII)	Amino acids	Yes	330	400	i	(9)
p-Nitrobenzyl methyl amine	Isocyanates	Yes	265		0.62 (254)	(18)
Phenanthrene boronic acid	1,2-, 1,3- or 1,4-diols	Yes	313	385	i	(18a)

[a] Roman numerals in parentheses refer to structures of Table 17.2.
[b] "Yes" means that the reacted sample cannot be injected onto the column without further workup.
[c] For photometric detection, refers to wavelength of maximum absorbance.
[d] Number in parentheses refers to wavelength λ_2 for ϵ value cited.
[e] Other phenacyl and naphthacyl derivatives have also been used.
[f] Supplied by Regis Chemical Co.
[g] Supplied by Applied Science Labs.
[h] Supplied by Pierce Chemical Co.
[i] Primary or secondary only
[j] Fluorescence detection.
[k] Note that this compound (X) is believed to be both mutagenic and carcinogenic.

Table 17.2 Structures of derivatizing agents and reaction products for compounds of Table 17.1. Roman numerals identify derivatizing agent of Table 17.1.

(I) $Br-\phi-\overset{O}{\overset{\|}{C}}-CH_2-Br + RCOOH \rightarrow Br-\phi-\overset{O}{\overset{\|}{C}}-CH_2-O-\overset{O}{\overset{\|}{C}}-R$

(II) $O_2N-\phi-CH_2-\overset{O}{\overset{\|}{C}}-O-N(\text{succinimide}) + RCOOH \rightarrow O_2N-\phi-CH_2-O-\overset{O}{\overset{\|}{C}}-R$

(III) $CH_3-\phi-N=N-NHCH_2-\phi-NO_2 + RCOOH \rightarrow RCOO-CH_2-\phi-NO_2$

(IV) $(O_2N)_2\phi-\overset{O}{\overset{\|}{C}}-Cl + ROH \rightarrow (O_2N)_2\phi-\overset{O}{\overset{\|}{C}}-O-R$

(V) $CH_3-\overset{\underset{COCl}{|}}{C}=N-N-\phi-(NO_2)_2 + ROH \rightarrow CH_3-\overset{\underset{COOR}{|}}{C}=N-N-\phi-(NO_2)_2$

(VI) $I-\phi-SO_2Cl + ROH \rightarrow I-\phi-SO_3R$

(VII) $\phi-COCl + ROH \rightarrow \phi-CO_2R$

(VIII) $O_2N-\phi-CH_2-O-NH_2 \cdot HCl + R_2C=O \rightarrow O_2N-\phi-CH_2-O-N=C\overset{R}{\underset{R}{<}}$

735

Table 17.2 (Continued).

(IX) $CH_2-O-N=CH(CH_3)_2$ with $C(CH_3)_2$

$+ R_2NH \rightarrow O_2N-\phi-CH_2-C(=O)-N(R)(R')$

(X) $(NO_2)_2-\phi-F + R_2NH \rightarrow (NO_2)_2-\phi-NR_2$

(XI) $SO_2Cl \cdots N(CH_3)_2 + R_2NH \rightarrow SO_2-NR_2 \cdots N(CH_3)_2$

(XII)   pyridine ring with CHO, CH_2OH, OH, CH_3

$+ R-CH_2-COOH$ with NH_2 $\xrightarrow{NaBH_4}$ $R-CH_2-COOH$, CH_2, N, pyridine ring with OH, CH_2OH, CH_3

reactions lack broad application at the present time. Other reactions not listed in Table 17.1 include those specific to a particular compound (e.g., oxidation of morphine to its fluorescent dimer), because these are not of general applicability. Still other reactions, such as the oxidation of phenols to fluorescent or colored products, are not sufficiently well characterized or specific to be predictable.

Finally, a large number of other derivatizing reactions can be considered in special cases. Examples are provided from such related areas as thin-layer-chromatography color development (*19*), colorimetric or fluorometric analysis (*20*), and classical organic qualitative analysis (Section 14.1).

In considering whether to apply sample derivatization before or after the column, several factors should be considered. In general, precolumn derivatization is easier to carry out; thus, reaction and cleanup (if required) of the sample can be done manually or off-line, which simplifies the overall LC system required. Furthermore, there are few restrictions on the way the reaction is carried out; any solvent or combination of solvents can be used, reaction times can be long, an excess of reagents can be used (and later removed from the sample before injection), and so on. As we will see in Section 17.4, this is not the case for postcolumn derivatization. On the other hand, there are also some disadvantages of precolumn sample reaction, as opposed to postcolumn derivatization. For example, some samples yield more than one reaction product for a given compound and, therefore, two or more bands for each derivatized sample component, which can complicate the resulting chromatogram. For quantitative analysis the reaction must either go to completion or proceed to a precisely determined percent of completion for each sample analyzed. Finally, the manual application of precolumn derivatization can be tedious and time-consuming, as opposed to the automatic mode of operation in postcolumn reaction detectors (but see Section 17.5 on automation).

Derivatization can be applied to a much wider range of compound types than are listed in Table 17.1, simply by chemically degrading samples of interest to form the reactive compound types listed. For example, esters can first by hydrolyzed, then either the acid or alcohol fragment can be derivatized for detection. In this way, Methyl-carbamate insecticides have been hydrolyzed to yield the characteristic phenol fragment, which in turn is derivatized with dansyl chloride, the sample separated by LC, and the dansyl phenol detected fluorometrically. Figure 17.4 shows the analysis of water samples for Carbofuran and Mesurol in this fashion, and it is apparent that detection limits are of the order of 1 ppb of insecticide. It should be kept in mind for all analyses such as that of Figure 17.4 that the resulting assay may be less specific than in the direct separation and detection of a compound of interest (or its derivative). Thus, in Figure 17.4, the phenol band can arise from other sources than a carbamate. Fortunately, in this specific case, confirmation of identity is provided by the

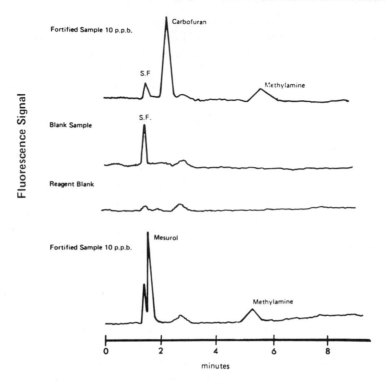

Figure 17.4 Analysis of water samples for carbamate insecticides (carbofuran and mesurol), following preparation of dansyl-phenol derivatives from hydrolysis products of sample. Column, 100 × 0.24 cm, Zipax plus 0.5%w/w BOP; mobile phase, 5%v ethanol/hexane; temp., ambient; 0.83 cm/sec; fluorometric, 360 (1°), 510 (2°); 1-10-μl samples. Reprinted from (*21*) with permission.

dansyl-methylamine peak that is also formed during the hydrolysis and dansylation of these carbamates.

Figure 17.5 shows the analysis of a crude plant extract used as an intermediate in the synthesis of various corticosteroidal drugs. The various sapogenins of interest in this sample are assayed by LC as the benzoyl derivatives. Of special interest in this application was the precision obtained: ±1% S.D., which includes errors in both the pretreatment and LC steps (*22*).

Some workers have reported that sample derivatization can provide improved separation of the sample, although this is much more common in GC than in LC. Generally, the derivatives formed in the reactions of Table 17.1 are less polar than the parent compounds, and this can have two advantages in LC. *First,* more polar compounds are often found to give tailing bands and lower column-plate numbers with many LC systems. Derivatization sometimes eliminates this problem, without the necessity of trying many different LC systems before successful separations result. *Second,* mixtures of polar compounds often exhibit a wide range of k' values, in turn necessitating the use of gradient elution. However, derivatization of such mixtures often reduces the spread in k' values and allows isocratic separation. Frei and coworkers (*22*) found this to be the case for separations of digitalis glycosides and aglycones on silica.

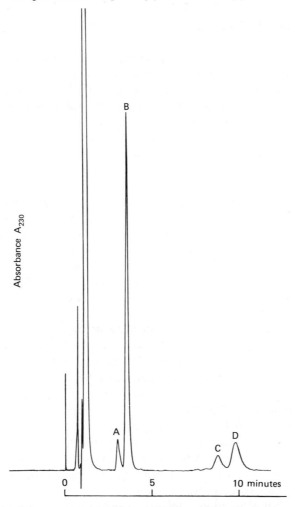

Figure 17.5 Analysis of crude hecogenin sample, following preparation of benzoyl derivatives. (*A*) 9(11)-dehydrohecogenin; (*B*) hecogenin; (*C*) unknown; (*D*) tigogenin. Column, 25 × 0.4 cm, LiChrosorb RP-8 (10 μm, C$_8$ bonded-phase); mobile phase, 80%v acetonitrile/water; ambient; 3.9 ml/min, P = 1800 psi; UV, 235 nm. Reprinted from (*13*) with permission.

Derivatization can also be used to create larger α values for selected pairs of compounds. However, this should be a last resort, after the various chromatographic parameters have been varied for maximum resolution.

For a discussion of the possible use of derivatization for enhancing compound sensitivity with electrochemical detectors, see (*18b*).

17.4 REACTION DETECTORS

Reaction detectors combine postcolumn reaction or derivatization of separated sample bands with a suitable detector, usually a photometer or fluorimeter. The object of postcolumn derivatization is to increase the detection sensitivity of sample bands or to enhance their sensitivity with respect to interfering bands that overlap bands of interest. A classical example of reaction detectors is the ninhydrin reaction used in conjunction with amino-acid analyzers: colorless amino acids are reacted with ninhydrin as they leave the column to form colored compounds absorbing at 570 nm. The latter can then be measured by means of a suitable photometric detector.

The adaptation of reaction detection to modern LC columns requires careful attention to the selection of the derivatizing reaction and the design of equipment, because extracolumn effects can be serious. For these reasons reaction detectors have so far found rather limited use in modern LC. Nevertheless, this approach can be quite powerful, and it should not be ruled out, particularly when the analyst is faced with the combination of a major problem in detection sensitivity and/or specificity for a given application, plus the need to analyze a large number of samples.

Some examples of reactions used in postcolumn derivatization are summarized in Table 17.3. All of these examples are compatible with the requirements of modern LC, at least in some applications. The reaction times for these examples are less than 20 min, which represents an upper limit beyond which unacceptable extracolumn band broadening usually results. Almost all these reactions require a major proportion of water in the final reaction mixture, and this limits most reaction detector systems to aqueous mobile phases or solvents miscible with water. For this reason most applications of reaction detectors have been in conjunction with ion exchange. Although reverse-phase LC is another possibility, it has not yet been widely exploited. Some of the reactions of Table 17.3 are also applicable for pre-column derivatization.

The general design of reaction detectors, including requirements for allowable extracolumn band broadening, has been discussed in detail elsewhere (*24-28*). The reaction-detector system must provide for the continuous addition of controlled volumes of one or more reagents to the column effluent, followed by mixing of the effluent/reagent mixture and incubation at some temperature T

Table 17.3 Some reactions that have been used in high-performance LC reaction detectors.

Reagent	Compounds detected	Comments	Refs.
Fluorescamine (Fluram)	Amines, amino acids, peptides	Fast, very sensitive, fluorescent detection	(9,24,32)
o-Phthaldehyde	Amines, amino acids	Fast, very sensitive, fluorescent detection	(9,32,33)
Ninhydrin	Amines, amino acids	1-min at 140°C; 440 and 570 nm	(33a)
o-Nitrophenol, sodium salt	Carboxylic acids, other acids	Fast, absorbs 432 nm	(9)
2,4-Dinitrophenylhydrazine	Aldehydes, ketones	3-min reaction time, absorbs at 430 nm	(9,27)
Ce^{+4}	Phenols, carbohydrates, carboxylic acids, other oxidizable organics	Reaction time and temperature vary from acceptable to marginal for different compounds; fluorescent detection	(34,35)
Enzyme plus substrate(s)	Enzyme inhibitors (e.g., organo phosphate and carbamate insecticides)	Can be very sensitive and specific	(9)
Substrates for a given enzyme	Any enzyme (e.g., LDH, CPK, alkaline phosphatase)	Can be very specific and sensitive (e.g., Figs. 9.13, 9.17 b,c)	(36)
Griess reagent	Nitrite, nitrosamides, nitroso-carbamates, alkyl nitrites	3-min reaction time, absorbs 550 nm	(37)
Photochemical reaction of effluent	Cannabinoids	Use of appropriate wavelength radiation to selectively convert sample compounds to fluorescing derivatives	(40)

Table 17.3 (Continued).

Reagent	Compounds detected	Comments	Refs.
Ferricyanide	Reducing sugars, other oxidizable compounds	Ferrocyanide product detected electrochemically	(29)
Neocuproin	Reducing sugars	3-15 min, 97°C	(29a)
5,5′-Dithio(2-nitrobenzoic acid)	Thiols, enzymes bearing —SH groups	Fast reaction, product absorbs at 412 nm	(26)
Ethylene diamine/hexacyanoferrate	Catechol amines	5-min reaction at 75° with fluorometric detection at 400 (I) and 510 (II) nm	(40a)
Kober reaction (H_2SO_4/hydro-quinone)	Estrogens	10-15-min reaction time at 120°C; fluorometric detection 535 (I) and (561) (II) nm	(40b)
9,10-Phenanthrene-quinone	Guanidino-compounds	2-min reaction time at 60°C; fluorometric detection 365 (I) and 460 (II) nm	(40d)
N-methylnicotinamide chloride	—CH_2—CO—groups	2-min reaction at 100° fluorometric detection at 380 (I) and 450 (II)	
isonicotinyl hydrazine	Δ^4—3—ketosteroids	2-min reaction at 70°; fluorometric detection at 360 (I) and 450 (II)	(40e)

for some time t. At the completion of reaction, the reaction mixture normally passes through a detector such as a photometer or fluorometer. To provide for incubation (for the interval t), between the column and the detector, three expedients have been employed:

- Use of a short length of narrow-diameter tubing; this allows incubation times of as much as 10-30 sec, and is hence limited to rather fast reactions (*22-24*).

- Use of a short column packed with glass beads, preferably comparable in terms of particle size and plate height to the preceding separation column; this allows incubation times up to about 3 min, and thus permits the use of somewhat slower reactions (*25, 26*).

- Use of a long length of narrow-diameter tubing, with air segmentation of the reaction stream; this provides incubation times of as much as 20 min, and hence allows still slower reactions (*27-29, 40b*); very long reaction times ($\geqslant$20 min) are possible with the technique of effluent storage (*30*).

Because of the specialized nature of reaction detectors, the present discussion will not go into the details of how these detectors are designed for optimum performance. The reader is referred to general references (*24-31*) for details, as well as to references (*9, 19, 20, 31*) for additional ideas on possible reactions suitable for such detectors. However, a few general comments are in order. With respect to the choice of the derivatizing reaction, this must obviously be compatible with the mobile phase used; reagents and mobile phase must be miscible, and the reaction must be fast under the given experimental conditions chosen. Extracolumn band broadening increases for more viscous reaction mixtures and for longer reaction times; therefore, elevated temperatures are often preferred for carrying out the sample reaction.

With fast reactions carried out in unsegmented flow through short lengths of tubing, the major practical problem is to ensure adequate mixing of the reaction mixture prior to its arrival at the detector (*24*). Usually this is achieved via specially designed tee connectors followed by tightly wound coils of tubing ("mixing coils"). With reactions carried out in packed beds, reduced extra-column band broadening is favored (up to a point) by shorter, wider beds that are packed with the smallest possible glass beads. However, the pressure drop across these beds is usually considerable, requiring high-pressure pumps for dispensing of reagents. The same care must be given to packing these beds as to preparing similar LC columns. Finally, for segmented flow of the reaction mixture through long lengths of narrow-bore tubing, the major consideration is reduction of extracolumn band broadening. This can be achieved by using small-bore (0.5 mm) tubing in the reaction coil, coupled with optimized flow rates and air-segmentation rates (*28*). For a good analysis of extracolumn band broadening in these three types of reaction detectors and comparison of the utility of each approach, see the discussion of (*38*).

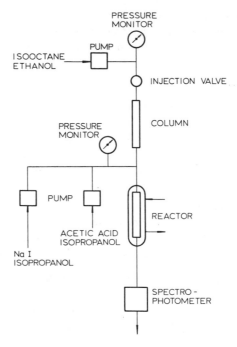

Figure 17.6 Reaction detector based on packed-bed reactor. Reprinted from (*25*) with permission.

Precision in assays based on reaction detectors depends on the usual chromatographic variables, as summarized in Chapter 13. It also depends on the ability to achieve constant proportioning of reagents and column effluent, and good mixing of the reaction solution before it reaches the reactor and detector. Under favorable circumstances, with attention to these variables, precisions of better than ±1% are achievable [e.g., see the study of (*32*)].

An example of an LC system plus reactor detector is shown in Figure 17.6. Two reagent streams, NaI/isopropanol and acetic acid/isopropanol, are metered into the column effluent via high-pressure pumps of the type used for LC, and the reaction mixture enters a reactor that consists of a column packed with glass beads. There a reaction between the iodide reagent and various hydroperoxide samples takes place to yield free iodine as chromophoric reaction product. The iodine formed, which is proportional to the concentration of each component hydroperoxide, is then measured in the spectrophotometer detector at 362 nm. The resulting chromatogram for a mixture of hydroperoxides is shown in Figure 17.7.

Other examples of reaction detectors based on packed, postcolumn beds are provided in Figures 10.13*b* and 10.17*b*, *c*. In each of these latter three separa-

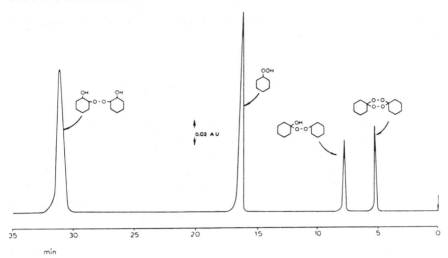

Figure 17.7 Separation of test mixture of peroxides, using reaction detector of Figure 17.6. Column, 30 × 0.4 cm, Merckosorb SI-60 (5 μm); mobile phase, 5%v ethanol/isooctane (50% water saturated); temp., ambient; 1.1 ml/min; UV, 362 nm. Reprinted from (25) with permission.

tions, enzyme bands are specifically detected as they leave the column, using reaction detectors based on the enzyme-catalyzed reaction. This approach allows both extreme selectivity in detection and potentially high detection sensitivity. Two examples of reaction detectors based on segmented-flow incubation coils are shown in Figures 10.14 and 10.16. Here the ninhydrin reaction is used to detect either amino acids or polyamines, respectively.

Still another approach to reaction detection should be noted: photochemical reaction as illustrated by Twitchett (39). In this case, cannabinoids were detected fluorometrically by passing the column effluent through a reaction coil that was illuminated by an intense beam of light of a given wavelength. The wavelength chosen permits specific photochemical reactions to be initiated, so that the resulting fluorescent reaction products allow both specificity and sensitivity in final detection.

A recent modification of the segmented-flow reaction detector has been described by Frei et al. (40). It has been applied to the detection of ionic (or ionizable) solutes in the effluent from reverse-phase BPC separations, although it could be used in other LC separations that employ aqueous or water-organic mobile phases. The aqueous column-effluent is segmented, and a fluorescent counter-ion is added. The next step is addition of an immiscible organic solvent to the effluent stream (plus counter-ion), with extraction of the ion pair. The organic phase containing the fluorescent counter-ion (as the ion pair) is then separated and run into a fluorometric detector. This approach seems quite general and widely applicable.

17.5 AUTOMATION OF SAMPLE PRETREATMENT

In many laboratories that use gas chromatography (GC) for the analysis of foodstuffs for pesticide residues or clinical samples for various biomedical constituents, sample preparation commonly requires more time and effort than the ensuing GC separation. The reason is simple: sample pretreatment is done manually, whereas GC separation can be automated to a high degree. Furthermore, data reduction is often achieved via an on-line computer system. Thus, the automation of sample pretreatment for these and other GC applications would be worthwhile.

A similar situation exists for LC also, although so far it has not been as apparent. The reasons are that LC generally requires less pretreatment of the sample than does GC, and high-volume analyses have been slower to develop in LC than in GC, because GC is 15 years older. Nevertheless, more and more laboratories are finding the need to automate various steps involved in an overall LC analysis.

The full automation of sample pretreatment in LC or GC has many advantages, apart from merely reducing the time and effort required for sample preparation. Thus, the precision of sample assays can be dramatically improved in many cases, inasmuch as manual pretreatment commonly introduces a large source of imprecision into an overall LC assay. Furthermore, where full automation is achieved in an on-line mode, the opportunities for sample mix-up (with results for one sample being reported for a different sample) are much reduced. Finally, in a fully automated procedure there is less reason to restrict the sample pretreatment procedure to one or two simple steps. Rather, several more or less complex procedures can be readily combined to yield a more reliable and/or more accurate overall assay.

The simplest form of automation in LC is automatic sampling. Automatic samplers permit several samples to be loaded into the system, following which each sample is automatically injected and run on the LC system, one by one. These devices have already been discussed in Section 3.8.

Various semiautomatic devices can be purchased for simplifying the repetitive steps of extraction, dilution, reaction, and so on, involved in typical pretreatment procedures for LC. These include automated or semiautomated diluters and dispensers (e.g., Micromedics), extractors (e.g., Du Pont), and so on. Although such equipment can certainly aid in processing large numbers of samples prior to LC analysis, they are for the most part batch-analyzers that must be used in an off-line mode. Thus, much of the benefit of complete on-line automation is lost.

For the complete, on-line automation of complex sample pretreatment procedures, the technique of continuous-flow analysis appears to offer many advantages. This proven procedure [see general reviews of (*31, 41*) and discussion of the continuous-flow-HPLC applications of (*42, 43*)] has now been in use in typical analytical laboratories for almost 20 years. It allows the automation of virtually every manual operation, as partially summarized in Table 17.4. For

Table 17.4 Operations that can be carried out in on-line and fully automated mode by continuous-flow analysis (Auto Analyzer) components.

Operation	Module
Dissolution and extraction of solid samples	Technicon Solid-Prep Sampler (*31*)
Evaporation to dryness and re-dissolution into second solvent	Technicon Evaporation-to-Dryness (EDM) module (*43*)
Dilution, buffering, addition of reagents	Standard Technicon components (*31*)
Solvent extraction, back extraction	Standard Technicon components (*31, 47*)
Incubation for reaction	Standard Technicon components (*31*)
Filtration	Agarose-coated tube (*48*)

example, continuous-flow techniques have recently been adapted to the fully automated analysis of fat-soluble vitamins in tablets (*44, 45*). This particular application features various pretreatment steps carried out on-line: high-speed dissolution and extraction of the vitamin tablet, evaporation to dryness of the extract, uptake of the extract into a second solvent, and chemical reaction (hydrolysis) of vitamin-A esters prior to final injection of sample onto an LC column via an automated sampling valve. Complete systems that combine auto-mated sample-pretreatment plus HPLC are commercially available (FAST-LC^{TM}; Technicon).

An in-depth discussion of the possibilities of automated sample pretreatment is beyond the scope of this book. For more details on the use of continuous-flow analysis for HPLC applications, see (*42, 43*). For a related example in-volving the analysis of pesticides in milk by automated pretreatment plus LC, see (*46*).

REFERENCES

1. G. Zweig and J. Sherma, eds., *Handbook of Chromatography,* Vol. II, CRC Press, Cleveland, Ohio, 1972, pp. 191-254.

2. *Pesticide Analytical Manual, Vol. I, Methods Which Detect Multiple Residues,* 2nd ed. 1968, revised July, 1975. U.S. Dept. of Health, Education & Welfare, Food & Drug Administration, Washington, D.C., Chap. 2.

3. L. R. Snyder, in *Techniques of Chemistry,* Vol. III, Part I, 2nd ed., A. Weissberger and E. S. Perry, eds., Wiley-Interscience, New York, 1978, p. 25.

4. B. L. Karger, L. R. Snyder, and C. Horvath, *An Introduction to Separation Science,* Wiley-Interscience, New York, 1973.

5. J. A. Schmit, R. A. Henry, R. C. Williams, and J. F. Dieckman, *J. Chromatogr. Sci.,* 9, 645 (1971).

6. E. R. Fett, D. J. Christoffersen, and L. R. Snyder, *J. Gas Chromatogr.,* 6, 572 (1968).

6a. H. V. Vliet, U.A. Th. Brinkman, and R. W. Frei, *J. Chromatogr.,* in press.

6b. R. G. Webb, "Isolating Organic Water Pollutants: XAD Resins, Urethane Foams, Solvent Extraction," U.S. Environmental Protection Agency, Corvallis, Oreg., 1975.

6c. P. Schauwecker, R. W. Frei, and F. Erni, *J. Chromatogr.,* 136, 63 (1977).

6d. J. F. K. Huber and R. R. Becker, *ibid.,* 142, 765 (1977).

7. D. D. Oehler and G. M. Holman, *J. Agric. Food Chem.,* 23, 590 (1975).

8. H. Guerro, E. R. Biehl, and C. T. Kenner, *J. Assoc. Off. Anal. Chem.,* 59, 989 (1976).

8a. L. A. Sternson, W. J. DeWitte, and J. G. Stevens, *J. Chromatogr.,* 153, 481 (1978).

8b. J. Goto, M. Hasegawa, S. Nakamura, K. Shimada, and T. Nambara, *ibid.,* 152, 413 (1978).

9. J. F. Lawrence and R. W. Frei, *Chemical Derivatization in Liquid Chromatography,* Elsevier, Amsterdam, 1976.

10. T. H. Jupille, Am. Lab., 85 (May, 1976).

10a. March issue, 1979 *J. Chromatogr. Sci.* (Vol. 17).

11. H. D. Durst, M. Milano, E. J. Kikta, Jr., S. A. Connelly, and E. Grushka, *Anal. Chem.,* 47, 1797 (1975).

12. R. W. Roos, *J. Chromatogr. Sci.,* 14, 505 (1976).

13. J. W. Higgins, *J. Chromatogr.,* 121, 329 (1976).

14. F. A. Fitzpatrick, M. A Wynalda, and D. G. Kaiser, *Anal. Chem.,* 49, 1032 (1977).

15. S. Selim, *J. Chromatogr.,* 136, 271 (1977).

16. F. Nachtmann and K. W. Budna, *ibid.,* p. 279.

17. M. J. Levitt, *Chem. Eng. News,* July 18, 1977, p. 30.

18. K. L. Dunlap, R. L. Sandridge, and J. Keller, *Anal. Chem.,* 48, 297 (1976).

18a. C. F. Poole, S. Singhawangcha, A. Zlatkis, and E. D. Morgan, *J. High Res. Chrom. Chrom. Commun.,* 1, 83 (1978).

18b. P. T. Kissinger, K. Bratin, G. C. Davis and L. A. Pachla, *J. Chromatogr. Sci.,* 17, 137 (1979).

19. E. Stahl, *Thin-layer Chromatography: A Laboratory Handbook,* 2nd ed., Academic, New York, 1969.

20. M. Pesex and J. Bartos, *Colorimetric and Fluorometric Analysis of Organic Compounds and Drugs,* Dekker, New York, 1974.

21. R. W. Frei, J. F. Lawrence, J. Hope, and R. M. Cassidy, *J. Chromatogr. Sci.,* **12**, 40 (1974).

22. J. F. K. Huber, K. M. Jonker, and H. Poppe, *Anal. Chem.,* **50**, (1978).

23. R. R. Schroeder, P. J. Kudirka, and E. C. Toren, Jr., *J. Chromatogr.,* **134**, 83 (1977).

24. R. W. Frei, L. Michel, and W. Santi, *ibid.,* **126**, 665 (1976); **142**, 261 (1977).

25. R. S. Deelder, M. G. F. Kroll, and J. H. M. van den Berg, *ibid.,* **125**, 307 (1976).

26. T. D. Schlabach, S. H. Chang, K. M. Gooding, and F. E. Regnier, *ibid.,* **134**, 91 (1977).

27. R. S. Deelder and P. J. H. Hendricks, *J. Chromatogr.,* **83**, 343 (1973).

28. L. R. Snyder, *ibid.,* **125**, 287 (1976).

29. J. C. Gfeller, G. Frey and R. W. Frei, *ibid.,* **142**, 271 (1977).

29a. M. H. Simatupang and H. H. Dietrichs, *Chromatographia,* **11**, 89 (1978).

30. L. R. Snyder, *J. Chromatogr.,* **149**, 653 (1977).

31. W. B. Furman, *Continuous Flow Analysis: Theory and Practice,* Dekker, New York, 1976.

32. D. L. Mays, R. J. Van Apeldoorn, and R. G. Lauback, *J. Chromatogr.,* **120**, 93 (1976).

33. G. Schwedt, *Anal. Clin. Acta,* **92**, 337 (1977).

33a. K. M. Jonker, H. Poppe, and J. F. K. Huber, *Chromatographia,* **11**, 123 (1978).

34. A. W. Wolkoff and R. H. Larose, *J. Chromatogr.,* **99**, 731 (1974).

35. S. Katz, W. W. Pitt, Jr., J. E. Mrochek, and S. Dinsomre, *ibid.,* **101**, 193 (1974).

36. F. E. Regnier, K. M. Gooding, and S. H. Chang, in *Contemporary Topics in Analytical and Clinical Chemistry*, D. Hercules et al., eds., Vol. 1, Plenum, New York, 1977, p. 1.

37. G. M. Singer, S. S. Singer, and D. G. Schmidt, *J. Chromatogr.,* **133**, 59 (1977).

38. R. S. Deelder, M. G. F. Kroll, A. J. B. Beeren, and J. H. M. van den Berg, *ibid.,* **149**, 669 (1978).

39. P. J. Twitchett, *ibid.,* **149**, 683 (1978).

40. R. W. Frei, I. Honigberg, L. Feenstra, J. F. Lawrence, and U. A. Th. Brinkman, *J. Chromatogr.,* in press.

40a. T. Seki, *ibid.,* **155**, 415 (1978).

40b. S. van der Wal, *High Pressure Liquid Chromatography of Steroid Conjugates and Other Compounds of Biomedical Interest,* Thesis, University of Amsterdam, 1977.

40c. C. R. Clark and M. M. Wells, *J. Chromatogr. Sci.,* **16**, 332 (1978).

40d. Y. Yamamoto, T. Manji, A. Saito, and K. Maeda, *J. Chromatogr.,* in press.

40e. N. Nakamura and Z. Tamura, *ibid.,* **168**, 481 (1979).

40f. R. Horikawa, T. Tanimura and Z. Tamura, *ibid.,* **168**, 526 (1979).

41. L. R. Snyder, J. Levine, R. Stoy, and A. Conetta, *Anal. Chem.,* **48**, 942A (1976).

42. D. A. Burns, "Automating HPLC Preparatory Techniques," *Res. Dev.,* 21 (April 1977).

43. J. C. MacDonald, *Am. Lab.,* 69 (August, 1977).

44. D. A. Burns, *Advances in Automated Analysis: 1976 Technicon International Congress,* Vol. 2, Mediad, Tarrytown, N.Y., 1976, p. 332.

45. J. W. Dolan, J. R. Gant, R. W. Giese, and B. L. Karger, *ibid.*, p. 340.
46. R. J. Dolphin, L. P. J. Hoogeveen, and F. W. Willmott, Philips *Tech. Rev.*, **36**, 284 (1976) (N. V. Philips' Gloeilampenfabrieken, Eindhoven, The Netherlands).
47. H. Alder, D. A. Burns, and L. R. Snyder, *Advances in Automated Analysis. 1972 Technicon International Congress,* Vol. 6, Mediad, Tarrytown, N.Y., 1973, p. 23.
48. J. W. Dolan and L. R. Snyder, *J. Chromatogr.*, in press.

BIBLIOGRAPHY

Sample Cleanup

Dell, D., "Sample Preparation in HPLC," *Methodol. Dev. Biochem.* (London), **5**, 131 (1976) (a brief review of procedures for biological samples).

Ko, H. and E. N. Petzold, in *GLC and HPLC Determination of Therapeutic Drugs,* Part 1, K. Tsuji and W. Morozowich, eds., Dekker, New York, 1978, Chap. 8 (good discussion of the isolation of compounds of interest from biological samples prior to LC).

Morris, C. J. O. R., and P. Morris, *Separation Methods in Biochemistry,* 2nd ed., Wiley-Interscience, New York, 1976 (detailed procedures for various separation steps).

References 1-4 of Chapter 17.

Derivatization

Blau, K., and G. King, eds., *Handbook of Derivatives for Chromatography,* Heyden, New York, 1977 (compares alternative derivatization schemes, includes discussion of optical resolution and ion-pair separations, plus fluorescent and nitrophenyl derivatives for LC).

Frei, R. W., "Derivatization as an Aid to HPLC," in *Res./Dev.,* 42 (February, 1977) (summary of LC derivatization plus some very current examples in pharmaceutical area).

Lawrence, J. F., and R. W. Frei, *Chemical Derivatization in Liquid Chromatography,* Elsevier, Amsterdam, 1976 (the definitive book on the subject).

Morozowich, W., in *GLC and HPLC Determination of Therapeutic Drugs,* Part 1, K. Tsuji and W. Morozowich, eds., Dekker, New York, 1978, Chap. 5 (an excellent review of additional derivitizing reagents as well as principles of derivatization; a "must" for anyone with an unusual derivatization problem).

N. Seiler, *J. Chromatogr.*, **143**, 221 (1977) (good review of derivatization procedures for amines, particularly those of biogenic origin).

van den Berg, J. H. M., *Improvement of the Selectivity in Column Liquid Chromatography: Design of Post-column Reaction Systems,* Thesis, Technische Hogeschool, Eindhoven, 1978 (good discussion of theory of reaction detectors).

March, 1979 issue of *J. Chromatogr. Sci.* (Vol. 17) (total issue devoted to review articles on derivatization for LC and GC; very current and complete).

Automation

Baker, D. R., in *GLC and HPLC Determination of Therapeutic Drugs,* Part 1, K. Tsuji and W. Morozowich, eds., Dekker, New York, 1978, Chap. 12 (good review of automation using less complex, off-the-shelf modules).

References 31, 41, and 42 of Chapter 17 (continuous-flow approach).

EIGHTEEN

SELECTING AND DEVELOPING ONE OF THE LC METHODS

18.1	Introduction	753
	Nature of the Sample	753
	Water-Soluble Samples	756
	Organic-Soluble Samples	757
	Separation Selectivity	758
	Experimental Convenience	759
	Experience with the Method	761
	Other Considerations	762
18.2	Developing a Particular Separation	762
	Hypothetical Problem: Organic-Soluble Sample	762
	Hypothetical Problem: Water-Soluble Sample	766
	Real Problem: Hydrogenated Quinolines	766
	Real Problem: Hot-Melt Adhesives	769
	Combination Methods	772
18.3	Special Applications	776
	Petroleum and Related Compounds	776
	Polynuclear Aromatic Hydrocarbons (PNA)	777
	Oligomeric Samples	777
	Carbohydrates	777
	Amino Acid Derivatives	777
	Clinical Applications	778
	Optical Isomers	778
	Isotopically Labeled Isomers	779
	Aliphatic Amines and Carboxylic Acids	779
	References	779

18.1 INTRODUCTION

Previous chapters have provided detailed discussions of each of the four basic methods (i.e., partition, size exclusion, adsorption, and ion exchange), plus important variations of these basic approaches. This chapter gives guidelines for selecting a particular LC method (or methods) for a given separation problem. Any discussion of how to choose the "best" LC method for a given separation is biased by the preferences of the individual worker, and often more than one approach will give satisfactory results. It should nevertheless be stressed that no one LC method is capable of handling all possible samples, and therefore the typical LC laboratory should have access to several of the basic methods.

Which LC method will work best is determined by factors such as

- The nature of the sample.
- The type of separation selectivity required.
- Experimental convenience.
- Experience with the method.
- Other considerations.

We discuss each of these areas in turn.

Nature of the Sample

The known physical and chemical properties of a sample are important guides for selecting the preferred LC method. Answers to certain questions can often rule out one method or suggest another, for example:

- Is the sample of high or low molecular weight?
- Is the sample water-soluble or oil-soluble?
- Is the sample ionic or nonionic?

Figure 18.1 provides a flowchart for the selection of an LC method, based on these and other questions about the sample. The methods shown here in dark-bordered boxes are generally preferred, based on the discussion of Chapters 7-12. The scheme shown in Figure 18.1 is necessarily oversimplified, but the general approach is applicable in most cases.

In choosing a particular LC method, the most important sample property is relative solubility in aqueous versus organic solvents. This in turn gives rise to two separate approaches, that of Figure 18.1a versus that of Figure 18.1b. If sample solubility is unknown, it should be determined experimentally, using water versus heptane or chloroform as solvents. The expected chemical structure of the sample constituents often suggests whether water or an organic solvent provides better solvency.

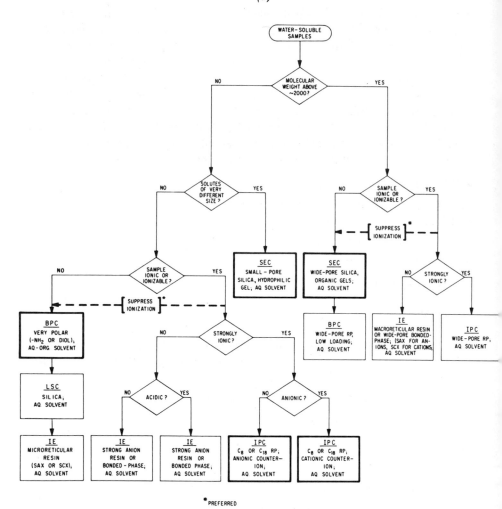

Figure 18.1 Simplified guide to selecting the LC method.

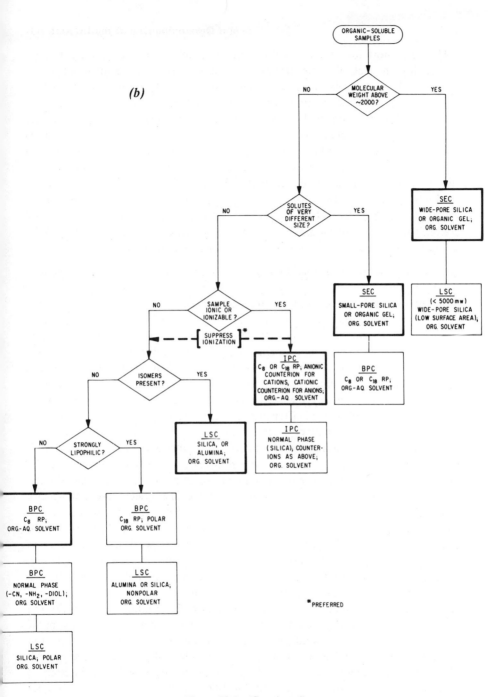

Figure 18.1 (Continued).

755

The next most important classification of sample properties for selecting an LC method is that of molecular size or weight. If no prior information on sample molecular weight is available, data can be quickly developed via a simple size-exclusion (SEC) separation carried out as described in Section 12.5. SEC can be used for both oil-soluble and water-soluble compounds, depending on the type of column packing. This separation indicates whether the sample is relatively simple and contains components of various sizes that can be resolved by SEC, or whether the sample is complex and/or contains several components that have little or no difference in molecular size. If the latter case is found, LC methods other than SEC must be considered. Where applicable, however, final separation by SEC should be considered because of its simplicity, speed, and convenience. In most instances the LC separation problem will not be completely solved by SEC, and one of the other LC methods must then be used.

Water-Soluble Samples (Figure 18.1a). Consider first the case of low-molecular-weight samples that are water-soluble. If the sample constituents are of quite different size, SEC separation with small-pore packings is probably the preferred approach. If sample molecules are of similar size, then we must determine whether the sample contains compounds that are ionic or ionizable. If all sample compounds are in the nonionic form, then BPC separation is usually preferred, and reverse-phase systems should generally be tried first. However, more-polar, water-soluble samples elute near t_0 in reverse-phase systems and are poorly separated. For these samples (e.g., carbohydrates), polar-phase BPC columns such as amino and diol phases are often useful—with organic/water solvent mixtures as mobile phase. Alternatively, some samples may respond better to LSC separation on silica, using mobile phases that contain water or methanol. Finally, macroreticular strong-anion or strong-cation ion-exchange resins can be used with water/alcohol solvents to separate small nonionic solutes by a partitioning mechanism. The best conditions for the last system are not usually predictable, but important separations have been effected in this manner.

If the sample is of low molecular weight and ionic (or ionizable), several approaches are possible. When dealing with relatively weak acids or bases as samples, the simplest technique is to suppress sample ionization completely, for example, with carboxylic acid samples, to lower the mobile-phase pH to a value of about 1-2. The separation can then be carried out as for nonionic samples. Ion suppression in this fashion generally yields satisfactory results and is more straightforward than coping with the special requirements and problems of ion-pair chromatography—and especially ion-exchange chromatography. Ion suppression is not applicable for strong-base samples (pK_a values above 6) with BPC systems, because the mobile-phase pH cannot be raised high enough to suppress ionization without attacking the BPC packing. Strongly basic samples should be handled in the same way as strongly ionic samples (Figure 18.1a);

by ion-pair chromatography. Anionic compounds require a cationic counter-ion, whereas cationic samples and strong bases require an anionic counter-ion. Ion-pair chromatography can be used for almost all ionic or ionizable solutes, but is especially preferred for more strongly ionic components (e.g., sulfonic acids, tetraalkyl ammonium compounds). Ion exchange is less convenient but is often satisfactory for weakly ionic samples (e.g., carboxylic acids, aromatic amines).

High-molecular-weight water-soluble samples (Figure 18.1*a*) are handled differently, depending on whether the samples are ionic (or ionizable) or nonionic. Ion suppression is also applicable for high-molecular-weight compounds and is often effective. SEC separation is usually preferred for all classes of compounds with molecular weights above 2-5000, but of course is limited to resolving constituents of differing molecular size. Wide-pore BPC packings with a low organic content are finding increasing favor for the separation of macro-molecules, particularly those in the 1-10,000 MW range.

Strongly ionic macromolecules that are water-soluble present a special problem in LC at present. Ion-pair chromatography with hydrophilic counter-ions (e.g. PO_4^{-3}) represents one approach, and special (still experimental) hydro-philic-matrix ion exchangers are another. At present such compounds as proteins and nucleic acids are still separated mainly on the classical, low-pressure, organic gels, but organic-modified silica packings show increasing promise.

Organic-Soluble Samples (Figure 18.1*b*). The initial approach in separating organic-soluble samples is similar to that used for water-soluble compounds. SEC separation should be used first to determine the size characteristics of the sample (if these are not known). High-molecular-weight samples can be further separated on appropriate wide-pore silica packings or organic gels, with organic solvents as mobile phase. For samples in the 1-5000 MW range, either SEC or LSC on silica can be used. SEC is preferred for samples containing a small number of compounds of significantly different molecular sizes. LSC (or BPC on polar-phase columns) is able to handle many synthetic and natural mixtures of oligomers, particularly those that are nonionic and with molecular weights under 2000.

For samples with molecular weights below 2000, the first question is whether the sample can be separated by SEC. If only a few compounds are present, and these are of quite different molecular size, SEC on a small-pore silica or gel column is the easiest and often the best approach. Alternatively, such samples can be handled in most cases by BPC (generally reverse-phase mode).

If the sample is more complex and/or involves compounds of similar molecular size, then various other LC methods become applicable. For these samples, the next question is whether the sample contains ionic or ionizable compounds. If the answer is yes, then the preferred options are either ion suppression (as for water-soluble compounds) or ion-pair (reverse-phase) chromatography. Normal-phase IPC and ion-exchange chromatography represent somewhat less desirable alternatives for most samples. Normal-phase IPC is useful when separating non-UV-absorbing solutes, because UV-absorbing counter-ions can be utilized to simplify the detection problem (Section 11.4).

For nonionic samples the next question is whether isomers are likely to be present in the sample. If the answer is yes, then the preferred technique is LSC, generally with silica as packing. If isomer separation is not a problem, we must determine whether the sample is strongly lipophilic. If it is, aqueous solvents should be avoided. This suggests either LSC or BPC with an organic mobile phase (e.g., a C_{18} column plus acetonitrile/chloroform).

Organic-soluble, less lipophilic samples that are moderately soluble in polar solvents are encountered fairly often. For such materials a reverse-phase column (e.g., C_8) with an aqueous-organic mobile phase is generally the preferred approach. Alternatively, polar-phase BPC packings (e.g., $-CN$, $-NH_2$ $-$diol) can be used with moderately polar organic solvents. Of equal utility for such samples is LSC (e.g., silica) with more polar organic solvents.

The scheme presented in Figure 18.1 suggests that the decisions involved in selecting a particular LC method are clear-cut and definable. Actually, this is not the case; particular separations problems often fall between or overlap the suggested decision patterns. Fortunately, this is usually not a serious problem, for alternate or adjacent LC methods listed in Figure 18.1 often provide satisfactory separations when appropriately optimized. Moreover, Figure 18.1 represents only one version of many possibilities for selecting an LC method in a given situation. In most cases, more than one LC method is satisfactory for a particular problem, and the method that is finally selected often depends on experimental convenience and experience with the method, as discussed later. Finally, a combination of two or more of the LC methods may be required for satisfactory solution of complex separation problems, as discussed in Section 18.2.

Separation Selectivity

Even very-high-efficiency LC systems may fail to provide the desired separation for mixtures of certain compounds. In these instances, optimization of separation selectivity offers the only chance for a successful separation. However, attempts to optimize separation selectivity for mixtures with a large number of components often are unsuccessful, because separation selectivity usually can only be simultaneously optimized for a few pairs of compounds. It is unlikely that one mobile-phase/stationary-phase combination will be optimum for all components in a complex mixture. In this case, the best approach is to select conditions that provide at least a partial separation for all components in the mixture.

(This will involve trials with a number of mobile-phase systems of different selectivity using the approaches described in Section 6.5.) The efficiency of the separating system is then increased by increasing column length or by using columns of smaller particles, to obtain the desired resolution.

Choosing a procedure that will be successful in separating very similar substances requires an understanding of the selectivity provided by different LC methods, and this in turn is based on a knowledge of the mechanism of the retention process for each method. Discussion of each of these two points, selectivity and mechanism, has been presented in the chapters on each of the LC methods. Some general aspects of selectivity have also been presented in Figure 18.1. For additional information, see Chapters 7-12 and the appended bibliographies. Finally, the separation of very complex mixtures may require combinations of LC methods designed to take advantage of the selectivity of each LC method (Section 18.2).

Experimental Convenience

The labor required in setting up a final LC separation, and in applying it routinely to samples of interest, is an important consideration in choosing one LC method over another. This fits into our general theme for the design of LC separations: They should provide *adequate* resolution of the sample with *minimum* overall effort. Some factors that affect experimental convenience are as follows:

- Sampling.
- Ease in determining experimental condition for satisfactory separation.
- Column stability.
- Column efficiency.
- Ease of packing efficient columns.

By *sampling convenience* we mean the use of an LC method that allows direct injection of the sample, with no pretreatment other than filtration. Thus, reverse-phase BPC is a logical choice for samples that arrive in the form of aqueous solutions. Such samples can usually be injected directly onto the column, without the need for removal of water (as would be the case for typical LSC applications). Also, relatively large sample volumes can usually be injected, thus avoiding the need for a preliminary concentration step (e.g., evaporation of water to give a more concentrated sample). Similarly, if the sample is an organic extract of an aqueous solution, it must be dried before injection onto an LSC column. Otherwise, column reproducibility is likely to be adversely affected. Again this suggests the use of some other LC method—for example, the various BPC procedures, which are usually insensitive to small amounts of water in the sample.

If the sample is initially dissolved in an organic solvent, on the other hand, LSC or normal-phase BPC are preferred to reverse-phase BPC. In short, it is generally desirable to consider first an LC method that allows direct sample injection without pretreatment.

The ease in determining adequate experimental conditions for a given LC separation has a great impact on setting up a given procedure. SEC methods are most attractive when applicable, because control of band migration rates is easily predicted in advance of the actual separations. In SEC all the sample constituents are eluted in a short time, which is not always the case for the other LC methods. With the other LC methods, gradient elution or some equivalent technique may be required for separating mixtures containing compounds with a wide k' range. Other than SEC, the selection of experimental conditions is probably most convenient with reverse-phase BPC. Satisfactory preliminary separations can often be achieved very quickly, using a water/methanol or water/acetonitrile gradient with a C_8 or C_{18} reverse-phase BPC column. Separations of many types of compounds are possible with such reverse-phase systems, and the appropriate isocratic mobile-phase strength can be estimated from the gradient run (Section 16.1).

Although reverse-phase BPC separations are easy to set up for many samples, LSC is often a more powerful separation procedure, and conditions for optimum resolution are readily predicted (Chapter 9). In addition, TLC systems can be easily and effectively used to pilot the conditions for LSC separations (Section 9.3).

The selection of experimental conditions for a successful ion-exchange or ion-pair separation is generally somewhat more tedious, for a variety of reasons. LLC, although used less frequently, has a good theoretical basis (Chapter 6) for guiding the search for optimum conditions.

Column stability is dependent on how well the column was initially packed and how it is subsequently used. Normally, columns for all of the LC methods should last for at least 3-6 months, which means that many hundreds of separations are possible with a single column before it must be replaced. Reverse-phase BPC columns are among the most stable, but columns of silica or packings for LSC and SEC also exhibit excellent durability when properly packed. Strong anion-exchange columns usually are among the least stable of modern LC packings, particularly when used above pH 7.5. LLC columns with mechanically held stationary phases also are subject to short life, unless the experimental conditions described in Section 8.6 are rigorously observed.

A major factor in the selection of an LC method could be the *efficiency of the columns,* particularly when complex mixtures must be separated. Fortunately, columns for all the LC methods now can be prepared for relatively high efficiencies, so this usually is not a major consideration. Columns of totally porous gels (e.g., ion-exchange resins and SEC organic gels) tend to be less efficient than silica-based particles, because of poorer mass transfer.

The *ease in preparing columns* of LC packings can have a strong influence on the selection of a particular LC method. Special equipment and skills are required for packing columns of small particles (<20 μm); therefore, many users prefer to purchase columns of these types, particularly because some manufacturers now provide columns with performance certified to be essentially equivalent to that reported in research studies. Pellicular packings may be preferred for separations of a more routine nature involving less complex separations, because these materials can be readily packed into useful columns by less-skilled operators.

Experience with the Method

Chromatographers—for good reasons—normally prefer to work with methods with which they have had direct experience. Likewise, in the area of liquid chromatography, many workers prefer to attempt all separations with one or two favored LC methods (e.g., reverse-phase BPC or SEC). A number of problems can in fact be successfully handled in this fashion, for we have seen that any given sample can usually be separated by more than one LC procedure. However, most laboratories will benefit by having several LC methods available. This allows the best LC method to be used in a given case, and it allows almost all separation problems to be handled by one method or another.

Table 18.1 Suggested columns for complete LC separation capability (in order of decreasing general utility).[a]

C_8 (or C_{18}) BPC column
Bimodal SEC silica column set
Silica (LSC) column
Polar-phase (amino, diol, or cyano) BPC column(s)
Bimodal SEC gel column set (e.g., μ-Styragel)
Bonded-phase ion-exchange columns (anion and cation)

---

[a]Columns should be packed with (e.g., 5- or 10-μm) particles of totally porous packings; it is desirable to have more than one column of the more popular types, for increased N values when needed; column diameter should be 0.4-0.5 cm, to allow both analysis and semipreparative recoveries.

Table 18.1 provides a list of LC columns (methods) arranged in decreasing order of general utility. Most workers should have access to at least the top three or four column types represented in this table. Additionally, laboratories with special needs or a preponderance of certain kinds of applications will need additional columns (e.g., prep columns, which are not listed in Table 18.1).

Other Considerations

The selection of an LC method involves other factors that can also be important. For example, in SEC all sample compounds elute quickly but only a limited number of solutes can be resolved in a single separation. Although SEC is usually inadequate for resolving complex mixtures, it is a useful method for screening a completely unknown sample. Much can be determined about the nature of the sample in a short time, because separations by SEC are rapid and the apparent complexity of a sample and the molecular-weight distribution of its components are revealed in the first separation.

The possible reaction of a sample during LC separations is usually not a problem. Of the various LC methods, SEC with organic gels probably is the least likely to lead to sample alteration; solutes that react with water or silanol groups should not be used with silica-based supports. Although LSC is more likely to lead to alteration or irreversible loss of sample, rapid separations allow little opportunity for changes in the sample to occur, and instances of problems with modern LSC are rare.

18.2 DEVELOPING A PARTICULAR SEPARATION

Different experimental approaches are required to optimize each LC separation. Therefore, it is impossible to set up rigid guidelines for the development of experimental conditions for a particular separation. However, there are some general steps that should be followed in each method development. Figure 18.2 presents a brief, systematic approach to LC separation problems. After proper selection of the analytical sample, the decisions involve (a) choosing an LC method, (b) selecting a specific column, (c) establishing the proper operating conditions, and (d) coping with any special situations, such as the general elution problem.

At this point it may be helpful to illustrate the steps involved in the development of a particular LC method by discussing several hypothetical and real LC separation problems of different kinds.

Hypothetical Problem: Organic-Soluble Sample

In the first hypothetical problem a routine, quantitative analysis is desired of a reaction intermediate that is partially soluble in moderately polar organic solvents. The first approach is to carry out a reverse-phase gradient elution separation, using a small-particle C_8 column and a gradient of 10-100% methanol in water. Figure 18.3a represents the gradient elution chromatogram obtained on the unknown sample. [This gradient elution separation was per-

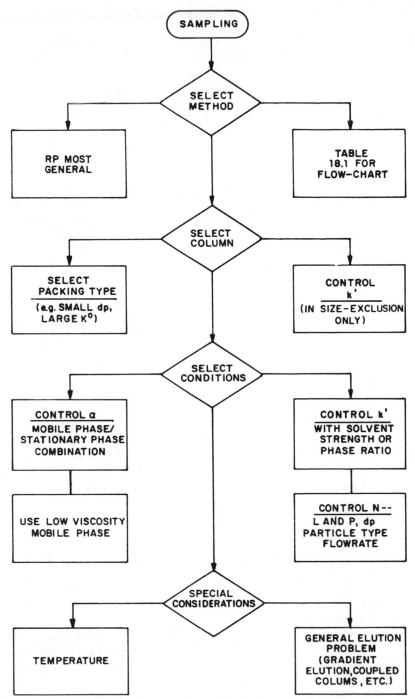

Figure 18.2 Systematic approach for developing the LC Separation.

formed with a small (10-μl) sample in methanol, because the sample was only sparingly soluble in methanol/water mixtures.] From this separation it is apparent that sample constituents elute in a relatively narrow k' range, indicating that a gradient elution separation is unnecessary and that an isocratic separation should be carried out (particularly because precise analyses are required). The separation is repeated using isocratic conditions of 40% methanol/water (Figure 18.3b). This particular solvent is suggested by the rule-of-thumb guide of Section 16.1: "Note the mobile-phase composition during elution of the bands of interest ($\sim$55%v MeOH), then extrapolate back by 2.5 t_0 (3 min $\times$ 4%/min = 12% methanol)," giving in this case a recommended isocratic composition of 40-45% methanol. This mobile-phase strength produces a satisfactory k' range (about 1-11 in Figure 18.3b) in a satisfactory separation time, but several peaks are not sufficiently separated (peaks 5, 6, and 11, 12 in Figure 18.3) to allow a precise quantitative analysis of the mixture.

An attempt is next made to change the separation selectivity of the isocratic separation, and 40% acetonitrile is substituted for 40% methanol to produce the isocratic separation in Figure 18.3c (methanol and acetonitrile have similar strengths in reverse-phase BPC; see Table 7.4).

Changing from methanol to acetonitrile results in peaks 11 and 12 being better separated, but peaks 5 and 6 (although reversed) still are not adequately separated. Peaks 7 and 8 now overlap and peaks 9 and 10 also are poorly separated. Thus, changing the organic modifier in this reverse-phase separation does change the separation but does not produce the desired resolution of all the peaks in the mixture.

At this point in developing the separation it appeared that the reverse-phase separation might not possess the selectivity required, perhaps because of isomer separation problems. Because it is unlikely that further change of organic modifiers in reverse-phase separation would be fruitful, the operator now has one of two alternatives. The first is to begin with a partial separation (Figure 13.3c) and increase column efficiency to produce the desired resolution. In most routine assays increased efficiency is best accomplished by increasing column length. It is estimated (see Figure 2.13) that R_s is 0.8 in Figure 18.3c, and we want it increased to about 1.0. For the given set of conditions in Figure 18.3c, it was estimated (Table 2.5) that column length should be increased 1.4-fold, with a twofold increase in separation time. Because this would result in a 40-min separation and routine application of the analysis was anticipated, this approach was abandoned.

The alternative approach is to consider another LC method that might provide a different selectivity. Because the reverse-phase separations previously carried out suggest that very closely similar structures are present (e.g., probably isomers), it appears that a LSC separation might provide a better approach. Accordingly, TLC tests were carried out with the sample, using butyl chloride as the basic mobile phase (because of sample solubility characteristics), with

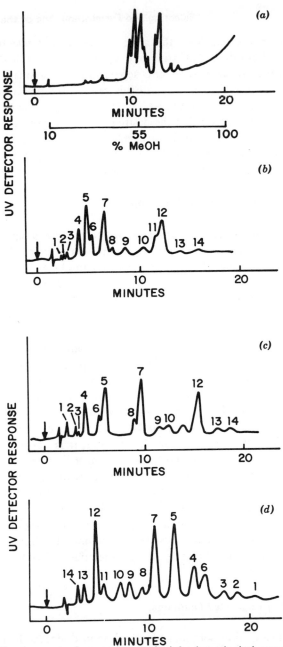

Figure 18.3 Development of separation method for hypothetical organic-soluble sample. Reverse-phase column, 15 × 0.46-cm small-particle C₈; 15 × 0.46-cm small-particle silica for LSC; mobile phase as shown; flowrates, 1.0 ml/min; temp., 50°C; detector, UV, 254 nm; sample, 10 μl. (*a*) Gradient reverse-phase separation, 10-100% methanol in water. (*b*) Isocratic reverse-phase separation, 40% methanol/water. (*c*) Isocratic reverse-phase separation, 40% acetonitrile/water. (*d*) Isocratic LSC separation, 3% methanol/butyl chloride.

modifying solvents selected from the apex of the solvent selectivity triangle from Figure 6.4. From these TLC studies, a 3% methanol/butylchloride mobile phase was subsequently used in an LSC separation with a small-particle silica column, as illustrated in Figure 18.3d. In this case all the peaks of interest are adequately separated for quantitative analysis by peak-height assay in a reasonable separation time (and in the desired k' range). Note that the elution order in the separation is essentially reversed from that shown for the reverse-phase separation, Figures 18.3a-18.3c, as expected. The separation of Figure 18.3d is now satisfactory for the desired routine quantitative analysis.

Hypothetical Problem: Water-Soluble Sample

In another hypothetical case the problem was to analyze a sample of water-soluble carbohydrates from the enzymatic hydrolysis of plant tissues. Because it was not known whether the hydrolysis would produce only monomers or yield higher-molecular-weight, partially hydrolyzed polymeric fragments, the sample was first separated by SEC (see Figure 18.4a). For this separation a bimodal column set (Section 12.2) was utilized to define the sample molecular-weight characteristics, because the molecular-weight distribution of the sample was initially unknown. Water was used as a very strong mobile phase, to ensure that all sample components eluted (not adsorbed) and that only the SEC process occurred. The separation in Figure 18.4a shows that the sample consisted largely of low-molecular-weight components (16-22 min), which are difficult to separate by SEC. There is also a small amount of very-high-molecular-weight, totally excluded solute (at 10 min), which presumably is starting material. After defining the sample in this fashion, it is apparent that a high-resolution method is needed for resolving the similar monomeric components. Accordingly, an NH_2-BPC column was utilized with an acetonitrile/water mobile phase to produce the chromatogram shown in Figure 18.4b. To ensure that this system remained suitable for the routine analysis of the monomeric carbohydrates, a short guard column of pellicular-NH_2 packing was inserted between the sampling valve and the analytical column. This removed the small amount of unchanged, high-molecular-weight material and other strongly held impurities of no interest.

Real Problem: Hydrogenated Quinolines

In an actual problem, dilute solutions of quinoline in dodecane had been hydro-genated catalytically, and it was required to determine qualitatively what products were produced and to set up a quantitative method for components of interest. Major interest centered on the aromatic, nitrogen-containing compounds in the samples. The total concentration of nitrogen in the hydro-

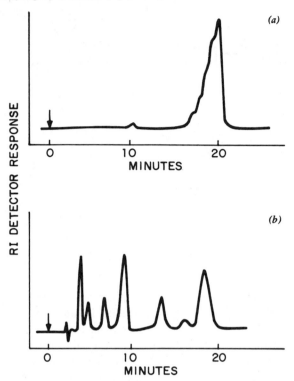

Figure 18.4 Development of separation method for hypothetical water-soluble sample. (*a*) Size-exclusion separation; column, bimodal pore-size set–25 × 0.62 cm each porous silica, 2-60 Å, 2-750 Å; mobile phase, 0.05 *M* Na$_2$SO$_4$; flowrate, 1.0 ml/min; detector, RI; sample, 50-μl solution. (*b*) BPC separation; column, 25 × 0.45 cm, −NH$_2$ bonded-phase; mobile phase, 60% acetonitrile/water; flowrate, 1.0 ml/min; temp., ambient; detector, RI; sample, 50 μl.

genated products ranged from 1000 ppm down to less than 1 ppm. Figure 18.5 shows the principal products of interest, although these were not known at the start of the project.

The first step was selection of an LC method. These basic, water-insoluble samples seemed amenable to LSC separation, and alumina was selected for its compatibility with basic compounds (Section 9.1). The next step was the qualitative analysis of a typical product sample to determine what compounds were present. Gradient elution is well suited to initial exploratory separation for determining the compositional range of a sample. Figure 18.6 shows the chromatogram obtained from the gradient elution of a typical product sample by LSC. The major components (numbered) occur in the middle of the chromatogram, with minor concentrations of hydrocarbons (i.e., nitrogen-free compounds)

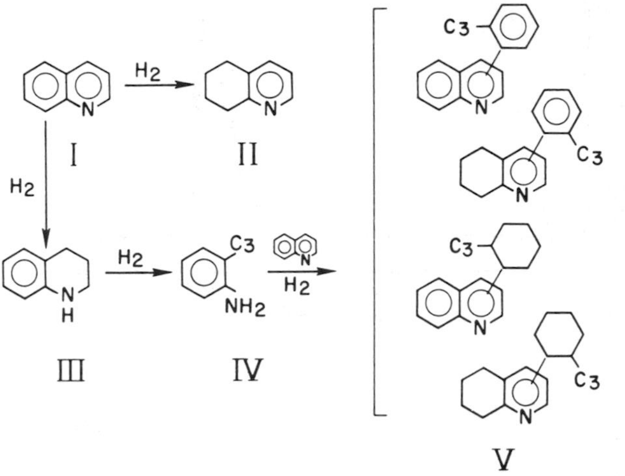

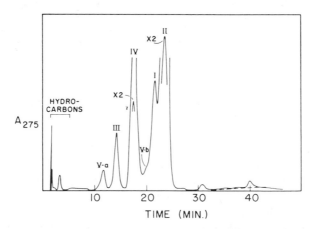

Figure 18.5 Products from the hydrogenation of quinoline in dilute solution.

at the beginning and trace amounts of more polar compounds at the end. The major bands shown in Figure 18.6 were readily isolated by scaling up this gradient elution separation, in much the same manner as that described in Chapter 15. The identity of each isolated peak was established by spectrophotometric analysis, as indicated in Figure 18.6. The later-eluting (minor) components were also characterized but were found to be highly rearranged compounds of little interest. Band V-b in Figure 18.6 occurs as an unresolved shoulder in

Figure 18.6 Gradient elution separation of hydrogenated quinoline sample on alumina. Column, 120 × 0.46 cm, 74-105-μm Alcoa F-20 alumina (4% H$_2$0); mobile phase, pentane/ dichloromethane to acetonitrile gradient; flowrate, 7.8 ml/min; detector, UV. Reprinted from (*1a*) with permission.

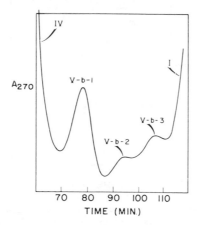

Figure 18.7 High-resolution isocratic separation of band V-b from hydrogenated quinoline. Same conditions as for Figure 18.6, except mobile phase, 35% dichloromethane/pentane (60% water saturated); flowrate, 0.9 ml/min. Reprinted from (*1a*) with permission.

the chromatogram. Because there was interest in the identity of this material, the resolution of the LSC system was increased by using a single solvent of optimum strength and a lower solvent flowrate (0.9 vs. 7.8 ml/min). In gradient elution, the elution of a given peak is a function of the strength of the mobile phase (in the column effluent) at the time of elution. Optimized single-solvent separation of the same band generally requires a solvent of somewhat lesser strength, and this approach was used in the present separation (see discussion of Figure 16.15). The resulting separation of band V-b is shown in Figure 18.7. Three distinct bands are now observed between bands IV and I. After the nature of the products formed from the hydrogenation of quinoline had been established, it was desired to analyze several hundred samples for compounds of interest: I, II, III, IV, V-a, and V-b. A detailed breakdown of band V-b into its three components was considered unnecessary. As a result, it was possible to use a single-solvent system (since bands I-V elute within a narrow range) of lower resolution than that shown in Figure 18.7. This simplified the routine analysis of the samples and reduced the necessary time per sample. The resulting separation is shown in Figure 18.8. The good resolution of all bands of interest permitted convenient quantitation.

Real Problem: Hot-Melt Adhesives

SEC is particularly useful for the rapid analysis of relatively simple mixtures that contain polymeric components. It was desired to analyze commercial hot-melt adhesives, which are blends of polymer, wax, and various resins. Prior to the application of SEC, these competitor samples were roughly characterized by

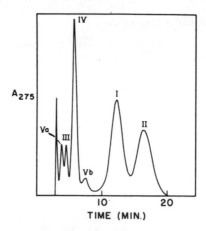

Figure 18.8 Medium-resolution isocratic separation of hydrogenated quinoline sample for routine analysis. Same conditions as for Figure 18.7, except flowrate, 10 ml/min. Reprinted from (*1a*) with permission.

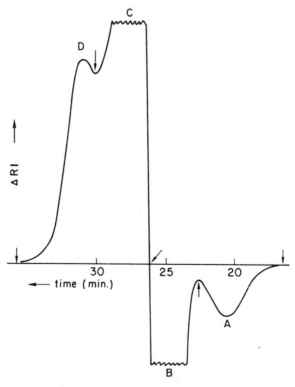

Figure 18.9 Separation and analysis of hot-melt adhesive. Columns, 120 × 0.8 cm, 2-Styragel-100 Å, 1-Styragel 10³ Å, 1-Styragel 10⁴ Å; mobile phase, toluene; temp., 100°C; detector, RI.

770

infrared spectrometric analysis of the total sample, followed by trial-and error blending experiments that attempted to duplicate the properties of the hot-melt adhesive sample of interest. In the example of Figure 18.9, a single SEC separation resulted in the separation and quantitative (gravimetric) assay of the four components present in the sample (bands A-D). In this case, infrared analysis provided an easy identification of the recovered fractions (arrows in Figure 18.9 show cutpoints). Finally, the average molecular weight of polymer (A) and wax (B) could be inferred from the chromatogram, to complete the total analysis of this particular sample. In this way, the SEC analysis was carried out in less than a day, versus more than the month required for the old blending approach based on infrared spectroscopy.

Another analysis by SEC of a competitor's hot-melt sample is shown in Figure 18.10. In this case, the initial separation yielded only two peaks, and it was subsequently shown by infrared spectrometry that the second band consisted of unresolved wax plus a resin. The refractive index of the resin was greater than that of the mobile phase (toluene), whereas the refractive index of the wax was less. The (net) wax band is shown as the dashed curve in Figure 18.10. By repeating the separation with a new solvent, a blend of chloro- and dichlorobenzene whose refractive index was matched to that of the resin, it was subsequently possible to obtain a chromatogram in which only the wax band appeared. This then permitted the quantitative analysis of the sample for the wax concentration, and also allowed the molecular weight of the wax to be determined.

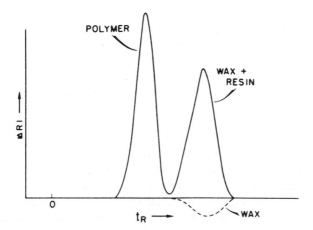

Figure 18.10 Analysis of wax in hot-melt adhesive with refractive index matching. Conditions same as in Figure 18.9, except mobile phase was a blend of chloro- and dichlorobenzene.

For an interesting application of LC in optimizing a process used to produce a coatings-industry intermediate, see the example described by Rehfeldt and Scheuing (*1c*).

Combination Methods

Complex mixtures that contain both similar and dissimilar materials can be resolved by combining different methods of LC in sequence. Because fraction collection and sampling handling are relatively simple in modern LC, we can readily move from one separating system to another by using sequential LC operations. Figure 18.11 illustrates one such technique of sequential analysis, in which the sample is first separated by SEC, and then various fractions are collected for further separation by a high-resolution LC technique (e.g., BPC, LSC). Separation methods can be devised so that sufficient sample can be isolated in the final separation for positive identification by supplementary techniques such as infrared or mass spectrometry.

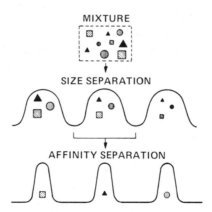

Figure 18.11 Representation of sequential LC analysis. Reprinted from (*1b*) with permission.

In certain cases, sequential techniques can be combined for *two-dimensional on-line chromatography* (*1*). An illustration of this approach is seen in the separation of senna glycosides from a complex plant extract, using the apparatus shown schematically in Figure 18.12. In this approach the first separating dimension is SEC, and the second separating dimension is reverse-phase BPC chromatography with a step gradient. The senna glycoside solution is first injected via loop 1 (Figure 18.12). The separation is then carried out on a SEC column at a low flowrate. Each fraction is collected via a loop, injected onto the

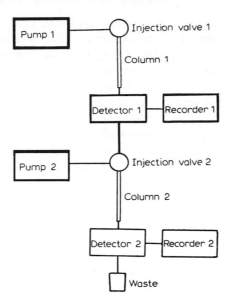

Figure 18.12 Schematic diagram of on-line two-dimensional LC. Reprinted from (*1*) with permission.

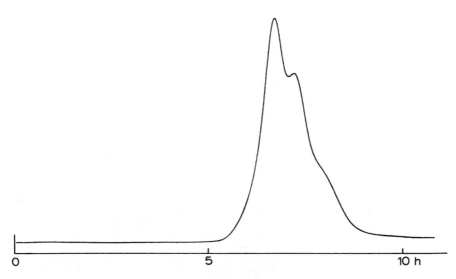

Figure 18.13 SEC separation of a senna glycoside extract (trace from recorder 1 in Figure 18.12). Column, 200 × 0.4 cm, 38-74 μm, controlled-pore glass, 88 Å/113 Å/170 Å/240 Å; mobile phase, titrisol buffer, pH 6; flowrate, 1.2 ml/hr; detector, UV, 254 nm. Reprinted from (*19*) with permission.

reverse-phase column, and then further separated with a step gradient. The general approach is such that the flow of pump 1 (SEC system, Figure 18.12) is adjusted so that loop 2 is just filled during the separation time on the second column. The flow of the reverse-phase system (pump 2) is such that complete flushing of loop 2 is possible within the injection time. The complex chromatogram that was obtained with this sample when only one dimension was used for the SEC run is shown in Figure 18.13, Figure 18.14 shows the similar separation for the step-gradient reverse-phase run with only one dimension. Figure 18.15 shows the two-dimensional approach with the apparatus of Figure 18.12: Seven fractions from the SEC run are separated by reverse phase

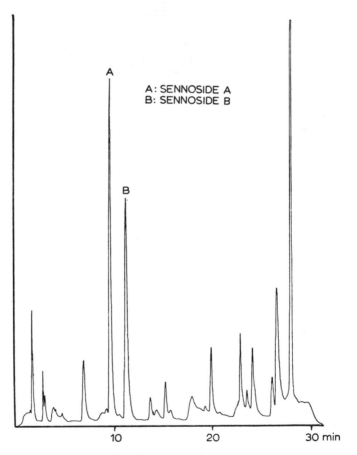

Figure 18.14 Reverse-phase separation of senna glycoside extract. Column, 25 × 0.4 cm, Nucleosil C_{18}, 5 μm; mobile phase, seven steps of acetonitrile-0.01 N NaHCO$_3$ in water (1.5-50% acetonitrile); flowrate, 2 ml/min; detector, UV, 254 nm. Reprinted from (*1*) with permission.

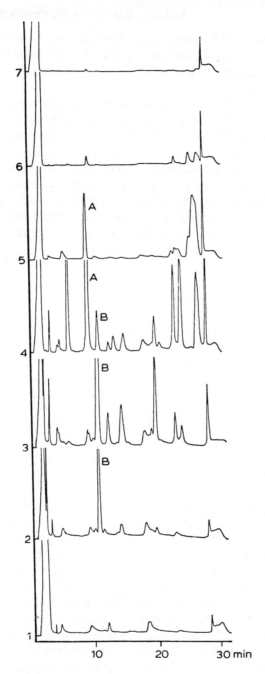

Figure 18.15 Reverse-phase chromatograms of seven fractions from SEC separations of senna glycoside. Conditions as in Figure 8.13 and 8.14, using apparatus of Figure 18.12. (*A*) Sennoside A, (*B*) Sennoside B. Reprinted from (*1*) with permission.

using a step gradient. The original very complex chromatograms (Figures 18.13 and 18.14) are now subdivided into a set of seven simplified and better-resolved chromatograms, which are more easily adapted to both qualitative and quantitative analysis.

Although in this study the principle is demonstrated for a combination of SEC and reverse phase, this same approach can easily be expanded to other separation methods. A limitation is that in the second separation, a preconcentration effect should be produced (i.e., use of gradient elution in Figure 18.14); otherwise, band broadening will eradicate any increase in resolution provided by the on-line sequential approach. A further development of this approach would be to utilize multidimensional techniques, that is, to use more than two columns with different separation principles. Of course, sequential methods can always be carried out separately (i.e., not on-line) in any appropriate combination, as long as the requirements given earlier are met.

18.3 SPECIAL APPLICATIONS

The preceding discussion allows the successful separation of most samples in a reasonably straightforward manner. However, some samples or analytical goals have distinctive features that require more or less special handling. In these cases, the experience of other workers who have faced similar separation problems can be invaluable. In this section we summarize a few such sample types and applications, with references to pertinent work in each case. The following listing is not intended to be exhaustive, merely helpful in a chosen few situations.

Petroleum and Related Compounds

Classical LC has been exhaustively applied to the separation of the different hydrocarbon and nonhydrocarbon compound classes in petroleum [e.g., see (2, 3)]. Typically, the resulting compound-type fractions include hundreds or thousands of individual homologs, isomers, and polycycloalkyl derivatives that must then be determined by another technique such as mass spectrometry [e.g., see (4)] or gas chromatography [e.g., see (5)]. In a few cases, modern LC has been adapted to these same separations, mainly to automate or speed up the separations (e.g., the separation of saturated, olefinic, and aromatic hydrocarbon classes as in Figure 16.25). For a recent review of the applications of both classical and modern LC to petroleum analysis, see (6).

Petroleum is typical of a number of complex samples (e.g., natural oils, coal extracts, etc.) that include too many individual compounds to allow complete separation of every sample consituent. In this case, compound-type elution bands are determined by the t_R values of individual compounds within the band, rather than by column efficiency. Therefore, increased N values generally have little effect on improving the separation of adjacent compound classes by LC. Because of the complexity of petroleum, "fingerprinting" of the sample may be a preferred alternative for some applications. Thus, the source of oil spills has been identified by comparing LC chromatograms of the offending sample with samples from various possible sources (7). The country-of-origin of *Cannabis*-resin samples has been confirmed in similar fashion (8).

Polynuclear Aromatic Hydrocarbons (PNA)

The analysis for compounds of the PNA class in water, air, or various foodstuffs is of increasing interest. Reverse-phase BPC is now used most commonly, because of its ability to separate both hydrocarbon classes and homologs, as well as for the good separation it provides of the two benzopyrene isomers. However, such PNA mixtures are usually quite complex, and overlapping bands are common. It is dangerous to assume the identity of a band in such a sample on the basis of t_R value alone, and the use of dual-wavelength detection is particularly advantageous. For a further discussion of the separation of these samples, see (9).

Oligomeric Samples

Polymeric samples are usually separated into a molecular-weight distribution (e.g., Figure 1.4) by SEC, without separation of individual oligomers. However, an analysis for the individual oligomers can be important for low-molecular-weight polymers (see Figure 9.4), for natural oligomeric samples such as carbohydrates or tannins (Figure 9.2), and for surfactants. For a recent review, see (10). In most cases, such separations use gradient elution from either an LSC or BPC column.

Carbohydrates

Mixtures of sugars and related compounds are of considerable importance, and they present special problems in their separation. This has led to the manufacture of special "carbohydrate columns," which are usually BPC columns with amino functionality. For one example of such applications, see (11).

Amino Acid Derivatives

Most separations of amino acids in protein hydrolysates are still done by the classical ion-exchange method with a polymeric cation-exchange resin. Alternatives based on modern LC have been presented [e.g., see (*12*)] but offer little in the way of concrete benefits (e.g., faster separation).

The use of LC for characterizing PTH amino acids from Edman sequencing has received considerable study. For a recent successful application, see the examples of Figures 9.23 and especially 7.30.

The separation of peptides and proteins by modern LC is receiving increasing attention. For a discussion of peptide separations by reverse-phase BPC, see (*13*). Most of the successful protein separations have been carried out on wide-pore ion exchangers (e.g., Figures 10.13, 10.17), but in a recent separation by reverse-phase BPC, phosphate ion was used to provide good retention characteristics for proteins such as insulin (*14*).

Clinical Applications

One of the major growth areas in LC analysis is in the assay of various trace-level substances in human serum or urine, for the purposes of diagnosing or treating diseased patients. Because of the complex nature of the sample matrix, as well as the need to measure analyte concentrations in the low ng to μg/ml range, such determinations present special problems. Additionally, rapid analysis times are essential to the clinical and hospital laboratories that make use of these tests. For a recent review, see (*15*).

Optical Isomers

Increasing interest is apparent in the separation of enantiomers by LC. In principle there are three different approaches: (1) use of a chiral stationary phase, (2) use of a chiral mobile phase, and (3) prereaction of the sample to form diasteriomeric derivatives of the two enantiomers of each compound to be separated. The use of optically active stationary phases (1) has so far proved relatively unsuccessful [but see (*15a*)]. A further difficulty is that a different stationary phase is required for each new class of optical isomers if good separation factors are to be achieved. For a good general discussion of this area, see the reviews in (*16, 16a*), and for a recent application see (*17*). The use of optically active mobile phases (2) is more flexible, but only rarely can the necessary separation factors for particular enantiomer pairs be achieved [e.g., examples in (*18*), (*18a*)] . The formation of diastereomers prior to separation (*3*) is the most popular approach, and effectively combines the known advantages of this classical technique with the power of LC resolution. In favored cases, α values as large as 3 for the diastereometers can be obtained. For recent good examples, see (*19, 20*) and Figure 7.29.

Isotopically Labeled Isomers

Compounds in which deuterium or tritium replaces protium, ^{14}C replaces ^{12}C, and so on, frequently show differences in retention, and give α values for isotopically labeled isomer-pairs of as much as 1.05-1.10 (but more often less than 1.01). For an early review, see (21), and for a detailed study in reverse-phase BPC systems, see (22). Good separations of such isomers are generally possible only when several atoms within the molecule have been replaced with the isotope, and especially when the molecule is small. However, the availability of very efficient columns (e.g., $N > 10^5$) should increase the prospects for this type of separation.

Aliphatic Amines and Carboxylic Acids

Compounds of this type have proved difficult to handle for several reasons: low detection sensitivities, variable ionization during separation, and strong interaction with many LC supports. The underivatized compounds are best separated by BPC with ion suppression, or by ion-pair chromatography. However, these compounds are easily derivatized (e.g., Tables 17.1 and 17.2) to give products that are simultaneously strongly UV-absorbing and nonionizable. The resulting derivatives are usually separable with minimal difficulty. The use of ion-pair chromatography with a UV-absorbing counter-ion (e.g., as in Figure 11.5) should also not be overlooked in this connection. See also the ion-exchange separation of Figure 16.3b,c.

REFERENCES

1. F. Erni and R. W. Frei, J. Chromatogr., 149, 561 (1978).

1a. L. R. Snyder, J. Chromatogr. Sci., 7, 595 (1969).

1b. J. N. Little, Am. Lab., 3, 59 (1971).

1c. T. H. Rehfeldt and D. R. Scheuing, Anal. Chem., 50, 980A (1978).

2. L. R. Snyder, ibid., 37, 713 (1965).

3. L. R. Snyder, ibid., 40, 1295 (1968).

4. L. R. Snyder, ibid., 41, 314 (1969).

5. E. R. Fett, D. J. Christoffersen, and L. R. Snyder, J. Gas Chromatogr. 6, 572 (1968).

6. K. H. Altgelt and T. H. Gouw, eds., Chromatography in Petroleum Analysis, Dekker, New York, 1979.

7. W. A. Saner, G. E. Fitzgerald, and J. P. Welsh, Anal. Chem., 48, 1747 (1976).

8. B. B. Wheals and R. N. Smith, J. Chromatogr., 105, 396 (1975).

9. L. R. Snyder and D. L. Saunders, in Chapter 10, Ref. 6.

10. F. P. B. van der Maeden, M. E. F. Biemond, and P. C. G. M. Janssen, *J. Chromatogr.*, **149**, 539 (1978).

11. J. N. Little and G. J. Fallick, *ibid.*, **112**, 389 (1975).

12. E. Bayer, E. Grom, B. Kaltenegger, and R. Uhmann, *Anal. Chem.*, **48**, 1106 (1976).

13. E. Lundanes and T. Greibrokk, *J. Chromatogr.*, **149**, 241 (1978).

14. W. S. Hancock, C. A. Bishop, R. L. Prestidge, D. R. K. Harding and M. T. W. Heaven, *ibid.*, **153**, 391 (1978).

15. L. R. Snyder, B. L. Karger, and R. W. Giese, in *Contemporary Topics in Analytical and Clinical Chemistry*, Vol. 2, D. Hercules et al., eds., Plenum, New York, 1978, p. 199.

15a. W. H. Pirkle and D. W. House, *J. Chromatogr.*, in press.

16. C. H. Lochmuller and R. W. Souter, *J. Chromatogr.*, **113**, 283 (1975).

16a. I. S. Krull, *Adv. Chromatogr.*, **16**, 175 (1978).

17. F. Mikes and G. Boshart, *J. Chromatogr.*, **149**, 455 (1978).

18. W. A. Pirkle and D. L. Sikkenga, *J. Chromatogr.*, **123**, 400 (1976).

18a. J. N. LePage, W. Lindner, G. Davies, D. E. Seitz and B. L. Karger, *Anal. Chem.*, **51**, 433 (1979).

19. R. W. Souter, *Chromatographia*, **9**, 635 (1976).

20. J. Goto, M. Hasegawa, S. Nakamura, K. Shimada, and T. Nambara, *J. Chromatogr.*, **152**, 413 (1978).

21. P. D. Klein, *Adv. Chromatogr.*, **3**, 3 (1966).

22. N. Tanaka and E. R. Thornton, *J. Am. Chem. Soc.*, **98**, 1617 (1976).

NINETEEN

TROUBLESHOOTING THE SEPARATION

19.1	Troubleshooting the Equipment	782
19.2	Separation Artifacts	791
	Band Tailing	791
	Column Overload	791
	System Mismatch	795
	Heterogeneous Retention Sites	795
	Mixed Retention Mechanism	798
	Extracolumn Effects	801
	Poorly Packed Columns	803
	Unbuffered Systems	804
	Slow Retention Kinetics	804
	Micelle Formation	804
	Sample-Solvent Effects	805
	Pseudo Tailing	807
	Other Band-Tailing Phenomena	807
	Sample Reactions	807
	Reaction of the Sample during Workup or Pretreatment	807
	Slow Reaction during Separation	809
	Fast Reaction during Separation	810
	Negative Peaks in LC	810
	Vacancy Peaks	812
19.3	Troubleshooting Analytical Errors	813
	References	823

In preceding chapters we have tried to cover all aspects of setting up an LC separation, including attention to the various problems that can arise during method development and subsequent application of the final LC procedure. However, most chromatographers spend significant time in dealing with various unexpected difficulties. These include equipment malfunctions and failure as well as unsatisfactory chromatograms and/or unacceptable analytical data. In this final chapter we try to focus on this kind of problem solving.

Troubleshooting the LC unit is normally discussed in the literature supplied by the equipment manufacturer. Because of the rapidly increasing diversity of LC systems and modules, it is not practical to repeat all this information here. And if it were possible, only a fraction of this material would be of use to any given reader at a particular time. Our approach in the following section is to provide general troubleshooting advice that should be applicable to a broad range of equipment. This discussion cannot substitute for information available from the manufacturer, but it does provide an organized approach to diagnosing and correcting what is wrong with a particular LC unit.

Assuming that the equipment is operating satisfactorily, troubleshooting the LC separation can take two different directions. First, we can assume that everything was done properly, but some unusual or unpredictable phenomenon is present. To that end, we provide here a section (19.2) on separation artifacts that were not stressed in previous chapters. Second, it more often happens that we have simply overlooked precautions set forth in the earlier part of this book. This latter problem was anticipated in earlier chapters, where we attempted to review and organize the individual steps involved in various phases of method development. In the last section of this chapter we give a final checklist of contributions to analytical error or imprecision, as yet another way of guiding the reader to the solution of his particular problem.

19.1 TROUBLESHOOTING THE EQUIPMENT

Table 19.1 is a simplified troubleshooting checklist that should provide some help in pinpointing general equipment malfunctions or certain column problems. Detector malfunctions mostly involve the optical detectors (e.g., UV, RI, fluorescence) that are widely used in modern LC. Remedies and correction procedures for restoring the LC system to its true capability are suggested, approximately in the order of ease and/or desirability. For additional information on this subject, the reader is referred to (1-3). The manual for the LC apparatus being used should also be consulted for help in identifying and correcting difficulties unique to that particular equipment.

Table 19.1 Diagnosing and correcting equipment problems.

Symptom	Diagnosis		Remedy	
1. Noisy baseline	1a	Contamination in sample or reference cells	1a	Flush cells with 1 M nitric acid, water, fresh solvent; remove detector cell, clean or replace windows
	1b	Bubble in sample or reference cells	1b	Suddenly increase flowrate to force out bubble; increase pressure on cell by restrictor at detector exit (if cell is designed for back pressure)
	1c	Recorder or instrument grounding problem	1c	Eliminate ground loop—ensure that recorder and instrument have common ground
	1d	Defective detector source (lamp bulb)	1d	Replace source
	1e	Leaking fitting or connector	1e	Tighten or replace fittings
	1f	Very small bubbles passing through detector cell	1f	Degas mobile phase; increase back-pressure on detector cell; check for leaks in system
	1g	Regular pulses from pump stroke	1g	Incorporate pulse dampener; use pulseless pump
	1h	Leaking septum	1h	Replace septum; preferably, use sampling valve
	1i	Particulate material passing through detector cell	1i	Clean cell; check column lower frit for leaks of packing

783

Table 19.1 (Continued).

Symptom	Diagnosis	Remedy
	1j Leaching of septum	1j Use recommended septum for mobile phase; preferably, use sampling valve
	1k Partial blockage of valve loop injector	1k Clean sampling valve loop
2. Drifting baseline	2a Contamination in detector cell	2a Flush cells with 1 M nitric acid, water, fresh solvent; remove detector cell, clean or replace windows
	2b Contaminated or "bleeding" columns	2b Regenerate column, or replace if regeneration is unsuccessful; saturate mobile phase with stationary phase and/or use a precolumn in LLC and normal-phase ion-pair chromatography; thermostat system in LLC
	2c Changes in detector temperature	2c Thermostat system
	2d Malfunctioning detector source	2d Replace source lamp
	2e Previous mobile phase not completely removed	2e Thoroughly purge systems with fresh mobile phase
	2f Contamination in solvent reservoir	2f Clean reservoir, refill with fresh mobile phase, purge column

Problem	Possible cause	Remedy
	2g Strongly adsorbed components being eluted from column	2g Elute all components from column prior to next separation using strong mobile phase; use solvent gradient to clean column
	2h Partial pluggage of injection port, sampling valve, or column inlet by particulates	2h Clean sample introduction system and column inlet frit
	2i Solvent demixing	2i Use compatible mobile-phase solvents
	2j Slow change in pump output	2j Check flowrate; operate in temperature-controlled room if pump output is temperature dependent.
3. Large spikes on recorder baseline	3a Bubbles passing through detector cell	3a Degas solvent and thoroughly purge system; check fittings for air leaks
	3b Other electrical (usually radio-frequency) equipment in laboratory (e.g., thermostatted ovens, other chromatographs)	3b Remove source of noise; make sure equipment is grounded; isolate apparatus with isolation transformer
4. Recorder baseline stair-stepping, flattop peaks, baseline does not return to zero.	4a Improper recorder gain and damping control	4a Adjust gain and damping controls; repair recorder
	4b Recorder or instrument not properly grounded	4b Ground equipment carefully
5. Negative peaks (also see Section 19.2)	5a Incorrect polarity of detector output	5a Reverse polarity of detector output
	5b With RI detector sample has lower re-	5b Select mobile phase with higher re-

Table 19.1 (Continued).

Symptom	Diagnosis	Remedy
	refractive index	fractive index; reverse detector output polarity
	5c Sample injection problem	5c Use sample valve injector; ensure sample loop free of air bubbles during sampling
	5d Impure mobile phase used	5d Change to purified mobile phase
6. Spurious or "ghost" peaks	6a Leak or contamination of sampling valve or septum; contamination of syringe	6a Repair leak in valve or septum; clean valve loop, septum inlet, or syringe
	6b Elution of sample solvent (unlike composition of mobile phase)	6b Dissolve sample in mobile phase; greatly reduce sample size
	6c Air dissolved in sample	6c Degas sample solution
	6d Impurities in solvents used during gradients	6d Used purified solvents (water a particular problem here)
7. Pump on, no column flow or column pressure	7a Air in liquid end of pump	7a Prime pump; disconnect pump outlet and operate at maximum flow until no bubbles appear
	7b Large leak between pump and column or in sampling system	7b Repair leak
	7c Solvent reservoir empty	7c Refill reservoir

Symptom	Probable Cause	Remedy
8. Pump on, pressure on but no flow from detector	8a Large leak in system	8a Repair septum, sampling valve, or lines and fittings between pump and detector
	8b Flow blockage	8b Remove particulates in injection port, sampling valve, or in capillary between column and detector, or in detector cell
	8c Pluggage in column inlet due to particulates	8c Clean or replace column inlet frit; replace column if required; filter all solvents and samples
9. Erratic column pressure	9a Leak in high-pressure system (septum inlet, sampling valve, inlet fitting, etc.)	9a Locate and repair leak
	9b Dirt in pump check valves	9b Disconnect liquid inlet and outlet lines to the pump; flush through 25-50 ml of 1 M nitric acid, followed by distilled water; disassemble liquid end and clean or replace check valves
	9c Air in liquid end of pump	9c Disconnect pump outlet and pump at maximum flow until no bubbles appear; prime pump if required
10. Column pressure increases, column flow decreases	10a Partial pluggage in column, pre-column, or "guard" column	10a Clean or replace column inlet frit; replace column if required
	10b Partial pluggage in detector cell or detector inlet tubing	10b Disassemble and clean cell and/or lines

Table 19.1 (Continued).

Symptom	Diagnosis	Remedy
11. Recorder will not balance at zero	11a Malfunctioning recorder	11a Repair
	11b Bubble in sample or reference cell	11b Increase mobile-phase flowrate to remove bubbles; place restrictor at detector exit to increase cell pressure; degas mobile phase
	11c Severe reduction of energy from sample or reference detector cells	11c Check light path, remove blockage; clean cell or replace cell window; (RI detector) flush reference cell with mobile phase
	11d Defective source lamp	11d Replace source lamp
	11e Poor electrical connections from detector to recorder	11e Check and secure electrical leads
	11f Excessive column "bleed"	11f Use different system; replace with fresh column
	11g Contamination with previous mobile phase	11g Thoroughly purge system of previous mobile phase
	11h Air in column packing	11h Thoroughly purge column with mobile phase
	11i Strongly absorbing mobile phase (with UV detector)	11i Replace mobile phase with UV-transparent solvent

Problem	Possible cause	Remedy
12. Non-Gaussian peak shape (tailing peaks)	12a See Section 19.2	12a See Section 19.2
13. Loss of column resolution	13a Column overloaded with sample	13a Reduce sample load or sample volume
	13b Degraded column	13b Replace with new column (see Section 5.4)
	13c Strongly retained materials on column packing	13c Regenerate column or replace if required
	13d Loss of stationary phase from column	13d (LLC) use packing with bonded stationary phase; replace column; insert precolumn or guard column for protection
	13e Degradation of column stationary phase	13e Regenerate or replace column
	13f Column packing not fully equilibrated	13f Thoroughly purge system with new mobile phase; readjust separating parameters for desired resolution
14. Increased solute retention time	14a Low mobile-phase flowrate	14a Readjust mobile-phase flowrate
	14b Column temperature low	14b Increase column temperature to proper level
	14c Improper gradient setting	14c Check gradient system; correct gradient setting

Table 19.1 (Continued).

Symptom	Diagnosis	Remedy
	14d Column activity increasing, system not equilibrated; mobile phase removing water from LSC column	14d Reequilibrate column with mobile phase having proper water or organic modifier content; flush column with 10 V_m of mobile phase; see Section 9.3
	14e Incorrect mobile phase	14e Reequilibrate column with correct mobile phase
15. Decreased retention time	15a High mobile-phase flowrate	15a Readjust mobile-phase flowrate
	15b Column temperature high	15b Decrease column temperature to proper level
	15c Improper gradient	15c Check gradient system; decrease rate of gradient
	15d Incorrect mobile phase	15d Reequilibrate column with proper mobile phase (10 V_m)
	15e Loss of stationary phase from column	15e Replace column
	15f Column activity decreased; system not equilibrated	15f See 14d
	15g Deactivation by strongly retained "garbage"	15g Purge column with strong solvent, then proceed as in 14d

19.2 SEPARATION ARTIFACTS

Band Tailing

In a properly designed LC system, all sample bands should appear symmetrical (Section 5.5), with no obvious signs of band tailing. Many examples of LC separation that meet this requirement can be found in Chapters 7-12. Under optimum conditions of separation (e.g., Figures 10.7, 11.10), bands in LC appear symmetrical over two or more orders-of-magnitude change in sample concentration—relative to the peak concentration of the band. This degree of band symmetry is rarely observed in gas chromatography, except for nonpolar compounds.

However, tailing bands in LC are a common complaint among practical workers. Fortunately, the causes of band tailing are well documented, and it is often possible to diagnose the reason for band asymmetry in a particular separation. In most cases, band tailing can then be corrected by systematic application of certain changes in separation conditions. In this section we look at different phenomena that can result in asymmetric or distorted sample bands and discuss the confirmation and cure of these different effects.

Table 19.2 summarizes a number of different phenomena that can lead to band distortion or tailing. Symptoms for each kind of band distortion are listed in this table, along with proposed changes in separation conditions that confirm and/or fix the particular problem. The following discussion follows the organization of Table 19.2. However, it should be stressed that the identification and cure of a particular cause of band tailing is not always straightforward. The following comments are meant to be suggestions, not absolute guides that are guaranteed to solve every problem involving band tailing.

Column Overload. Symmetrical bands are normally observed only for samples that do not exceed some maximum size (e.g., see Section 9.4). When the amount of sample injected is larger than this, retention times and plate numbers begin to decrease, and various kinds of band distortion become evident. These changes in band shape can be further magnified by nonlinearity in the detector at high sample concentrations. Chapter 15 provides several examples of this effect; specifically, see Figures 15.12 and 15.27.

Column overload should be suspected whenever the amount of sample injected onto a column exceeds about 1 mg/g (100 μg/g for pellicular packings). This effect can be confirmed by the simple expedient of decreasing sample size (and detector attenuation) three- to five-fold. If a significant increase in retention time(s) is observed, with some improvement in band shape, it is almost certain that column overload is the problem. The solution in this case is to decrease sample size to the point where retention times for all bands become constant.

Table 19.2 Diagnosing and correcting band tailing and related problems.

Phenomenon	Symptoms	Confirmation and/or correction
Column overload	One or more distorted or tailing bands	Reduce sample size and detector attenuation by fourfold; retention should increase and tailing decrease; finally, decrease sample size until retention is constant
System mismatch	Severe band tailing, particularly at large k', with low N values and (in some cases) low recoveries	Change to another column packing (e.g., avoid cationic or basic samples on silica); try ion-pair chromatography
Site heterogeneity	Band tailing increases with in-crease in k' for LSC or ion exchange	Tailing should decrease with decreased sample size; addition of water or polar modifier to mobile phase will correct for LSC; gradient elution will solve in all cases
Mixed retention	Band tailing increases with in-crease in k', especially for SEC, BPC, and LLC	See Table 19.3
Extracolumn effects	Band tailing decreases for larger k'	Replumb LC system to reduce extracolumn band broadening; use lower volume detector, faster response detector, larger volume column

Poorly packed column	Band tailing similar for all bands, all samples	Test standard sample/mobile phase suggested by manufacturer; return or discard column
Unbuffered system	Tailing of acidic or basic samples, unbuffered mobile phase.	Buffer mobile phase or add formic acid or triethylamine to mobile phase for ion suppression
Slow retention kinetics	Severe band tailing	Tailing decreases with decrease in mobile-phase flowrate, or increase in temperature
Micelle formation	Ionic samples in SEC; less polar mobile phase; retention changes with sample size	Use more polar mobile phase
Sample-solvent effects	Sample dissolved in solvent that is stronger than mobile phase	Decrease volume of injected sample; re-dissolve sample in weaker solvent
Pseudo tailing partly resolved bands	One or two tailing bands in middle of chromatogram	Increase column plate number (e.g., by two-fold) as in Section 2.5, under controlling resolution
polymer bands in SEC	Later bands do not tail and are narrower	Ignore

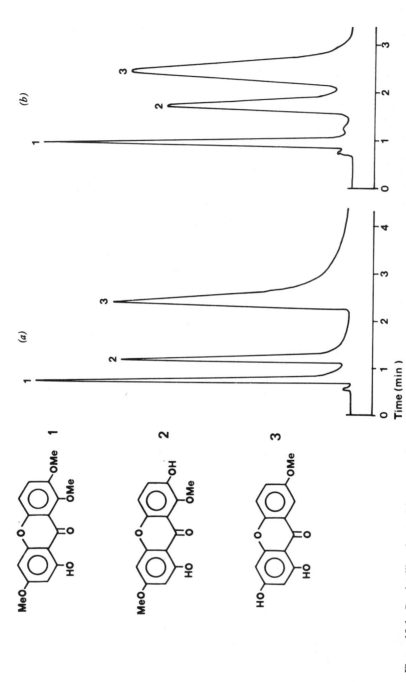

Figure 19.1 Band tailing due to (*a*) system mismatch corrected by (*b*) change in column packing. (*a*) Column, 25 × 0.22 cm, Micropak CN (10 μm); mobile phase, 80%v chloroform/cyclohexane; temp., ambient; flowrate, 1.33 ml/min (1000 psi); detector, UV, 254 nm. (*b*) Column, 25 × 0.22 cm, Micropak NH₂ (10 μm); mobile phase, 85%v chloroform/cyclohexane; temp., ambient; flowrate, 1.16 ml/min (900 psi). Reprinted from (*4*) with permission.

Column overload in ion-pair chromatography can occur at lower sample sizes than for other LC methods. Increase in the concentration of the counter-ion is one technique for increasing column capacity and decreasing resulting band tailing. Alternatively, a long-chain alkylammonium compound can be added to the organic phase when separating amine samples (*3a*), or a long-chain carboxylic acid can be added (*3b*) when separating acidic samples.

Sytem Mismatch. Many band-tailing problems can be attributed to the wrong combination of sample and column packing. In fact, it is generally good advice when band tailing is encountered to first try a different column (and different stationary phase) for the sample in question. Examples of sample/column mismatch have been discussed in earlier chapters; for example, cationic or strongly basic samples with silica as packing; very polar and/or multiply ionized samples in LSC or BPC, and so on. The symptoms of system mismatch include (a) markedly tailing bands, particularly for larger k' values, as in Figure 9.3a; (b) much lower column N values for bands with $k' > 2$; (c) in some cases low recovery of later-eluting compounds—particularly for smaller samples. Where system mismatch is suspected, the best procedure is to try a column of different type; an example is shown in Figure 19.1 for the separation of three naturally occurring xanthones. Tailing occurred on a BPC-nitrile column (Figure 19.1a) but was corrected on a BPC-amino column (Figure 19.1b). Alternatively, the technique of ion-pair chromatography is ideally suited for many of the sample types that are poorly adapted to the other LC methods; that is, compounds that are multiply ionized and/or strongly basic.

Heterogeneous Retention Sites. Problems involving heterogeneous retention are encountered most commonly with LSC or ion-exchange systems. For these packings, retention of the sample occurs on active sites (e.g., discussion of Figures 9.5c and 10.2). Usually these retention sites in LSC and IEC are not exactly equivalent within a given packing, which results in sites of varying retention affinity: strong sites that yield greater retention and larger k' values (on those sites), and weaker sites that give smaller k' values. In the usual case a plot of the concentration of these sites within the stationary phase versus site energy E (or log k') yields a roughly Gaussian distribution, as in Figure 19.2a. At very low sample concentrations, the stronger sites are preferentially combined with sample molecules, but these sites are not overloaded and normal elution of sample bands is observed. However, for more practical sample sizes the strong sites quickly become saturated or overloaded—as in Figure 19.2b. The overloading of these strong sites results in pronounced band tailing, as in Figure 9.9a for separation of a steroid mixture on silica or Figure 16.3a for separation of organic acids by ion exchange. For more strongly retained (later-eluting) sample bands, the site-energy distributions as in Figure 19.2 broaden; a greater fraction of the sample band is present in the stationary phase to over-

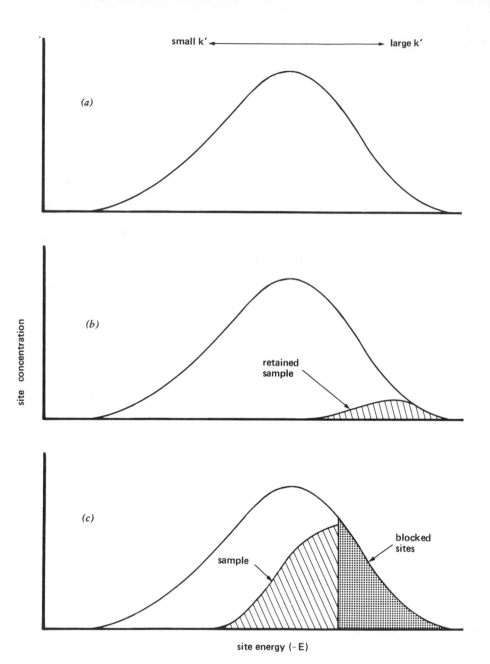

Figure 19.2 Distribution of sorption sites in adsorption or ion-exchange chromatography by site energy E (equivalent to log k' for a given solute and mobile phase). (*a*) Site distribution in original (underactivated) packing. (*b*) Rapid overloading of high-energy sites at small sample concentrations; cross-hatched area refers to sites covered by sorbed sample molecules. (*c*) Effect of sorbent deactivation on site-energy distribution; shows larger sample size possible (versus *b*) before overloading of sites occurs.

load retention sites, and band tailing becomes more pronounced. Band tailing resulting from site heterogeneity always increases with increase in k', and tailing also becomes less severe for smaller samples.

One approach to the correction of tailing caused by site heterogeneity is to partially deactivate the stationary phase, so as to selectively remove the stronger sites. In LSC this is most commonly achieved by addition of water or other polar modifier to the adsorbent (or mobile phase) (e.g., discussion of Section 9.3). The result is the truncation of the site-energy distribution curve, as in Figure 19.2c. It is seen here that the relative concentration of strongest sites that remain after deactivation is larger than in Figures 19.2a or b, thus allowing larger sample sizes without band tailing. The beneficial effect in LSC of water deactivation on band tailing was illustrated in Figure 9.9b versus Figure 9.9a.

Some suppliers of LSC adsorbents provide so-called deactivated silicas (e.g., Vydac). These materials are claimed to be chemically modified so as to eliminate strong sites and solve the related problem of band tailing. Many workers have encountered various problems in using these deactivated silicas. Furthermore, tailing is not eliminated for many samples, requiring addition of water or polar modifier to the mobile phase, as for conventional LSC adsorbents. However, the concept of modifying LSC or ion-exchange stationary phases so as to eliminate strong sites permanently is fundamentally the best approach to solving the heterogeneous-site problem. It is to be hoped that research in this area will eventually result in adsorbents and ion-exchange packings with minimal site heterogeneity.

An alternative solution to the problem of site heterogeneity is reduction in sample k' values by a change in mobile phase or temperature. Where the sample bands cover a range in k' greater than about 5 for isocratic elution, the most effective expedient is gradient elution, as illustrated by Figure 16.3b versus Figure 16.3a. Finally, tailing resulting from site heterogeneity can generally be minimized by a significant reduction in sample size. Often this results in marginal detection sensitivity, so it is a less practical approach. Alternatively, an increase in the concentration of stationary phase within the column may achieve correction of tailing, without loss in detection sensitivity. Or these two expedients may be used in combination: increase in amount of stationary phase plus a small reduction in sample size. Thus, in Figure 19.3 the tailing of the aniline band (no. 7) on this pellicular cation-exchange column is probably due to site heterogeneity. In this case, a small reduction in sample size could be combined with substitution of either a porous-polymer or bonded-phase porous-silica cation exchanger for the original pellicular packing (plus use of a stronger mobile phase). We would then expect decreased tailing with no loss in detection sensitivity, particularly if the column N value was not reduced for the second column.

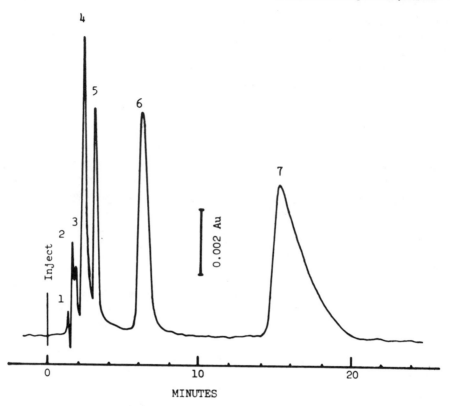

Figure 19.3 Band tailing due to site heterogeneity in ion-exchange chromatography. Sample: *p*-aminophenol (4), *m*-aminophenol (5), *o*-aminophenol (6), and aniline (7), Column, 100 × 0.21 cm, Zipax SCX (pellicular cation exchanger); mobile phase, 0.1 M H$_3$PO$_4$/0.1 M KH$_2$PO$_4$ in water (pH = 2.9); temp., 25°C; flowrate, 0.8 ml/min; detector, UV, 254 nm; samples, 10 μl. Reprinted from (5) with permission.

Mixed Retention Mechanism. Probably the most common band tailing results from stationary phases that allow more than one kind of retention. An example would be a bonded-phase packing in which the silica surface is only partly covered by the organic phase. The resulting column could then allow retention by both the normal BPC mechanism with the organic coating and by adsorption onto the bare silica surface.

Two retention mechanisms do not necessarily lead to band tailing. In the usual case, however, the two stationary phases available for retention (organic coating and bare silica above) are present in greatly different concentrations. Thus, one retention process predominates, and the other retention mechanism quickly overloads the small amount of stationary phase available for that process.

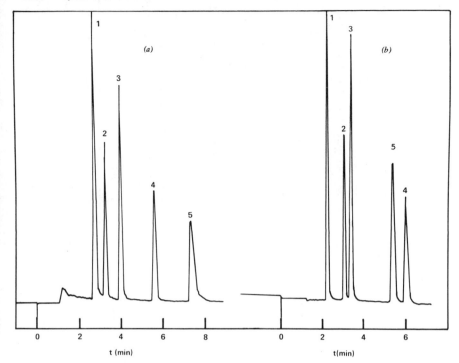

Figure 19.4 Band tailing due to mixed-retention mechanism in separation of xanthines by reverse-phase BPC. (*a*) Tailing of band 5 due to adsorption on incompletely silanized silica. (*b*) Tailing corrected by complete silanization of support. Column, 15 × 0.46 cm, 5-μm C$_8$ BPC packing; mobile phase, 20%v methanol/0.01 M phosphate in water (pH 3.2); temp., 30°C; flowrate, 1.5 ml/min; detector, UV (254 nm). Solutes are theobromine (1), dyphylline (2), theophylline (3), ethosuximide (4) and caffeine (5).

The net result is somewhat similar to the preceding case of site heterogeneity: tailing bands, primarily at higher k' values. An example is shown in Figure 19.4*a* for an incompletedly reacted reverse-phase BPC packing. Here several xanthines are being separated by reverse-phase on a C$_8$ bonded silica. All the bands except No. 4 show some tailing, but tailing is more pronounced for the last band, No. 5. This reverse-phase packing was then further reacted to complete the coverage of the surface by organic groups, and the separation was repeated in Figure 19.4*b*. Now all the bands are more symmetrical, particularly the No. 5 band, which is also seen to elute earlier in (*b*) than in (*a*)

Some important differences exist between mixed retention and site heterogeneity. Mixed retention is found primarily in stationary phases of weaker retention activity: SEC, BPC, LLC, and (occasionally) LSC. Site heterogeneity is found primarily in ion exchange and LSC. Site heterogeneity always results in

Table 19.3 Different types of mixed retention: classification, diagnosis, and cure.

Column packing	Sample type	Secondary retention	Cure
SEC (silica)	Polar	Adsorption	Add high-molecular-weight polar compound to mobile phase, e.g., 0.1% Carbowax 20M; addition of salt may also help
BPC/LLC	Polar (normal-phase)	Adsorption	Add polar modifier to mobile phase, e.g., 0.5% methanol
	Ionic	Ion exchange	Add salt to mobile phase to compete with ion-exchange sites; e.g., ammonium carbonate (reverse-phase), tetramethyl ammonium nitrate, etc. Buffer system to suppress sample ionization, or add acetic acid (acidic samples) or ammonia (basic samples)
LSC	Ionic	Ion exchange	Same as above

increasing band tailing with increase in k', whereas this is not necessarily true in mixed retention. Mixed-retention problems can generally be inferred from the nature of the sample and column packing used, and several practical fixes can be suggested for each case. Table 19.3 summarizes solutions for the more common examples of mixed retention.

Adsorption often occurs as a secondary retention process in SEC or normal-phase BPC or LLC systems. It often can be suppressed by adding 0.1-0.5% methanol or isopropanol to the mobile phase. Similarly, ion exchange can occur in these same systems, as well as in LSC. In these cases, where ionic or ionizable sample compounds are involved, addition of a soluble inorganic or organic salt to the mobile phase suppresses ion exchange as a secondary retention process, and decreases band tailing. For samples involving the ionization of an acid or base, ion suppression can be employed: the use of high pH for basic samples and low pH for sample acids. This is commonly achieved by adding acetic or formic acid to samples of acids (e.g., RCOOH), and triethylamine or ammonia to samples of bases (e.g., R_3N). These polar modifiers also suppress any adsorption of the sample onto the silica surface.

It is important to recognize that mixed retention in BPC systems can arise either from incomplete silanization of the support or from particularly strong interaction between certain samples and silica. Thus, in many cases band tailing can be blamed on a poorly made support, and the column should be returned to the supplier. However, this is not always the case, and for some samples a good (fully bonded) column requires addition of methanol, acetic acid, triethylamine, and so on, to the mobile phase to suppress band tailing. Incomplete coverage of a BPC packing can be checked as discussed in Section 7.2.

Mixed-retention effects that lead to band tailing are less often encountered in ion-exchange chromatography, although tailing in IEC systems often occurs for other reasons. Traces of heavy metals in the mobile phase can lead to ligand exchange as a secondary retention mechanism for some IEC systems. In such cases, addition of a chelating agent such as EDTA to the mobile phase has proved beneficial in reducing tailing. An example is that of the ion-exchange separation of tetracyclines (6). The separation of large, multiply ionized molecules such as proteins by ion-exchange presents special problems, which have already been discussed in Section 10.2. Here the main problem is usually one of incomplete recovery of sample, rather than tailing bands.

Extracolumn Effects. It has been noted that an increase in the volume of the sample injector, detector cell, or tubing between these units and the column can lead to unwanted band broadening and an apparent decrease in column plate number (see discussion of Section 2.3). These extracolumn effects also lead to band tailing. A typical example is shown in Figure 10.1. In every such case, early bands show pronounced tailing, with later bands becoming more symmetrical. Thus, this problem is easy to diagnose. The problem can be solved by

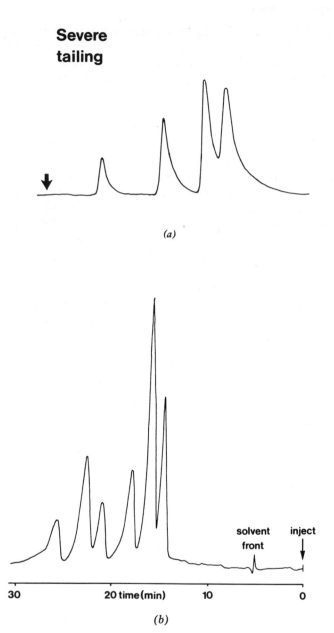

Severe tailing

(a)

solvent
front

inject

30 20 time(min) 10 0

(b)

Figure 19.5 Band tailing due to poorly packed column. (*a*) Possible void at column inlet. (*b*) Separation of urobilin dimethyl esters (bile pigments) on silica. Column, 30 cm, µ-Porasil; mobile phase, 72/25/0.01 benzene/ethanol/diethylamine; temp., ambient; detector, UV 450 nm. Reprinted from (7) with permission. (*c*) Inhomogeneous bed structure and/or channeling. Reprinted by permission of Spectra-Physics.

replumbing the LC unit to reduce the various extracolumn volumes, using a smaller-volume detector flowcell or faster-response detector or recorder, and so on. This type of tailing becomes more pronounced for smaller-volume, higher-efficiency columns. The use of 10-15-cm lengths of narrow-diameter columns (e.g., i.d. less than 0.4 cm) packed with 5-μm particles is especially demanding in this respect, and it may not be practical to totally eliminate tailing of early-eluting ($k' < 1$) bands in such systems.

Poorly Packed Columns. Columns that are poorly packed, including those that develop voids, cracks, and so on, as a result of handling or shipment, often exhibit band tailing. Figure 19.5a is a typical example of this kind of problem, but the degree of tailing can be greater or less for a given column. The characteristic feature of this type of tailing is that *all* bands in the chromatogram show significant asymmetry. Furthermore, the amount of band tailing is more or less independent of sample size and sample type. Figure 19.5b is an actual example taken from the literature, which suggests this type of problem. In some cases, severe band distortion, and even splitting of the band for a single compound into an apparent doublet is observed. Figure 19.5c illustrates this effect. In each of these various cases, the column should first be rechecked using the particular test mixture and mobile phase recommended by the manufacturer to check

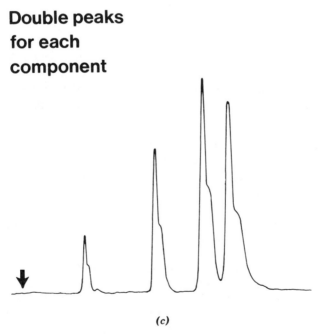

Double peaks for each component

(c)

Figure 19.5 Continued.

column performance (see Section 5.5). If band tailing is confirmed in this system, the column should be returned to the supplier—assuming that the column was not abused and has not outlived its useful life.

Tailing as in Figures 19.5a-c can also be caused (less often) by poor technique in injecting the sample onto the column or by use of very viscous samples (viscous "fingering").

Unbuffered Systems. When acids or bases are separated in an LC system, either the mobile phase should be buffered or ionization should be suppressed by working at low or high pH. If these precautions are not followed, an individual acidic or basic compound in the sample can exhibit partial ionization under the conditions of separation. In the absence of buffering, the degree of sample ionization varies with sample concentration. Because sample concentration varies between the band peak and band tail, this means that the degree of sample ionization then also varies. The result can be either a fronting or tailing band, depending on conditions. The severity of band asymmetry in this case is greatest for bands that are about half-ionized (i.e., pH $\approx$ pK_a) under the conditions of separation, and, therefore, band tailing can vary randomly with k'. It is good general practice to buffer the mobile phase, whenever ionizable sample compounds are to be separated. For an example of ion suppression at low pH, see Figure 7.15.

In ion-pair chromatography, another type of buffering problem can be encountered. For very low concentrations of the counter-ion, the concentration of sample ions during the separation can be comparable to that of the counter-ions, leading to change in free counter-ion concentration and a variation of k' with sample concentration (e.g., see Eq. 11.2a). This results again in band tailing, which can in this case be corrected by increasing the concentration of the counter-ion. This can be accompanied by adding neutral salt to maintain k' values constant (Section 11.4).

Slow Retention Kinetics. In principle, slow mass transfer between mobile and stationary phases is possible, as a result of slow breaking of bonds between sample molecules and functional groups within the stationary phase. This effect is not commonly encountered, but can be expected in some LC systems using complexing ions that concentrate into the stationary phase. Tailing of this type decreases as the flowrate of mobile phase through the column is decreased, or as separation temperature is increased.

Micelle Formation. When separating ionic surfactants or other compounds that tend to form micelles in nonpolar solvents, band tailing may be observed in SEC systems. Typically, for this case retention also varies with sample size, which is rarely true in SEC (except for very large samples). The correction of the problem is generally achieved by using a more-polar solvent, particularly one with a higher dielectric constant (see Table 6.1).

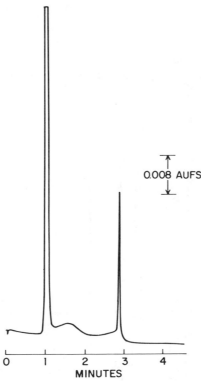

Figure 19.6 Result of injecting sample (unknown) in a solvent stronger than the mobile phase. Column, 100 cm × 0.21 cm, Corasil II; mobile phase. 0.5% dioxane in isooctane; detector, UV, 254 nm; sample, 10 μl in dioxane, 0.1 mg/ml. Reproduced courtesy of A. Vatvars, E. I. duPont de Nemours & Co.

Sample-Solvent Effects. If the sample is dissolved in a solvent that is stronger than the mobile phase, resulting bands can be markedly distorted—particularly in the case of early-eluting bands, as illustrated in Figure 19.6. Here a poorly soluble sample was dissolved initially in dioxane for injection into a column operated with a much weaker solvent as the mobile phase. The result was distorted bands and an indication of solvent demixing. Such a problem often can be solved by dissolving the sample at a 10-fold lower concentration in the mobile phase, and injecting a 10-fold larger sample volume. With this approach the final separation for the Figure 19.6 sample showed two completely separated symmetrical bands. This sample-solvent effect also can be tested by reducing sample volume, which should also reduce band distortion. Thus, this type of problem can be corrected either by replacing the solvent containing the sample with a weaker solvent, as just indicated or (sometimes) by using sample volumes no larger than 5 μl.

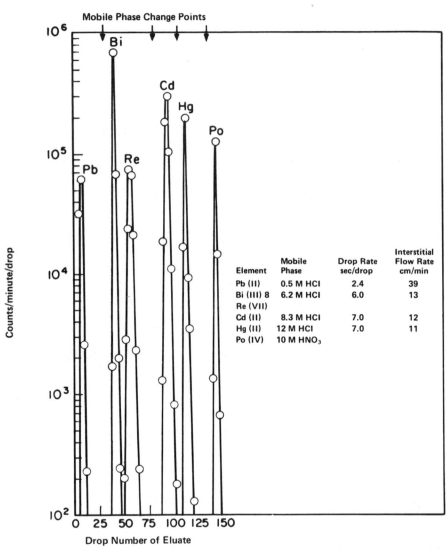

Figure 19.7 Virtually complete suppression of band tailing on a well-packed column. Separation of radioisotopes by ion-exchange chromatography. Column, 5 × 0.28 cm, Zipax coated with 1%(w) tricaprylmethylammonium chloride; mobile phase, HCl gradient in water; temp., 60°C; flowrate, 2-7 sec/drop; detector, radioactive counter. Reprinted from (8) with permission.

Pseudo Tailing. In addition to "real" examples of band tailing, *apparent* band tailing is observed in some separations. In Section 2.5, the discussion of how large R_s should be, we mentioned the case of two poorly resolved bands that appear as a single, tailing band. A good example is seen in Figure 9.20 for the middle "anti" band. This band tails noticeably, but it is seen that the bands immediately preceding and following the "anti" band do not tail. Furthermore, the chemical structures of "cis" and "anti" bands are virtually the same. Under these circumstances, one should suspect incomplete resolution of a compound eluting just after the "anti" band, rather than "true" tailing. This possibility can be tested by increasing the column plate number for increased resolution, for example, as discussed in Section 2.5 under controlling resolution. If pseudo tailing is involved, two bands should become apparent as resolution is increased. If the apparently tailing band shows no change in band shape as N is increased, then some form of "real" tailing is involved.

Apparently tailing polymer bands are often noted in SEC separations (e.g., Figure 12.14). When the polymer band appears to tail, but later bands are symmetrical, the probable cause is an asymmetrical molecular-weight distribution of the polymer, not tailing of individual-compound bands.

Other Band-Tailing Phenomena. As in gas chromatography, LC bands are never perfectly symmetrical, because of the inherent nature of the chromatographic process. This often becomes apparent if very sensitive detectors are used with larger samples (without attenuation). Then the extremes of the tail on each side of the band are magnified, and the later-eluting tail is much more pronounced. When it is necessary to reduce this type of band tailing as much as possible, careful attention must be paid to the preparation of the column and to sample injection [as in (8)]. An example of what can then be achieved is shown in Figure 19.7, for the separation of various radionuclides. Here the various bands show no evidence of tailing, over as much as 4 decades of solute concentration. The reduction of band tailing to this degree, however, is usually a tedious undertaking; the effort should be restricted to those applications where extreme band symmetry is required.

Sample Reactions

The possibility of reaction or degradation of the sample during its workup or actual LC separation has been discussed in previous chapters. Here we look briefly at certain consequences of such sample alteration.

Reaction of the Sample during Workup or Pretreatment. In some cases a single compound can be isomerized or reacted to yield two or more products during sample handling prior to the LC separation. Then a single compound present in the original sample will appear as two or more bands in the final chromatogram.

An example is shown in Figure 19.8 for the assay in urine of ingested LSD. In all assays by this procedure, including that of control urines spiked with LSD, it was found that a later-eluting "iso-LSD" band was observed in addition to the LSD band (9).

Derivitization of a sample prior to LC separation sometimes leads to two or more products from a single starting compound. An example of the latter is shown in Figure 16.4, for the LC assay of carbamate-insecticide residues as their dansyl derivatives. In this case, hydrolysis of the sample during its workup leads to the

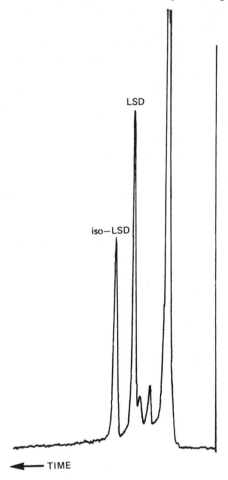

Figure 19.8 Isomerization of LSD to iso-LSD during sample pretreatment. Determination of LSD in urine by LSC. Column, 25 × 0.46 cm, Partisil (6 μm); mobile phase, 55%v methanol/water plus 0.2% ammonium nitrate; temp., ambient; flowrate, 1.0 ml/min (2000 psi). fluorescent detection, 325-nm excitation, 430-nm emission: 1-μl sample, representing 1 ml or urine. Reprinted from (9) with permission.

formation of a dansyl phenol plus dansyl amine product; that is, there are two bands for each original insecticide.

The formation of multiple bands during sample workup need not represent a serious problem for the analyst. It may be necessary to quantitate both bands (unless the ratio of products is constant for all samples) and then to combine the results for each band to obtain a result for the original compound of interest. Although the formation of such multiple bands from a single compound is an obvious complication and should be avoided if possible, there is one redeeming feature. The presence of two bands for a given compound provides an additional check on peak identity, because now two retention times must match the standard, rather than just one.

Slow Reaction during Separation. The slow reaction of the sample during separation typically leads to another type of band distortion or tailing, as illustrated in Figure 19.9. In this example the shoulder on the major band (arrow) is due to the reaction of the initial compound during the separation, with formation of a

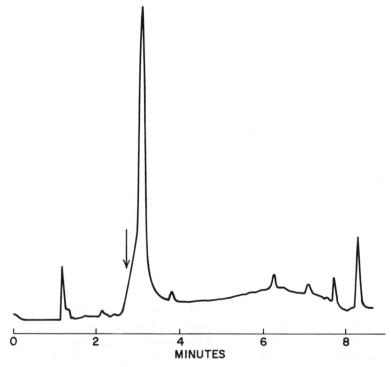

Figure 19.9 Reaction of major sample component during LSC separation (unknown sample). Column, 25 × 0.32 cm, Zorbax-Sil; mobile phase, gradient from 12%v dioxane/hexane plus 0.1% water to 100% dioxane plus 1% water; temp., ambient; detector, UV, 254 nm. J. J. Kirkland, unpublished studies, 1975.

product that moves with a smaller k' value. Because the reaction continues throughout separation, two distinct bands are never obtained. Sample artifacts of this type can be confirmed by changing the flowrate and residence time of the reacting band. Faster flow should result in a shorter reaction time, and decrease in the apparent band tail. Slower flow should enhance the degree of band distortion.

When reaction of the sample during separation has been confirmed, various means exist to slow down or eliminate this unwanted effect. Specific cases have been discussed in earlier chapters (e.g., Section 9.5). Generally, a change in experimental conditions (pH, mobile-phase composition, temperature, addition of antioxidants, etc.) will suffice.

Fast Reaction during Separation. Rapid reaction of a sample component during separation will generally mean the final elution of a single product band, at a different place in the chromatogram. Some workers have asked whether the rapid equilibration of a single compound between two forms will give rise to two bands: one for each compound. An example would be the reversible ionization of an acid $HX \rightleftharpoons H^+ + X^-$. The answer in this case is that only one band for the compound will be observed. The reason is that each molecule of HX ionizes and recombines with H^+ many times during its migration through the column. The final retention of an injected solute molecule is therefore an average of retentions for the two forms of the molecule (HX and X^-).

Negative Peaks in LC

Negative peaks, as for band 7 in Figures 19.10a and b, are occasionally encountered in LC separations and are a frequent cause of confusion to both beginning and experienced chromatographers alike. These negative bands can arise from different causes. The most obvious origin for a negative band is a greater detector response for the mobile phase versus the band of interest. This commonly occurs with refractometer detectors, where sample compounds can have refractive indices either greater than or less than that of the mobile phase. An example is that of Figure 18.9, where bands A and B go below baseline, whereas bands C and D are above baseline. In this case, compounds A and B have refractive indices less than that of the mobile phase, and bands C and D have higher refractive indices. Negative bands of this type can occur with other detectors, but are less common for photometric detectors—because nonabsorbing mobile phases are generally employed and sample components of interest have high absorptivity (except below 220 nm). Sharp, negative peaks or "spikes" are sometimes observed at the beginning of the chromatogram or at t_0 (e.g., Figures 2.24 and 10.5). These are generally found when the detector is set at lower attenuations (higher sensitivity setting); negative spikes can occur with photometric detectors. Negative peaks at time zero can result from pressure

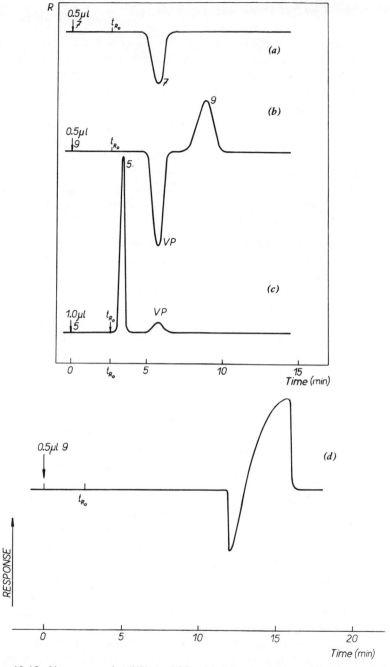

Figure 19.10 Vacancy peaks (VP) in LSC with binary-solvent mobile phases. Column, 50 × 0.18 cm silica; mobile phase, 2%v ether/cyclohexane (*a-c*), 0.5% ether/cyclohexane (*d*); temp., ambient; refractive index detection; sample size indicated on figure. Samples are benzene (5), ether (7) and nitrobenzene (9). Reprinted from (*10*) with permission.

fluctuations that accompany sample injection. As a result, there is a momentary change in flow of mobile phase through the detector, and usually a corresponding small change in the temperature of the mobile phase (due to differences in temperature between ambient and the detector). These temperature changes can result in a change in detector response.

Negative peaks at t_0 in the case of photometric detectors are also due to Schlierren effects, when the sample-solvent and mibile phase are different. The mixing of the sample solvent, which elutes generally at t_0, with the mobile phase leads to a change in refractive index of the mobile phase leaving the column at t_0. This in turn causes changing refraction of the light passed through the detector flowcell, which leads to fluctuation in the amount of light lost via refraction at the walls of the flowcell.

Vacancy Peaks. This phenomenon is well known in gas chromatography, and its occurrence in LC has recently been discussed in detail (*10*). An example of a vacancy or "ghost" peak is shown in Figure 19.10*b* and *c*, for band 7 (or VP) in each case. In these two examples, compound #7 was *not* injected as a sample, but is a component of the mobile phase. Vacancy peaks are associated with the use of mobile phases that contain small concentrations of a strongly retained moderator (e.g., 2%v ethyl ether/cyclohexane in Figure 19.10*a-c*, for silica as stationary phase). A second factor required for the observation of vacancy peaks is the use of rather large amount of sample, for which reason these bands are more often observed with refractometer detectors than with photometric devices. Vacancy peaks can be either positive or negative.

The origin of these vacancy peaks is as follows. First, consider a sample that is less strongly retained than the mobile-phase moderator *B*. This is illustrated in Figure 19.10*c* (*B* = 2% ether), where compound 5 (benzene) is injected and elutes with a t_R value less than that of the moderator *B* (ethyl ether, compound 7 in Figure 19.10*a*). In this case, injection of compound 5 simply dilutes the mobile phase at the column inlet, lowering the concentration of *B*. Because compound 5 elutes prior to *B*, it cannot displace adsorbed *B* from the stationary phase. The resulting volume element of diluted mobile phase then moves through the column as would an injected volume of *B*, except that the concentration of *B* is *less* in this volume element, rather than higher as for injected *B*. The result is a vacancy peak that elutes at the same time as *B*, but with an opposite displacement of the peak from baseline—compared to injected *B*. Thus, the negative peak in Figure 19.10*a* for injection of *B* is replaced by a positive peak in Figure 19.10*c* for injection of benzene.

When a sample is injected that is more strongly retained than *B* (e.g., nitrobenzene, no. 9 in Figure 19.10*b*), it displaces *B* from the stationary phase on entering the column. This quantity of displaced *B* then elutes in normal fashion, giving a vacancy peak equivalent to that for injection of pure *B*.

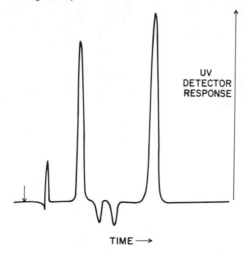

Figure 19.11 Vacancy peaks with a UV detector. Peaks below baseline represent non-UV-absorbing compounds eluting with a partially UV-absorbing mobile phase.

Vacancy peaks are readily confirmed when their presence is suspected. For example, increasing the concentration of B in the mobile phase leads to a reduction in the size and t_R value of the vacancy band. Similarly, the vacancy peak will be variously positive or negative, depending on whether an injected compound elutes before or after the vacancy peak, and whether the normal band for injection of B (as sample) is positive or negative.

Vacancy peaks that are negative (or other negative peaks) can lead to another type of peak distortion. As illustrated in Figure 19.10d, a negative vacancy peak partially overlaps an adjacent "real" band, leading to the effect noted.

Vacancy peaks can also occur when using a UV detector, as illustrated in Figure 19.11. In this case the mobile phase contains UV-absorbing impurities that equilibrate with the column and become part of the detector background that is electronically balanced to zero before the separation. When non-UV-absorbing impurities elute during the separation, the transmission of the solvent background *increases* (less background impurity) during peak elution, resulting in a negative peak, as illustrated for the two peaks in the middle of the separation in Figure 19.11.

19.3 TROUBLESHOOTING ANALYTICAL ERRORS

In Table 19.4 we present a checklist of problems involving analytical errors or imprecision, together with possible solutions to these problems, as still another

Table 19.4 Diagnosing and correcting assay and trace-analysis problems.

Effect	Situation or Symptom	Cause		Cure	
Problems with assay-type analyses					
1. Poor peak height reproducibility	1.1 With all peaks	1.1a	Irreproducible sample volume	1.1a	Use valve injector; check sample injector and column inlet for leaks
		1.1b	Improper detector response	1.1b	Repair detector
		1.1c	Improper recorder response	1.1c	Adjust or repair recorder
		1.1d	Column or detector is sample-overloaded	1.1d	Reduce sample size
		1.1e	Calibration and sample solutions decomposing	1.1e	Make up fresh solutions, run promptly; change separation system
		1.1f	Internal standard decomposing or reacting	1.1f	Use different internal standard; use calibration without internal standard
		1.1g	Irreproducible retention times	1.1g	See Table 19.1, items 14, 15
	1.2 With symmetrical peaks	1.2a	As in 1.1	1.2a	See 1.1
		1.2b	Sample decomposition in column (fast)	1.2b	Change separation system
	1.3 With unsymmetrical peaks (see Section 19.2)	1.3a	See 1.1	1.3a	See 1.1

		Cause	Remedy
2. Poor peak-area reproducibility	2.1 With all peaks	1.3b Undesired interaction of solute with stationary phase	1.3b Change separation system
		1.3c Poorly packed column	1.3c Replace column
		1.3d Sample decomposition in column (slow)	1.3d Change separation system
		2.1a See 1.1	2.1a See 1.1
		2.1b Improper integrator or computer response	2.1b Repair area-measuring device
		2.1c Internal standard decomposing or reacting	2.1c Use different internal standard; use calibration without internal standard
	2.2 With symmetrical peaks	2.2a See 1.2	2.2a See 1.2
	2.3 With unsymmetrical peaks (see Section 19.2)	2.3a See 1.3	2.3a See 1.3
3. Poor analytical accuracy	3.1 Analytical results higher than theoretical (e.g., for known mixtures of pure compounds)	3.1a Improperly made standards	3.1a Repeat calibration carefully
		3.1b Calibration and/or sample solutions decomposing	3.1b Make up fresh solutions, run promptly; change separation system
		3.1c Unknown interference with peaks for desired component(s)	3.1c Increase resolution (particularly α) to eliminate, identify, and eliminate overlap; change separation system

Table 19.4 (Continued).

Effect	Situation or Symptom	Cause	Cure
Problems with assay-type analyses			
		3.1d Internal standard reacting or decomposing	3.1d Use different internal standard; use calibration without internal standard
		3.1e Component of interest or internal standard too close to t_0	3.1e Increase k' (e.g., decrease strength of mobile phase)
	3.2 Analytical results lower than theoretical (e.g., for known mixtures of pure compounds)	3.2a See 3.1a, 3.1b	3.2a See 3.1a, 3.1b
		3.2b Unknown interference with internal standard peak	3.2b As in 3.1c, or use different internal standard
4. Peak-height or peak-area calibrations do not intercept zero	4.1 No peak seen for very small sample	4.1a Desired component decomposing	4.1a Change system; run samples immediately
		4.1b Improper peak-size measurement	4.1b Repair detector; adjust or repair recorder, integrator, or computer
	4.2 Strange peak seen at or near retention time of component of interest	4.2a Peak overlap	4.2a Increase separation resolution; use peak-height instead of peak-area measurement; run sample "blanks" or "control"
	4.3 Peak(s) seen in control or "blank" run	4.3a Contaminated injector	4.3a Clean sample injector

Problem	Cause		Remedy
5. Variable analytical accuracy as function of concentration	5.1 Nonlinear calibration plot or calibration factor S (Chapter 13)	5.1a Nonlinear detector response	5.1a Reduce sample size; repair detector or data readout system
		5.1b Sample component or internal standard decomposing	5.1b Change system; change internal standard; run samples immediately
6. Analytical precision insufficient for need	6.1 Analytical variability too large, even though proper procedure followed (see 1-5)	6.1a Less precise calibration method used	6.1a Use peak-height method with frequent calibration; run replicate analysis and average; use automatic sample injector if required
		6.1b Lower precision with internal standard	6.1b Eliminate internal standard (if possible), follow 6.1a
		6.1c Equipment deficiencies	6.1c Repair or obtain better equipment
		6.1d Gradient elution used	6.1d Use isocratic elution, coupled-column method if k' range is great
7. Analytical speed insufficient for needs	7.1 Higher analysis speed required for routine analysis, process analysis	7.1a Low column plate count	7.1a Increase N by using column of smaller particles (constant L), then increase u by increasing P
		7.1b Low column selectivity	7.1b Increase α (by optimizing mobile phase/stationary phase combination), reduce column length and/or increase u by increasing P

Table 19.4 (Continued).

Effects	Situation or Symptom	Cause	Cure
Problems with assay-type analyses			
		7.1c Resolution of components too good	7.1c Decrease column length or increase u by increasing P
8. Difficult to determine which peak is to be measured	8.1 Irreproducible sample profiles, peak sequences, or retention times	8.1a Poor separation reproducibility	8.1a Repair defective equipment; obtain better equipment; thermostat equipment
		8.1b Column irreproducibility	8.1b Purge thoroughly with strong solvent, then fresh mobile phase; replace if degraded
		8.1c "Ghost" peaks present	8.1c Clean sample injector, check for leaks, thoroughly purge column with strong solvent; change septum or use sample valve
		8.1d Spurious peaks in calibration solutions due to sample decomposition	8.1d Make up fresh solutions
		8.1e Peak retention affected by large concentration of unknown (sometimes undetected) "garbage"	8.1e Use method of standard additions (Section 13.5) for quantitation

Problem	Cause	Sub-cause	Remedy
9. Variable analytical results	9.1 Changing calibration factor S	9.1a See 1.2	9.1a See 1.2

Unique problems with trace analyses (see preceding entries for general analytical problems)

Problem	Cause	Sub-cause	Remedy
10. Insufficient sensitivity for needs	10.1 Poor detectability for peak of interest	10.1a Insufficient detector response	10.1a Increase detector sensitivity; derivatize sample (Sections 17.3, 17.4)
		10.1b Improper detector	10.1b Change to more sensitive, selective detector
		10.1c Sample volume too small	10.1c Increase sample volume
		10.1d k' too large	10.1d Decrease k' by increasing strength of mobile phase (if R_s is still adequate); use bonded phase with shorter chain (C_8 instead of C_{18}); use lower surface area stationary phase
		10.1e Column too long (more resolution than needed)	10.1e Shorten columns to provide adequate resolution
		10.1f Value for α too small	10.1f Increase α by mobile phase/stationary phase change, decrease L and/or decrease k' (as in 10.1d)
		10.1g N too small	10.1g Increase N, preferably by decreasing d_p; use as short a column as required resolution allows

Table 19.4 (Continued).

Effect	Situation or Symptom	Cause	Cure
Unique problems with trace analyses (see preceding entries for general analytical problems)			
		10.1h Value for u too large	10.1h Decrease u to u_{opt} (or H_{min})
		10.1i Separation method not optimum	10.1i Change to method that elutes peak of interest rapidly, but with good selectivity (e.g., reverse-phase instead of LSC)
	10.2 Poor detection when sample volume is limiting	10.2a Column internal diameter is too large	10.2a Decrease column i.d.
11. Poor trace analysis accuracy or precision	11.1 Overlap of trace peak with interferences	11.1a Use of peak-area method	11.1a Use peak heights
		11.1b Insufficient resolution	11.1b Increase resolution (by increasing α if possible)
		11.1c Interference from spurious unknown peaks	11.1c Run "blank" or "control" samples; increase resolution if required; automate sample pretreatment (Section 17.5)
	11.2 Poor recoveries with sample pretreatment	11.2a Variable loss of component(s) of interest	11.2a Introduce appropriate internal standard into sample before workup
	11.3 Change in calibration S factor	11.3a See 1.2	11.3a See 1.2

Symptom	Possible Cause	Remedy
11.4 Baseline drift makes measurement of peaks difficult	11.4a Detector drift	11.4a Repair detector; clean cells
	11.4b Solvent contamination	11.4b Fill reservoir with proper fresh solvent, thoroughly purge column before use
	11.4c Broad peaks from previous runs are eluting during the analysis	11.4c Purge column with strong solvent, reequilibrate with mobile phase (see below)
	11.4d Flowrate change	11.4d Check pump; reestablish proper flowrate
	11.4e Gradient used at high detector sensitivity	11.4e Use isocratic separation, with column switching, if required
	11.4f Column "bleed"	11.4f Replace column; use pre-column; use different separation system
12. Analysis time too long		
12.1 As in 7.1	12.1a See 7.1	12.1a See 7.1
12.2 Components eluting long after peak of interest	12.2a Improper separation system	12.2a Change separation system (e.g., from LSC to reverse-phase), so that "garbage" peaks elute first.
	12.2b Simple isocratic system used	12.2b Increase strength of mobile phase after peak of interest elutes, to purge column of "garbage"; use column-switching technique, or column back-flushing

Table 19.4 (Continued).

Unique problems with trace analyses (see preceding entries for general analytical problems)

Effect	Situation or Symptom	Cause	Cure
		12.2c Sample too complex	12.2c Use preliminary clean-up step (e.g., size exclusion) to eliminate strongly retained components

aid to the reader in troubleshooting LC difficulties. Because many problems in quantitative analysis are the result of interrelated equipment, technique, and procedural factors, the hints in Table 19.4 are only a simplified guide; detailed discussion of these factors are presented in appropriate sections throughout this book. For example, Chapter 13 discusses quantitative methods in detail, including the preferred approaches for various types of analytical needs and the effect of various separation parameters on quantitative results. As with all analyses, best results are obtained when good laboratory techniques are observed.

REFERENCES

1. J. Q. Walker, M. T. Jackson, Jr., and J. B. Maynard, *Chromatographic Systems: Maintenance and Troubleshooting*, 2nd ed., Academic, New York, 1977.

2. N. Hadden et al., *Basic Liquid Chromatography*, Varian Aerograph, Walnut Creek, Calif., 1971, Chap. 11.

3. R. P. W. Scott, *Liquid Chromatography Detectors*, Elsevier, New York, 1977, Chap. 3.

3a. I. M. Johansson, K.-G. Wahlund, and G. Schill, *J. Chromatogr.*, 149, 281 (1978).

3b. J. Hermansson, *ibid.*, 152, 437 (1978).

4. K. Hostettmann and H. M. McNair, *ibid.*, 116, 201 (1976).

5. H. Sakurai and S. Ogawa, *J. Chromatogr. Sci.*, 14, 499 (1976).

6. A. G. Butterfield, D. W. Hughes, N. J. Pound, and W. L. Wilson, *Antimicrob. Agents Chemother.*, 4, 11 (1973).

7. M. S. Stoll, C. K. Lim, and C. H. Gray, in *High Pressure Liquid Chromatography in Clinical Chemistry*, P. F. Dixon et al., eds., Academic, New York, 1976, p. 97.

8. E. P. Horwitz and C. A. A. Bloomquist, *J. Chromatogr. Sci.*, 12, 200 (1974).

9. J. Christie, M. W. White, and J. M. Wiles, *J. Chromatogr.*, 120, 496 (1976).

10. K. Slais and M. Krejci, *ibid.*, 91, 161 (1974).

APPENDIX I

SUPPLIERS OF LC EQUIPMENT, ACCESSORIES, AND COLUMNS

The following table gives some suppliers of modern LC instruments, accessories, and columns. Although the manufacturers and suppliers listed are primarily located in the United States, some are also given for other countries. Many major manufacturers have marketing outlets in different parts of the world; for brevity only one listing is given. This table is presented as a convenient reference, but is by no means comprehensive. For more complete and up-to-date listings, the reader is referred to the most recent edition of laboratory buyer's guides (e.g., those of *Analytical Chemistry* and *American Laboratory*).

Supplier	Complete LC instruments, pumps, detectors, etc.	Accessories, including valves, fittings	Columns, column packings
Ace Glass 1420 N.W. Blvd. Vineland, N.J. 08360, U.S.		X	
Alltech Associates, Inc. 202 Campus Drive Arlington Heights, IL 60004, U.S.		X	
Altex Scientific, Inc. 1780 4th Street Berkeley, CA 94710, U.S. (Now Division of Beckman Instruments)	X	X	X
American Instrument Co. 8030 Georgia Avenue Silver Spring, MD 20910, U.S.	X	X	

Supplier	Complete LC instruments, pumps, detectors, etc.	Accessories, including valves, fittings	Columns, column packings
Amicon Corp. Scientific Systems Div. 21 Hartwell Ave. Lexington, MA 021073, U.S.		X	X
Analabs, Inc. P.O. Box 501 North Haven, CT 06473, U.S.		X	X
The Anspec Co., Inc. P.O. Box 7044 Ann Arbor, MI 48103, U.S.	X	X	X
Antek Instruments, Inc. 6005 N. Freeway Houston, TX 77076, U.S.	X		
Applied Chromatography Systems, Ltd. Concord House, Concord St., Luton, Beds., England			X
Applied Science Laboratories, Inc. P.O. Box 440 State College PA 16800, U.S.		X	X
Baird-Atomic, Inc. 125 Middlesex Turnpike Bedford, MA 01730, U.S.	X		
Baird & Tatlock (London) Ltd. P.O. Box 1, Romford, Essex RMl 1HA, London, England	X		
Beckman Instruments, Inc. 2500 Harbor Blvd. Fullerton, CA 92634 U.S.	X	X	X
Bioanalytical Systems, Inc. P.O. Box 2206 W. Lafayette, IN 47906, U.S.	X	X	X
Bio-Rad Laboratories 2200 Wright Avenue Richmond, CA 94804, U.S.	X	X	X
Buchler Instruments 1327 16th Street Fort Lee, NJ 07024	X		

Supplier	Complete LC instruments, pumps, detectors, etc.	Accessories, including valves, fittings	Columns, column packings
Burdick & Jackson Labs, Inc. 1953 S. Harvey St. Muskegon, MI 49442, U.S.		X	
Carlo Erba Scientific Instruments P.O. Box 4342 20100 Milan, Italy	X		
Cecil Instruments Ltd. Trinity Hall Industrial Estate Green End Road Cambridge, Great Britain	X	X	
Chromatography Services Ltd. Carr Lane Industrial Estate Hoylake, Wirral, Merseyside, England			X
Chrompak, Nederland B.V. P.O. Box 3 Middelburg, The Netherlands			X
C. Desaga G.m.b.H., Nachf. Erich Fecht MaaBstr. 26-28, P.O. Box 101969, 6900 Heidelberg 1, West Germany			X
Dionex Corp. 1228 Titan Way Sunnyvale, CA 94086, U.S.	X		
E. I. du Pont de Nemours and Co. Instrument Products Div. Wilmington, DE 19898, U.S.	X	X	X
Durrum Instrument Corp. 1228 Titan Way Sunnyvale, CA 94086, U.S.	X	X	X
Electro-Nucleonics Inc. 368 Passaic Avenue Fairfield, NJ 07006, U.S.			X
EM Products 2909 Highland Ave. Cincinnati, OH 45212, U.S.			X

Supplier	Complete LC instruments, pumps, detectors, etc.	Accessories, including valves, fittings	Columns, column packings
Farrand Optical Co., Inc. 117 Wall Street Valhalla, NY 10595	X		
Finnigan Instruments 845 W. Maude Avenue Sunnyvale, CA 94086, U.S.		X	
Fischer & Porter Co. (Lab Crest) Country Line Road Warminster, PA 18974, U.S.		X	
Gilson Medical Electronics Inc. P.O. Box 27 Middletown, WI 53562, U.S.	X	X	
Glenco Scientific Inc. 2802 White Oak Houston, TX 77007, U.S.	X	X	X
Gow-Mac Instrument Co. P.O. Box 32 Bound Brook, NJ 08805, U.S.	X	X	
Hamilton Company P.O. Box 17500 Reno, NV 89510, U.S.		X	X
HETP 34 Gonville Ave., Sutton, Macclesfield, Cheshire, SK 110EG, England		X	X
Hewlett-Packard Avondale, PA 19311, U.S.	X	X	X
ICN-Paris-Labo 49 Rue Defrance 94300 Vincennes, France			X
Instrumentation Specialties Co. (ISCO), 4700 Superior Ave. Lincoln, NB 68504, U.S.	X	X	X
Jasco, Inc. 218 Bay Street Easton, MD 21601, U.S.	X		

Supplier	Complete LC instruments, pumps, detectors, etc.	Accessories, including valves, fittings	Columns, column packings
Jobin-Yvon 16-18 Rue Du Canal 91160 Longjumeau, France	X		
Johns-Manville Ken Caryl Ranch Denver, CO 80217, U.S.			X
J-Y Optical Systems 20 Highland Avenue Metuchen, NJ 08840, U.S.	X		
Kipp Analytica, B.V. Phileas Foggstraat 24, 7821 Ak Emmen P.O. Box 620 Emmen, The Netherlands	X	X	X
Kipp and Zonen 390 Central Avenue Bohemia, NY 11716	X		
KG Dr. Knauer Strasse 635 1000 Berlin 37, West Germany	X	X	X
Koch-Light Laboratories, Ltd. 2 Willow Road Colnbrook SL3 OBZ Bucks., England	X	X	X
Laboratory Data Control P.O. Box 10235 Interstate Industrial Park Riviera Beach, FL 33404, U.S.	X	X	X
Lachat Chemicals Inc. 10,500 N. Port Washington Rd. Mequon, WI 53092, U.S.	X		X
L. C. Co., Inc. 619 Estes Avenue Schaumburg, IL 60193, U.S.			X
Lion Technology, Inc. P.O. Drawer 898 Dover, NJ 07801, U.S.			X

Supplier	Complete LC instruments, pumps, detectors, etc.	Accessories, including valves, fittings	Columns, column packings
LKB Instruments, Inc. 12221 Parklawn Drive Rockville, MD 20852, U.S.	X	X	
Machery-Nagel & Co. 5160 Duren Werkstrasse 6-8, West Germany			X
Mandel Scientific 395 Norman Street Montreal, Canada	X	X	X
E. Merck Darmstadt, West Germany			X
Metrohm Ltd. CH-9100 Herisau, Switzerland	X		
Micromeritics Instrument Co. 5680 Goshen Springs Road Norcross, GA 30093, U.S.	X	X	X
Millipore Ashby Road Bedford, MA 01730, U.S.		X	
Negretti & Zambra Hawthorn Road Willesden Green London, England		X	
New England Nuclear 549 Albany St. Boston, MA 02118, U.S.	X		
Optilab P.O. Box 138 S. 16212 Vallingley, Sweden	X		
Orlita KG Max-Eyth-Strasse 10, 63 Giessen, West Germany	X		
Packard-Becker BV P.O. Box 519 Vulcanusweg 259 Delft, The Netherlands	X	X	

Supplier	Complete LC instruments, pumps, detectors, etc.	Accessories, including valves, fittings	Columns, column packings
Packard Instrument Co. 2200 Warrenville Road Downers Grove, IL 60615, U.S.	X		
Perkin-Elmer Co. Main Avenue Norwalk, CO 06856, U.S.	X	X	X
Pharmacia Fine Chemicals 800 Centennial Avenue Piscataway, NJ 08854, U.S.			X
Phillips Electronic Instruments Inc. 750 S. Fulton Ave. Mount Vernon, NY 10550, U.S.	X		
Pierce Chemical Co. P.O. Box 117 Rockford, IL 61105, U.S.		X	X
Princeton Applied Research Corp. P.O. Box 2565 Princeton, NJ 08540, U.S.	X		
Pye Unicam Ltd. York Street Cambridge, Great Britain	X		
Rainin Instrument 94-T Lincoln Street Brighton, MA 02135, U.S.	X	X	X
Regis Chemical Co. 8210 N. Austin Ave. Morton Grove, IL 60053, U.S.			X
Rheodyne 2809 10th Street Berkeley, CA 94710, U.S.		X	X
Schoeffel Instrument Corp. 24 Booker Street Westwood, NJ 07675, U.S.	X		
Scientific Glass Eng. Pty., Ltd. 111 Arden St. N. Melbourne Victoria, Australia 3051		X	X

Supplier	Complete LC instruments, pumps, detectors, etc.	Accessories, including valves, fittings	Columns, column packings
Separations Group 8738 Oakwood Avenue Hesperia, CA 92345, U.S.			X
Shimadzu Scientific Instruments 9147 Red Branch Road Columbia, MD 21045, U.S.	X	X	X
Siemens Aktiengesellschaft Geschaftsbereich MeB-und ProzeBtechnik, Abt. AnalysenmeBtechnik, E 63, Postfach 211080 D 7500 Karlsruhe 21, West Germany	X	X	X
Spectra-Physics 2905 Stender Way Santa Clara, CA 95951, U.S.	X	X	X
Supelco Inc. Supelco Park Bellefonte, PA 16823, U.S.		X	X
Technicon Industrial Systems 511 Benedict Avenue Tarrytown, NY 10591, U.S.	X	X	X
Tracor, Inc. 6500 Tracor Lane Austin, TX 78721, U.S.	X	X	X
Unimetrics Corp. 1853 Raymond Ave Anaheim, CA 92801, U.S.		X	
Varian Associates 611 Hansen Way Palo Alto, CA 94303, U.S.	X		X
Valco Instruments Co., Inc. 7812 Bobbitt La. Houston, TX 77055 U.S.		X	
Waters Assoc., Inc. Maple Street Milford, MA 01757, U.S.	X	X	X

Supplier	Complete LC instruments, pumps, detectors, etc.	Accessories, including valves, fittings	Columns column packings
Whatman Inc. 9 Bridewell Place Clifton, NJ 07014, U.S.		X	X
Wilks Scientific Corp 140 Water St. S. Norwalk, CT 06856	X		
M. Woelm Adsorbenzien Abteilung 344 Eschwege, West Germany			X
Carl Zeiss, Inc. 444 Fifth Avenue New York, NY 10018, U.S.	X	X	

APPENDIX II

MISCELLANEOUS TABLES USED BY WORKERS IN LC

The following tabulations have been selected for their value in assisting certain more-or-less common calculations or decisions required in different aspects of LC. A brief discussion is given of the potential application of each table.

Table II.1 The Gaussian or Error Function

The Gaussian function describes the shape of normal LC bands. If the band is plotted in x-y coordinates, then the mathematical description of a *band of unit area* is given as

$$y = \frac{1}{\sqrt{2\pi}} e^{-x^2/2}. \tag{II.1}$$

In terms of time t (sec) for the x axis, x is replaced by (t/σ_t), where t is measured from the band center ($t = 0$) and σ_t is the standard deviation of the band in seconds. The Gaussian curve is plotted in Figure II.1 along with some important relationships of this function.

The Gaussian function can be used to synthesize partially resolved LC bands, as in Figures 2.11-16. In some cases more detailed information on such overlapping bands is required for a particular application. For example, in very-high-precision assays involving R_s values lower than 1.5, one might be concerned with the exact error of either band-area or peak-height measurements. Such calculations are easily made with the data of Table II.1, for varying ratios of the two over-lapping compounds.

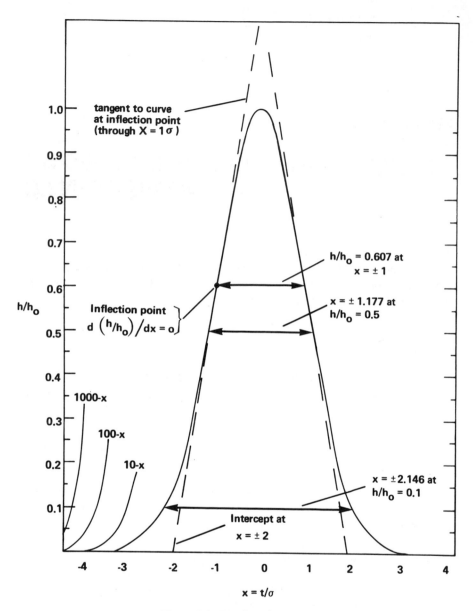

Figure II.1 The Gaussian curve.

Table II.1 The Gaussian function.

$x = t/\sigma_t$	Area (band center to x)	y (band height at x)	y/y_0 (relative band height)
0.0	0.000	0.399	(1.000)
±0.1	0.040	0.397	0.995
0.2	0.079	0.391	0.980
0.3	0.118	0.381	0.956
0.4	0.155	0.368	0.923
0.5	0.191	0.352	0.883
0.6	0.226	0.333	0.835
0.7	0.258	0.312	0.783
0.8	0.288	0.290	0.726
0.9	0.316	0.266	0.667
1.0	0.341	0.242	0.607
1.1	0.364	0.218	0.546
1.2	0.385	0.194	0.487
1.3	0.403	0.171	0.430
1.4	0.419	0.150	0.375
1.5	0.433	0.130	0.325
1.6	0.445	0.111	0.278
1.7	0.455	0.094	0.236
1.8	0.464	0.079	0.198
1.9	0.471	0.066	0.164
2.0	0.477	0.054	0.135
2.2	0.486	0.036	0.084
2.4	0.492	0.022	0.056
2.6	0.496	0.014	0.034
2.8	0.497	0.008	0.020
3.0	0.499	0.004	0.011
3.2	0.499	0.002	0.006
3.4	0.500	0.001	0.003
3.6	0.500	0.0006	0.0015
3.8	0.500	0.0003	0.0007
4.0	0.500	0.0001	0.0003

Table II-2   Reduced Plate-Height Data for "Good" Columns

We have discussed in Section 5.7 the concept that reduced plate-height/velocity plots for "good" columns are independent of particle size d_p. This allows us to estimate the performance of a "good" column as a function of experimental conditions. The important relationships are

$$\text{reduced plate height,} \quad h = \frac{H}{d_p} \qquad (\text{II.2})$$

$$\text{reduced velocity,} \quad \nu = \frac{u d_p}{D_m} \qquad (\text{II.3})$$

$$\text{column plate number,} \quad N = \frac{L}{H}. \qquad (\text{II.4})$$

Reduced plate heights h for "good" columns can be estimated from the Knox equations:

$$\text{totally porous particles,} \quad h = (2/\nu) + \nu^{0.33} + 0.05\nu \qquad (\text{II.5})$$

$$\text{pellicular particles,} \quad h = (2/\nu) + \nu^{0.33} + 0.003\nu \qquad (\text{II.6})$$

Values of h versus ν from Eqs. II.5 and II.6 are tabulated in Table II.2.

The data of Table II.2 were used in calculations for Tables 2.2-2.7 and 5.25-5.27. They allow the calculation of the best possible LC efficiencies as a function of various parameters, as well as the change in efficiency with change in one or more experimental conditions. Values of N for many commercial columns are about half as large as those for these "good" columns.

Table II.3   Viscosity of Solvent Mixtures

The viscosity η of solvent mixtures is often of interest in LC. Table II.3 gives values of η versus %v composition for two common solvent mixtures that have anomalous values of η: methanol/water and acetonitrile/water. The viscosity of "normal" solvent binaries is given approximately by the relationship (1)

$$\eta = (\eta_a)^{X_a} (\eta_b)^{X_b}. \qquad (\text{II.7})$$

Here η_a and η_b refer to viscosities of pure solvent components A and B, and

Table II.2 Values of h versus v for "good" columns.

Reduced velocity, v	Values of h	
	Porous	Pellicular
0.1	20.5	20.6
0.3	7.35	7.33
1.0	3.05	3.00
2.0	2.36	2.27
3.0	2.25	2.11
5.0	2.35	2.12
7.5	2.59	2.23
10	2.84	2.37
20	3.79	2.85
30	4.57	3.16
50	6.18	3.83
75	7.93	4.41
100	9.59	4.89
200	15.8	6.36
300	21.6	7.47
500	32.8	9.28
750	46.4	11.1
1000	59.8	12.8

X_a and X_b refer to their respective mole fractions:

$$X_a = \frac{\phi_a(d_a/M_a)}{\phi_a(d_a/M_a) + \phi_b(d_b/M_b)}, \qquad (\text{II.8})$$

$$X_b = 1 - X_a. \qquad (\text{II.8a})$$

The quantities d_a and d_b refer to the densities of pure solvents A and B (Table 6.1), and M_a and M_b refer to their molecular weights. The terms ϕ_a and ϕ_b are the volume fractions of A and B.

Solvent viscosity values are of use mainly in determining column pressure drop, by means of Eqs. 2.12 or 2.12a. Table II.3 also lists estimated values of D_m for these different mixtures, assuming the molar volume V_a of the solute is 300 ml. The latter values are useful in connection with Eq. II.3. Values of D_m for others solutes are equal to these values times $(300/V_a)$.

Table II.3   Viscosities of solvent binaries: methanol/water and acetonitrile/water.

	Methanol		Acetonitrile	
%v Organic-water	η^a	$D_m \times 10^{5\ b}$	η^a	$D_m \times 10^{5\ b}$
0	0.89	0.55	0.89	0.55
20	1.26	0.43	0.91	0.58
40	1.42	0.37	0.98	0.54
60	1.40	0.38	0.76	0.68
80	1.01	0.53	0.58	0.79
100	0.57	0.95	0.43	1.07

Source: From (2).
[a] 25°C, values in centipoise; for η values at other temperatures, see (2a).
[b] V_a = 300 ml.

Table II.4   Particle Size Expressed as Mesh Size

The particle size d_p of LC packings is commonly measured in terms of micrometers. Larger particles are sometimes expressed in terms of "mesh size" (e.g., 200-400 mesh). The relationship between mesh size and d_p in micrometers is given in Table II.4, for the commonly used Tyler screens. Thus, 400 mesh refers to particles that have passed through a 200-mesh screen (74 μm) and been retained on a 400-mesh screen (37 μm), and therefore are nominally within the 37-74-μm range for d_p. For a brief further discussion of particle screening and mesh sizes, see (3).

Table II.4   Mesh size versus particle diameter d_p that is retained on the screen.

Screen mesh number	Retains particles with d_p larger than:
20	833 μm
28	589
35	417
48	295
65	208
100	147
150	104
200	74
270	53
400	37

REFERENCES

1. W. R. Gambill, *Chem. Eng.*, 151 (1959).
2. L. R. Snyder, *J. Chromatogr. Sci.*, 15, 441 (1977).
2a. H. Colin, J. C. Diez-masa, G. Guiochon, T Czajkowska and I. Miedziak, *J. Chromatogr.*, 167, 41 (1978).
3. B. L. Karger, L. R. Snyder, and C. Horvath, *An Introduction to Separation Science*, Wiley, New York, 1973, pp. 543-545.

LIST OF SYMBOLS

The following list is divided into commonly used and less commonly used symbols, for convenience in referring to the symbols cited in this book. Units for each symbol are given in parentheses. Letters $(A, B, C, \ldots, x, y, z)$ are variously used to describe solvents, sample compounds, and so on, and the meaning of these particular symbols is generally clear from the context of the pages where they appear. One possible exception is the general use of A and B to refer to the weak (A) and strong (B) components of a mobile-phase binary mixture.

COMMON SYMBOLS

d	Tubing inside diameter (cm); Eq. 3.1.
d_p	Particle diameter of column packing material (generally cm in various equations, μm elsewhere).
d_c	Column internal diameter (cm).
D_m, D_s	Solute diffusion coefficient in mobile and stationary phases, respectively (cm^2/sec).
F	Mobile-phase flowrate (ml/sec, unless otherwise specified)
h	Reduced plate height, equals H/d_p (dimensionless)
h_r, h_1, h_2	Heights of valley between two bands, band 1 and band 2, respectively (Figure 2.19)
h'	Height of a band (Figure 13.1)
H	Height equivalent of a theoretical plate, equal to L/N (cm)

k'	Solute capacity factor; equal to total amount of solute in stationary phase divided by total amount of solute in the mobile phase within the column, at equilibrium (Section 2.2)
K	Solute distribution coefficient (Eq. 2.5).
K_0	Distribution constant in SEC, equal to $(V_R - V_0)/V_i$; Eq. 12.1.
K^0	Column permeability (cgs units); Eq. 2.12.
L	Column length (cm).
L_m	Molecular length (Å); Eq. 12.8 for n-alkanes.
n	Refractive index (Table 6.1).
N	Column plate number; Eq. 2.6 (also Eqs. 5.8-5.9a).
P	Pressure drop across column (usually in psi, but dynes/cm^2 for equations involving P, unless otherwise specified) (1 atm = 14.7 psi = 1.01×10^6 dynes/cm^2).
P'	Solvent polarity parameter (dimensionless); Section 6.4 and Table 6.1.
P_a, P_b, P_c, P_w	Values of P' for pure solvents, A, B, C, and water.
R_s	Resolution function; Eqs. 2.9, 2.10 and Fig. 2.9.
R_1, R_2	Values of R_s in (1) initial and (2) final separations.
pK_a	Acid dissociation constant, defined by Eq. 10.10.
t	Time required for completion of separation, equal (approximately) to t_R for last eluted band (sec); Eq. 2.11; also, time after sample injection (sec).
t_R	Retention time of a given band (sec); Figure 2.4.
t_w	Baseline band width in time units (sec); Figure 2.4.
t_0	t_R value for mobile-phase molecules injected as sample; also (except in SEC) t_R for unretained sample molecules.
u	Mobile-phase velocity (cm/sec).
V_m	Total volume of mobile phase within column (ml); equal to $t_0 F$.
V_R	Solute retention volume (ml); Eq. 2.4.
α	Separation factor; Eq. 2.10; equal to k' for second band divided by k' for first band.
ϵ	Dielectric constant; Table 6.1.
ϵ^0	Solvent strength parameter in LSC; Section 9.3.
η	Solvent viscosity (poise in equations) (centipoise in tables and in Eqs. 5.16, 5.3a).

$\theta_{0.1}$	Column linear capacity; weight of sample/weight of column packing, such that k' decreases by 10% from its constant value at small sample sizes; Figure 2.6.
ϕ	Flow-resistance factor; Eq. 2.12.
ϕ_a, ϕ_b, etc.	Volume fraction of component A, B, etc., of solvent binary.
ν	Reduced velocity, equal to ud_p/D_m
ν_1	Value of ν for initial separation; Table 2.3.

LESS COMMON SYMBOLS

A	Absorbance; Eq. 4.3.
$A, B, C \ldots$	See comment at beginning of List of Symbols.
$C_d, C_e, C_m, C_s, C_{sm}$	Plate height coefficients; Section 5.1 and Table 5.1.
$C_{\max}, C_0$	The concentration of a sample component at the band maximum and in the injected sample, respectively; Eq. 13.2.
d_f	Thickness of stationary-phase layer (cm).
E	Ion-pair extraction constant; Eq. 11.2.
f	Column-packing parameter; Eq. 2.12a.
I, I_0	Transmitted and initial light intensity, respectively; Eq. 4.2.
k_1, k_2	Values of k' for bands 1 and 2.
k_t	k' value at some time t during gradient elution for a given compound; Figure 16.1.
$M_{\text{peak}}, \overline{M}_w, \overline{M}_n$	Sample "average" molecular weights: respectively, for molecules eluting at band "peak" in SEC, weight-averaged and number-averaged.
n_s, n_m	Total moles of solute present in stationary phase (s) and mobile phase (m), respectively.
N_{eff}	Column effective plate number; Eq. 2.13; value of N corrected for effect of k' on resolution.
pK_1, pK_2	pK_a values for a given compound referring to first proton dissociation (1), second proton dissociation (2), and so on.
R	Fraction of total sample molecules in the mobile phase, equal to $1/(1 + k')$.

R_f	In TLC the fractional movement of a solute band, relative to the distance moved by the solvent front; ideally, $R_f = R$.
R', R^-	A charge-bearing sample ion (Chapter 6) or a charged functional group attached to an ion-exchange packing (Chapter 10).
S	Solvent or mobile-phase molecule; Figure 2.2.
S	Calibration factor for quantitative analysis; Section 13.3.
t_1, t_2, t_X	Values of t_R for bands 1, 2, X, and so on (sec).
t_{w1}, t_{w2}	Values of t_w for bands 1 and 2 (sec); Figure 2.9.
$t_{w\frac{1}{2}}$	Value of t_w at band half-height (sec).
u_x	Velocity of sample molecules as they move through the column (cm/sec); Eq. 2.1.
V_c	Volume of a detector cell (ml).
V_{ec}	Extracolumn band broadening (ml, baseline) that would occur if the column were removed from the LC system; width of resulting (columnless) band t_w F.
V_0, V_i	Volume of mobile phase (within column), respectively, outside and inside of SEC particles; Figure 12.2.
V_p	Baseline bandwidth of a compound in the absence of extracolumn effects (ml).
V_s	Volume of sample injected; also, volume of stationary phase within column.
V_w	Observed baseline bandwidth of a peak, including extracolumn effects (ml); equal to t_w F.
x	Peak-height ratio for two overlapping bands; Section 2.5, discussion of how large R_s should be.
X	In Figure 7.2, a $-Cl$ or $-OCH_3$ group.
$(x)_i$	Concentration of component x in phase i (moles/l).
α_i	A group-increment-factor; addition of a substituent-group i to a molecule in a given LC system will change k' for the parent compound by the factor α_i.
δ	Hildebrand solubility parameter; Section 6.4.
E	Molar absorptivity; Eq. 4.2.
λ_{max}	Wavelength of maximum light absorption for a given compound.

LIST OF ABBREVIATIONS

AAX	Anion-exchange-resin designation.
alc.	Alcoholic
API	Atmospheric pressure ionization.
aq	Aqueous (usually subscript).
AUFS	Absorption-units full-scale; i.e. a full-scale recorder deflection is equivalent to the AUFS value in o.d. units.
BHC	Benzene hexachloride (a germicide).
BHT	Butylated hydroxytoluene (an antioxidant).
BOP	Oxydipropionitrile (a liquid stationary phase).
BPC	Bonded-phase chromatography (Chapter 7).
CI	Chemical ionization
CK, CPK	Creatine kinase (an enzyme).
CPG	Controlled-pore glass (a porous silica).
C_1, C_2, C_8, C_{18}	Refers to various reverse-phase packings, where the substrate is substituted with $-O-Si(CH_3)_2C_n$ groups (n refers to the carbon number of the longest group, equal 1, 2, 8, or 18).
DEAE	Diethylaminoethyl; $-C_2-N(C_2H_5)_2$; a common ion-exchange functional group for anion exchange (under conditions where the DEAE group is protonated).
DMSO	Dimethylsulfoxide
ec	Extracolumn.
EC	Electrochemical (pertaining to detectors).
EDTA	Ethylenediamine tetraacetic acid (a metal chelator).
EO	Ethylene oxide unit
FTIR	Fourier Transform infra-red
GC	Gas chromatography.

GC-MS	Gas chromatography-mass spectrometry.
GE	Gradient elution.
GFC	Gel filtration chromatography
GPC	Gel permeation chromatography.
HCB	Hexachlorobenzene.
HPSEC	High-performance SEC
IC	Ion chromatography (Section 10.6).
i.d.	Internal diameter.
IEC	Ion-exchange chromatography (Chapter 10).
IPC	Ion-pair chromatography (Chapter 11).
IR	Infra-red
LC	Liquid chromatography
LDC	Laboratory Data Control (a company).
LDH	Lactate dehydrogenase (an enzyme).
LLC	Liquid-liquid chromatography (Chapter 8).
LSC	Liquid-solid chromatography (Chapter 9).
LSS	Linear-solvent-strength (Section 16.1, Eq. 16.1).
MS	Mass-spectrometry
MW	Molecular weight
ODS	Octadecylsilyl; equivalent to a C_{18} reverse-phase packing.
org	Organic (usually subscript), referring to the nonaqueous phase.
PIC	Paired-ion chromatography (a Waters Associates trademark), equivalent to IPC.
PS-DVB	Polystyrene-divinylbenzene; a polymer used for certain SEC packings.
psi	Pounds per square inch.
PTFE	Polytetrafluoroethylene (Teflon).
RI	Refractive index.
RP	Reverse-phase.
RSMN	Rapid-scanning multiwavelength (Section 4.3); a type of LC detector.
SAS	Short-alkyl-silyl; a type of reverse-phase packing equivalent to C_1.
SAX	Strong anion exchanger.
SCX	Strong cation exchanger.
SEC	Size-exclusion chromatography (Chapter 12).

TBA^+	Tetrabutylammonium ion.
TBA, TEA	Tributylamine, triethylamine
THF	Tetrahydrofuran.
TLC	Thin-layer chromatography (Section 9.3).
TMA, TMA^+	Tetramethylammonium ion.
Tris	Tris(hydroxymethyl)aminomethane; a buffer
UV	Ultraviolet; however, broadly used in LC to include visible measurement as well
ϕ	A phenyl group; $-C_6H_5$
2,4-DNP	2,4-dinitro phenyl

INDEX

Abate^{TM}, 342
Abbreviations, 844-6
Absorbance, 131
Accuracy, 542
  peak height versus area, 545
Acid, carboxylic, 779
  separation by LSC, 362-3
Acidity constants (pK$_a$), 425
Adhesives, hot-melt, 769
Adsorbents (for LSC), 184-5, 196, 200-1,
    361-5. *See also* Column packings;
    Liquid-solid chromatography, *individual*
    materials e.g., Silica
Adsorption chromatography, *see* Liquid-sol-
    id chromatography
Adsorption (in SEC), anion effect, 502
  polymer modifier, 503
  prevention, 489, 501
  silanization, 525
  unmodified silica, 502
Aflatoxins, 135, 147, 352, 606
Alcohols, derivatization of, 733
Aldehydes, derivatization of, 733, 741
Alkali metals, 428
Alkaline phosphatase, 437
Alkaloids, 400
Alkanes, detector response, 150
Alumina, 362-3
Amines, 779
  derivatization of, 733-4, 741
Amino acids, 437, 777
  derivatives, 311, 777
  derivatization of, 734, 741
  PTH-derivatives, 311

Aminophenols, 798
Amperometric detectors, *see* Electrochemi-
    cal detector
Analysis, 542-612
  trace, 560-72, 778
Anilines, 645
Anthraquinones, 53, 678
Antiarrythmia drugs, 475, 478
Anticonvulsants, 305
Antioxidants, 342, 704-5
Antipyrine, 433
Apparatus, *see* Equipment
Aromatic hydrocarbons, *see* Hydrocarbons
Artifacts, separation, 791-813
Arylhydroxylamines, 731
Association, molecular, 524
Association factor, 237
Asthma tablet, 412
Asymmetric peaks, 222, 295, 791-807
Asymmetry factor, 222, 224
Automated method development, 80-1
Automation, for sample pretreatment, 746-
    7
8-Azaprostaglandins, 654

Back-extraction, 728
Backflushing, 226, 709-10
Balanced-density slurry packing, 208
Band, 20-1
  baseline width, 21
  half-height width, 222, 834
  identification of, 575-614
  shape, 220
  tailing, 222, 295, 791-807

volume, 87
Band area, *see* Peak
  error in, 46-7
Band broadening, 27-34
  contributions, 169
  extra-column, 31-3, 86-8, 114, 130, 140,
    289-92
  processes, 169
  in SEC, 508-9
  see also Column efficiency; Plate number N
Band migration, 16-21
  in gradient elution, 663-5
Band width, 21
  in SEC, 508
Baseline interpolation, 545
Bases, separation in LSC, 362-3
Baths, temperature controlling, 118
B-concentration effect, 370-2
Bed consolidation, 207, 213
Beer's law, 130
Benomyl, 436, 596-7
Benzodiazepams, 431
Benzologs, 271
  retention versus n, 579-83
  separation of, 355-7, 359
Benzopyrenes, 777
Bergapten, 406
Bimodal pore-size, 494
Biogel P-2, 488
Blanks, sample, 551
Blood (serum), 303
Blowdown water, 443
Boiling point, of solvents, 248-50
μ-Bondagel, 489
Bondagel-E, 492
Bonded-phase chromatography (BPC), 22,
  269-322
  applications, 301-16
  band-tailing, 295, 792, 799-801
  column packings, 186, 188, 197, 272-
    80
  column selection, 316
  gradient elution, 674, 684
  ion-suppression, 294
  mobile phases, 281-9, 317
  operating variables, 319
  problems, 294-301
  retention mechanism, 281
  retention prediction, 587-8
  sample size, 289-92

selectivity, 271
  solvent selectivity, 283
  solvent strength, 283
  solvent viscosity effects, 293
  temperature effects, 293
  TLC scouting, 389
  see also Normal-phase BPC; Reverse-phase
    BPC
BPC, *see* Bonded-phase chromatography
BPC packings, 186, 188, 197, 272-80
  bulk volume effects, 278
  chain-length effects, 278, 280
  characterization of, 278
  covalent, 273
  deterioration of, 297
  hydrophilic, 277
  hydrophobic, 277
  in-situ preparation, 277
  lifetime, 296
  packing problems, 214
  polymerization of, 275
  pore effects, 279
  preparation of, 276
  preparative reactions, 272
  reaction stoichiometry, 274
  regeneration of, 297
  sample size effects, 280
  silicate esters, 273
  siloxane-type, 273
  stability, 281, 296
  surface concentration, 278
  suspending liquids for, 210
  temperature effects, 281
  testing, 278
  variability, 294
  wettability, 298
Bread extract, 725
Broad range samples, 494, 663
Bromoseltzer^{TM}, 432
Brushes, 273
Bubbles, in detector, 89
  elimination of, 130
Buffers, 425
  effect on tailing, 804
Bull shit, *see* Manure, bovine
Butter, saponified, 136

Calculations, quantitative, 550, 571
Calibration factor, 549
Calibration frequency, 551

Calibration methods, 549-60
  external standard, 549
  internal standard, 552
  non-linear, 551
  peak normalization, 552
  selection of, 556
  spiking, 571
  standard addition, 571
Calibration mixtures, 551
Calibration plot, 550-1
  in SEC, 489-99
Cannabinoids, reaction detector for, 741
Cannabis-resin, 777
Capacity factor, 23-5
  effect on R_s, 36, 51-6
  liquid-loading effect, 331
  optimum value of, 49-54
  preparative separation, 644
  temperature effect on, 293
Capping, 277
Carbamates, 653, 738
Carbohydrate polymer, 510
Carbohydrates, 777
  reaction detector for, 741
Carbon-paste electrode, 154
Carbonyls, 342
  derivatization of, 733, 741
Carbowax, 503
Carboxylic acids, 666-7
  derivatization of, 733, 741
Carboxymethyl cellulose, 531
Carrier, see Mobile phase; Solvent
Cassia oil, 396
Catechol amines, reaction detector for, 742
Cathode-ray tube, 121
Cells, see Detector cell
Cellulose hemi-formal, 8
Cellulose hydrolysate, 307
Chemical effects, and resolution, 79
Chiral compounds, see Optical isomers
Cholesterol, 138
Cholesteryl ester, 646
Chromatogram, 3, 21
  caption format, 12
Chromatography, see Gas chromatography;
    Liquid chromatography; etc.
Chitosan, 511
Classical LC, 4-5
Cleanup, of sample, 722-31
  by SEC, 533

Clinical LC, 778
Coating methods, for LLC, 328
Collection port, 594
Colloids, 535
Color reactions, for qualitative analysis, 593,
    601-2. See also Derivatization
Column, 168-245
  backflushing, 226, 709-10
  bed consolidation, 207, 213
  bed stability, 216
  blanks, 203, 212, 625
  carbohydrate, 777
  for column switching, 710-17
  compression, 625
  connecting of, 204, 227, 696
  cost, 172, 203
  deactivation, 486
  degradation, 226
  diameter effects, 205, 624
  end-fittings, 204
  equilibration in LSC, 378-80
  evaluation of, 218
  flexible wall, 625
  geometry, 204
  glass, 203
  glass-lined, 203
  "good," 225, 234-40, 836-7
  guard, 182, 228
  handling of, 225
  hardware, 203
  infinite-diameter, 230
  limiting performance, 240-3
  linear capacity, see Linear capacity
  longitudinal compression, 626
  maximum length, 215
  optimization, 240
  overloading of, 26-7, 181, 292, 617, 791-
    5
  pellicular versus porous, 182
  performance specifications, 219, 225
  permeability, 37, 219, 221, 240, 293
  pH range, 226
  plugged, 112, 227
  poorly packed, 170, 803
  pre-, 227, 337
  preparative, 624, 626, 631, 634
  radially compressed, 625
  recommended practice, 226
  regeneration, 691-2
  repairing, 227

replacing, 225
specifications, 218-25
stability, 227, 760
storage, 226
suggested, 761
tapered, 625
terminators, 204
testing, 220
thermostatting, 118
troubleshooting, 226
voids, 226-7
wall effects, 232
wall smoothness, 203
see also Column packings
Column efficiency, 28
  injection effects, 86, 112
  internal diameter effect, 624
  pump pulsation effect, 117
  see also Plate height H; Plate number N
Column length, 204
  effect on R_s, 56-8, 60, 65-71
  maximum, 215
Column-packing methods, 202-18
  balanced-density, 208, 215, 217
  bed consolidation, 213
  column length effect, 215
  down-flow, 212
  dry-fill, 206
  dynamic packing, 218
  hard gels, 217
  hardware, 203
  ion-exchange resins, 218
  machines for, 206
  particle sizing, 206
  pressure effects, 212
  slurry-packing, 207-18
  soft gels, 218
  up-flow, 216
  vibration effects, 206
  weight of packing, 215
  wet-fill, 207-18
  see also Column packings
Column packings, 183-202
  bed structure, 177
  characteristics, 178
  coating for LLC, 328
  diatomaceous earth, 176
  dissolution of, 227
  hard gels, 173
  for IEC, 190, 414-22

inter-lot variation, 183
for IPC, 186, 457-8
for LLC and LSC, 184, 327-32, 361-5
mass transfer in, 170
for normal-phase BPC, 188, 272-80
optimum size, 240
particle size, 171, 179, 180, 182, 207, 631
particle-size range, 176
pellicular, 170, 196-9, 229, 243
physical properties, 173
pH effects, 226
porous-layer, 170
porous versus pellicular, 182, 236
preparative, 176, 200
for reverse-phase BPC, 186, 272-80
rigid, 173
for SEC, 193, 487-500
sedimentation, 207
separation characteristics, 179
settling velocity, 208
shape, 177
sizing, 176
soft gels, 174
stability, 227, 427-8
superficially porous, 170
suppliers, 192
thick layer, 170
see also Ion-exchange chromatography;
  Liquid-liquid chromatography, *etc.*;
  Particles; *and* Supports
Column performance, 219-225, 234-43
Column plate height, *see* Plate height H
Column pressure, *see* Pressure
Column switching, 694-711
  versus gradient elution, 775-7
  packings for, 710-11
Compressibility of liquids, 95-6
Computers, 548
Conditions, in figures, 12
Conductivity detectors, 161, 440-1
Connectors, 204
Continuous-flow analysis, 743, 747
Control samples, 551
Convenience, of LC method, 759
Corasil, 175
Corrosive solvents, 266-7
Corticosteroids, 335, 739
Cosolvent, 334
Cough syrup, 400
Coumarin, 326

Counter-current distribution, 324
Counter-ion, in IPC, 459-64
  UV-absorbing, 470-2, 779
Coupled-columns, 707-9
CK (enzyme), 440
Cross-check, chromatographic, 591-2
Cutpoint, equal-purity, 45
Cut-and-weigh, 547
Cyclosporins, 565
Cytidine, UV spectrum, 134

Data handling systems, 119
Degassing, of mobile phase, 89, 109
Deproteinization, 724
Derivatization, 568, 731-46
  for qualitative analysis, 602-3
  pre- versus post-column, 737
  see also individual compounds; e.g.,
    Alcohols; Ketones; etc.
Detection, 125-67
  dual-wavelength, 777
  preparative anomaly, 638
  routine, 140
  trace analysis, 139, 145, 148, 155
Detection sensitivity, 181
  dilution effects, 562
Detector, 125-67
  amperometric, 153
  anomaly, 638
  bulk property, 127
  cells, 129
  conductivity, 161
  drift, 128
  electrochemical, 153
  fluorometer, 145
  ideal, 126
  infra-red, 147
  less-used, 163
  linear range, 129
  noise, 127
  overloading, 637
  preparative, 637
  radioactivity, 158
  refractometer, 140-5
  response, 544, 556
  response time, 130
  selective, 127
  sensitivity, 129, 162
  specifications, 162
  time constant, 130

trace analysis, 569
ultraviolet, 130-40
  UV-scanning, 138
  see also Reaction detectors
Detector cell, carbon paste, 154
  design, 129
  electrochemical, 154, 157
  infra-red, 151
  radiometric, 160
  refractometer, 141-2
  ultraviolet, 133
  volume, 129
Detector electrode, carbon paste, 154
  mercury drop, 157
Deuterium labeled compounds, 779
Diastereomers, 311, 654, 778
Diatomaceous earth, 176
Dieldrin, 699
Dielectric constant, of solvents, 248-50
Dielectric interaction, 257
Diffusion, longitudinal, 20
Diffusion coefficient, calculation of, 237
Dimerization, of solutes in SEC, 524
2,4-Dinitrophenylhydrazine derivatives, 271,
  733, 741
Diols, derivatization of, 734
Dipole interactions, 256
Dipole moments, 256
Disc integrator, 547
Dispersion interactions, 256
Distribution coefficient, in SEC, 491
Distribution constant, 25
Distribution head, 641
Dodecanoic acid, 296
Donor, proton, 257
Double-peaking, 803
Drugs, 285, 287, 302-5
  SEC calibration for, 518
Dry-fill packing, 206
Dyes, 346, 474-5, 477, 479

Eddy diffusion, 18-9
Edman sequencing, 778
Effective plate number, 54
Electrochemical detector, characteristics,
  153
  differential pulse mode, 158
  oxidation, 155
  reduction, 156
Electrode, carbon-paste, 154

glassy-carbon, 158
mercury drop, 157
Electrolyte, supporting, 154
Elemental analysis, by IPC, 442, 445
Eluotropic series, *see* Solvent, strength; *and
    individual methods* (Bonded-phase
    chromatography; Ion-exchange chro-
    matography; *etc.*)
Emulsions, breaking, 723-4
Enantiomers, 778
Enzyme, 437, 438, 440, 533
    reaction detectors for, 741
Enzyme inhibitors, reaction detector for,
    741
Equal-purity-cutpoint, 45
Equipment, 83-167
    criteria, 85
    data-handling, 119
    detectous, 125-67
    gradient elution, 103
    gradient formers, 105-6
    integrated, 119
    micro, 122
    microprocessor-controlled, 119
    miscellaneous, 117
    preparative, 636-41
    process control, 122
    pumps, 90
    sample injection, 110, 233
    safety, 122
    schematic, 86
    selection of, 85
    specialized, 119
    slurry-packing, 211
    troubleshooting, 782-91
Error, analytical, 542, 813-23
Error function, 833-5
Estrogens, 77, 306
    reaction detector for, 742
Ethylene oxide oligomers, 271
Exclusion chromatography, *see* Size-exclu-
    sion chromatography
Exclusion limit, 490
Exponential peak tailing, 224
Extra-column effects, 31-3, 86, 183, 225,
    242
    connecting tubing, 87
    preparative, 639
    recognition of, 87
    sample volume, 289-92

tailing, 801-3
Extraction, 306
    back-, 728
    of sample, 722-25
Extraction constant E (IPC), 455

Fabric finisher, 679
False bands (in SEC), 522
Fatty acids, phenacyl esters, 300
Figure captions, format, 12
Filters, line, 117
Fingerprinting, 777
Fish extract, 647
Flammability, of solvents, 253
Flow programming, 712-3
Flowrate, control of, 92-100, 559
    effect on R_s, 56-8, 60, 65-71
    measurement, 118
Flowrate variation, 119
Flow resistance $\emptyset$, 37
Flowtube, 119
Fluoroalcohols, 654
Filtration, sample, 226
Fines, 212
Fittings, "zero-dead-volume," 204
Flumethasone pivalate, 341
Fluorometers, 145-7
Fluorometric reagents, 316, 733-4, 741-2
Food, 340
Food additives, 309
Fourier transform, *see* Infra-red; Mass spec-
    trometry
Fractionation range (SEC), 490
Fraction collector, 118
Freeze-drying, 647
Frits, 204
Functional-group constants (α_i or ΔR_M),
    583-4
Fused-ring aromatics, *see* Hydrocarbons

Gallotannins, 353
Gas chromatography, versus LC, 2-3
Gauges, 117
Gaussian band, 20-1, 833-5
Gaussian function, 833-5
Gel chromatography, Gel filtration chroma-
    tography, Gel permeation chromatogra-
    phy, *see* Size-exclusion chromatography
Gels, organic, 488
General elution problem, 54-5, 663

Glassy-carbon electrode, 158
Glue, wood, 528
Glycophase, 489
Gradient, selection for gradient elution, 668-86
Gradient delay, 680
Gradient elution, 55-6, 663-94
  advantages, 663-8
  applications, 693-5
  column regeneration, 691-2
  versus column switching, 715-17
  design of separation, 691
  equipment, 103-10
  equivalent isocratic separation, 693
  gradient shape, 669-70, 676-80
  gradient steepness, 669-70, 680-6
  high-pressure mixing, 105
  low-pressure mixing, 106
  LSS gradients, 668, 674
  mechanism, 663-5
  precision, 559
  preparative, 646
  reproducibility, 110
  R_s versus sensitivity, 681
  "scouting" with, 693
  solvent demixing, 675-6
  solvent program, 667-86, 688-91
  varying N, 686-8
  varying α, 688-90
Gradient formers, 103
Grapefruit oil, 396
Grapejuice, 445
Grating, 137
Grignard reaction, 273
Group constants (α_i or ΔR_M), 583-4
Guanidines, reaction detector for, 742
Guard column, 182, 228, 297

Halogenated solvents, corrosion by, 266-7
Heart-cut, 621
Hecogenin, 739
Height equivalent of a theoretical plate (HETP), see Plate height H
Hemoglobins, 440
Hexachlorophene, 395
High performance liquid chromatography (HPLC), see Liquid chromatography
High-performance TLC (HP-TLC), 8, 389
Hildebrand solubility parameter, 259

Homologs, retention versus n, 579-81
Hot-melt adhesives, 769
Hydraulic capacitor, 96
Hydrocarbons, 776
  aromatic (PNAs), 6, 299, 301, 340, 500, 671-2, 681-2, 771, 777
Hydrochlorothiazide, 303
Hydrodynamic radius, 491
Hydrogenated quinolines, 766
Hydrogen bonding, 257
  in LSC, 372-3
Hydroperoxides, 745
  reaction detector for, 744
5-Hydroxy-indole acetic acid, 571

IEC, see Ion-exchange chromatography
Indole acetic acid, 155
Infinite-diameter column, 230
Infra-red (IR), detectors, 147-53
  on-line detection, 607-12
  qualitative analysis, 593, 597-600, 607-12
Injection, Injector, see Sample Injection; Sample injector
Insect attractants, 364
Insect growth-regulator, 730
Insecticides, carbamate, 783. See also Pesticides
Insoluble samples, 722
Instruments, see Equipment
Insulin, 778
Integrators, Integration, 547
Interactions, intermolecular, 255-8
Internal standard, 552-5
Interparticle volume, 490
Intrinsic viscosity, 515
Ion chromatography, 438-45
Ion-exchange capacity, 417-8
Ion-exchange chromatography (IEC), 21-2, 410-52
  applications, 429-45
  buffers, 425
  column, 445-6
  column packings, 190-2, 198-9, 414-9
  column stability, 427-8
  complexation effects, 412
  gradient elution, 674, 684
  ionic strength effects, 420-1
  ligand exchange, 413-4
  mobile phase, 419-26, 447, 450

organic solvent effects, 426
pH effects, 411, 417-8, 423-6
problems, 427-9
protein separations, 416, 418, 433, 435,
    438, 440
resin structure, 415
retention mechanism, 412
retention prediction, 588
sample size, 418-9, 426-7, 450
solvent selectivity, 421-6
solvent strength, 420-6
tailing, 667, 795-7
temperature effects, 426, 450
Ion exchangers, 414-9
Ionic strength, in IEC, 420-1
    in SEC, 500-3
Ionization constants (pK$_a$), 425
Ionized solutes, selectivity for, 286
Ion-pair chromatography (IPC), 22, 454-
    82
    applications, 473-9
    column packings, 457-8
    counter-ion, 454-7, 459, 463-5
    design of separation, 476-81
    gradient elution, 674, 684
    mobile phase, 458-70, 477-80
    normal versus reverse-phase, 454-5, 458-
        462
    partitioning phases, 458-60
    pH effects, 467-9
    problems, 471, 473
    retention mechanism, 454-6
    retention prediction, 589
    sample size, 470
    secondary-ion effect, 467
    selectivity, 467-70
    solvent strength, 459, 463-9
    temperature effect, 470
    UV-absorbing counter-ion, 470-2
Ion-pair extraction, 731
Ion suppression, 294, 332, 756
IPC, see Ion-pair chromatography
IR, see Infra-red
Isocratic, 663
Isocyanates, derivatization of, 734
Isoenzymes, 440
Isohydric solvents, 377-81
Isolated fractions, 646
Isomer separations, 356-60, 638
    endo-exo, 521

isotope-labeled, 779
    optical, 778
    preparative, 649, 653-4
Isotopes, 301, 779

Ketones, derivatization of, 733, 741
Ketosteroids, reaction detector for, 742
Kinetics, of retention, 804
Knox equation, 235, 836
Knox-Parcher ratio, 220

Lannate, 553
Laser fluorometer, 147
Latices, 534
LC, see Liquid chromatography
LDH (enzyme), 440
Length, see Column length
LiChroma, 204
LiChrosorb, 175
Linear capacity, 26-7, 180, 616
Linear-solvent-strength (LSS) gradients, 668,
    674
Lipids, 647
Liquid, see Solvent
Liquid chromatography (LC), 1-8
    abstracts, 10-1
    advantages, 2-8
    band broadening, 16-21
    basic concepts, 16-37
    books, 13-4
    classical, 3-4
    control of separation, 37-81
    history, 9
    literature, 9-12
    meetings, 10
    methods, 21-2, 753-80
    modern versus traditional, 3-8
    retention, 16-8, 22-5
Liquid crystals, 281
Liquid-ion-exchangers, 416, 430
Liquid-liquid chromatography (LLC), 21,
    323-48
    advantages, 324, 327
    characteristics, 324
    coating of supports, 328
    column packings, 184, 198, 327-32
    column stability, 334
    comparison with BPC, 327
    gradient elution, 674, 684
    k' optimization, 332

liquid-phase loadings, 331
mobile phases, 258, 260, 332-6
normal-phase, 325
partitioning phases, 333
preparative, 649
problems, 336
retention prediction, 585
reverse-phase, 327
sample size, 336
selectivity, 332
solvent strength, 258, 260
solvent systems, 333
temperature effects, 336
Liquid-solid chromatography (LSC), 21,
   349-409
adsorbents, 184-5, 196, 200-1, 361-5
applications, 398-405
band tailing, 394, 795-7, 800
column packings, 184-5, 196, 200-1,
   361-5
column regeneration, 393-4
design of separation, 405-7
gradient elution, 674, 684
history, 351
mobile phases, 365-89, 406-7
open columns, 351
organic modifers (to replace water),
   381-3
preparative, 626
retention mechanism, 356-61
retention prediction, 585-7
versus reverse-phase LC, 355-60
sample size, 389-90
selectivity, 355-60
silica versus alumina, 361-3
solvent selectivity, 370-4
solvent strength, 248-50, 258-60, 365-74
temperature effects, 390-2
TLC scouting, 383-9, 585-6
water deactivation, 374-83, 393
LLC, *see* Liquid-liquid chromatography
Loading limit, 620
Lobar, 634
Longitudinal diffusion, 20
Low-pressure column chromatography,
   725-7
LSC, *see* Liquid-solid chromatography
LSD, 402, 808
Lyophilization,
   647

Manure, bovine, 730
Mark-Houwink constants, 512
Martin rule, 579-84
Mass spectrometry (MS), on-line detection,
   603-7
   qualitative analysis, 593, 595-7
Mass transfer, 18-20, 170
Material balance, 647
Mercaptans, derivatization of, 733, 742
Mercury drop electrode, 156
Mesh sizes, 838-9
Metal, by LSC, 405
Metal-β-diketonates, 340
Methomyl, 553, 570
Methyl amines, 446
Methylene blue, 354
Micelle formation, 804
Micro-LC, 122
Microprocessor control, 119
Microsampling valves, *see* Valves
Milli-Q-system, 265, 298
Mixed retention, 798-801
Mobile phase, 17
   for BPC, 281-9, 317
   column testing, 221
   compressibility, 95, 98
   effect on R_s, 73-80
   filtration, 101
   frictional forces, 240, 346
   fractionation, 89
   for gradient elution, 668-86, 688-91
   for IEC, 419-26, 447, 450
   for IPC, 458-70, 477-80
   for LLC, 258, 260, 332-6
   for LSC, 365-89, 406-7
   mass transfer, 20
   microbiological inhibitors, 226
   pH and R_s, 76
   presaturation for LLC, 333, 337
   purity, 265-7, 298
   repurification, 640
   reservoirs, 88
   reuse, 90
   for SEC, 500
   ternary, 285
   TLC selection, 383-9
   velocity, *see* Velocity
   see also Solvent
Modern liquid chromatography, *see* Liquid
   chromatography

Molar volume, 517
Molecular association, 524
Molecular conformation, 509
Molecular hydrodynamic radius, 491
Molecular length, 516
Molecular permeation, 485
Molecular shape, 491
Molecular size, 485
Molecular weight, weight and number
  averaged, 509
Molecular weight calibration, 489, 509-18
Molecular weight effects, 237
Multidimensional separation, 776

NARP, 301
Niacin, in bread, 725
Nickel complexes, 517
Nicotinic acid, 456, 700
Nitrites, 315
  reaction detectors for, 741
Nitroso-compounds, 156, 340, 570
  reaction detectors for, 741
NMR, 593, 600-1
Noise, detector, 127
Nonapeptides, 289, 314
Normal-phase BPC, 270, 283-5
  applications, 301
  column packings, 188, 197
  mobile phases, 284
  problems, 294
  retention mechanism, 281
  selectivity, 283
  solvent strength, 283
  see also Bonded-phase chromatography
Nucleic acids, 314
Nucleosides, 314
Nucleotides, 438
Nujol, 608

Oil, lubricating, 526
Oil spills, 777
Olefins, 340, 776
Oligomers, 316, 529, 679, 777
  ethylene oxide, 271
  retention versus *n*, 579-83
On-column enrichment, 565
Opium extracts, 306
Optical brighteners, 354-5
Optical isomers, 778
Orange juice, 445

Organization of book, 12
Overload, column, 617
  sample, 642
Overloading BPC columns, 291
Oxygen, removal, 90

Packing, *see* Column packings
Paper chromatography, 4-5
Paraffins, 340
Parameter sets, stored, 122
Particles, 169-83
  advantage of rigid, 488
  aggregation, 209
  charge effects, 209
  electron micrographs, 175-6
  fracturing, 212
  mesh size, 838-9
  optimum size, 240
  physical properties, 173
  segregation, 208
  size and permeability effects, 179
  size range, 176
  sizing, 176, 207
  types, 171
  see also Column packings
Particulates, 243, 535
Partition chromatography, *see* Liquid-liquid
  chromatography
Partitioning phases, 332
Peak, 20-1
  area, 546-7
  asymetry factor, 222
  broadening, *see* Band broadening
  capacity, 486, 519
  dilution, 562
  double, 803
  height, 545
  measurement, 545
  moments, 221
  normalization, 552
  overlapping, 38-41, 622
  shape, 220, 791-810
  skew, 224
  vacancy, 812-3
  volume, 87, 242-3
Peak size, flowrate effect on, 558
Peanut butter, aflatoxins in, 352
Pellicular packings, 170, 182, 243
  control of R_s, 60-1
  k' values, 25

Knox equation, 836
  loading capacity, 27
Penicillin ester, 343
Peptides, 289, 314, 778
  derivatization of, 734, 741
  protected, 652
Perisorb, 175
Permaphase, 275
Permeability, column, 37, 179
Permeation, molecular, 485
Permeation limit, 490
Permeation volume, 490
Pesticides, 282, 315, 342, 554, 647, 699,
    712
Petroleum, 776
pH, effect on R_s, 76
  in IEC, 411, 417-8, 423-6
  in IPC, 467-9
  in reverse-phase BPC, 286
  variation for extraction, 723
Phenols, 78, 291, 307
  derivatization of, 733-4
Phospholipids, 81
Photometers, 131
  cells, 133
  dual-channel, 132
  multi-wavelength, 131
  rapid-scanning, 139
PIC^{TM} reagents, 475-6
pK_a values, 425
  retention versus, 588, 590
Planimetry, 547
Plant tissues, 534
Plasma, human, 503
Plasticizers, 342
Plate count, see Plate number N
Plate height H, 28, 169-73
  calculation, 235
  coefficients, 169
  column diameter effect, 205, 624
  dependence on conditions, 28-30
  dependence on u, 30, 169-72
  equations, 170, 172, 234-5
  liquid-loading effect, 330
  particle-size effects, 172, 180, 234-5
  reduced, 234, 236, 836
  sample size effect, 292
  stationary phase effect, 330
  versus velocity, 29, 171, 180, 234-5
  see also Plate number N

Plate number N, 27-8
  additivity of, 228
  calculation, 221
  to control R_s, 56-73
  effect of guard columns on, 229
  effective N_{eff}, 54
  effect of permeability, 240
  error, 223
  in gradient elution, 686-8
  in SEC, 499
  maximum, 238
  measurement, 30-1, 221-3
  particle-size effect, 631
  preparative separation, 631
  sample-size effect, 33-4
  tailing peaks, 223
  typical, 238
  viscosity effect, 242
Plugged columns, see Column, plugged
Polar solvents, 257-60
Polyacrylamide gels, 488
Polyamines, 439
Polyaromatic hydrocarbons, 6, 299, 301,
    340, 500, 671-2, 681-2, 771, 777
Polychlorobiphenyls (PCBs), 350
Polydextran, 488
Poly (ethylene terephthallate), 317
Polymer additives, 526, 527
Polymer extract, 530
Polymer-film extract, 525
Polynuclear aromatics (PNAs), see Polyaro-
    matic hydrocarbons
Polysaccharide, acetylated, 634
Polystyrene, 679
  gels, 488
  oligomers, 359
Polyvinylalcohol, 501
Polyvinylchloride, 527
Porasil, 175
Pore size, 485
  selection, 493
Pore structure, for LLC, 327, 329
Porosity, total, 220
Porous frits, 204
Porous silica, for SEC, 488, 496
Postcolumn derivatization, see Reaction de-
    tector
Precision, 542
Precolumn, 227, 337
Precolumn venting, 698-701

Prefractionation, heart-cut, 621
  in prep-LC, 533
  sample cleanup, 720, 724-31
Preparative LC, 615-61
  column choice, 631
  column packings, 628
  conditions for, 660
  effect of, 73-4
  extra-column effects, 639
  flow-programming, 645
  fraction purity, 48
  heart-cut, 621
  loading-limit, 620
  material balance, 647
  maximum yield, 620
  purity estimation, 45
  sample injection, 639, 641
  sample weight, 617
  separation goals, 619
  solvent removal, 646
  strategy, 617
  throughput, 617, 643
  typical conditions, 624
  yield estimation, 48
PrepPak^{TM}, 634
Presaturation, 337
Pressure, 36-7
  versus column heating, 240
  effect on R_s, 56-8, 60, 65-71, 238-41
  feedback, 95
  monitors, 117
  optimum, 238
  practical limits, 240
  transducers, 117
Propoxyphene hydrochloride, 399
Proteins, 416, 418, 433, 435, 438, 440,
    463-4, 506, 778
  adsorption of, 503
  diffusion coefficients, 237
  ionic strength effects, 503
  separation, 532, 533
Proton acceptors, 257
Proton donors, 257
PTH amino acids, 403-4
Pulse dampers, 117
Pulse damping, 94
Pumps, 90-103
  accuracy, 91
  comparison, 101
  compressibility effects, 93

  constant flow, 92
  constant pressure, 100
  design approaches, 101
  diaphram, 97
  displacement, 98
  drift, 91
  dual-head, 94
  feedback-flow-control, 96
  flowrate compensation, 94
  Haskel, 211
  hydraulic amplifier, 99
  maintenance, 101
  materials of construction, 90
  noise (pulsation), 91
  operation, 103
  output patterns, 95
  pneumatic amplifier, 100
  precision, 91
  preparative, 636
  pressure equilibration, 98
  programmed, 105
  pulsation, 94
  pulseless, 98, 100
  reciprocating, 92
  requirements, 91
  resettability, 91
  specifications, 91
  three-head, 97
Purification of solvents, 265-7
Purity, 647
  estimation of, 45
P' values, 248-50, 283
Pyrazoline isomers, 401
Pyrethins, 649
Pyridines, 456, 555

Qualitative analysis, 576-614
  chromatographic cross-check, 591-2
  color reaction, 593, 601-2
  elemental analysis, 593
  IR, 593, 597-600, 607-12
  MS, 593, 595-7, 603-7
  NMR, 593, 600-1
  on-line detectors, 603-12
  peak isolation, 593-5
  relative detector response, 592
  resolution required, 43-5
  using t_R values, 576-89
  UV, 593
Quantitative analysis, 541-73

accuracy, 542
band area method, 46-7, 546-52, 556-9
band height method, 45-6, 545-6, 556-9
collect-and-weigh, 47-8
isocratic versus gradient, 559
precision, 542
recommendations, 560
separation variables, 556
trace analysis, 560-73
Quartering, 543
Quinolines, 766
  hydrogenated, 694

Radial compression, 625
Radioactivity detector, 158
Radioisotopes, 428-30, 806
Radiolabeling, 647
Reaction detector, 433, 437, 439, 568, 740-
  6. *See also* individual compounds; *e.g.*
  Aldehydes; Ketones; *etc.*
Reactive solvents, 252, 266-7
Reactivity index, 530
Recorders, 119
  response time, 130
Recycle chromatography, 622
  equipment, 519-20
Reduced parameters, 234, 836
  to control Rs, 62-73
Refractive index, dispersion interactions,
  256
  matching, 771
  of solvents, 248-50
Refractometer detectors, 140-5
  selective detection, 771
Reservoirs (solvent), 88
  column packing, 216
  preparative, 637
Resin, phenolic, 529
Resin degradation, 528
Resolution R_s, 34-6
  for accurate t_R, 43-4
  compromise, 616
  control of, 49-81
  estimation of, 38-43
  flowrate effects, 645
  versus k', 51-6
  versus N, 56-73
  one band or two?, 43
  preparative separation, 48-9, 618
  qualitative analysis, 43-5

quantitative analysis, 45-8
sample volume, 289
sample weight, 291, 618
scheme to control, 62-73
versus selectivity α, 73-80
standard curves, 38-41
see also Separation
Retention, 16-21
  accurate comparison of, 577
  linear isotherm, 26
  mixed mechanism, 795-8
  liquid-loading effect, 332
  prediction of, 578-89
  for qualitative analysis, 576-89
  time, 21, 23-4
  volume, 24-5
Retention mechanism, *see individual LC*
  methods (Bonded-phase chromatography;
  Liquid-solid chromatography; *etc.*)
Retinal isomers, 351
Reverse-phase BPC, 270-1, 285-9
  addition of salts, 287
  applications, 301
  column packings, 186, 197
  mobile phases, 285
  packing stability, 281
  pH effects, 286
  preparative, 631
  problems, 294
  retention mechanism, 281
  selectivity, 286
  solvent selectivity, 264-5
  solvent strength, 286
  temperature effect, 293
  wetting of packing, 298
  see also Bonded-phase chromatography
RI, *see* Refractive index

Safety, 122, 123, 637
  of solvents, 253, 255
Salting-in effect, 288
Salting-out effect, 288
Sample, concentrating, 729-30
  convenience, 759
  decomposition, 394-5, 543, 807-10
  derivatization, 721, 731-42
  dilution, 562
  injectors, 110, 230
  insoluble, 722
  loss, 552

mass, 642
organic-soluble, 762
preparation, 543, 720-50
recovery, 593-5, 646
solubility, 635
throughput, 644
viscosity and SEC, 507
volume, 289, 641
Sample cleanup, 722-31
Sample concentration, in SEC, 507
Sample injection, 230, 543
  automatic, 640
  concentration effects, 292
  mobile phase effects, 291
  normal, 231
  overload, 642
  point (center), 231-3
  preparative, 639-41
  solution effects, 298
  syringe versus valve, 115
  volume, 564, 642
Sample injector, 233
  automatic, 116
  septumless syringe, 110
  stop-flow, 112
  syringe, 112
  valve, 113
  see also Sample injection
Sample pretreatment, 720-40
  automated, 746-7
  cleanup, 722-31
  concentration, 729-31
  derivatization, 731-46
  extraction, 722-4
Sample size, 26-7, 33-4, 180, 183, 289, 564
  BPC, 289-92
  effect on N, 33-4
  effect on separation, 26-7
  IEC, 418-9, 426-7, 450
  IPC, 470
  LLC, 336
  LSC, 389-90
  quantitative analysis, 544
  SEC, 506
Sampling, see Sample injection
Scale-up, 616
Screens, 204
  mesh-size, 838-9
SEC, see Size-exclusion chromatography
Sedimentation, 207

Selectivity, see Separation selectivity
Selectivity groups, of solvents, 263-4
Selenium, 315
Semi-preparative LC, 625, 631
Senna glycosides, 773
Sensitivity, 181
  dilution effects, 562
  optimization, 562
  see also Detector; Sample pretreatment
Separation, 16-82, 753-81
  capacity, 496
  control of, 37-81
  general approach, 37-8
  organic-soluble samples, 762
  purity of fractions, 45
  water-soluble samples, 766
  see also Resolution R_s
Separation efficiency, 36. See also Plate number N
Separation factor α, 35. See also Separation selectivity
Separation selectivity, 36, 73-80, 758
  alumina versus silica, 362
  BPC, 271
  to control R_s, 73-80
  gradient elution, 688-90
  IEC, 421-6
  IPC, 467-70
  LLC, 332
  LSC, 370-4
  LSC versus reverse-phase, 355-60
  mobile phase, 73-80, 260-5
  pH, 76, 411, 417-8, 423-6, 467-9
  silica versus alumina, 362
  stationary phase, 76-9
  switching, 706-7
  temperature, 79-80
Separation time, 36-7, 239
  decrease via α, 79-80
Sephadex, 488
Sep-Pak^{TM} columns, 728
Sequential separation, 772
Serum, 778
Shear forces, 346
Silanes, 274
Silanols, 272
  measurement, 278
  residual, 275
Silica, 272, 361
  chlorination of, 273

dehydration, 277
dissolution, 501
drying of, 276
Grignard reaction, 273
hydrolysis of, 276
ion-exchange effect, 501
in LSC, 361-5
modified, 533
preparation, 176
suppression of ionization, 537
surface silanols, 272, 799
wide-pore for LLC, 327
Silicate esters, 273
Silver salts, in BPC, 80
in LSC, 364-5
Sil-X, 175
SIMPLEX, 80-1
Size-exclusion chromatography (SEC), 22, 483-575
additive retention volumes, 494
advantages, 485
column packings, 193-5, 487-500, 536
control of separation, 493
design of separation, 534
flowrate, 538
mobile phases, 504
problems, 522
quantitative analysis, 530
retention prediction, 509-18, 585
sample types, 484
small-molecule separations, 530
temperature, 538
Site heterogeneity, 795
Skew, 223-4
Slurry-packing, 207
Soap chromatography, 414, 457, 458, 462
Soft drinks, 307
Solubility, 635
P' versus ϵ', 635
salt effect, 288
Solubility parameter, 259
Solubility problems, 298
Solutes, test, 221
Solutions, sample, 543
Solvent, 247-68
availability, 251
binary, 52-3
boiling point, 248-50, 252
BPC, 281-9, 317
concentration gradients, 388

corrosive, 266-7
degassing, 524
demixing, 386-7, 395-7
detector capability, 251-2
dielectric constant, 248-50
flammability, 123, 253
gradient elution, 668-86
IEC, 419-26, 447, 450
IPC, 458-70, 477-80
leaks, 123
LLC, 258, 260, 332-6
LSC, 365-89, 406-7
preparative LC, 634
metering system, see Pumps
miscibility, 253-4
presaturation, 337
polarity, 258-9
properties, 247-55
purification, 265-7
P' values, 248-50
reactivity, 252
refractive index, 248-52
removal, 646
SEC, 504
selection, 247, 251
selectivity, 248-50, 260-65, 283
slurry, 214
slurry-packing, 209
solubility of water in, 248-50
strength, 36, 52, 257-60, 283
strong, 52
suppliers, 248-51
μ-Styragel compatability, 523
ternary, 285
TLC selection, 383-9, 585-6, 657
UV absorption, 248-52
velocity, 22-3, 29
viscosity, 248-50, 252-3, 836-8
wesk, 52
see also Mobile phase
Solvent demixing, 386-8, 675-6
Solvent selectivity, Solvent strength, see individual LC method (Bonded-phase chromatography; Ion-exchange chromatography; etc.)
Spectrofluorometer, 147
Spectrophotometer, 135
Spiking, 571
Stagnant mobile phase mass transfer, 20
Standard addition, 571

Stationary phase, 17, 324
  bleed, 346
  effect on R_s, 76-9
  mass transfer, 20
  polymeric, 174
  programming, 700-6
  see also Column packings
Steroids, 341, 377, 702, 713
  preparative isolation, 653
  sulfates, 476
Stream splitter, 639
Styragel, 489
μ-Styragel, 496
Sugars, 307, 777
  reaction detector for, 742
Sulfonamides, 465-6
Sulfonic acids, 78, 475
Supports, hydrophilic, 344
  for LLC, 327-31
  hydrophobic, 334
  see also Column packings
Surface area, effect in LSC, 363-4
Surfactants, 777
Sweetener, 307
Swinney filter, 226
Symbols, 840-3
Syphon counter, 119
Syringes, 112
System mismatch, 795

Tailing bands, 220, 791-807
  pseudo, 807
Tap-fill procedure, 207
Temperature, effect on R_s, 79-80
  programming, 714-5
  see also individual LC methods (Bonded-
    phase chromatography; Ion-exchange
    chromatography; etc.)
Ternary LLC systems, 334
Testosterone, 377
Tetraalkylammonium salts, 501
Tetracyclines, 473-4
Theophylline, 304
Theoretical plate number, see Plate num-
  ber N
Thin-layer chromatography (TLC), 4-5,
  352-4
  high performance (HP-TLC), 8
  preparative scouting, 657
  scouting for LSC, 383-9

t_0, 23
  measurement of, 24
Tobacco constituents, 307
Total porosity, 220
Toxicity, of solvents, 255
Trace analysis, 560-73, 778
  calibration, 570
  detection, 139, 145, 148, 155, 566
  flowrate effect, 563
  k' effect, 563
  preferred conditions, 572
  sample injection, 564
  sample pretreatment, 572, 722-31
  selectivity effect, 564
Trace enrichment, 565
Triangulation, 547
Tridigitoxosides, 302
Triglycerides, 151
Tritium-labeled compounds, 779
Troubleshooting, 781-823
  equipment, 782-91
TSK-gels, 489
Tubing, blanks, 203
  coiling, 87
  connecting, 87

Ultraviolet detectors, 130-40
  cells, 133
  characteristics, 139
  light source, 131
  low-wavelength, 139
  operation, 140
  qualitative analysis, 593
  rapid-scanning, 139
  solute structure, 134
  spectrophotometers, 139
Urea herbicides, 282, 315
Urine analysis, 7
UV, see Ultraviolet detectors

Vacancy peaks, 812-3
Valves, automated, 114
  sampling, 113
  switching, 115
Variance, 33
Velocity (mobile phase), 22, 28-30
  effect on column efficiency, 28-30, 56,
    170-172, 234-5
  effect on pressure drop, 36-7
  reduced, 234

Ventilation, 123
Viscosity, of binary solvents, 836-8
  of sample, 507
  of solvents, 248-50
Vitamin A, 136, 529
Vitamin B-12 intermediates, 652
Vitamins, 345, 725
  A, 136, 529
  B-12 intermediates, 652
  in cake, 434-5
  D_2, D_3, 79-80
  fat-soluble, 345
  riboflavin, 146

Water, contamination, 226,
    297
  deactivation of LSC adsorbents, 364-5
  LSC modifier, 374-83, 393

purification, 265
  solubility in organics, 248-50
Wavelength, solvent cut-off, 251-2
Wavelength "detuning," 639
Wax, 771
Wetting of BPC packings, 298
Wilke-Chang equation, 237
"Wings," 622

XAD resin, 730
Xanthines, 799
Xanthones, 794

Yield, in preparative LC, 48

"Zig-zag" classifiers, 176
Zorbax-PSM, 499, 537
Zorbax-Sil, 175